Larousse
Science of Life

Distributed in America by Tudor House Inc.

Published by the Hamlyn Publishing Group Limited
London   New York   Sydney   Toronto
Hamlyn House, Feltham, Middlesex, England
Translated by Delano Ames from *La Vie* by
Jean Rostand and Andrée Tétry, first published
in France in 1962
Copyright © 1971 Augé, Gillon, Hollier-Larousse,
Moreau & Cie., Librairie Larousse, Paris, and
the Hamlyn Publishing Group Limited

ISBN 0 600 02364 8

Printed in Czechoslovakia by TSNP Martin
51627

Larousse
Science of Life

a study of biology
sex, genetics, heredity
and evolution

Jean Rostand and Andrée Tétry

HAMLYN
London·New York·Sydney·Toronto

Contents

Preface

The Importance of Biology

One of the most remarkable features of our age is the sudden advance which biology — the science of life — has made during the past fifty years. Having lagged far behind sciences which deal with inert matter such as physics and chemistry it is at present rapidly gaining ground and will soon have drawn level, not only in theoretical grasp of biological phenomena but in practical application of biological knowledge.

It is true that the basic nature of life still eludes us and, for the moment, we cannot reasonably envisage a laboratory synthesis of a living protoplasm which would be the supreme triumph of biology; but, for that matter, can physics explain the real nature of electricity or gravitation?

As for the fundamental problems of heredity, of sex, of the development of the individual and the evolution of species, they have by now been sufficiently clarified to permit replies to a host of questions eagerly asked by the layman, who is all too often unaware that science has already supplied the answers.

Why, for example, do we resemble our parents, and why do we differ from them? Why do children of the same parents differ from each other? What accounts for individuality — the organic uniqueness of the

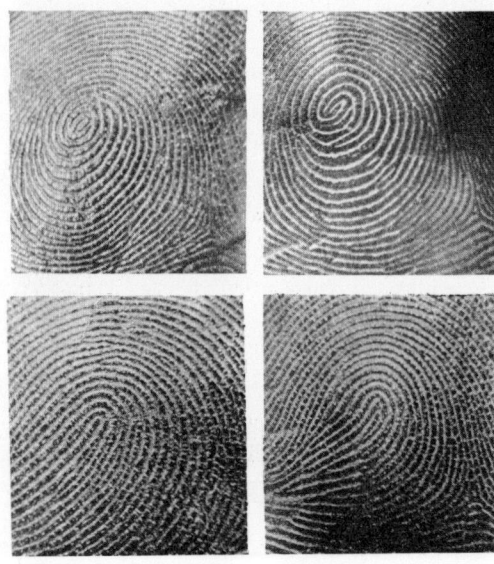

Fingerprints of two thumbs and two forefingers. Everyone's fingerprints are different and can thus be used for purposes of identification.

Larousse.

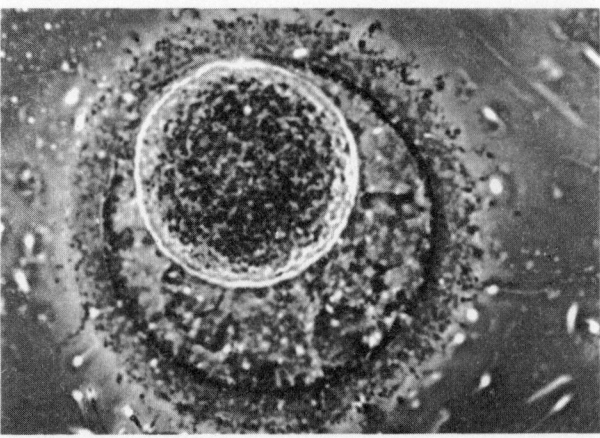

For the first time in history man has witnessed the actual conception of a human being, or rather, the exact moment of a human ovum's fertilization by a spermatozoon. This remarkable achievement is due to Dr. Landrum B. Shettles of New York, who during an operation removed the ovum from the ovary of a patient and put it in contact with semen. Cinemicrography then revealed the phases of human fertilization.

P. Popper.

human individual, demonstrated, as everyone knows, by the possession of fingerprints peculiar to himself? In the formation of personality what part is played by heredity, and what by circumstances, by environment? Are the sins of the fathers visited on their children? What are we to think of 'hereditary alcoholism' or of the 'children of the old'? How are twins produced? Why are certain twins as alike as two peas in a pod while others bear no more resemblance to each other than ordinary brothers or sisters? Why do certain people have one blue eye and one brown eye? Where do 'birthmarks' come from?

How does it happen that in a normal family a deformed child or a monster is born, and are monstrosities hereditary? How are Siamese twins formed? How is sex determined. Why do women, on an average, live six or seven years longer than men? Why do women almost never go bald? Is it true that certain people change sex during the course of their lives, and are there individuals, neither male nor female, of intermediate sex? How do the father and the mother collaborate in the formation of a child? In what fashion is the future living creature represented in the germ from which it springs? Where does the human species come from, and in what particulars does one race differ from another?

Applications of biology to animal breeding and agriculture

Biology is not only in a position to satisfy much of our legitimate intellectual curiosity about life itself, and in particular about human life; but in many cases it can effectively intervene in the course of nature. Biology has been applied with fruitful results to those plants and animals of particular interest to the human race, and also to the human race itself.

Agriculture and animal breeding have already greatly benefited from recently acquired knowledge of heredity, selection, mutation, cross-breeding, the mechanics of fertilization and the physiology of reproduction in general. To cite a few typical examples, it has been possible to increase the fertility of domestic mammals by the employment of certain *hormones* which stimulate the ovarian glands, so that multiple births have been induced among animals such as ewes, cows and mares, which normally produce their young one at a time. Artificial insemination permits the hereditary qualities of selected males to be widely exploited for the rapid improvement of livestock, since, by the use of this technique, a single male can sire hundreds or even thousands of descendants in the course of one year. Artificial insemination thus provides the most effective means of applying speedily and accurately the principles of selective breeding, and in the more advanced parts of the world is in the process of completely modifying the rearing of livestock.

Wheat-rye hybrid between its two parent plants. Its four chromosome sets comprise fifty-six chromosomes (4N = 56). On the left, *Triticum vulgare* (2N = 42). On the right, *Secale cereale* (2N = 14).

After Dorsey.

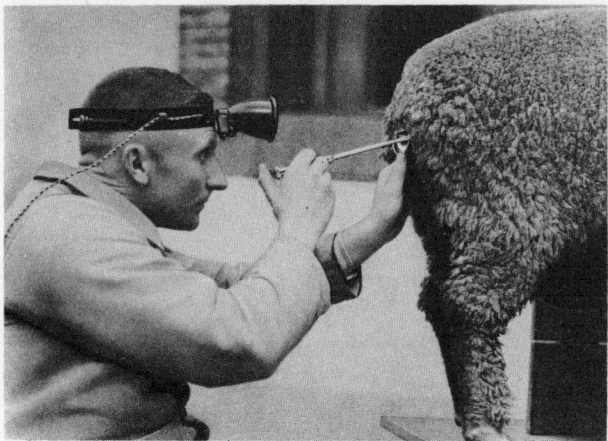

Artificial insemination of a Merino ewe at the French National School of Sheep-breeding at Rambouillet.

Bergerie nationale.

7

Thanks to this method, the production of milk, the percentage of cream, the quantity of wool, of meat, and so on has already been strikingly increased.

Generally speaking, the principles of selective breeding, directly inspired by biological findings, have in application proved extremely effective in the case of cattle, sheep, horses, pigs and poultry. In the realm of agriculture the economic value of applied biology is perhaps even more obvious. By utilizing the principles of genetics new varieties of cereals have been created which combine the most desirable features of those previously known: for example wheats which combine the high yield of the English Squarehead with the quality and ability to resist cold of certain Swedish varieties. In the cultivated varieties of these plants early ripening, quality of grain and quality of straw, have been systematically improved. Hybrids of high yield such as *Triticale* (wheat-rye) have been created.

In Sweden, where these studies are particularly advanced, yield has increased in a few years by twelve per cent in the case of oats and summer wheat, and by twenty-five per cent in the case of winter wheat.

In the same way improvements have been made in sugar beet, clover and cotton. New varieties of plants have been created which are immune or more resistant to infection — wheats, for instance, which are resistant to twenty-two kinds of 'rust', and many varieties of new ornamental flowers. Giant vegetables

and fruits have been grown, and at the moment there are hopes of creating a variety of tobacco which will be welcomed by smokers, since it will no longer contain the carcinogens which produce lung cancer.

One of the most spectacular successes of genetics applied to agriculture is perhaps the creation of the famous hybrid maize or corn which plays such an important role in the agricultural economy of America. This hybrid corn might well be called the major triumph of applied biology during the first half of this century. It saved more lives during that period than any of the great biological discoveries made in the field of medicine.

Insulin and penicillin have certainly saved thousands of human lives during the last quarter of a century, but the abundance of nourishment supplied by hybrid corn saved millions. Indeed, thanks to hybrid corn the American people were able not only to dispense with food restrictions during the second world war but could provide the armies with not inconsiderable supplies of foodstuffs and, later, send provisions to the undernourished peoples of Europe. Hybrid corn has moreover served in the distillation of alcohol, the manufacture of synthetic rubber and of other products.

Since the creation of this plant the average crop of corn in the United States has been increased by fifty per cent, although only three-quarters of the corn grown is of the hybrid type. This *agricultural revolution* has already spread to Italy, Southern and Eastern Europe, and above all to Mexico where the new plant, introduced in 1943, has entirely transformed the economy of the country to the extent that Mexico eventually found it unnecessary to import maize. In view of the fact that the world's population continues to increase and the problem of food supply is one of the most serious which faces the human race, it is impossible to overestimate the importance of such an innovation. Obviously, hybrid corn cannot by itself solve the general problem of feeding the world's population; but, with the aid of other scientific methods it contributes to the solution. In any case, this example demonstrates very clearly the value, scope and practical consequences of biological progress.

Single maize hybrid.
J. Vincent, J. de Beaupré.

The general public, and for that matter most governments, scarcely suspect the material importance of the science of life. The world has need of biologists.

Applications of biology to the human race

The technique of artificial insemination has also been applied to human beings, the semen of the husband being employed in some cases and in others that of an anonymous 'donor'.

At the present moment thousands of children, colloquially referred to as 'test-tube babies', owe their existence to this method which, because of its slightly mechanical or even 'veterinary' principle, first aroused violent protest. Human artificial insemination has nonetheless become fairly extensive since the war. Although only very approximate figures are available it is estimated that over 100,000 children have been engendered by this method in the United States, and that from 10,000 to 15,000 are added yearly. The population of Australia includes about 5,000 to 6,000 test-tube babies, that of England about 10,000, that of France a few thousand.

Since 1939 a special technique of preserving semen in a glycerine medium and refrigerating it has been perfected, and it is now possible to produce babies from semen preserved in this way for several months.

A well-known application of the biology of reproduction is the early diagnosis of pregnancy by means of a urine test of the woman presumed to be pregnant.

From the first weeks of gestation the pregnant woman's urine contains special hormones which are connected with the development of the embryo. These can very easily be identified by the manner in which they stimulate the sexual glands of certain animals. This was the chief means of diagnosis at one time. The animal employed for this purpose was usually a mouse, male or female, or a male frog or toad, whose testicular gland reacted with speed and certainty when the animal was injected with the urine of a pregnant woman. A few hours after injection the animal would eject its semen which mingled with its own urine, and microscopic examination then revealed the wriggling spermatozoa. More recently pregnancy diagnosis has become more refined, and there are now simple, accurate microscopic tests for which no experimental animals are required.

Biology and medicine

It is hardly an exaggeration to say that all branches of medicine, including surgery and hygiene, are dependent on biology. Much of our understanding and treatment of infectious diseases is based, after all, on the brilliant researches of Louis Pasteur, through which he established the origin of fermentation and the existence of bacteria.

In addition, a great number of diseases arise from a functional disorder of certain glands. These are known as the *endocrine* or *ductless* glands because they secrete certain essential chemical substances called hormones directly into the blood stream.

In this rapid inventory of the chief consequences of biology we shall take into account only some of its direct contributions; for example, those made to that vast body of medicine recently gathered together under the name of *medical genetics*.

A host of diseases and human afflictions arise from hereditary factors and are connected with the constitution of *genes*. These are invisible elements contained within the *chromosomes* of cell nuclei which play a dominant role in the phenomena of heredity. Such diseases are comparable in every respect with hereditary variations, or *mutations*. Mutations occur in all animal and plant species and have been the subject of profound biological research. Brittle bones, Daltonism, congenital cataract, various forms of blindness and deaf-mutism, haemophilia, certain nervous and muscular degenerations, certain types of idiocy, of dementia, of epilepsy, and so on, are all produced by deficient genes which are transmitted by parents to their offspring with predictable and often almost mathematical regularity.

Haemophilia, or the failure of the blood to clot, is among the best known of these defects by reason of the role it has played in the history of royal families, a gene producing haemophilia being present in the descendants of Queen Victoria.

In medicine the concept of bad or morbid genes is constantly borne in mind. Apart from the considerable number of illnesses directly determined by such genes, they also control the organism's predisposition or resistance to various other maladies such as tuberculosis, cancer, and asthma. The importance of the individual's hereditary background has been incontestably established.

To realize the importance which genetics has

Human chromosomes. Photomicrograph of the nucleus of a normal cell.

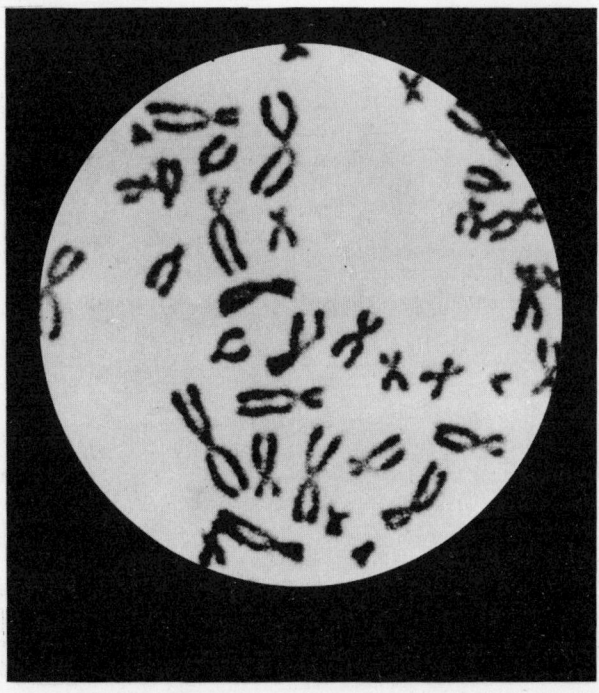

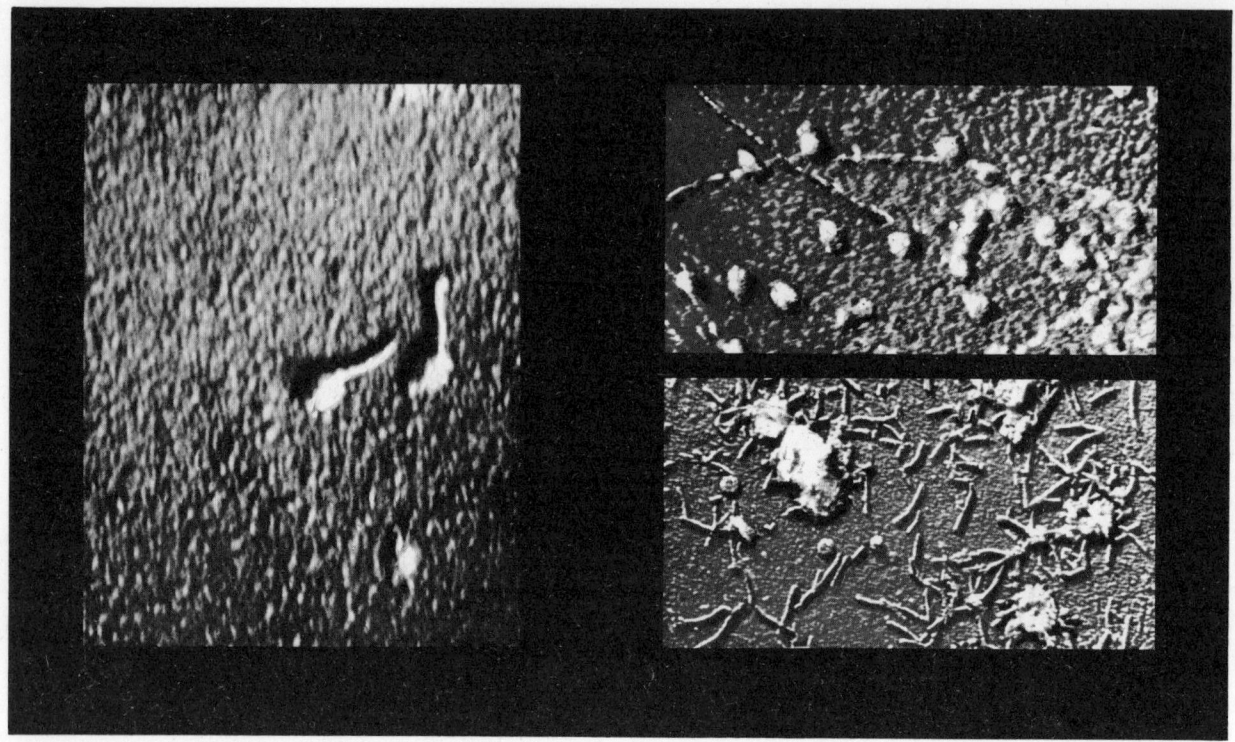

Corpuscles of a vaccine virus, magnified 2,000 times.

Inst. Pasteur.

Top, the virus of influenza seen under the electron microscope. *Above*, bacteriophages magnified 5,500 times by electron microscope.

Inst. Pasteur. Giuntini.

assumed in medicine one has only to glance through the monumental work (876 octavo pages) which Professor A. Touraine has dedicated to *Heredity in Medicine*. A knowledge, at least elementary, of medical genetics is indispensable to the practitioner, and can save him from grave and costly mistakes in diagnosis.

As Touraine points out, the practitioner, faced with an illness which persists, 'must always seek for and evaluate the elements which are due to heredity... To keep its role constantly in mind is often to avoid an error of diagnosis and, in consequence, of prognosis and treatment.'

Genetics also aids the doctor in dealing with anomalies and deformities which are at times of genetic origin but may also be caused by an accident which occurred during the development of the embryo: for example, a disturbance of the hormones or the effect of a virus, such as the virus of German measles. The distinction between genetic deformities and accidental deformities is essential, for the latter cannot be inherited. Nevertheless their study remains within the province of biology which can throw light by means of experimental research on the conditions which produce such accidental abnormalities.

Genetical investigation of blood characteristics has led not only to an explanation of certain grave mishaps which occur during pregnancy, but has found means to remedy them.

'There are good marriages,' said La Rochefoucauld, 'but there are no delightful ones.' To this disillusioned observation, the biologist adds his own warning that there are 'dangerous' marriages. Such are those which unite two individuals whose blood — containing

a substance, an antigen, known as the Rhesus factor — is mutually incompatible.

These marriages can result in children who are born with serious anaemia, children who until recently were fated to die. Since the causes of the illness have been discovered they are no longer so condemned. One successful form of treatment for the newborn baby consists of completely draining away its blood and substituting blood of appropriate type. This is known as an *exsanguino-transfusion*. Throughout the world thousands of babies are annually saved by this method.

Mention must be made of the contribution of the study of genetics to our knowledge of microbes, the variations in their virulence and the variations in their resistance to antiseptics and antibiotics. Once commonly referred to as 'miracle drugs', penicillin, streptomycin, etc. must not be abused, but should be reserved for genuinely serious cases. The reason for this is that every time an antibiotic is employed there is a risk of giving rise to *mutant* strains of bacteria which are resistant to the antibiotic. If these refractory bacteria became too numerous the antibiotic would lose its efficacy. Genetics has, however, enabled us to improve the varieties of fungi which secrete antibiotics and thus endowed us with more powerful weapons in the war against disease.

Finally there is a vast field to which genetics or, more broadly speaking, biology has much to contribute: namely, the study of those sub-microscopic entities known as *viruses*.

The importance of viruses appears to be continually growing: they are undoubtedly responsible for numerous infectious diseases such as German measles, smallpox, chickenpox, poliomyelitis, encephalitis, rabies, and trachoma, and they are suspected of playing their part in many other diseases. These viruses seem, in many respects, to be closely related to hereditary elements, or genes. Indeed, it has even been freed from intra-cellular discipline and subsequently acquired pathogenic powers. This speculation suggests a return to a kind of morbid 'spontaneous generation' which, though very different from the 'spontaneous generation' disposed of by Pasteur, could nevertheless cause havoc with classic doctrine. The disease of cancer is perhaps produced by a virus. In any case it is the task of biology, in the first place, to solve the problems posed by the cancerous cell, endowed with invading and malignant properties which set it apart as being the product of a race of cells unlike all others.

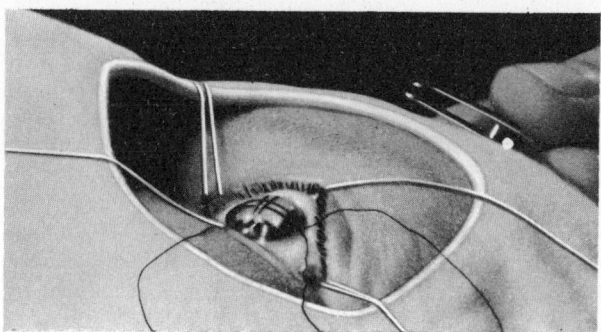

Keratoplasty or graft of the cornea. Final stage of the operation.

L. Rollet.

While we know a great deal about cancer and are able to cure some types of the disease with radiation or drug treatment, biologists and physicians have much research to complete before we fully understand the nature of the disease.

Great strides, however, have been made recently in the grafting of various organs of the body. It was believed by earlier biologists that grafting as a general rule was only possible with tissue removed from the person grafted (autografting) or from a person whose genetic constitution was identical as in the case of an identical twin. It was thought that genetically even a mother and a son were too dissimilar for success in such an operation to be counted on. While to date the failure rate has been high in attempts to transplant organs such as the heart or kidney, much progress has been made. In the simpler matter of cornea grafting in eye surgery the success rate is much higher. In practice it is the most successful kind of operation in which tissue from one individual is successfully transferred to another, and the success of the operation is probably due to the special category of the graft.

It is a 'dead graft', so called because it does not involve cellular survival, the cornea of the donor being reinhabited by the cells of the receiver.

The analysis of what occurs in grafting and a search for methods capable of extending its possibilities are today among the chief preoccupations of biological research. The time is already foreseen when, by ingenious techniques, the obstacle of genetic individuality will be surmounted, with all the practical consequences which such progress will entail.

Top, Penicillin, *Penicillium chrysogenum*, greatly enlarged.

Cultivated moulds used in the preparation of penicillin.

U. S. I. S.

A piece of skin to be grafted is removed by means of Padgett's device, the dermatome.

J. Boyer.

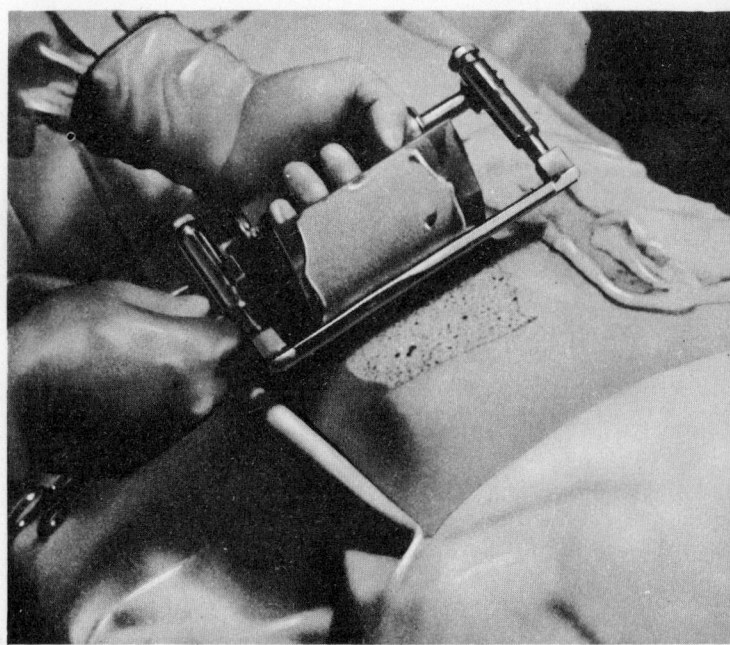

11

The atomic threat

Foremost among the problems of today is the atomic danger caused by nuclear explosions made by leading world powers. The governments of such nations consider such experiments necessary for the preparation of their 'non-conventional' armaments.

Violent debates rage on the subject. Many, including those with authority to speak, have demanded the instant suspension of these experiments. Confusion has been increased by political passions with their attendant array of ignorance, illusions and lies. Meanwhile the insecurity, to say nothing of the anxiety, of the general public, faced by a menace which is all the more disquieting because it remains mysterious, continues to grow.

In order to understand the issue a little more clearly we must again consult the biologist.

Setting aside the thought of a full scale atomic war and the monstrous destruction and massacre it would bring in its train and, for the moment, considering only the present 'armed atomic peace', or, in other words, the bombs which are now being tried out, we must recognize that each such explosion releases into the upper atmosphere radio-active elements which can afterwards fall to earth. Certain of these elements are long-lived; in twenty-eight years, for instance, the radio-activity of strontium-90 is only half destroyed, while the half-life of caesium-137 is thirty-three years. Both contaminate plants and thereby plant-eating animals; hence a minute but certain pollution reaches all our food supply. The doses of radiation which in this way reach our tissues are, for the moment, extremely small, and possibly negligible in comparison with the natural radio-activity of the universe to which we are constantly submitted. But if nuclear explosions continue, the doses will necessarily increase until they become a real danger.

Radiation is capable of causing leukaemia and cancer and also of increasing the frequency of *mutations* or hereditary variations.

It is here that the biologist can clarify the debate. He will point out that the effects of radiation are cumulative, which means there is no dose so feeble that it cannot, operating over a sufficient period of time, exert its power to cause mutations. Furthermore mutations almost always bring about changes which are undesirable, or those which produce the defects, diseases and deficiencies which have already been mentioned.

In brief, every increase in radiation tends to raise the mutation rate, and thus contribute to drawing upon the 'genetic capital' of the species. Here and now the evil produced by nuclear explosions is not quite nil. Here and now it can be stated with certainty that within two or three generations individuals, tainted with deficiencies, will be born because of these explosions.

Even the peaceful exploitation, therefore, of atomic energy is not without its dangers. Until satisfactory means of assuring the elimination of radio-active waste have been found biologists will with reason remain uneasy. A further source of their alarm arises from the excessive employment of radioscopy and radiography.

'At the moment,' said the eminent biologist L' Héritier, 'a source of irradiation much greater than that created by the bombs, and which ought to worry the

The underwater explosion of an atomic bomb of the 'Hiroshima' type at Bikini in 1946.

public even more, is that produced by the too frequent use of X-rays in medicine. Patients must learn not to submit to these rays so lightly.'

X-rays, however, are not the sole agents of genetic deterioration: a number of chemical substances, among which are products used therapeutically, can bring about undesirable mutations.

As Professor Lamy states: 'From now on we must be alert to the dangers to the unborn involved in the use of certain chemicals, administered professionally or medically. It is by no means certain that a patient can be submitted over a course of years to chronic intoxication or to repeated medical injections without harm.' The disastrous results of the recent use of the drug thalidomide on pregnant women underlines the validity of this warning.

The use of biology in the universe

Biology, as we have seen, can be applied with valuable results in the fields of agriculture, of medicine, of bacteriology. It has, in addition, supplied salutary lessons on the vulnerability of human hereditary factors, which are of vital importance to all men and to the future of the human race.

And yet biology is only in its early days. If, setting forth from the results already obtained, we give a little rein to our imagination, either by assuming that what has been achieved in animal biology is equally possible in human biology, or by logically prolonging certain lines of research already engaged on, we can, without falling into the extravagances of science fiction, foresee in the more or less near future the advent of very startling innovations.

Rejuvenation of the individual — or at any rate the prolongation of youth — voluntary sex determination of the unborn child, parthenogenesis or generation without male parent, the suppression of paternal heredity, the artificial preservation of an entire individual, artificially induced twins, human propagation by methods similar to plant-cuttings, pregnancy in a glass jar, transformation of sex, transformation of superior individuals, and so forth would all seem to be biologically possible.

An interesting experiment was conducted in 1957 by Benoit, Leroy, and C. and R. Vendrely. They injected ducks with DNA or desoxyribonucleic acid derived from cellular genes. It was discovered in this way that the hereditary characteristics of a living creature are capable of being altered even when treated after birth.

Peking ducks, from the age of eight days, were given a series of injections taken from another breed of duck, Khaki Campbells. When they reached the adult stage they revealed important modifications. The skin of the bill, instead of being yellowish-orange, like a Peking's, was spotted with black or purplish-black (the Khaki Campbell's bill is greenish-black). The head was of an

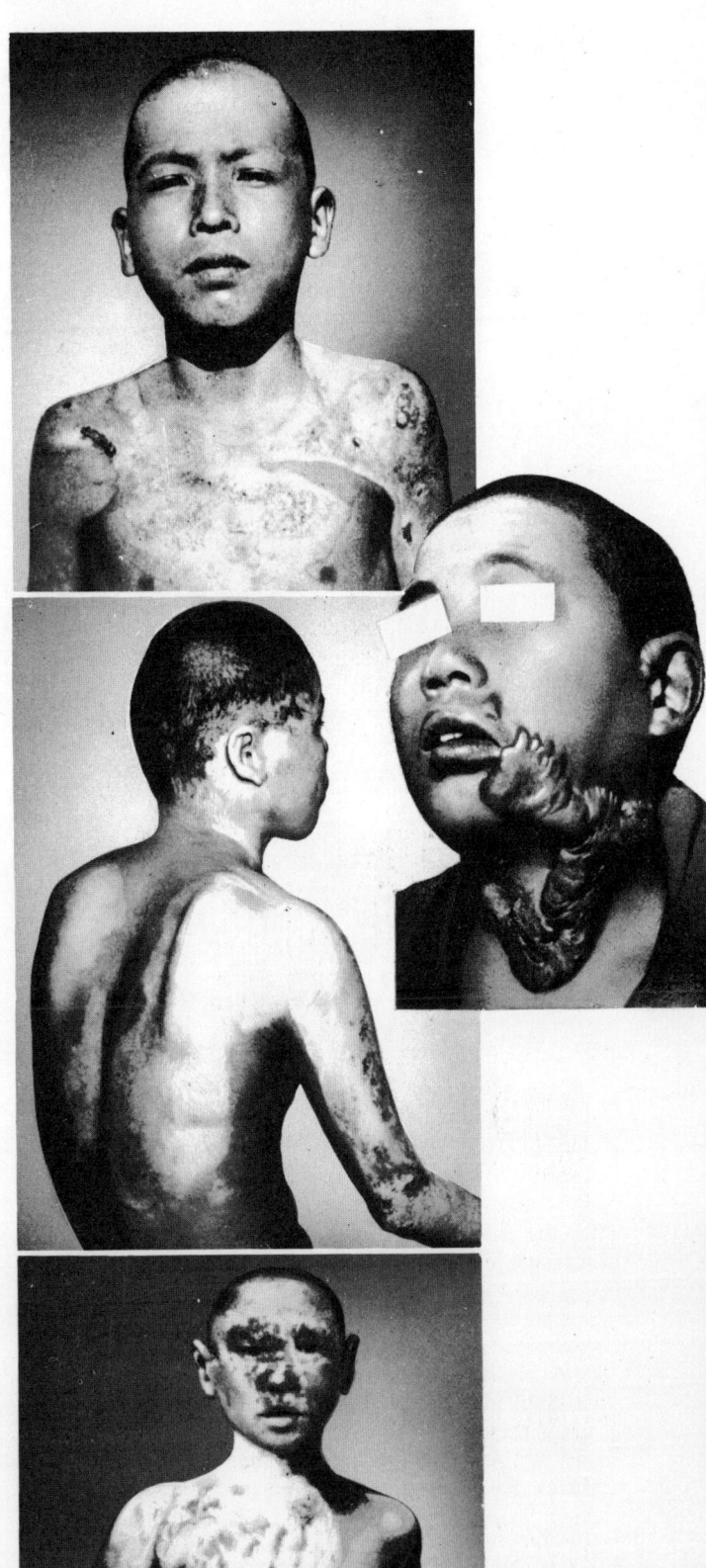

Burns caused by the explosion of the atomic bomb at Hiroshima.

'L'Arme atomique', Lt.-colonel P. Genaud, Dunod.

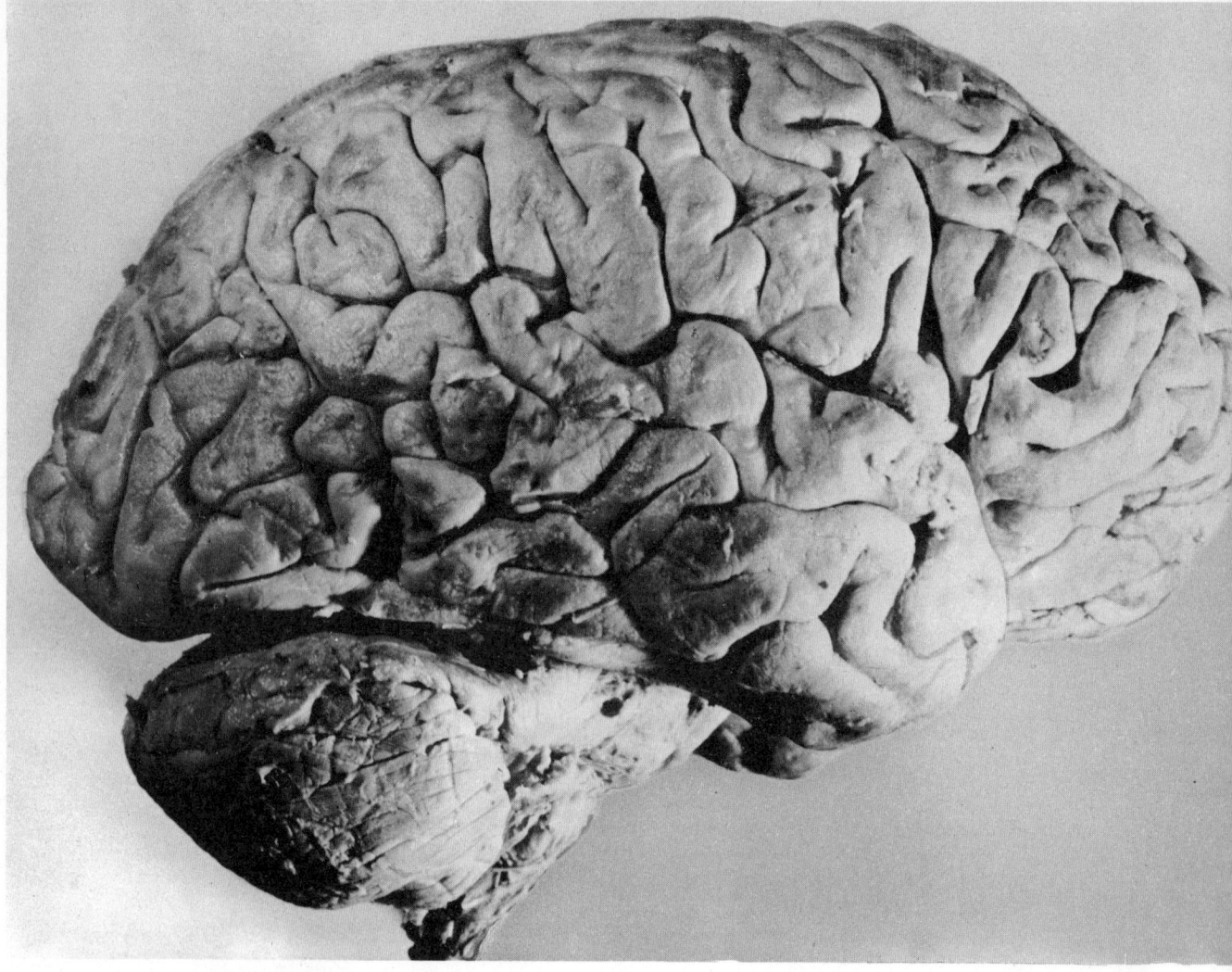

The human brain as seen from the side.

Larousse.

abnormal shape; the plumage, instead of being sulphurous-white, was snow-white and silky (in the case of the Khaki Campbell it is brown). The stature, carriage and gait had all changed.

Thus the injection of foreign DNA had modified the young ducks to such a degree that they no longer resembled either Peking ducks, Khaki Campbell ducks, or hybrids of the two breeds; but instead presented an original mixture of their racial peculiarities. In addition, these new features proved to be inheritable: seventy per cent of the ducklings produced by these ducks have revealed the aberrant characteristics of their progenitors.

These experiments have not unnaturally aroused mingled emotions. One need only imagine their results applied to the human species to feel a certain uneasiness at such a biological triumph. By means of this extraordinary substance DNA, the 'heredity drug', the biologist could conceivably transfer the hereditary traits of one individual to another and, as it were, perform a 'chemical hybridization'. For example, one individual could 'borrow' from another certain characteristics of pigmentation or structure, a facial feature, or even a predisposition to intelligence or artistic

ability. And not only would the individual himself have been thus modified: his offspring would be *different* from those he would have produced without the intervention of science. To the normal human being there are certainly grounds for alarm in such a notion of 'chemical adultery'!

At the moment we do not know if the results in question can be considered definite, but it is highly probable that biology, sooner or later, will succeed in producing such 'controlled' variations of hereditary traits — and thus open the startling perspective of the voluntary modification of mankind by man. Indeed, the 'best of possible worlds' would already seem to be taking shape around us.

It is not difficult to imagine for the future specialized institutions supplying a kind of *Standard DNA*, to be systematically administered to future parents in order to normalize or rectify their progeny. The question indeed arises: is it DNA which will reveal the secret of creation and allow us at will to produce superior beings and geniuses? If the problem of outstanding intellectual superiority is a purely biological one or, in other words, if, as Carrel believes, genius depends essentially on 'a chemical and structural state of the tissues' it is to be supposed that such a state could be artificially produced. Perhaps the moment has come for men of science 'to seek to alter the quality of the

14

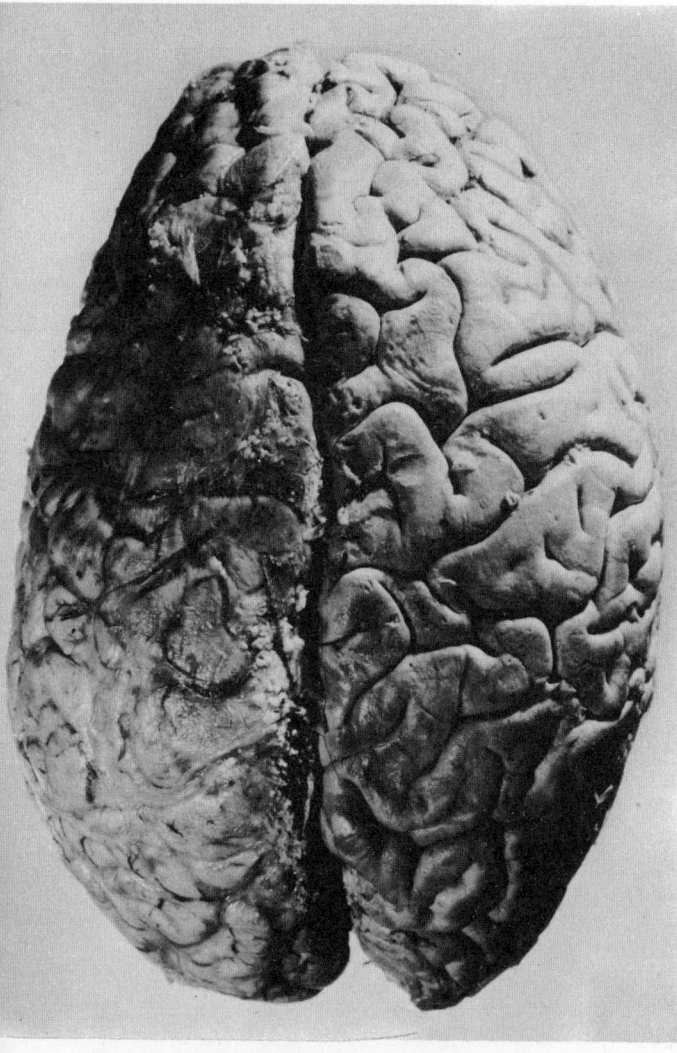

The human brain as seen from above.

Larousse.

cerebral substance and the endocrine glands and thus improve the mind?'

It is already well known that there are 'chemical idiots' due to defective metabolism and, more generally, that the exercise of mental activity depends on chemical reactions which take place in the cells of the cerebral cortex. 'Psycho-chemistry', as Georges Dumas has pointed out, has only just been born and as it matures we may expect it to supply us with 'psychogenes' which will be to our present day 'pep pills' what cortisone is to aspirin. It will perhaps also furnish us with pills which can change our feelings, or even our moral qualities: why not drugs to induce devotion, courage, virtue, brotherhood?

Biological speculation is not confined to the human race as it now is. Biologists contemplate, and not without seriousness, the creation of a super-human being, a sort of metanthropus, who would ascend a new rung of the evolutionary ladder. In theory, the simplest way of doing this would be to bring about structural alterations in the brain. There are several thousand million cells in the human cerebral cortex. If, in the course of this organ's formation, we could induce an additional or supplementary division the result would

be a doubling of the total number of cells. Probably the thought of thus creating a double-brain is somewhat naive.

Certain biologists contend that Man is a retarded, a *backward,* animal in comparison with his ape-like ancestor. In many ways he bears more resemblance to the foetus of the great ape than to the adult great ape. It is true that his most specifically human characteristics, such as the size of his brain compared to the total size of his body, are foetal or infantile characteristics. On such facts the famous theory of foetalization is based, from which the following corollary is derived: if man owes his superior brain to the phenomenon of organic backwardness, to evolve further, to become more humanized or super-humanized, he must accentuate his backwardness and undergo increased foetalization. It has even been suggested that Superman will be the human foetus which is capable of reproducing itself . . . In such an event the problem would be to 'super-foetalize' the human being. This could, presumably, be achieved by means of certain hormones.

The creation of a superman is, perhaps, beyond the realm of possibility, but it is highly significant that contemporary biologists allow themselves to think about it.

Biology and the law

Any serious change which man may bring about in his organic status will necessarily have serious moral, social and even legal consequences. It is important to note that even now the progress of biology has begun to have repercussions in the law courts, either because it can, in certain cases, cooperate in the establishment or clarification of justice, or because new biological techniques have rendered certain laws obsolete.

Thanks to knowledge supplied by genetics, and especially the study of blood-groups, we can to-day establish negative paternity. In other words, though it cannot be proved that a certain child is the son of a certain man, it can sometimes be proved that he is not the son of that man. Blood analysis is being constantly improved and the time is foreseeable when proof of exact paternity can be established.

By the use of the same methods the biologist can also prove that a certain child cannot belong to a certain couple. Newspaper readers will be familiar with dramatic cases of baby substitution which have occurred in maternity hospitals. This is just one instance of the way medical science can assist in social and moral problems. In some spheres it can even lead to actual changes in the legal code. Since artificial insemination allows the sexual act and fertilization to be dissociated, the birth of a child need not necessarily mean that its two progenitors have had physical relations. Jurists therefore agree that it would be desirable to amend laws of affiliation which authorize the husband to repudiate a child when he can prove that he was not with his wife at the time the child was conceived.

There is much controversy at the present time regarding the legal aspects of transplant surgery. Already the technical mastery of replacing a diseased human heart or kidney with a healthy one has been accomplished technically, although complete success depends on overcoming the body's rejection of the foreign organ. The increasing practice of transplant surgery makes it necessary to establish firm legislation to safeguard the individual, to stipulate, for example, the period of time imposed by law before organs and tissues may legally be removed from corpses. At the same time it can be questioned whether the prolongation of a useless or unhappy existence can be justified on either religious or legal grounds.

On this subject of grafting Professor Savatier, an eminent French specialist in medical ethics, wrote: 'The ministry which society gives to the doctor, the only person permitted to mutilate the human body for curative purposes, is given him solely for the benefit of the patient: he has no right to mutilate his patient for the benefit of another's health, even with the patient's consent.'

It has been suggested that the individual is not free to dispose of his own body, and that one of the functions of the law is precisely to protect him from damage to himself of his own volition. Hence the very principle of grafting, which tends to mingle individual bodies and treat living organs and tissue as though they were things, could be considered as basically opposed to the legal conception of the individual.

The jurist Aurel David wrote: 'Removable living organs are objects which circulate on the market like any other commercial commodity. It is extremely difficult to assign a legitimate limit to grafts and to the interchange of 'living' organs, as well as the replacement of organs by mechanical devices such as an artificial heart or kidney. Certainly the latter are merely objects and therefore impersonal and marketable. But what becomes of the individual in the midst of all these exchanges? The real identity to which we affix a proper name seems to disappear in the confusion. Paul's kidney has gone to John, while Paul walks with the aid of John's leg. What has happened to John and Paul themselves as a result of such chopping and changing about?'

Aurel David would also be justified in asking what would become of the individual identity in the event of biology mastering the technique of altering a man at will by injection of D N A.

He sees only one method of overcoming such difficulties, difficulties which will increase with biological and technical progress. He suggests that a sharp distinction be made between the central person and his surrounding body; the latter, according to this thesis, would be only a 'physiological puppet', a robot and, like his other material possessions, the property of the person. For, unless the body and the person are one and the same, the legal principle of the uniqueness and continuity of the individual could be undermined and legal identification destroyed — a situation obviously unacceptable to a lawyer. Without subscribing to the extreme views of Aurel David we must be grateful to him for having laid emphasis on the inevitable conflict between certain applications of biology and the theoretical demands of the law.

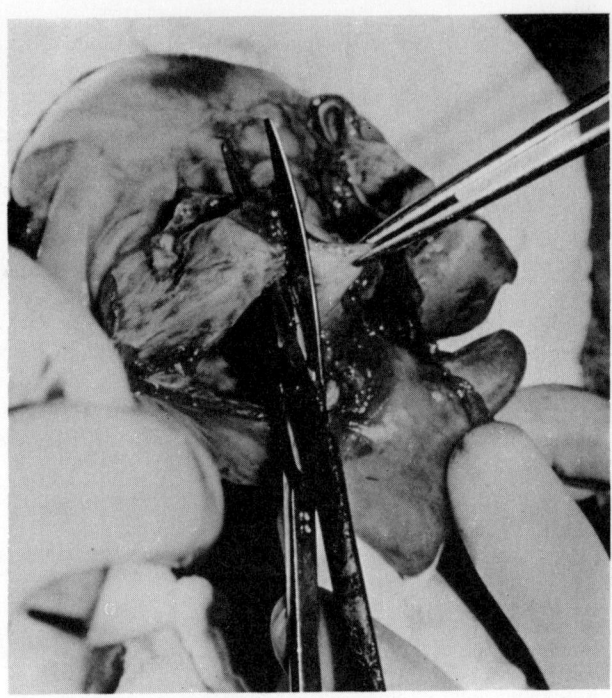

A heart transplant operation being performed in the United States by Dr Arthur Cooley.

P. Popper

The problem of eugenics

As long ago as 1929 Charles Richet saw in human selective breeding the supreme goal of biology. 'If,' he wrote, 'in the biology of the future, only one question remained to be studied this would be it. The

The distinguished physiologist, Charles Richet (1850—1935), an enthusiastic advocate of eugenics.

Gerschel.

greatest duty which awaits biology is the study of human selection.' Indeed, genetics has taught us that a vast number of diseases, blemishes and defects are transmitted to our offspring with mathematical regularity. An individual who suffers from Daltonism or congenital cataract will inevitably transmit this affliction to half of his descendants. Is it, from the point of view of the species, admissible to allow children thus gravely defective to be born? In order to prevent

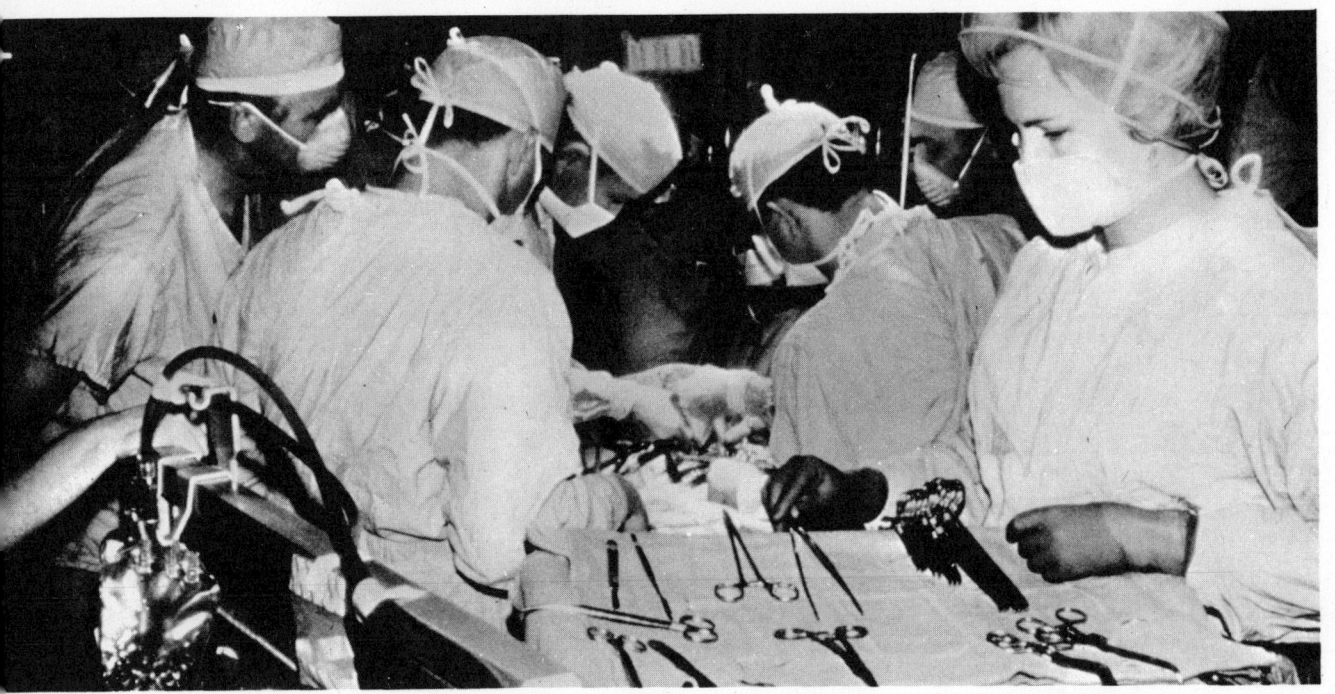

The operating theatre during a heart transplant operation. Here the doctors are waiting by the side of the donor to remove the heart.

P. Popper.

their birth is it, on the other hand, admissible to have recourse to methods which violate the right, until now inalienable, of reproduction? One side of the argument is urged by social interest and by charity towards the unborn invalid; the other side is defended by respect for personal liberty and the dignity of the individual. The progress of biological knowledge thus brings us face to face with a highly delicate choice on which authorities of equal eminence disagree.

In 1939 a *Geneticist's Manifesto* was signed by the most distinguished biologists of America and England — Haldane, Muller, Needham, and others. It resolutely advocated *eugenic selection,* planned and controlled, which would be able not only to prevent the genetic degradation of the human species but also, within a relatively few generations, raise its level.

It would, the manifesto stated, be regarded as an honour and a privilege, if not a duty, for a mother to produce the best possible children, not only the best educated but the best genetically, even though this latter point implied artificial (though always voluntary) control of procreation. But, the statement pointed out, before the people, or the governments which are supposed to represent them, could submit their reproduction to a rational policy a knowledge of biological problems must be widespread and the truth be universally recognized. This truth is that heredity and environment together constitute the dominant and decisive factors of human well-being, factors which can both be controlled by mankind, whose progress would then know no limits.

This immense problem of eugenic selection, which is primarily a social and moral problem, will perhaps arise in a more acute form in a few decades if the genetic patrimony of mankind is dissipated too recklessly as a consequence of military or civil exploitation, or of atomic radiation. In any event, and even if society refuses to introduce a rational control of reproduction, one can only hope for an ever wider diffusion of knowledge about heredity. In this way, presumably a sense of the responsibility of parenthood will develop. Such a sense may already be awakening, for more and more often biologists receive letters from unknown people asking for enlightenment or advice before marrying a blood relation, a person with an hereditary disease, and so on.

The immediate creation of a Bureau of Eugenic Information would appear to be necessary. It is vitally important that the universal dissemination and scientific development of more and more efficient methods of birth control is achieved. Methods both positive and negative can be employed at all stages of the reproductive processes: voluntary permanent or temporary sterilization, contraception, abortion, control of fertilization and of the sexual cycle.

Contraceptives have posed serious social and moral problems. For example, in July 1957 it was announced that G. Pincus, the great biologist who had already succeeded in breeding unsired rabbits had, with his colleagues, perfected a contraceptive which was both effective and simple, since it could be administered in the form of pills. These pills, it seemed, were soon to be freely sold throughout the United States, but in many parts of the world they could be procured only by doctor's prescription.

Connected with the question of contraception is that of medical abortion. To take a simple example, German measles is normally a harmless disease, but when it

17

is contracted by a mother during the first weeks of pregnancy it very often results in the malformation of the foetus. Should we, then, allow the child to come into the world, knowing that it will almost certainly be born with some deformity? Again there are two opposed opinions which can both advance powerful arguments, founded on moral precepts, which are equally worthy of respect.

Biology and religion

There have been great battles between biology and religion in the past, especially in the days when the theory of evolution could be attacked by Monseigneur Dupanloup in such terms as these:

'Fathers of families demand assurance against this ignominious science which today attempts to substitute the superior Ape for the sublime word by which the Bible teaches the child the sublime origin of Man.' Numerous and distinguished ecclesiastics such as Teilhard de Chardin, the Abbé Breuil and many others have since stoutly defended the theory of evolution and accepted most of the theoretic conclusions of contemporary biology. But it has not always been easy to reconcile the teachings of biology with those of revelation. In reality, apart perhaps from certain questions of secondary importance, such as that of human monogenesis and polygenesis (the origin of life from one or many cells, there is no necessary clash between science and religion, which move in independent spheres. In fact some men of religion subscribe not only to the theory of organic evolution, but also to the positivist arguments which science advances to explain it.

On the other hand most clerics are extremely reserved in their attitude towards the action of Man on mankind, gravely condemning certain experiments on human beings. Secular opinion often agrees with them in censuring certain technical interventions which, though they might appear physically advantageous to men, nevertheless seem *unlawful* to Man as a moral creature.

J. B. Montini wrote: 'Recent discoveries in biology and psychology are the basis of future technical study of the problems of health and like all scientific progress, merit the highest esteem. But they must not be applied to human beings without discernment or without weighing the superior demands of natural and Christian morality.'

No doubt nature can be given a helping hand; organs and functions can be restored. But according to this view one must, in doing so, always respect the laws of 'human nature'. It is respect for such laws which has given rise to much criticism on religious grounds of birth control and the sterilization of the unfit. Indeed, all interference with the seed of man has been prohibited by many and artificial insemination absolutely condemned. The argument forwarded is that to be legitimate all procreation between man and wife must be the fruit of physical union, accomplished according to the laws of nature. (See, for example, the Papal allocution of the twenty-ninth of September

Because of serious overpopulation in India contraception is being encouraged and men are paid to undergo sterilization.

David Channer. Camera Press.

1949 to members of the Sixth International Congress of Catholic Medicine). Artificial insemination by means of a 'donor' has been proclaimed to be a very much graver fault, for husband and wife have undertaken the exclusive duty of mutual parenthood: 'The ova of the wife can only be fertilized by the sperm of the husband, and the latter must fertilize only the ova of the wife.' The beliefs are not held by all religious faiths, but the morality of new biological practices has yet to be established.

The Encyclical letter 'Humanae Vitae' of Pope Paul VI in 1968 condemning the practice of artificial birth control again raised a conflict between medical science and religious belief:

' .... new life is not the result of each and every act of sexual intercourse. God has wisely ordered the laws of nature and the incidence of fertility in such a way that successive births are already naturally spaced through the inherent operation of these laws. The Church, nevertheless, in urging men to the observance of the precepts of the natural law, which it interprets by its constant doctrine, teaches as absolutely required that any use whatever of marriage must retain its natural potential to procreate human life.'

18

Biology and ethics

Identical twins, in other words those who have exactly the same hereditary patrimony, bear a striking resemblance to each other not only physically but mentally. This fact in itself serves to indicate that biology can contribute to the understanding of psychology.

Though it may seem excessive to think, as Darlington suggested, that all the elements of intelligence and character are imprinted in advance on the germ-cell from which the individual derives, there can be no doubt that the biological diversity of human beings is partly and perhaps very largely responsible for their intellectual and emotional development.

However the concept of *biological determinism* must always be borne in mind when attempting to judge human conduct or ascertain the degree of the individual's responsibility for his behaviour.

'For the noblest aspirations of the soul to vanish it is sufficient,' wrote Carrel, 'that the blood plasma be deprived of certain chemical substances. When the thyroid gland, for example, ceases to secrete thyroxin into the bloodstream there is no longer either intelligence, sense of evil, sense of beauty or religious sense.'

'Furthermore,' he points out, 'the pituitary, the thyroid, the sexual glands, the suprarenal gland, make possible love, hatred, enthusiasm and faith. The Christian virtues are more difficult to practise when our endocrine glands are deficient.'

Hence there is a possibility of curing vicious conduct biologically — the *therapeutics of perversity* that Carrel foresaw. This is in fact being practised in many of todays prisons and hospitals. 'Science', he wrote, 'is capable of giving man the means of acquiring a disposition to do good and to avoid evil.'

We already know, for instance, the powerful effects of the 'fertility drug', and that among animals a simple lack of manganese can abolish the maternal instinct and, furthermore, that the instinct can be awakened by the action of certain hormones such as progesterone. We are also aware that a lack of male hormones can extinguish courage to such a point that a fighting cock, so deprived, becomes that timorous fowl, the capon.

Many biologists and philosophers have, in the footsteps of Herbert Spencer, tried to draw moral lessons from biology. According to Bergson all moral pressure or aspiration is in essence biological, provided one uses the word biology in its very broadest sense.

Carrel, emphasizing the concept of the 'sin against life', states that the morals of biology are far more strict than those of the Ten Commandments.

'Good,' he writes, 'consists of that which conforms with the basic tendencies of our nature: in other words, of things, thoughts, sentiments and acts which tend to preserve life, to propagate the race, to promote the mental ascent of the individual and of the spirit. Evil, on the other hand, is that which is opposed to life, to the multiplication of life and to its spiritual advance. In reality the supreme good is identical with the success of life itself under its specifically human aspect.'

Gas-powered limbs help this four-year-old child to lead a normal and independent life.

John Drysdale. Camera Press.

G. G. Simpson attempted to found an ethical system on man's sense of evolution. Man is the animal which possesses by far the most knowledge and thinks the most. He is, therefore, the 'responsible animal'. Hence he must strive to develop his aptitudes as a superior animal and rise still higher, primarily by enlarging his knowledge, not only through the acquisition of new truths, but by communicating them to other men. He must, moreover, accept his responsibilities in all circumstances, and loyally, courageously, assume all their consequences. Thus, according to Simpson, biology implies a moral system based on knowledge and individual responsibility.

As for Paul Chauchard, he insists on the essential unity of the human species, the genetic equality of the different human races and the historic fraternity of all men. From this he derives a system of ethics founded on emancipation and solidarity: 'In a healthy ethical system based on biology, which has no illusions about human nature but knows that man is a work still to be accomplished, evil is all that which in me and in all men opposes freedom and the rising tide of consciousness; that which drags us back towards the unconscious, the automatic, the animal; that which separates us

The philosopher Henri Bergson in 1939 at the age of 85.
X Photography.

from others and makes us the privileged, persons apart.'

Marcel Boll has sharply criticized the very principle of a moral system founded on biology, and it is obvious that such a system could never claim absolute or categoric validity. It can tell us in what direction man must proceed if he wishes to continue his evolution and pass beyond the stage he has reached. But the individual still remains free to refuse, if he wishes, to cooperate in the ascent of his species, and to ignore the rules which encourage the ascent.

Biology and politics

In as far as biology is the concern of ethics it is of necessity the concern of politics, using also the word 'politics' in its widest sense. Although biological moralists do not always arrive at the same conclusions as to the ideal social structure they do, to a remarkable extent, agree that such a structure should be broadly based on equality and respect for the individual.

'There will always,' said Alexis Carrel, 'be biological inequalities of stature, sex, vitality, intelligence, aptitudes. In an organic community, individuals are, like the organs of the body, unequal in structure and potentiality, but equal in that all alike are needed for the proper functioning of the whole. Thus social classes are abolished. If biological classes cannot be abolished we can at least provide everyone with the possibility of improving his life and of raising his spiritual level, both mentally and morally.'

P. Chauchard is even more equalitarian: 'All men's equality of rights, then, is founded on a factual equality, and every man must have the same possibility of inheriting mankind's best qualities.'

Simpson, while admitting that democracy is bad in many of its current aspects, sees in it the only political ideology which can produce a morally good society according to the findings of biology, since it can prevent the selfish exploitation of certain individuals by others and, at the same time, the intolerable regimentation of us all.

It may be remembered that in about 1950 a wordy battle broke out between Western biologists and Russian biologists, disciples of Michurin. Among other points under dispute was that of the inheritance of acquired characters.

The Soviet biologists sustained with extreme passion the thesis that acquired characters were transmissible. It is obvious that this purely biological question, which can only be solved by experiments conducted and interpreted without preconceived ideas, is of far-reaching importance to the sociologist or the politician whose attitude will naturally differ according to whether he believes that hereditary traits are fixed and permanent or can be changed by external conditions.

If acquired characters were heritable we could, of course, hope that the improvement of social conditions would in the long run bring about the physical and mental improvement of the human race. If, on the contrary, acquired characters are not transmissible, we must renounce this 'sociological optimism' and, although continuing to do our best to improve social conditions, realize that we have nothing to expect as far as the genetic progress of the species is concerned. Such progress could only be achieved by selective breeding of our more highly gifted individuals, and would require that voluntary control of reproduction advocated by eugenics, a control which is not widely practised in our civilized societies.

It is also conceivable that in the future biology will supply us with the means of directly modifying our hereditary factors or genes, for example by use of D N A, which has already been mentioned. If that day arrives there will be no further need for selection: humanity will be master of its genetic destiny, with all the hopes and fears which that implies.

Biology and everyday life

Without falling into the error of making of biology 'the supreme and universal science, which embodies our only possible approach to knowledge', we must, at least, agree that the science of life has contributions to make in every sphere of human activity and that it throws its own special light on all intellectual problems.

It helps us to comprehend the position we occupy in the universe; it gives us information about the differentiation of individuals, of sexes, of races. According to Darlington it can renovate the study of history, and its advice should be sought even in our choice of a mate.

What appears indubitable is that the farther we progress the more need we shall feel to consult the findings of biology, and the more frequently they will influence our opinions, our judgments and our decisions. There is no human problem which has not its biological aspect, and it will become more and more necessary for everyone to 'think in biological terms'. In the words of C. Darlington: 'At the present moment, in every country in the world, men of learning who pride themselves on instructing their contemporaries, are completely ignorant of the great problems which surround the mystery of life, a subject on which their ancestors professed contradictory opinions. They are totally unaware that the great questions of heredity,

A technician at the National Cancer Institute feeds cancerous tissue cultures for research study.

Camera Press.

Nepalese students on a training course before being assigned to dispensaries that are opening all over Nepal.

P. Popper.

of development or of evolution, can influence the conduct of public affairs, education, the preservation of health or the administration of justice. The moment has come, we believe, to reform this state of things . . . That which we can do is to peer closely at the scientific data, at the facts, the methods and concepts which science reveals. They have great bearing on the mystery of life. And, when we have studied and understood the facts, it will be up to us to use them for our own salvation — before it is too late.'

Biology and education

In spite of the growing importance which biology has assumed, the man in the street could be better informed on the subject. Gradually however, improvements in education and more widespread knowledge about the human body — the importance of factors such as diet, sleep, exercise and fresh air — is making the general public aware of its importance. Accessible to all, demanding no special mental aptitude to be understood, able to supply answers to questions everyone asks, biology ought to be considered as a subject essential to human culture.

Biology, which strengthens our sense of observation and ability to make comparisons, which develops our subtlety of mind, which puts us into direct contact with visible and concrete reality, and can thus act as a corrective to the cold abstractions of mathematics; biology which, as Henri Doffin pointed out, must, in tightening the bonds between animate nature and academic learning, re-animate our knowledge of the world; biology which, in Boris Rybak's words, is in part, 'inductive and psycho-sensory'; biology must occupy a privileged position in contemporary humanism — that of an intellectual discipline which is both supple and too alive to be reduced to dry figures and diagrams.

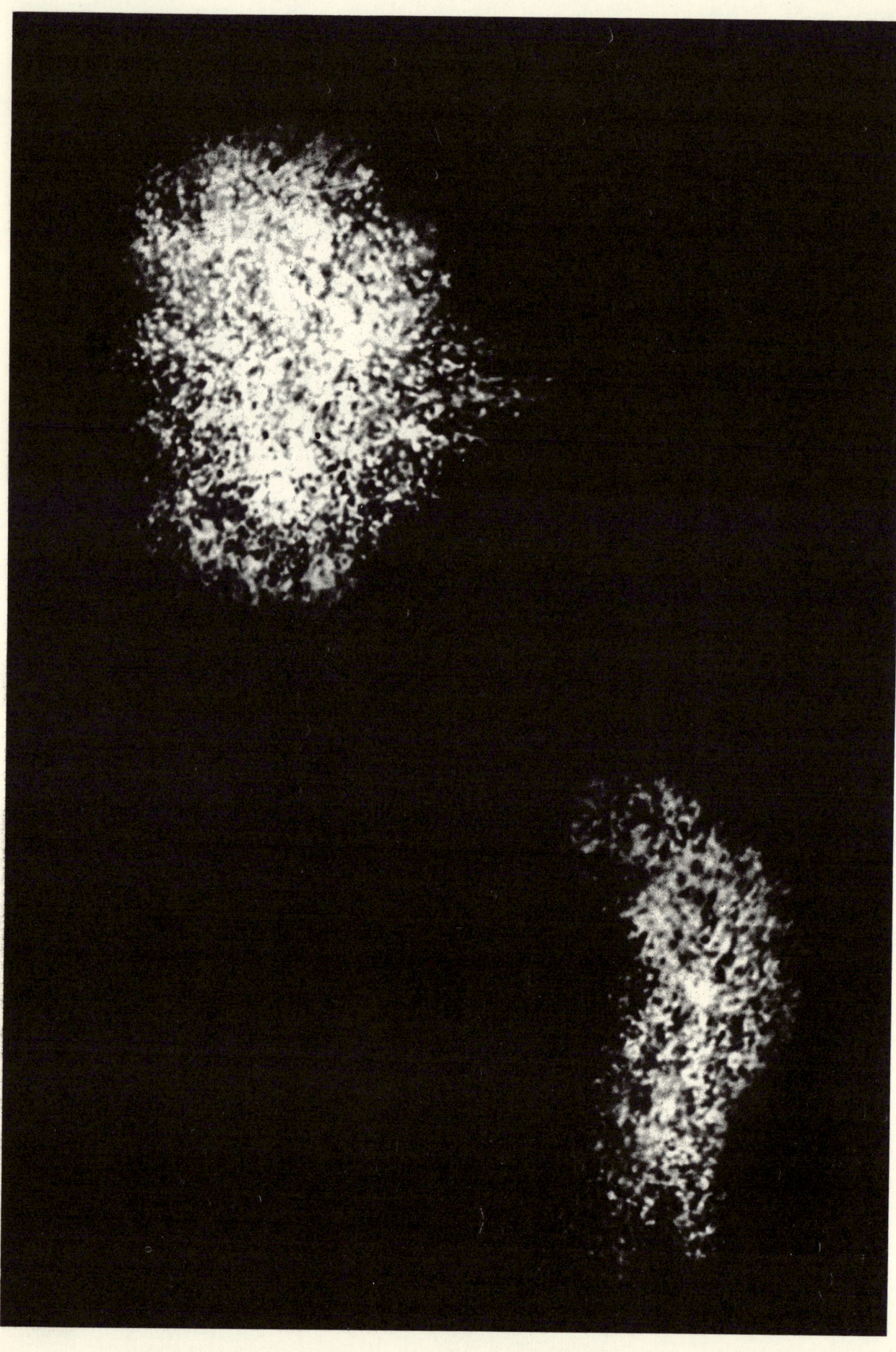

The Cell

Research on the Biological Unit

It is generally agreed that what is called life begins with the cell, a collection of matter which is already highly complex both structurally and in the nature of the elements which compose it. If the atom is the natural unit of inanimate matter, the cell by analogy may be called the natural unit of living matter. It constitutes the vital element which is present in every organism, whether plant or animal. The minute yeast grain, the blade of grass, the age-old tree, the amoeba, the butterfly, are all composed of one or of many simple or specialized cells.

The idea of the cell was not understood until about 1839 when, after much work by earlier investigators, two German biologists, Schleiden and Schwann, established the cell theory. The cell, from the Latin *cellula* (little room), was discovered in 1665 by the English botanist, Robert Hooke. Examining a thin lice of cork under the microscope, Hooke observed ts honeycombed or porous structure. He found similar alveoli in other plant tissue and gave them the name 'cells'. What Hooke had seen of the cell was mainly its walls of thickened cellulose, and he did not grasp its true nature and significance.

Malpighi in 1672 and Grew in the same year, without using the word cell, mentioned that certain parts of plants were made up of minute elementary organisms — utricles, sacs, and vesicles. During the eighteenth century and the beginning of the nineteenth similar ideas recur in the 'fibres' of Haller (1757) and of

Living *Amoeba proteus* (printed from the negative).

H. Mugard.

Bonnet, in the 'cylinders' of Fontana (1781), in the 'utricles' of Brisseau-Mirbel, and in the elementary 'vesicles' of Oken.

In 1824 Dutrochet recognized the individuality of cells, and wrote: 'Undoubtedly the organs of animals are formed by agglomerations of utricles which themselves contain smaller utricles, as in the case of plant cells or utricles. These observations leave no doubt about the utricular nature of the globules which compose the tissues of most animals. We thus see that nature possesses a uniform plan for the inner structure of organic beings, whether animal or vegetable.'

At about the same time Schleiden and Schwann recognized the cell as the basic unit of all living things. They built up the cell theory and profoundly influenced their contemporaries.

Schleiden (1804—81) was a botanist and professor at the University of Jena. In 1838 he published a thesis in which he advanced certain ideas about the structure of plants which he compared to 'polyparies of cells': he insisted on the independent life of the cells which compose the organism, and on the role played by the nucleus.

Schwann (1810—82) was a zoologist and professor at the Universities of Louvain and Liège. A close friend of Schleiden, Schwann himself recounted how, during a chance conversation, the two were led to recognize the connection between plant and animal structure. In the following year, 1839, Schwann published his famous monograph on the analogy between plant and animal structure: the cell, the elementary unit of life, was, he wrote, the point of departure of individual development. 'The elaboration of the proposition that there exists a general principle for the production of all organic bodies, and that this principle is the formation of cells, together with the conclusions which may be drawn from the proposition, comprises the cell theory'. Schwann himself thus proposed the expression 'cell theory'.

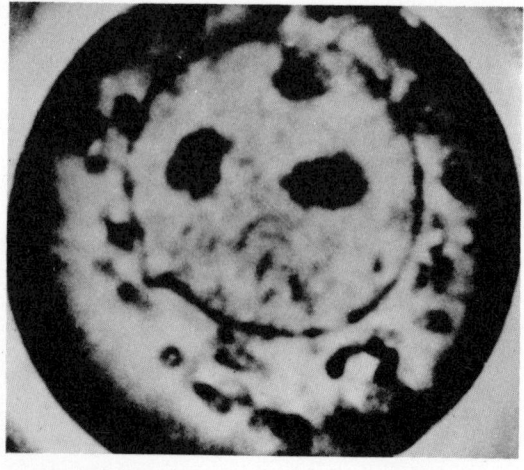

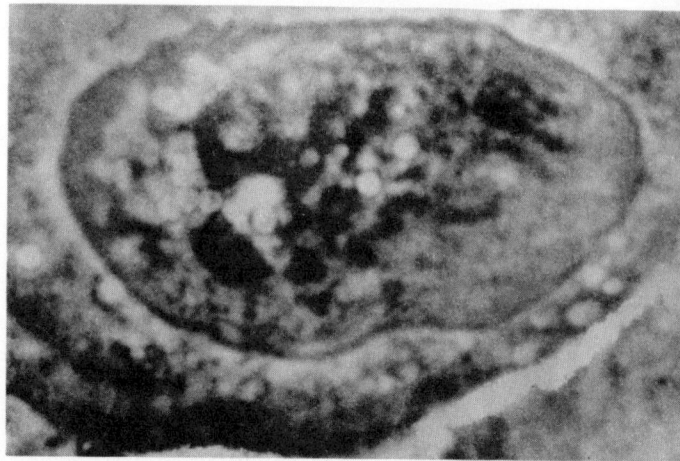

Red blood-cell. Photograph taken in the Paris Blood Transfusion Centre.

Bessis.

Living cell of the caudal epidermis of *Pleurodeles waltl*, photographed under a phase-difference microscope.

H. Mugard.

Epidermal cells of an onion bulb showing the cytoplasm, the nucleus, and the membrane.

L. Plouvier.

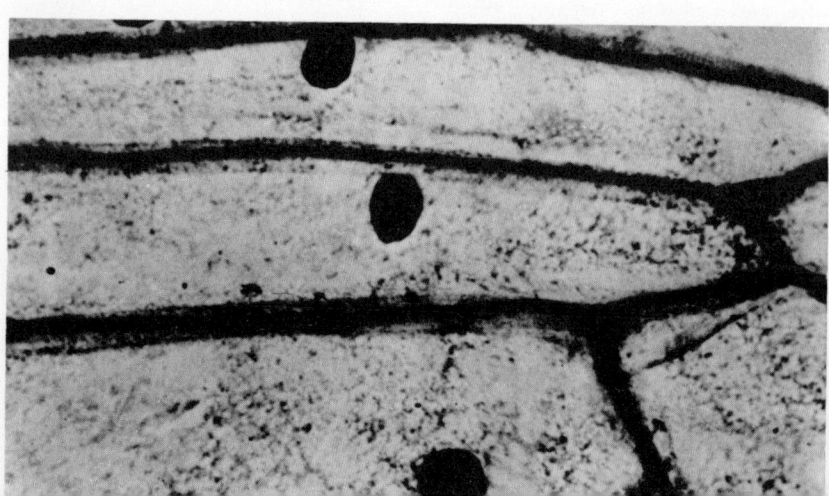

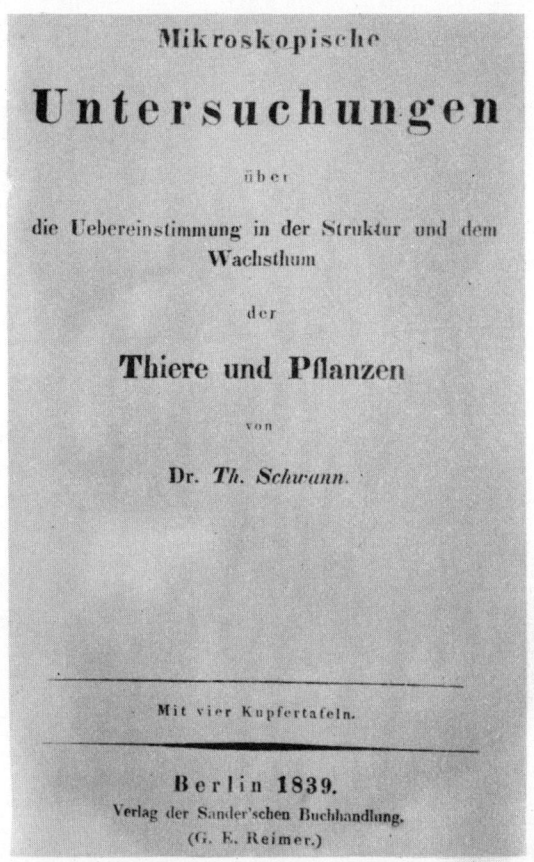

Above left, the zoologist, Théodore Schwann (1810—1882).

Library of the Muséum national d'histoire naturelle. Larousse.

Above right, the botanist, Mathias Jakob Schleiden (1804—1881).

Library of the Muséum nationale d'histoi e naturelle. Larousse.

Title page of the *Microscopic Research on the Resemblances of Structure and Growth in Plants and Animals*, 1839, by Théodore Schwann.

Library of the Muséum national d'histoire naturelle. Larousse.

Title page of the *Elements of Botanical Science*, 1842, by Dr. M. J. Schleiden.

Library of the Muséum nationale d'histoire naturelle. Larousse.

Title page of Robert Hooke's *Micrographia, or some physiological descriptions of minute bodies*, 1665.

Palais de la Découverte. Larousse.

The cell theory enjoyed immense success and was at once adopted by zoologists and botanists. It explained the then known facts and supplied a vast synthesis which was intellectually satisfying. For all that, it was not exempt from error. Schwann, for instance, believed that cells could be spontaneously generated by a process analogous to crystal formation. In the following thirty years the errors in the theory were corrected by the work of cytologists of many nationalities: Remak, Virchow, Henle, Purkinje, Von Mohl, Max Schultze, Ranvier, Nageli. The repercussions of the cell theory were great in every branch of biology: pathology, the study of generation and reproduction, embryology, evolutionary theory.

Unity in diversity

The cell, though microscopic, exists in a great variety of shapes and sizes. A cell can be spherical, oval, crescent-shaped, rod-like, polyhedral, parallelepipedic, globular, cylindrical, spindle-shaped, star-shaped or branched. Certain cells, such as the amoeba, are globules of no fixed form, changing shape as they move. In general the form of animal cells is more supple and varied than that of plant cells which are enclosed in a rigid, cellulose membrane. Cells vary in size even more than in shape. Animal cells measure on an average some ten thousandths of a millimetre — or 10 microns. The smallest, found in certain protozoa, are only half this size while human red corpuscles measure 7—8 microns. The human ovum with its 200 microns

is among the largest cells, while the ovum of the hen — 2 centimetres — is an example of a giant animal cell.

The size of plant cells, generally greater than those of animals, is in the order of 20—50 microns. But many yeast cells measure only 5—6 microns. The male gametes of one of the Cycadales (primitive seed plants of the order of Gymnospermae) reach 500 microns, while the largest are the fibre cells of ramie or China-grass which measure more than 20 centimetres.

The general organization of the cell has been known since the second half of the nineteenth century. It is the subject matter of a special science of fundamental importance to cytology, a study in which revolutionary changes are currently taking place. Improvements in microscopic technique and the perfection of new and more powerful instruments have allowed us to examine the cell more and more closely. Quite recently the infrastructure of cellular elements, previously unknown, has been discovered. For this reason cytology has again become descriptive. New findings in this domain

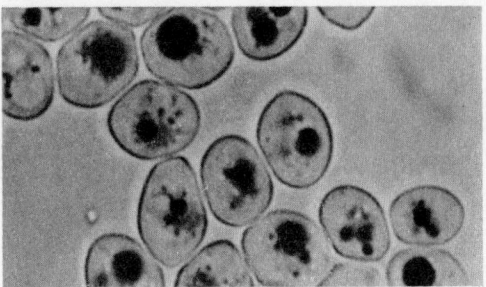

Yeast: each cell contains one or more rounded vacuoles.

R. J. Gautheret.

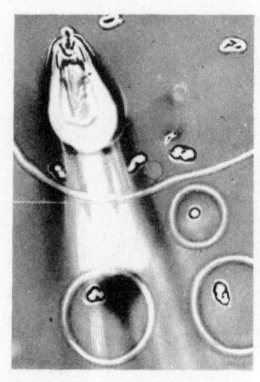
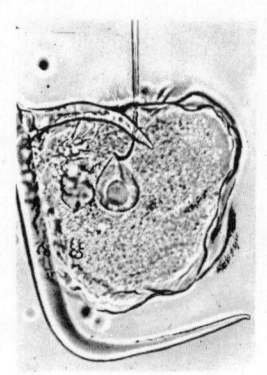

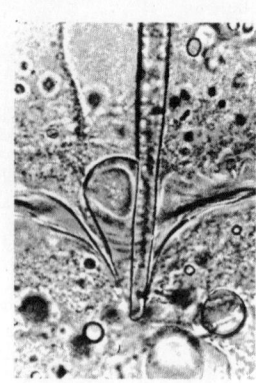
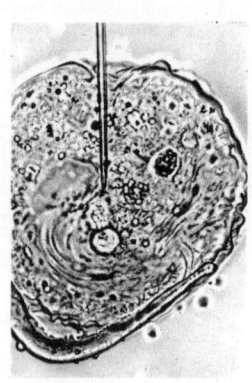

Top left, micro-pipette (diameter 2 microns) used to separate microbes. In the photograph two microbes have already been removed from a drop of culture, while a third is about to be sucked in by the micro-instrument.
Top right, an amoeba is immobilized with the aid of a glass fork: the nucleus is then extracted with a minute hook. The overall diameter is 35 microns.

Above left, an amoeba, *Amoeba sphoeronucleus,* is grafted with a nucleus. The two amoebas are kept in contact; then, with a glass splinter, the nucleus of one is pushed towards the other into which it will penetrate by a break in the distended cellular surface.
Above right, micro-injection of colchicine (the alkaloid of autumn crocus) in the region of the nucleus.

Yves Durand.

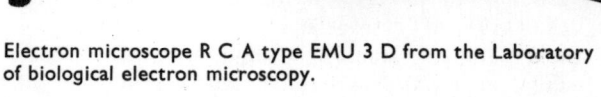

Electron microscope R C A type EMU 3 D from the Laboratory of biological electron microscopy.

Larousse.

continue at an accelerated rate, and they are of capital importance in the comprehension of cellular mechanism.

Methods of studying the cell

A cell can be observed either alive or after it has been killed. Transparent cells in particular can be conveniently examined alive: colouring matter which does not alter the cell's structure permits the chief elements which constitute it to be identified.

But as a rule a living cell observed under an ordinary microscope does not reveal very much. The minute entities which compose it vary too little in their index of refraction to be easily distinguished. For this reason the examination of living cells has been largely abandoned.

The dead cell, on the other hand, can be studied in a variety of ways. To the classical method of cutting and staining sections for microscopical examination new techniques have been added: notably cytochemical, cytophysical and histochemical techniques, and micro-dissection. Cytochemical methods employ special reagents which enable the chemical nature of the cell to be identified. Radioactive isotopes serve to mark particular molecules which can then be followed in their courses through the cells of the organism. Cytophysical methods provide clues to the structure of cellular elements: absorption spectroscopy, X-ray spectrography, ultrafiltration, the ultra-centrifuge, electrophoresis and paper chromatography, have all become current practice. Histochemical methods aim at revealing the chemical 'image' of the tissues by means of quantitative analysis, colour reactions, etc. By such means it has been possible to discover, for example, which cells in the gastric glands secrete the hydrochloric acid, or to estimate the degree of

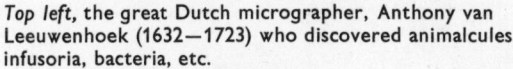

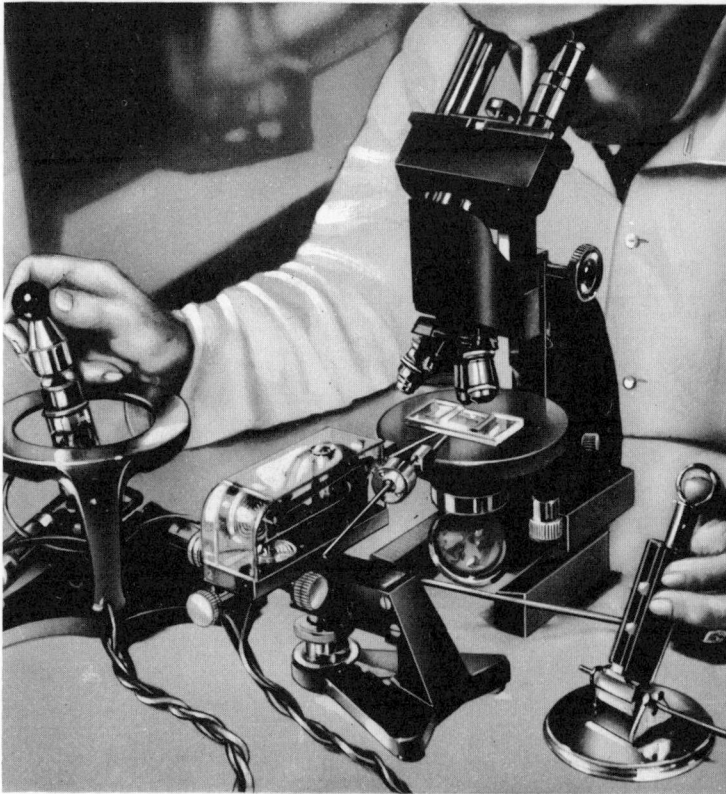

The micromanipulator of Fonbrune. The micro-instruments, contained in an oil chamber under the microscope, are manipulated by the right hand while the left hand controls the micro-injection pump.

Yves Durand.

Top left, the great Dutch micrographer, Anthony van Leeuwenhoek (1632—1723) who discovered animalcules, infusoria, bacteria, etc.

Top centre, Leeuwenhoek's microscope. A lens is mounted between two thin copper plates. It faces a detachable stand to which a thin sheet of mica bearing the object is affixed. The object to be viewed is then adjusted by a screw.

Larousse.

Above, the microscope presented to Buffon by his pupils in 1758. The optical tube is attached to a grooved scale which slides along a similar scale and is adjusted by a rack and pinion.

Nachet collection. Larousse.

The microforge of Fonbrune with which micro-instruments are made from splinters of glass which are fused, drawn out, and broken to the requisite size.

Yves Durand.

27

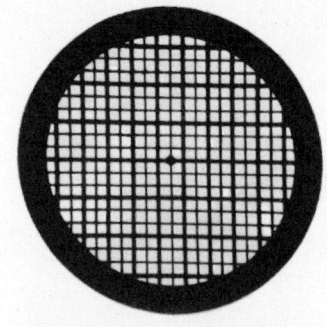

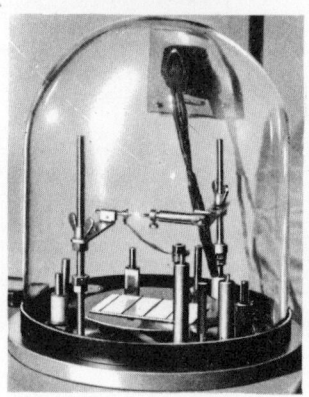

respiratory exchange or the production of lactic acid.

Microscopy, too, has made extraordinary progress. Here we shall consider only two remarkable instruments: the electron microscope and the contrast phase microscope. The electron microscope provides magnifications of from 50,000 to 500,000 diameters, which enable us to see the most minute structure of living matter and even the relationship among cells themselves. The electron microscope differs from the ordinary light microscope in the nature of the rays it utilizes. Ordinary light waves are replaced by beams of electrons projected at great speed. In addition to the source of the electrons, which is a cathode ray tube, the electron microscope comprises a *lens* which concentrates the beam on the specimen, an objective lens, and an eyepiece. The lens is not of glass but of mild steel surrounding a magnetizing coil called a magnetic lens. Sometimes the electrons, being charged particles, are refracted by an electrostatic field.

The high speed of the electrons, which is indispensable, is considerably reduced when the electrons pass through the specimen. If this is more than 0.2 thousandth of a millimetre in thickness, they are unable to penetrate it. Hence a first necessity is to obtain specimens sufficiently thin to be successfully examined.

Certain cells can be flattened so that they are less than 100 millimicrons in thickness. Other methods of procedure rely on the partial absorption of the cellular contents or on the destruction of the cells. But in numerous cases the only possible technique consists of cutting ultra-thin sections which are from 100 to less than 500 Å thick. (Å is the angstrom, the unit for expressing wave-lengths of light; it is equal to one ten-thousandth of a micron, or 1×10^{-8} cm.) The thickness normally employed varies between 150 and 300 angstroms. To obtain such thin sections or slices requires a special technique of inclusion and the use of special cutting machines or microtomes.

Far left, microtome for cutting ultra-thin sections for the electron microscope; *top left,* grille supporting sections to be examined under the electron microscope; *above left,* evaporator for metallic plating and coating in electron microscopy; *above right,* the knife of the microtome, a sliver of bevel-edged glass, and the block which will be converted into ultra-thin sections.

Larousse.

The phase-difference or contrasting phase microscope.

Nachet collection. Bazaine-Publicité.

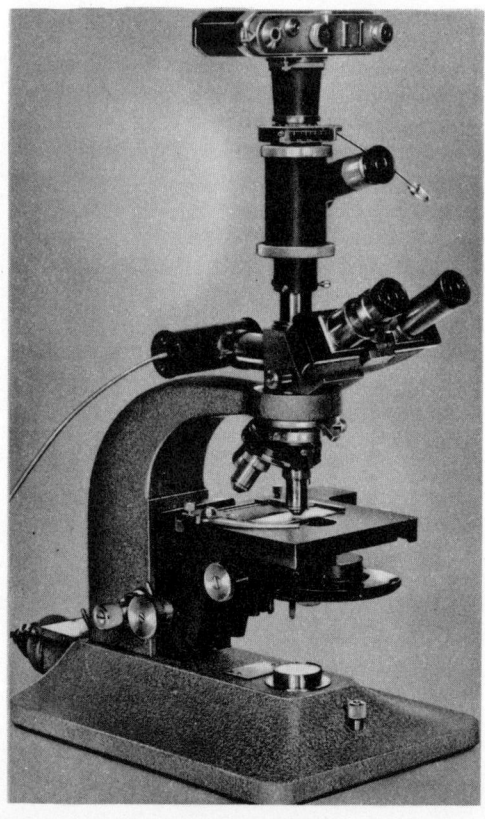

Formulae of the twenty amino acids.

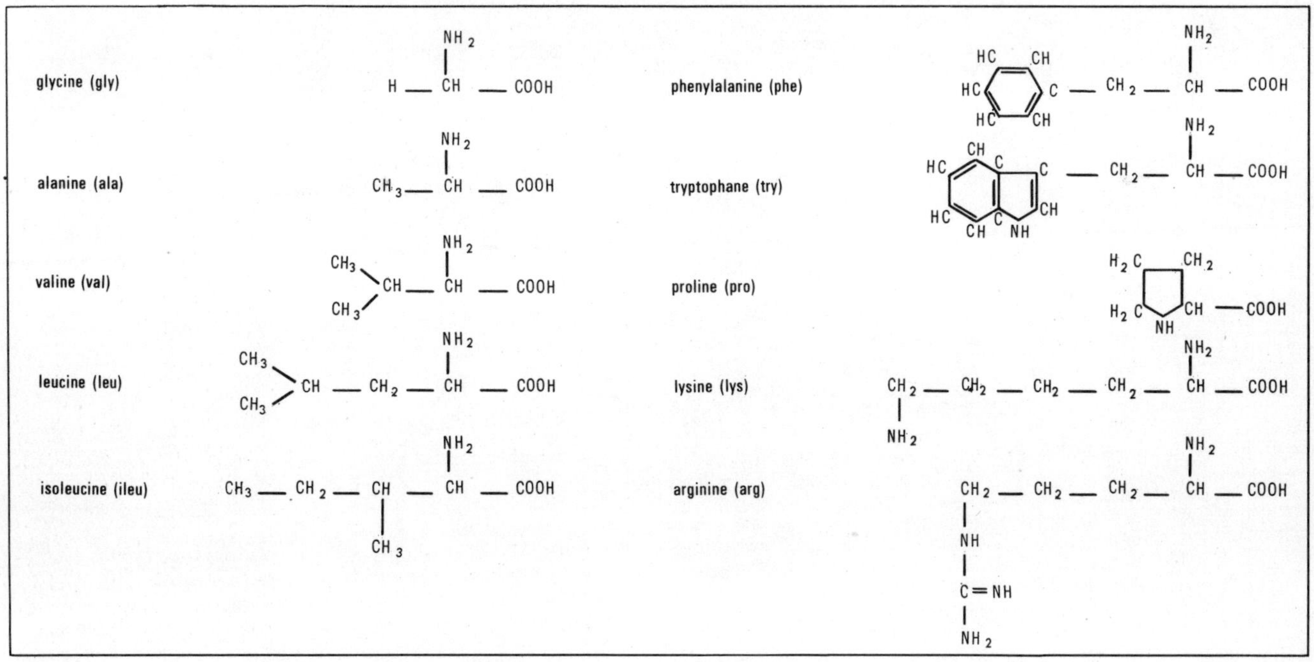

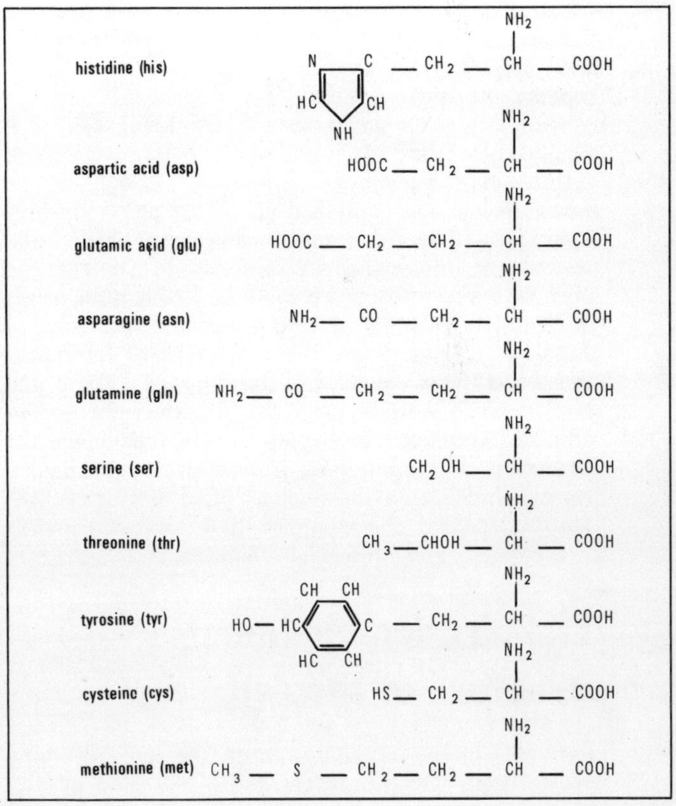

glycine (gly)	$H — \overset{NH_2}{\underset{}{CH}} — COOH$	phenylalanine (phe)	(benzene ring) $C — CH_2 — \overset{NH_2}{\underset{}{CH}} — COOH$
alanine (ala)	$CH_3 — \overset{NH_2}{\underset{}{CH}} — COOH$	tryptophane (try)	(indole ring) $C — CH_2 — \overset{NH_2}{\underset{}{CH}} — COOH$
valine (val)	$\overset{CH_3}{\underset{CH_3}{}}CH — \overset{NH_2}{\underset{}{CH}} — COOH$	proline (pro)	(pyrrolidine ring) $CH — COOH$
leucine (leu)	$\overset{CH_3}{\underset{CH_3}{}}CH — CH_2 — \overset{NH_2}{\underset{}{CH}} — COOH$	lysine (lys)	$\underset{NH_2}{CH_2} — CH_2 — CH_2 — CH_2 — \overset{NH_2}{\underset{}{CH}} — COOH$
isoleucine (ileu)	$CH_3 — CH_2 — \underset{CH_3}{CH} — \overset{NH_2}{\underset{}{CH}} — COOH$	arginine (arg)	$CH_2 — CH_2 — CH_2 — \overset{NH_2}{\underset{}{CH}} — COOH$; NH ; $C=NH$; NH_2

histidine (his)	(imidazole ring) $C — CH_2 — \overset{NH_2}{\underset{}{CH}} — COOH$
aspartic acid (asp)	$HOOC — CH_2 — \overset{NH_2}{\underset{}{CH}} — COOH$
glutamic acid (glu)	$HOOC — CH_2 — CH_2 — \overset{NH_2}{\underset{}{CH}} — COOH$
asparagine (asn)	$NH_2 — CO — CH_2 — \overset{NH_2}{\underset{}{CH}} — COOH$
glutamine (gln)	$NH_2 — CO — CH_2 — CH_2 — \overset{NH_2}{\underset{}{CH}} — COOH$
serine (ser)	$CH_2OH — \overset{NH_2}{\underset{}{CH}} — COOH$
threonine (thr)	$CH_3 — CHOH — \overset{NH_2}{\underset{}{CH}} — COOH$
tyrosine (tyr)	$HO — $ (benzene ring) $C — CH_2 — \overset{NH_2}{\underset{}{CH}} — COOH$
cysteine (cys)	$HS — CH_2 — \overset{NH_2}{\underset{}{CH}} — COOH$
methionine (met)	$CH_3 — S — CH_2 — CH_2 — \overset{NH_2}{\underset{}{CH}} — COOH$

Inclusion consists of embedding the section in an easily hardened supporting substance. Use is made of a plastic material, butyl methacrylate to which the harder methyl methacrylate is added. Specimens to be examined are placed on a special support, a fine film of collodion, silicon, beryllium, polystyrene, parlodion or carbon.

Ordinary microtomes have been improved; the mechanism of adjustment by a cogged wheel has been replaced by a thermal device. The prepared section is fixed on to a metallic shaft which lengthens in proportion to the heat to which it is subjected. In this way the finest adjustment can be obtained. An arrangement assures that the specimen passes only once under the cutting-edge as it descends. The steel blade, which must be perfectly sharpened, is often replaced by a sharp-edged splinter of plate glass which, being of minute dimensions, can conveniently be employed. Since collisions with air molecules interfere with the path of the electronic beam, the microscope must be evacuated and, as water absorbs electrons, the cells to be examined must be dried.

The greater the mass of atoms the projected electrons encounter, the more quickly their course is halted: thus the differences in density between the various elements of the specimen under examination are revealed in the magnified image produced. These contrasts can be further accentuated by certain technical devices. The phase contrast microscope, invented in 1938 by Zernike, a Dutchman, is, in fact, a perfected form of the normal microscope and employs ordinary light waves. It has revived interest in the examination of living cells for it allows us, without staining the cells, to perceive faint differences in thickness and refractive index, always provided that the cells examined are transparent. In this apparatus slight differences of refractive index in the specimen, imperceptible to the eye, are transformed into differences

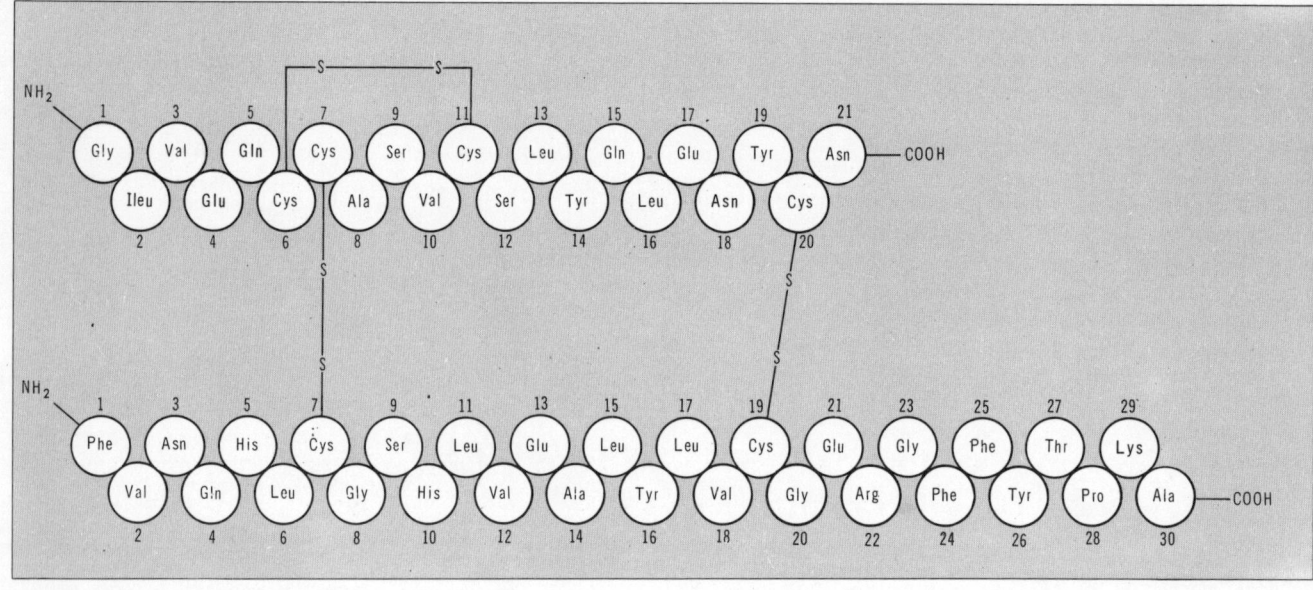

Primary structure of the insulin of beef.

of intensity, that is to say into differences of brightness which the eye can perceive. The contrasts are so sharp that the minute components of the living cell, unstained and practically invisible to the ordinary microscope, become clearly discernible.

Thus it is possible to observe the living cell as it functions, to study its pathology and its reactions to various environmental conditions, to poisons, and to medicaments. The combined use of the phase contrast technique and the cine camera enables us to photograph cellular and intra-cellular movements. Movements too slow for the eye to perceive can be accelerated, while those too fast can be slowed down and analysed. In this way cytology has been able to describe cell movements and the movements of its component parts: cell displacements, surface puckering, contractions, extension, the spontaneous display of white globules, the formation of pseudopods, the oscillation of nuclei, the expulsion from the nucleus of erythroblasts, the dance of the mitochondria, the rhythmical movements of the centrosome, and other phenomena.

The physico-chemical structure of the cell

The study of the cell is, in a sense, the study of living matter itself. The word *protoplasm* was used by von Mohl in the 1840s to replace the word *sarcode* proposed earlier by Dujardin to designate the jelly-like substance which composes the cell of the amoeba. Protoplasm is essentially a colloidal system made up of large molecules, highly impregnated with water. Inflated by this water they are in contact with each other and constitute a *hydrogel*. But this 'water-jelly' is peculiar in that it has the property of changing very rapidly, according to conditions, from a viscous, jelly-like state to a fluid, almost liquid state. In certain conditions, when acted on for instance by heat or acids, protoplasm flocculates, then coagulates.

Biochemistry, a relatively new science, has already made immense progress. Biochemists have discovered the basic chemical composition of living matter and, further, the original characteristics which distinguish it from inert matter.

Some thirty of the 105 elements known to contemporary science are found in living matter. Of these thirty a dozen comprise the major elements necessary to life, and in themselves make up 99.9 per cent of the total weight of living matter. Oxygen, carbon, hydrogen, and nitrogen alone account for ninety-six per cent of the bodies of animals. Besides the major elements there are some twenty others, metals or metalloids, some of which are found in only infinitesimal quantities.

From these elements the living organism synthesizes enormous and complicated molecules: sugars, fats, and proteins. While the relatively simple molecules of common mineral salts are composed of some twenty atoms at most, the glucose molecule contains twenty-four.

But the proteins, which are the most important constituents of protoplasm, represent infinitely more complex molecules. Among the proteins with high molecular weights are: lactalbumin (17,500); ovalbumin (38,000); haemoglobin (horse — 69,000); serumglobulin (horse — 150,000); thyroglobulin (pig — 350,000); haemocyanin (snail — 5,000,000).

For comparison, we remember that the molecular weights of polyamides such as nylon are between 6,000 and 20,000; those of polystyrenes and polyvinyls are about 300,000. The molecular chains of plastics are relatively simple beside those which are manufactured by living things.

These substances with very high molecular weights (proteins, cellulose, starch) are polymers or macromolecules; in other words they are complexes of smaller units joined by co-valent bonds; these may be

30

all identical or different and form a repetitive series. These macro-molecules, of which the molecular weight is over 10,000, have a definite structure, stable and specific, which is frequently reproduced with great accuracy.

Molecular biology, a fairly recent science (the name was used for the first time by Astbury in 1950), consists of research into the structure of the macro-molecules which are responsible for the various cell processes and consequently for the functioning of all organisms. Because of their extreme importance to cellular physiology and genetics, we will deal with the proteins in detail.

Proteins

Composed of carbon, hydrogen, oxygen, nitrogen, sulphur, and phosphorus, the proteins are the essential macro-molecules of living matter; free or associated with sugars, lipids, nucleic acids, they form between fifty and eighty per cent of the dry weight of cells. All the enzymes (or catalysts) of cell reactions are proteins; more than 1,000 enzymes exist in the cell. The molecular weight may be from 6,000—500,000. They are concerned not solely with the structure of organisms, but also with their defence; one foreign protein, introduced into an organism by a bacterium for example, immediately causes the elaboration of a defensive protein, known as an *antibody,* which combines with the foreign protein, now known as an *antigen,* and neutralizes it; this is exactly the principle of immunization.

Whatever their origin, proteins are constructed from a series of twenty amino acids. As their name indicates, amino acids consist of an acid radical and an amine, which together determine their chemical

properties. The general formula is
$$R-\underset{\underset{\displaystyle NH_2}{|}}{C}H-COOH,$$

R being a different radical for each amino acid. The twenty amino acids are: glycine (Gly), alanine (Ala), valine (Val), leucine (Leu), isoleucine (Ileu), phenylalanine (Phe), tryptophan (Try), proline (Pro), lysine (Lys), arginine (Arg), histidine (His), aspartic acid (Asp), glutamic acid (Glu), asparagine (Asn), glutamine (Glu), serine (Ser), threonine (Thr), tyrosine (Tyr), cysteine (Cys), methionine (Met). The names in brackets denote the conventional abbreviation of each name. Apart from these amino acids, the special amino acids ornithine, citrulline, β-alanine may be present in cells.

Two or more amino acids may react together with the loss of water, forming a dipeptide, a tripeptide, or a polypeptide. Polypeptides contain a maximum of fifty amino acids and can pass through semi-permeable membranes. Proteins are macro-molecules comprising at least 100 amino acids which, in solution, do not pass through a semi-permeable membrane and behave as colloids.

Like other macro-molecules, proteins possess four types of structure. The primary structure corresponds to a chain of amino acids; the same molecule of protein always has the same number of amino acids connected in a stable sequence. The secondary structure is composed of a helical chain of polypeptides. The tertiary structure is formed by more complicated combinations of primary and secondary structures. Finally, the quaternary structure found among various proteins comprises many polypeptide chains, amounting to an enormous molecular edifice.

The structure of many proteins is now known, since the work of Sanger (1944) on insulin. Insulin, a pancreatic hormone, controls the level of blood glucose. It is composed of fifty-one amino acids including 777 atoms of carbon, hydrogen, nitrogen, oxygen, and sulphur in varying quantities. Bit by bit, Sanger arrived at a solution for the structure; insulin is formed of two polypeptide chains joined by two disulphide bonds. Sanger finally arrived at a model of the sequence of all the amino acids in the insulin molecule. This important work won him the Nobel Prize for Medicine in 1958. The sequence of amino acids is as yet known in a very small number of proteins, hormones, and enzymes.

The secondary structures are most often of the α-helix type proposed by Pauling; some rare proteins have an accordion-like structure.

The chain is eventually folded back on itself and forms compact and globular shapes which are in fact tertiary structures. The repetition of many polypeptide chains corresponds to quaternary structure.

Tertiary and quaternary structures are still barely understood, although two models have been constructed of the molecules of myoglobin and haemoglobin. From these it appears that it is the sequence of the amino acids which determines the tertiary structure. As with the hormones, if the sequence of amino acids is changed it may disrupt normal functioning. For example, we know that if the amino acid valine is substituted in a certain position for glutamic acid, such a change will cause the hereditary disease known as drepanocyte anaemia.

Nucleic acids

Nucleic acids were discovered in 1869 by Miescher, who called them ‘*nucleine*’. The name ‘nucleic acids’ was given by Altonan in 1889. During a century of intensive work this name has been established. For a long time only two nucleic acids were known: one which is extracted from yeast (zymonucleic acid) and one extracted from the thymus gland (thymonucleic acid). Admittedly the first characterized plants and the second animals, but soon both of these acids were found in both plant and animal cells.

Today the two main types of nucleic acid are recognized according to the type of sugar they contain. Ribonucleic acids (RNA) contain ribose; desoxyribonucleic acids (DNA) contain desoxyribose.

Desoxyribonucleic acid (DNA)

DNA has a basic chain structure and has in fact a backbone of desoxyribose sugar and phosphoric acid with four purine and pyrimidine bases (A, T, C,

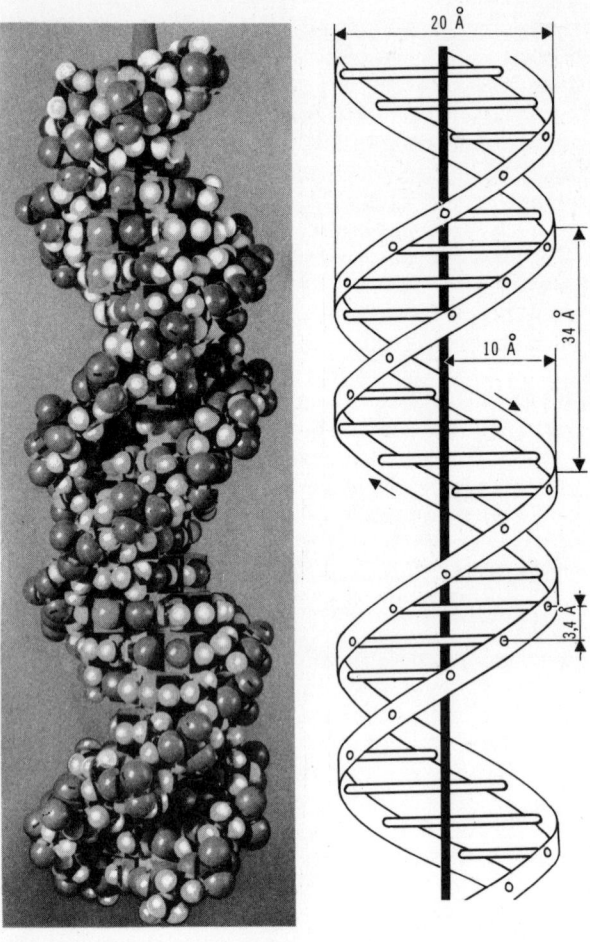

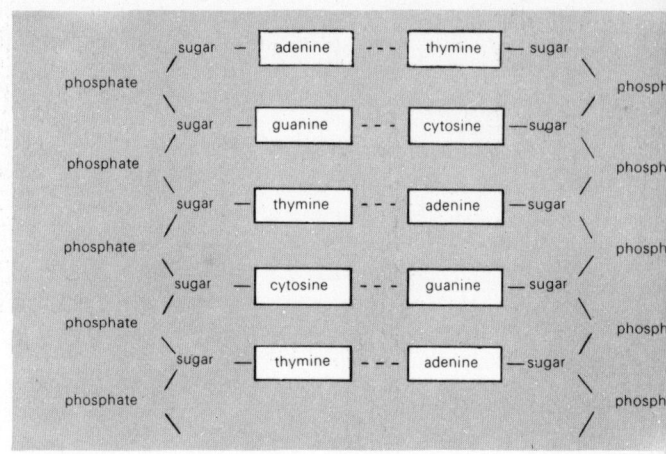

Left and above, **Structure of a molecule of desoxyribonucleic acid.**

Larousse.

Biosynthesis of DNA by self-replication: 1. Initial molecule of DNA; 2. New molecule of DNA locking with half the initial molecule. 3. Another new molecule of DNA locking with the other half of the initial molecule.
The two synthesized halves of the new molecules are shown in bold face.

Formation of a nucleoside and a nucleotide of RNA.

G) arranged along this backbone. The polynucleotides differ between themselves in the sequence of bases along the backbone, itself a stable structure. Experiment has shown that the sequence (the order of bases) is not fixed; probably one DNA differs from another DNA in the sequence of the four bases. This sequence can be of the utmost importance in determining all the information necessary for protein synthesis.

Chargaff has shown ingeniously that the different molecules of DNA contain as many adenine molecules as thymine ($A = T$) and as many molecules of cytosine as guanine ($C = G$); but two DNAs from two different animals can be distinguished by the proportions of the proportions of the pairs of bases $A+T$ and $C+G$.

A number of observations seem to support the view of a secondary structure for DNA; from X-ray diffraction pictures, it appears that the bases are separated from one another by a regular distance of the order of 3.4 Å (1 Å equals one ten-thousandth of a micron). The bases are therefore piled up rather like a stack of plates.

Watson and Crick (1958) formulated a basic hypothesis giving a helical structure to DNA. This model consists of two polynucleotide chains helically coiled around an axis. The two chains are arranged with their bases opposite, along the inside of the helix; one base of one chain is linked to one base of the other chain by hydrogen bonds; but the coupling is only possible through two particular bases, one purine (adenine) and the other pyrimidine (thymine) (A-T), or the pair guanine-cytosine (G-C). The pairs of bases are perpen-

dicular to the axis; the sugar is in a plane parallel to the axis. The chain revolves once in 34 Å; phosphorus atoms are arranged at 10 Å from the axis, which gives the double helix a diameter of 20 Å. This model can be compared with a ladder; the uprights are represented by the backbone (polydesoxyribose phosphate) and the rungs by pairs of bases; the ladder is twisted and appears as a double spiral.

The pairing of the bases is only compatible with the four following arrangements: A-T, G-C, T-A, C-G; an adenine is always opposite a thymine and a cytosine opposite a guanine. The two chains are thus complementary but their base sequences are not identical; the following pairs give bases for each of the chains shown:

$$2A, 2C, G, T \left\{ \begin{array}{c} A-T \\ C-G \\ G-C \\ T-A \\ A-T \\ C-G \end{array} \right\} A, C, 2G, 2T$$

The two chains are also orientated so that one base is opposite its partner on the other chain.

The number of nucleotides and their sequence characterizes each DNA, with a practically infinite number of possibilities. A chain of 100 nucleotides with these four possible base combinations would permit 4,100 different combinations; this number is many times larger than the number of atoms in the solar system. We know that DNA in a human sex-cell contains 109 pairs of bases, giving the double helix a length of one yard; it has to be very compressed to fit into the cell-nucleus, unless there is a tertiary structure not yet known.

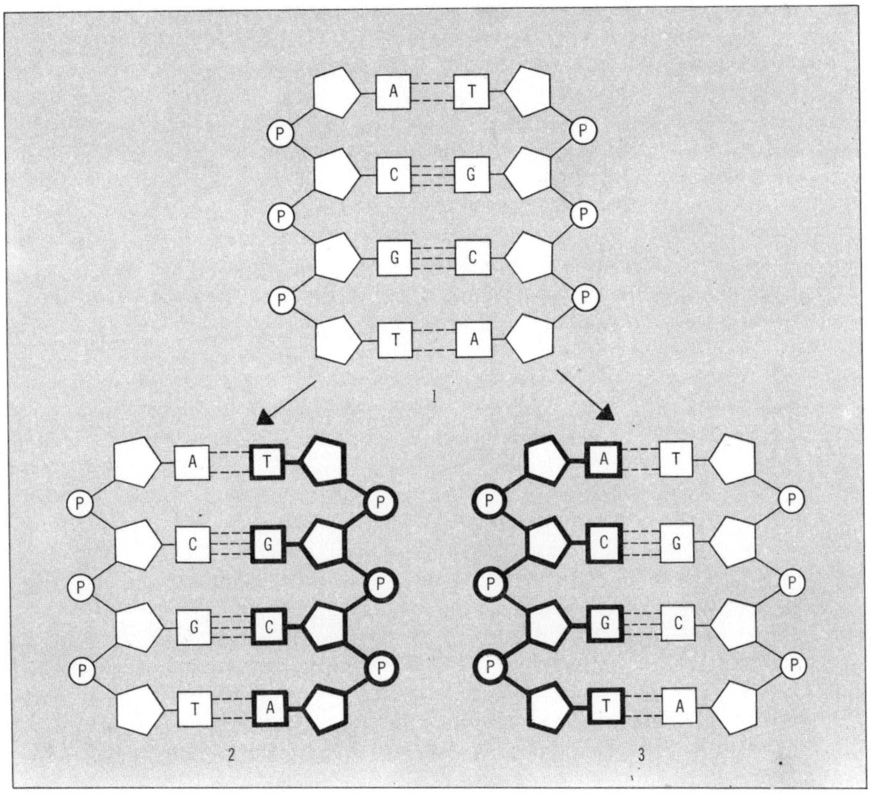

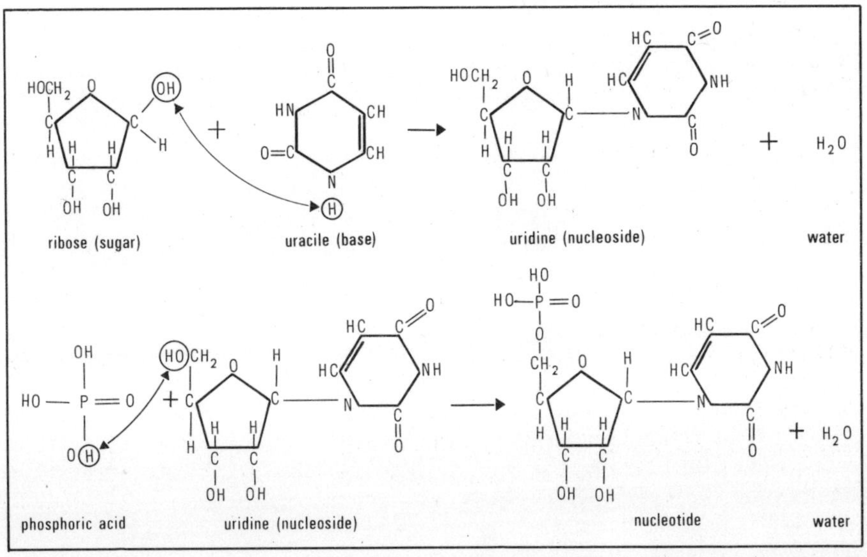

The problem is all the more complex since in all organisms, except the simplest, DNA in the chromosomes forms combinations with histone proteins, giving desoxynucleohistones which are important elements of chromosomes.

The original model of the secondary structure of DNA which won the Nobel Prize for its discoverers accounts for all the various properties of DNA; it is compatible with the self-replication of DNA and it represents a four-unit code capable of transmitting genetic information.

The mechanism of self-replication is as follows: the two polypeptide chains shear apart; each one is capable of rebuilding its opposite chain. After the parting of the two chains, the bases are free and a free nucleotide from the cell solution becomes attached to each; thus a thymine nucleotide becomes attached to the adenine base and a guanine nucleotide to a cytosine base. In other words, A and B separate; a B chain is formed alongside the A chain and an A chain is formed along the B chain. The initial AB molecule gives rise to identical AB and BA molecules each containing half of the original molecule.

Ribonucleic acid (RNA)

Ribonucleic acids are mostly found in the cytoplasm but the nucleolus of the cell is an active centre of the

Primary structure of transfer RNA from baker's yeast (after Holley, 1965). A, adenine; C, cytosine; G, guanine; d-G, dimethyl-guanine; m-G, methyl-guanine; I, inosine; U, uracile; d-U, dihydroxy-uracile; p-U, pseudo-uracile.

synthesis of RNA.

RNAs are composed of nucleotides; we have already mentioned that the sugar is ribose and that one of the nitrogenous bases is different from DNA — uracil in the place of thymine. RNAs also contain a small proportion of other nitrogenous groups, similar to or derived from the four bases (inosine, methyl inosine, pseudouracil, dimethyguanine, dimethyluracil).

Ribonucleotides become linked together to form polyribonucleotides; the primary structure of RNA is similar to that of DNA. The shorter or longer chains are always composed of a skeleton of polyribose phosphate and the bases are distributed in sequence along the skeleton. This sequence carries information derived from that supplied by the DNA.

Three types of RNA have been identified, the ribosome RNA, messenger RNA, and soluble or transfer RNA.

Ribosome RNA, the most abundant (about eighty per cent of the total RNA), makes up part of the ribosomes of the cell. The molecular weights of these stable compounds vary from 500,000 to 1,200,000. Their primary structure is not yet completely known. Each ribosome contains two molecules of RNA, each one associated with characteristic proteins. The sequence of bases is complementary to that of DNA. The ribosome RNAs appear to be a single chain but may in some places form a double helix.

Messenger RNAs are free, and their molecular weights are a little lower, from 200,000 to 1,000,000. They break down rapidly and represent only five per cent of the total RNA; they play an important part, however; the Nobel Prize for Medicine in 1965 was presented to A. Lwoff, F. Jacob, and J. Monod, mostly for the discovery of messenger RNA. The structure of messenger RNA is now known with certainty; theory and experiment have shown that the base sequence is complementary to that of DNA in the nucleus of the same cell. Synthesized in the nucleus, messenger RNA is 'sent' into the cytoplasm to the ribosomes where protein synthesis takes place, with its series of bases complementary to the DNA series, which provides the information necessary for protein synthesis. DNA thus appears to store the information which the messenger RNA carries to the ribosomes.

We know that it forms hybrid helices, one helix being a polydesoxyribonucleotide while the other is a polyribonucleotide; the complementarity of the bases is always in the same pattern by which the pair A-T is replaced by the pair A-U.

Transfer RNA or soluble RNA is composed of 100 nucleotides and has a molecular weight of about 25,000. Transfer RNAs form five to ten per cent of the total RNA in the cell. As their name indicates, they are free and soluble in the cytoplasm. They combine in a reversible fashion with amino acids. The chain of a transfer RNA is shaped like a hairpin and the two branches are associated in a helix by the pairs of bases (A-U, G-C). The bend of the two branches of the hairpin carries three free bases capable of attaching to a particular complementary region on messenger RNA.

Holley and his colleagues (1965) succeeded in working out the sequence of bases of a transfer RNA from baker's yeast which specifically fixes an amino acid, alanine, in the presence of a particular enzyme. The molecule is composed of seventy-seven nucleotides and has a molecular weight of 26,600. The terminal series of the bases C-C-A is common to all transfer RNAs, the amino acid fixes on the terminal adenine.

To summarize, the RNAs, whatever their type, are polyribonucleotides with a single chain of variable length and capable of folding to form sections of double helix.

Nucleoproteins

These molecular complexes formed by association of nucleic acids with proteins comprise the desoxyribonucleoproteins and ribonucleoproteins.

In the cell-nucleus, DNA is combined with proteins; these desoxyribonucleoproteins form part of the structure of the chromosome. The nature of the combinations is not understood. DNA combines to a certain extent with the protamines to give desoxyribonucleoprotamines and also with histones to form desoxyribonucleohistones. The first are particularly abundant in the sperm heads of various animals. Histones assume a role of regulation of function in the various region of the DNA molecule.

Cellular RNA associates with proteins in variable

proportions and forms cytoplasmic particles, the ribosomes.

Mucoproteins

These result from the union of proteins with carbohydrates; very diverse and widespread, mucoproteins are found in connective tissues, milk, eggs, red blood corpuscles, insect chitin. Their composition varies greatly with the nature of the carbohydrate, and with the proportions of protein and carbohydrate contained.

Lipoproteins

These result from the combination of proteins with lipids, particularly cholesterol and phospholipids. They take part in the formation of all cell membranes and organelles; they constitute about fifteen to twenty per cent of the protein of human serum (globulin); the vitelline membrane of eggs is rich in lipoproteins.

The animal cell

The animal cell is composed of a thin film or *plasma-membrane* and of *cytoplasm*, which surrounds a more or less spherical body, called the nucleus. In the nomenclature of cellular constituents a regrettable confusion exists. Cytoplasm designates the cellular substance with the exclusion of the nucleus. The term protoplasm is applied to the whole corpus of living matter.

The plasma-membrane

The plasma-membrane, or cell membrane, is simply an external differentiation of the cytoplasm. It permits exchanges between the cell and its surroundings, and to this effect has the special property of letting certain substances enter the cell while denying passage to others. It thus behaves as a semi-permeable membrane. Its permeability is, however, selective and varies according to the type and functions of the cell.

The cytoplasm

Cytoplasm is composed of a fundamental substance or hyaloplasm which holds in suspension various inclusions. Under the ordinary microscope cytoplasm appears transparent and homogeneous, though now and again differentiation is observed. In the amoeba the peripheral hyalin cytoplasm is called the *ectoplasm*, the granular internal cytoplasm being the *endoplasm*.

Cytoplasm is more or less viscous. Age and the surrounding conditions of temperature and acidity affect its viscosity. It also has a certain elasticity. Its chemical composition is very complex. Thirty per cent of its content consists of proteins, carbohydrates, fatty acids, enzymes. The seventy per cent is water.

Cytoplasmic inclusions

Chondriosomes, the Golgi apparatus, ergastoplasm, and metaplasm all occur in the cytoplasm.

Chondriosomes

Chondriosomes were discovered in 1890 by Altmann. They appear in many shapes, being sometimes granular (mitochondria), sometimes granular strings (chondriomites), and sometimes the rod-shaped flexuous chondriocontes. The mitochondria measure from a half to one micron, while chondriocontes reach a length of from 3-6 microns. In spite of this variety of shape and size, the electron microscope reveals that they have the same structural plan. Mitochondria are double-walled vesicles, the interiors of which are divided by complete or incomplete partitions, also double, which are more or less totally inserted into the outer wall of the vesicle. These internal partitions are usually perpendicular to the principal axis of the mitochondria, though they are sometimes parallel to it. They are short and in plants they are disposed radially.

Golgi apparatus

The Golgi apparatus was discovered in 1898 by the Italian cytologist Golgi who found it in nerve-cells as an irregular and complex network around the nucleus. Later it was observed not only in neurons but also in other animal cells. Sometimes the network was replaced by shuttle-shaped bodies, known as dictyosomes. The dictyosomes are composed of two parts, one chromophilic, or 'colour-loving', shaped like a shell or a shuttle, and the other chromophobic, or 'colour-fearing' — forming a droplet lodged in the cavity of the chromophil area.

Because of difficulties of fixation and direct observation the Golgi apparatus gave rise to bitter controversy: its existence was queried and even denied. The question of whether plant cells possess a Golgi apparatus was disputed for thirty years and the conclusion was reached that they did not.

In fact they do, and the argument has now been settled by the clear and repeated images supplied by the electron microscope. These reveal that the Golgi apparatus is composed of independent dictyosomes, accompanied by a more or less dense following of *osmiophilic* vesicles. The network originally described was simply an artifact which resulted from fixation. Each dictyosome is composed of minute sacs, flattened and piled one upon the other.

The Golgi apparatus is not unique to certain metazoan cells, but exists in many of the Protozoa, notably the parasitical Flagellata, where it is often referred to as the parabasal apparatus. Plant cells too possess a Golgi apparatus which has the same morphology and the same ultra-structure. The Golgi apparatus, then, is a distinct cellular element, independent of other inclusions especially of the vacuoles.

The ergastoplasm

Ergastoplasm was the term which Garnier applied in 1897 to the basophilic or easily stained filaments found in gland cells. Frequently merged with the chondriosomes, the ergastoplasm appeared to have no further independent existence of its own. The electron microscope has, however, revealed its structure. It is composed of long laminations with double-walls some 200 to 250 angstroms thick which are situated towards the base of the cell. The external surfaces of the laminations are dotted with granules which are between 150 and 200 angstroms in diameter and probably composed of R N A.

Ergastoplasm is found in the cells of vertebrates, of invertebrates (insects) and in plant cells. It is not a permanent substance and disappears when cells are in repose. Its presence must be associated with the accomplishment of some cellular activity, perhaps the formation of certain proteins.

Metaplasm

Metaplasm consists essentially of elements resulting from cellular activity, of substances absorbed from the environment to be stored by the cell, and of waste matter. It is a heterogeneous assembly of emulsified fat droplets, glycogen or animal starch, albumin, pigments and minute corpuscles, mineral or half-mineral and half-organic (crystals of silica and of calcium urate and calcium carbonate). The vacuoles are enclaves which contain substances about to be excreted or food in various stages of assimilation.

The nucleus

The nucleus was discovered in 1831 by Robert Brown in the cell of orchids and subsequently in animal cells. All cells possess a nucleus, though the nucleus of certain Cyanophyceae (blue-green algae) and of bacteria is rudimentary. Cells lacking a nucleus — such as the red blood corpuscles of mammals — are cells which have lost the nucleus they originally had. Normally each cell contains a single nucleus. Hepatic cells, however, have two, while the myeloid tissue of bone marrow and cancerous cells possess several. Cells with two nuclei occur much more frequently in plants and can be observed during certain phases of development.

The shape of the nucleus varies with the nature and age of the cell. In spherical cells the nucleus is spherical, while in elongated cells it stretches into an ellipsoid with its major axis parallel to that of the cell. In old white blood-cells and in many secretory cells the nucleus becomes polymorphous, lobed, reticulated or branched. Its size depends on the size of the cell.

Examination of the nucleus under the electron microscope presents serious obstacles which have not yet been overcome. Its fixation is difficult. Our most reliable information about the nucleus concerns its surrounding membrane. The nuclear membrane, clearly visible, is of double thickness and in certain cells has the appearance of a fine network or reticulum, the mesh of which suggests the presence of pores through which nuclear fluid is in direct communication with the cytoplasm.

At all times during cell-division the nucleus contains independent units whose number remains constant. These are permanent elements called *chromosomes*. The appearance of the chromosomes varies immensely according to whether the cell is dividing or in a state of repose. During cell-division, called mitosis, chromosomes steadily contract by coiling up into a spiral. As the phases of mitosis proceed they uncoil and elongate, altering in dimensions and appearance. Uncoiled, the chromosome of a snail measures at least 240 microns; at its maximum contraction it is only 1.5—2 microns long. A comparison of these figures demonstrates the importance of spiralization in the morphology of the chromosomes. The contraction is further accomplished by a motion comparable to accordion pleating.

The thread-like chromosome is at one or more points constricted: one such point appears as a delicate granule, called the *centromere*, by which the chromosome is attached to the *spindle* during cell division. In multicellular organisms the centromere is never terminal, that is, the chromosome extends on either side of it in sections known as arms. Hence there are V-shaped, J-shaped and U-shaped chromosomes. One of the arms can also be so short that the chromosome has a rod-like appearance. A secondary constriction of the chromosome sometimes corresponds to the point from which the *nucleolus* is secreted.

The classical diagram of the chromosome shows that its two arms are composed of a stain-resistant sheath enclosing an easily stained filament of homogeneous appearance. Depending on the organism this 'thread' is or is not 'strung' with minute 'beads' or granules. The sheath is called the matrix, the filament the chromatid, and the granules chromomeres.

The electron microscope has as yet given us little fresh information about the inner structure of chromosomes, although study of the formation of spermatozoa has thrown some light on the subject. In the sperm of the snail the chromosomes appear as extremely minute threads some 70—100 angstroms in diameter. (The diameter remains the same over the length of the chromosome.) The thread is homogeneous: no structural differentiation appears. No bulges reveal the chromomeres, points of particular interest and especially rich in D N A. The matrix is absent. The visible chromosome is thus reduced to its chromatids.

The homogeneous structure of the chromosome makes it impossible to advance direct evidence of particles which could be identified as *genes*.

To sum up, a chromosome is a filament of complex structure in which D N A, in a highly polymerized form, is strung on a fibrous thread of nucleoproteins of various kinds. Its two fundamental elements are the filament and the centromere. The matrix is an unstable and can disappear and reappear according to the phases through which the chromosome passes. The chromomeres are found when the chromosome coils up, and occupies fixed and invariable positions.

The chemical composition of the chromosome is

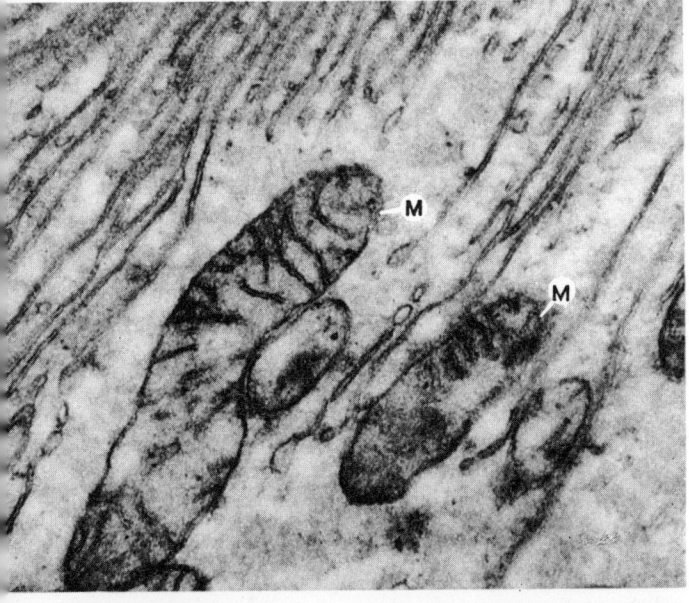

Mitochondria (M) in an excretory nephridium cell of a leech. The crests which project from the mass of the mitochondria can be seen, as well as the numerous folds of the cellular membrane. (Magnified 30,000 times by electron microscope.)

Laboratoire de microscopie électronique appliquée à la Biologie. C. N. R. S. Mlle. Gourvest.

The Golgi apparatus (enlarged 50,000 times) of a snail's semen. It is formed of dictyosomes, four of which are visible in the section. *Above,* part of one of the mitochondria can be seen.

Laboratoire de microscopie électronique appliquée à la Biologie. C. N. R. S. P. P. Grassé.

Epidermal cells of *Pleurodeles waltl.* By a special technique (impregnation with silver) the nucleus and the boundaries of the cell are made clearly visible.

H. Mugard.

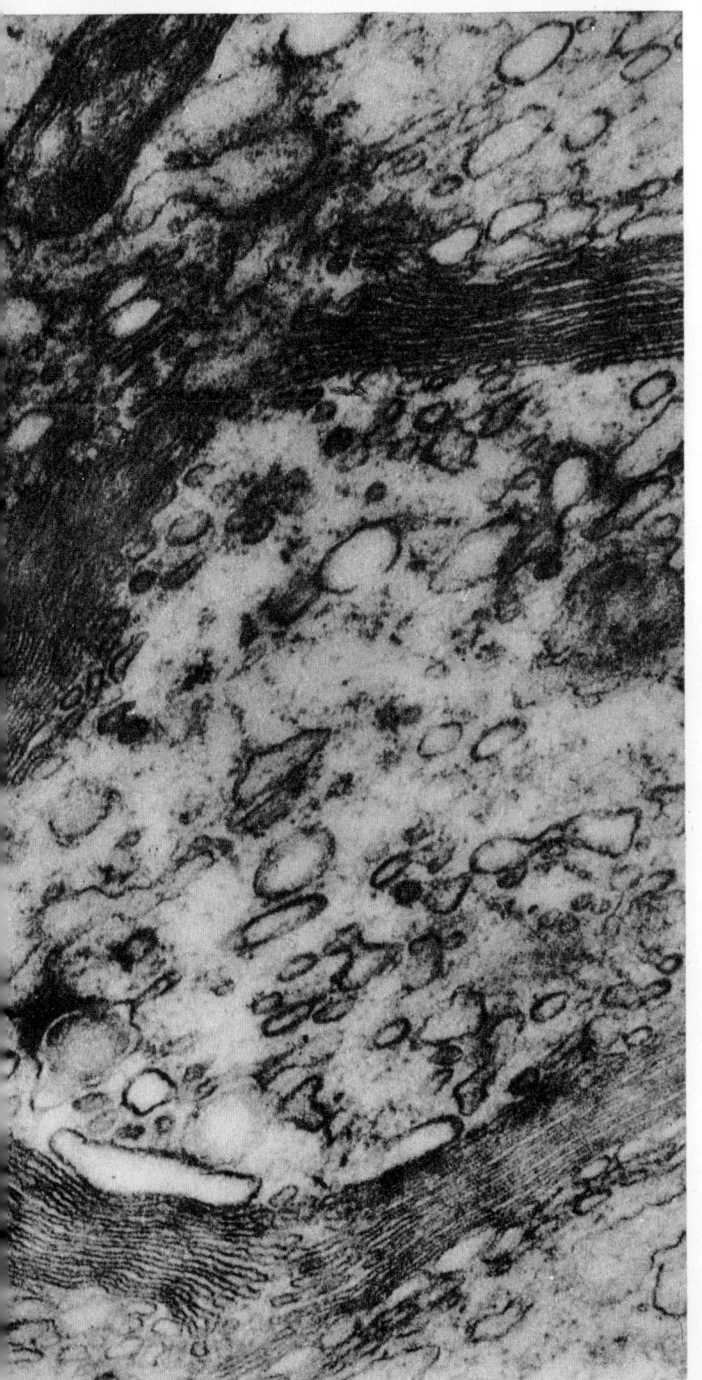

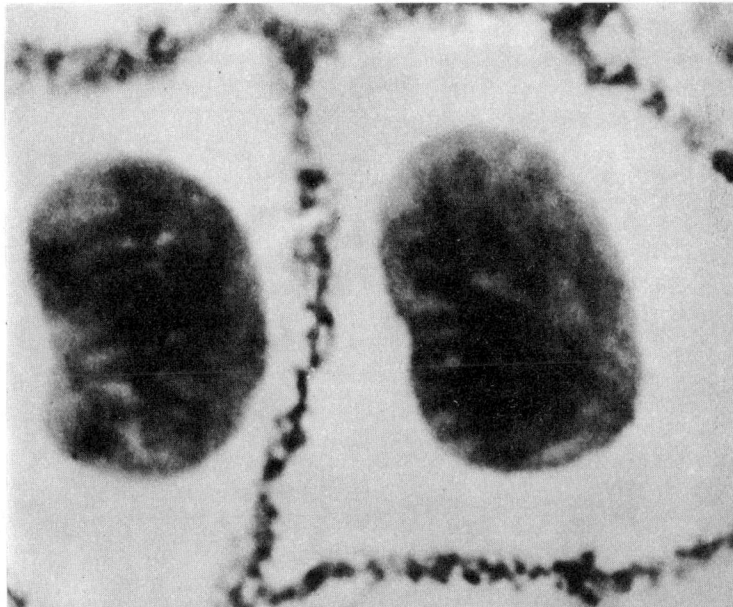

complicated, its chief ingredient being D N A. The percentage of D N A present in any given class of cell remains constant and depends on the number of chromosomes the cell contains. Experiment has proved that, in the same species, a tetraploid cell with its double complement of chromosomes contains twice as much D N A as the corresponding diploid cell. The gamete, which has half the number of chromosomes found in ordinary somatic cells, also contains half the amount of D N A. Thus the percentage of D N A per nucleus is exactly proportional to the number of chromosomes the nucleus contains.

As well as chromosomes the nucleus contains a varying number of minute spherical bodies which, like the cytoplasm, can be stained by acid dyes; these are the *nucleoli* which are produced by the chromosomes. They were discovered in maize and usually vary in number between one and four. In cells where metabolism is intense, such as egg cells and glandular cells, the number of nucleoli increases and can reach the neighbourhood of one hundred, the extra nucleoli being the result of nucleolar activity. The nucleolus is basically composed of histone in association with R N A. The biggest nucleoli sometimes contain enclaves. The nucleolus has a life of its own, and is capable of budding and secreting.

The centrosome

Near the nucleus of many cells there is a mass of clear and homogeneous cytoplasm known as the *centrosome,* the middle of which is occupied by a minute granule called the *centriole.* Around the centrosome radiate filaments of varying length which constitute the *asters* which develop during cell division. The centrosome is absent in the cells of higher plants, and in those of certain protozoa. The line which connects the centrosome with the centre of the nucleus defines the axis of the cell. The centrosome attains its greatest development during the course of cell-division.

The work of the cell

The cell is not only a morphological unit but also a physiological unit. True life is only manifested by the complete cell: a cell without a nucleus dies and a nucleus isolated from its cytoplasm is not viable. The nucleus is indispensable to cellular life, but its function is scarcely known. It plays a principal part in the transmission, maintenance and carrying into effect of hereditary characters. The role of the nucleolus is undoubtedly secretory. It would seem to supply the cytoplasm with products used in the synthesis of proteins and establish a kind of equilibrium between its own R N A content and that of the cytoplasm.

The chondriosomes appear to be above all the bearers of enzymes and vitamins, that is of the catalysts indispensable to biochemical reactions.

The Golgi apparatus and the ergastoplasm play an important part in cellular secretions: both possess enormous surfaces in relation to their dimensions, a factor which facilitates such activity. In this connection the electron microscope has revealed the extraordinary development of the surfaces displayed by cellular inclusions. Surface activities certainly play a principal role, not only externally such as that performed by the pleated membrane of the excretory cells, but also within the cell itself.

The plant cell

The description of the animal cell given above could be applied to the plant cell, if we add to it the pecto cellulose membrane (or cell-wall) and the plastids. The plastids are minute inclusions in the cytoplasm of plant cells which prepare various substances essential to the plant. Thus chloroplasts manufacture chlorophyll, to which green plants owe their colour; amyloplasts elaborate starch; protoplasts produce protein; chromoplasts pigments and oleoplasts fats.

The cell-wall of plants is composed chiefly of cellulose and the pectic substances, pectose, pectin and calcium pectate. As the plant grows the wall undergoes profound modifications, being impregnated with varied substances like suberin, cutin and lignin. The centrosome is, as we have noted, absent in higher plants, existing only in certain groups such as the algae, fungi, and the gametes of mosses and vascular cryptogams.

The electron microscope gives evidence that the animal cell and the plant cell are structurally almost identical — a fact of great evolutionary significance. It adds weight to the hypothesis that all forms of life have a common origin.

The function of living matter

The molecular structure of living matter is complex, but no more complex than the work which living

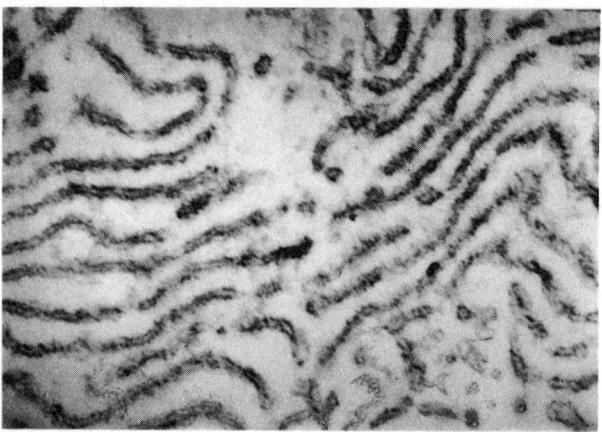

Ergastoplasm in a salivary cell of a larval *Chironomus* (a small fly or midge) enlarged 31,350 times.

Laboratoire de microscopie électronique appliquée à la Biologie. C. N. R. S. Mme. Pochon.

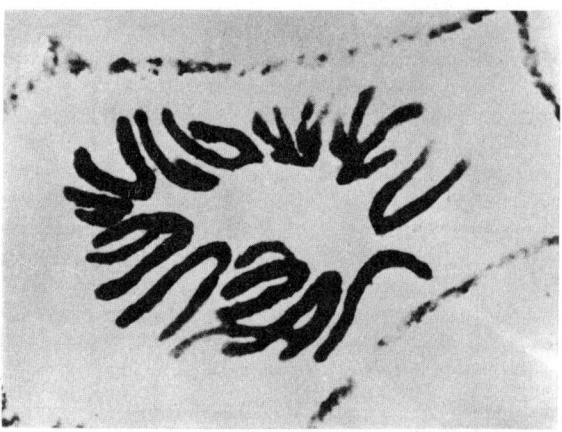

Metaphase. The cell is impregnated with silver and derives from the caudal epidermis of *Pleurodeles waltl.*

H. Mugard.

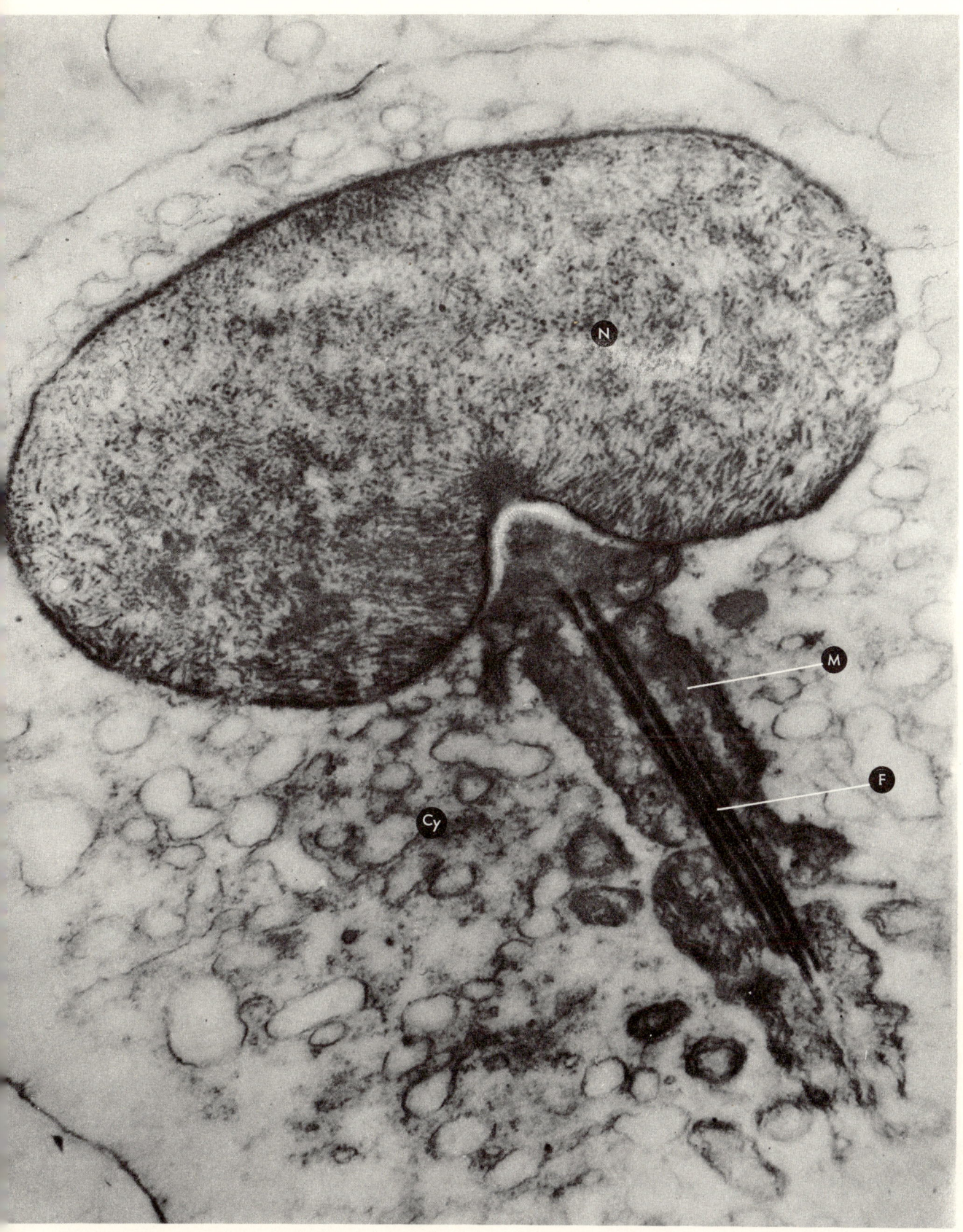

Spermatid of a snail *(Helix pomatia)* as seen by the electron microscope, 43,000 times enlarged. A spermatid is a cell which develops into a spermatozoon. N, the nucleus, is the future head of the spermatozoon. The chromosomes can be distinguished in the form of very fine filaments (chromonemata) polarized by the centrosome and in the remainder of the

nucleus extremely wavy. M, mitochondria, form a sleeve around the flagellum. F, the flagellum, shows three of its contractile filaments. Cy, represents the cytoplasm of the cell.

Laboratoire de microscopie électronique appliquée à la Biologie. C. N. R. S.
P. P. Grassé.

matter performs. The cell is the centre of astonishing activity which continues without cease throughout the life of the organism. Food is absorbed, reduced to its basic constituents and then synthesized into proteins, fats and carbohydrates. For example, proteins derived from foodstuffs are broken down so that the amino-acids they contain can be used to build up the new proteins which the organism requires. The energy needed to bring about these reactions is supplied by the combustion of sugars and fats. As in the laboratory, all these reactions necessitate the presence of specific catalysts: vitamins, hormones, amylases (diastases) and metallic atoms. There are countless enzymes and the appropriate enzyme is available for every reaction such as the isomerases, decarboxylases, esterases and transaminases. Some ten enzymes are indispensable for the combustion of glucose, while a further ten participate in the synthesis of a fatty acid. Every cell contains an extraordinary enzyme equipment.

Indeed the cell is like a factory in which constant balance between all working elements is maintained by a system of automatic control which, within the limits of the possible, corrects any deficiency or trouble which may arise. In this way *homeostasis* or the main-tenance of constancy of the internal environment is achieved.

The analysis of living matter and the study of its behaviour suggest that it could be thought of as inert matter whose molecular constituents are arranged in an orderly, definite and precise manner: this special organization would thus be the unique property of living matter, which could then be described as inert matter on which structure had been imposed. Without organization, without architecture, without differen-tiation, matter would remain lifeless.

One-celled organisms

The cell, the morphological and physiological unit, is self-sufficient, and is capable of leading a free and independent life as it does in the case of unicellular animals and plants. All one-celled organisms are composed of a membrane, of cytoplasm with inclusions and of a nucleus, but they display great diversity of structure.

Perhaps the simplest of these organisms is the amoeba (*Amoeba proteus*), a protozoan measuring from 200—500 microns, widely distributed in damp soil and still water. The amoeba is a typical cell, a little mass of protoplasm of indefinite form which puts out transitory projections of cytoplasm known as *pseudo-pods*. By means of these powerful 'false feet', which can be blunt or filament-like, the amoeba moves. Amoeboid movement proceeds as the pseudopod adheres to a support and the main body of the cell flows forward to join it. A new pseudopod then appears, finds further support, draws the amoeba after it, and so on. The pseudopods also serve in the capture and ingestion of the animal's prey. The amoeba is dependent on its environment, drawing from it what is indispensable to its existence and eliminating into it what it cannot digest. The animal reacts to external stimuli and will move away from danger.

Naked cells, or cells unprotected by a rigid envelope, are also found in the vegetable kingdom, being observed in the male sex-cells of numerous plants. The spermatozoa of certain mosses are protoplasmic spirals about fifty microns in length. Those of *Zamia* (of the Cycadaceae family) are of great size, being sometimes more than 300 microns in diameter and 330 in length.

Yeasts are unicellular fungi belonging to the Asco-mycetes. Yeast cells show no differentiation: they are composed of a skeletal membrane surrounding cytoplasm which holds in suspension the nucleus, the chondriosomes and a few vacuoles.

Yeasts and amoebas have nothing which could be called a head or tail end. Many single-celled organisms do, however, possess polarity or a symmetrical axis.

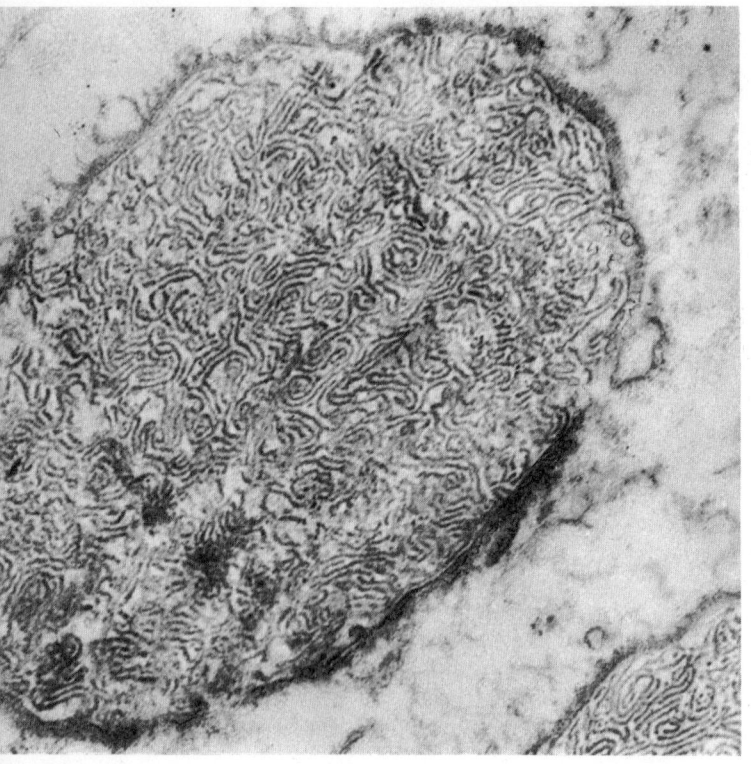

Section of an older spermatid of a snail, enlarged 50,000 times by the electron microscope. Here the chromonemata appear in the form of fairly thick, wavy filaments.

Laboratoire de microscopie appliquée à la Biologie. C. N. R. S. P. P. Grassé.

Among the green unicellular algae of the genus *Chlamydomonas,* which often colour sea-water bright green, the forepart of the cell is distinguished by a rostrum, or a thickening of the membrane, which bears two filaments, the flagella, by means of which the alga moves through the water. In the anterior third of the cell are found the nucleus and the pulsating vacuoles which expel waste matter. An enormous chloroplast, responsible for food synthesis, occupies almost all the cell. On its anterior edge can be seen an orange-tinted lenticular corpuscle, the eye-shaped stigma coloured by carotene. It affects the alga's orientation by following the direction of light rays. Towards the posterior extremity of the cell, embedded in the chloroplast, is the *pyrenoid*, a small protein granule surrounded by a starch sheath. The anterior region of the cell is thus chiefly sensory and locomotive while the posterior region is concerned with the assimilation of food. Thus in the *Chlamydomonas* there are the faint beginnings of functional differentiation. This differentiation becomes more striking among the Ciliophora such as *Paramecium*.

Paramecium are very common in ponds or wherever decayed matter collects in stagnant water. They measure two-tenths of a millimetre and are oval in form. The supple cellular membrane of *Paramecium* is covered with short vibratile 'hairs' or *cilia*. Embedded in the outer cytoplasm, between the cilia, there are small oval bodies called trichocysts which can be discharged like minute arrows to attack an enemy. On the left side of the under or ventral surface there is an oblique groove slanting to the right. At the end of

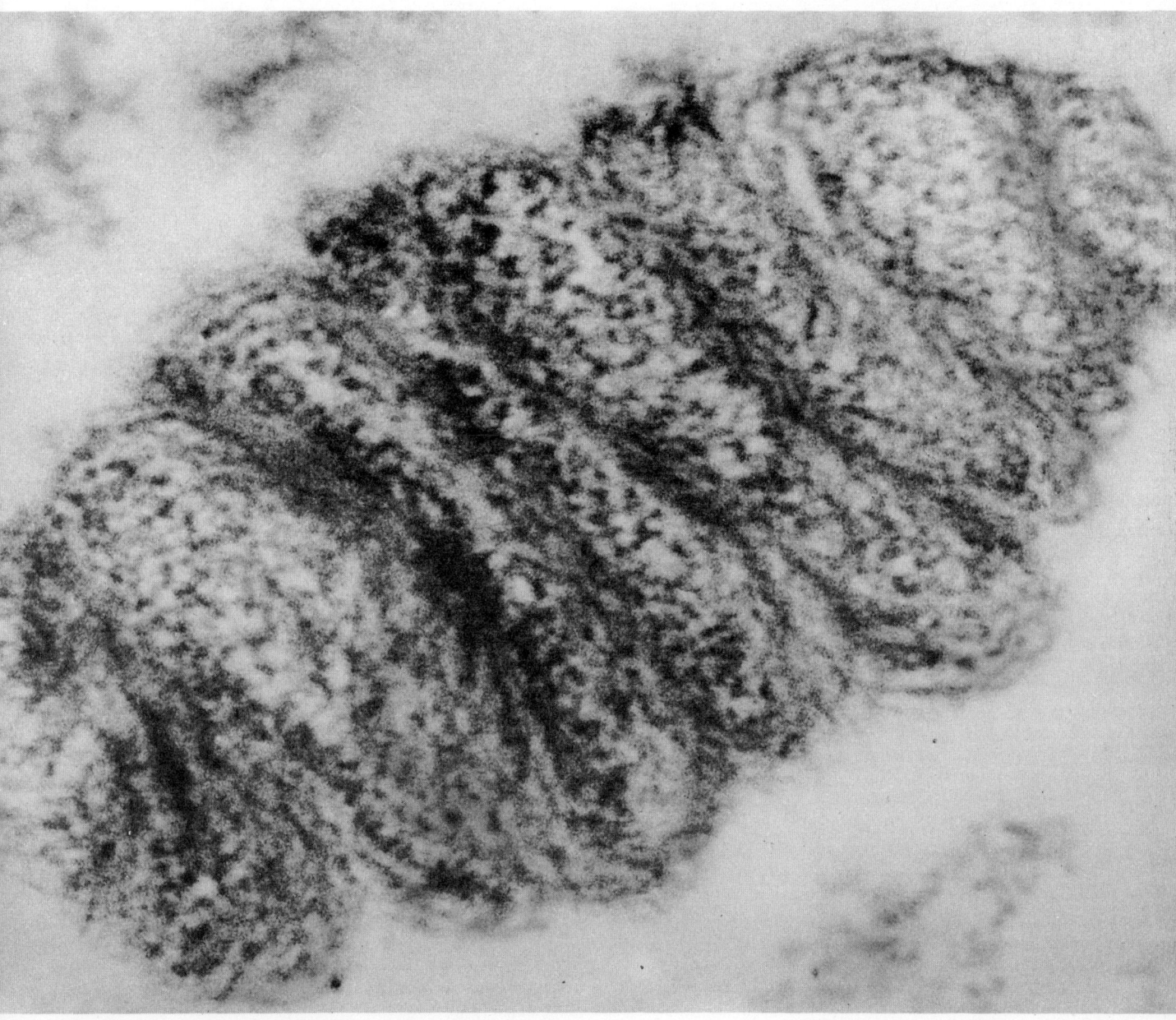

Single chromosome of the protozoan *Amphidinium* species as seen 200,000 times enlarged by the electron microscope.

Laboratoire de microscopie appliquée à la Biologie. C. N. R. S. P. P. Grassé.

this concavity there is an opening, the mouth pore, which leads into the cytopharynx or gullet and is supplied with an undulating membrane. Food particles — bacteria and other minute organisms, are drawn into the oral groove and enter the mouth and the cytopharynx where they are concentrated into a ball and form a food vacuole. When digestion has been completed the vacuole moves to the surface — always to the same point — bursts, and eliminates the indigestible waste matter. At each extremity of the cell there is a contractile vacuole, composed of a central reservoir and a circle of minute radiating canals. When one of these contracts (systole) the other relaxes (diastole) and vice versa.

Paramecium, like all other Ciliata, contains two nuclei, one of which is large and concerned with the cell's ordinary functions, while the other, which is small, plays the essential reproductive role.

A system of muscular fibres cooperates with the beating cilia to enable *Paramecium* to swim, to curve inwards and to contract. Although unicellular, *Paramecium* has thus achieved a certain complexity, it possesses locomotive, defensive, digestive, and excretory systems.

Diatoms are brown unicellular algae, each cell being covered by a carapace of silica, variable in form, but always composed of two overlapping halves, finely sculptured. Through the centuries the siliceous cell-walls of countless fossil diatoms have formed deposits of friable rock known as kieselguhr.

The simple yeast cell, which offers no visible signs of differentiation, and the complicated cell of *Paramecium* represent two extremes of unicellular structure. In the latter, cytoplasmic differentiations already foreshadow specialized organs: digestive vacuoles carry out digestion; contractile vacuoles eliminate waste; cilia and flagella assure locomotion, and the stigma behaves like a sensory organ.

The coenocytes

Some inferior plants, notably certain fungi and algae, are composed of a mass of protoplasm containing numerous nuclei but no cell-wall divisions. This rather odd structure is called *coenocytic.*

For example, Myxomycetes, or slime moulds, consist of a naked viscous and often voluminous mass of multi-nucleate protoplasm. Such masses are known as *plasmodia* and change continually in appearance, sending forth prolongations and moving in forests on the ground or up tree-trunks rather like a giant amoeba at the speed of 1-10 centimetres an hour.

In the same way the Saprolegniaceae, mould fungi which live in fresh water on decaying animal or vegetable matter, possess a thallus formed of tubes or of branched siphons. Each siphon comprises a large axile vacuole surrounded by a band of cytoplasm with numerous nuclei, mitochondria, and oily droplets.

Certain green siphonal algae, the Vaucheriaceae, found growing in ditches, and the Caulerpaceae, display the same structure. The cytoplasm, which has no cell-walls, is full of nuclei. One species, the compara-

tively large *Caulerpa prolifera* which frequents the Mediterranean coast, resembles a superior plant and has a kind of rhizome which affixes itself with crampons and sends green blades up through the water. The plant consists of a rigid membrane enclosing an unwalled protoplasmic mass rich in nuclei. In all parts of the alga this mobile protoplasm is similar, although the green blades contain a greater abundance of chloroplasts. The plant's pronounced morphological differentiation is accompanied by very slight internal differentiation.

All of these coenocytic organisms are reduced, during one period of their life cycle, to typical cells with a membrane isolating a nucleus. In them all, then, the cell virtually exists during the course of their vegetative life. This 'virtual' cell, enclosed by imaginary walls, comprising a nucleus and the cytoplasm which surrounds it, is called an *energid.* The plasmodia of Myxomycetes and the filaments of the Vaucheriaceae are formed by a multitude of energids. In fact coenocytes are organisms composed of an assembly of energids. Similar structures are also found in the animal kingdom. For instance, the parablast of the eggs of certain fish, the early placenta of mammals, striped muscle and hepatic cells are partly coenocytic in structure.

From the physiological point of view the coenocytes behave as a cellular whole. Instead of plasmodium the word *syncytium* is sometimes employed. The two terms are synonymous, though botanists generally use the first while zoologists prefer the second.

Multicellular organisms

Unicellular organisms and the coenocytes form only part of the Earth's population. By far the most important of its inhabitants are multicellular. Nevertheless these more complex organisms all began as a single cell, the fertilized ovum or *zygote*. The zygote, by its successive divisions, gives rise to a great number of cells which are at first alike. Then they differentiate and acquire a recognizable form and structure. At the same time they undergo functional differentiation and become specialized for the role they are destined to play.

Cells of the same form, the same structure and the same function group together to form a *tissue*. By this is meant a region consisting mainly of cells of the same sort and function, a histological unit, histology being the study of tissues. Animal tissues include the epithelial tissue; the nervous tissue; the supporting tissues (reticulated, connective, elastic, cartilaginous and bone); *muscular* tissues; blood tissues and *glandular* tissues.

Tissues in their turn form organs which are defined by their structure and function. The organ is an anatomical unit. A division of physiological work is then achieved with the specialized organs assuming their various functions. The entire physiology of the multicellular organism results from the elementary task performed by each of its cells. Certain cells specialize in the absorption of oxygen and others in its fixation; others produce ferments while others eliminate waste matter. As animal organization

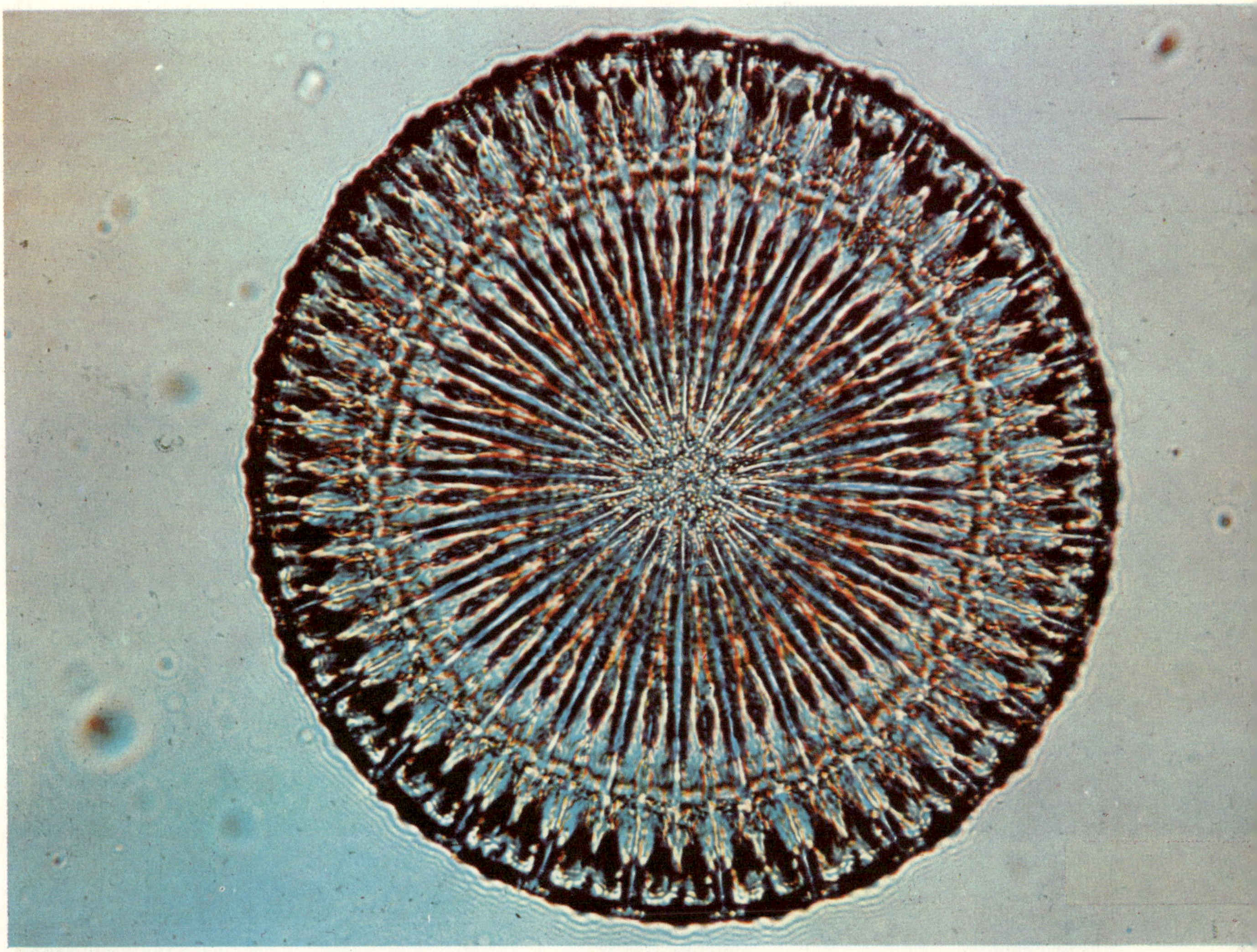

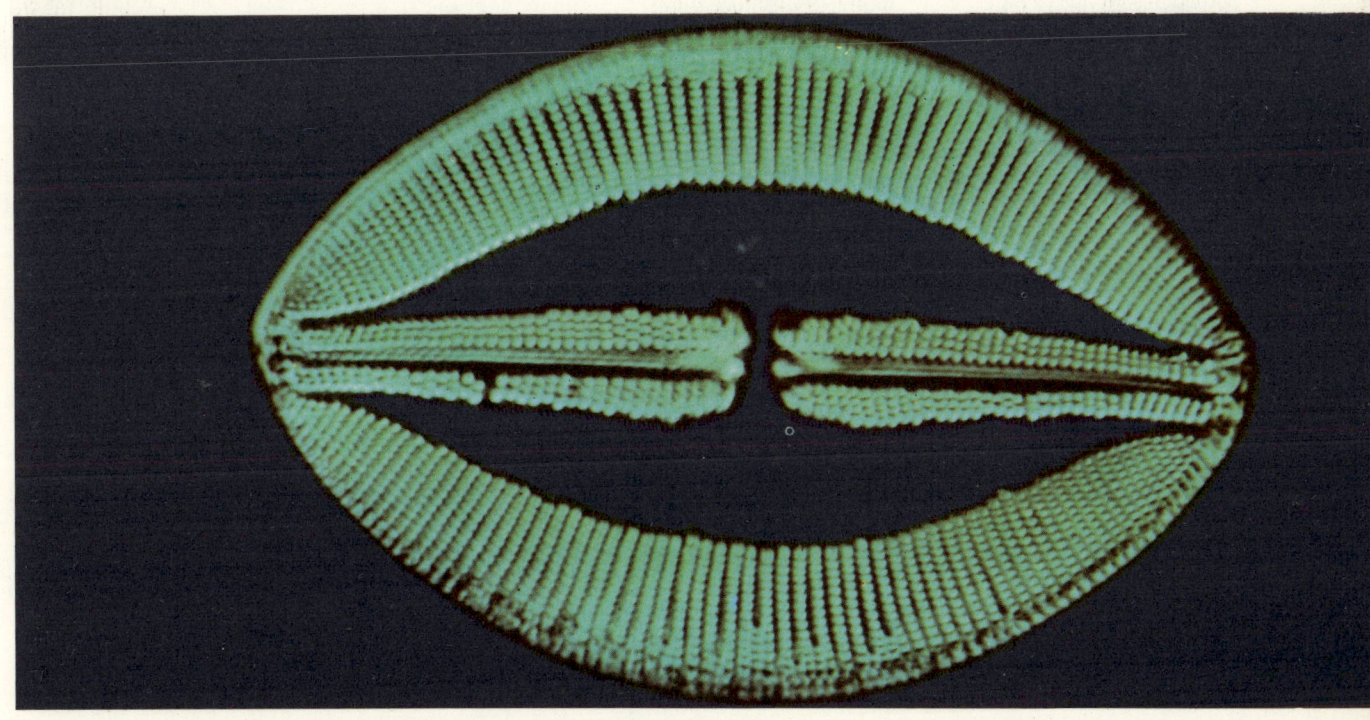

Above, Lepidodiscus elegans from Singhewsky, U.S.S.R.
Magnification X 980.

Below, Navicula Hennedyi from Yokohama, Japan.
Magnification X 980.

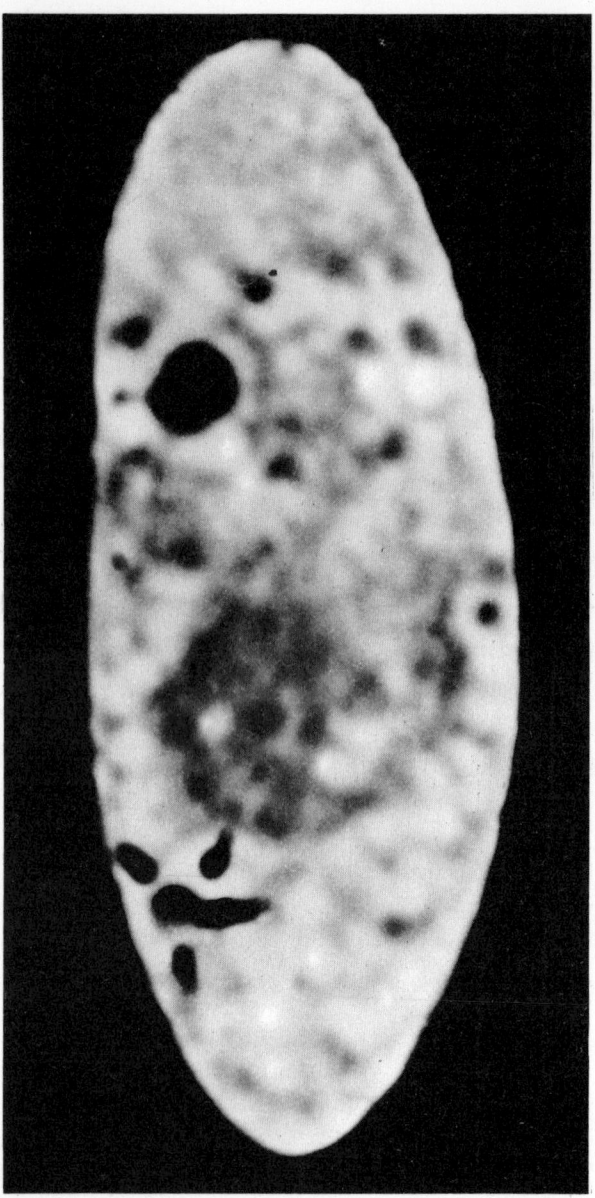

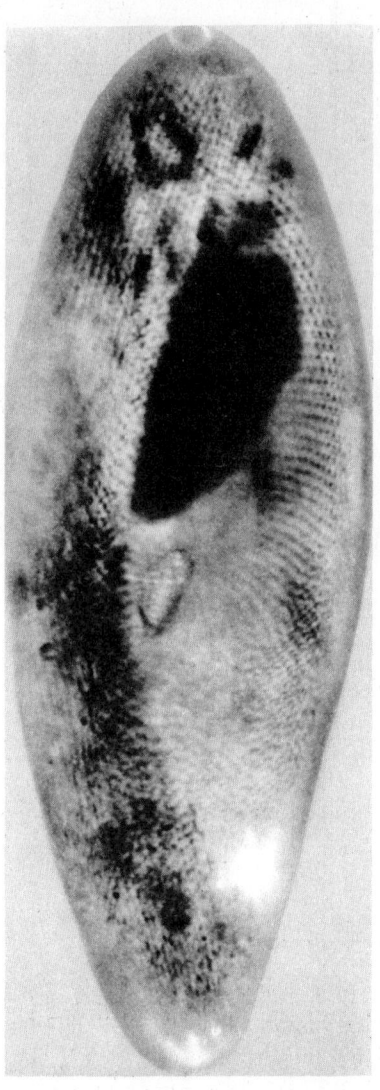

A living *Paramecium caudatum.*

H. Mugard.

Paramecium caudatum stained with silver.

H. Mugard.

becomes more complex the cells are separated from each other. The spaces left between them are occupied by the nutrition the animal has absorbed. These substances dissolve in the interstitial fluid, and in this manner reach the various regions of the organism. This primitive kind of lacunary circulation is replaced, in higher forms of life, by a closed circulatory system of canals and vessels. Still other cells are sensitive to outside stimuli which they interpret by responses useful to the organism. Finally some cells retain their original potentialities: these are the germ-cells, the 'totipotent', potentially immortal cells, which are responsible for the creation of new life and the continuity of the species.

All cells, then, with the exception of the germ-cells, are subject to differentiation and group themselves into organs which accomplish the many tasks indispensable to the maintenance of life. Each individual cell leads its own life and, at the same time, cooperates in the total life of the organism.

Several organs can work together to achieve a single purpose. Such collaborating organs compose special systems. Each function has its own system: respiration, the respiratory system; digestion, the digestive system; elimination, the excretory system; and so on. The connection is purely physiological and does not imply structural identity. The various respiratory systems of animals, for example, bear little resemblance to each other. The trachea of insects, the gills of fish, the lungs of mammals possess different structures though they have one feature in common, namely the ability to allow atmospheric oxygen to pass through them.

In order that the various functions of the organism should work together harmoniously some means of correlation is indispensable. Among animals this is assured by the nervous system and, in the case not only

Amoeba proteus (printed negative).

H. Mugard.

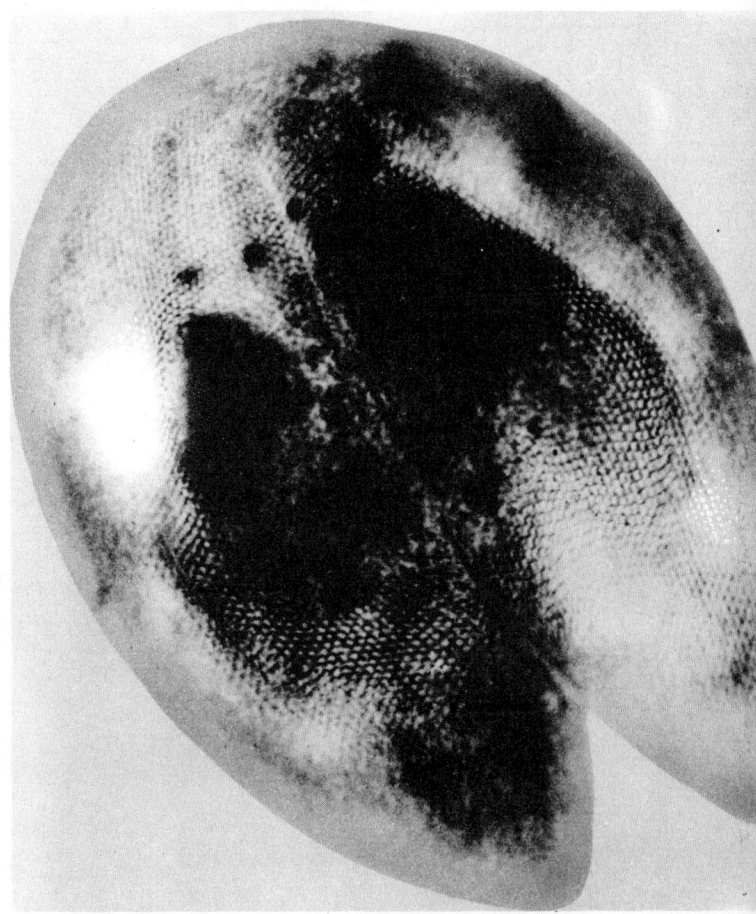

Paramecium caudatum in conjugation (stained with silver).

H. Mugard.

of vertebrates but of many invertebrates, by the hormones as well. Among the invertebrates the hormones of metamorphosis and of pigmentation are the best known. The hormone and nervous systems function synergistically, that is, their combined effect is greater than the sum of each acting alone. The higher in the evolutionary scale the organism rises, the greater will be its anatomic and functional differentiation, and the more need it will have for coordination.

The cell, nonetheless, remains the fundamental element of all living things, microscopic or macroscopic, whether it remains in isolation or combines with countless other cells, its daughter cells, granddaughter, great-grand-daughter cells . . . for all alike were derived from one unique cell, the zygote.

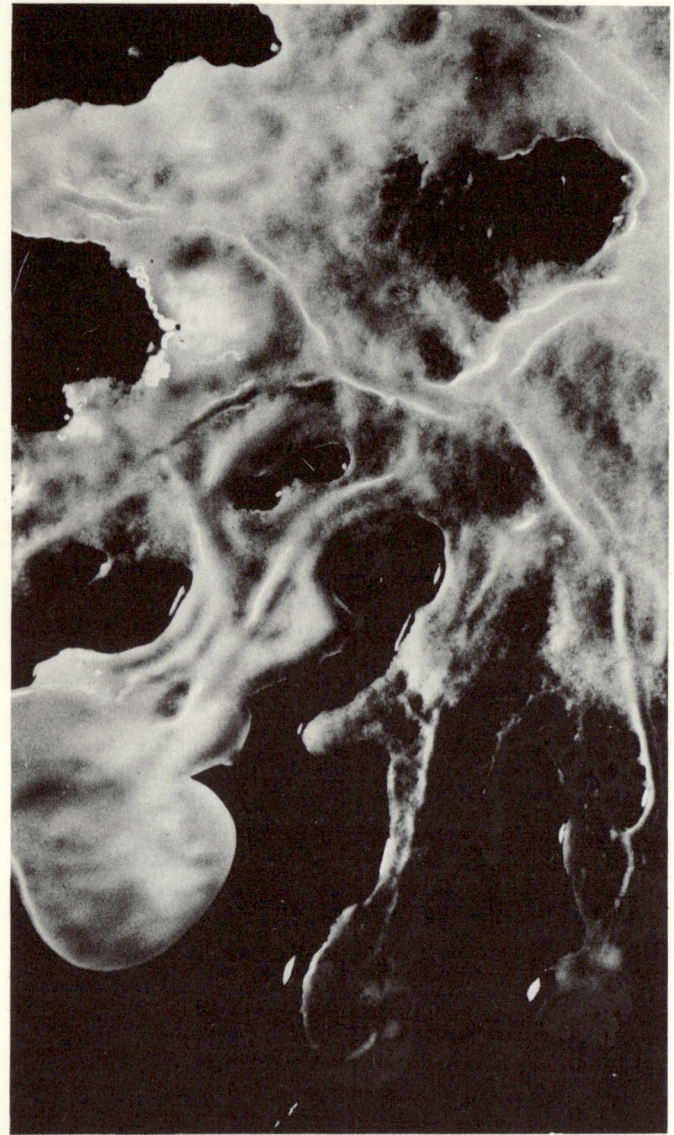

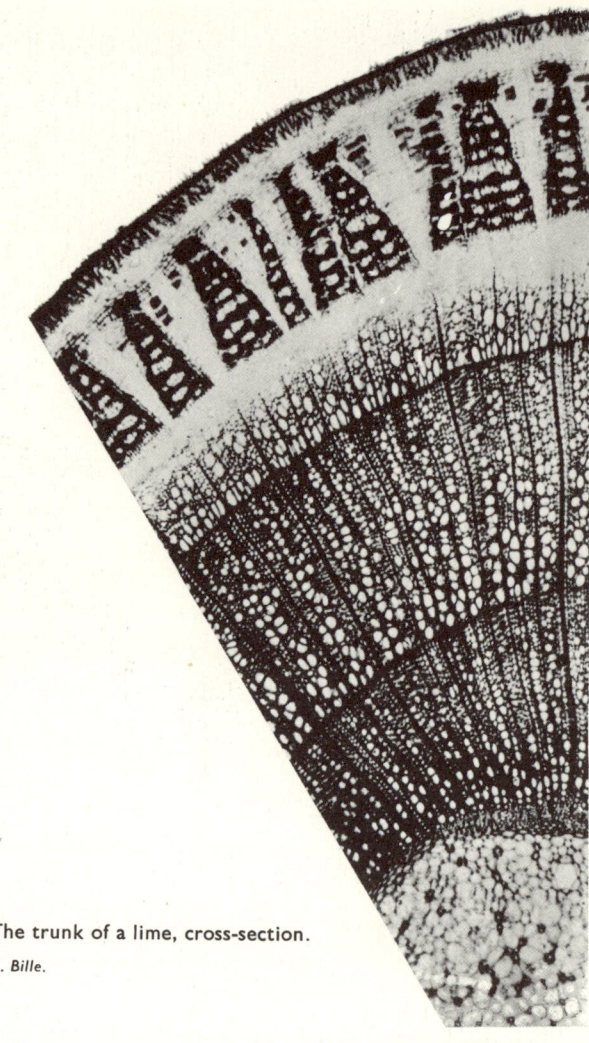

The trunk of a lime, cross-section.

C. Bille.

Plasmodium — the vegetative stage — of a slime mould (Myxomycetes).

Dragesco.

Human striped muscle.

M. C. Noailles.

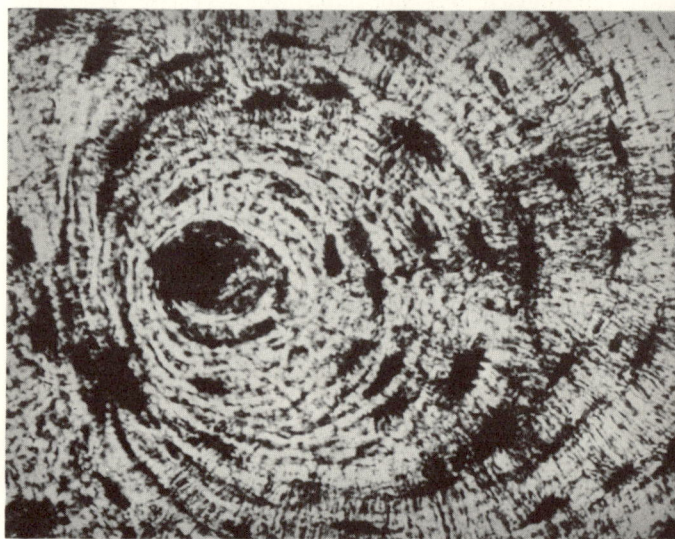

Cross-section of a dry bone showing the Haversian canals and bone cells.

M. C. Noailles.

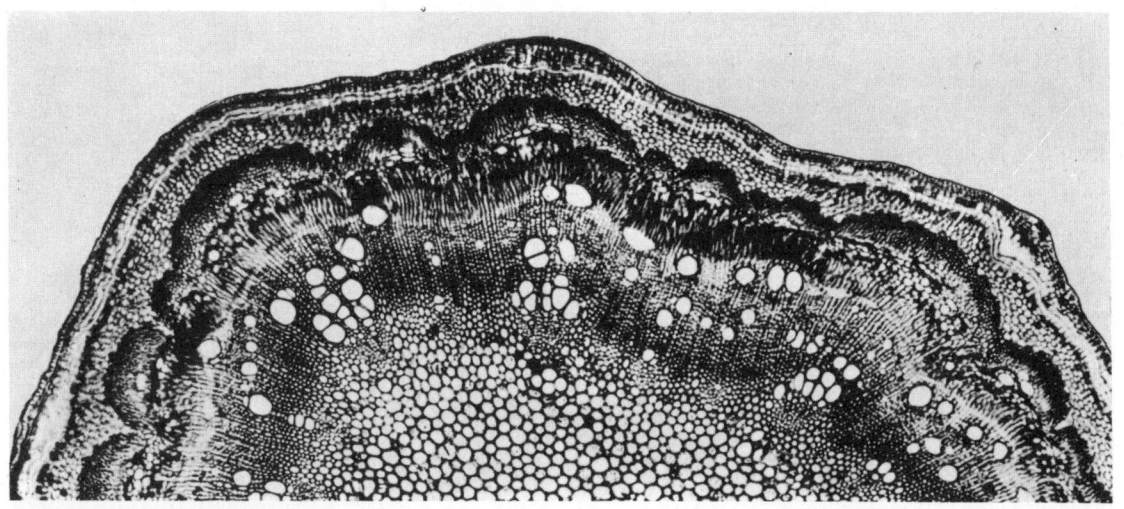

Cross-section of wistaria.

C. Bille.

Cell-division
and Multiplication

The reproduction of single-celled organisms and the building up of multicellular life has already indicated the vast importance of cell-division.

Every cell derives from a pre-existing cell. *Omnis cellula e cellula:* such is the celebrated axiom which Virchow enunciated in 1850. Today this truth seems to be generally accepted, although certain schools of thought affirm that living cells can be created from simple proteins. The experimental verification of this affirmation remains to be made.

The discovery of cell division is relatively recent. For many years the *blastematic theory* was thought to be true. This held that cells sprang from amorphous living matter called 'formative blastema or cytoblastema', the cells of which were supposed to have arisen by crystallization. Cell-division, suspected by Raspail, was discovered by Remak, Mohl, and Virchow. Numerous cytologists then analysed and defined its various manifestations. Even today, however, we cannot pretend to know everything about the inner mechanism of cell-division. Cell-division takes place in two different manners. One process, which is very uncommon, is direct division or *amitosis*. The other, almost universal, is indirect division, *mitosis* or *karyokinesis*. Every division occurs first in the nucleus, then in the cytoplasm and its inclusions, and finally in the membrane. The two daughter cells which result from the division of the mother cell possess all the elements indispensable to cellular life.

Amitosis

Amitosis is a simple process comparable to cutting an apple into two appreciably equal parts. In amitosis near identity is achieved but not total identity. Formation of daughter cells which in all details are exactly alike, and exactly like the mother cells, does not result, and amitosis is, therefore, rare. The development of the organism and the building up of tissues demand an exact division of all the cell's potentialities, a division which is only achieved by mitosis.

Mitosis

Mitosis, a complex phenomenon which proceeds in successive stages, takes place in an almost identical manner in both the animal and vegetable kingdoms. The essence of mitosis is the separation of the duplicated nuclear chromosomes into two rigorously equal sets, one set going to each daughter cell. The nuclei of both daughter cells then have a complement of chromosomes identical to that of the original cell. The four successive stages into which the process is divided are known as the prophase, metaphase, anaphase, and telophase.

1. *Prophase:* within the nucleus double chromosomes appear, double because they have split longitudinally in two. They shorten and, in consequence, thicken. Just outside the nuclear membrane in the cytoplasm the centriole forms a radiating system of striations — the aster. Centriole and aster divide and the two new centrioles, each with its aster, move apart to opposite poles of the cell where they remain throughout the

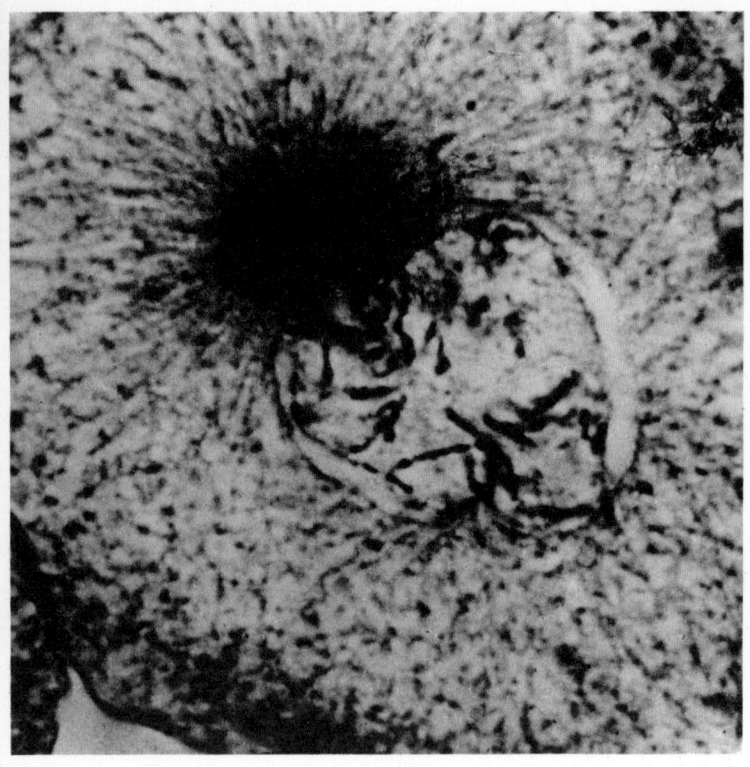

1

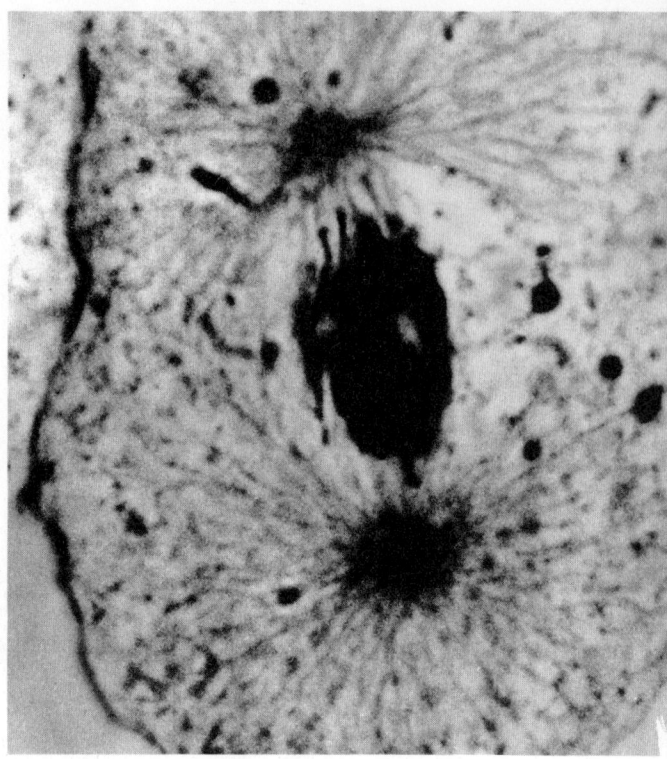

2

Mitosis of the cell of a fish. 1. Beginning of the prophase: division of the aster and appearance of the chromosomes. 2. Prophase more advanced: the spindle forms between the asters; the chromosomes are visible. 3. Beginning of the metaphase: front view of the 'equatorial plate'. 4. Metaphase more advanced: the chromosomes are joined in an equatorial plate at the 'equator' of the spindle between the asters. 5. Anaphase: the duplicate of each chromosome (chromatids) move towards the two poles of the spindle.

Gallien. Larousse.

division. Between the two poles the spindle develops. The spindle plays an important part in the subsequent stages of mitosis. The second phase begins.

2. *Metaphase:* at an equal distance between the two poles of the cell the equatorial zone of the cell is situated. The nuclear membrane dissolves and the nucleus loses its individuality. The equatorial zone is occupied by the duplicated chromosomes, one strand of which is called a chromatid. These bend, generally into a V, the angle being its centromere or spindle-attachment. The two arms of the V may be of unequal length. The third phase begins.

3. *Anaphase:* anaphase is characterized by the polar ascension of the chromosomes. The paired chromosome strands — or chromatids — separate and move towards opposite poles, following the fibres of the spindle. When each set has arrived in the neighbourhood of the centriole the fourth and last phase begins.

4. *Telophase:* at this stage the cytoplasm divides. At the same time in each of the daughter cells a nuclear membrane forms and the nucleus is reconstituted. The

centriole either disappears or persists, single or doubled, already prepared for the next mitosis.

Mitosis has finished. It has lasted, in the case of certain unicellular organisms, for half an hour. In the case of plant cells it can take more than five hours. The duration varies according to whether the cell is animal or plant, and also according to the kind of cell in question. External factors too, such as temperature, can hasten or delay the speed of mitosis.

In the cells of embryo chickens cultivated *in vitro* the following duration of the various stages was observed: prophase, five to ten minutes; metaphase, less than five minutes; anaphase, ten minutes; and telophase, more than twenty minutes. In all observations telophase, during which the most complicated processes take place, was of longest duration. Mitosis thus comprises two fundamental operations: the division of the nuclear chromosome complement into two sets, each identical with that of the original nucleus, and the division of the cytoplasm.

The division of the nucleus can occur without being followed by cleavage of the cytoplasm. In that case the result is a voluminous cell with multiple nuclei, such as the cells of striped muscle-fibre and the polykaryocytes of bone marrow, which can, as we have seen, be considered as a syncytium or a plasmodium or a coenocyte. The division of the nucleus, then, appears to be the chief element in mitosis and it is by this operation that the number and quality of the chromosomes are exactly maintained. Mitosis is a method of cell-division which is practically universal. Every cell which reaches a suitable size divides into two cells by mitosis, but what determines mitosis itself still remains obscure.

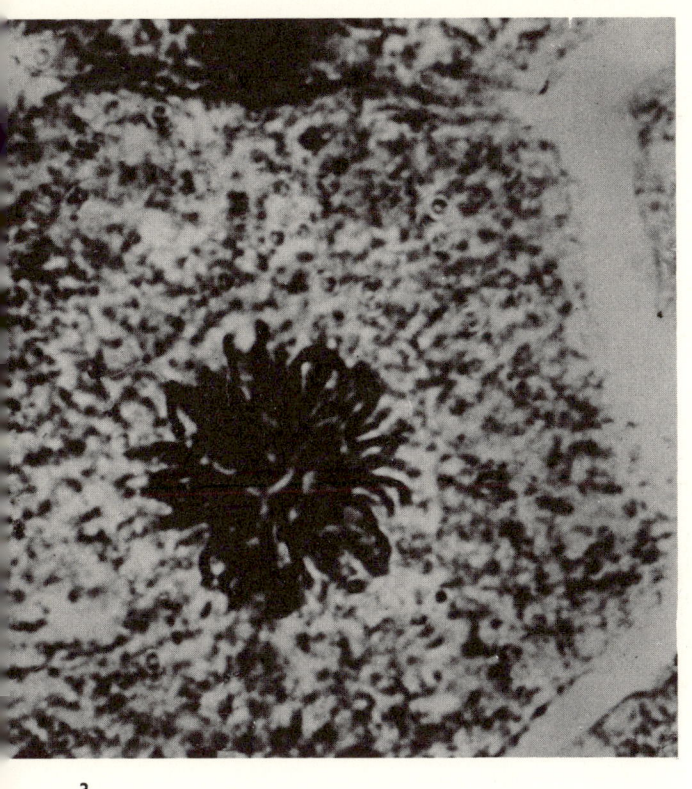

3

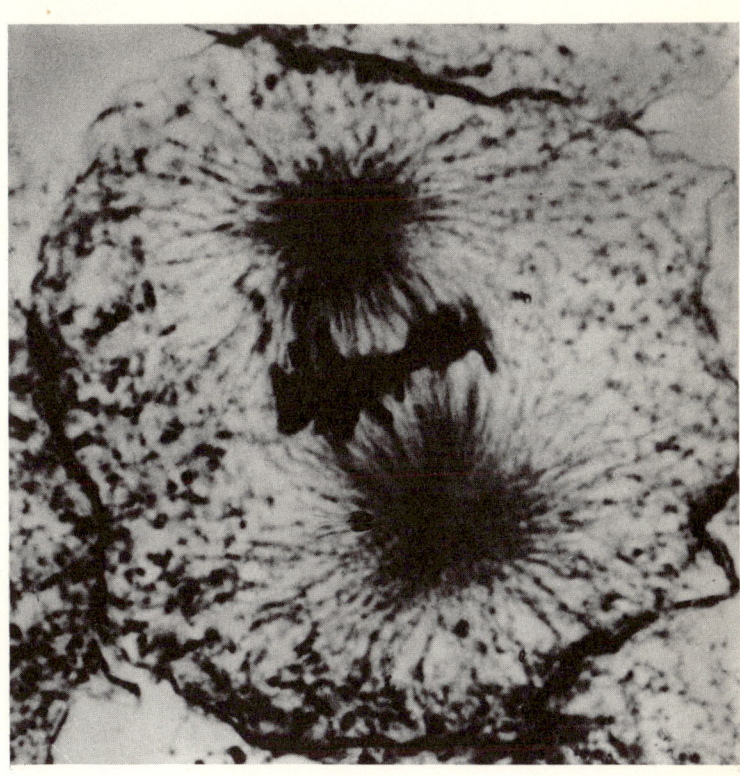

4

5

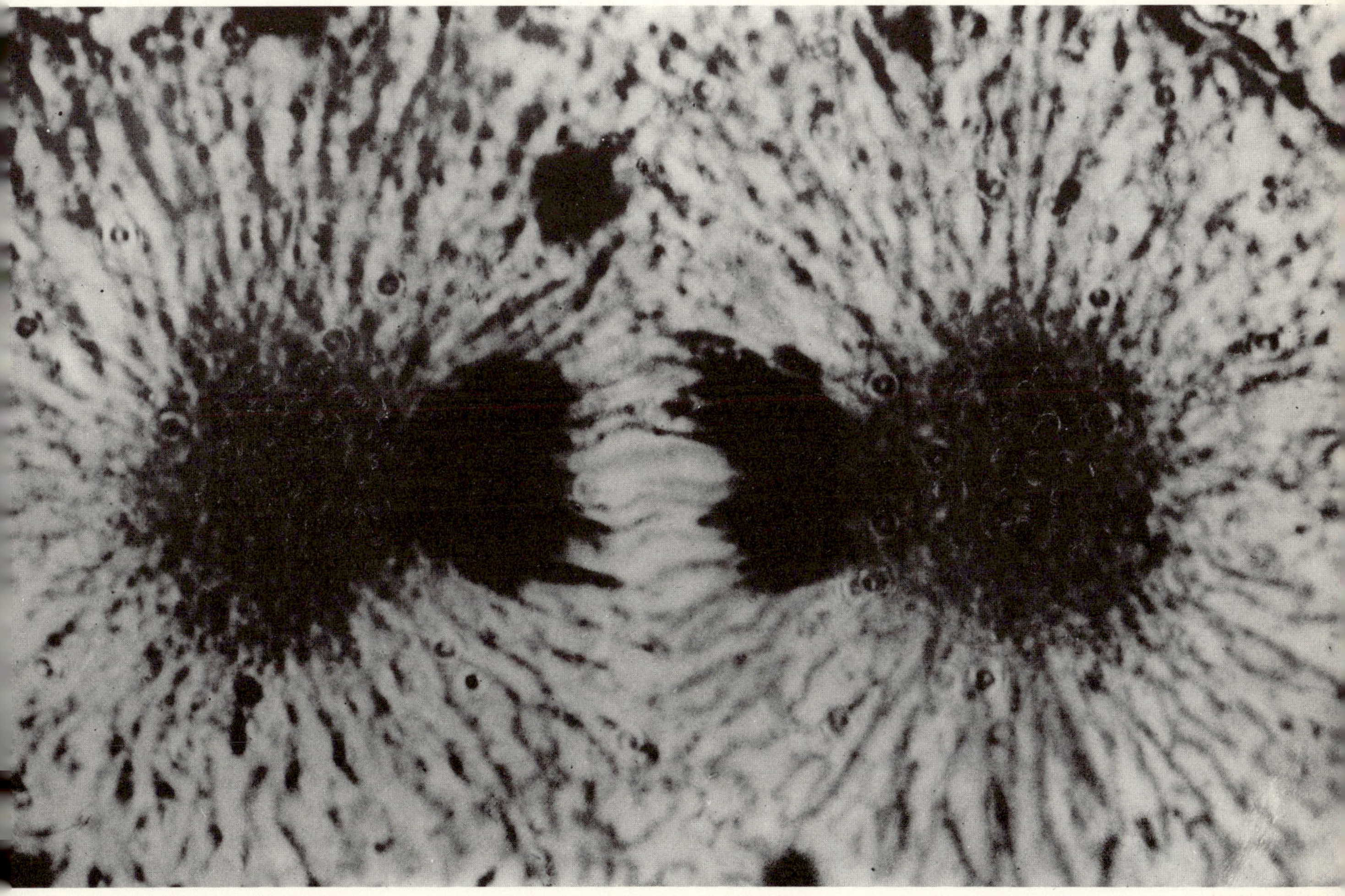

Meiosis

There exists, however, a special kind of mitosis which is confined to the creation of male and female sex-cells. This is *meiosis* or *reduction division.* All the cells of the organism are *diploid,* that is, their chromosomes are in pairs. The germ-cells or *gametes* are the sole exception, being *haploid,* that is, having in each nucleus only a single set of unpaired chromosomes. Now the cells which produce the gametes are themselves diploid and have the normal double set of chromosomes. During the division which gives rise to the gametes the daughter cells or gametes receive only a single set of chromosomes and, in consequence, only half the mother cell's chromosome potentialities. Meiosis is the cell-division which results in the formation of gametes. Among men and women, for example, every cell in the body has forty-six chromosomes. Their gametes, ovum and sperm, contain only twenty-three.

Meiosis consists of a combination of two successive mitoses: the first is the 'reduction division' which reduces the number of chromosomes in the two daughter cells to one half, cells which are haploid instead of being diploid — like the mother cell. The second mitosis maintains the haploid number of chromosomes in the four resulting 'grand-daughter' cells. Reduction division, like normal mitosis, includes a prophase, a metaphase, an anaphase and a telophase which we shall not describe in detail.

Meiosis is characterized by the phenomenon of chromosome reduction which was discovered in 1883 by van Beneden. As we have seen, it results in the male and female gametes possessing only half the normal number of chromosomes. During fertilization, when the male gamete unites with the female gamete to produce a fertilized egg or *zygote,* the normal diploid number of chromosomes is restored. The zygote thus contains both the maternal and the paternal haploid sets of chromosomes.

Expulsion of the polar body by the human ovum. This phenomenon which occurs during meiosis or reduction division reduces the number of chromosomes in the ovum from 46 to 23. These 23 with the addition of the 23 contributed by the spermatozoon make up the total 46 chromosomes which are characteristic of the human species.

Landrum B. Shettles. P. Popper.

Microbe Biology

Until now we have considered only the cell, the basic unit of all living organisms, the single cell leading a free and autonomous life in the case of the Protozoa and the Protophyta; cells in groups, differentiated specialized, composing the more or less complicated, multicellular organism of the Metazoa and the Metaphyta. No mention has yet been made of the bacteria, the *Rickettsia* and viruses, which are often grouped together under the general name of 'microbes'. In spite of their infinitesimal size these micro-organisms are of enormous importance.

Bacteria

In a drop of sea-water or a speck of manure, microscopic examination always reveals the presence of bacteria, varying in size between a few tenths of a micron to perhaps 2 microns. The largest bacteria are still smaller than the smallest of cells. Their forms are simple but varied. The micrococcus is a little sphere, the bacillus a small straight rod, the vibrio, spirillum and treponema are wavy or spiral in shape. Certain bacteria are provided with long, fine filaments or flagella. These are organs of locomotion and allow such bacteria to move. Those having no flagella are normally immobile and are subject to the movements of the liquid in which they are in suspension.

The structure of bacteria is not yet fully known. For many years it was believed that the bacterium was a rather exceptional kind of cell devoid of a nucleus; its cytoplasm contained glycogen, sometimes fatty inclusions, and in the case of the Thiobacteria even sulphur. It was shown to have a membrane, but the nature of the membrane is still disputed.

Later, bacteria were recognized to possess some-

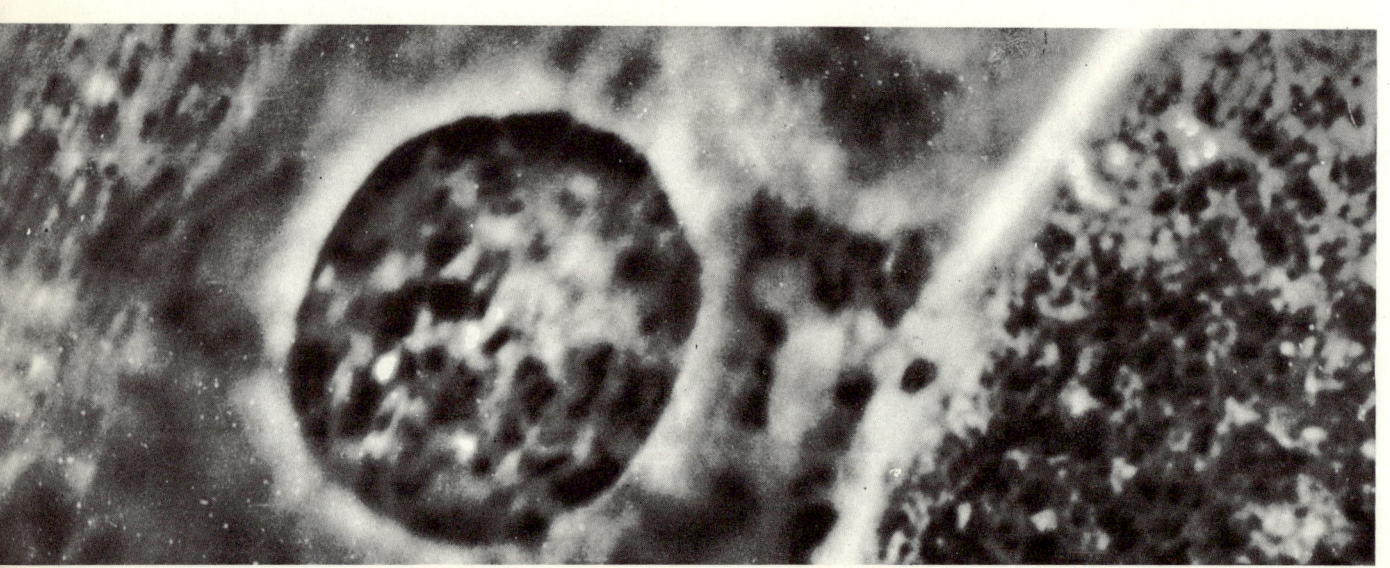

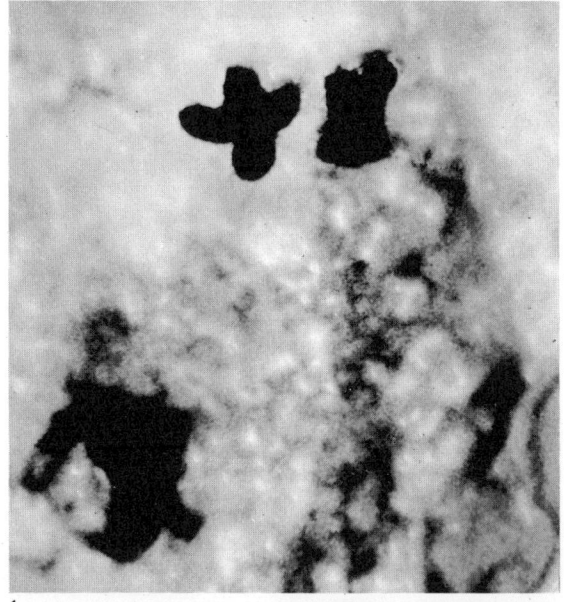

1

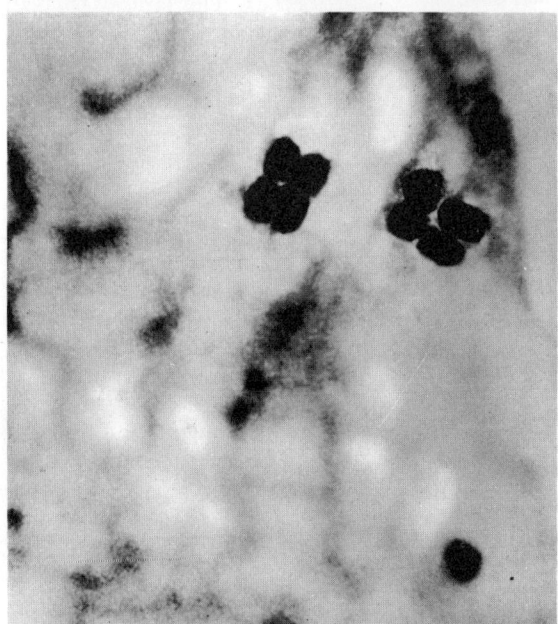

2

The successive stages of meiosis.
1. The two tetrads are formed of four chromatids each.
2. The two tetrads: separation of the chromatids by reduction division.
3. Emission of the first polar body. Two dyads remain parallel to the wall.
4. The first polar body is expelled, and the two dyads have turned 90°. Achromatic spindle of the second division.
5. Emission of the second polar body. Two chromatids remain.

H. Mugard.

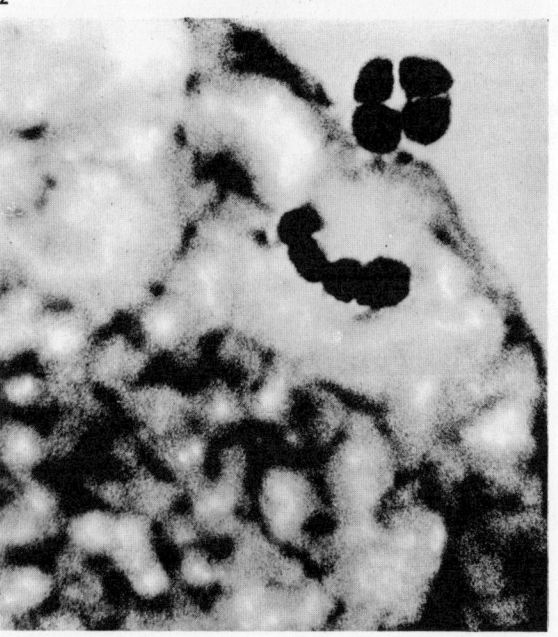

3

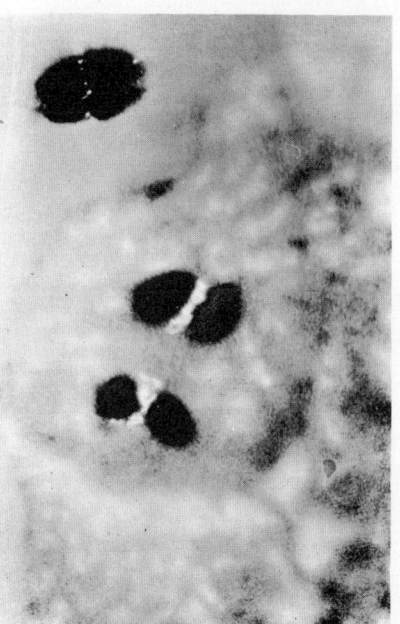

4

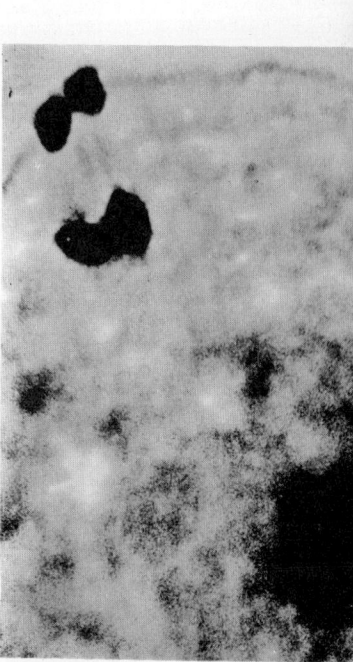

5

thing which suggested a nucleus. Nuclear stains were found to colour certain irregular filaments in the cytoplasm and also certain granules of chromatin which have been named chromidium. This chromidium, representing a diffuse nucleus, may well be chemically equivalent to a nucleus without, however, having the form of a nucleus. For this reason bacteria are sometimes considered as 'Prokaryotes' — with a primitive nucleus, in contrast to the 'Eukaryotes', namely, plants and animals which have a normal nucleus. Harmless or non-pathogenic bacteria live freely in the ground and in water where their activity is of great importance in the fermentation and decay of plant and animal tissues. Pathogenic bacteria live either freely or install themselves in superior organisms, animals and men, where they cause serious diseases such as diphtheria, tetanus, typhoid fever, tuberculosis, cholera, or syphilis. Bacteria are thus both valuable and disease-producing parasites.

All bacteria multiply by simple fission. Many form cysts, which are called spores.

Rickettsia

Rickettsiae are micro-organisms which occupy an intermediate position between bacteria and viruses. In size they are about a half-thousandth of a millimetre. Normally they inhabit the intestines of arthropods. Some are parasitic in man and animals and are responsible for exanthematic typhus and certain related diseases. Among animals they cause psittacosis, conjunctivitis in cattle, rickettsioses in dogs, sheep, and oxen. These diseases are transmitted either directly or by arthropods like lice. The pathogenic species of rickettsia live in the cells of the host and cannot be grown on the usual bacterial culture plates. They multiply only in the presence of living cells. Their precise nature is not fully understood.

Viruses

Viruses are disease-producing agents, invisible under the ordinary microscope and generally so small that they pass through the fine porcelain filters which retain bacteria. For this reason the term filterable viruses is sometimes employed. In size they are much smaller than bacteria. The biggest virus, the smallpox virus, is only a few hundred millionths of a millimetre or thousandths of a micron, while the smallest, the virus of foot and mouth disease, is not more than ten thousandths of a micron. In size large viruses are not very far from bacteria, while small viruses are somewhere between the biggest and the smallest molecules of certain proteins, the albumin molecule being between two and three thousandths of a micron and the haemocyanin molecule measuring twenty thousandths of a micron.

Viruses are of necessity totally parasitic: they can

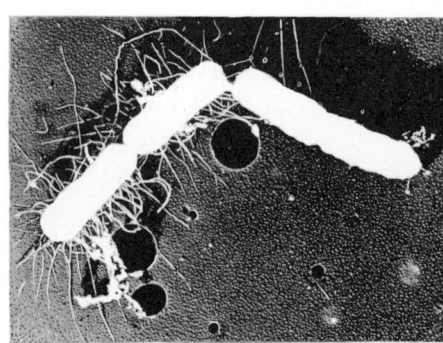

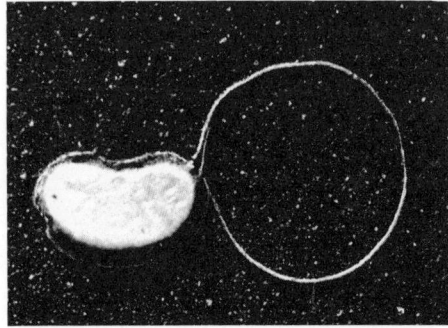

Conjugation of two bacteria. The female *(on the left)* is provided with numerous cilia and has begun to divide. The division of the male bacterium *(on the right)* has scarcely started. The two bacteria are united by a bridge of cytoplasm.

Institut Pasteur. Electron microscope photograph by Th. F. Anderson, E. Wollmann, and F. Jacob.

Vibrio metschnikovi, which causes cholera.

Delft Institute of Electron Microscopy.

multiply only within the living host cells they invade. Unlike bacteria they cannot be cultivated in nutritive broths. They attack human beings and cause diseases. These include the common cold, influenza, measles, lethargic encephalitis, poliomyelitis, smallpox, chicken-pox, shingles, yellow fever, and rabies. Among animals they are responsible for cattle plague, foot and mouth disease, swine fever, encephalomyelitis in horses, sheep-pox, distemper in dogs, and fowl pest. In plants viruses are responsible for numerous diseases: the mosaic disease of tobacco and other plants, *bushy stunt* in tomatoes, *curly top* in beetroots, *bunchy top* in banana trees, and many more. Certain viruses, known as *bacteriophages* or 'phage*s*', attack bacteria causing their 'lysis' or dissolution.

Among the all too many viruses which attack plants, the tobacco mosaic virus has been longest studied and merits particular attention. In 1898 Beijerinck began to suspect that the causative agent of tobacco mosaic revealed characteristics which were neither entirely those of living nor of inert matter, but something in between. In nature this virus attacks certain members of the Solanaceae family, principally tobacco and tomatoes. It causes more or less large patches of discoloration in the leaves as the result of destroying the chloroplasts. The virus can be transmitted through fine breaks in the plant's epidermis or by grafting. The juice obtained by crushing the diseased plants contains the virus which can then be extracted by a high speed differential centrifuge.

In 1935 in the United States Stanley, by adding ammonium sulphate, succeeded in precipitating purified and concentrated solutions of the tobacco mosaic virus. They appeared as microscopic paracrystalline, that is, having two-dimensional symmetry — needles. This achievement caused great excitement at the time. Rather too hastily it was believed that the virus would fill in the hiatus between living and inert matter.

Other viruses, including those of tobacco necrosis and tomato *bushy stunt,* were in their turn crystallized as faceted crystals which approached three-dimensional symmetry. These viruses are rounded particles while the particles of tobacco mosaic virus are rod-shaped, being fifteen thousandths of a micron thick and of a length which varies and can attain 100—200 thousandths of a micron when several particles join end to end.

These particles are composed of a nucleoprotein which contains six per cent ribonucleic acid or R N A. Each virus has its own specific ribonucleic acid. The proportions of nitrogenous bases, purine and pyrimidine, vary from one to the other. The protein content of viruses is chemically known and the various amino acids which compose it have been identified. A structural diagram has even been made of the rod-shaped particle, according to which it comprises a protein spiral enveloping a molecule of nucleic acid.

A partial 'synthesis' of a virus was achieved in 1955 by Fraenkel and Conrat. Under very special conditions they mixed the protein fraction and the nucleic acid

Treponema minutum.

Institut Pasteur. Giuntini-Moureau.

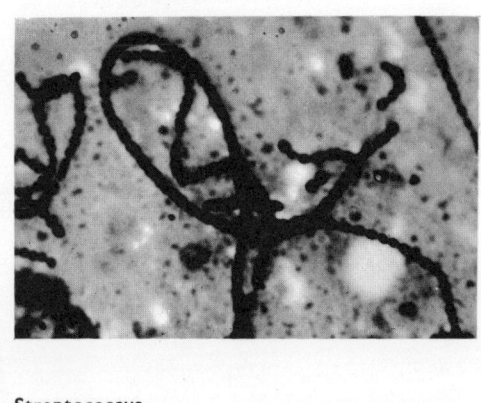

Streptococcus.

Institut Pasteur.

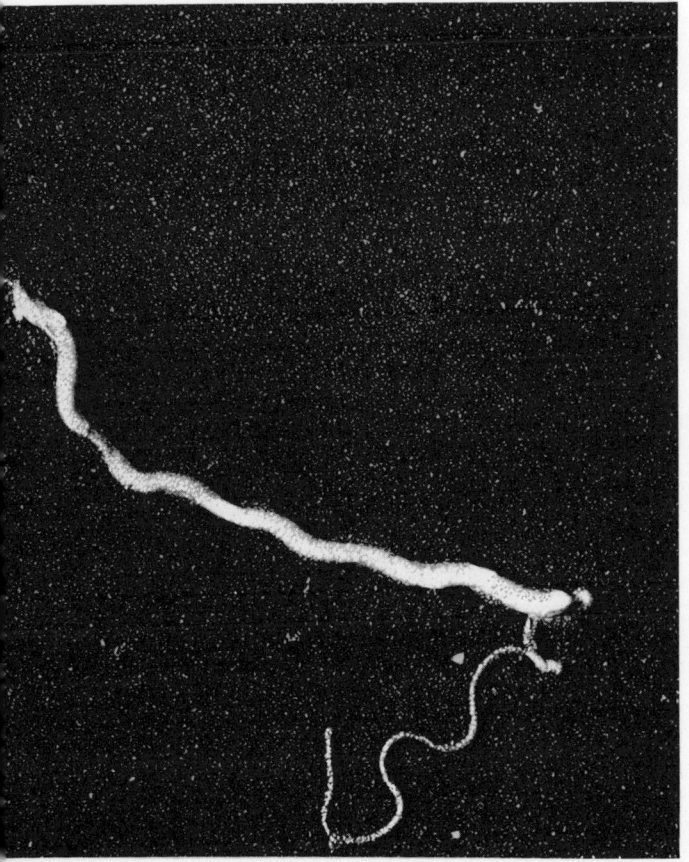

1. Virus particles: avian variola.
2. Twelve-sided crystals of the virus *bushy stunt*, a parasite of the tomato.
3. *Salmonella enteretidis* and bacteriophages D 4.
4. Bacteriophages D 4 of *Salmonella enteretidis*.

Institut Pasteur. Giuintini-Anastasin.

After Bawden and Pirie.

Institut Pasteur. Giuntini-Barbu.

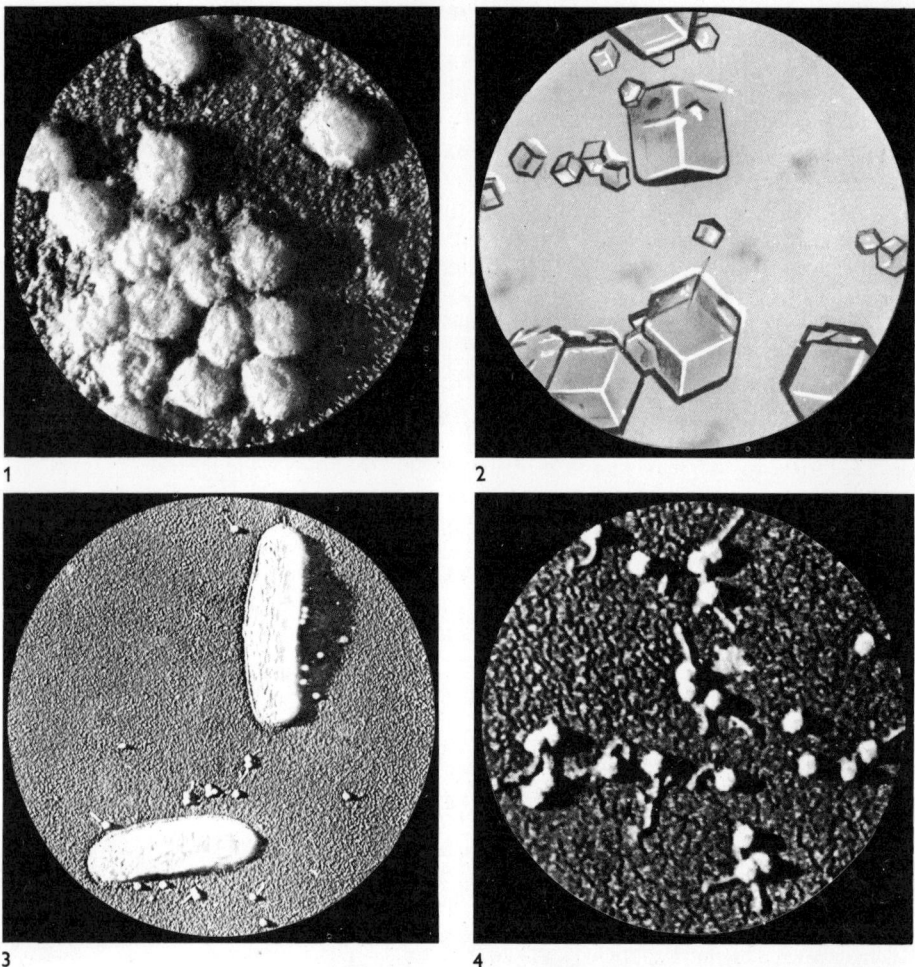

fraction — previously separated — and obtained rod-shaped particles which were identical to those of the virus and endowed with the same infectious properties. It seems that the nucleic fraction exercises the greater influence and must be intact, while the protein fraction need not be. Work by Bawden in 1956 on a relation of the tobacco mosaic virus which attacks a certain tropical leguminous plant showed that the protein fraction could in part be interchanged. The transfer was made from the virus of a variety of bean to the virus of tobacco and vice versa. Such successes, though important, must not mislead us. An authentic synthesis of living matter, though sometimes proclaimed, has not been achieved. In practice the so-called synthesis of virulent particles was attained by separating two constituents of the living virus and then putting them together again.

The virus multiplies in a very special manner. It seems that the host cell itself synthesizes the virus particles at the same time that it synthesizes normal proteins. The virus would thus appear to be devoid of means to reproduce itself, and is duplicated by the host cell from the model of the first particle introduced. Various techniques have enabled the stages of this multiplication to be followed. At first the quantity of virus injected into the host plant diminishes, then a period of latency begins. During this period it has been technically impossible to observe the virus particles which have already undergone a reduction in length. At the end of this period they reappear, having enormously multiplied.

During multiplication the virus may mutate, that is to say undergo sudden hereditary changes. Such mutations of the tobacco mosaic virus can affect either

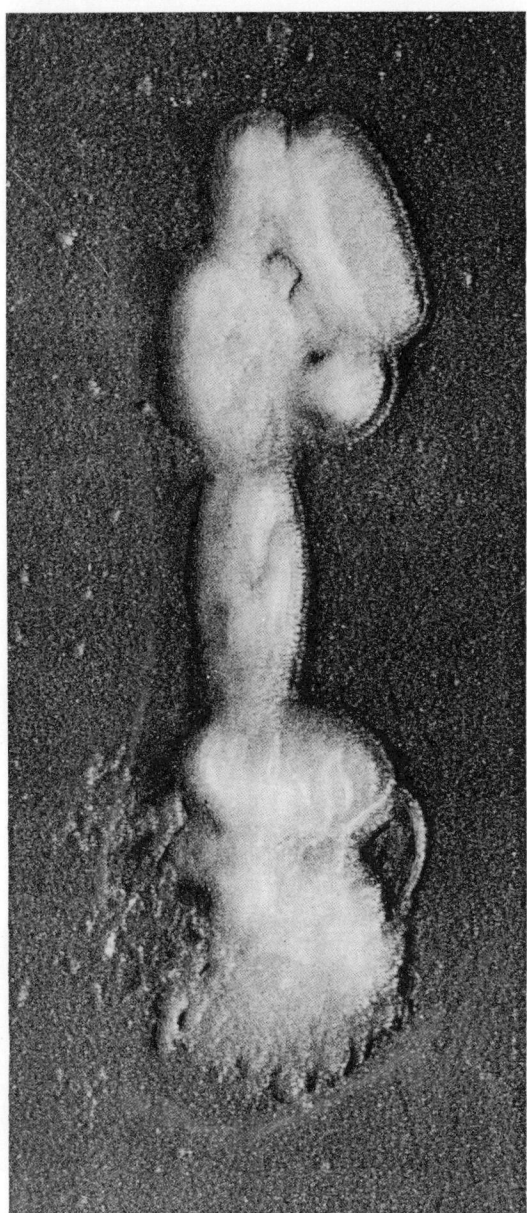

Rickettsia as seen by the electron microscope.

Institut Pasteur.

Particles of vaccine virus absorbed by the stroma of a chicken's red blood corpuscle.

Institut Pasteur. Giuntini-Vieuchange.

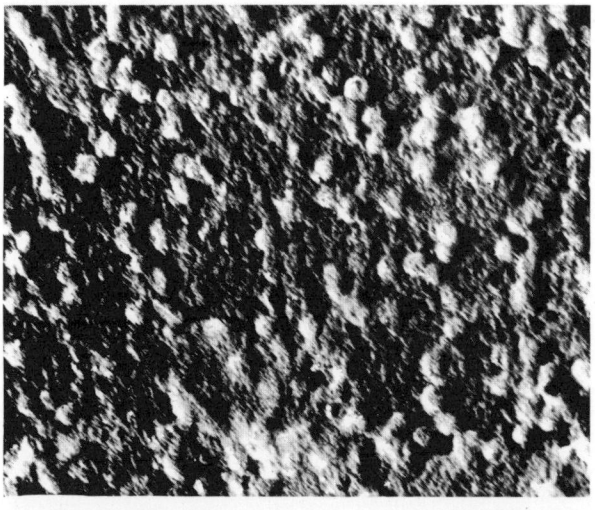

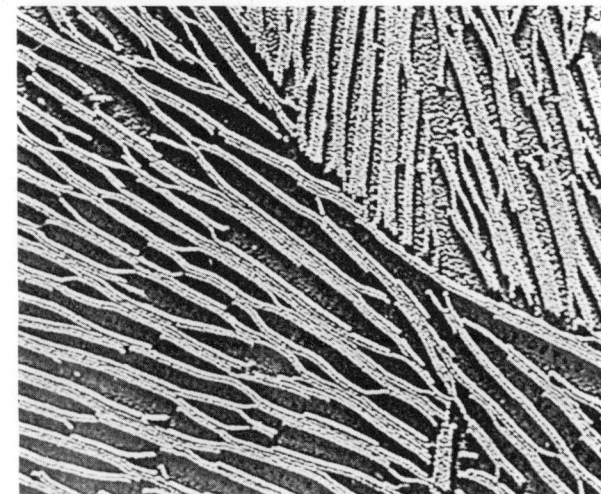

Particles of the tobacco mosaic virus magnified 70,000 times by the electron microscope. The photograph shows a concentrated solution in which the particles lie parallel to each other.

Institut Pasteur.

its degree of virulence, or the colour or intensity of the mosaic it produces. These various mutations are independent of each other.

Mutation of the bacteriophage viruses is analogous to the mutation of the tobacco mosaic virus. The bacteriophage of colibacillus is composed of a nucleoprotein and D N A. The bacteriophage penetrates the colibacillus and some twenty minutes later the colibacillus dissolves, releasing hundreds of bacteriophages. Thus the bacteriophage has multiplied, which of necessity implies that more desoxyribonucleic acid has been formed. Now on one hand the bacillus contained *no* D N A, its nucleic acid being R N A. On the other hand the virus (bacteriophage) is itself totally incapable of synthesis!

The only possible explanation of the two cases, that of the tobacco mosaic virus and that of the bacterio-

phage, relies on the synthesizing faculty of specific proteins and nucleoproteins: in other words, on their faculty of *auto-reproduction* or rather of *auto-duplication,* the mechanism of which still remains obscure. It is conceivable that scattered chemical elements, on contact with the invading parasite nucleoprotein, should regroup in such a manner as to induce the formation of a similar molecule, the necessary energy for the process being provided by metabolism.

The inevitable question arises: is the virus itself alive? But to make an attempt to answer this question as objectively as possible we must first try to define life itself and discover exactly what characterizes living organisms.

What is Life

All living things have in common certain recognizable characteristics: they all have a definite shape and structure; the ability to absorb and transform oxygen and food into substances which become an integral part of themselves, namely, the faculty of assimilation; irritability, the power to survive certain changes in environment; reproduction; ageing and death. Growth, metabolism, and reproduction are three characteristics which are common to all living things. But as we descend the scale of animate beings it becomes increasingly difficult to apply the usual criteria and our conclusions will be different according to the importance we attach to one or other of these characteristics.

If the essential characteristic is auto-reproduction, then the tobacco mosaic virus could be a living thing, even though its method of reproduction is highly exceptional. But does it really reproduce itself? Is it not, rather reproduced by the host cell, therefore making it an example of auto-synthesis rather than of auto-reproduction? If, on the other hand, the essential characteristics of life are assimilation and metabolism, the virus, totally devoid of enzymes and incapable of synthesizing anything, is not a living thing but a nucleo-protein. It does not breathe, and contains no water, salt, or sugars.

The same contradiction arises when the virus is considered from other angles. For instance, the physiopathologist, studying the tobacco mosaic disease, discovers all the symptons of a parasitical disease. The virus, he finds, spreads from one plant to another and when the infection is established, it obviously multiplies, for the disease becomes more intense. Superficially the virus behaves like bacteria and seems to possess all the attributes of a living parasite.

The biochemist examines the virus in his laboratory employing the techniques of physics and chemistry. He purifies the virus and obtains paracrystals which, in solution, retain all their power to cause disease to the tobacco plant. These paracrystals are composed of a protein rolled as a spiral around a molecule of nucleic acid. Taking a solution of purified virus it has been possible to separate these two constituents. When they are brought together again the electron microscope shows that the nucleic acid resumes its position in the centre of the protein helix. Not only is the virus reconstituted, but it again possesses all its original properties including its pathogenic power.

This 'partial synthesis' and the multiplications of the virus, or more exactly its synthesis by the host cell, gives biochemists strong grounds for considering that the virus particles are authentic molecules. If so, the virus would no longer be a parasite but a mutant cytoplasmic protein, endowed with the property of being transmissible from one organism to another. The student of plant diseases and the biochemist thus reach contradictory conclusions, the former recognizing those characteristics in the virus which are common to all living organisms.

Some scientists have tried to find a compromise. The tobacco mosaic virus, in its purified state, amorphous or crystallized, is not living; *in vitro* it is lifeless. But plunged into the living environment of the host cell it behaves like various elements of the cell, elements which are endowed with genetic continuity such as, the mitochondria, the chloroplasts, and the chromosomes. Such cellular particles are not alive: they merely participate in the life of the cell which itself is alive and capable of autonomy. The virus, with its vital properties, also participates in the life of the cell, but under special conditions. With other cell elements having properties complementary to its own, it shares in the 'infra-lives' which when integrated make up, as it were, the total cellular life.

For other scientists the difficulty lies in the lack of a definition of life which establishes a precise frontier between the realm of the inert and that of the living.

Does, or can, such a line of demarcation exist? If we consider the total series of living organisms in descending scale, from the most complex to the most simple, we arrive at single-celled animals and plants, then at bacteria and imperceptibly to the viruses, the proteins, the big molecules. All are linked into what appears to be a continuous series. Where in it does life begin? Is life, perhaps, already a potentiality of inert matter? An answer to this question cannot be found by observation or by experiment. Thus the question lies beyond the province of science.

The origin of life

The origin of life has been explained by many theories which have become progressively incompatible with the progress of scientific knowledge: spontaneous generation; cosmic fertilization, according to which the Earth was sown with germ-cells brought by meteorites or by interstellar cosmic dust; creation by pure chance. None of these suggestions stand up to serious critical examination.

The French physicist Dauvillier, in 1958, considered the problem of the origin of life in terms of its cosmic, geo-chemical and energetic aspects, in other words, in connection with the evolution of our planet. He attributed the origin of life to a photochemical process. The appearance of organic matter, then, is a problem for cosmic physics. What requires study is the transparency of the primitive atmosphere with its ultraviolet light of distant origin, the photosynthesis of ternary and quaternary chemical compounds, of asymmetrical molecules, and the appearance of an atmosphere containing free oxygen with its dual role in respiration and in affording protection against distant ultraviolet radiation.

Since organic matter is endothermic (heat-absorbing) its synthesis requires energy from outside, terrestrial or cosmic. Only ultraviolet light is capable of furnishing this source of energy. Today these rays are absorbed by the Earth's screen of atmospheric oxygen. This oxygen, characteristic of life, is, from the cosmic point of view, *totally* derived from the activity of chlorophyll in plants, so that atmospheric oxygen is contemporary with life. The primordial atmosphere contained no oxygen and offer no shield against ultraviolet light.

In 1910, with ultraviolet light provided by a mercury arc, Berthelot and Gaudechon succeeded in synthesizing ternary and quaternary compounds from carbon dioxide, steam, and ammonia. Thus formaldehyde was produced from carbon dioxide and vaporized water with the release of oxygen:

$$CO_2 + H_2O \rightarrow H.CHO + O_2.$$

Formaldehyde $H.CHO$ (or $C.H_2O$) is easily polymerized to produce a chain of monosaccharides, or simple sugars which have the general formula $C_n(H_2O)n$ derived from $n(C.H_2O)$. For example, we pass from glucose to cellobiose, and from cellobiose to cellulose, essential to plant cells.

Just as carbon dioxide broken down by ultraviolet rays yields formaldehyde, so it can combine with ammonia to produce formamide:

$$CO_2 + NH_3 \rightarrow NH_2.CHO + O.$$

The two reactions occur in the same place and under the same conditions, and the combination of formaldehyde and formamide gives rise to glycine, an aminoacid of importance to living matter. From glycine it is possible to synthesize the polypeptides. All these chemical reactions could have occurred at the surface of primeval seas. During the reactions oxygen would be released, and in this way free oxygen would remain dissolved in the waters.

But how can we picture the origin of asymmetrical molecules, which are basic to life?

Today the enzymes, themselves asymmetric, act as agents which catalyse the asymmetrical molecules characteristic of life. The presence of quartz crystals or of calcite in the neighbourhood of a brackish lagoon could have stimulated the formation of an asymmetric molecule. For crystallization and life are two manners in which matter is organized, crystallization representing a static coordination and life a dynamic coordination. Just as a crystal plunged into a supersaturated solution at once determines how the solution will crystallize, so the primitive asymmetrical molecule might determine the asymmetrical structure of its entire organic environment. The phenomenon may have happened only once and inverse symmetry had exactly the same chance of occurring. Asymmetrical syntheses of this kind can take place only photochemically, which is an important argument in favour of the photochemical theory of the origin of life.

Owing to its salinity, its alkalinity, and its temperature, sea-water was propitious to chemical syntheses, and thus there appeared immense banks of gelatinous formations, engendered by ultraviolet light. These formations, moreover, could combine with sulphur, phosphorus, magnesium, iron — all present in sea-water. Such photosyntheses could have occurred more than 4,000 million years ago since the oceans had by then sufficiently cooled down. In this first stage, organic matter, then, arose from mineral matter. In the second, and much more complex, stage, organic structure was elaborated.

Dauvillier envisaged the phenomena as occurring in this order: at first molecules of adenosine triphosphate would appear, together with the enzymes indispensable to anaerobic fermentation. These fermentations of organic matter would create a living medium or milieu, still devoid of organisms. The adenosine triphosphate would allow the formation of nucleoproteins endowed with the faculty of auto-reproduction and corresponding to free genes or perhaps to viruses. These particles, being in a living medium, would reproduce at the expense of the proteins in the medium.

Next, plasma membranes would appear, made up of molecules forming networks or webs, the two surfaces of which possessed different properties and constituted catalytic surfaces. We could imagine such membranes folding into minute closed sacs about a micron in diameter. Their internal surfaces could conceivably be capable of secreting fats, starch, glycogen, while their outer surfaces secreted enzymes. Other similar sacs would enclose macromolecules of desoxyribonucleic acid. Thus the vague preliminary outlines of heterotrophic bacteria or of mitochondria could appear. The heterotrophic bacteria would derive their nourishment from the organic matter formed by photosynthesis and would breathe the oxygen dissolved in the sea. Living slightly below the surface of the water they would be sheltered from the destructive action of the ultraviolet light.

Then the appearance of chlorophyll would completely revolutionize the living world. This green pigment contains carbon, oxygen, hydrogen, nitrogen, and magnesium. Its molecule, which is composed of 137 atoms, is not a macromolecule endowed with genetic continuity. The production of chlorophyll requires the successive action of eleven genes. By means of chlorophyll, living organisms could build up carbohydrates from water and carbon dioxide by photosynthesis, absorbing the energy of quanta of visible light and, in the process, releasing oxygen.

Only the ultraviolet light of outer space made this synthesis possible. Life, which until then had been heterotrophic, that is, requiring organic material as food from its environment, became autotrophic, could manufacture its own organic matter and release oxygen in abundance. Oxygen rose from the waters of the sea; a layer of ozone was created and soon an atmosphere of molecular oxygen made surface life possible by forming a screen to arrest the distant ultraviolet rays and counter their deadly action. From then on life could multiply and spread, while the breathing of air became possible. The biosphere had been created. Nearly 3,000 million years had been necessary to rise from the 'living medium' to the birth of the Cyanophyceae, blue-green algae which are the most primitive of autotrophic organisms.

Finally that heterogeneous and complex structure, the cell, was born. The Protista, single-celled organisms, were the origin not only of plants (which are generally autotrophic), but also of animals (which are heterotrophic).

In this photochemical hypothesis the appearance of life on earth seems of necessity to depend on the simultaneous fulfilment of a number of strictly defined conditions, notably that the surface of the oceans should be at a certain temperature. If, by accident, all life disappeared from the globe presumably the oxygen and ozone in the atmosphere would also automatically disappear. The atmosphere would once more be unable to shield the Earth from ultraviolet light from outer space. Ultraviolet photosynthesis would then again begin at the surface of the oceans, rich in ammonium, bicarbonate, and phosphates, and a new cycle of life would begin.

Possibilities of life on other planets

The many strict conditions essential to the appearance of life would help to explain why it seems to have appeared only on a single planet in the solar system. The giant planets with temperatures as low as minus 120 °C and an atmosphere composed of ammonia and methane could not support life. This leaves us our near neighbours, the planets Venus and Mars, to consider.

Venus is comparable in mass to the Earth and it could have oceans and an atmosphere. But its tropical oceans must be boiling while in its polar regions the atmosphere, rich in carbon dioxide, contains no free oxygen. Hence the atmosphere affords no protection against ultraviolet light, and if the photosynthesis of organic matter took place in the polar regions only marine life, utilizing dissolved oxygen, would be possible.

Due to its small size Mars possesses a hydrosphere reduced to a polar cap of frost. Its sparse atmosphere, mainly of carbon dioxide, is devoid of oxygen. The climate is harsh and no photosynthesis requiring water in liquid state is possible.

But the absence of life on the other planets of our solar system does not imply that there are no planets in the entire galaxy which could be inhabited by living beings. The galaxy contains about a thousand million planetary systems. Some millions of planets, then, could belong to systems like our own and provide the multiple conditions which are compatible with life. Living beings on such planets could differ from terrestrial living beings in a variety of ways, notably by an inverse molecular asymmetry.

In Dauvillier's words, 'As the partisans of cosmic generation will have it, life and mind are indeed perpetual in the universe, but only in a statistical manner. Life arises independently, evolves and disappears on every planet where the physio-chemical conditions of its appearance are momentarily satisfied.'

All of this is, of course, hypothesis. But the hypothesis is admissible and even probable. It provides a rough idea of the processes by which in the distant past life may have arisen. The first essential steps can be made in the laboratory, and audacious biologists envisage the laboratory creation of organized living particles, that is, of particles with a heterogeneous structure. The total synthesis of adenosine triphosphate and its enzymes has not yet been achieved in the laboratory. When it is, and if the synthesis of nucleic acids proves possible, biologists can perhaps create particles endowed with genetic continuity, in other words, viruses. But the complexity of more advanced organisms seems to render their synthesis highly improbable.

Sex

Reproduction

An organic species, animal or plant, can continue to exist only by reproduction, the function which assures that the individuals that compose the species are perpetually renewed. Ceaselessly, new beings draw life from parents, only to supplant them, the young and intact organisms replacing organisms which have aged and deteriorated.

The link between mortality and the ability to reproduce is so close that the philosophical biologist can argue that this ability is the consequence of mortality or, equally, that mortality is the consequence of this ability. In either event the function of reproduction is a direct result of the function of growth. By whatever process the new organism is created it is always a product of parental growth.

There are two fundamental ways in which living organisms reproduce to assure the perpetuation of their species. Sometimes the new individual derives its existence directly from the parent, which divides into two or in some cases many. The older organism can produce buds from which completely new individuals are formed. This method of reproduction, by division, segmentation, or budding, is known as asexual or vegetative reproduction. It is found in all unicellular organisms such as bacteria and Ciliophora, among certain worms, among the tunicates and among myriad plants. Asexual reproduction is unknown among higher animals, who procreate in another fashion, namely that known as sexual reproduction.

In the case of sexual reproduction the new individual is derived from a single cell, the egg-cell, which is itself generally formed by the conjunction of two specialized cells the gametes (from the Greek *gamos,* marriage) which have both been detached from a parent organism. The two gametes, always more or less different from each other, are the ovum and the spermatozoon or sperm. The ovum, or female germ-cell, is relatively large. Its nucleus is surrounded by an abundance of cytoplasm which contains reserves of nourishment destined to supply the first needs of the new being. The ovum is roughly spherical in shape and is immobile, hence the spermatozoa, in order to reach the ovum, must possess the ability to move. The sperm is fragile, thread-like and almost devoid of cytoplasm; but it is furnished with a means of propulsion — a whip-like flagellum which allows it to move with agility through the medium, always liquid, in which the meeting of the two gametes takes place.

The gametes are formed by special organs, the genital glands or gonads. Glands which produce the ova are called ovaries while those which produce spermatozoa are called testes or testicles.

Among certain animals known as hermaphrodites the same individual is provided with both ovary and testis. The snail is one example of this. In the majority of cases, however, the species is split up into two sections, from which we derive the word sex, from the Latin *sexus* meaning section or separation. The male section has the testes, the female the ovaries, and this separation of the sexes is called gonochorism.

Sexual reproduction exists not only among all animals which are fairly advanced in the organic hierarchy but also among many lower animals. Early beginnings of sexual behaviour are found even among unicellular organisms, and phenomena of conjugation between cells of different shape and form can often be observed. Wherever a small cell attaches itself to a large cell it is agreed to call the former masculine and the latter feminine. When conjugation takes place between apparently identical cells it is presumed that there are invisible differences between the two which could be qualified as sexual. In any case the very existence of cellular union can be considered as an adumbration of sexuality, and it is interesting to note that even microbes have been observed to mate, while

'The Kiss' by Auguste Rodin, at the Tate Gallery.

Mansell Collection.

Coupling of crane flies. The male is on the right.

Pesson.

virologists have begun to suspect that some form of conjugation takes place among viruses.

In many species and in a variety of ways the two methods of reproduction can be combined. At times asexual and sexual reproduction have an alternating pattern. At other times sexual reproduction only occurs, it would seem, by accident, under the influence of unfavourable environmental conditions such as cold, confinement, or lack of food.

It was formerly supposed that from time to time a phase of sexual reproduction must of necessity interrupt the series of repeated asexual reproduction. It was thought that sexuality had a rejuvenating effect on the line, and that in its absence degeneration must inevitably ensue. That asexual reproduction can in reality continue indefinitely is demonstrated by numerous examples supplied by the Ciliophora, the Turbellaria and the Oligochaeta.

Secondary sexual characteristics of vertebrates

Among a great number of animal species the two sexes differ one from the other not only in having different kinds of sexual glands with the ovaries of one producing ova and the testes of the other sperm, but also in certain anatomical structures which serve to deliver and receive the sexual cells. They often differ in a number of both internal and external characteristics, which at times play only an indirect part in the generating act, or even no part at all. The primary sexual character is, of course, the gonad or sexual gland; other differences between the sexes are known as secondary sexual characters.

The degree of sexual dimorphism, or lack of resemblance, between the sexes varies greatly. In some species it is very difficult if not impossible to distinguish the sexes by their external appearance. In others, on the contrary, the distinction is at once apparent, the male and female differing in shape, size or colour.

These sexual differences can appear as soon as the animal is born or, on the other hand, become noticeable only at some later stage of its growth. They can be permanent, from the moment of their appearance, or be manifested only at certain seasons.

Among mammals a few examples of well-marked sexual dimorphism will be noted. Attributes characteristic of male animals are the mane of the lion, the antlers of the stag, the horns of the ram, the calloused rump of the mandrill (an African baboon), the vocal apparatus of the howler monkey, the protuberant trunk of the elephant seal, the fragrant glands of the musk deer and the tusk-like canine teeth of the walrus, the wart-hog, and the babiroussa. The two sexes of the nylghau antelope *(Portex pictus)* differ in the colour of their coat, reddish with a white patch on the neck in the female, in the male slate-grey with a tuft of black hair on the throat.

Female marsupials are provided with an abdominal pouch in which their young are reared. In all mammals the female mammary glands are more highly developed than those of the male. Among mice the submaxillary glands differ greatly according to sex.

The relatively slight differences between men and women chiefly concern the figure, weight, hair growth, the larynx, and so forth. There are marked differences between the weight of the long bones in men and in women: the femur, tibia, humerus, radius, and cubitus. In particular the bones of a man's forearm can weigh twice as much as those of a woman.

The vocal chords, which are 10 millimetres long in a child of six, begin to display a certain sexual difference towards the age of twelve, when the voice breaks. At twenty a man's vocal chords are about 24 millimetres long, as against a woman's 16. At thirty the difference in length has increased to 30 as against 20.

Snails preparing to couple.

R. H. Noailles.

'Sleeping Hermaphrodite', Greek replica of a third century B. C. original, in the Louvre.

Giraudon.

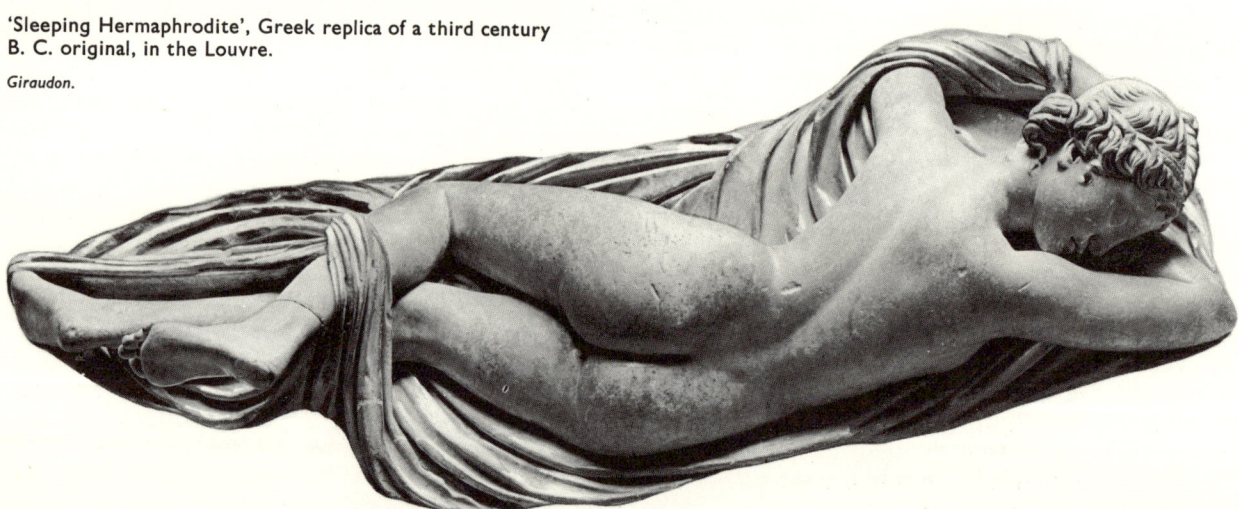

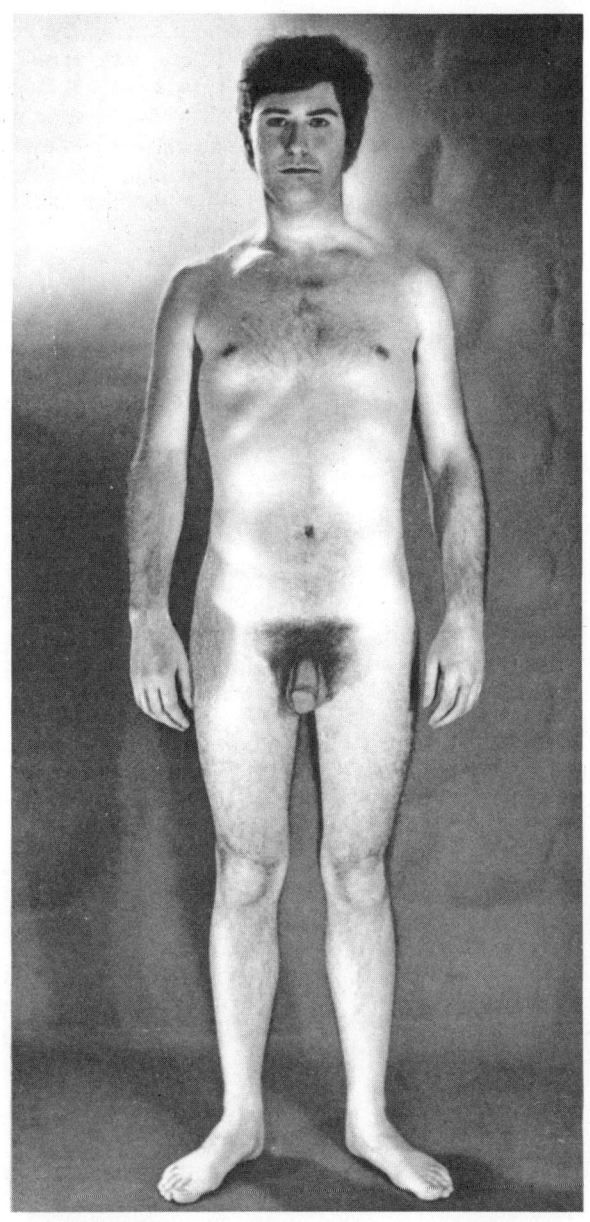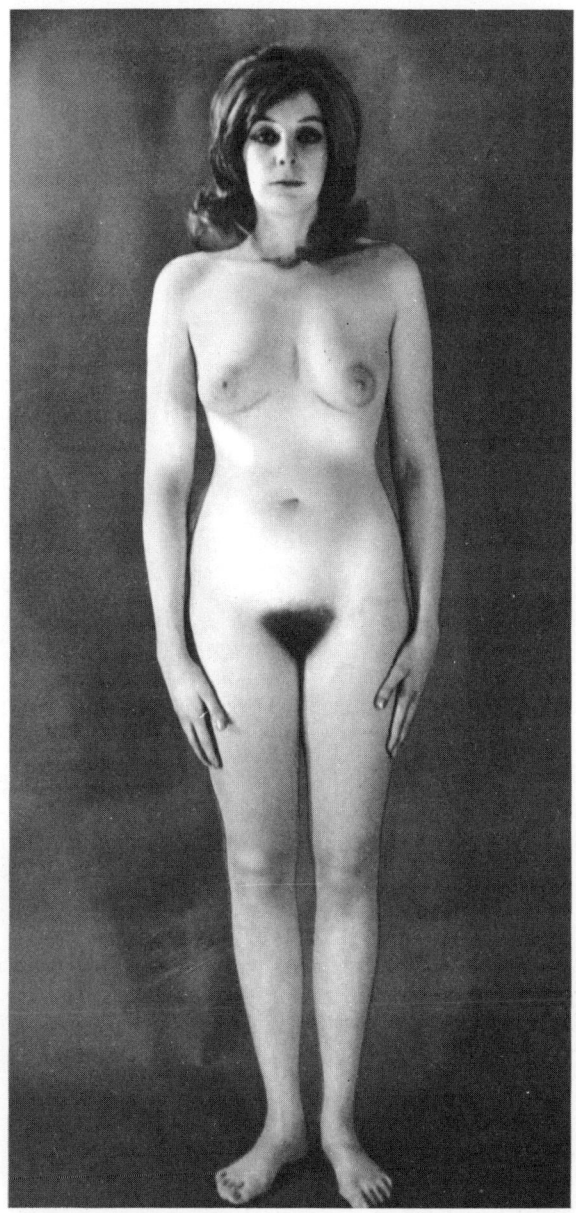

Man and woman.

Among birds, the range of sexual dimorphism is very extensive and even practically absent in the case of pigeons and canaries. On the other hand, sexual differences are obvious in the case of such fowls as cock and hen, peacock and peahen, male and female pheasants, as well as among birds-of-paradise, lyre-birds, and certain hummingbirds.

A list of the masculine attributes found in amphibians includes the vocal sacs of the common frog, the natterjack *(Bufo calamita)*, and the tree-frog; the digital callosities of frogs and toads; the dorsal crests of tritons; the webbed toes of the alpine triton, etc. In fish: masculine attributes include the incubating pocket of the seahorse *(Hippocampus)*; the caudal fin — or sword — of the sword-fish, etc.

A particularly striking case of sexual dimorphism was discovered by the English naturalist Tate Regan among certain deep-sea fishes of the Ceratiidae family

(related to angler fish) in which the male, a dwarf in comparison with the female, lives on the female as a permanent parasite.

Even more interesting from the biological point of view is the fact that this parasitical situation is accompanied by parabiosis or a Siamese graft. In the region of the mouth the male is completely fused to the female. The cutaneous tissues of the female have actively proliferated around the jaws and tongue of the male in such a manner that the two animals coalesce. Their circulatory systems are in direct communication so that they literally share the same blood.

Most of the dwarf male's internal organs have atrophied. He no longer has teeth or a digestive tube, and is almost without a heart. He is fed by the female that carries him in the same way that a foetus mammal is fed by its mother. The only functions which he retains are those of breathing — for he still has gills —

63

and of reproduction, assured by an enormous testis.

In this case, which is unique among vertebrates, the male is virtually absorbed by the female and, being little more than an appendage, has lost all individuality and physiological independence.

He is no longer more than an organ in the service of the species.

The cellular diagnosis of sex

The remarkable work of R. L. Moore and M. L. Barr revealed, first among cats and then among several other vertebrates including the human species, a rather unexpected sexual difference which could not have been observed without the aid of the microscope, since it concerns the structure of the cell nucleus. In female cells, after appropriate fixation and staining, they distinguished a minute mass of chromatin, a chromatic spot on the membrane which surrounds the nucleus or the nucleolus. It should be emphasized that this difference, which is a genuine secondary sexual character on the microscopic scale, must not be confused with the difference which exists between the chromosome equipment of the male and the female — which we shall examine later.

A lioness.
Reichler.

A lion.
Reichler.

Turkey cock. The wattles are characteristic of the male.
Ergy Landau-Rapho.

Stag and hind.
Kenneth, Rödle.

Meanwhile, with this test the diagnosis of sex by examining the cells is not difficult. It is sufficient to have a few cellular elements of an individual to be able to state whether they came from a male or from a female. The diagnosis can be made from globules of blood, from biopsies of skin, from smears of buccal mucus, vaginal smears, roots of hair, from varied cellular debris, and can even be taken up to four weeks after death from fragments of bone or teeth. The test has been applied to diagnosing the sex of the foetus. If, with a fine surgical trocar a few drops of amniotic liquid are removed from the embryonic sac, cells will always be found which come from the desquamation or peeling of the foetus. Microscopic examination of these cells will reveal with certainty the sex of the child to be born. Only the risk entailed and the inconvenience of the process detracts from its great scientific value.

White peacock spreading its tail with, in foreground, a white peahen.

Claire.

Male bird-of-paradise with, *on its right*, the female.

X Photography.

Male alpine triton.

R. H. Noailles.

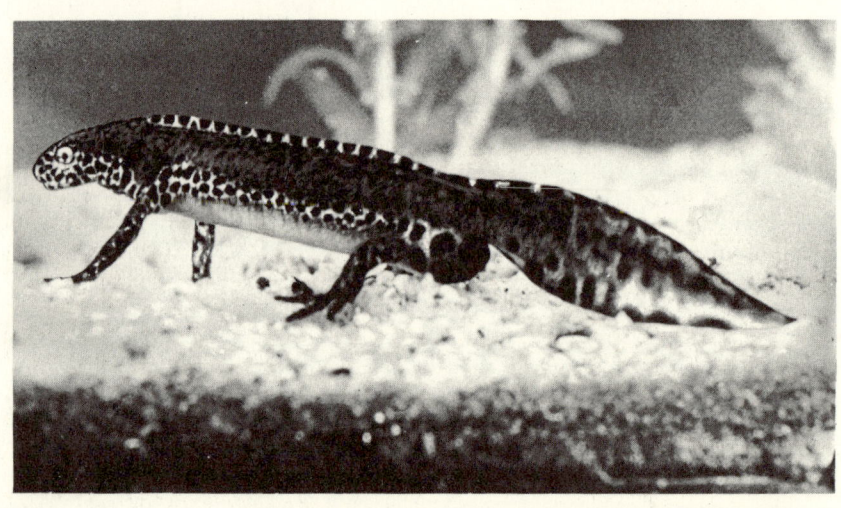

Male *Pleurodema*, an amphibian of the Andes, which floats by means of inflating its voluminous vocal sac.

Dorst.

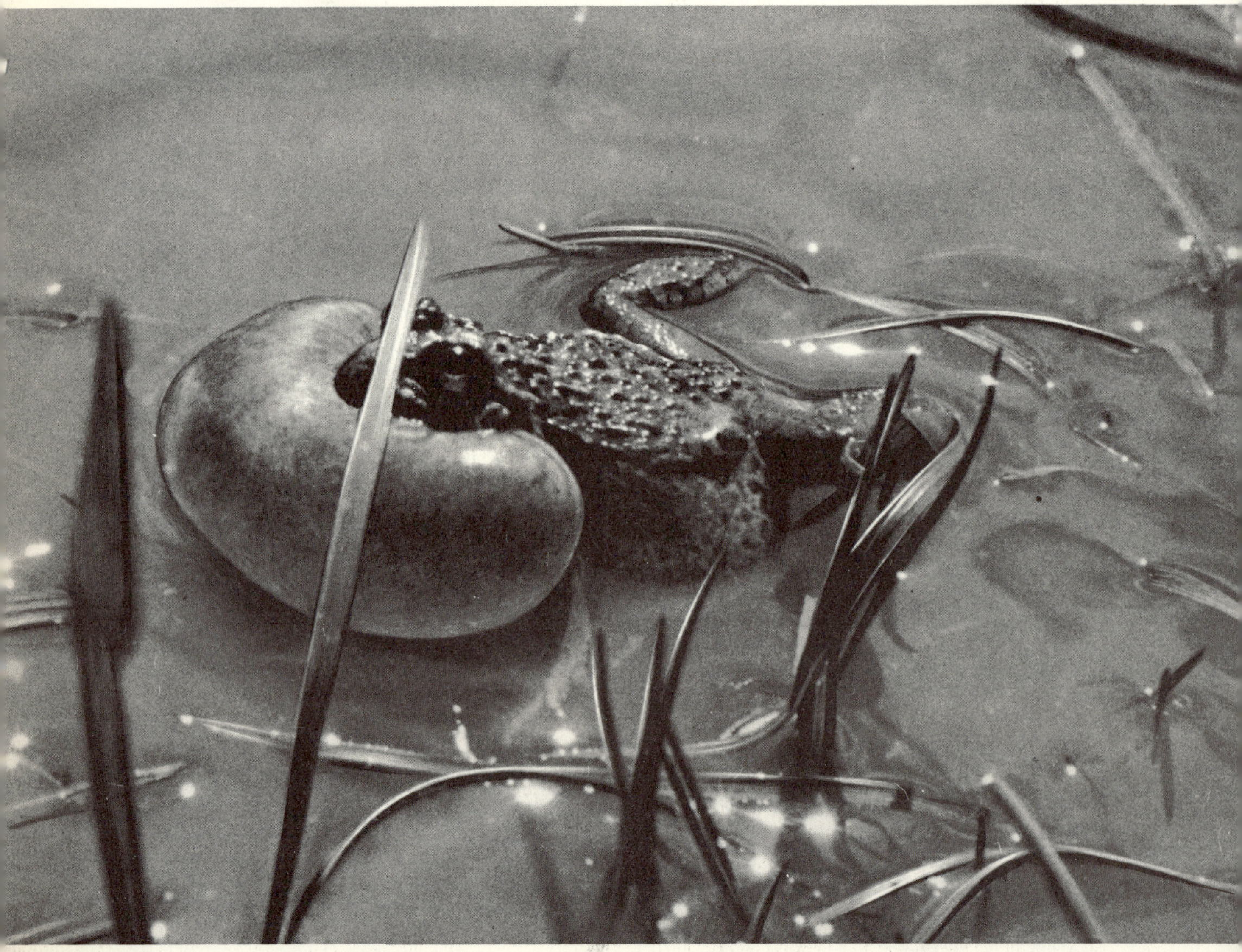

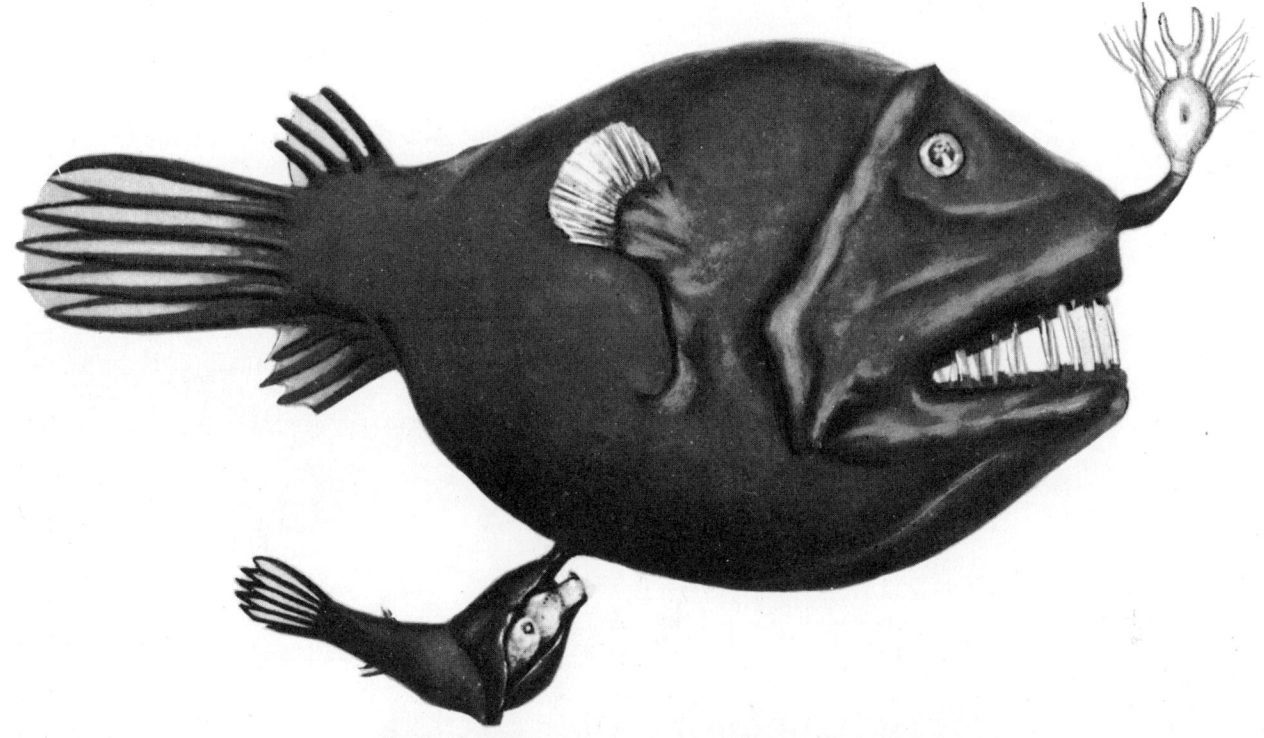

A female ceratiid with its 'lantern' and the little male parasite attached to its belly.

X *Photography.*

The coupling of toads *(Bufo bufo).*

J. Vanden Eeckhoudt.

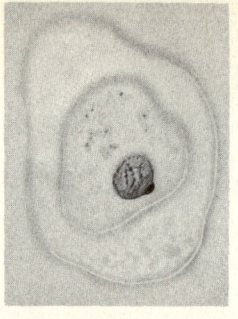

Smear of buccal mucus from female mouth. The arrow indicates the speck of chromatin.

Gefroy.

Position of the chromatin spot against the nuclear membrane. The cell is from the tissues of a female cat. Graham and Barr found that this 'sex-chromatin' exists in the tissues of all the female cat's organs except the liver and the exocrine pancreas: from 50 to 65 per cent of the cellular nuclei reveal such a spot.

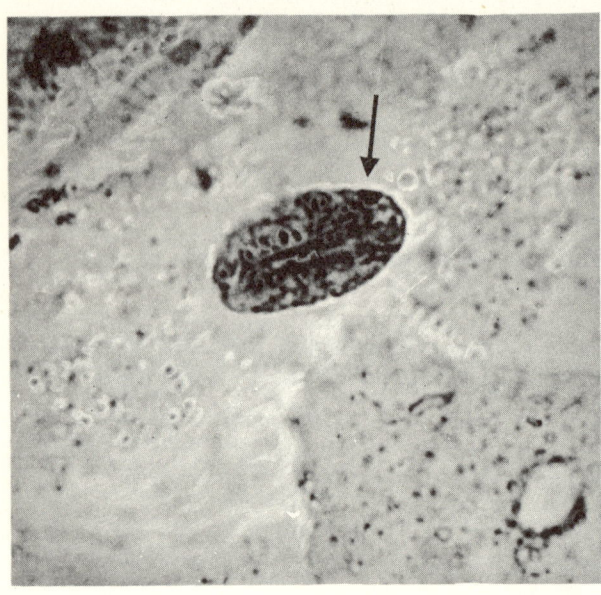

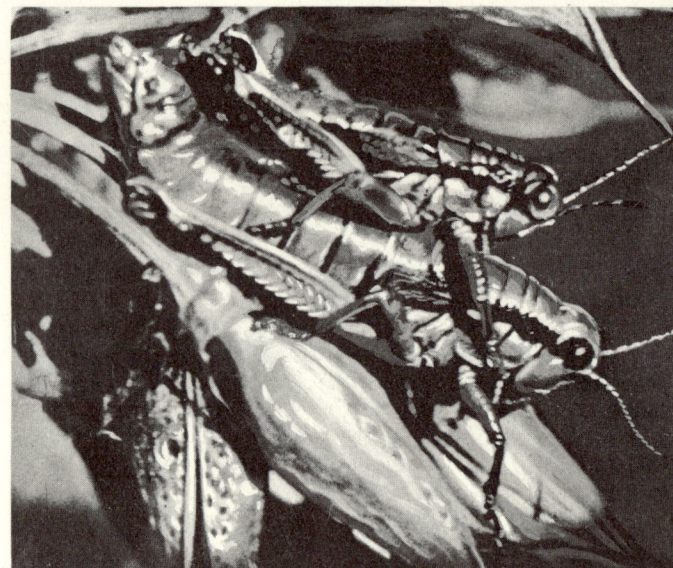

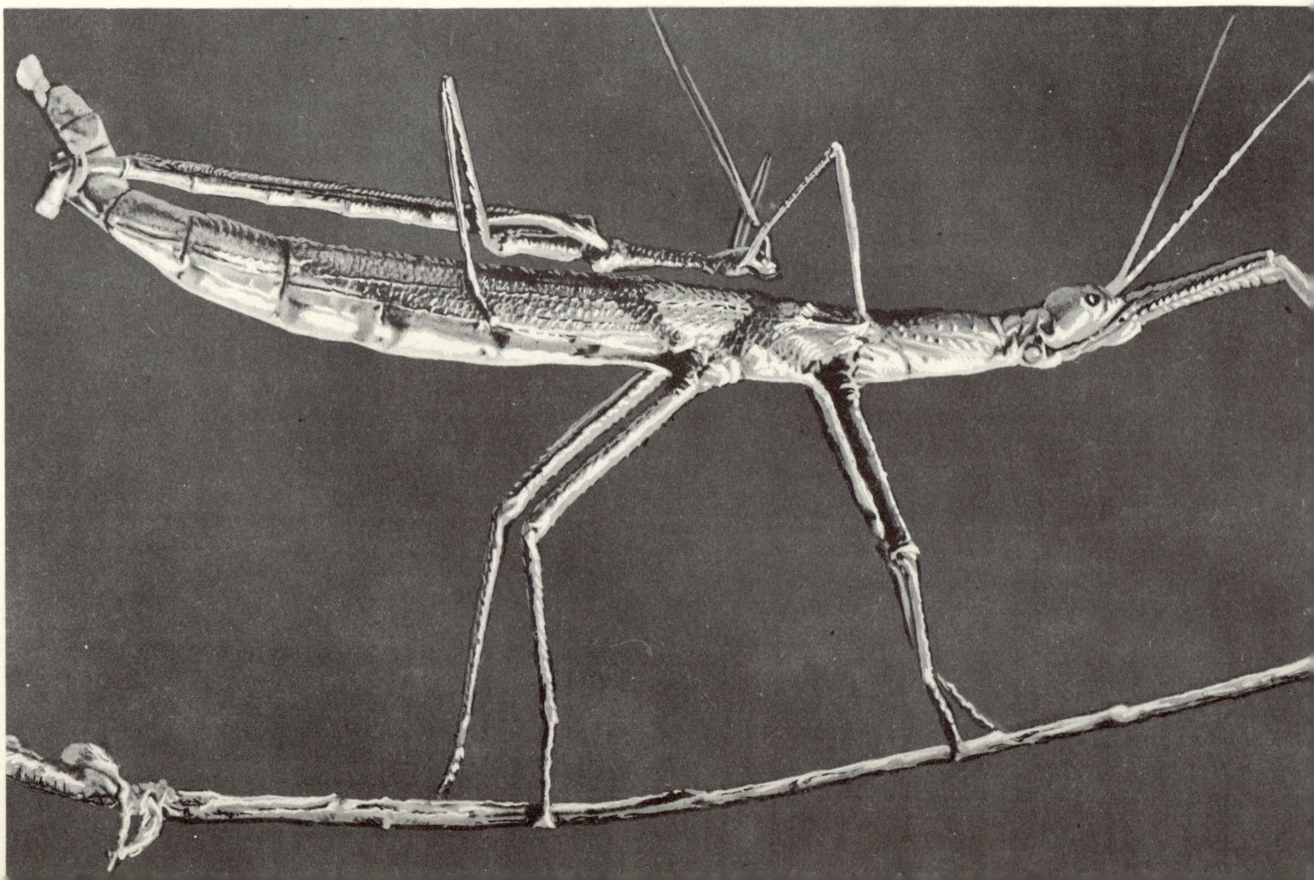

The coupling of cyphocranes. The male, *above*, is much smaller than the female.

Foucher.

The coupling of grasshoppers.

R. H. Noailles.

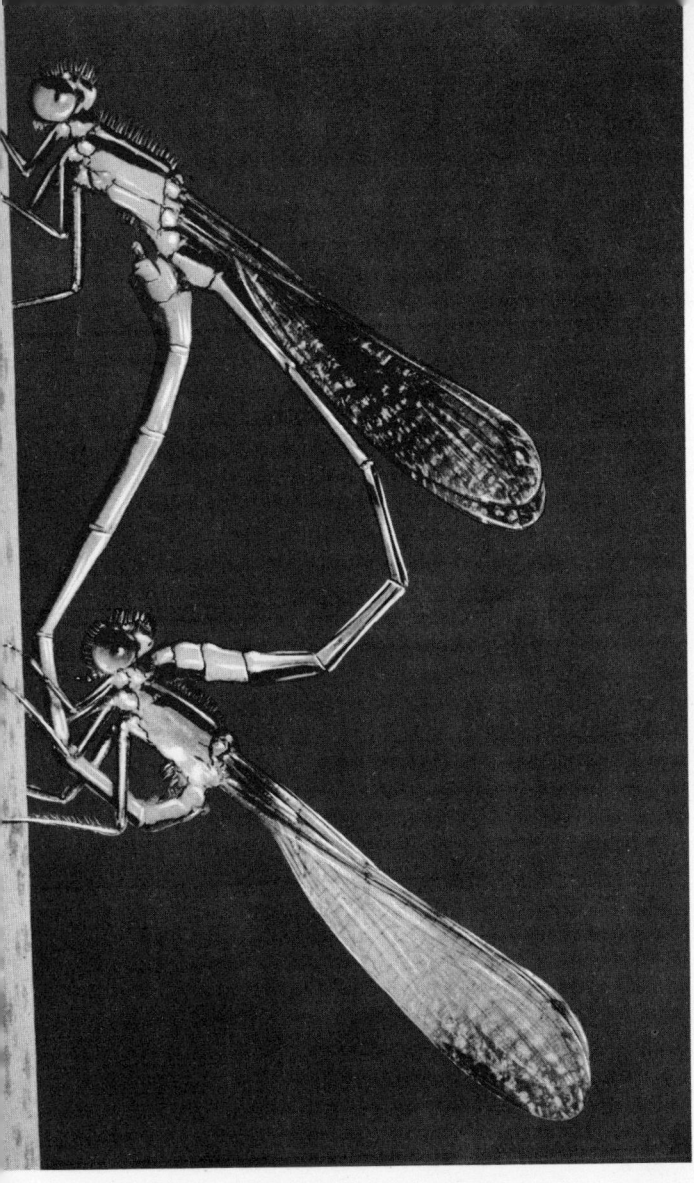

Secondary sexual characteristics of invertebrates

Dimorphism among the invertebrates is bewildering in its variety: dwarf males among the rotifers and many crustaceans, the copulating 'arm' of male cephalopods, the keel-like shell of the female argonaut, the giant claw of certain male crabs and the sexual variations in shape and colour among numerous butterflies, spiders etc.

In colour the male beetle *Hoplia coerulea* is a very beautiful metallic blue, while the female is reddish and dull. The male dragonfly *Calopteryx virgo* is green with brown wings, while the female is brown with transparent wings. The fly *Bibio hortularius* is entirely black while the female has a red corselet and a yellowish-red abdomen. The male butterfly *Lymantria dispar* is brown, the female off-white. Among the exotic butterflies of the genus *Morpho* the male is famous for the pearly blue of his wings, the male of the *Ornithoptera* for his velvety black and green and that of the *Urania* for his sumptuous coppery bronze.

Among such insects as the mosquito, the feather midge *(Chironomus plumosus),* the cockchafer, and the emperor moth, the male antennae are generally much more developed, while in many diptera such as ants the male's eyes are more richly faceted.

Features characteristic of the male include the stag-beetle's mandibles which are transformed into powerful weapons and, perhaps, serve in seizing the female during copulation. The two long horns of the *Dynastes hercules,* one inserted in the head and the other in the corselet, form pincers. A pair of horns is

The coupling of dragonflies, with the male above.

J. Vanden Eeckhoudt.

The coupling of common flies, *Musca domestica.*

J. Vanden Eeckhoudt.

The antennae of the male *Bombyx mori*, or silkworm moth.

R. H. Noailles.

Abdomen of the female *Bombyx mori* showing at its extremity
the odoriferous bladders which attract the male.

R. H. Noailles.

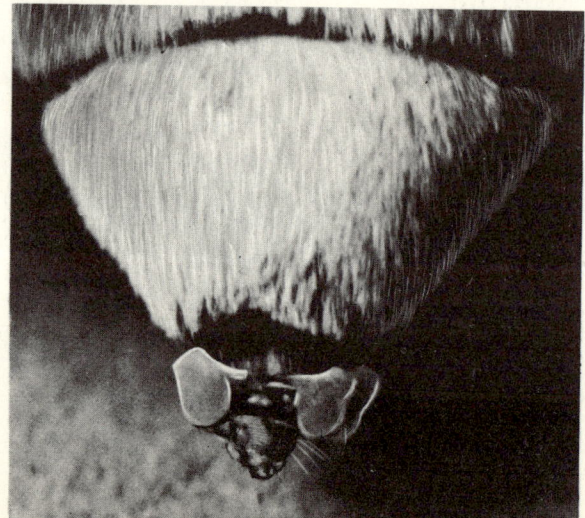

Opposite, The coupling of *Melitaea varia*, a species which lives in the Valais at an altitude of 1½ miles.

J. Vanden Eeckhoudt.

The coupling of emperor moths: *on the right*, the male.

A. Buston.

borne on the lower surface of the thorax of *Onitis furcifer*. The left mandible of the male *Taphroderes distortus* is longer than the right, while the tarsi of the forefeet spread out like a palette and are provided with little suckers. The pincers of the male earwig (*Forficula*) are larger than that of the female. The mole-cricket has a special gland between its thorax and abdomen which the female licks greedily during coupling. Male cicadas, locusts, crickets, and grasshoppers possess a musical apparatus which remains rudimentary in the females. The male scorpion fly has three long salivary glands while the female has only two and those very small.

Characteristic of the female are the basket and brush which the honeybee carries on her hind legs. These work tools serve to gather pollen and are lacking in the male bee who of course lives the life of a drone. Among many females of the Hymenoptera the abdomen is armed with a sting and poisonous glands. The female glow-worm bears a luminous apparatus and has no wings, while the male, which does not glow, is winged.

Scent in insect-mating

Perfumed females

Among certain insects the female gives off a particular odour by which she attracts and stimulates the male. The female emperor moth, for example, emits no scent perceptible to human nostrils, yet she must give off an extremely powerful perfume to which only the males of her species are susceptible. That the sense of smell is involved is proved by the fact that the female receives no further visits if she is enclosed in a well-sealed box or glass jar. If, however, the slightest opening is left, the males will cluster round the jar or box.

In addition, it has been observed that any object which has been in fairly prolonged contact with the female, a piece of cloth or a plant for example, will acquire the power to attract the male and can retain it for several days, or even several weeks. It is impossible to mask this mysterious female odour, or to disguise it by other odours such as naphthaline, paraffin, or lavender, strong as they may seem to our sense of smell. To join the female, moths will if necessary make complicated detours. They have been known to have penetrated a closed room in which a female was kept, by way of the chimney.

It is agreed that the seat of the male's sense of smell is in the antennae, appendages situated in the fore-part of the head and generally well developed in species with odoriferous females. Males deprived of antennae are incapable of finding the female at a distance. In many species of Lepidoptera, including the peacock butterfly, the oak eggar moth, the tussock moth, the gypsy moth *(Lymantria dispar),* and the Indian moth *(Actias selene),* the female also attracts males from a distance. The peacock butterfly, which is diurnal, visits the female not during the evening but towards midday. Sometimes the female can attract even before emerging from the cocoon, while still bound up in the chrysalis: sometimes she becomes attractive only a day or two after emergence. Normally her power to give off scent lasts the few days of her brief existence.

The sexual attraction of the silkworm moth is only effective at a short distance — at half a yard or at most a yard — but her role in the insect's reproduction is nonetheless great. Separated from the females, male silkworm moths maintain complete repose and if a female is brought within proper range they at once beat their wings in a sort of frenzy. In this case the female attracts by secreting a substance from certain glands which from time to time project from her posterior extremity and resemble translucent bladders. If this extremity of the female is amputated she loses all power to attract the males who, on the other hand, will beat their wings excitedly around the severed parts where the glands are situated.

Remarkable facts concerning sexual attraction by odour have been discovered among certain Coleoptera of the family of the Scarabaeidae. The female of *Pachypus cornutus* is wingless and ordinarily lives underground, coming to the surface only to meet the male. A host of suitors, having perceived her odour, await her emergence with lowered heads.

Female *Actias selene* drying its wings after emerging from the cocoon.

Le Charles.

One entomologist has related how, during the course of a journey through Corsica, he captured a female *Pachypus* which squirted a jet of liquid secretion on to his sleeve. For several days, even in the middle of towns, he was pestered by swarms of male beetles, attracted by his sleeve.

In the case of a certain cotton parasite American entomologists have protected crops by making use of the odour of the female as a bait to lure away the males.

In the same way, for *Lymantria dispar,* which devastates forests, parks, and orchards, recourse has been had to 'lures' obtained by preparing sticky paper coated with amputated female abdomens preserved in alcohol. These papers are placed in the trees and if male moths become stuck to them it is a certain indication that there is a community of *Lymantria dispar* within less than 4 kilometres. In this way small communities can be spotted and exterminated.

The chemical study of the attracting substance, or rather substances for they certainly differ according to the species, is still only in its infancy. Though there have been successful attempts to isolate the attracting substance from many species, the substance, which is soluble in fats, may be related to an alcohol. In any case it contains carbon, oxygen, and hydrogen, but not nitrogen.

The action, exercised over great distances, of substances responsible for sexual attraction, raises theoretical problems which are not yet entirely solved. We have noted that the male emperor moth is attracted from a distance of more than a mile. The female of another moth, *Actias selene,* can signal her presence at a distance of seven miles. When we remember that odour is propagated by means of material particles it becomes necessary to imagine a sphere with a seven-mile radius filled with 'feminine particles' capable of attracting males. The possibility of this enormous dilution has been seriously questioned. Fabre took refuge in the hypothesis of 'olfactory waves'. Odour, he suggested, could be propagated in two ways. By substituting undulations for simple emission the problem of the emperor moth could be solved.

'Without losing any of its substantiality a luminous point sets the aether vibrating and fills an orb of indefinite amplitude with its light. The female moth's system of signalling her presence must function in much the same way. It does not give off molecules but sets waves vibrating which are capable of penetrating to distances which actual diffused matter could never reach. There would, then, be two ways in which odour is propagated: by particles dissolved in the air and by waves in the aether. Only the first method is known to us.'

It would seem, however, that Fabre's hypothesis must be discarded. Today it is generally agreed that the attracting substances produced by female moths act, like all odorous substances, through the medium of material particles. Male moths must therefore possess a sensitivity to these substances which is out of proportion to anything we yet know about the sense of smell. It has been claimed that a *single molecule* of the attracting substance has the power to excite the male!

Perfumed males

Until now we have been concerned only with scents which are emitted by females. But in certain species of moth it is the males which secrete the sexually stimulating perfumes. These perfumes are only effective over short distances, at most a few dozen centimetres, and can often be perceived by human nostrils. The range of such scents varies greatly according to the species. It may be of orange blossom or of snuff, of mignonette, violets, geranium, musk, heliotrope, chocolate, cloves, strawberries, primroses, broom, syringa, meadowsweet, sandalwood, chloroform, pineapple . . . . . Among butterflies of the genus *Satyr* the glands which secrete scent are situated in the forewings of the male who flutters around the female and, using his wings, fans her with a current of perfumed air. Finally he imprisons the antennae of the female with his wings and impregnates her with his scent.

Among other Lepidoptera, in *Hepialus,* the attracting odour emanates from the legs of the male: the tibia of the third pair of legs, swollen into a knob, contains glands which secrete a substance smelling of ripe pineapples. In the course of the fluttering flight which characterizes courtship, the male shakes his perfumed tibias around the female, and when she is won the couple swoop down to the ground.

Among the *Amauris* of the family Danaidae the

A pair of *Cheimatobia brumata:* the male is winged while the female is wingless.

Le Charles.

aphrodisiac odour is found in hairs which grow in a special region of the back wing. On the abdomen there are brushes which, when they stroke the hairs, produce a spray of perfume, rather like an atomizer. Nevertheless, while the attracting substances produced by the females play a role of unquestioned importance in bringing the two sexes together, the function of substances of masculine origin is less evident.

Unisexual polymorphism

Among many species of butterflies, and also among certain dragonflies and earwigs, several types of females are known. There are two distinct types of female *Colias philodice,* one yellow like the male and the other white. There are three different female *Papilio memmon* (swallowtails) which have been given the distinct names of *achates, agenor,* and *laomedon. Papilio polytes* of Ceylon also produces three types, one of which, *cyrus,* resembles the male, while the other two, *polytes* and *romulus,* are plainly different. In the case of *Parasemia plantaginis,* conversely, it is the female which is unique, while the male exists in various forms. These cases of polymorphism in both male and female are conditioned by hereditary factors, or genes, whose diversifying effect is only apparent in one of the two sexes.

Beauty and the sexes

The male can always be distinguished from the female by his noble, virile, and handsome appearance. This is true of all animals but is particularly evident among fowl. It is impossible to imagine anything more beautiful than the dazzling hues of the male pheasant, for example. But he has no female counterpart. In her there is no trace of the brilliant golds, purples, and azures that adorn her partner.

What is the purpose of, or reason for, this 'beauty', this masculine brilliance, or at least that of certain males? Today we know that sexual dimorphism arises from differences in *present* hormone conditions, a subject which will be discussed in a later chapter. Knowledge of this physiological cause does not, however, explain why secondary sexual characteristics are often more striking in the male than in the female. Is it pure chance, a kind of epiphenomenon without consequences, that the male displays more vivid colours, that he is adorned with coxcombs, crests, plumes, aigrettes, ruffs, wattles, fantails? Or must we, on the contrary, assume that the greater ostentation and beauty of the male corresponds to some deep necessity of nature?

The opinion which Charles Darwin expressed towards the end of the nineteenth century is well known.

Male and female pheasants.

Larousse.

'Bourbonnaise' cock and hen.

Ministère de l'Agriculture, France.

According to the great naturalist the ornamental characteristics of male animals — like all other organic characteristics — appeared progressively during the course of the species' evolution. But they would have had a special origin, due to the power of seduction which they conferred on those who possessed them: the most pleasant, the most 'attractive' males would have been chosen by the females in preference to the less attractive. They would therefore leave more offspring, and, by the law of heredity their descendants would be provided with those same characters to which their male progenitors owed their sexual success. Thus, in Darwin's view, the beauty of the male would be essentially the result of selective pressure ceaselessly exerted by generations of females.

This theory clearly postulates the existence of a kind of aesthetic sensibility among animals, assuming that animals are capable of showing preference for this or that form or colour. Darwin cited the case of an old peacock of gaudy plumage with whom all the peahens were 'in love' to such an extent that they ignored every other male of their species. One year when he was captive but on view they would constantly gather by the trellis which formed the wall of his prison, refusing to permit another peacock, a young one with black wings, to approach them. The following year the seducer was locked up in the stable and then all the peahens turned towards his rival.

Among the Padaung tribe living in the central jungles of Burma it was considered beautiful for a woman to have a long neck. It was a custom from the time they were babies to wear brass rings around their necks until the vertebrae stretched to fit them. Gradually more and more rings were added until some women were wearing more than twenty brass necklaces. Although this custom has died out now, there are still some women left who cannot lie down to sleep because of the rings on their necks.

Keystone.

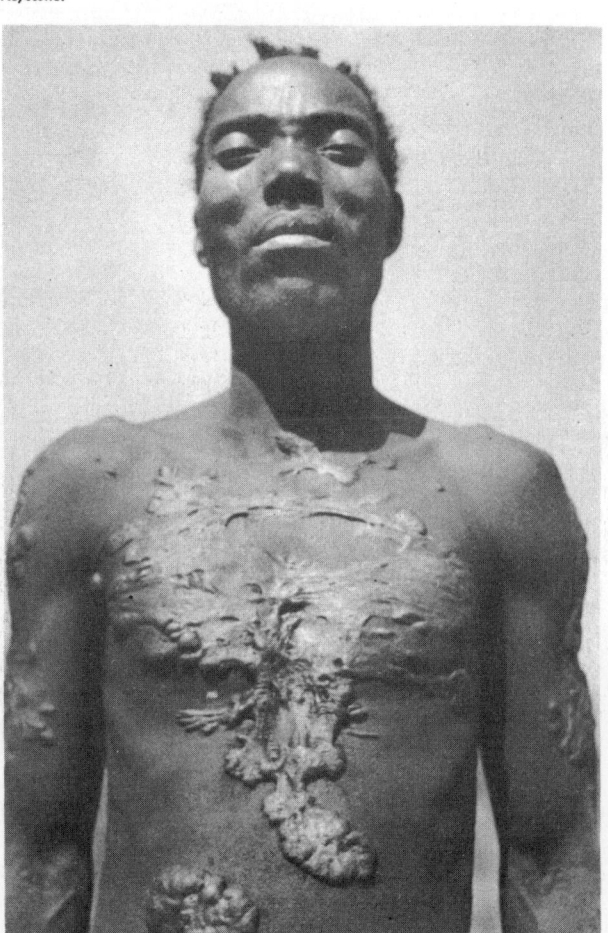

A Bushman in Tanzania considers that his beauty is enhanced by these cicatrices on his back and chest. They are gained by cutting the flesh in innumerable places and then stitching the wounds which form prominent scars in elaborate patterns.

P. Popper.

75

Efforts have been made to establish Darwin's theory on firmer ground by more precise studies in different species of the behaviour of females towards the males offered to them. Among insects these experiments have produced only negative results.

The American moth *Callosomia promethea* is a large bombycid with pronounced sexual dimorphism. Mayor has amputated the wings of the female and glued them to the male: males thus disguised proved quite as successful with females as were normal males. One can also paint the male's wings green or red without in the least diminishing his chances of success.

Analogous results were obtained with another moth, *Lymantria dispar*. In the case of moths, then, the question seems to have been settled: moths are not sensitive to the beauty of their mates. But it would appear otherwise among certain female birds.

However, among ruffs, common European and Asian sandpipers, the chosen male is often the one who has the most elegant ruff. It has been observed that two males of remarkable beauty monopolized the favours of the feminine population while the majority of the other males suffered enforced celibacy. In the course of three and a half hours ten out of twelve females yielded to a male provided with a bronze ruff of brilliant iridescence.

According to Noble and Wurm, if the feathers of the male black-crested nighthawk are plucked the female is not slow in abandoning him to go in quest of a more attractive partner. The careful observations of Mrs. Cinat-Tomson, the Riga biologist, revealed the remarkable powers of discernment which the female budgerigar displays when choosing a mate. The male of the species is distinguished from the female by the amount and colour of its breast plumage, which is bright yellow, spotted in the centre with black, and forms a little collar which frames the bird's head. The female budgerigar, when offered a choice of mates, almost always chooses the male with the best developed and most prettily coloured collarette. The more marked the difference is between her suitors the more promptly she shows her preference. If the male budgerigar is deprived of the feathers which form his collar he will be ignored from then on by all the females. If, on the contrary, his collarette is artificially enhanced by adding an extra circle of feathers, borrowed from another male, his power to attract is singularly increased: the percentage of success obtained by this 'supermale' in borrowed plumage is much higher than that of his normal rivals.

There is, then, little doubt that at least among some birds the beauty of the male contributes to his sexual success. Nevertheless the majority of contemporary naturalists think that masculine behaviour — activity, eagerness, assiduity, sexual ardour — is more important in the competition for mates than physical appearance alone. In many cases when the female appears to show a preference for a certain kind of physical adornment it is found that this adornment is more or less closely associated with special qualities of behaviour. Thus, the reason why peahens in the London Zoo show a decided preference for white peacocks has little to do with their snowy plumage; it is because the white peacocks are given to continual sexual par-

ades, eroticism being much more pronounced among them than among normal peacocks.

In other words, where the most handsomely adorned males supplant their rivals we can presume that the development of sexual adornment has kept pace with the development of more passionate behaviour, since physical characteristics and psychic characteristics depend on the same hormones.

Sexual selection in the human species

In the human species it is obvious that reason and reflection have played their part in encouraging sexual selection, and that this selection occurs in two ways: the choice of males by females, and the choice of females by males. Thus each sex would, as it were, help to shape the other, stimulating the opposite sex to evolve in the image of its own tastes — its own sexual ideal.

It is also interesting to note that one effect of the development of civilization has been to promote the female sex to the rank of the 'fair sex' by giving it the privilege of richer adornment and apparel. While in the animal kingdom it is ordinarily the males who are better adorned than their mates, the human male has become more and more content to dress soberly, dully and without imagination. This is a curious inversion of the tendency which nature has previously shown. Darwin remarked, with that dry humour peculiar to him, that to decorate their hats women borrow from male birds those feathers which served to fascinate females of their own species.

Edmond Perrier emphasized the truly revolutionary innovation which men have introduced into the evolution of life by abandoning elegance and foppery in dress. 'In our civilization, as the desire of women for adornment increases, so that of men subsides. Men have gradually given up all effort to achieve personal elegance. Little by little they have renounced jewellery, gold braid, ribbons. This sharp division of a taste for adornment between the two sexes, apparent in the evolution of costume, is entirely characteristic of the human species and sets it apart from animal species more distinctly than any one psychic trait.'

Lucien Cuénot amusingly qualified those tempting artifices which women use in an attempt to enhance their sex appeal as 'tertiary sexual characteristics'.

'Contrary to that which we observe in the animal kingdom, it is the female of our species who ornaments herself in a striking, though sometimes subtle, fashion, who heightens her colour with rouge and dyes, who anoints herself with alluring scents, and even augments the volume of her hair, already luxuriant, by strange artefacts.'

But as in all fields of life, as women become the equals and the competitors of men it is foreseeable that differences in clothes and adornment between the sexes will become less and less marked.

Prelude to the coupling of the praying mantis, *Mantis religiosa*. Acquaintance is made by the titillation of the antennae.

J. Vanden Eeckhoudt.

The Early Determination of Sex

For many years it was believed that the sex of the individual was determined during the course of his or her development. The first cell or egg was thought to be sexually neuter, and would give rise to a male or a female according to the influence of external conditions on the embryo.

The sex of the offspring, and especially of the human child, was once believed to depend on the physiological state, the strength, and the diet of the mother. Among animals numerous experiments on caterpillars, the larvae of flies, tadpoles, and others, were alleged to have proved that diet, temperature and light could influence sex. All of these theories have today been abandoned since we now know that sex is irrevocably determined at the moment the ovum is fertilized. This was discovered in the first place by the study of twins. While fraternal twins, born of two distinct ova, can be of the same or different sex, identical twins, born of one and the same ovum, must always be of the same sex.

Polyembryony

The above observation which applies to the human species is all the more evident in species in which polyembryony, or formation of more than one embryo per fertilized ovum, is normal. Thus the litter of the small South American armadillo, *Dasypus,* is always produced from a single egg-cell. Its offspring are thus identical twins (or rather identical quadruplets) differing in this respect from the offspring of other mammals of multiple gestation such as dogs, cats, and mice, which bear fraternal or false twins in the same litter.

The offspring of the armadillo are always of the same sex, as the natives of Paraguay noticed many years ago. Newman and Patterson studied *Dasypus novemcinctus* of Texas which in each litter bears identical quadruplets while *Dasypus hybridus* of the Argentine, studied by Fernandez, has a litter of from eight to twelve. H. von Hering in 1885 was the first to advance the idea that the armadillo always produced litters of the same sex because all the offspring derived from the same ovum.

Certain insects carry the principle still farther. The small parasitic Hymenoptera such as *Encyrtus fuscicollis* are polyembryonic to an extraordinary degree. The fertilized ovum is deposited by the female in the egg of a moth and fission begins. After a period of inactivity which corresponds to the winter pause of the host, the ovum of the parasite stretches and elongates into a kind of sausage which finally breaks into hundreds of tiny embryonic segments. These develop into hundreds of individuals, all identical twins and all of the same sex.

J. de Beaupré.

Sex and the chromosomes

If sex is determined in the zygote or fertilized ovum, how does it occur and what is the mechanism which assures, in species with separate sexes, that two types of zygote are formed, one destined to become a male, the other a female?

The mechanism of sex determination depends on the behaviour of the chromosomes, later to be described. The mechanism is very simple and depends on the formation, by one of the sexes, of two types of gametes. The sex which produces these two types of gametes is called the heterogametic sex, and in the case of the human species and indeed, in the majority of species, the heterogametic sex is the male. The male, then, forms two kinds of spermatozoa, one kind carrying a certain chromosome called the X-chromosome, the other kind a smaller chromosome called the Y-chromosome. The female sex (in humans) is homogametic, the ova being alike in carrying the same kind of chromosome. This is always an X-chromosome.

In fertilization there are thus two possibilities: the ova (X) can be fertilized by an X-sperm or by a Y-sperm. In the first case the zygote will have two X-chromosomes; in the second it will have one X-chromosome and one Y-chromosome.

XX zygotes become daughters. XY zygotes become sons. If we recall what has been previously said about meiosis, or reduction division, it will be easily understood how females (XX) must produce ova which contain one X-chromosome, while males (XY) can produce spermatozoa carrying either an X-chromosome or a Y-chromosome.

The sexual gamble

In brief, a man's semen contains two sorts of germ-cells, one which produces boys and the other girls. Since the two sorts are equally numerous, and since the encounter of the male and female cell occurs at random, for there is no reason why the ova should be penetrated by one sperm rather than another, the sex of the child is itself determined at random. In other words, the chance of producing a boy or a girl is what is called in gambling games an even chance. It

Nine-banded armadillo, *Dasypus novemcinctus.*

Camera Press.

can be safely predicted that boys will be very nearly as numerous as girls.

Sex determination

Among numerous animals, including mammals and many insects, the mechanism of sex determination is similar to that we have just described. But it can occur that the Y-chromosome, instead of being smaller, is larger than the X-chromosome, as in the case of the fruit fly *Drosophila,* or it could be that the Y-chromosome is entirely lacking, as in the case of the insect *Protenor belfragei.* The male *Protenor belfragei* has one chromosome less than the female: instead of the XY chromosome pair he possesses only a solitary and unpaired chromosome XO and produces X-sperm and O-sperm.

In birds, certain amphibians, Lepidoptera, and others the situation is reversed and it is the female who is heterogametic. This means that she forms two types of ova while the male forms spermatozoa which are all alike in their sexual potentialities. In such animals the chromosome pair is sometimes designated by the letters Z and W. The female (ZW) forms Z-ova and W-ova, while the male (ZZ) forms only Z-sperm. Here again fertilization has two possibilities: either a Z-ovum is fertilized by a Z-sperm and forms a ZZ zygote which produces a male, or a W-ovum is fertilized by a Z-sperm and produces a WZ zygote which becomes a female.

Among some moths the W-chromosome, proper to the female, is entirely lacking. The female then carries one less chromosome than the male: instead of the WZ pair she has an unpaired chromosome (ZO) and produces Z-ova and O-ova. This mode of sex determination where the female is heterogametic is known as *Abraxas,* a word derived from the name of the moth *Phalaena* in which it was first observed.

The other mode of sex determination where the male is heterogametic is known as *Drosophila,* from the name of the small fruit fly in which it was originally encountered.

It must be noted that within the same group two related species can differ in their pattern of sex determination. Thus there is every reason to think that, among tailless amphibians, frogs follow the example of *Drosophila* while *Xenopus* follows that of *Abraxas.*

Sex-linkage and heredity

The demonstration that sex is determined by the chromosomes is primarily supported by a host of cytological observations. In many cases the difference between male and female chromosomes can be seen under the microscope. Where it cannot, or can scarcely be perceived, which is the case with most vertebrates except mammals, it is reasonably certain, and generally agreed, that the difference nonetheless exists and that sex determination depends on a chromosome disparity in the gametes, whether masculine or feminine.

But the determination of sex by the chromosomes does not in any way imply a visible difference in the shape and size of this or that particular pair of chromosomes. What is essential is that in one of the sexes, the heterogametic sex, one chromosome pair should be composed of dissimilar chromosomes. In this pair the genes carried by each chromosome will differ, so that when the pair divides in meiosis two different kinds of gametes will result. Genes, as we shall see later, are the factors by which hereditary characters are transmitted.

The results of cross-breeding unequivocally betray the heterogametic nature of one or the other sex, for the sex chromosomes (X or Y, Z or W) carry not only the genes connected with sex determination but also genes which can determine other, more general characters such as coloration or the form of certain organs. These genes would obviously not be transmitted according to the rules which apply to genes carried on ordinary chromosomes. There are genes which are shared in common by the X and Y chromosomes, and also by the Z and W chromosomes, but they are transmitted more or less like the genes on autosomes, an autosome being a chromosome which is not a sex chromosome. Genes carried by ordinary chromosomes are transmitted in accordance with laws of ordinary Mendelian heredity, which is described later. Genes carried by sex chromosomes, at least those carried by the X, Y, Z or W chromosome are transmitted in a special way, being, of necessity sex-linked and inherited, as we should expect, in two different manners according to whether the organism's sex determination follows the example of *Drosophila* or of *Abraxas.* In the former event, as in man, the mother has of course two X-chromosomes and the genes they carry, so that they are transmitted to all her offspring, sons and daughters alike, while the father transmits the genes of his X-chromosome only to his daughters, those of his Y-chromosome going only to his sons. In the latter event it is the father who has a double dose of Z-chromosomes and thus transmits their genes to all his offspring of both sexes, while the mother transmits only Z-chromosome genes to her sons and W chromosome genes to her daughters. By observing in the offspring the evidence of various sex-linked hereditary characters it is possible to say with certainty to which of the two modes of sex determination *(Drosophila* or *Abraxas)* a given species conforms.

In the fruit fly *(Drosophila melanogaster)* more than 100 genes are known which are carried by the X- chromosome and transmitted as sex-linked hereditary factors: white eyes, red eyes, yellow body, rudimentary wings, abnormally shaped abdomen, etc. In men too there are numerous morbid characters which are sex-linked and transmitted in this manner: hemeralopia or day blindness, Daltonism, haemophilia or failure of the blood to coagulate, and many others.

Independently of the very revealing facts supplied by sex-linkage in heredity, further light is thrown on the subject of sex determination by studying the offspring of animals which have undergone a change of sex — described more fully in a later chapter.

As we shall see, it sometimes happens that an organism belonging genetically to a given sex (XX or XY) will produce the functional gametes of the opposite sex. This may occur as the result of a constitutional anomaly or be artificially induced by the biologist. In either case the organism retains its genetic sex, so that in uniting with a normal individual the union of two individuals belonging to the same genetic sex is achieved. A proportion of the descendants of such a union will of necessity be sexually aberrant.

The sex ratio

Since, in the *Drosophila* mode of sex determination, the (Y) spermatozoa which produce males are theoretically equal in number to the (X) spermatozoa which produce females, and since, in the *Abraxas* mode, the (W) ova which produce females are theoretically equal in number to the (Z) ova which produce males, the two sexes should be more or less equally represented among fertilized eggs or zygotes, providing that the conjunction of the two gametes, that is, sperm and ova, takes place strictly by chance, and that the conditions of fertilization are in no way advantageous to one or other of the two kinds of gametes.

This sex ratio at conception, or primary ratio, cannot be known by direct observation, as sex statistics become a practical possibility only at birth or hatching. If, then, there is any variation in the embryonic mortality rate it will affect the sex ratio at birth, and this ratio, the secondary ratio, may differ appreciably from the primary ratio. The sex ratio is the number of males per 100 females. The secondary sex ratio of the human species is, taken for all countries, approximately 105, that means about 105 males per 100 females. Thus the fact which emerges from the statistical study of human births is that everywhere there are a few more boys born than girls.

Among 77,000 miscarriages Russel found ratios of 129 for the seventh, eight, and ninth months; of 142 for the fifth and sixth months, and of 375 for the fourth.

It is thus obvious that during intra-uterine life mortality acts selectively to the detriment of the male sex, so that the excess of males at conception must be still greater than at birth. It is estimated that at conception there are about 120 males or more for every 100 females. This raises two questions: why, at conception, are more males than females formed, and why, during the course of pregnancy, do they die more frequently than females?

It would seem, in brief, that the male foetus, or embryo, is weaker and more vulnerable than that of the female. The superior resistance of the female continues to be apparent after birth, for in the first year of life 161 boys die to every 131 girls. Due to the higher mortality rate among males the number of females, at first inferior, soon rises so that it is equal to and even appreciably greater than the number of living males. If statistics are taken at different ages we find that at maturity there are 115 women to 100 men, and in old age that there are twice as many.

An effort has been made to explain this apparent biological weakness of the masculine sex genetically by taking into account the fact that the male bears the chromosome pair XY, while the female bears the XX pair. As a result of this all the genes carried by the X-chromosome are, in a woman, doubly present,

while in a man the only genes which are doubly present are those which are carried by both the X-chromosome and the Y-chromosome. Those genes which are characteristic of the X-chromosome only, or of the Y-chromosome only, are present singly: they are, so to speak, odd or unpaired genes.

It is known that certain genes can undergo sudden changes or mutations and that these mutations can result in the appearance of unfavourable characters, ranging from those serious defects which are incompatible with life to those slight imperfections which merely result in slightly enfeebling the organism. Let us examine the effect of one of these mutant, more or less deleterious, genes on first one and then the other sex.

If one of the woman's X-chromosomes carries the harmful gene it is probable that her other X-chromosome will carry the corresponding gene in its normal state and be able to compensate for, or neutralize the effect of, the mutant gene. But the case of the man is different, or at least it is if the mutant gene is unpaired. In that event the harmful mutation will inevitably produce its effect, and deterioration or debilitation will ensue.

It is for this reason that certain defects and hereditary diseases, all of those which depend on genes carried on the X-chromosome, are observed much more frequently among men than among women. Thus Daltonism, or the inability to distinguish green from red, is found in four per cent of the male population of Europe while among women it is very much rarer.

Thus the biological inferiority of the masculine sex can be connected with harmful mutations which affect his unpaired genes. Masculine fragility could be attributed to the fact that men are genetically unbalanced organisms whose hereditary factors are imperfectly doubled.

The fact that in women every gene has its corresponding gene is an insurance against the unfavourable mutation of one of them. A woman receives the maximum benefit from the advantage of having two parents from each of whom she derives a full set of matching genes. A man, and this would seem to be the cause of his organic frailty, is of bi-parental origin only in respect of the majority of his genes; for *some* of his genes, those carried on the X-chromosome come only from his mother, while others, those carried on the Y-chromosome come only from his father.

It is very likely that this genetic explanation of the weakness of the 'stronger sex' contains part of the truth, though it is perhaps not entirely satisfying. In 1957 R. and C. Vendrely suggested an interesting hypothesis which brings D N A into the picture. They suggested that a supplementary reason lay simply in a few extra molecules of D N A in the nucleus of feminine cells, from which could result greater harmony between the D N A of the nucleus and the cellular mass it controls. Some biologists believe that in addition to genetic considerations one should take into account differences between the sexes introduced by their internal secretions, or hormones, which the genital glands produce even in the embryonic state. A more unlikely theory postulates that among mammals every individual male has accomplished his embryonic development within the body of an organism of the opposite sex and that it is therefore conceivable that he receives, from the maternal hormones, some foreign and harmful influence which affects his entire life.

In any case, from the fact that the male is more frail than the female, it would follow that he is less able to support unfavourable conditions in his prenatal environment. Hence both economic and social factors could have their influence on the sex ratio at birth.

As for the other problem of why more males than females are conceived, this remains even more obscure. Must we assume that the normal conditions of insemination, including temperature, acidity, alkalinity and so on, favour a sperm with a Y-chromosome? Or is it that such sperm, bearing the smaller chromosome, are lighter and more agile than sperm bearing the bigger X-chromosomes? One can for the moment only invent theories without foundation.

Families of daughters and families of sons

It is often asked if certain couples are more or less predisposed to engender boys or girls. The question cannot at the present time be answered categorically. It is extremely likely that the vast majority of couples have an almost equal chance of procreating either a boy or girl, although certain genetic anomalies can upset the sex ratio. For instance, a woman with a lethal gene on one X-chromosome will produce only one boy to two girls, since of the four possible genetic combinations only three are viable. Even in very exceptional cases when in a family half a dozen or even a dozen children of the same sex are born successively it is simpler to assume that the phenomenon is caused by pure chance and has no more significance than red or black turning up the same number of times in succession at roulette. Nevertheless, it is not impossible that there can be a family tendency to produce children of a given sex. In the 'family of daughters' studied by Lienhart and Vermelin the tendency is certainly marked, for in three generations not one male was born. They reported:

'The grandparents of Madame B. had six daughters, all of whom married. Their children, the second generation, were all girls. The eldest of the six had eight daughters, the second two, the third two, the fourth four, the fifth two and the sixth, who is Madame B's mother, had nine. This makes a total

of twenty-seven girls in only one generation.

The third generation comprising children of Madame B. and of her sisters, is again composed exclusively of girls. The eldest sister remained unmarried, the second sister had twelve children, the third nine, the fourth five, the fifth four, the sixth three, the seventh two, the eighth two, and the ninth (Madame B) two. The present count for this third generation is thus thirty-nine daughters.'

Six plus twenty-seven plus thirty-nine equals seventy-two: out of seventy-two births over three generations seventy-two were girls! Since the various fathers of the girls were unrelated to each other, one is inclined to believe that the tendency to give birth to females must arise from some hereditary factor in the mothers. But this is difficult to reconcile with what we know about the mechanism of sex determination. It could be argued that among women of Madame B's family the ova, as a result of some constitutional abnormality, refused to be fertilized except by X-chromosome sperm. But in favour of this hypothesis there is not the slightest positive evidence to offer: no such example of 'selective fertilization' among animal species is known.

Another argument, more probable but still without evidence, is that among these mothers who produced only daughters the vaginal secretions were so composed that they had an adverse effect on Y-chromosome sperm, or, finally, that in these women the hormone conditions during gestation were so unfavourable to male foetuses that they were very rapidly eliminated. Whatever the reasons may be, it would seem that they do not depend on some peculiarity of the chromosomes, but rather on some special quality of the cytoplasm which is transmitted from mother to daughter.

In addition to this 'family of daughters' a 'family of sons' has been studied. Harris, in 1946, reported that in this family only two daughters, as against thirty-three sons, had been born during ten generations. Of the two daughters one was of such indefinite sex that at birth there had been some hesitation in declaring whether she was male or female. Nothing is known of the other, except that she died at the age of two. The 'family of boys' was distinctly less fertile than the 'family of girls'.

At the moment no satisfactory suggestion can be made to explain the transmission from father to son of a tendency to engender males. It remains one of those extraordinary facts which may one day throw light on other commonly observed phenomena.

In the animal kingdom certain observations have

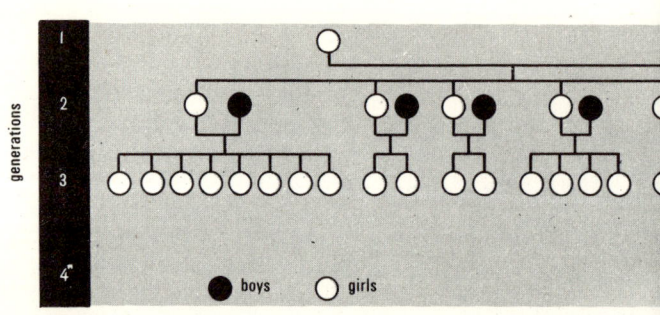

generations

boys   girls

been made which appear to concern males with a tendency to engender females.

Genealogical tree of a family of boys. Not until the tenth generation did two girls — white in the diagram — appear. One died in infancy and the other was an intersex and could leave no descendants.

After Harris.

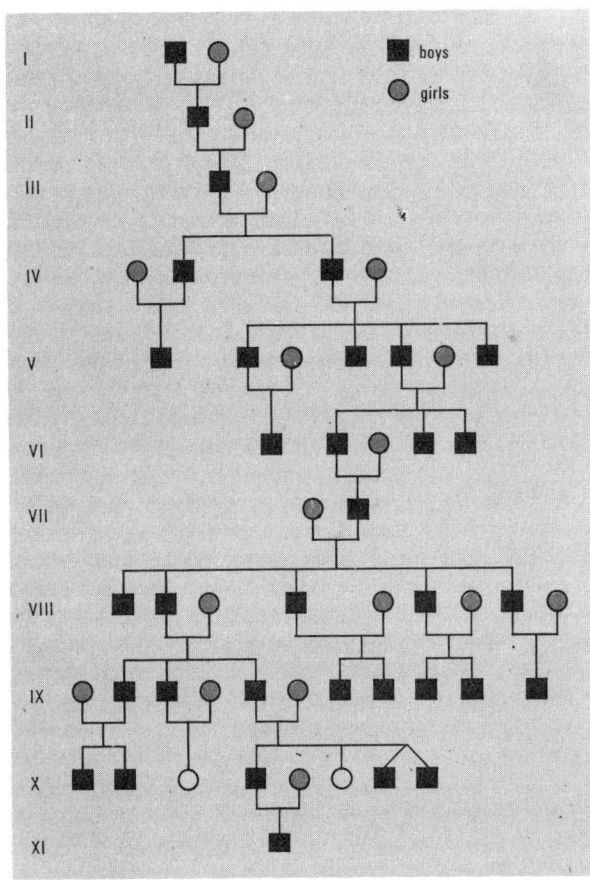

Genealogical tree of a family of girls, after Lienhart and Vermelin. In three generations there were 72 births, all female.

been made which appear to concern males with a tendency to engender females. Sanson reports the case of an ass which, with mares, sired only she-mules and was in consequence much sought after by owners of mares, in spite of a defect in shape, because the marketable value of she-mules is much greater than that of he-mules.

The voluntary determination of sex

Few problems have taxed the imagination of men more than that of voluntary sex determination. It is certainly one of the most ancient and persistent dreams of mankind to acquire the power to produce at will a son or a daughter. Innumerable recipes, strange and ingenious, were formerly concocted for this purpose by doctors and naturalists who in fact knew no more about the real causes of sex determination than did the credulous parents themselves.

According to one learned authority such and such a position should be adopted while the seed was implanted, since males were engendered from the right and females from the left. It was all a matter of timing, replied another: what was required was to choose the hour of conception wisely and arrange matters so that the egg was fertilized when fully mature. A third medical expert would recommend the father to follow a strengthening diet and thus imprint his, the stronger, sex on the child to come, while a fourth advised him to weaken himself since it was well known that the child inherited the sex of its weaker parent. Other grave medical men have prescribed suitable diets and various drugs for the mother.

In recent years Dr. Kleegman, a New York gynaecologist, has claimed a success rate of seventy per cent in the sex determination of families. This is based on her observation that children conceived thirty-six or more hours before ovulation seemed to be predominantly female and those conceived two to twenty-four hours before ovulation were predominantly male. Of course the difficulty in establishing the exact time of ovulation in many cases has hindered confirmation of this claim.

From all that is now known of sex determination it would appear that the voluntary choice of sex is in theory very simple. Among animals belonging to the *Drosophila* type, which includes our own species, it should be sufficient to separate the spermatozoa into

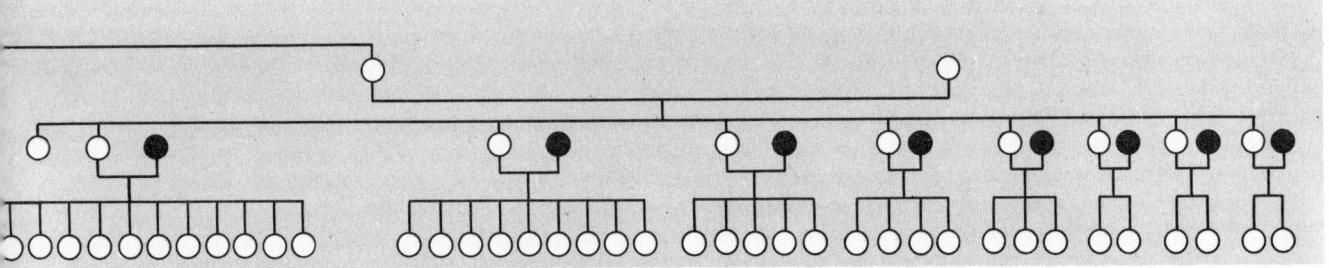

two categories, male-producing and female-producing. Then, with purified semen, artificial insemination would result in the sex desired. It is also reasonable to suppose that if we could change the female's chemical reactions to fertilization by one or other type of semen the survival of the chosen semen, or at least its chances of success, could be assured. But all efforts to do these things experimentally have until now produced only vague and unsatisfactory results.

Taking as a point of departure the assumption that there may be a difference in size and weight of the two kinds of gametes, Lush and Lindahl have tried to separate them mechanically, by centrifugation or by filtration of the semen. These attempts have proved ineffective. Equally disappointing have been attempts to discover selective survival among X or Y sperm when semen has been allowed to 'age' at normal temperature or in a refrigerator.

It is, however, claimed that sperm preserved by refrigeration produces an excess of females. It is thought that rats, deprived of vitamin B, produce more female offspring. Among mice, females are also said to be more numerous when the male is warmed, and among rabbits when the sperm is itself directly warmed.

Again among rabbits, sexual fatigue appears to have an effect on the sex ratio. If the male rabbit is encouraged to couple repeatedly the proportion of his male offspring decreases little by little. Expressed in percentage the first copulation results in 129 per hundred females and the twentieth in as little as twenty-eight.

When male mice are given alcohol, caffeine or yohimbine the proportion of male births is said to increase; but these results have been contested.

It has been claimed that the percentage of male births rises when the medium of fertilization has been made alkaline, either by introducing an alkaline solution into the vagina of the female or by mixing it with the semen before insemination. All these experiments are, however, open to criticism for inefficiency of method or insufficiency of care.

Among mice the sex ratio may depend on the pH value of the male's blood. The pH value is a measure of a solution's hydrogen-ion concentration, that is its acidity or alkalinity. Males of certain strains having a pH value of 7.42 produced sixty per cent females; others with a pH value of 7.46 produced only forty per cent females. The pH value of the mother's blood appears to have no influence on the sex ratio of the offspring.

According to certain authors it should be possible with cattle and other animals of economic importance to select males of appropriate pH value, and also to alter this value by means of chemical agents, feeding and so forth. In the classic experiments of Miss King with rats in which by parental selection she isolated lineages which departed from the normal either by producing an excess of males or an excess of females, it is possible that paternal pH was the important factor.

Mention should be made of the much debated work of Koltzoff and his pupil V. Schröder who claimed that by submitting the semen of rabbits to the action of an electric field the sperm which produces males can be separated from that which produces females: the

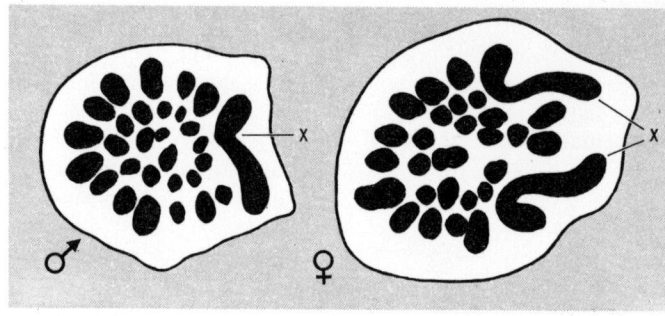

Chromosome content of the orthopteran *Leptophyes punctatissima. On the left,* the male has one X-chromosome; *on the right,* the female has two X-chromosomes.

After O. L. Mohr.

female-producing sperm moving towards the positively charged anode, and the male-producing sperm towards the negatively charged cathode. The different behaviour of seminal elements would thus seem to depend on a difference in their biochemical nature, itself linked to their difference in chromosome content, whether, for instance, they are X or Y.

Koltzoff and Schröder themselves admitted that electrophoresis does not always give absolutely constant results. When sperm from the anode or from the cathode is used for insemination not only is it far from producing 100 per cent of the predicted sex, but also changes in the sex ratio are influenced by auxiliary factors like temperature, reaction of the seminal fluid, and the physiological condition of the male donor. Schröder proposed an ingenious variation of the method, which consisted of immunizing the male progenitor against one or other type of spermatozoa. It is well known that when an organism is injected with microbes or poisonous substances it will form 'antibodies' which are capable of destroying the microbes or of neutralizing the poisons: vaccination, or active immunization, is based on this phenomenon. In a similar way if an organism is injected with cells of a certain type, red or white blood-cells, epithelial cells, sperm-cells, and so on, specific 'cytolysins' are produced which can destroy the specific cells introduced.

Schröder, therefore, injected male rabbits with semen which had collected at the anode or at the cathode, that is, with female-producing and male-producing sperm respectively, hoping to induce the formation of specific sperm-destroying antibodies or spermolysins which would destroy not all spermatozoa but only the specific X or Y sperm injected.

The experiments fulfilled the expectations of the

84

investigator: the offspring of males injected with anode-sperm were predominantly male, while those of males injected with cathode-sperm were predominantly female. It took some ten days to immunize the male, with intervals of from three to five days. In all, the animal received from 1—4 cubic centimetres of spermatozoa in serum. Two weeks after the end of the treatment he was placed with the female.

Finally Schröder employed a third procedure which consisted of injecting the male progenitor with the serum of a rabbit which had itself been previously immunized against either anode or cathode sperm. By means of this passive immunization a higher percentage of males or, according to the case, of females was obtained.

There have been numerous experiments using X-rays to cause an increase in the number of females. This appeared to be successful when applied to rabbits, but had the opposite effect when used on mice.

In the offspring of populations submitted to atomic bombardment statistics seem to suggest that the percentage of male births is increased by one powerful dose of paternal irradiation, while feebler, repeated doses may possibly cause a decrease in this figure.

Lejeune and Turpin supported the theory that the percentage of boys diminishes in proportion to the advancing age of the father, and a possible explanation of this phenomenon could be the long-term influence of natural radio-activity.

Finally, various investigators hold that artificial insemination appreciably raises the proportion of boys, which if their statistics are valid, would seem to indicate that when the encounter of gametes is thus facilitated the Y-chromosome sperm somehow benefits.

Voluntary sex determination in heterogametic females

Among those animals whose sex is determined in the *Abraxas* manner the problem of voluntary sex determination is obviously different. In this case we can try to modify the primary sex ratio by influencing the ovum at the moment it is formed, that is, at the moment of meiosis or reduction division when it expels either the Z-chromosome or the W-chromosome to one of its poles. If by technical means it could be brought about that one of these chromosomes, say the Z-chromosome, was retained by the ovum with more than normal frequency, the number of Z-ova would, of course, be proportionately greater, and the sex ratio of the organism would be altered.

This is precisely what the biologist Seiler succeeded in doing in the series of elegant experiments he made with the moth *Talaeporia tubulosa*. Seiler found that when the external temperature was 18 °C the proportion of females was 136 per 100 males. If the temperature was raised to 35—37° the ratio dropped to sixty-two per 100 males, while it again rose to 155 when the temperature was lowered to 3—5 °C. By cytological observations Seiler was able to verify these findings, that heat and cold act in an opposite way on the

retention by the ovum of the Z-chromosome.

An ovum which is over-matured behaves like one submitted to higher temperature, producing seventy-two females for 100 males. With other moths — *Lymantria dispar, Bombyx mori* — J. Rostand obtained an increase in the number of males by artificially postponing the time when the female lays her eggs. Among fowl that also follow the *Abraxas* pattern it seems that when egg-laying begins the composition of the egg is such that it encourages the retention of the Z-chromosome — whence males are produced in greater abundance at the beginning of the season.

Perhaps there is sometimes an ovular dimorphism connected with sex. The yolk of a pigeon's egg which produces a female may be larger than that of an egg which produces a male. It is thought that the heavier eggs laid by hens produce a larger proportion of females, and the same may be true of silkworms. But further research is needed to settle the question.

The Sex Hormones

We have just seen that sex is determined in the fertilized ovum by the nature of the chromosomes it contains. But it is not, at least not among vertebrates, the chromosomes themselves which control sex differentiation. In other words, the female chromosome formula XX or ZW does not directly determine the appearance of feminine characters in the organism, any more than the male chromosome formula XY or ZZ directly determines the appearance of masculine characters.

Sexual characteristics, masculine or feminine, are conditioned by the action of the genital glands which are themselves either masculine or feminine as a result of the kind of chromosomes present at the beginning of the organism's development. Thus a female chromosome formula will lead to the formation of female glands, or ovaries, which will then feminize the animal. A male chromosome formula will lead to the formation of male glands, or testes, which will then masculinize the animal.

The comb of a cock compared with the comb of a capon.
Ylla-Rapho.

Golden Leghorn cock and hens.
Larousse.

Japanese Golden Pheasant. The male is above and the female on the left.
Larousse.

Secondary sexual characters in poultry

The glandular determination of secondary sexual characters has been studied with particular care among birds, and especially among poultry which, because of their marked sexual dimorphism, have furnished scientific investigators with striking results.

The cock is outwardly distinguished from the hen by its more brilliant and varied plumage and the particular form of certain feathers such as sickle-feathers; by the development of its erectile organs — comb and wattles; by the presence of spurs; by its ability to crow, by its sexual ardour and aggressive behaviour.

Now if a young cock is castrated, its development will be considerably altered: it will, in fact, become a special animal, a capon, with the following characteristics: plumage resembling a cock's; comb and wattles, like a hen's, reduced; spurs similar to a cock's; inability to sing cock-a-doodle-doo and, again like the hen, lack of sexual and combative ardour.

What occurs if, on the other hand, a young hen is deprived of its ovaries? Again the bird's development is considerably changed. As an adult it displays traits almost identical with those of the capon, having, notably, the spurs and the plumage of a cock.

The capon and the caponized hen are thus the neuter form of the species, that is to say the form which develops in the absence of any sex glands, testes or ovaries, quite a different animal from either the male or the female and different from the animal it would have become had it undergone the natural changes which the retention of its natural parts would have

produced. Such experiments conclusively demonstrate that the chromosome differences between male and female are not alone sufficient to create sexual dimorphism. Furthermore they show that this dimorphism is intimately linked with the presence of genital glands, ovaries or testes.

The genital glands act through the intermediary of certain secretions or hormones which they pour into the bloodstream. In fact it is possible to prevent the capon and the caponized hen from assuming the neuter form by grafting on to them a testis or an ovary. Even more striking is the fact that the same result can be obtained by injecting them with extracts made from testes or ovaries.

The secretions which have this powerful effect are the sex hormones, chemical substances of which the chief are testosterone for the male, oestrone or oestradiol for the female. The male hormone, testosterone, has a positive or stimulating action on comb and wattles, on the ability to crow and on sexual ardour. The female hormone, oestrone, has a negative or inhibitory action on cock-like plumage and on the growth of spurs. By the use of these hormones a total inversion of secondary sexual characters can be artificially brought about. With testosterone a caponized hen can be made to grow a comb, or with oestrone the plumage of a hen can be made to appear on a capon.

The somatic or body-cells of the capon and the caponized hen are different in their chromosome formulae, but in spite of this the capon will take on the aspect of a hen if supplied with female hormones, while the caponized hen will take on the aspect of a cock if supplied with male hormones. During one spectacular demonstration hen's ovaries were removed and its neck, breast, back and rump were half-plucked.

Leghorn cock turned into a hen by castration followed by the graft of an ovary.

After Caridroit.

A few months later new feathers grew over the plucked areas. Since the new feathers had grown when the hen no longer had ovaries, and hence in the absence of oestrone, they were masculine in type, while the original, unplucked feathers, grown while oestrone was still secreted, were those of a normal hen. The plumage of the bird was thus male on one side and female on the other: the experiment had created a chicken which was half-hen, half-cock, a so-called 'bipartite-gynandromorph'.

A variation of the experiment can be made: a cock castrated, grafted with an ovary and then entirely plucked on one side. A few months later new feathers will appear on the plucked area, feathers which, having grown in the presence of oestrone, will be of feminine type. The old feathers will remain masculine in type and again the result will be a bird which is half-cock, half-hen.

If the bird's hormone supply is changed while its first feathers are still growing, each feather will at maturity be half-male, half-female in type, the division between the masculine and the feminine zones being straight and clearly defined.

In the case of a capon which has been injected with oestrone the tip of the feather (which grew before the process of feminization began) will be male in type, while the base of the feather (formed after the injections) will be female in type. In the case of a hen whose ovaries have been removed the tips of the feathers will be of female type and the lower, newly-grown part of male type. It is a fact, known since the days of Aristotle and perhaps even earlier, that old female birds sometimes develop masculine traits which must be attributed to the eventual failure of their ovaries to function. Biologists and naturalists have cited cases of this kind among hens, turkeys, pheasants, and peahens. Everard Home reported having seen an old duck, thus masculinized, pursuing young females and even mounting one of them, simulating the actions of a genuine drake. Toussenel, a confirmed feminist, wrote: 'It is so true that among birds the female is in every way superior to the male that she has only to deteriorate in order to borrow his most gaudy costume and his most melodious song. Every day one meets old hens, old peahens, or pheasants who, grown weary of maternity, amuse themselves by assuming the livery of their males.'

Leghorn hen in process of being masculinized by treatment with male hormones.

Leghorn hen being turned into a cock by removal of ovary and graft of testis.

After Caridroit.

89

Hormones and sexual characters in mammals

The influence of hormones on sexual characteristics has been analysed as thoroughly among mammals as among birds. All the experiments we have mentioned can be repeated with rats, mice, guinea-pigs, and so on: the suppression of sexual characters by castration, the creation of a neuter form in both sexes, the restoration of feminine traits in the neutered female by grafting ovaries or injecting female hormones, the restoration of masculine traits in the castrated male by grafting testes or injecting male hormones, the experimental inversion of sexual characteristics by injecting the castrated male with female hormones or alternatively by injecting the neutered female with male hormones.

The nuptial callosities of the male frog will disappear when the frog is castrated and reappear when a testis is grafted. In male newts castration brings about the regression of the dorsal crest which can be made to grow again by grafting a testis. By the same graft a female newt whose ovaries are removed can be made to grow a crest.

In the human species castration has in the past been practised for social, religious, or other reasons, and its effects, especially in the masculine sex, are well known. If the operation takes place when the male is a child he will not achieve full manhood: his genital tract, penis and prostate, remains under-developed, his skin hairless, his larynx infantile, and his voice, remaining high, does not break.

The chemistry of sex hormones

We have said that sex hormones are produced by the sex glands, without stating in which part of the glands the hormones originate. They do not, in fact, originate in the germinal tissues where the gametes are formed, but in a particular tissue with secretory properties — the interstitial tissue in the case of the testes, and special glandular cells in the ovaries.

The arrangement of these secreting tissues differs according to the animal species. The only basic and invariable resemblance is the essential duality of the sex gland which always comprises two parts endowed with very distinct functions. The function of the strictly *genital* part is to form and ripen the reproductive cells, ova or spermatozoa. The function of the secretory or *endocrine* part is to secrete the substances or hormones which, poured into the bloodstream, insure the sexualization of the entire organism.

The sex hormones like the other hormones of the pituitary, thyroid and adrenal glands are, at least among vertebrates, devoid of zoological specificity. In other words, a sex hormone can be borrowed from a mammal and used to sexualize a bird or an amphibian. Inversely, a sex hormone borrowed from a bird or an amphibian can be used to sexualize a mammal. It is because of this lack of zoological specificity that sex and other hormones can be employed in human therapy.

The principal sex hormones have been identified by biochemists and can be prepared in a pure state. The hormone secreted by the mammalian ovary is generally believed to be oestradiol, a hydrogenated derivative of oestrone or folliculine. Its empirical formula is $C_{18}H_{24}O_2$, while that of oestrone is $C_{18}H_{22}O_2$. Both are derivatives of phenanthrene and belong to the family of the sterols. Oestrone is found in great abundance in the urine of pregnant mares: as much as 25 grams can be obtained from 52 metric tons of the liquid. It was when this practically unlimited source of female hormone was discovered that chemical research could begin in earnest. The male hormone or testosterone — $C_{19}H_{28}O_2$ — is closely related to oestrone.

Today the sex hormones can be synthetically produced from cholesterol. The female hormone (oestradiol) is by preference employed in the form of benzoate; the male hormone (testosterone) in the form of propionate. Both are widely used in human therapeutics, either to compensate for a lack of sex hormones — during the menopause or senility, for instance — or to exert an inhibitory action in the case of certain tumours. Oestradiol is used, for example, in treating cancer of the prostate.

Chemists have synthetically created sex hormones which are in some respects more powerful than those made by nature, as is the case with diethylstilboestrol. Diethylstilboestrol is as active when swallowed as when injected for, unlike natural female hormones, it is not destroyed by the liver. Particular care is taken to prescribe sex hormones in precise amounts, and for this purpose biological tests are made of certain functional reactions which consistently accompany the administration of a given dose of the hormone.

In the case of the female hormone use is made of the property it possesses of bringing about characteristic changes in the vaginal lining of rodents when on heat — copious desquamation with the appearance of large keratinized epithelial cells in the vaginal smears. The minimum hormone dose which causes this reaction in a mouse under the age of puberty or in a mouse with its ovaries removed is known as a mouse-unit. One milligram of oestradiol is equal to about 10,000 mouse-units, an indication of the feminizing power of the substance.

Oher tests are sometimes used: opening and keratinization of an immature rodent's vagina, increase in weight of the uterus, which can be almost immediate by oedematous swelling or else slower by cellular proliferation. In the case of the male hormone, its effect on the growth of a capon's comb, or on the genital tract of an immature rat, is observed. The capon-unit is that which enlarges the surface of the capon's comb by twenty per cent, and corresponds to the administration of 15 ten-thousandths of a milligram of testosterone in oily solution over a period of six days. Since the male hormone has a local action the comb of a capon or of a cockerel can be daubed with testosterone and after six days of the treatment cut off and weighed. By these tests it is possible to measure quantities of testosterone well under one thousandth of a milligram.

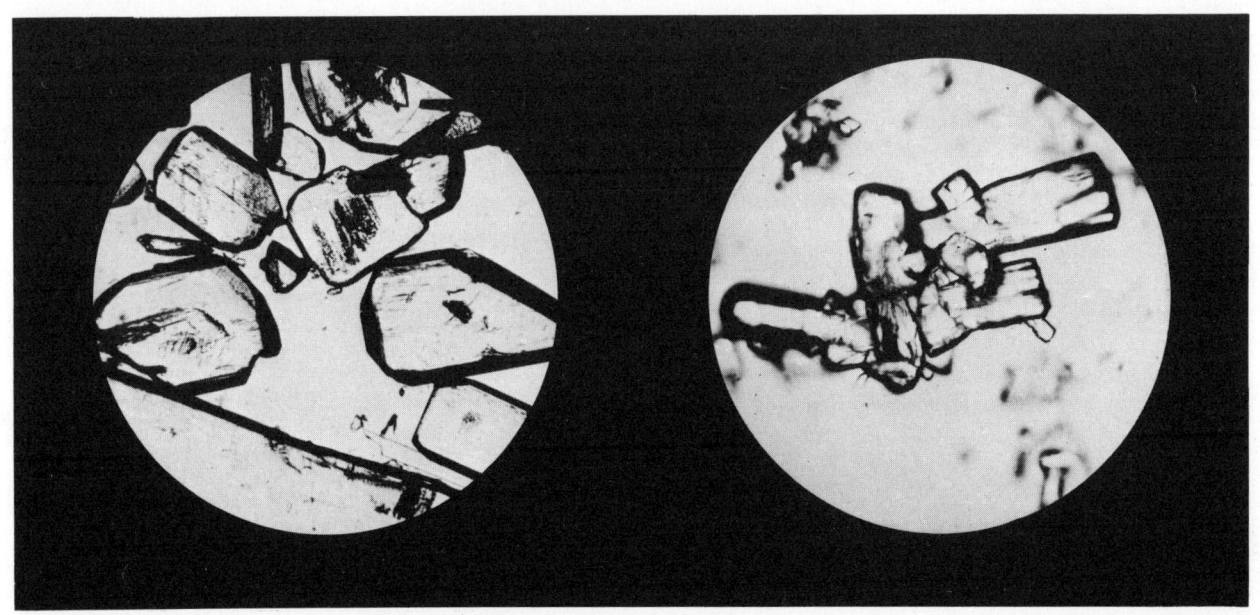

Crystallized sex hormones. *Left,* testosterone, *right,* oestrogen.
Doc. Lab. Pharlon.

Chemical castration

It is well known by all poultry breeders that capo-
nizing or removing the testes of a six or seven week old
male chick results not only in the complete sexual
neutralization of the bird but also in fattening it and
making its flesh more tender. For some time poultry
breeders have tended to replace surgical castration
by 'chemical castration' which consists of treating
the young birds (seven to eleven weeks old with synthetic female hormones such as diethylstilboestrol.
These are preferably implanted as pellets under the
animal's skin in the region of the neck.

The operation, taking thirty seconds, is very rapid
and the bird can be sold in good condition a few weeks
later, while surgical castration requires several months
before its full effects are apparent. The chief effect
of the treatment is to increase the layer of subcutaneous
fat; furthermore the birds become calmer, more resistant
to infections, and a greater yield is obtained for less
expenditure in feed.

This chemical castration is, it would seem, without
toxic effect on the poultry. The hormone does not
appear to accumulate in the edible tissues of the bird,
which can be eaten without risk of feminization to the
consumer! Chemical castration has also been practised
with turkeys, and even with cattle.

The action of the sex hormones on the embryo

The examples of hormone action which we have
discussed affect the secondary sexual characters of
adult organisms or at least of organisms which have
already reached a stage of advanced development. But
at birth, or on hatching, organisms have already
developed certain secondary sexual characters, notably
everything connected with the anatomic structure of
the genital apparatus. The male has formed *vas deferens,* penis, and prostate; the female oviducts or
Fallopian tubes, uterus, and vagina.

What is the origin of these sexual characters which
are formed so early? Are they, like later characters,
determined by the action of hormones? If so, must we
assume that from the beginning the sex glands themselves exercise an endocrine function and secrete
hormones which differentiate the sexes?

For many years it was believed that the sex glands
remained totally inactive until the age of reproductive
maturity was reached, and that this inactivity applied
not only to their reproductive function but also to their
endocrine function. We now know that active hormones are in fact produced and secreted by the ovaries
and by the testes during embryonic life.

When ducklings are hatched they reveal very definite sexual dimorphism: the males have a 'genital tubercle' and their voice organ, or *syrinx,* is much bigger than that of the females. It is also clearly dissymmetric, being dilated towards the left and enlarged in front. Now if young embryo ducks are castrated by means of X-rays — which totally destroy their sex glands, it is observed that all the ducklings hatched have a syrinx which develops according to the masculine type. Among embryos thus castrated there will be more or less one half who genetically were certainly destined to be females, so we may reasonably conclude that it is the presence of the ovarian gland in the female embryo which determines the development of the female syrinx.

The neuter, sexless type of syrinx, that which is produced in the absence of all sex hormones, is thus the masculine type, as we have already found to be the case with the neuter, sexless type of plumage in poultry.

There is a complementary proof: when strong doses of female hormone, oestradiol benzoate, were injected into duck eggs (between the fifth and the ninth day) *all* the embryo ducks developed the feminine type of syrinx. A further experiment consisted of removing the syrinxes from young embryos of ducks before any differentiation had taken place. They were then cultivated *in vitro* in an artificial nutritive medium which contained no trace of sex hormones: all developed in the male manner. When, however, female hormone was added to the culture medium a diametrically opposite result was obtained: all the

syrinxes developed in the female manner. These experiments, so neat and instructive, teach us in addition that the hormone acts directly and immediately on the syrinx, without any organic intermediary, for *in vitro* the syrinx was isolated form the rest of the organism and so uninfluenced by outside factors.

Similar studies have been made of the differentiation of the genital tubercle. When duck embryos are castrated by X-rays it is observed that in all of them the genital tubercle develops as though the animal were male. Again it is the secretion of the embryonic ovary which in females determines the regression or atrophy of this organ. If genital tubercles are cultivated in a medium deprived of all hormones they all develop in the way characteristic of the male. Thus, as in the case of the syrinx, the neuter type is the masculine type.

The differentiation of sexual passages

The differentiation of sexual ducts or channels which serve to deliver the reproductive cells is also controlled by sex hormones produced by the glands of the embryo. Every embryo duck possesses a double assortment, male and female, of these passages: the future *vas deferens* or Wolffian duct, and the future oviducts or Mullerian ducts. If embryo ducks are submitted to X-ray castration all retain the feminine or Mullerian ducts. Similarly, if these channels are cultivated in

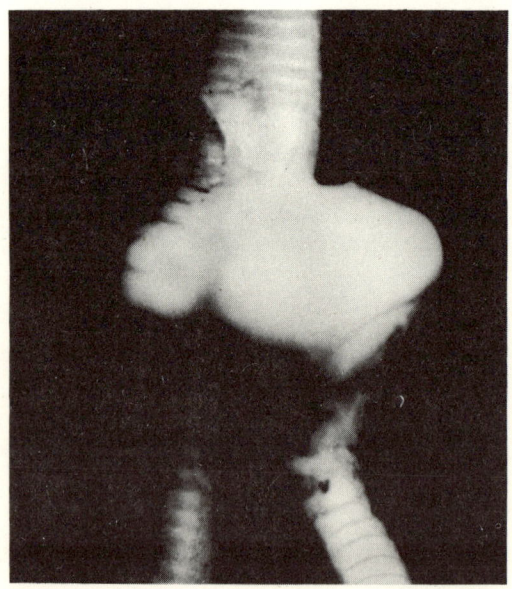

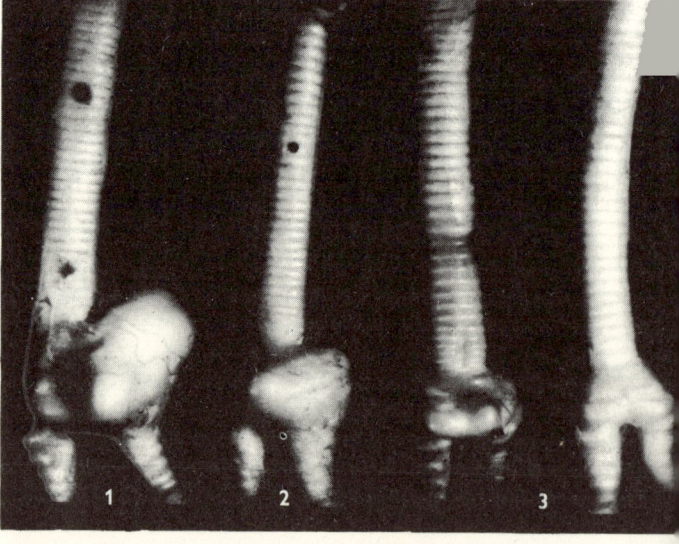

Influence of embryonic castration on the development of the syrinx in the embryo duck. The embryos are castrated at a very early stage.
1. The syrinx of the castrated subject, male or female, has exactly the same form as the normal male syrinx.
2. Two cases in which the ovary of the female embryo was partially destroyed, showing incomplete inhibition of male characters. (All three cases show that the hormone produced by the ovary of the embryo prevents the development of the male form — which is thus of neuter or asexual type.)
3. Syrinx of normal female.

After Et. and Em. Wolff.

Syrinx of a castrated embryo developing male characteristics in accordance with the above conclusion.

After Em. Wolff.

a medium without hormones, it is observed that they develop normally. But if male hormone is added to the nutritive medium the channels regress and atrophy, as they do in the normal male embryo. Therefore, in this case the neuter type corresponds to the feminine type.

How do the sex hormones act?

We have just seen how, in one case, the female hormone in the embryo duck prevents the syrinx from evolving into the male type — which is the neuter type — and how in another case the male hormone in the same embryo causes the atrophy of the female or Mullerian ducts, channels which in the neuter type are retained. Is it not possible for the biologist to pursue the analysis of these matters a little farther and discover how the sex hormones are able to bring about such phenomena of prevention, inhibition and regression? The problem of the manner in which the sex hormones, and in general all the hormones, act is one of the most important which physiology has to solve.

With regard to the inhibitory effect exercised by female hormone on the syrinx of the duck, Wolff and his colleagues have, with their technique of organ culture, been able to show that certain changes in the culture medium can bring about effects which are rather anologous to those caused by female hormone.

If the organ is no longer cultivated in the ordinary nutritive medium, which contains embryonic chicken extract, but in a medium composed exclusively of amino acids in a pure state, it is observed that the syrinx, although able to survive and develop in this artificial medium for several days, develops constantly towards a type which is transtitional between the male type and the female type, and even shows a tendency to be more female than male.

In other words, the specific effects of the female hormone can be at least partly imitated, 'mimed', simply by an impoverishment of the nourishing medium. In this experiment an organ was 'feminized' merely because it had been cut off from certain alimentary materials. One is therefore led to suspect that the normal effect of the female hormone may be to bring about a kind of food shortage by preventing the tissues from fully exploiting the sources of nutrition at their disposal. It must be added that Wolff has also obtained an effect of feminization by submitting the syrinx, in culture, to very strong doses of male hormone, testosterone propionate. This paradoxical result may be due to the hormone's toxic action on the organ, reducing it to a state of inanition.

With regard to the destructive effect of the male hormone on the Mullerian ducts, Wolff and his colleagues also made important discoveries. In the first place, the effect, which is relatively easy to observe *in vitro,* is very remarkable in that it is, unlike ordinary toxic action, selective. When other organs are exposed to the same concentrations of male hormone they undergo no regression. We are thus led to conclude that the Mullerian ducts are somehow particularly sensitive and vulnerable to the male hormone. From

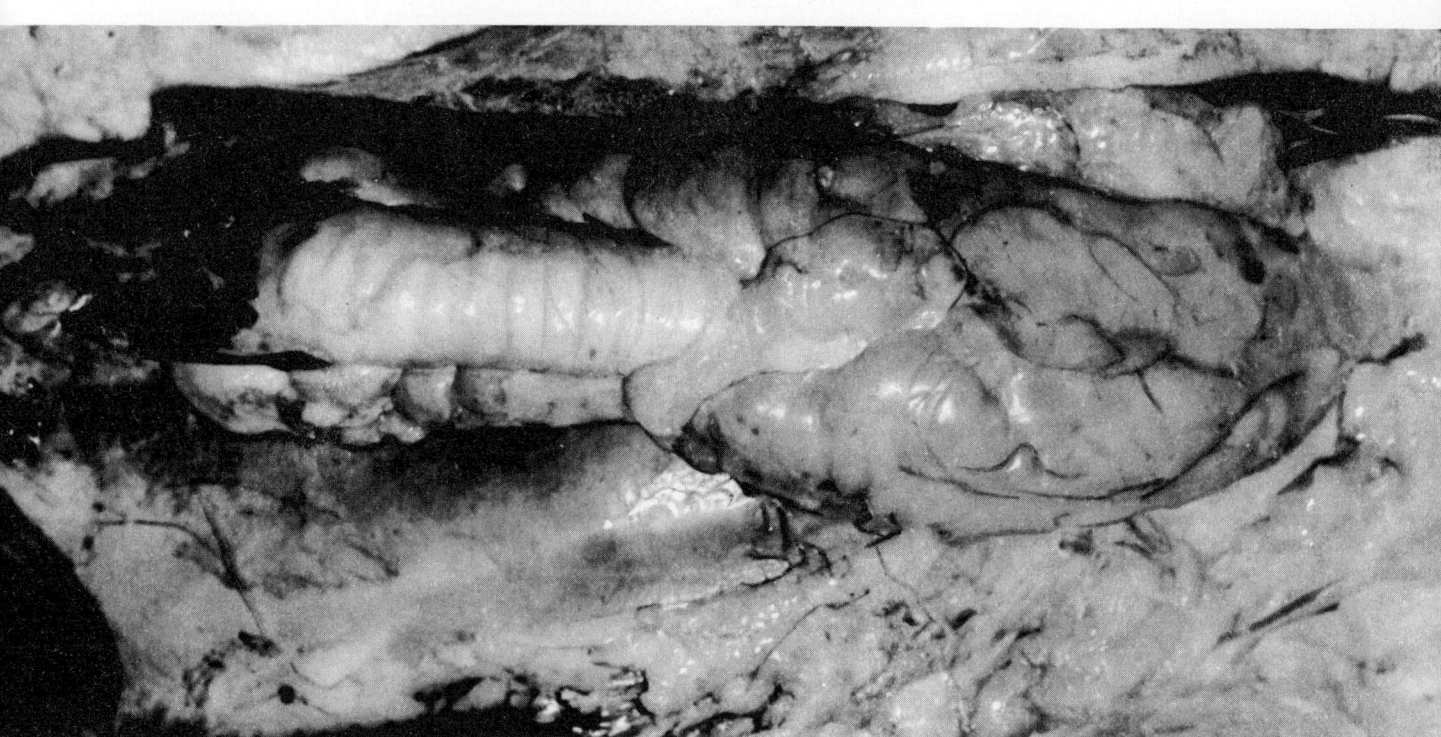

Thymus of a calf.

Larousse.

the moment that they are submitted to the action of male hormone they are irretrievably condemned to atrophy and necrosis. Everything done to save them is unavailing. It is as though the hormone has set about a process of destruction or 'lysis' which once begun cannot be arrested.

Again Wolff's method has given us a deeper understanding of the phenomenon. The male hormone brings about the destruction of the Mullerian ducts by stimulating the liberation of special ferments in the tissues of the ducts by means of which the ducts consume themselves.

The existence of these ferments has been demonstrated in this way: from Mullerian ducts which have undergone the action of male hormone an extract is made. This extract is next placed with foreign tissue — embryonic pancreas. We then see that the tissue is digested, but it is not digested when exposed to an extract identically prepared from Mullerian ducts which have not undergone the action of male hormone.

Finally, a new fact has been recently added to those revealed in this beautifully executed and logically interpreted series of experiments. H. Lutz and Yvonne Lutz-Ostertag have shown that ultra-sounds (high-frequency pressure waves above the limit of audibility) can cause the Mullerian ducts of an embryo male to persist. To explain the phenomenon they assumed that the ultra-sounds have a destructive effect on the ferments responsible for the ducts' regression, and they have been able to verify this hypothesis experimentally.

The researches which we have been discussing have, of course, made invaluable contributions to the analysis of sexual differentiation; but they have also opened the way to further studies of the phenomena of regression, necrosis, and atrophy so frequently encountered in the course of animal development, and of which some at least seem to be closely connected with the production of certain hormones. Thus among mammals the thymus, or sweetbread, is reabsorbed before the age of puberty, probably under the influence of sex hormones. Again, among Batrachians, the tail is reabsorbed as a result of the action of the hormone secreted by the thyroid gland.

Finally, among men, there is a well-known phenomenon, namely the loss of hair, which must to some extent be connected with the male hormone, since eunuchs are never bald and women very rarely so. It is therefore reasonable to assume that the male hormone in some fashion encourages the production or the liberation of lysis-causing ferments which attack and destroy the roots of the hair. The problem is to prevent this destruction. Loss of hair among men may one day be prevented in the same way that the regression of the Mullerian duct in embryo birds has been prevented, and ultra-sounds be employed in the treatment of baldness.

Experiments on mammals

The series of experiments on poultry, just described, has been paralleled by similar experiments on the embryos of mammals. These have led to equally significant results and confirmed the importance of embryonic hormones in early sex differentiation.

A. Jost, in particular, was successful in castrating the young embryos of rabbits in 1947. The rabbit's period of gestation is about twenty-eight days. In the genital gland sex differentiation is apparent towards the fourteenth day, and in the somatic organs towards the twentieth day. During the interval which separates these two stages the foetus is castrated. To do this the abdomen of the pregnant female is opened, the womb incised and one of the embryos exposed. Its genital glands are then extracted, and the embryo is re-enclosed in the womb. Finally the womb and the rabbit's abdomen are stitched up.

In spite of the technical difficulties of such an operation (its delicacy may be grasped when we remember that the foetus of a rabbit at this age measures only 3 centimetres) a good number of the embryos thus castrated were born alive, and it could be noted that the removal of the sex glands had prevented all sexual differentiation of the soma or body. When operated upon, both male and female foetuses evolve in strictly the same way, and their development follows the feminine pattern. Therefore, in this species, the neuter type is the feminine type. Both male and female embryos form a uterus and Fallopian tubes but no seminal channels, no prostate, no penis. The external sexual equipment is of feminine type. The genital gland in the male foetus must then exercise a double action: on one hand it brings about the formation of masculine equipment, and on the other it prevents the formation of feminine equipment. In the female foetus the sex gland is not required for the formation of feminine organs, which evolve in its absence, and its function is limited to stimulating their development.

The question is how the genital gland of the foetus so powerfully affects somatic differentiation. Undoubtedly the answer is that it acts through the intermediary of the soluble hormones which it secretes into the surrounding blood vessels.

If a foetal testis is grafted on to a female foetus whose ovaries have been removed the ensuing development will, in fact, be masculine. Furthermore, with a male foetus, it is possible to replace the masculinizing action of the testis by implanting a crystal of male hormone, testosterone propionate. This last experiment proves, moreover, that the hormone secreted by the testis of the foetus must be similar if not identical to the hormone secreted by the same gland in the adult organism. For, with the compensating hormone, even castration fails to prevent normal masculine development.

It may be added that by injecting pregnant females or their embryos with sex hormones a whole series of sexual anomalies can be brought about, of which the most extreme result is the complete transformation of the somatic sex.

In this way females, under the influence of testosterone, can be made to grow seminal ducts, a prostate and a male-type urethra, while the male, under the influence of oestradiol will develop a vagina and a uterus.

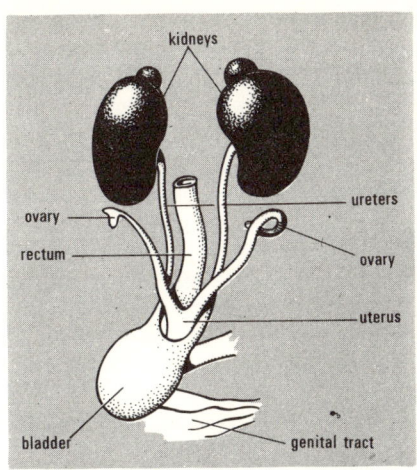

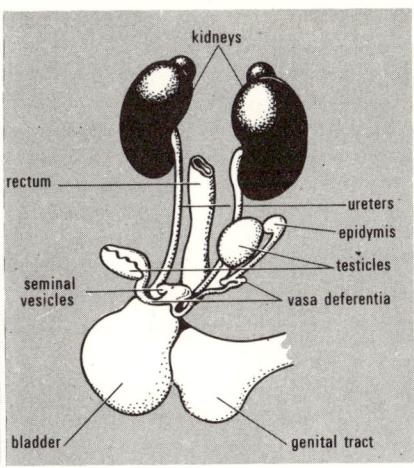

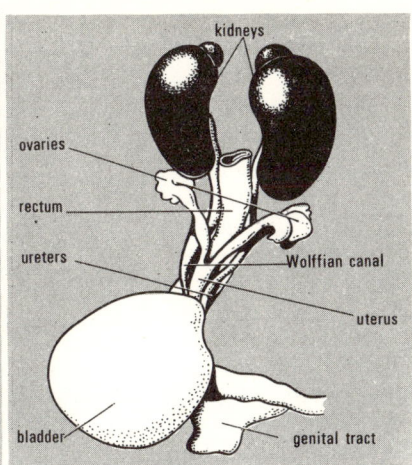

1   **2**   **3**

Diagrams of position at birth of the genital apparatus of 1. a normal female mouse, 2. a normal male mouse, and 3. a female intersex. The female intersex was produced by injecting the pregnant mother with male hormones.

After A. Raynaud.

Battery of hens' eggs in an experiment devised by Professor Etienne Wolff. Through the transparent cover of certain eggs the dark mass of the embryo and its attached blood vessels may be seen.

Larousse.

Modifications of the sex gland

By the use of these powerful drugs, the sex hormones, can we not influence the sexual evolution of the individual still more profoundly? If before the sex gland has been formed these hormones are introduced into the embryo the question arises: are they capable of changing the nature of the sex gland itself, in other words, of altering the primary sexual character of the organism? Experiments to discover this were made as long ago as 1939 by Willier, Vera Dantchakoff and above all by E. Wolff and Gingingler, with birds, and the positive results obtained exceeded all expectations.

Hens' eggs are taken between the third and fifth day of incubation and a small circular opening made in the region where the embryo is situated. On the chorio-allantoic membrane a few drops of female hormone (oestrone) in oily solution are deposited. The circular opening is then covered with a thin sheet of glass or cellophane and the edges sealed with wax. Through the transparent glass or cellophane window the progress of the experiment and the development of the embryo can be easily followed.

If a considerable dose of hormone, that is, more than 500 mouse-units, is introduced into the egg all male embryos are almost completely turned into females: they develop the typical ovary with its left oviduct. Normal hens, of course, have only one ovary, the left, and a single oviduct. This dissymmetry of the genital tract permits the sex of an embryo more than ten days old to be diagnosed at a glance. By using weaker doses males of intermediate sex are obtained with gonads of intermediate type: the degree of inter-sexuality depends on the quantity of hormone injected. Anything over twenty units (over two-thousandths of a milligram) is enough to bring about a distinct change in the male gonad: a very slight feminization can be noted with as little as ten to twenty units. Whatever the dose may be, it must, to be effective, be administered before the eighth day of incubation, for it is then that the chicken's sex glands are differentiated.

Comparable results are obtained by using a synthetic substance, diethylstilboestrol, which, though quite different in chemical constitution from the female hormone, closely resembles it in its biological effects.

Wolff made his 'change of sex' experiments even more apparent to the eye by crossing a hen of white plumage (a Light Sussex) with a cock of red plumage (a Rhode Island Red). In this kind of cross the differential characteristics depend on genes carried on the Z-chromosome and are transmitted in accordance with the laws governing sex-linked heredity, so that all the males are white, like the mother, and all the females are red, like the father. Now if oestrone is introduced into the hybrid eggs the result is pullets with white plumage: their colour at once betrays the fact that they are really genetic males turned into females.

Attempts have been made to bring about the inverse transformation, that is, of female into male, by introducing male hormone into the embryo. By this process, and then only when massive doses of the hormone are administered, sling and wery imperfect masculinization of the female gonad has been observed. Among chickens thus masculinized the oviduct remains incomplete and does not communicate with the cloaca, while the conchitic or shell gland is missing, so that the eggs cannot be laid and instead accumulate inside the bird.

What, it may be asked, finally becomes of cockerels who have been changed into pullets by hormone treatment? Their change of sex is impermanent. At the end of a few months they recover their original sex. They become genuine cocks who, from their feminine phase, retain only a useless oviduct.

Even if the treatment begun in the embryo is continued after hatching by renewed injections of two-tenths to a milligram of female hormone every two days, the bird's ovary remains rudimentary. It never becomes a true hen capable of producing eggs. Until now science has not succeeded in making a cock lay an egg.

Must we then assume that the chromosome constitution of the male opposes his complete and permanent feminization, or would it not be more plausible to argue that the ovary, in order to attain its state of functional maturity, must receive substances other than female hormone which biologists have not yet found.

Whatever the answer may be, these experiments of sex changes are far-reaching in their implications. In the first place they suggest the idea of the genetic bi-potentiality of the sexes. In every egg, whether it is hatched as a male or as a female, there exists the means of evolving a sexual apparatus of both masculine and feminine type. Which of the two is developed depends on a difference in hormones only. If, moreover, the sex hormones of the adult have such a powerful effect on the genesis of the gonads it is reasonable to suppose that in normal conditions differentiation of the sex glands is produced by substances very similar to these hormones, if not identical.

The situation, then, might be summarized rather cursorily in this fashion: a given chromosome formula leads to the production of a given hormone, male or female, and the hormone then controls the differentiation of the gland which will continue to pour this same hormone into the bloodstream during the life of the animal. It would seem certain that the same substances which determine the sexual characters of the adult also determine the differentiation of the embryo's genital glands: the ovarian hormone would thus form ovaries and the testicular hormone would form testes.

Eggs soaked in hormones

It has recently been claimed that the sex of chicks can be changed simply by soaking the eggs in an oily solution of male or female sex hormone, the solution being maintained at a temperature lower than that of the egg. The hormone filters through the pores of the eggshell and is deposited on the conchitic membrane whence it is gradually absorbed by the embryo. The smaller end of the egg should be placed downwards, and the soaking is said to be effective until the sixth day of incubation. Such a simple procedure would obviously be of the greatest convenience both to the poultry breeder and to the research biologist, but until now it has not been confirmed.

The embryo of a chicken with its blood vessels. (Mosaic by
Nelly Vandel which decorates the entrance to the Laboratory
of Experimental Embryology at Nogent-sur-Marne.)

Larousse.

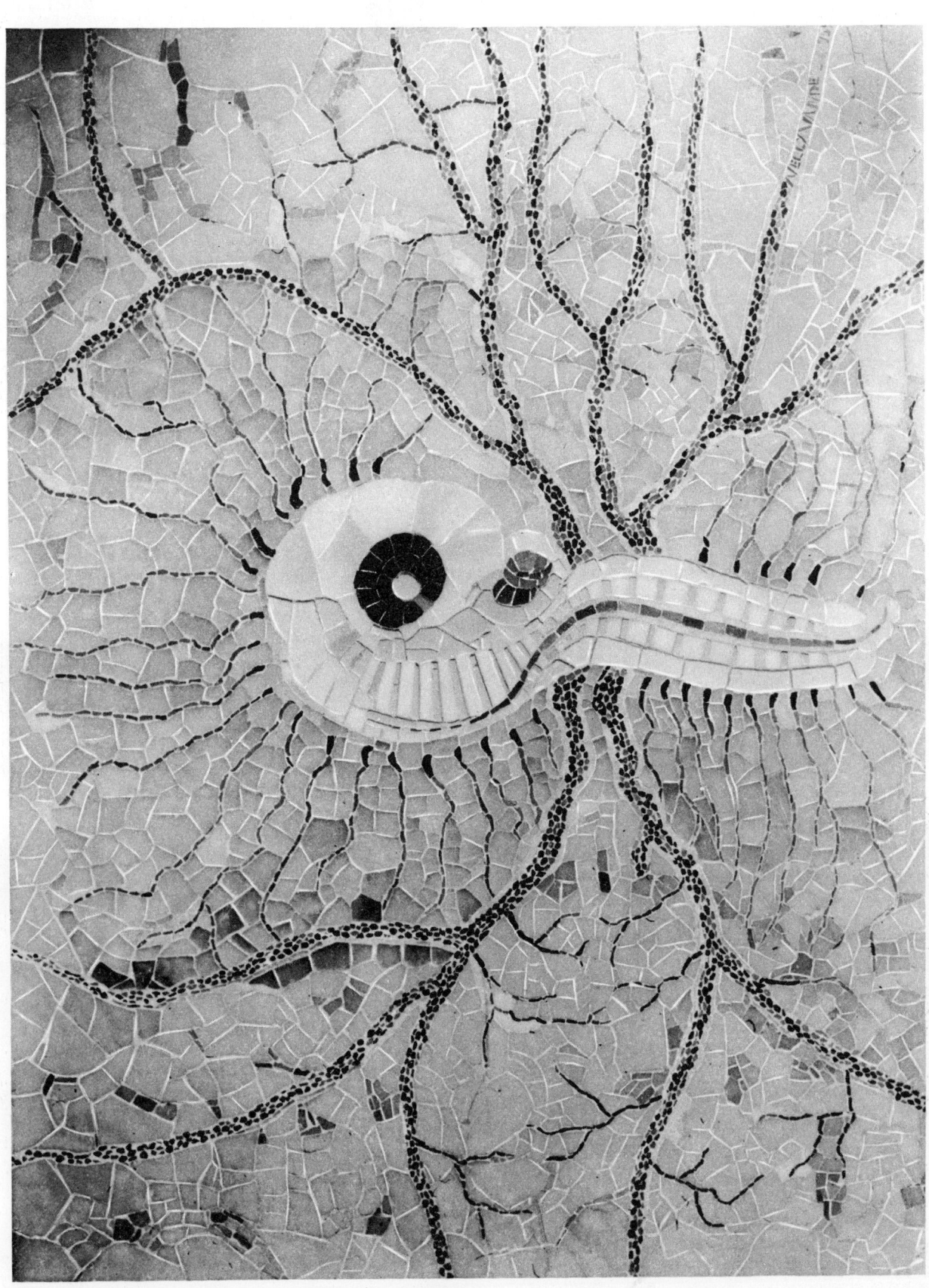

Sex transformations in amphibians

Although inversion or change of sex does not, as we have seen, go so far as to produce fertile individuals among birds, and the new sex types created by hormone treatment are incapable of reproduction, it is possible, and even relatively simple, to bring about the total, permanent and *functional* inversion of sex among amphibians.

If the larvae of the common frog are reared in hormone solutions of suitable composition they will all acquire the same sex: exposed in this way to female hormones all develop ovarian glands; exposed to male hormones all develop testicular glands. As, in such experiments, normally laid eggs are employed, the two sexes must statistically be roughly equal in number, from which the conclusion is inevitable that all the males have been turned into females or alternatively that all the females have been turned into males. This change from male to female is not permanent, while that from female to male appears to be lasting and final. At the end of many months young frogs which have been masculinized maintain their acquired sex and seem in no way prepared for a new inversion.

The offspring of two fathers

Because of the practical difficulties which attend the rearing of young frogs, individuals thus transformed into males cannot be kept until the age of reproductive maturity. But the experiment can be successfully made with another amphibian, *Xenopus* or the clawed frog, a native of South Africa. Because of its habits *Xenopus* can easily be reared in the laboratory. *Xenopus* is exclusively aquatic and unlike the frog does not demand living prey but can be nourished with small pieces of flesh or liver. Its evolution is rapid and if the surrounding temperature is high enough it will hatch and reach reproductive maturity within less than a year.

With *Xenopus* Gallien has not only brought about a complete change of sex but also succeeded in causing males feminized by female hormone to reproduce as though they were in fact females. This result is obviously of considerable interest since, as the change of sex has plainly not affected the genetic sex of the individual, we can by uniting feminized males with normal males, bring about the union of two genetic males and produce offspring which are really the offspring of two fathers.

The experiments which gave these fascinating

Groups of *Xenopus*, amphibians of South Africa.

Larousse.

results took place in the laboratory of L. Gallien in this manner: on hatching, *Xenopus* larvae were reared in water which contained several milligrams of oestradiol benzoate per litre. All these larvae produced toads which were apparently female and had typical ovarian glands. It was, however, reasonable to assume that a good proportion of these females were false females, in other words, males feminized by the hormone. The question was how to distinguish the false females from the true, and the answer was simply by letting them reproduce.

From the animals treated, nine females were chosen at random and united with normal males. The sex of their progeny was then carefully noted, and it was seen that the sex ratio of the offspring produced by the nine was utterly dissimilar. Five of them produced offspring of both sexes in roughly equal numbers. Out of a total of 280 there were 137 females and 143 males. The four remaining 'females' produced only males. One produced seventy-one, and the other three produced 135, 101, and 111 males respectively. In other words, there were 418 males out of 418 births. There can be no doubt about the interpretation of this result: it is abundantly clear that those females which produced males exclusively were the false females whose existence had been foreseen, females who resulted from the feminization of genetic males by the

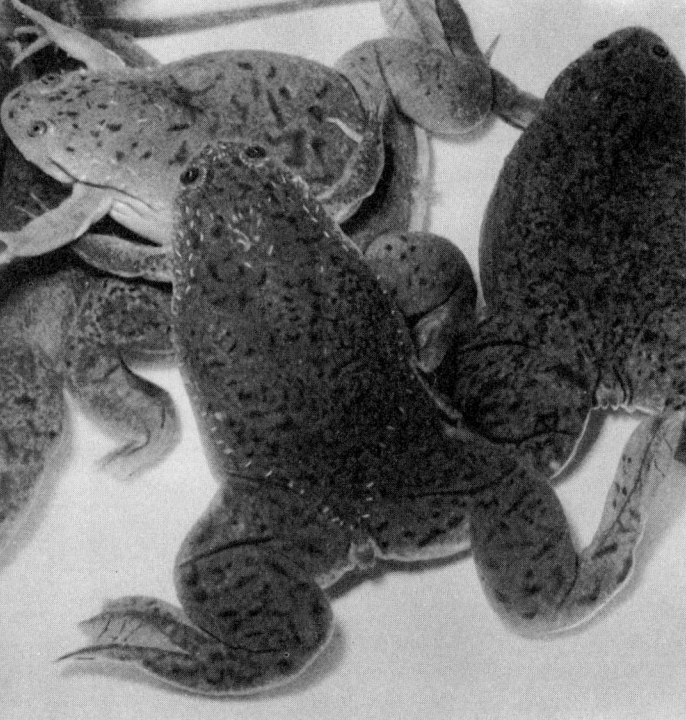

female hormone.

This spectacular experiment also revealed another fact which had been unknown until then, namely that the sex determination of *Xenopus* follows the *Abraxas* pattern, the chromosome formula of the male being ZZ and that of the female ZW. This would explain why males (ZZ) even when transformed into females can with normal males (ZZ) produce only ZZ offspring, in other words male offspring.

Chang and Witschi have successfully performed the same experiment in America, and state the case well when they say that with a few males who have undergone an inversion of sex it is easy to produce exclusively male descendants by the hundreds of thousands; in the case of *Xenopus* the genetically male type can be propagated indefinitely and, if wished, the heterogametic female completely eliminated.

For the experimental biologist it can be of great importance to know as soon as possible, in the embryo stage or even in the egg, whether he is dealing with a male or a female. By beginning his researches with two male parents he can know with certainty that all the offspring obtained will be genetically male.

A total inversion of sex together with the production of offspring had previously been obtained with an amphibian of the Urodela order, one of the salamanders (genus *Ambystoma*). This splendid achievement was the work of the American scientist Humphrey in 1945, and was done not with a hormone bath but by transplanting or grafting. By grafting a testis taken from a male embryo on to a young female larva he brought about a complete change of sex, thus creating a false male. To be certain that the grafted testis should not itself influence reproduction Humphrey had removed the testis from a *black* salamander and grafted it on to a (female) *white* salamander. When the white salamander had thus been converted into a male he coupled it with a normal female, also white. Since all the offspring were white it was possible to affirm that the grafted testis had had no effect on their genesis, since the character 'black' is dominant to the character 'white'.

The progeny obtained in this experiment were born to two genetically female parents and were thus the offspring of two mothers. What was their sex ratio? Among 2,097 offspring 509 were males and 1,588 were females; that is to say there were approximately three times as many females, a ratio which is obviously aberrant. The only plausible interpretation of this result is that the salamander in question follows the *Abraxas* pattern of sex determination. In this case the union of two genetic females (ZW) can produce four chromosome combinations: ZZ, ZW, WZ and WW. The first only (ZZ) will produce males, the remaining three producing females. In other words, statistically there will be three females to one male, which was confirmed by the experimental data.

But Humphrey did not stop at this point. Among the females obtained (ZW, WZ, WW) there should in theory be two categories: one with the ZW or WZ chromosome complement of normal females, and the other (WW) having two W-chromosomes. The first category should, of course, be roughly twice as numerous as the second. In order to distinguish the females

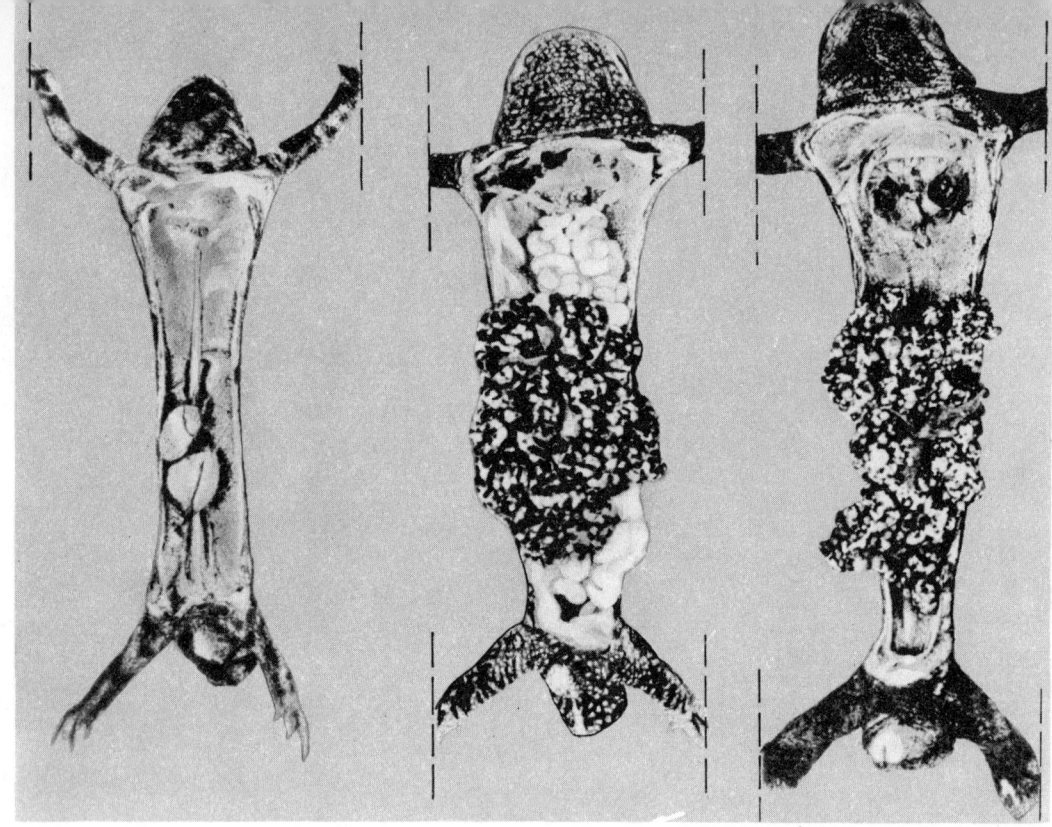

Pleurodeles dissected to show the sex organs. *Left*, adult male of eighteen months. *Centre*, adult, female of eighteen months. *Right*, two-year-old male transformed into female by treatment with female hormone. The ovaries are well developed.

After Gallien.

Two *Pleurodeles*. Gallien succeeded in inverting the sex of this species of triton by administering oestradiol, the female hormone.

Larousse.

Pleurodeles waltl.

After Duméril and Bibron.

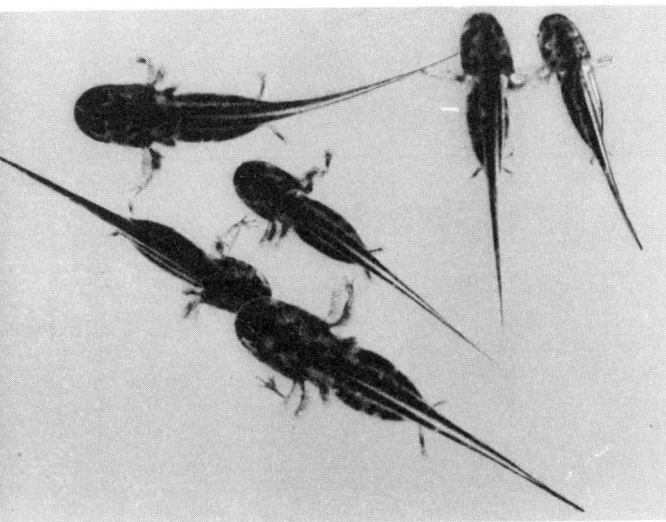

Pleurodeles larvae.

Larousse.

Pleurodeles waltl. The specimen above is a normal male, while below is a male, ten years old, which has been changed physiologically into a female by hormone treatment undergone during its larval stage.

Larousse.

with the normal chromosome formula from the abnormal WW females, the familiar process was employed: they were made to breed. Applying the rules of sexual algebra it is evident that the females of WW formula will be betrayed by their offspring, since when united to normal (ZZ) males all their offspring must be WZ, that is female. In practice this was verified by uniting female salamanders (daughters of two mothers) with normal males, when it was found that a certain proportion of them bore females only. In addition it was found that these exceptional females, the aberrant females incapable of producing males, were, as in theory they should be, only half as numerous as the remaining females who bore offspring of both sexes. A more striking confirmation of the chromosome theory of sex determination can hardly be imagined. It is interesting to note that by uniting two genetic females Humphrey created a category of females (WW) which does not exist in nature: females who could, strictly speaking, be called hyper-females, and who can procreate offspring of their own sex only.

But our sexual algebra can be carried even further. Having achieved a reversal of sex among salamanders let us suppose that instead of changing females into males we change males into females. Since the male has two Z-chromosomes, the male which has become a female can only produce ova bearing a Z-chromosome. If this false female is then united with a true male — which can produce only Z-sperm — all the fertilized eggs must be ZZ, and therefore male.

In other words, the union of two genetic males must give birth to a purely masculine line of descendants. This experiment, complementary to Humphrey's work, has been successfully made by the French biologist Gallien, using not the same species of salamander Humphrey worked with, but another tailless amphibian, *Pleurodeles waltl.*

The results obtained were those which had been logically foreseen, and have thus completed the important researches made by Humphrey. The offspring of two genetically male *Pleurodeles* were exclusively male — 940 out of a total of 940. The experiment was, in brief, identical to that which Gallien had made with *Xenopus,* and to bring about the feminization of male *Pleurodeles* he reared the young larvae in a solution of female hormone — oestradiol benzoate.

Partial changes of sex in mammals

More or less complete transformations of sex have been obtained among fish and reptiles by means of hormones. In mammals the action of hormones is confined to secondary sexual characters. The primary sexual character (that is the genital gland) is never affected, no matter how intense the hormone treatment is or how early it is administered. Even among marsupials, whose method of reproduction lends itself to direct action on the embryo which, in the marsupial pouch, can be exposed to sex hormones, only modifications of secondary sexual characters have been attained.

Nevertheless nature itself has experimentally demonstrated that the sex gland of a mammal can, in certain circumstances, be structurally deviated under the influence of hormones. It has long been known that when a cow drops two calves of different sex (false or fraternal twins) it commonly occurs that the female calf reveals characteristics which are intermediate between those of a female and of a male. She is a kind of 'sexual monster' which English breeders have named a free-martin, a term which is now internationally adopted by scientists.

In free-martins the external genital members are female in type, but the ovaries are often replaced by rudimentary testes, and the internal organization can resemble that of a young bullock. It is reasonable to suppose that this masculinization of the female twin is due to the influence of the male twin and effected through the intermediary of some hormone or other factor present in the bloodstream. This hypothesis is substantially confirmed by the fact, discovered by F. R. Lillie, that between the two twin calves there is almost always a connection of the umbilical arteries which allows their blood to mingle. This connection is lacking in those rare cases in which the female twin escapes masculinization.

Free-martins are always sterile, but as K. Ponse suggests, it would be most interesting to try to stimulate the development of their rudimentary testes: if in this way the testes could be made to produce semen it might be possible to bring about the union of two genetically female mammals.

Many biologists have set about reproducing the phenomenon of free-martinism among amphibians, especially *Ambystoma*, by grafting two larvae together — parabiotic, or artificial 'Siamese' twins. If a sufficient number of these parabiotic twins are reared some of them will inevitably comprise a male and a female. In this event, do we in fact observe any modification of the sex glands? In practice the male-female twins constantly reveal anomalies. A masculinization of the female's ovary almost always occurs, though the feminization of the male's testis is very rare. These findings have been confirmed by Humphrey and extended by Witschi to tailless amphibians, namely frogs and tree-frogs.

The free-martin effect can also be brought about by grafting an embryonic sex gland on to a larva of the same species. If the larva grafted and the larva from which the graft is taken are of different sex both will frequently develop sexual anomalies. Etienne Wolff switched the embryo sex glands of chickens and observed, according to the case, a masculinization of the ovary or a feminization of the testis.

The prediction of sex before birth

From the work of Jost and of Wolff, and from the study of free-martinism, it is abundantly clear that the embryo or the foetus secretes sex hormones. These hormones must pass into the tissue-fluids of the mother. Hence, in the case of human beings, we may

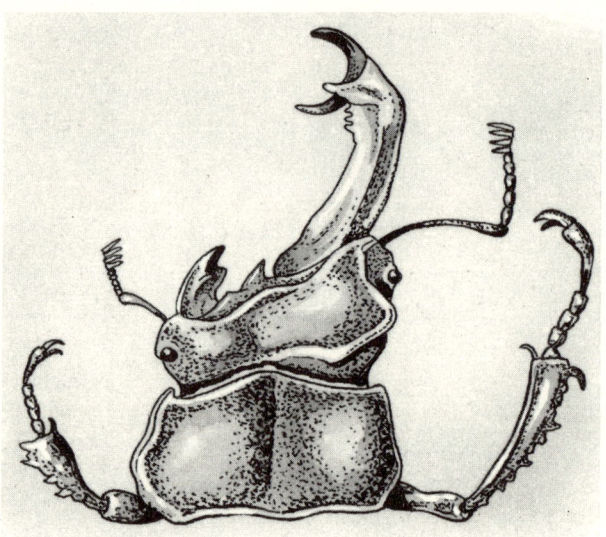

Lucanus cervus, a stag beetle or bipartite gynandromorph. The characters on the right are male, those on the left, female.

After Dudich.

suppose that their presence should be detectable, either in the maternal bloodstream, or else in such maternal secretions as saliva, excrement, or urine.

One of the first methods to apply this supposition was that of Dorn and Sugarman, and consisted of injecting a young male rabbit with the urine of the pregnant woman. Two days later the rabbit's testes were removed and submitted to thorough examination.

If the foetus was female the testes were swollen and congested and revealed a maturative growth of germinal cells; if the foetus was male the testes showed no modification. The test would seem to rely on the presence in the urine of a hormone substance produced by the female foetus, unless of course it was produced by the mother herself under the influence of the foetus. The prognosis is claimed to be correct in eighty per cent of the cases when this test is made without recourse to microscopic examination, and in more than ninety per cent when the microscope is used.

Other experimenters have, however, reported much less satisfactory results, and the fact remains that the test has never become current practice, either because it has been considered too inaccurate or too delicate to perform. Nor has more success attended another test which consists of injecting the urine of the pregnant woman into certain fish whose sexual organs are supposed to react differently according to the sex of the foetus.

More credit has been gained by a method perfected by Rapp and Richardson which was hardly out of the research laboratory before it was used commercially. It has the advantage of being both simple and of

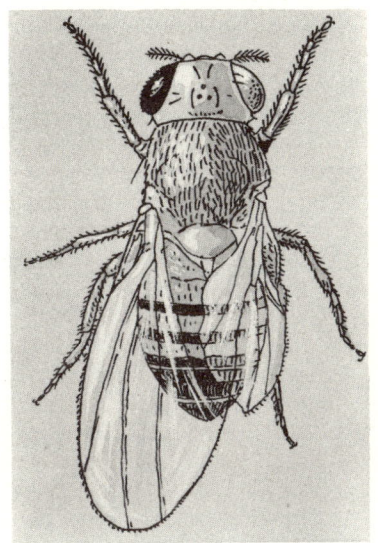

Gynandromorph *Drosophila melanogaster* obtained by crossing. The left side, with normal characters, is female; the right side with mutant characters — eosine eyes, miniature wings — is male.

After Morgan and Bridges.

Female beach flea *Orchestia.*

R. H. Noailles.

giving quick results. A little saliva, supplied by the future mother, is treated with a reagent — a derivative of phenylhydrazine — which turns brown if the foetus is male and remains colourless if the foetus is female. The same result is obtained if a 'pastille' is impregnated with the reagent and kept in the mother's mouth for a few minutes.

A change of colour, the positive reaction, is presumed to be connected with the presence of a sex hormone which, from the sixth month of pregnancy, passes from the male foetus into the bloodstream and, thence, into the saliva of the mother.

According to Rapp and Richardson the reaction is specific and will practically never be obtained with the saliva of a woman who is not pregnant or who bears a female foetus. Of 225 positive reactions, 218 correctly foretold the birth of boys; of 151 negative reactions, 148 that of girls. Such a high percentage of success would amply justify the vogue of the new test were it not that, again, contradictory reports arouse our scepticism. According to a paper published by John Rieger, the Rapp-Richardson test is totally devoid of significance, and predictions made by it are of no more value than predictions made at random.

The statistics published by this author are indeed disappointing. In 176 cases the prediction was correct eighty-seven times, that is a little less than fifty per cent, and incorrect eighty-nine times. Of the eighty-five males predicted, forty-seven were born (55.3 per cent) and thirty-eight (44.7 per cent) were girls. Of the ninety-one female predictions, forty (44 per cent) were correct, fifty-one (56 per cent) incorrect.

In an earlier chapter the 'cellular diagnosis of sex' was described, by which it is today possible to predict sex by microscopic examination of the amniotic liquid.

This problem of sex prediction has aroused human curiosity and parental impatience since the earliest times. Prejudice, superstition, and fantasy have, as always, outstripped the contributions of positive knowledge. The Bérol papyrus, dating from 1,350 B.C., mentions a method of pre-natal sex diagnosis which consisted of having the pregnant woman urinate on two sacks containing grains of barley and wheat respectively. If subsequently only the barley seeds germinated the birth of a male child could be foretold, while the exclusive germination of the wheat seeds announced the birth of a girl. If neither one nor the other germinated the pregnancy was an illusion.

According to Hippocrates the male foetus was revealed by its movements from the third month of gestation while the female foetus, less precocious, did not stir until after the fourth month.

In modern times it has been claimed that the sex of the foetus can be known by the frequency of its heart-beats, or by certain radiological signs, or again by various modifications undergone in the mother's organism, alterations in facial expression, insistent nausea, changes in the colour of the iris, etc. And of course we also have our twentieth-century alchemists who will undertake to foretell the sex of the unborn child by swinging a pendulum over the maternal abdomen.

103

SEXUAL DIMORPHISM
Left, pair of *Labeotropheus quelltorni* (family Cyprinidae). Male (blue). Female (mottled).

Six.

Below left, male European common toad *(Bufo bufo)* with its nuptial callosities.

Lauros. Atlas-Photo.

Below right, male capercaillie *(Teatro urogallus).*

Bel-Vienne.

Bottom left, nuptial callosities on two fingers of the male European common toad greatly magnified.

Lauros. Atlas-Photo.

Bottom right, female capercaillie.

Bel-Vienne.

104

SEXUAL DIMORPHISM. *Ornithoptera urvilleanus* (Solomon Islands). *Above*, the male; *below*, the female.

Muséum national d'histoire naturelle. Larousse.

Sex hormones among the invertebrates

Everything which has so far been said about sex hormones has concerned the part they play among vertebrates. The problem now arises of how far our findings can be applied to less highly developed animals, for example to insects and crustaceans.

Among insects hardly any facts, natural or experimental, are known which demonstrate the existence of sex hormones. If we deprive the larva of a cricket or of a caterpillar of its sex glands the adult cricket or butterfly thus castrated nonetheless acquires its (normal) secondary sexual characters. The castration of a male insect followed by the grafting of ovarian glands, or the removal of a female insect's ovaries followed by the grafting of testicular glands, cause no apparent sexual inversion of the insects. According to Kopec, a castrated male butterfly grafted with five ovaries was perfectly normal in its secondary sexual characters and coupled normally with a female whose ovaries had been replaced by five grafted testes!

The phenomena of *mosaic development,* or development when determination is complete before functional differentiation, are also incompatible with hormone determination.

Among moths, dragonflies, and fruit flies it is not uncommon to encounter individuals which are half-male and half-female. Such individuals are known as *gynandromorphs.* They arise from the accidental loss of a chromosome during the first cleavage of the egg cell. The ovum of a fruit fly destined to produce a female has two X-chromosomes. If, by accident, one of the two first embryonic cells receives only one X-chromosome instead of two it will produce the male half of the fruit fly while the other, normally endowed, cell will produce the female half. The fly will then be male on one side and female on the other. The loss of the X-chromosome may take place during the course of a later cell-division and in this case instead of one half only a small portion of the insect will present male characteristics.

In all cases of gynandromorphism there is, in the same insect, a juxtaposition of male and female parts. And yet these parts exercise no visible influence on each other, as they undoubtedly would if hormones played a part in the insect's sex differentiation.

On the other hand crustaceans — or more precisely sand shrimps — have recently been the subject of delicate experimental research which leaves no doubt of the existence among them of true sex hormones which act in a very similar way to sex hormones in vertebrates.

In the male *Orchestia gammarella* Hélène Charniaux-Cotton has discovered a minute gland which no-one had previously noticed: an organ of lobular appearance, attached to the vas deferens and completely independent of the testis. The gland has been well named *androgenous,* for it secretes hormones which masculinize and it is the gland which controls the appearance of secondary sexual characters in the male *Orchestia.*

If an androgenous gland is removed from a male *Orchestia* and grafted on to a female of the same species the female is seen to become slowly masculinized while at the same time her ovaries are changed into testes. The female thus masculinized behaves in every way like a male; like him she seizes normal females with whom she goes through the customary preliminaries of courtship. By means of artificial insemination she can even be made to procreate as a male and in this way offspring are obtained which, as among amphibians, are produced by two genetic females.

Organs analogous to the androgenous gland of *Orchestia* have been found in crabs, hermit crabs, etc. It would thus appear that the above findings are valid for a number of crustaceans, and possibly for all.

The Analysis of Sex Determination

We have already seen that sex determination is assured by a very precise genetic mechanism which leads to the formation of two kinds of fertilized ova, differing in their chromosome content and producing, according to the case, males or females. We have also seen that these chromosome differences control the formation of the sex glands, testes or ovaries, which, at least in all vertebrates and in certain invertebrates (crustaceans), secrete hormones which masculinize or feminize.

Experiments have, moreover, shown us that these hormones possess not only the power to sexualize the body but also that of influencing the course of the sex glands' evolution. We have assumed that such and such a hormone will be produced in greater quantity in the embryo according to the chromosome formula of the fertilized ovum: where follicle-stimulating hormone predominates female glands or ovaries will appear, but where testosterone predominates male glands or testes will develop.

The question now arises of why a given chromosome formula should bring about the predominance of a given hormone. 'Why,' as Wolff puts it, 'does a genetically female organism secrete a greater quantity of female hormone than an organism which is genetically male? What is the order of the reactions which occur in certain cells, the series which begins with the sex genes carried on the X-chromosome and ends with the secretion of sex hormones?'

Here we approach the fundamental problem of genetics, of how genes act, which will be examined in detail later. Though the problem is far from being solved, biologists have, from the fruit fly *Drosophila,* gathered data of great significance on the manner in which the chromosomal genes intervene in sex determination.

In the first place we know that in some species, but not in all, of the two sex chromosomes (X and Y) only the X-chromosome plays a part in sex determi-

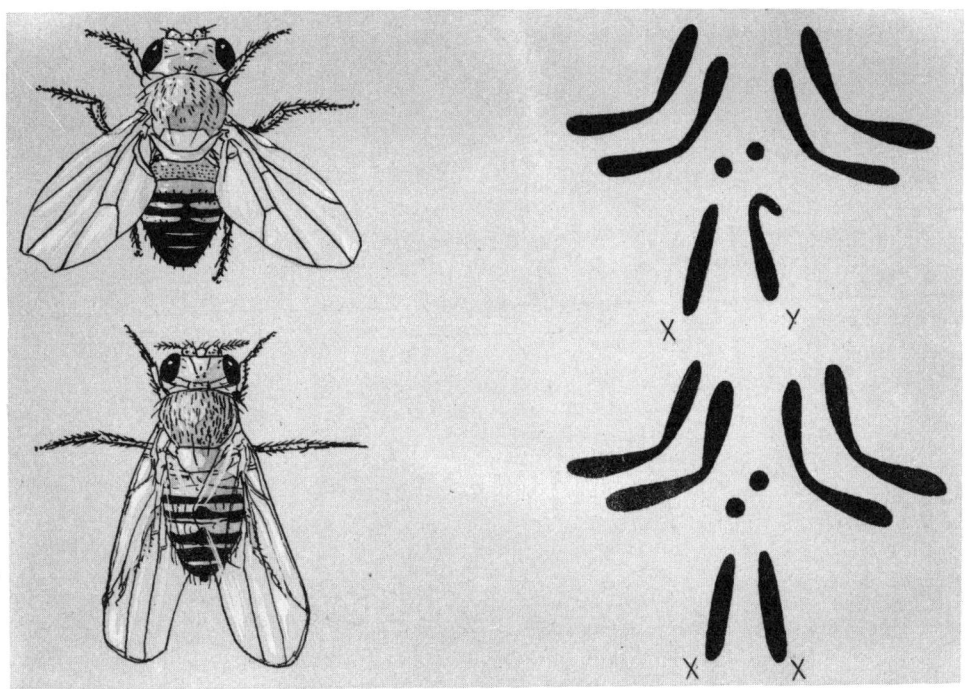

nation. An individual of XY formula is of male sex not because he has a Y-chromosome, but because he only has a single X-chromosome. Actually males are known who are unprovided with a Y-chromosome and whose formula is XO. Although sterile their sexual characteristics appear to be perfectly normal.

When, on the other hand, the fertilized ovum of *Drosophila* contains two X-chromosomes and two sets of autosomes (that is, non-sexual chromosomes) it

Diagram showing the four pairs of chromosomes of the male *(above)* and the female *Drosophila melanogaster.* The hook of the Y-chromosome is characteristic of the male. Chromosomes forming a pair are normally found in proximity.

After Morgan, Sturtevant, Muller, Bridges.

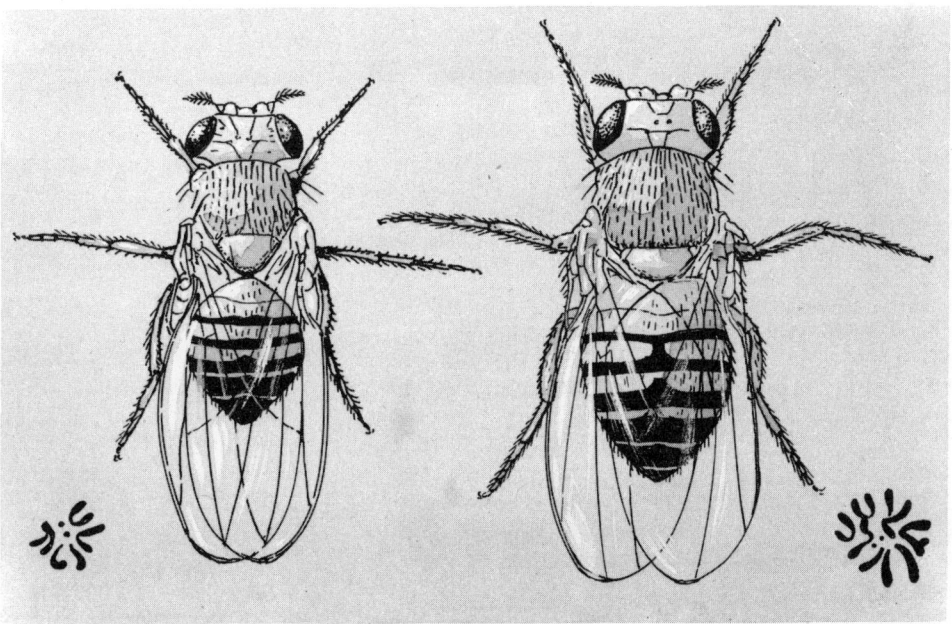

Triploidy in *Drosophila melanogaster. Left,* normal, or diploid, female with eight chromosomes. *Right,* triploid female with twelve chromosomes.

After an original sketch by Miss Wallace.

Swordfish. *Above* the male; *opposite*, the female.

will give birth to a female. If it contains only one X-chromosome and two sets of autosomes, it will give birth to a male. The normal female thus has an X-chromosome for each set of autosomes, while the normal male has only one X-chromosome for two sets of autosomes. As long as these proportions remain constant no sexual anomalies occur. Thus a triploid female fly who exceptionally carries three complete sets of chromosomes, in other words three X-chromosomes and three sets of autosomes, will be larger than an ordinary fly but her sexual characters will be perfectly normal because she has an X-chromosome for every set of autosomes. If, however, a fly has only two X-chromosomes for three sets of autosomes it will be sexually abnormal. Its sexual characters will be intermediate between those of a female fly and of a male fly, because it has too few X-chromosomes in proportion to its number of autosomes. Instead of the normal female's one to one proportion or the normal male's one to two, it has a ratio of two to three, which is intermediate.

Similarly if a fly has three X-chromosomes and only two sets of autosomes sexual abnormality will result. Like the 2/3 intersex just described, it will be sterile but for the opposite reason, namely, too many X-chromosomes per autosome set. The ratio is 3/2, instead of the normal female's 1/1. Such flies, which are as it were too feminine, are called 'super-females'. Finally if a fly carries only one X-chromosome for three sets of autosomes it will be equally sterile, because its ratio, 1/3, is smaller than the 1/2 of the normal male. Such too masculine flies are called 'super-males'.

From these valuable findings, from the research of the American geneticist Calvin Bridges, we learn that the autosomes also participate in sex determination. Since the sex of the organism depends on the

ratio of X-chromosomes to sets of autosomes, there would appear to be an antagonism between the sex determining tendencies of X-chromosomes and autosomes. From this it has been concluded that the X-chromosome carries genes of femininity, or at least a surplus supply of them, while the autosomes carry genes of masculinity or, similarly, at least a surplus of such genes.

In the light of these results sex no longer appears to us as a physiological absolute, but rather as an expression of proportion. Between the male organism and the female organism there is, in brief, only a quantitative difference. Both sexes carry both genes of femininity and genes of masculinity. As Guyenot has expressed it, 'A male is more male than a female, and reciprocally a female is more female than a male.'

Researches of a similar inspiration were made by Richard Goldschmidt on the gypsy moth, *Lymantria dispar*. The sexual dimorphism of the gypsy moth lends itself especially to this type of study. The female gypsy moth has whitish wings, a squat and woolly abdomen and thread-like antennae. The male, smaller and more frail, has brown wings and feathery antennae. Since the moth is found in many parts of the world, strains from widely separated regions can be crossed. Richard Goldschmidt not only found that such cross-breeding can result in the birth of intersexes, he also found that in such cases the type of sexual aberration differs according to the region of origin of the father or the mother.

Thus if a European female is crossed with a Japanese male a great number of masculinized females will appear in the first generation. In the second generation all the males are still normal, but half the females reveal masculine characteristics. The opposite cross, a Japanese female with a European male, produces in the first generation moths which are all normal as

Sexual dimorphism in the moth *Lymantria dispar*. Above, the female.

After Goldschmidt.

far as sexual characters are concerned but, in the second generation, all the females remain normal while half the males are feminized. In both cases, of masculinized females and feminized males, the intersexuality can go as far as the total inversion of sex: in that event further offspring will sometimes be 100 per cent female or 100 per cent male, fifty per cent deriving from the intersexes.

In a long and penetrating analysis of these results, the details of which we cannot discuss here, Richard Goldschmidt comes to the conclusion that among gypsy moths there exist local or geographical races which differ in sexual force or *valency*, though their morphological differences are indiscernible. Masculinized females appear if, in the original cross, the male belonged to a stronger race than the female. On the other hand, if the male belonged to a weaker race the result was the appearance of feminized males. The degree of intersexuality is proportional to the difference in sexual valency between the parent moths.

Sexual strength and weakness are purely relative, one race being strong in comparison to a second, and weak in comparison to a third. By multiple crossings between moths of various geographical origins Goldschmidt has been able to arrange a great number of races in order of strength and thus to predict with astonishing accuracy the results of new crosses.

According to Goldschmidt every fertilized ovum, male or female, of the gypsy moth contains a double assortment of sex factors: factors which masculinize (M) and factors which feminize (F). Normal sex differentiation is assured by the prevalence of M factors in the egg destined to produce a male, and by the prevalence of F factors in the egg destined to produce a female.

Intersexuality results when, following a cross between races of unequal sexual valency, the feminine factors are too powerful in the male egg, or the masculine factors are too powerful in the female egg. In these cases the organism will begin to develop in harmony with its true sex, but when a certain stage is reached it will evolve towards the opposite sex; the earlier this turning point is reached the more accentuated intersexuality will be.

A number of facts, drawn from the study of both vertebrates and invertebrates, are in accord with this idea of an original bi-sexuality of the organism, giving rise to an early conflict between masculinity and femininity. Ordinarily the predominance of one of the sexual tendencies is sufficiently accentuated to impose a definite choice from the beginning of development. The male or female gland is then formed at once and will persist throughout the life of the individual. But in certain animals the issue of the struggle remains in doubt for some time. The organism appears to hesitate, it goes through a period of hermaphroditism, or even adopts the opposite sex.

In the case of the mollusc Crepidula every individual begins as a male. It attaches itself to a congener (community of like individuals), which is itself attached to an oyster shell. It then changes sex. In this way chains are formed in which the oldest members are females, the youngest males, and those in between hermaphrodites.

In certain undifferentiated races of frogs and toads all the young possess a sex gland of feminine type, since female tendencies always begin by being the more powerful. The predominance of these tendencies is maintained by half the young which become definite females, while in the other half, the definite males, masculine tendencies little by little gain ground and at last triumph. All male eels pass through a hermaphrodite stage, and it is possible that the same occurs with trout. Phenomena of this order are encountered even in mammals: in the course of their development certain female opossums go through a masculine period. In many species of fish sexual equilibrium is so unstable that adults may spontaneously change sex. Female swordfish are often found who, as they age, become males. They take on masculine characteristics and acquire the caudal sword. Their masculinization is not simply a superficial phenomenon like that observed in old hens or old pheasants: their sex gland undergoes structural changes and can emit functional sperm so that they are able to fertilize the eggs of normal females and produce fish which are the offspring of two genetic females.

The toad offers an equally remarkable example of rudimentary hermaphroditism and was thoroughly studied and expounded by K. Ponse of Geneva. Above its sex gland every male toad has a small organ of reddish tint which is called the Bidder organ, and is, in reality, a rudimentary female gland. If the toad's male gland is removed the Bidder organ is, as it were, set free. It develops and turns into a true ovary that produces normal eggs which can be fertilized. At the same time the oviducts, which are present in the male in a rudimentary state, become sufficiently enlarged to allow the passage of ova. It takes a few years for the feminization to be completed. The male characters regress, and the feminized male can then couple like a female. By coupling feminized males with normal males tapdoles have been produced which are the progeny of two genetic males.

Poultry furnish us with another example comparable to that of the toad, though in this case it is the female and not the male which undergoes a reversal of sex. The domestic hen has a single left ovary; on the right it has an atrophied sex gland which has the structure of a rudimentary male gland. If the ovary is removed before the female chick is thirteen days old the rudimentary male gland will develop into a functional male gland and even be capable of producing active sperm. For purely anatomical reasons, however, the animal is unable to emit its semen, so that it is impossible to couple such masculinized hens with normal hens and in this way obtain progeny of two genetic females.

The experiment was nevertheless made in 1931 by Christian Champy who is believed to have successfully crossed a Phoenix hen with a golden Leghorn whose left ovary had been removed during the second month of life and who had been externally changed into a cock. The four offspring of the two hens were all females. Though the experiment is interesting the results were clearly too few in number to allow genetic conclusions to be drawn from it.

It would seem that certain infections, tuberculosis for instance, can bring about the masculinization of hens by transforming the rudimentary right gland into a testis. Crew has reported a case of spontaneous sexual inversion in an adult hen which had already laid fertile eggs. This hen, after it became a false cock, fertilized a true hen which gave birth to two chicks, a male and a female.

The testicular gland can also illustrate the toad's sexual bi-potentiality. If a fragment of its testis is grafted under the toad's skin the abnormality of the situation will in itself cause the animal's rudimentary ovary to develop. Conversely, if a fragment of a hen's ovary is grafted under the hen's skin, the animal's testis will tend to evolve.

Sex determination by environment

During the extreme youth of some animals sexual uncertainty is so pronounced that the choice of sex may be determined by the environment. Such is the case of certain crustaceans where the males are dwarfs and parasites of the females. The young crustacean will evolve into a female or into a male according to whether or not it attaches itself to an adult female. Experimentally the process of sexual differentiation can be reversed: young females will turn into males if they are attached by the biologist to adult females, and conversely young males will turn into females if they are detached from the female to which they were previously affixed.

Among amphibians the genetic determination of sex is not so hard and fast that the evolution of their sex glands is rigidly fixed; it can be altered by factors unrelated to hormone action. Thus, with frogs, Witschi and Jeanne Piquet obtained clear inversions of sex simply by changing the temperature. An

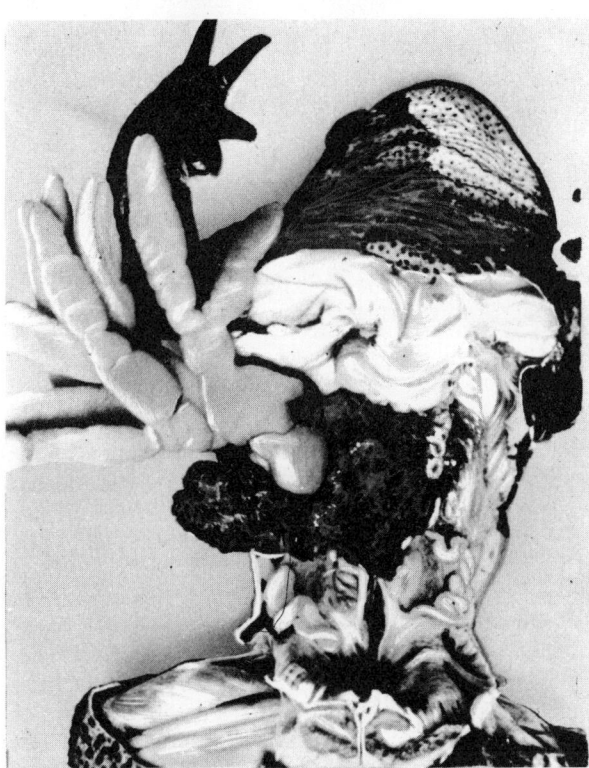

Inversion of sex in the toad. Experiment made by K. Ponse.

Palais de la Découverte.

In the castrated male the ovary develops from Bidder's organ.

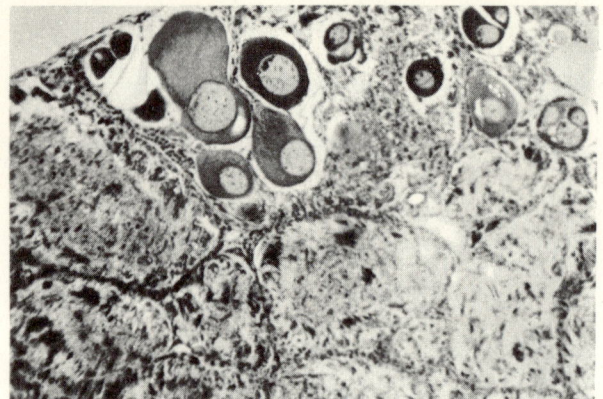
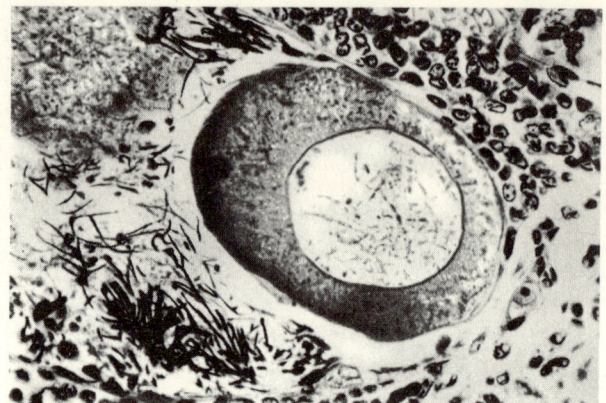

Section of the testis of an adult toad *(Bufo bufo)* showing the presence of ova in the seminiferous canals. *Left*, magnified 50 times; *right*, 100 times.

Dr. Elkan.

excessive number of females resulted from ova kept at a temperature of 10 °C while a temperature of 27 °C produced a great surplus of males.

The over-maturity of the egg-cell also considerably increases the proportion of males. By fertilizing ova which had waited more than four days in the uterus an almost exclusively masculine progeny was obtained. Just how these various factors influence sex determination is not fully understood. It is possible that they somehow upset the hormone mechanism, or it may be that in the tissue they directly bring about physico-chemical changes for which the hormones are normally responsible.

Among eels, too, there seems to be a certain sexual flexibility. The proportion of male births will increase simply by raising the temperature.

Among worms, the male *Bonellia* does not exceed 3 millimetres in length and might be mistaken for one of the larger ciliophorans while the female's plum-shaped body reaches a length of 7 centimetres and is itself prolonged by a proboscis measuring more than 1 metre.

The organization of the microscopic *Bonellia* is extremely rudimentary, and it lives as a parasite in the oviduct or in the anterior intestine of its giant female. Within a single female as many as eighty-five males are sometimes found. As the Swiss biologist Baltzer has shown in the course of a series of remarkable experiments, the worm's sex determination is connected with the behaviour of the newly hatched larvae. If young larvae are placed in sea-water containing appropriate nourishment they will all become females. If, however, they are put in contact with a large female they soon attach themselves to her proboscis and, living thus as parasites, all acquire male sex. If larvae are detached from the proboscis at the end of a relatively short time they will evolve into intersexes, that is, they will develop an intermediate structure between that of the typical male and that of the typical female. The masculinizing action of the proboscis is due to a chemical substance which is not unlike a sex hormone. The effect of masculinization can be obtained by submitting the larvae to extracts of proboscis, or of the digestive tube,

diluted in sea-water. If the solution is too strong these extracts prove toxic to the larvae. Similar effects of masculinization have been obtained with very dilute solutions of carbonic acid and hydrochloric acid.

Accidental hermaphrodites

Now and then we find frogs which are functionally hermaphrodite, that is, frogs in which both sex glands develop and even mature. One such hermaphrodite, discovered at Fribourg-en-Brisgau, furnished the biologist Witschi with a subject for a very curious experiment. He succeeded in fertilizing the ova of the hermaphrodite frog with its own sperm. All the offspring were female. When he fertilized the ova of a normal female with the sperm of the hermaphrodite the progeny which resulted was also exclusively female. On the other hand by fertilizing the hermaphrodite's ova with the sperm of a normal male, offspring of both sexes were obtained, roughly half males and half females.

These varying results are easily explained if we assume that the hermaphrodite was genetically female; and also that the female common frog, *Rana temporaria,* produces only one kind of ovum, with an X-chromosome, while the male produces two kinds of spermatozoa, one with an X-chromosome, the other with a Y-chromosome.

No cases of functional hermaphroditism are known among birds or mammals although intersexes or individuals of mixed sex are not particularly rare. For centuries pigs and goats of this sort have been observed. On the island of Espiritu Santo in the New Hebrides, for example, intersexed pigs are treated as sacred animals, and the natives have noticed that greater numbers of them are bred by pigs of a certain pedigree. Intersexed goats are frequently the result of inter-racial cross-breeding on the lines of the moths studied by Goldschmidt.

In the human species Petersson and Bonnier both reported several families of intersexes. The intersexed individuals look and behave like normal women, but their internal structure is abnormal: no womb, a closed vagina, and ovaries replaced by sterile testes which have half-descended.

Masculinization can take place under the influence of adrenal hormones which are very similar to male hormones. The presence of a tumour, benign adenoma, may often cause an excessive secretion of these hormones. When the tumour appears early in life a more or less complete inversion of sex may result; when it appears later it produces only superficial symptoms of masculinity, for instance women with beards.

Due to hypertrophy of the clitoris masculinized females are sometimes legally registered as males. K. Ponse cited the case of a young 'man' who married a woman when in reality he was himself a woman; his true sex was only detected at the autopsy. Probably the origin of the masculinization was adrenal, and perhaps the same cause was involved in the case carefully described in 1942 by Doctors Costanti and Toreilles which is quoted by Wolff in his book, *Changes of Sex*. In this case the subject was twenty-five years old, of androgenous appearance and predominantly masculine. The external genital apparatus was of male type, but there was only one testis, the left. The face was beardless, the features plump and the breasts abnormally developed. Sharp abdominal pains made operation necessary, and on operation the patient was found to possess, on the right side, a complete female genital apparatus including ovary, Fallopian tube, and uterus without outlet. The two sex glands, female as well as male, were capable of functioning, for the ovary showed unmistakable signs of ovulation and the testis contained mature spermatozoa which were identified by puncturing the organ. The individual was thus capable of simultaneously producing fertilizable ova and the sperm to fertilize them. He (or she) could, at least in theory, be both father and mother, like Witschi's frog.

The popular press often features athletic champions who change sex, for example, Claire Bresoles, the jumper, who in September 1948 became Pierre Bresoles and Zonia Koubhova, the runner, who in 1934 became Mr. Koubak.

Here, too, we may presume that the subjects changed into men were genetically feminine, but we cannot be certain without applying the Barr test (the cellular diagnosis of sex). Better still would be a direct examination of their chromosome equipment in a tissue culture.

It was the latter method which recently clarified the origin of certain sexual disturbances such as the Turner-Albright syndrome (which is related to the XO chromosome formula) and Klinefelter's syndrome (which is related to the XXY chromosome formula).

Such chromosome aberrations arise from an accident in the distribution of sex chromosomes which occurs when the reproductive cells are maturing.

Needless to say sexual anomalies can pose difficult questions of conscience to the doctor or surgeon. Sometimes, during the course of a surgical operation on a patient who believes he belongs to the male sex, a tumour of the uterus or of an ovary is discovered. In such a case the surgeon's position is peculiarly delicate and it is often hard to decide what his attitude should be.

April Ashley, the model, who was born George Jamieson but underwent a sex-change operation.

Keystone.

113

Greenfly or aphid.
R. H. Noailles.

Virginal Reproduction – Parthenogenesis

Natural Parthenogenesis

We have seen that in the great majority of animal species reproduction is accomplished only by the collaboration of two organisms belonging to different sexes. There is a male parent and a female parent, each of which furnishes one cell. The fusion of the female cell or ovum and the male cell or spermatozoon produces the fertilized egg, the zygote, which is the new individual's point of departure.

Nevertheless it is far from rare for a single cell to be sufficient to produce a living being. In such cases the cell is, as we should expect in view of the greater size of the ovum, always female. Though the two reproductive cells are more or less equal as far as hereditary factors are concerned, that is, factors represented by the chromosomes in the nuclei, only around the nucleus of the ovum is there an abundant supply of cytoplasm, which possesses the organization necessary to form the embryo and also contains the reserves required to nourish it during the early period of its formation.

Reproduction by means of the ovum alone was named parthenogenesis (from the Greek *parthenos,* virgin), by Richard Owen in 1848. The term has, of course, the same etymology as the word Parthenon, or temple of the virgins. Parthenogenesis is closely related to sexual reproduction, from which it is a kind of deviation. Probably species which reproduce in this manner are derived from species which, earlier, practised sexual reproduction. Although parthenogenesis cannot be considered as evolutionary progress it is assumed that it must bring some advantage to those species in which it is maintained.

In all events it occurs sporadically in almost all the fundamental groups of the animal kingdom: echino-dermata (sea-urchins, starfish), Nematoda, Trematoda, Annelida, Mollusca, Crustacea, Myriapoda or milli-pedes, Acarina (minute Arachnida), and Insecta. It is among insects that the great majority of parthenogenetic species are found. Natural parthenogenesis assumes very different aspects, according to the species which practises it. It can be obligatory or optional; produce males or females; be seasonal or permanent.

One of the best-known cases is that of the domestic bee, *(Apis mellifera)*. In this species of hymenopteran parthenogenesis is optional in the sense that the egg can develop in two ways, namely, with or without being fertilized. The female, or queen, couples during the course of her nuptial flight, unforgettably, but erroneously, described by Maeterlinck in his famous *Life of the Bees.* Until recently it was believed that the queen bee was fertilized by a single male but Soviet biologists have shown that she receives the sperm of several males. According to Woyke, a queen can be fertilized as many as sixteen times during the course of the mating flight, the average being between six and ten times. During the flight the queen receives a provision of sperm which, stored in a small receptacle or *spermatheca,* maintains its activity during the three or four years of the insect's reproductive life. The spermatheca is a little sac with contractile walls which communicates with the channel through which the eggs pass. If the spermatheca is opened while the egg passes, the egg comes into contact with the sperm and is fertilized. Otherwise it remains unfertilized. If fertilized, the egg will produce a female bee. This offspring will become a queen if, while young, she receives the 'royal jelly' and a worker if she does not. If unfertilized, the egg will produce a male or drone. In the case of the bee, therefore, parthenogenesis produces males: it is arrhenotokous (from the Greek *arrhenotokia,* male-bearing). Eggs which have been fertilized, that is to say those which will become females, are deposited in ordinary alveoles or worker cells, while those which

115

become males are deposited in larger cells.

The existence of partial parthenogenesis among bees was discovered by a Silesian priest, Johann Dzierzon (1811—1906). His theory aroused much controversy and even today some specialists in apiculture hold that it does not take all the observed phenomena into account. The theory rests, however, on a very coherent body of demonstrable facts. In the first place, when for one reason or another a queen has not been fertilized she gives birth to males only, thus producing a hive of drones. The same thing occurs when a fertilized queen has been exposed to temperatures near freezing point: it would seem that her reserve of seminal fluid has been destroyed by the cold. An ageing queen produces fewer and fewer females and often, towards the very end of her life, gives birth only to males, as though her store of semen had been exhausted. When a queen couples with a male of another race — for example, a *Black* queen with a *Yellow* male of Italian race — the workers, having two parents, are of hybrid type; but the drones, having no father, are of purely maternal type. Occasionally in the hive certain workers become fertile. When this happens they procreate males exclusively, which is what we should expect, since they cannot have been fertilized. Finally, according to various authorities, the presence of spermatozoa can be detected in eggs which occupy the worker cells, while sperm is never found in eggs contained in the male (drone) cells.

Males, being the issue of unfertilized eggs, should in theory possess only one set of (maternal) chromosomes. It appears, however, that this chromosome deficiency is made good by a doubling of maternal chromosomes, which takes place at the very beginning of the insect's development.

Though the existence of optional parthenogenesis among bees is now well established we still know absolutely nothing of how it comes about that the queen deposits female eggs in one type of cell and male eggs in another. The general opinion is that we must explain it by some kind of reflex action caused by the sight or contact of the cell: the reflex would, according to the case, provoke the opening or closing of the seminal receptacle and hence the fertilizing or non-fertilizing of the egg.

Some observers have pointed out that the queen bee's abdomen lengthens when she lays an egg in a worker's cell, and suggest that this lengthening would cause a delay in laying the egg which would give it time to be fertilized. Lienhart emphasized the relative distension of the queen's abdomen as a clue to the mystery, while Flanders believed that the fertilization of the egg requires the emission of an alkaline fluid which by neutralizing the normal acidity of the semen would activate it.

Such a large number of different hypotheses is a testimony to their inadequacy.

Analogous to the case of the honeybee is that of *Osmia*, often called the 'mason bee'. A solitary bee, it builds two types of cells in the hollows of reeds, first forming large cells well provided with nourishment and then smaller cells with reduced rations. In the first the female eggs are laid and in the second the male eggs.

According to Fabre, the mother mason bee 'knows' the sex of the egg she is about to lay and, keeping in mind the superior dimensions of the female larva, apportions the size of the lodging and the quantity of provisions required by its inhabitant accordingly. But how can such foresight be attributed to an insect? Again it would seem that we must fall back on reflex action. Armand Descy, who has made many experiments with *Osmia*, believes that it is the presence or absence of sperm in the seminal sac which incites the mother to build large or small cells for her eggs.

Numerous insects belonging to various orders — Hymenoptera, Hemiptera, and certain Acarina (mites and ticks) — practise parthenogenesis comparable to that of the social and solitary bees, that is to say, optional parthenogenesis which produces males.

Among certain Acarina in particular it has been observed that the male, hatched from an unfertilized ovum, carries half as many chromosomes as the female, hatched from a fertilized ovum. In this case, unlike that of the bee, there is no doubling of the egg-cell's chromosomes.

A totally different type of parthenogenesis is found among other insects, notably thrips, tiny creatures which inhabit the corollas of flowers. In their case parthenogenesis is obligatory, for the egg is never and can never be fertilized. It is matured in a special way and, contrary to the rule, no reduction in the number of chromosomes takes place; the new organism receives all the mother's chromosomes. As it has the same chromosome equipment as the mother it must of necessity be feminine in sex. Thus this type of parthenogenesis produces nothing but females and is called thelyotokous (from the Greek *thelyotokia*, female-bearing).

As we shall see later the two types of parthenogenesis — female-bearing and male-bearing — are sometimes combined in the same species.

In connection with obligatory parthenogenesis, which produces females only, a type of parthenogenesis known as *polyploid* may be discussed. In this form of parthenogenesis the species includes two very distinct races: one which reproduces asexually and is composed exclusively of females, the other which reproduces sexually and is composed of both sexes in roughly equal number.

The two races, as Vandel has shown in the case of the woodlouse *Trichoniscus*, differ in chromosome content. While the sexual race has a double set of chromosomes, the parthenogenetic race has a triple set. The addition of this supplementary set leads to various degrees of gigantism. Here too the eggs of the parthenogenetic female undergo no reduction in the number of chromosomes and in consequence produce only females. In certain regions the two races coexist, while in others only the virginal race is found. Even where they coexist they do not mingle.

The association between asexual reproduction and an increase in the number of chromosome sets has also been observed among certain butterflies *(Solenobia)*, among the curculionid *Otiorynchus*, among Orthoptera, Diptera, water fleas, and many others.

The queen bee, with her 'court' of worker bees.

R. H. Noailles.

Alternating parthenogenesis

The little hymenopterans called gall wasps (family Cynipidae) are unknown to the general public though everyone is familiar with the tumours or *galls* which they cause on the leaves or buds of trees. One of these gall wasps is called *Neuroterus lenticularis,* and the species is represented, in spring, by females that reproduce asexually. They lay their eggs on the leaves or on the catkins of oaks. Around the eggs are formed little galls, shaped like gooseberries, in the interior of which the larvae develop. During the course of summer the adults are born, males and sexual females. These females, unlike their mothers, are incapable of reproducing without being fertilized. In appearance they are so unlike their mothers that it was long thought that the two generations of *Neuroterus* formed two distinct species. After they have coupled, the new females lay their eggs on the undersurface of young oak leaves where galls are again formed, but this time shaped like lentils. With the dead leaves these lenticular galls fall to earth, where they pass the winter. From them, during the following spring, asexual females emerge, females like the ones with which we began.

Thus the parthenogenetic *Neuroterus* females have two parents, while the males and the sexual females have only one. Each generation resembles not the one which produced it but the preceding one: the offspring resembles not its parents but its grandparents, and heredity as it were skips a generation. It is believed that the two annual generations of the insect originally comprised both sexes, and that the spring generation later evolved the ability to reproduce asexually. Other members of the Cynipidae reproduce only asexually, the sexual generation having completely disappeared from the annual cycle.

Galls formed on the twig of an oak tree.

P. Popper.

Artichoke-shaped oak-apples.

X Photography

Galls on an oak leaf. (Oak-apples.)

R. H. Noailles.

118

A queen bee, unfertilized. She has just left the cell in which she developed.

R. H. Noailles.

Apis mellifera. Above, the fertilized queen; *centre,* the sterile worker; *below,* the drone.

After Grenier.

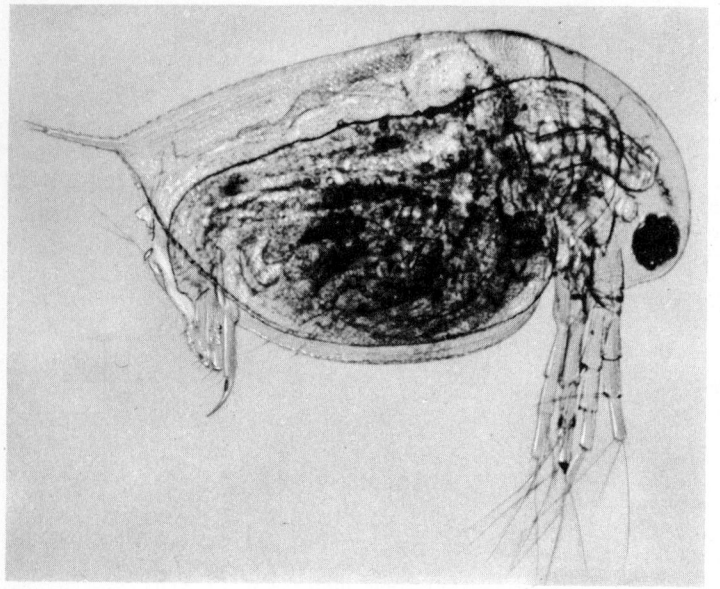

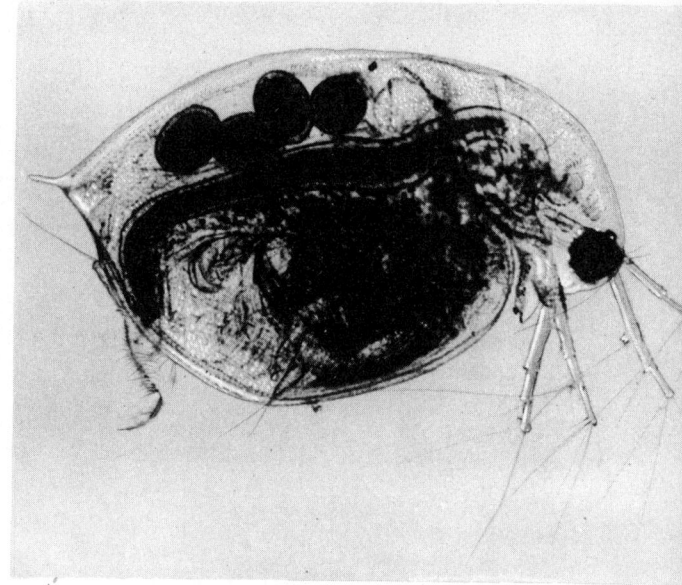

Daphnia, a small fresh-water crustacean. The eggs develop in an incubating cavity situated in the dorsal part of the animal.

Daphnia, whose eggs develop by parthenogenesis (without being fertilized).

Rotating reproductive cycles

The phenomenon of parthenogenesis becomes even more complicated among the rotifers, microscopic animals which live in fresh water and owe their name to the disc-shaped cilia-bearing lobes or 'rotors', by means of which they move.

Some rotifers practise sexual and asexual reproduction at the same time, but among them there is no regular alternation of generation as there is among the Cynipidae. A long period of parthenogenesis covering several generations is followed by a brief period of sexual reproduction: together the two periods form a reproductive cycle.

Let us briefly examine the cycle of *Hydatina senta*, a rotifer which has been much cultivated and studied in biologists' laboratories. The fertilized ovum produces a so-called *virginiparous* female, which gives birth by obligatory parthenogenesis to other females of the same type, that is to say females capable of asexual reproduction. These continue to reproduce for several generations. Then, at a certain moment — the timing depends both on internal and external conditions — appear so-called *sexuparous* females which are exactly like the females that engendered them except in their mode of reproduction. Sexuparous females practise optional parthenogenesis like the queen bee: if they remain unfertilized they produce males, if they are fertilized they produce females. In the latter case their eggs, instead of hatching at once, are surrounded by a thick shell and therefore take a certain time to dev-

elop. These delayed eggs give birth to virginiparous females, and a new cycle begins.

It is interesting to note that the kind of food given to the rotifer can play a decisive part in determining the moment when asexual reproduction gives way to sexual reproduction. If the rotifer is fed on a colourless flagellate, parthenogenesis continues more or less indefinitely; if it is given green flagellates (which contain chlorophyll) the appearance of sexuparous females will soon be observed, sometimes in the following generation.

In the case of the rotifer *Hydatina senta* there is normally only one sexual period during the course of the year, but among other rotifers there are two and even several such periods. Among still others, on the other hand, sexual reproduction has totally disappeared, having been replaced by constant or exclusive parthenogenesis. Such species are composed only of females and in them males never appear.

Among *Daphnia*, small crustaceans sometimes called water fleas and well-known to fish breeders, the reproductive cycle is very similar to those we have just discussed. Again parthenogenesis normally alternates with sexual reproduction, but the purely asexual stage can be prolonged by manipulating external conditions. In this way *Daphnia* have been kept in a state of constant parthenogenesis by Mortimer for sixteen months with a total of eighty-eight generations and by von Dehn for three years or 180 generations with 53,560 descendants! The change to sexuality can be brought about at will by starvation, cold or confinement.

120

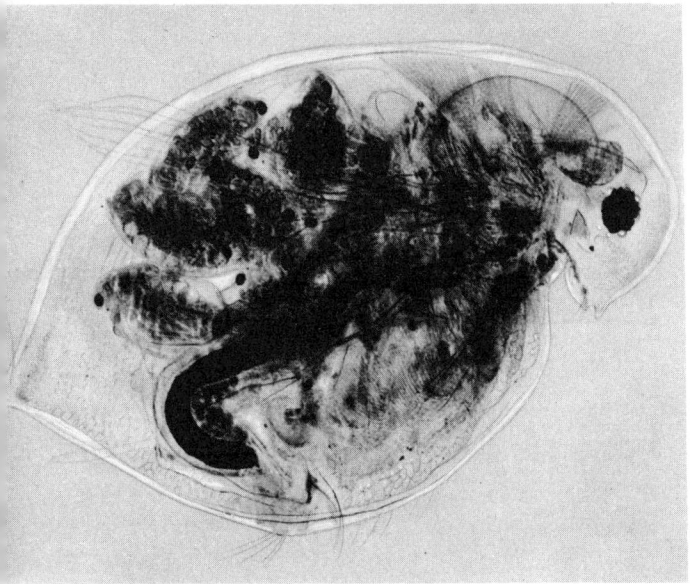

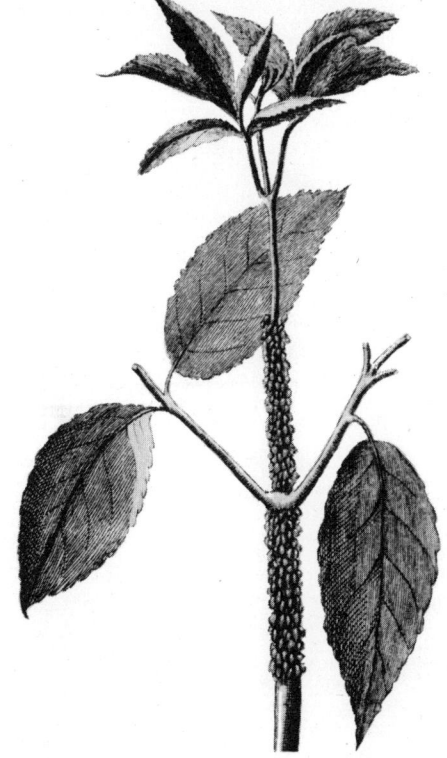

The incubating cavity of *Daphnia* contains several embryos of parthenogenetic origin. The small black dots are the eyes of these embryos, on the point of being liberated.

Aphids of the alder with which Charles Bonnet experimented during the course of his famous researches in natural parthenogenesis.

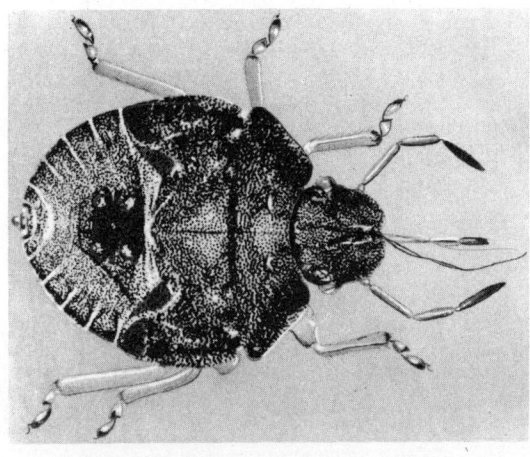

Hornbeam aphid.

Plouvier.

The aphid or plant louse

This familiar creature exists in a wide variety of species and colours: green, pink, yellow, brown, black. Almost every plant has its appointed plant louse, specialized in sucking its sap.

The aphid, apart from its considerable importance as an enemy to horticulture, occupies an outstanding position in the history of biology, for it was in the aphid that the phenomenon of parthenogenesis was first observed. This discovery was made in 1740 by the Swiss naturalist Charles Bonnet who was then hardly twenty years old.

Bonnet, having placed a new-born aphid in solitary confinement, was astonished some ten days later to see that it had produced a very much alive juvenile baby aphid. During the next twenty-one days no fewer than ninety-five further insects were produced by the aphid, 'all much alive and the majority come into the world under my own eyes,' he wrote.

The discovery caused a sensation. The possibility was first suggested that a single coupling might be sufficient for two successive generations, but Bonnet, patiently pursuing his researches, obtained as many as ten successive generations without union of the sexes. As he pointed out, it would indeed be much to swallow, that the descendants should be 'fertilized by their great-great-great-great grandfather!'

Cycles of reproduction exist among the aphids as they do among rotifers and *Daphnia*. During spring and summer only female plant lice are found, females which asexually engender other females. They are

The phasmid, *Carausius morosus*, reproduces exclusively by parthenogenesis and the species is represented by females only.

R. H. Noailles.

A cluster of aphids.

Pesson.

Rosebud aphid.

J. Vincent.

122

viviparous, that is to say their young are born fully formed. At the end of several generations, usually towards autumn, both males and sexual females appear, females which cannot procreate without being fertilized. After coupling they lay eggs which, when winter has passed, hatch out in the spring to produce parthenogenetic females, known as 'founders', who found the new reproductive cycle.

Here again the transition from asexual to sexual reproduction depends both on intrinsic factors and environmental conditions. Appropriately fed and kept at a suitable temperature, rosebush aphids have been kept by Kyber for as long as fifty parthenogenetic generations. On the other hand the arrival of the sexual stage can be hastened by cold, desiccation, darkness and even by the action of certain chemicals.

Among other species of aphid the insect does not accomplish its full reproductive cycle on the same plant, which leads to more or less complicated migrations. Since the winter egg is deposited by the sexual female in a plant, usually woody, it is on this plant, called the principal host, that the 'founding' female is hatched. But the majority of asexual generations take place on another plant, generally herbaceous, known as the 'intermediate host'. The passage from principal host to intermediate host in the spring is assured by females of a special type, provided with wings. In autumn the return journey is assured by other females, also winged. It can, however, happen that a certain number of individuals remain on the intermediate host where they continue indefinitely to reproduce by parthenogenesis. These asexual non-migratory strains assure the survival of the species in regions where the principal host is lacking. It is in this way that the redoubtable woolly aphid can in Europe accomplish its entire reproductive cycle on an apple tree, while in America, the country of its origin, it exploits two different trees, the apple tree and the American elm.

The reproductive cycle of *Phylloxera,* the dreaded vine louse which devastates vineyards, is in many ways like that of plant lice, although it is less dependent on external conditions. The migration of *Phylloxera* takes

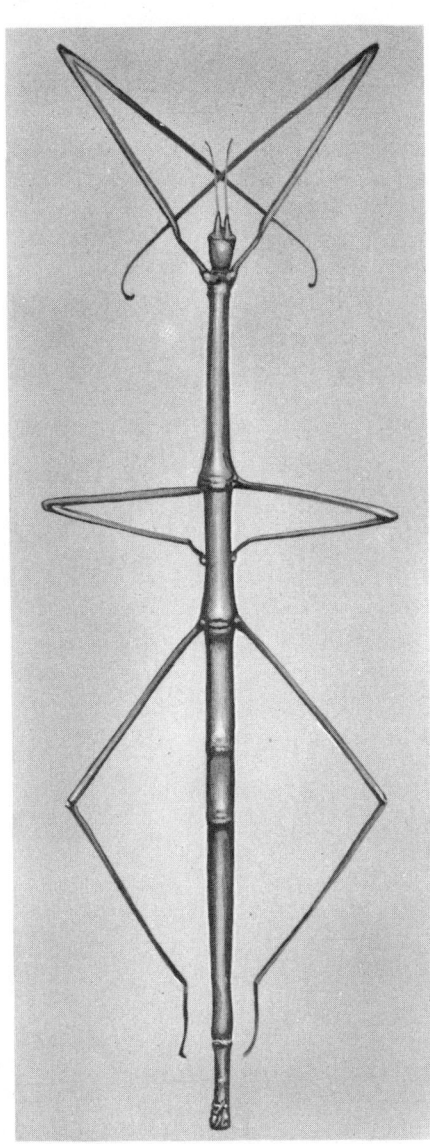

Baculum artemis, a parthenogenetic female.

After Cappe de Baillon.

Phylloxera — the winged form.

Poilpot.

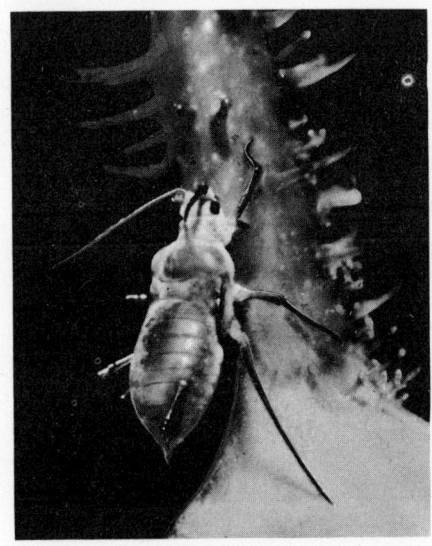

Aphid, many times enlarged.

J. Vincent.

123

place on the plant itself, from leaves to roots. The most injurious generations are those which attack the roots. For this reason the fight against *Phylloxera* consists of grafting European vines on to American plants whose roots are much more resistant to the pest.

Suppression of males

We have noted the existence of exclusive parthenogenesis with constant suppression of the males. To the examples already cited may be added that of the stick insects (Phasmida) — elongated orthopterans shaped like a little stick which, when the legs are folded against the body, can often be mistaken for a twig of dry wood.

One species, *Carausius morosus,* is reared so commonly in laboratories that it has been called the 'guinea-pig' of insects. Among the hundreds of thousands of the insects which have been examined all may be said to have been females, the rare males encountered being most probably masculinized females, incapable of fertilizing others.

The fact that these male-less species can be indefinitely perpetuated is of capital importance and worth underlining, for it demonstrates that sexuality, contrary to the beliefs of certain naturalists including Darwin and Maupas, is not at all necessary for the survival of the species. As we shall see later, the culture of Protozoa and segmented worms, as well as the culture of tissues taken from superior animals, yield facts which lead to the same conclusion.

Larval parthenogenesis — paedogenesis

All the cases of parthenogenesis so far examined have concerned adult females. But among certain animals parthenogenesis begins in the larval stage. It is then called *paedogenesis,* a word which literally means 'reproduction by the child'. Paedogenesis occurs among very small flies such as *Miastor* and in *Oligarces* whose larvae live in colonies under the bark of trees, in cultivated mushrooms, in the stercoral balls of dung beetles, etc.

While in the larvae of insects the sex glands usually remain in the rudimentary and inactive state, they reach maturity in the larva of *Miastor,* a singular phenomenon of accelerated development. Some twenty mature eggs are released in the interior of the larva where they develop without being fertilized. Since there is no natural orifice by which these daughter larvae can emerge they can only be born by bursting out of the skin of the mother. The celebrated biologist Elie Metchnikov saw in this explosive delivery, which always costs the life of the parent organism, a striking example of those 'discords of nature'.

Nonetheless, several succeeding generations of larvae are born in this fashion. Then, at a certain moment, larvae of a different type appear. These are called *imaginal* larvae and are unable to procreate asexually. Instead, as in the case of ordinary maggots, meta-

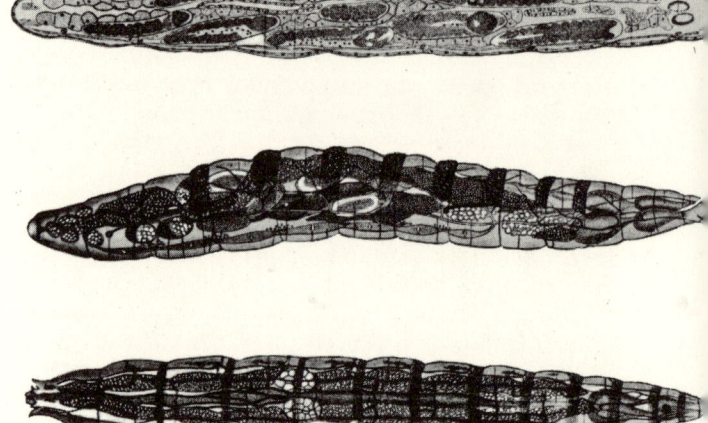

Larvae of *Miastor.* Within the organism several embryos can be seen in course of development.

morphosis takes place and they become flies. The adults, male and female, then couple and produce a few eggs of large size from which emerge new larvae, again capable of this extraordinary form of parthenogenesis called paedogenesis.

In paedogenesis, as in cyclic parthenogenesis, environmental conditions such as nourishment, humidity and so on, contribute to delay or hasten the arrival of the sexual generation.

Another example of larval parthenogenesis is found in a North American coleopteran, *Micromalthus debilis.*

Among certain insects related to mosquitoes parthenogenesis occurs not in the larval stage, but when the insect is a *nymph,* the nymph being intermediate between the larva and the adult, and corresponding to the chrysalis of the butterfly. Parthenogenesis during the nymph stage forms a curious transition between genuine paedogenesis and adult parthenogenesis.

Accidental parthenogenesis

In all the cases previously examined, whether of adult parthenogenesis or of larval parthenogenesis, asexual reproduction formed part of the *normal* cycle of the species. But in certain species where sexual reproduction is the rule it sometimes occurs that a virgin egg will prove capable of ordinary development. Such 'accidental parthenogenesis' is found in grasshoppers, cockroaches, butterflies, *Bombyx* or the moth of the silkworm, some species of the fruit fly, and so on.

Drosophila parthenogenetica owes its very name to the fact that the ratio of virginal births it produces is eight per 10,000, and can rise to fifteen per 10,000 if special females are selected.

Not long ago Olsen and Marsden revealed, to the

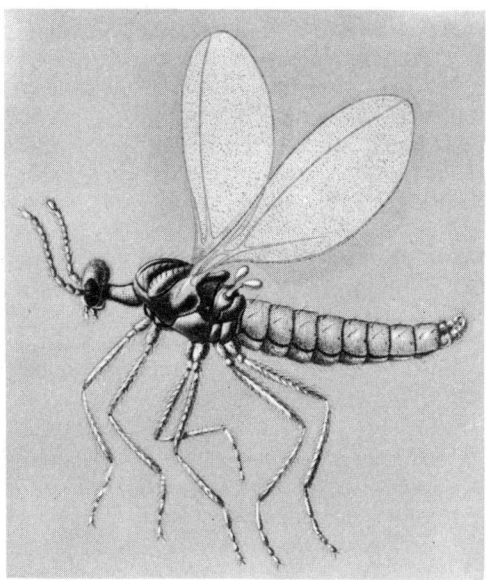

The fly *Miastor* in its adult male form.

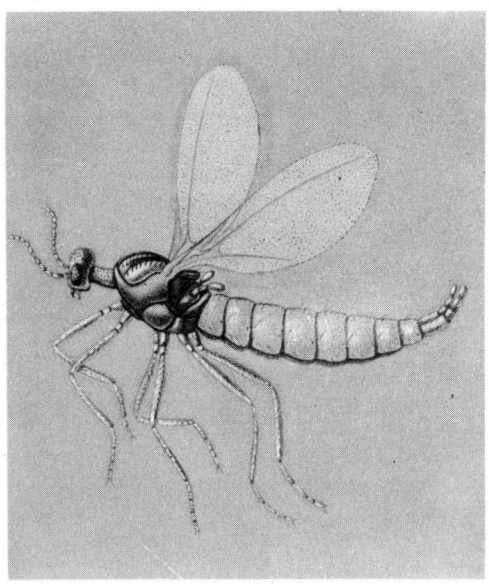

Miastor in its adult female form. The fertilized female lays eggs from which emerge larvae that reproduce by 'larval parthenogenesis' or 'paedogenesis'.

After Kahle.

extreme astonishment of poultry breeders and biologists alike, that the unfertilized eggs of a turkey hen could very exceptionally be hatched. They worked with a breed of turkey called Little White Beltsville. The hens had been separated from the males from the age of four weeks (their sex having been identified by means of a small abdominal incision). The eggs of these isolated females, examined before incubation, sometimes revealed faint signs of development which could be prolonged by incubation. Embryonic evolution was a little slower than normal, the germinal disc becoming visible only at the end of three or four days instead of the normal eighteen to twenty-four hours. Generally speaking, the process stopped there, but in exceptional cases extra-embryonic membranes were formed which covered the entire surface of the 'yolk', blood patches appeared, and even in some cases (twenty-seven out of 278 eggs) embryos of normal or almost normal structure took shape. Out of thousands of virgin eggs two embryos developed completely and were actually hatched. These 'fatherless' turkeys were both male and slightly smaller than those born of normal parentage.

The proportion of unfertilized eggs which develop varies according to the female, from 0 to 29.4 per cent. The average is ten per cent. Among eleven females two were clearly distinguished from the remainder by a marked aptitude for parthenogenesis.

More recent experiments have shown that the percentage of such eggs to develop is appreciably raised if the females, though separated from the turkey cocks, can hear them gobbling. From this it would seem that nervous reactions of a sensory origin can encourage parthenogenesis in turkey hens. Such nervous mechanisms would doubtless bring about changes in the bird's hormone equilibrium, perhaps stimulating the pituitary gland which secretes the substances that control the maturing of the eggs. It is possible or

probable that by selecting those turkey hens most given to parthenogenesis the percentage of unfertilized eggs to mature would be appreciably increased.

In any case it is very remarkable, and most unexpected, to discover the phenomenon of natural parthenogenesis in a superior vertebrate. It proves that there is little truth in what has often been held: namely that parthenogenesis becomes increasingly rare as we ascend the evolutionary scale. If a systematic search for natural parthenogenesis is made among birds, and even among mammals, it is possible that further surprises are in store for us.

Robert Matthey, an authority on chromosomes, has already suggested that natural parthenogenesis may exist among rodents of the genus *Ellobius,* in which the testicular glands of the males are never found in a functional state.

Rudimentary parthenogenesis

In connection with accidental parthenogenesis rudimentary parthenogenesis must be considered, for, besides the virgin eggs which mature completely, there are other and more numerous eggs which abort at all stages of development. Furthermore, in many species in which complete virginal development has never been reported, the beginnings of development can be observed in the unfertilized ovum. Such parthenogenesis is called partial, abortive, or rudimentary.

It is found among sea-urchins, roundworms, and insects; among birds such as peacocks, pigeons, turtle-doves, ducks, and above all chickens, in which the phenomenon was carefully studied first by Lécaillon, then by Igor Kosin.

When the eggs of an un-mated hen pass through the oviduct, segmentation always begins, and may continue until a little colony of embryonic cells is formed. By the time the egg is laid the embryonic cells have already begun to degenerate and, unlike what occurs in the case of the turkey's egg, it is impossible, by incubation, to prolong the evolution of the germ even for an instant.

Rudimentary parthenogenesis has long been known among mammals like sows, rabbits, guinea-pigs, hamsters, ferrets, etc. The unfertilized ovum can begin to undergo cleavage either when it is still enclosed in the ovarian follicle or when, having left the ovary, it travels through the Fallopian tube towards the uterus.

In the ovaries of virgin guinea-pigs under the age of puberty Leo Loeb, Robert Courrier, and Oberling have observed the presence of genuine small embryos: one showed the vague outlines of a neural tube and chorioplacental formations.

Like Lams, we might well wonder 'how far the development of an egg whose cleavage begins in the ovary would proceed if it reached the uterine wall where conditions would be more favourable.'

Even in the human species abortive parthenogenesis has been reported. In the case of a girl Hoche and Morlot have observed the segmentation of a virgin egg, and as early as 1864 the histologist Morel reported cases of 'cleavage without fertilization' among women who had died of puerperal peritonitis eight to ten days after childbirth. 'We have,' he writes, 'encountered several such ova in which cleavage was as clearly visible as in fertilized ova. Only the cells of the pseudoblastoderm had already undergone fatty metamorphosis ... The segmentation of the yolk, then, is possible without previous fertilization. Besides, the phenomenon of cleavage is not in itself in any way abnormal, for the ovum is only a cell and daily we find that cells of an organism, under the influence of some irritant, also undergo cleavage or a proliferation of the nuclei, and in consequence the most varied pathological tissues are produced.'

Finally all doctors know the dermoid cysts of the ovary which many authorities attribute to abortive parthenogenesis. On this subject Witschi cites a very curious observation made by Neumann: when two distinct cysts were removed from an ovary one was found to contain a wisp of black hair and the other a wisp of blond hair. It has been suggested that in this case the patient carried two different genes for pigmentation in her hereditary patrimony, one of which had gone into the germ-cell that produced the first cyst, and the other into the cell that produced the second cyst.

Ovarian cysts are often multiple and as many as fifteen or twenty are sometimes found in the same ovary. Their genesis seems to depend on hereditary or racial predisposition: they are most frequent among Japanese and the black races. There may be some connection between this predisposition and a hereditary predisposition to abortive parthenogenesis which we have observed among turkeys, *Drosophila*, etc. The parthenogenetic theory of ovarian cysts was propounded by Mathias Duval and his pupil Répin who in 1891 published a remarkable thesis on the *Parthenogenetic Origin of Dermoid Cysts of the Ovary*.

Although the theories of Albert Peyron are contested they should, in passing, be mentioned. Peyron connects the origin of certain tumours of the testes with a male parthenogenesis, complicated by polyembryony. Under the influence of inexactly understood morbid conditions the testicular cells are supposed to produce minute but genuine embryos which themselves break into fragments. In a single tumour thousands of these embryos are formed and, according to Peyron, should furnish incomparable material for the study of the earliest stages of human development.

Peyron distinguishes between these parthenogenetic tumours and parasitic tumours which have a completely different origin since the latter result from 'a foetal inclusion' and never contain formations which are genuinely embryonic. It was evidently to a foetal inclusion that Peyron would have ascribed the case of 'male parthenogenesis' which in 1953 was reported to the French Academy of Medicine by Lombard, Ferrand, and Legenissel, and concerned a four months foetus found in the abdomen of a boy twenty months old.

Artificial Parthenogenesis

If in species where natural parthenogenesis exists the unfertilized ovum has the power to develop of its own accord, why in all other species does it require the aid of spermatozoa? The problem is far from being solved. Some biologists hold that the parthenogenetic egg begins to develop because within it a stimulating substance is formed which is not formed in an ordinary egg. Others hold the opposite view that an inhibitory substance is not formed in the parthenogenetic egg which is formed in the ordinary egg.

Whatever the mechanism of natural parthenogenesis may be, science has succeeded by artificial means in bringing about in a great number of animal species the development of virgin eggs which in normal circumstances would never develop.

The first attempts to produce parthenogenesis artificially were made by Spallanzani towards 1780. Hoping to find a substitute for the stimulant action of sperm he submitted frogs' eggs to electrical discharges and to liquid irritants.

In 1847 Boursier claimed that a female *Bombyx,* or silkworm moth, could be made fertile by sunlight. He observed that some females among the Lepidoptera, moths especially, hurried to lay their eggs when captured by collectors and fixed by pins through their bodies. Eggs laid in this manner usually produced little caterpillars. He had previously assumed that the females had been fertilized by a male of their species prior to capture, but this idea was disproved by capturing a female silkworm moth as she emerged from the cocoon and had therefore not communicated with any male. The female was exposed to direct sunlight for several hours, then placed in the shade. The following

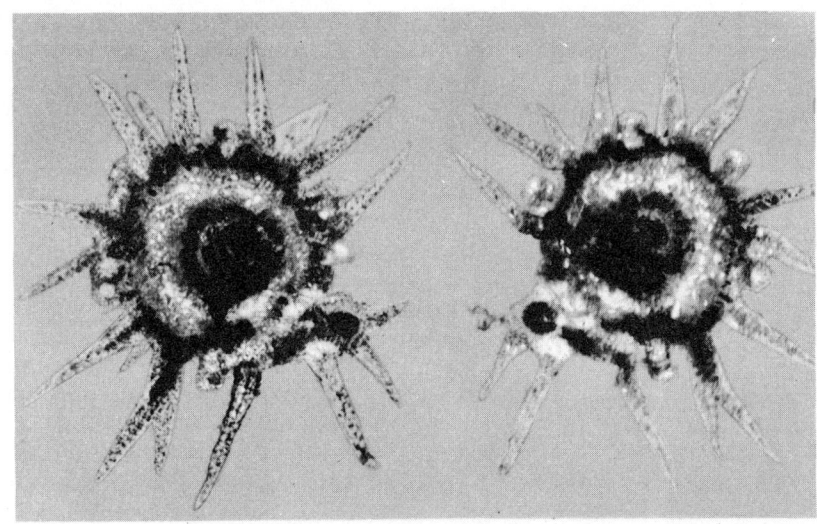

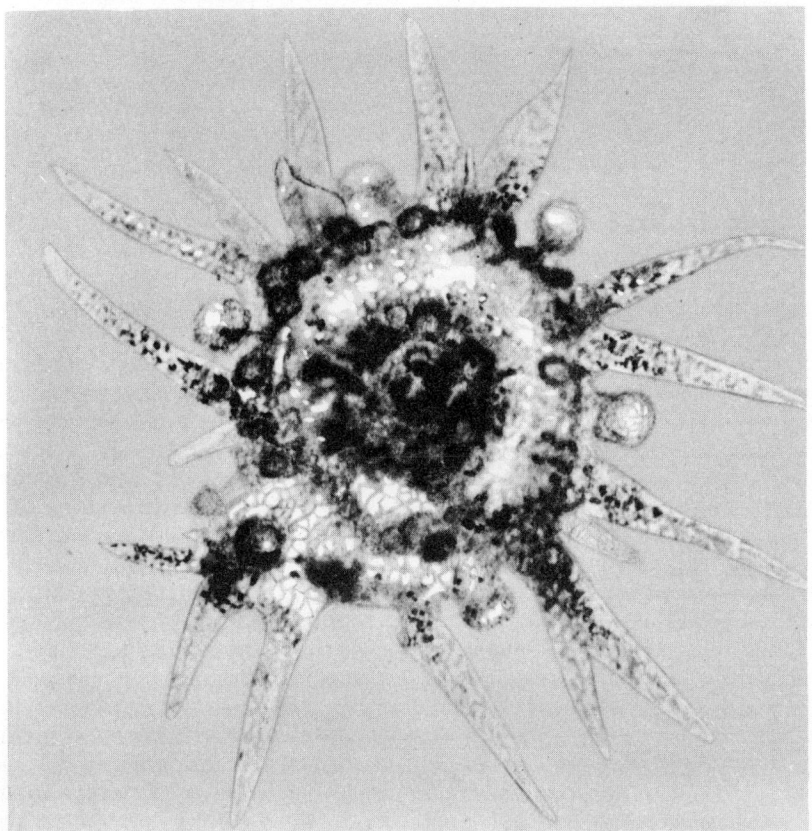

day she laid some forty eggs which hatched into small caterpillars, ate, and thrived. Unfertilized eggs of the silkworm moth were again used in 1886 by Tichomirov who caused their development to begin by rubbing them between two pieces of cloth. This is a procedure employed by breeders to hasten the development of eggs which after being laid normally pass through a long period of repose called the diapause.

However, the first great triumph of artificial parthenogenesis occurred in 1899 when the American biologist Jacques Loeb succeeded in 'chemically fertilizing' the sea-urchin by plunging its virgin eggs into a solution of magnesium chloride of appropriate strength.

This achievement had of course been prepared by much patient and less dramatic research, notably the

Young sea-urchins of parthenogenetic origin obtained in 1910 by the biologist Yves Delage.

After Yves Delage.

work of Richard and Oskar Hertwig, of Herbst, of Thomas Hunt Morgan, and others. But until 1899 only vague suggestions of segmentation had been obtained by exposing unfertilized eggs to chemical agents, while Loeb had succeeded at the first attempt in producing normally constituted embryos and even larvae which swam, products which in every way resembled those of natural fertilization. It is scarcely necessary to emphasize the great significance of this discovery. For the first time man had imitated in the laboratory the initial process in the creation of a living creature.

Eugène Bataillon, who invented traumatic parthenogenesis in 1910, by introducing a cellular element into the unfertilized ovum.

Loeb's discovery not unnaturally caused a sensation. Like all great novelties it was at once contested by other learned men who refused to admit that the fertilizing action of male semen could be replaced by a simple salt solution.

'When I published these results,' 'the great biologist said, 'the almost unanimous opinion was that I had been the victim of an illusion.' He had himself, he confessed, at first been afraid he was mistaken. But his astounding results were not long in being confirmed by other investigators. In France, for instance, Yves Delage, employing other techniques, succeeded in parthenogenetically reproducing sea-urchins and star-fish.

In his first experiments Loeb obtained 'chemical fertilization' by plunging virgin eggs into sea-water rendered hypertonic by the addition of magnesium chloride. Later, in 1906, he devised a method which gave even better results. The principle is briefly this: sea-urchins' eggs are first treated for a minute or two with sea-water to which butyric acid has been added. Then they are washed in normal sea-water and plunged for twenty to thirty seconds into a solution enriched with salts. Finally they are again put into normal sea-water where they develop as successfully as eggs which have been normally fertilized.

This procedure has the advantage of very faithfully imitating the action of male sperm, for after treatment by the acid the egg becomes surrounded by a 'fertilization membrane' exactly as it does after being penetrated by a spermatozoon. According to Loeb, the formation of this membrane indicates a beginning of destruction, or cytolysis, in the superficial layer of the egg. As a result oxidation increases and would inevitably lead to the death of the egg if it were not plunged into the second, salt-enriched solution which by neutralizing excess oxidation averts the danger.

Many other processes have proved effective in bringing about parthenogenesis in the sea-urchin's egg, including treatment with tannin and ammonia employed by Delage, and ultraviolet rays. For starfish carbonic acid (soda-water), shaking, cold, desiccation, chlorinated mixtures, and other methods have been successful.

With the unsegmented worm *Urechis caupo* Tyler employed diluted sea-water and even distilled water. In silkworms complete parthenogenesis can easily be caused by treating virgin eggs with hydrochloric acid or by exposure to warmth. In this species the same agents not only bring about parthenogenesis in the egg but interrupt the embryonic diapause.

Among fish the opening phase of development has been brought about by the action of saline or sweetened solutions, and in some cases even complete development has resulted from the action of heat combined with that of distilled water.

Among frogs and toads physico-chemical factors such as salts and thermal or electric shocks cause only the first phase of development, that is, activation. To obtain complete development recourse must be made to traumatic parthenogenesis. This is a special process invented in 1910 by Bataillon which consists of introducing a cellular element into the unfertilized ovum.

If, as Bataillon demonstrated in his experiments, virgin eggs of the frog are pricked with a fine glass stylet they are all activated, that is to say they all begin to develop. Nearly all of them, however, soon die, and only a few continue their evolution to produce embryos and larvae.

The question was: what constituted the difference between those eggs which achieved only abortive parthenogenesis and those in which parthenogenesis was complete?

The answer proved to be that some eggs had, without the experimenter having noticed it, been inoculated by the glass stylet with a minute globule of blood. This accidental inoculation was explained by the fact that the gelatinous matrix surrounding the eggs was frequently soiled by blood or lymph unless the eggs were extracted from the belly of the animal with particular care. The blood corpuscle, which is a cell, takes no active part in the formation of the embryo; it simply brings about parthenogenetic development. There is, moreover, no need for the blood-cell to be that of a frog. The frog's unfertilized ovum will develop with the aid of any other vertebrate's cell, reptile, bird, or mammal, and even the cells of certain invertebrates like the earthworm or the snail.

The cellular factor which assures the development of the egg is called the 'regulation factor' and its nature remains obscure. It is lacking in all insects, in the sea-urchin, in all plants, in microbes and in viruses. It can be separated from the cell itself and seems to be connected with the ribonucleoprotein granules of the cytoplasm. It will resist fairly high temperatures and can, as demonstrated by Jean Rostand, be completely dried and kept for long periods in glycerine or sugar syrup without losing its potency.

The phenomenon of artificial parthenogenesis by means of chemically undetermined cellular substances is one of particular interest, for it constitutes a transition between chemical and natural fertilization.

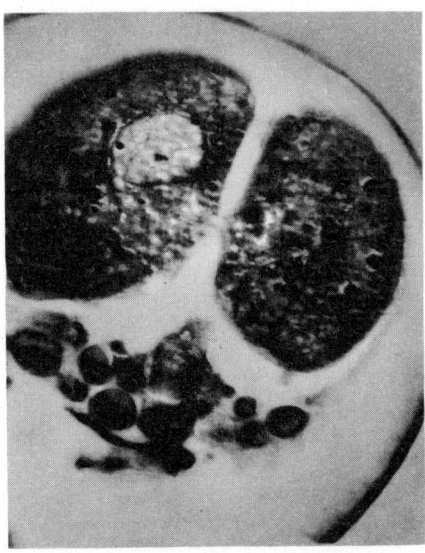

Parthenogenetic segmentation — stage of two blastomeres — obtained by chilling the unfertilized egg of a rabbit.

After Ch. Thibault.

Mammals

The eggs or ova of mammals are so small that they are practically invisible to the naked eye. They are, moreover, few in number. Even in prolific species like the rabbit and the mouse the female produces only seven or eight at a time. Finally the ovum is very fragile and its development takes place within the interior of the maternal organism. For these reasons attempts to bring about parthenogenesis in mammals are particularly laborious. Investigators, far from being discouraged by such obstacles, have skilfully surmounted them and, in recent years, have achieved successes which are incontestable.

As long ago as 1927 Christian Champy observed the cleavage of a rabbit's unfertilized ovum in an ovarian culture. The culture took place in rabbit plasma to which embryonic chicken juice had been added, and the cleavage occurred some fifty days after it had begun. Though the experiment was full of promise it led to nothing and more than ten years passed before the American, Gregory Pincus, successfully produced parthenogenesis in rabbits.

Pincus obtained virgin eggs by puncturing the ovarian follicle and by taking them directly from the uterine tube. These he submitted to a stimulant treatment — to a salt solution or, by raising the temperature to 45 °C, to a thermal shock. He then replaced them in the uterine tube of a rabbit which had been prepared to receive and nourish the ova by suitable hormone injections. Out of 615 ova thus treated he obtained three which developed completely and at the proper time were born as baby rabbits.

The percentage of success was no doubt low, but the theoretical results were of the highest importance. There was now no longer any doubt that artificial parthenogenesis could be extended to the superior animals.

In later experiments Pincus, with the collaboration of Shapiro, has produced parthenogenesis in rabbits by chilling the unfertilized ova. To do this he circulated iced water for a few minutes in a hollow casing which surrounded the uterine tube at the level where the eggs were situated. This method was applied to sixteen rabbits, one of which subsequently produced a live baby rabbit. In other words, there was one complete development out of about 200 chilled eggs. The percentage of success is again low, but it is probable that it can be raised by improvements in the technique.

Still working with rabbits Thibault, in France, chilled the eggs, as Pincus had done, by means of a small casing through which cold water ran around the uterine tube or by applying a small piece of ice to the desired part of the tube and leaving it there for several minutes. More than half the ova treated in this way were activated, that is to say they showed the first signs of development and three young embryos — in the *morula* stage of development — were formed and studied. In 1954 M. C. Chang kept the ova of rabbits in a serum medium at a temperature of 10 °C for a day, hoping to bring about parthenogenetic development. Fourteen per cent of the ova began regular development, but when they were replaced in the uterine tubes they did not evolve into little rabbits, having failed to become implanted in the uterine wall. It was as though the parthenogenetic embryo, the *blastocyst,* lacked some 'factor' necessary for settling down in the normal way (nidation).

Attempts to produce artificial parthenogenesis in sheep, rats, mice, and guinea-pigs have only resulted in abortive segmentations. Reimann and Muller reported in 1939 that the beginning of a cleavage had been observed in a human ovum which had been pricked *in vitro*.

As we have seen, the treatments by which artificial parthenogenesis can be induced are extremely varied: strong salt solutions, distilled water, traumatic and thermal shocks, acidity, alkalinity, and others. A given agent may prove most effective on the eggs of one species and totally ineffective on those of another. Generally, we do not know the reasons for these differences in sensitivity, nor why different ova react differently. As Vandel has said, the very variety of the agents which cause parthenogenesis shows that it is impossible to construct a general theory founded on the specific action of the factors employed.

Morgan came to much the same conclusion when he stated that virgin eggs could be induced to produce embryos by artificial agents of the greatest diversity. He suggested that such was the capital result of the work done in this field and was more important than all the hypotheses proposed to explain the phenomenon.

Although we are, at the moment, not sure of the detailed mechanism involved in artificial partheno-

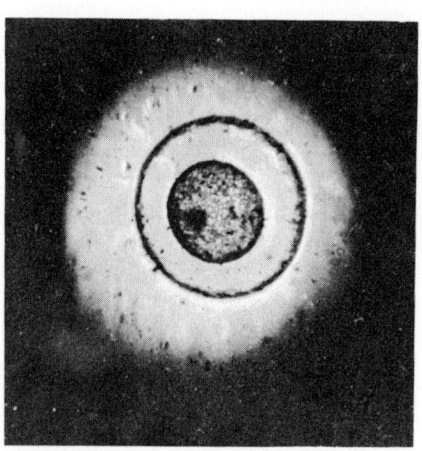

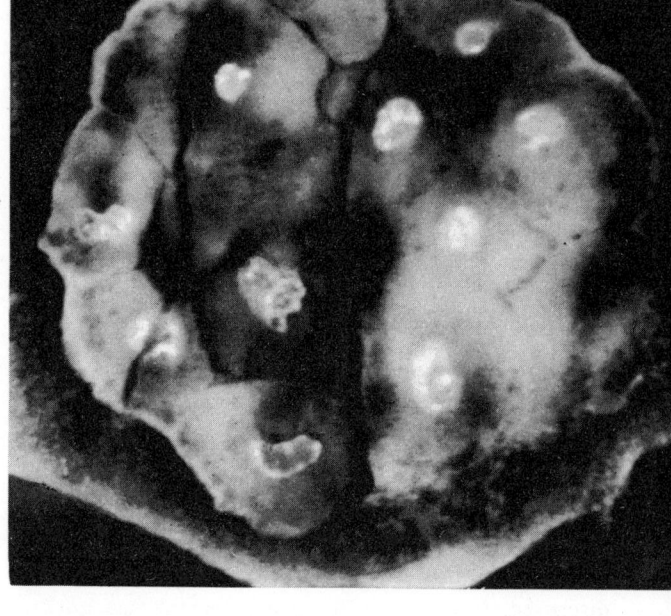

Blastocyst (embryonic vesicle) of a rabbit, obtained by chilling an unfertilized egg; it has already penetrated into the uterus.

After Ch. Thibault.

Parthenogenetic 'morula' of rabbit (52 hours) obtained by chilling an unfertilized egg.

After Ch. Thibault.

genesis we do, on the other hand, know why parthenogenesis gives birth sometimes to normal or almost normal organisms and sometimes to organisms which are feeble, malformed, or unable to live.

The difference lies essentially in the manner in which the chromosomes of the unfertilized ovum behave. In principle, the number of chromosomes in the virgin egg is reduced to half the total number which the mother possesses, since reduction division takes place at the moment the ovum is matured. If this ovum subsequently develops parthenogenetically it must produce an individual with a halved — or haploid — number of chromosomes. But certain forms of artificial parthenogenesis sometimes result in bringing about a doubling of the chromosomes or else — which comes to the same thing — in preventing their reduction in the ovum. In this case parthenogenesis will lead to the birth of an individual which possesses the normal, or diploid, number of chromosomes.

The disparity of the results of artificial parthenogenesis are caused by this disparity in chromosome number. Individuals which bear the reduced haploid number are feeble and deformed and usually they do not reach the adult stage. On the other hand, those that bear the normal diploid number are themselves normal and live.

The existence of these two types of parthenogenetically born individuals is noticeable among amphibians who sometimes give birth to vigorous tadpoles, capable of evolving into adults, and sometimes to dwarf larvae, incapable of feeding themselves, attacked by oedema, and condemned to perish before the end of the third week. Among silkworms Astaurov obtained perfectly normal subjects by the thousands, all capable of procreating in their turn and carrying the normal number of chromosomes. The same is true of the few rabbits which Pincus obtained.

The question of sex in parthenogenesis is slightly complicated by the fact that we must distinguish between species whose sex determination follows the *Abraxas* pattern and those which follow the *Drosophila* pattern.

In the former case the female forms W-ova and Z-ova, and the offspring can be WW or ZZ or, if reduction division has been prevented, even ZW. They can therefore be of both sexes. In the case of *Drosophila* the female forms only X-ova and all the offspring are of necessity XX, that is, females.

Human parthenogenesis

There is no *a priori* reason for assuming that the ova of women should be especially refractory to the agents of artificial parthenogenesis. Indeed the fact of abortive parthenogenesis inclines us rather to the opposite opinion. But until now no serious experiments of the kind have been made, with the possible exception of ova *in vitro* obtained from the operating theatre.

Such experiments might, like experiments with rabbits, consist of removing the ovum and after treatment replacing it in the Fallopian tube, or else in treating it directly *in situ*. Possibly other methods may be invented which will allow surgery to be dispensed with: for instance, hibernation, irradiation, or chemical treatment. If so, there will be more chance for human parthenogenesis to be seriously contemplated.

Bearing in mind what we know about sex determination in our species, namely that it follows the *Drosophila* pattern, we can see that all offspring must be female. But would such daughters be identical to their mothers, be, as it were, their mothers' younger twins? That would depend on how the chromosomes behaved. If the daughter received all the mother's chromosomes she would be an exact reproduction; if she received only half of them, doubled, then she need not resemble her mother at all.

It is hardly necessary to add that parthenogenesis in our species could serve little useful purpose. Quite apart from dangers of a biological nature, such as the more frequent appearance of anomalies, it would arouse the deepest moral and psychological objections. It would perhaps be acceptable if the husband were sterile or if he had a grave genetic defect which would be dominant in his offspring. In other words parthenogenesis would be indicated in those cases where artificial insemination by a donor is now employed, and the husband would no doubt prefer to owe his child to his wife alone than to the collaboration of a stranger.

Gynogenesis

Almost everyone today has heard parthenogenesis discussed, but few except specialists in biology are aware of gynogenesis which is, in fact, a variation of parthenogenesis since it too results, in a roundabout way, in producing offspring without male parent.

In some species of roundworms *(Rhabditis)* and of flatworms *(Dugesia),* etc., the sperm penetrates the egg and stimulates development, but soon degenerates without having taken any part in the formation of the embryo. The case, in brief, is one of 'parthenogenesis by sperm' and is known as gynogenesis, pseudogamy or partial fertilization.

Such a phenomenon is possible because in the egg the spermatozoon has a double function. In the first place it brings about the development of the ovum by introducing certain substances which have a stimulating action: in the second place it supplies the new being (zygote) with the chromosomes which make up its hereditary patrimony.

A dissociation between the two functions of fertilizing and heredity — can occur by accident. Thus in the course of crossbreeding salamanders Fankhauser found that a white female and a black male produced offspring which were all white. Their colour betrayed the fact that the male parent had not genetically participated in their conception and hence that they were products of spontaneous gynogenesis.

The phenomenon of gynogenesis can be caused at will by submitting the fertilizing cells to certain treatments. In this way artificial gynogenesis was first achieved some thirty years ago by the celebrated German embryologist Oskar Hertwig. His experiments, today classic, were made with frogs and toads, animals which are particularly suitable for this type of research because of the abundance of their eggs and the ease with which they can be artificially inseminated.

If the semen of the frog is exposed for a few minutes to the effects of radium or of a radio-active preparation and then used to fertilize normal eggs, the offspring will present certain anomalies or defects. It is as though the irradiation has acted on the hereditary substance in the male germ-cell — more precisely on the chromosomes in the nucleus — and caused a permanent lesion, the effects of which are apparent in the defective development of the embryo.

As irradiation is prolonged the anomalies produced will become more and more marked, and when a certain duration of time is exceeded the embryos can no longer be hatched. There is nothing very astonishing in this perhaps, but what is astonishing is that if the irradiation is continued for an even longer time, for several hours or more, the malformations produced, instead of becoming still greater, begin to diminish and a point is reached when again almost normal embryos appear, capable of evolving to the larval stage.

This paradoxical result has been named the Hertwig effect and it can be explained by the fact that beyond a certain point of irradiation the paternal chromosomes are so badly damaged that they can no longer play any part in the formation of the embryo. They have, as it were, been annihilated and cannot join the maternal chromosomes of the ovum as they do at the beginning of every normal development. Or, if they do enter the ovum, the junction is of short duration; they remain inert, idle, taking no part in the first divisions of the maternal nucleus, and soon degenerate in the cytoplasm of the ovum where they are absorbed and digested like any other foreign substance. If in these conditions embryos are formed which are more or less free from defects it is because they have been spared irradiated chromosomes and carry in their cells only healthy chromosomes derived from the maternal germ-cell. Thus, by the indirect agency of damaged semen, gynogenesis has been experimentally achieved.

Agents other than radium are equally capable of destroying the hereditary elements in male semen while leaving intact its capacity to bring about development. Thus X-rays can be used, ultraviolet rays, and even certain chemicals like methylene blue, trypaflavine, toluidine blue, pyronine, and yperite or mustard gas. Dalcq and Simon demonstrated that the use of ultraviolet rays is an excellent method of inducing gynogenesis in amphibians.

Gynogenesis can also be brought about by an extremely simple process which consists of sharply chilling the egg immediately after it has been fertilized. J. Rostand has shown that in the development of eggs thus treated only the maternal chromosomes are involved.

Experimental gynogenesis in higher animals, namely birds and mammals, has not yet been successful. R. G. Edwards tried to discover the Hertwig effect in mice, but the results obtained were only partially satisfactory and relatively uninstructive.

Gynogenesis by false hybridism

As a general rule the union of two individuals belonging to different species is found to be quite sterile. There are, however, interesting exceptions to the rule which are, in fact, the results of gynogenesis.

The female toad and the male tree-frog are two amphibians which are very unlike in shape, structure, colour, and habits, but by using artificial insemination they can be crossed. Surprisingly, a good proportion of the inseminated eggs develop and give birth to larvae. This is all the more surprising since no offspring result when much more nearly related species are crossed — for example, the common frog and the green frog.

The explanation of the phenomenon resides in the very fact that the two species are so distantly related, for the tree-frog's sperm can only partially fertilize the toad's egg, thus leading to gynogenesis. Here again the larvae born owe their existence to a father from whom no hereditary characters have been received. Put in another way, they are the result of a false hybridization. The larvae produced are of purely maternal race: a toad's larvae, without the slightest admixture of tree-frog. If the 'false' cross had been the other way around, that is, a female tree-frog with a male toad, the larvae would be of pure tree-frog breed with no admixture of toad.

Other amphibian combinations give rise to gynogenesis through false hybridization: the female tree-frog and the male green toad, the female tree-frog and the male *Pelobates,* the female toad and the male *Pelobates,* the female *Rana pipiens* and the male *Pseudacris,* etc. False hybridism is also known among certain lower animals: sea-urchin and mollusc, sea-urchin and worm. But it has never been observed in birds or mammals.

Whether they are produced by foreign sperm or by damaged sperm the offspring of gynogenesis carry only chromosomes derived from the maternal ovum. In theory, then, they should have only half the number of chromosomes of a normal organism which receives chromosomes from both female and male parent. In practice this is usually the case and from lack of chromosomes the gynogenetic organism is weak and sickly and, in the case of frogs especially, usually dies before the end of the third week. In any event it never reaches the adult stage.

Very occasionally, however, gynogenetically born organisms are found with a complete number of chromosomes. In their case the maternal chromosome set was doubled at the beginning of their development. Having benefited from this doubling of chromosomes such individuals appear to be quite as robust and as able to reach maturity as normal individuals, although it has been pointed out that they may reveal certain anomalies as a result of the doubling of harmful recessive genes.

It should be added that when eggs fertilized by damaged or foreign sperm are submitted to lower or higher temperatures the proportion of vigorous gynogenetic offspring is appreciably raised. The experiment has been made many times by J. Rostand with the toad. The thermal shock encourages the chromosomes to double, and by this method we are able to obtain at will a great number of gynogenetic products capable of becoming adults and differing in no visible way from their normally-engendered fellow creatures.

Human gynogenesis

As long ago as May 1913 the biologist Yves Delage envisaged the possibility of human gynogenesis in an

One-year-old toads obtained by gynogenesis.
After J. Rostand. Feher.

article that caused a sensation at the time. In his *Can Parthenogenesis exist in the Human Species?* he examined the possibility that spermatozoa could be so damaged by toxic agents like alcohol, morphine, cocaine, nicotine, and perhaps also the virus of syphilis, that, on penetrating the ovum they might be able to cause gynogenesis. (Delage used the term parthenogenesis, since at that time a clear distinction was not made between the two phenomena). 'Hence,' he concluded, 'it is not impossible that in the human species there exist parthenogenetic individuals, people whom perhaps we pass in the street without suspecting the extraordinary singularity of their origin — because this singularity does not reveal itself by easily interpreted or obvious characteristics.'

'Only by attentive observation of cases that might reveal such origin could an opinion on the subject be arrived at. The task would prove highly fascinating and of the greatest biological interest.'

As possibly responsible for parthenogenesis, Delage also considered 'those very rare sexual acts which nonetheless indubitably occur between human beings of one or the other sex and animals.' This same hypothesis was later advanced by the biologist L. Bounoure to explain the birth of a certain human 'monster'. The case concerned an anencephalous or headless child, born on the sixteenth of January 1897 in a maternity hospital at Vichy, to a girl of sixteen who lived in a gypsy caravan with her father and a monkey — probably a macaque, *Macaca mulatta.*

132

A queen bee emerges from the royal cell or alveolus. This was an ordinary cell which the workers have enlarged during the course of the queen's growth.

Daphnia, carrying in its incubating pouch eggs which develop without being fertilized, that is, by parthenogenesis.

Ephippial female *Daphnia*, having formed viable eggs after fertilization.

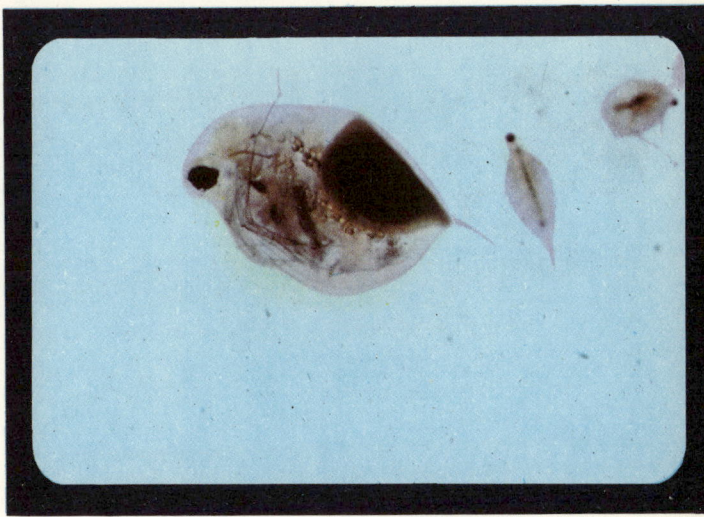

View of a terrarium containing some forty young toads
obtained by gynogenesis. They are of purely maternal origin,
like toads obtained by parthenogenesis.

After J. Rostand. J. Vincent.

Young toad, more than a year old, of gynogenetic origin.

After J. Rostand. H. de Segonzac.

Young albino toad, obtained by gynogenesis. Due to the animal's purely maternal origin, a 'recessive' hereditary character (albinism) was able to make its unmasked appearance.

After J. Rostand. H. de Segonzac.

The girl had never associated with outsiders, and in the district where the caravan was stationed public rumour insinuated, though without basis, that incestuous relations existed between father and daughter.

L. Bounoure suggested that the origin of the monster should be sought in sexual relations between the girl and the monkey, relations which had given rise to the phenomenon of pseudogamy or gynogenesis. However, it is believed that the hypothesis of incest is much less improbable than that of false hybridization between a human being and a monkey.

Androgenesis

A phenomenon can also be artifically induced which is in a sense the inverse of gynogenesis: namely, making a fertilized egg develop from the paternal nucleus instead of from the maternal nucleus. This is what is called androgenesis.

In androgenesis all the nuclei, and therefore all the chromosomes of the individual derive from the paternal gamete, though obviously the maternal role cannot be totally dispensed with, since the ovum contributes its cytoplasm, the portion of the protoplasm which surrounds the nucleus, to the development of the embryo. The problem thus arises of deciding whether the products of androgenesis, in other words products which derive from a cell with paternal chromosomes and maternal cytoplasm, should be considered as of purely paternal or of mixed origin.

From the theoretical point of view the question is of great importance, since it touches on the part played by the cytoplasm in determining hereditary characters. Before examining it, however, let us examine the various methods by which the research biologist can suppress the maternal nucleus and thus produce androgenesis.

In the first place he can, just before the egg is ferti-

lized, destroy the maternal nucleus, or at least so damage it that it takes no part in the development of the embryo. This is done by submitting the ovum to a radio-active preparation or to the action of X-rays. This experiment, which is of course a replica of Oskar Hertwig's method of producing gynogenesis by irradiated semen, has been successfully made with amphibians by Gunther and Paula Hertwig, and with moths by the Russian scientist Astaurov.

Again, the nucleus of the ovum can, after fertilization, be destroyed by touching it with a heated needle, or it can be got rid of by suction with the aid of an extremely fine pipette. Both processes have been employed with success with amphibians (of orders Anura and Urodela).

A third method which gives much less consistent results relies on exposing the egg to extreme temperatures, high or low. When submitted to such thermal shocks the female nucleus can be excluded from the cell's development which then takes place by androgenesis. By this process Hashimoto with silkworms and Humphrey and Fankhauser with the Mexican salamander *Ambystoma* have been able to obtain a certain number of products with purely paternal chromosomes. Even in the natural conditions of fertilization androgenetic embryos are exceptionally produced by salamanders.

Finally, there is a very efficacious method of suppressing the maternal nucleus: namely, to remove it by amputating that part of the egg which contains it. This special manner of producing androgenesis is known as andro-merogony, and has been successfully applied to a good number of aquatic invertebrates (sea-urchins, starfish, *Ascidiella, Cerebratulus,* tooth shell molluscs, etc.) and even to vertebrates belonging to the class Amphibia.

For example, in the case of sea-urchins the virgin egg is cut in two and the fragment without the nucleus is fertilized. The upper surface of the egg can also be removed with the aid of a fine needle, or, again, separation can be induced by shaking.

Among amphibians, especially water salamanders or tritons, the egg can be separated by means of a ligature. Shortly after fertilization the egg is bound with a very fine hair — preferably the hair of a child — which is slowly and cautiously tightened. In this way the egg can be divided into two parts, one of which contains the maternal nucleus and the other the paternal nucleus. The first fragment does not develop or at least develops only abortively. The second, which contains the male nucleus, often gives birth to an androgenetic larva or, more precisely, an andromerogon. Experiments of this kind have been brilliantly performed by the school of Spemann, by Baltzer, Fankhauser, and others.

In theory androgenetic organisms, or andromerogons, should carry in their cells only half the normal number of chromosomes, since, of course, they receive paternal chromosomes only. Very exceptionally, however, it occurs that a doubling of the paternal chromosomes takes place at the beginning of development. This doubling may be produced by the fusion of the first two nuclei to undergo cleavage or by the fusion of the paternal nucleus with the nucleus of a supernumerary spermatozoon. The latter phenomenon occurs especially among tritons whose eggs are normally penetrated by several spermatozoa, a phenomenon known as polyspermy.

Androgenetic organisms with a single set of chromosomes display the defects which are always associated with chromosome deficiency: dwarfing and early death. Those with a double set of chromosomes are normal in appearance, can develop into adults, and eventually reproduce.

In the same way that a double set of chromosomes of maternal origin suffices in parthenogenesis and gynogenesis to assure the normality of the organism, so in androgenesis normality is assured by the double set of chromosomes of paternal origin.

The sex of androgenetic organisms depends on the pattern of sex determination the species follows. Androgenetic silkworms, for instance, are always male, because the male is ZZ. Androgenesis has not yet been induced among higher animals, birds, and mammals, but in theory there is no reason why it should not be.

Hybrid androgenesis

The problem now arises of an egg deprived of its nucleus, or a fragment of egg without nucleus, which has been penetrated by the sperm of a species different from its own: what will happen to such an egg in which the nucleus and the cytoplasm belong to different species? If an embryo develops what will be the characteristics of the resulting individual? Will it be a hybrid organism with mixed paternal and maternal characters, or an organism of pure breed, with purely paternal characters? The answer to this question is obviously of the utmost importance in discovering the respective roles of the nucleus and the cytoplasm in the determination of hereditary characters.

The first attempts to produce hybrid androgenesis were those of the Austrian biologist T. Boveri in 1889. He strove to fertilize fragments of the ova of *Sphaerechinus* with the semen of *Echinus,* and his technique was not above criticism. Analogous efforts have since been made by Hörstadius under irreproachable conditions, but they have not fully answered the question of the respective parts played by nucleus and cytoplasm. Nor have the androgenetic hybrid products obtained among tritons by Paula Hertwig, by Curry, and by Baltzer lived long enough for us to draw final conclusions from their examination.

On the other hand Baltzer succeeded magnificently in 1941 with an experiment of this kind on the Mexican salamander *Ambystoma.* Two quite distinct races of this animal are known, one pigmented and the other an albino race totally devoid of pigmentation. These racial differences arise from a difference in a certain gene on one of the chromosomes. Now Baltzer succeeded in fertilizing eggs of the black race from which the nuclei had been removed with spermatozoa of white, or albino, males. The result was what we might have expected: from these eggs with black-race cytoplasm and white-race nuclear chromosomes were born androgenetic larvae which were totally white, in conformity with the racial type determined by the paternal nucleus.

Still more instructive are the researches of Astaurov on the moth of the silkworm. The Russian biologist took two different species of this moth *(Bombyx mori* and *Bombyx mandarina).* He crossed these and submitted the hybrid eggs to the thermal shock of raising the temperature, thus eliminating the maternal chromosomes. In this way he obtained moths which bore paternal chromosomes only. In spite of the presence of maternal cytoplasm their characters were all of the paternal race.

It may be asked how these insects, provided only with paternal chromosomes, can live, since a double set of chromosomes is necessary for normal development. They are viable because, in silkworms, the egg is penetrated by several spermatozoa, two of which fuse to produce a nucleus with a double set of chromosomes. In this case the androgenetic product will have been diploid since its origin, exactly like the product of normal sexual generation. The hereditary role of the chromosomes could not be better demonstrated.

Androgenesis is sometimes referred to as 'male parthenogenesis', a term which is not altogether accurate, since the spermatozoon makes use of the maternal cytoplasm. No artifice is known — or is likely to be known — by which a spermatozoon can be made to produce life without the collaboration of an ovum.

While there is no theoretical reason why women should not one day become mothers without the aid of men, men themselves will always have need of women in order to become fathers. This biological inequality is inherent in the dimensional inequality of the reproductive cells. No matter what progress science makes, the male sex will remain the producer of small cells, incapable of being sufficient in themselves.

Fertilization

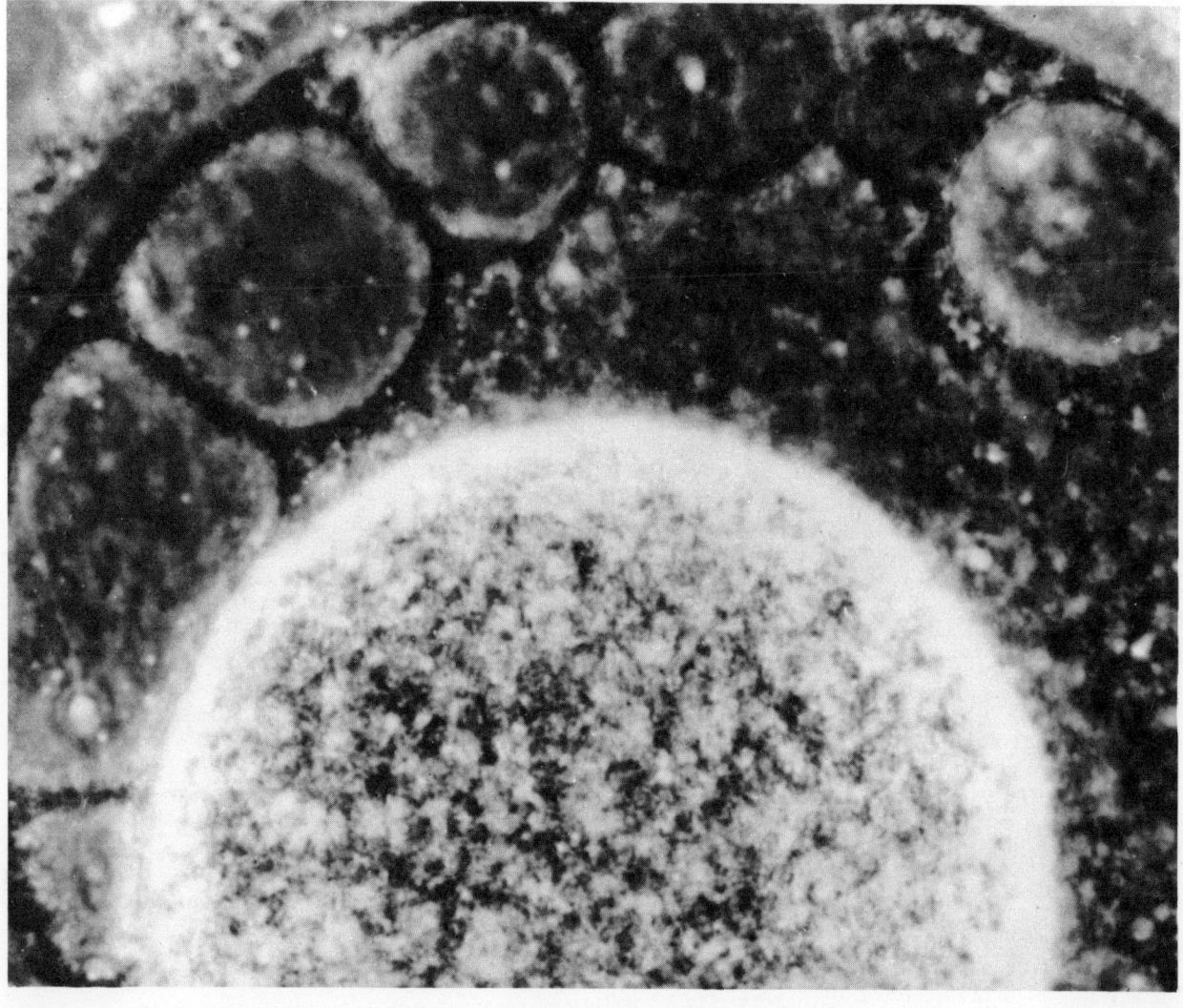

Human ovum, at the moment of fertilization.

Dr. Shettles, with a contrast phase microscope.

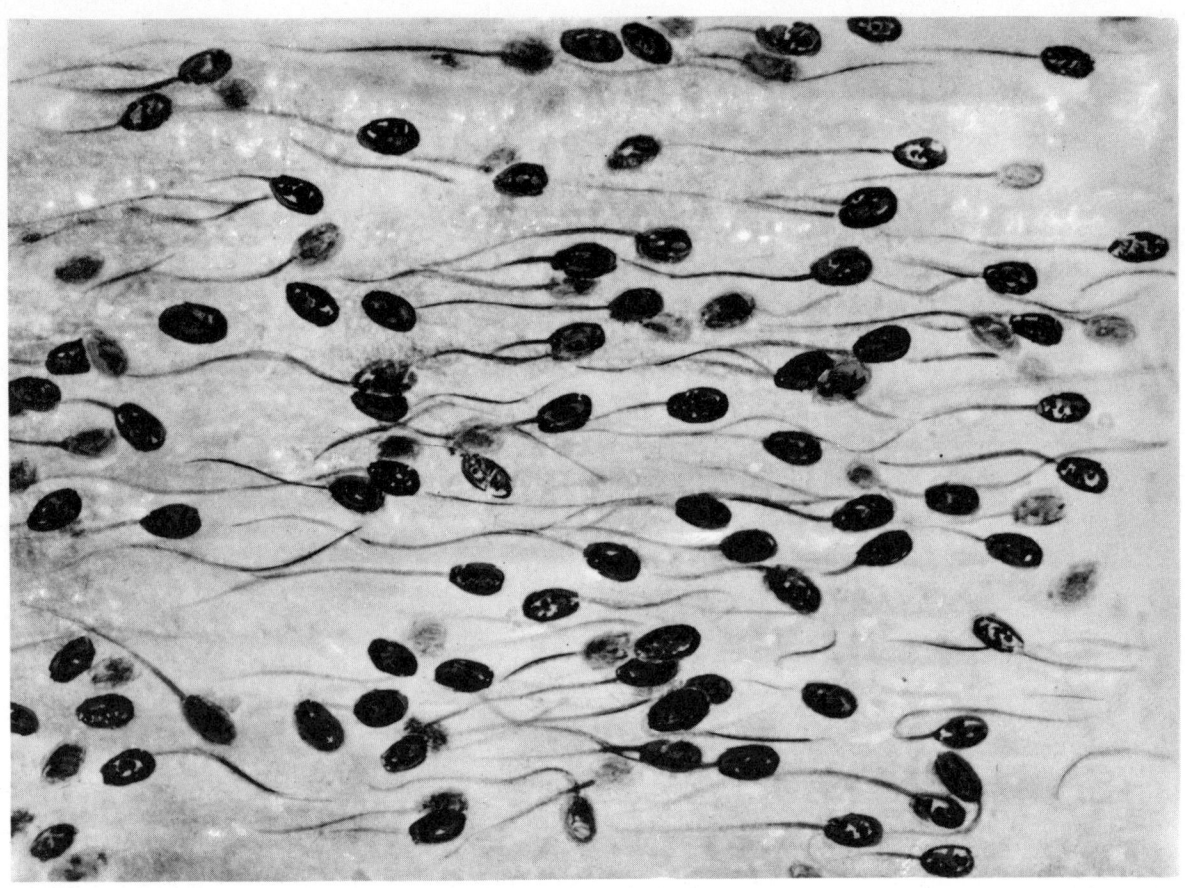

Spermatozoa of a ram.

Atlas-Photo.

As we have several times emphasized, in all organisms which reproduce sexually the new individual arises from a first cell — the fertilized ovum — itself formed by the conjunction of two reproductive cells, the gametes. The female gamete is the ovum, the male gamete the spermatozoon. Since the ovum is much larger than the spermatozoon it is the latter which penetrates the former in order to fertilize it.

The shape of the ovum is more or less spherical or ellipsoid and its size, which depends on the amount of food reserves it contains, varies greatly from species to species. The hen's ovum, which is identical with the egg yolk, measures several centimetres in diameter and weighs some fifteen grams which, for a single cell, is enormous. The ovum of the ostrich is still more voluminous, while the egg yolk — that is to say, the ovum — of the great fossil bird *Aepyornis* measured no less than 17 centimetres in diameter.

On the other hand the ova of mammals are very small: those of mice and rabbits do not exceed two or three tenths of a millimetre, dimensions which are hardly surpassed in the human species, or even among huge animals like the whale.

The annual number of ova produced by the female also varies greatly according to the species: certain fish produce millions, frogs and toads produce several thousands, the hen a few dozen, the majority of mammals only a few, women thirteen.

In general spermatozoa look like simple filaments. Those of crayfish, however, are star-shaped, others are amoeboid, while still others are furnished with undulating membranes, explosive capsules, and so forth.

Almost always very small compared with the ovum, they never exceed some 50 microns in diameter, but attain great length in amphibia of the family Discoglossidae (2 millimetres), in the aquatic *Notonecta* (12 millimetres) of the order Hemiptera, and in the ostracods, minute crustaceans, whose spermatozoa are six times as long as the entire body of the animal.

In all species spermatozoa are produced in great numbers: there are several tens of thousands, if not more, per cubic millimetre. The complexity of their structure seems to be connected with their need to penetrate the interior of the ovum. Roughly speaking a head containing the cell-nucleus can be distinguished, and a tail or flagellum which serves as an organ of propulsion. The encounter of the two reproductive cells can take place either externally — in fresh water or sea-water — or internally, within the genital tract of the female. Ordinarily the ripe ovum cannot await the

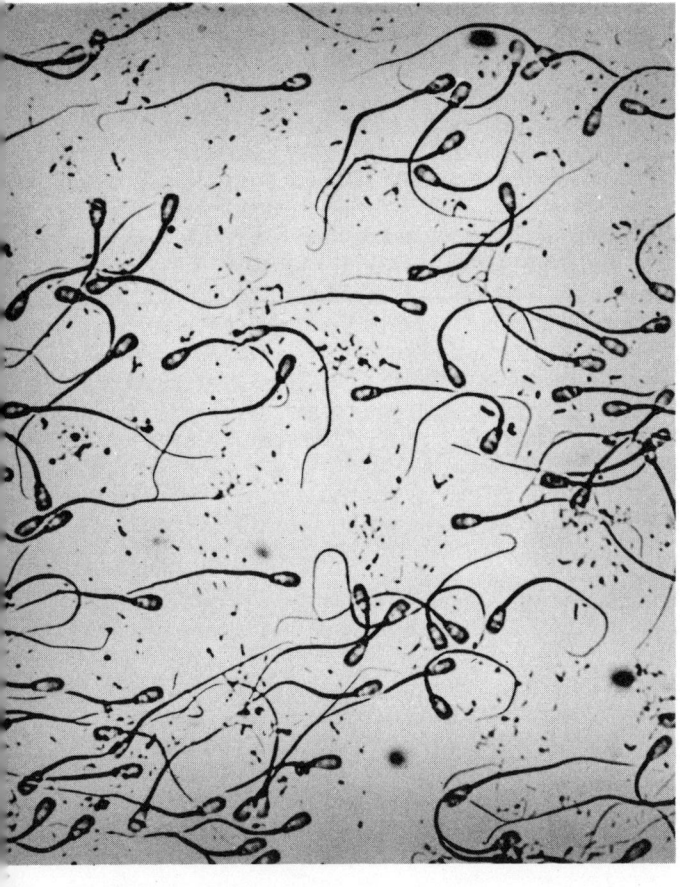

Spermatozoa of a bull.

Bureau of Dairy Industry.

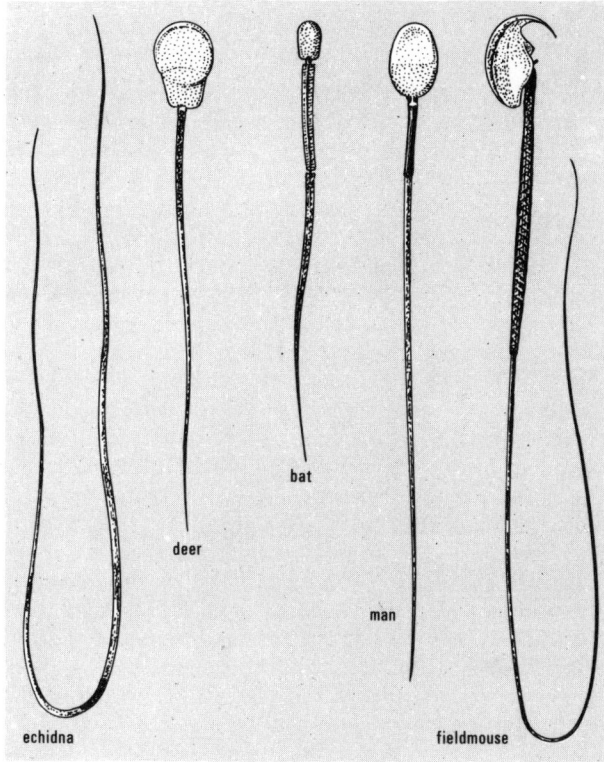

Spermatozoa of various mammals.

After Retzius.

spermatozoon for more than a few hours and, unless joined within the proper time, will die. The life-span of the spermatozoa is extremely variable, ranging from a few seconds in fish to several years in, for instance, bees. The sperm of the cockerel can survive as long as three weeks in the genital tract of the hen; among bats several months elapse between mating, which takes place in the autumn, and fertilization, which only occurs in the spring. The sperm of mammals undergo a sort of 'maturing' in the uterine tubes before they are competent to fertilize. In certain species fertilization is encouraged by secretions of the ova (gamones) which attract or stimulate the spermatozoa.

As a rule a single sperm penetrates the ovum (monospermy) and as soon as it does so the ovum becomes a fertilized egg and cannot be penetrated by any other spermatozoon. The action of the sperm on the ovum is twofold: on the one hand it supplies the fertilizing impulse which stimulates development and on the other hand it contributes the paternal hereditary factors contained in its chromosomes.

Shortly after being penetrated by the sperm the fertilized egg seems to emerge from a kind of torpor. Undergoing numerous physico-chemical modifications it becomes more permeable, increases oxidation, gives

off liquid, etc., and in most cases a fine membrane — the fertilization membrane — forms at its surface.

Meanwhile the spermatozoon loses its tail, and its head swells in a characteristic fashion: it assumes the aspect of a nucleus (the male nucleus) which soon joins the nucleus of the ovum (the female nucleus) to form a nucleus of mixed origin which becomes the nucleus of fertilization or the nucleus of the egg. While the nuclei of the two gametes — sperm and ovum — carried only a single set of chromosomes, the nucleus of the egg or zygote carries a double set. A few hours later the nucleus of the egg divides, thus initiating the first of the many cell divisions which culminate in the formation of the new individual. At the same time that the egg prepares for its first cleavage a rearrangement of its inner structure takes place, and the vague outlines of the future organism take shape. This phenomenon has been studied with particular care in the egg of the frog, in which indications of the embryo's bilateral plan of symmetry can be seen shortly after the entrance of the spermatozoon. Before this time the egg has revealed only 'polarity', its upper hemisphere being characterized by a relative poverty in food materials. An hour and a half after fertilization it not only possesses an upper and lower half but a right and a left side.

139

Polyspermy and fertilization

We have said that the fertilization of the ovum is achieved by a single spermatozoon. If, however, the ovum is submitted to abnormal conditions (too high or too low a temperature), or if the semen is treated with salt solutions (chlorides or bromides), the ovum can be made to accept several spermatozoa: artificial polyspermy has then been induced. In spite of the fact that several sperm have entered the ovum, the nucleus of only one of them can unite with the nucleus of the ovum. The nuclei of the remaining spermatozoa remain isolated and unmated, each dividing on its own account. Thus, in the embryo of polyspermic origin two kinds of nuclei are formed: those derived from the nucleus of fertilization and those derived from the nuclei of the supernumerary spermatozoa. The latter have only a single set of chromosomes, the paternal set, instead of the double set, maternal and paternal, carried by the former. Because of this the chromosome constitution of the new being is heterogeneous. Furthermore, since the spermatozoa of a single individual are all genetically different, the sperm nuclei themselves differ one from another. Thus in certain regions of the embryo a genetic heterogeneity will result.

The very early death of all organisms of polyspermic origin is most probably due to this double heterogeneity, quantitative and qualitative. Among frogs, in particular, polyspermic larvae cannot live for more than about ten days. The greater the number of supernumerary sperm involved in its genesis the sooner the larva dies.

Apart from artificially induced polyspermy, natural or physiological polyspermy is known to exist among various animals: insects, fish of the order Selachii, amphibians of the order Urodela, and reptiles. In natural polyspermy, as in the laboratory variety, a single sperm nucleus unites with the nucleus of the ovum, but the extra nuclei introduce no complications since they take no part in subsequent development. Instead they remain inert in the protoplasm, do not divide, and little by little are dissolved and digested by the egg to which they contribute only nourishment.

Since in artificial polyspermy the embryo contains nuclei derived from supernumerary sperm 'polyfertilization' — as Brachet called it — may be said to have occurred. In natural polyspermy nothing of the sort takes place, because the nuclei of the supernumerary sperm do not participate in the formation of the embryo. No kind of 'polyfertilization' is known, at least among vertebrates, which is compatible with normal development. No-one has ever seen the birth of an individual in whose genesis several spermatozoa have cooperated.

Human sperm and ova

The human spermatozoon consists essentially of a head, a short intermediate section and a long tail. The head is 5 microns long and the tail some 50 microns. Thanks to the rhythmical beating of its tail the sperm can, in an appropriate environment, swim at the rate of about

1-5 millimetres per minute. Spermatozoa are extremely numerous: about 300 million are discharged at a time varying from sixty to 100 million per cubic centimetre. When the age of genital maturity is reached sperm is formed and ceaselessly renewed in the testicular glands.

Physiologically, it is not a matter of indifference that these glands are externally situated; for the production of spermatozoa would, in fact, be impossible within the abdominal cavity where the temperature is too high for them. If, by chance, the testes — which in the embryo are intra-abdominal — do not descend into the scrotum, the seminal part of the gland degenerates and only the secretory part persists. The individual, known as a cryptorchid, will then be devoid of sperm and hence quite sterile, but due to male hormone he will acquire a masculine appearance and instincts.

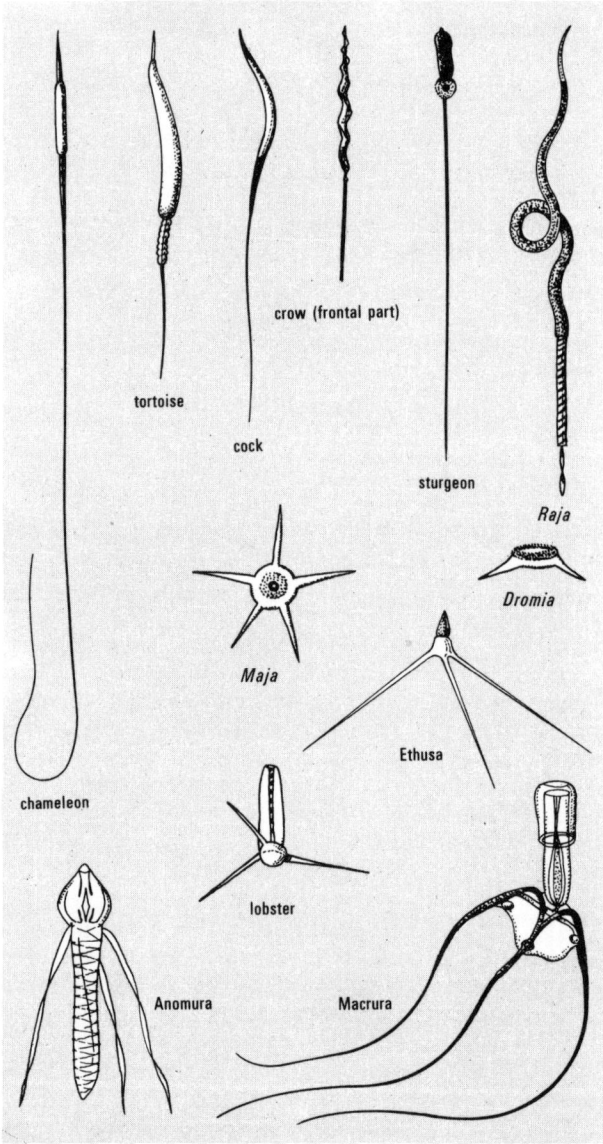

Spermatozoa of birds, reptiles, fishes, and invertebrates.
After Wilson.

The human ovum is 10,000 times greater in volume than the human spermatozoon. From the age of puberty until the menopause the ovary is, each month, the centre of a regular cycle, comprising the evolution of a Graafian follicle, a fluid-filled spherical vesicle which contains an oocyte, or immature ovum. When the follicle has reached a certain point in its growth it bursts and releases the mature ovum. The rupture of the follicle and the discharge of the ovum (ovulation) coincide with the middle of the intermenstrual period — that is, about fourteen days before menstruation begins.

When the ovum leaves the ovary it enters the Fallopian tube; it is here that it will be fertilized if it meets a spermatozoon. Once fertilized the egg descends into the uterus, to the wall of which it becomes fixed during *nidation,* and where all its subsequent development takes place. The first division or cleavage of the human ovum has been estimated to occur about thirty-six hours after it has been penetrated by the spermatozoon, but the exact time is difficult to establish. Menkin and Rock tried to fertilize, *in vitro,* ova taken from ripe follicles and believe that they have succeeded with four ova out of 138 so treated.

At the same time as the follicle is evolving important changes take place in the hormone equilibrium of the female organism. The follicle itself secretes a hormone — oestradiol — which causes the swelling and congestion of the uterine lining. One of the following alternatives is then possible:

1. If the ovum has been fertilized and if the egg has accomplished its nidation the ruptured Graafian follicle turns into the *corpus luteum,* a yellow tissue or body which secretes progesterone, the hormone that maintains the uterine lining in the state required for gestation, stimulates the development of the placenta, and at the same time prevents a new follicle from maturing until gestation has finished.

2. If, on the contrary, the ovum has not been fertilized, the *corpus luteum* formed in the place of the Graafian follicle soon degenerates. In this case a new follicle can develop during the following month, while the regression of the *corpus luteum* brings about the collapse of the uterine lining with subsequent discharge of blood — or menstruation.

By no means all ovarian follicles mature and emit their ovum. Of the hundreds of thousands which each foetal ovary contains, only a few hundred mature during the reproductive life of a woman, usually one each month for roughly thirty-five years.

The menstrual cycle, the average duration of which is twenty-eight days, is controlled by the activity of a small endocrine gland, the pituitary body, situated at the base of the brain. The pituitary gland or *hypophysis cerebri* is sometimes called the 'leader of the endocrine orchestra' and produces various hormones of considerable importance, many of which are concerned with sexual life, some determining the evolution of the Graafian follicle or the *corpus luteum,* others stimulating the development of the mammary glands. It is not, however, the pituitary gland which, in the event of gestation, determines the maintenance of the *corpus luteum* in a functional state, but one or several hormones secreted by the embryo itself.

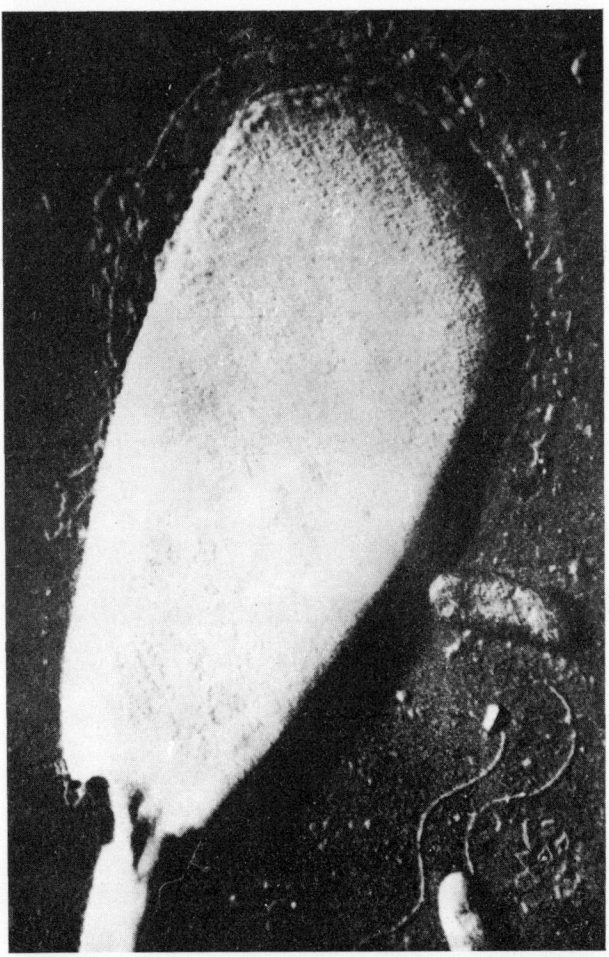

A spermatozoon of a bull. Above is the head and below the extremity of the tail.

After H. L. Bretxlmeider and W. Van Sterson, 1947. Electron microscope negative.

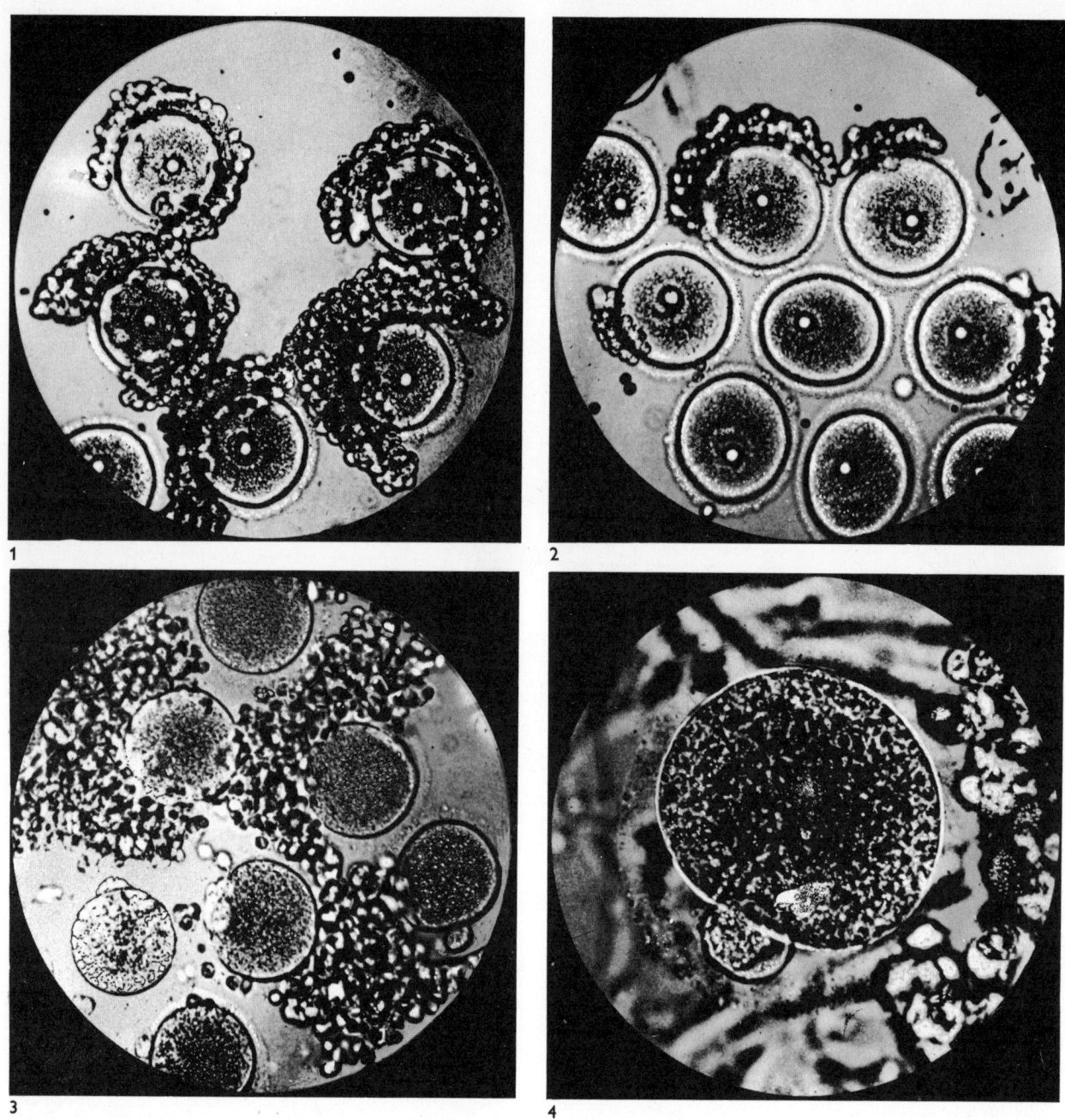

1

2

3

4

The gonadotrophic hormones

Among the hormones of the pituitary, those which stimulate the growth of the ovarian follicles are secreted by the anterior lobe of the gland. They are called the gonadotrophic hormones — or gonadotrophins — to indicate their general action on the sex glands, or gonads, of the two sexes.

They intervene in the sex life of all vertebrates, and it is very simple to demonstrate their stimulating power by injecting a male or female subject with a small amount of the appropriate pituitary hormone. The resulting superactivity of the sex gland is unfailing and

characteristic. By these means animals which have not reached puberty can be made to ovulate. From very young mice mature ova have been produced which, when fertilized and transferred to the uterus of adult females, developed normally.

The administration of a pituitary extract to tailless amphibians has effects which are very instructive. If two or three pituitary glands of the frog are crushed in a few drops of saline and the liquid thus obtained injected under the skin of a female frog in a state of sexual repose (in other words during the autumn and winter months) she will, within thirty-six hours, produce ova which, on leaving the ovaries, accumulate in the uterine sacs from which they can be made to

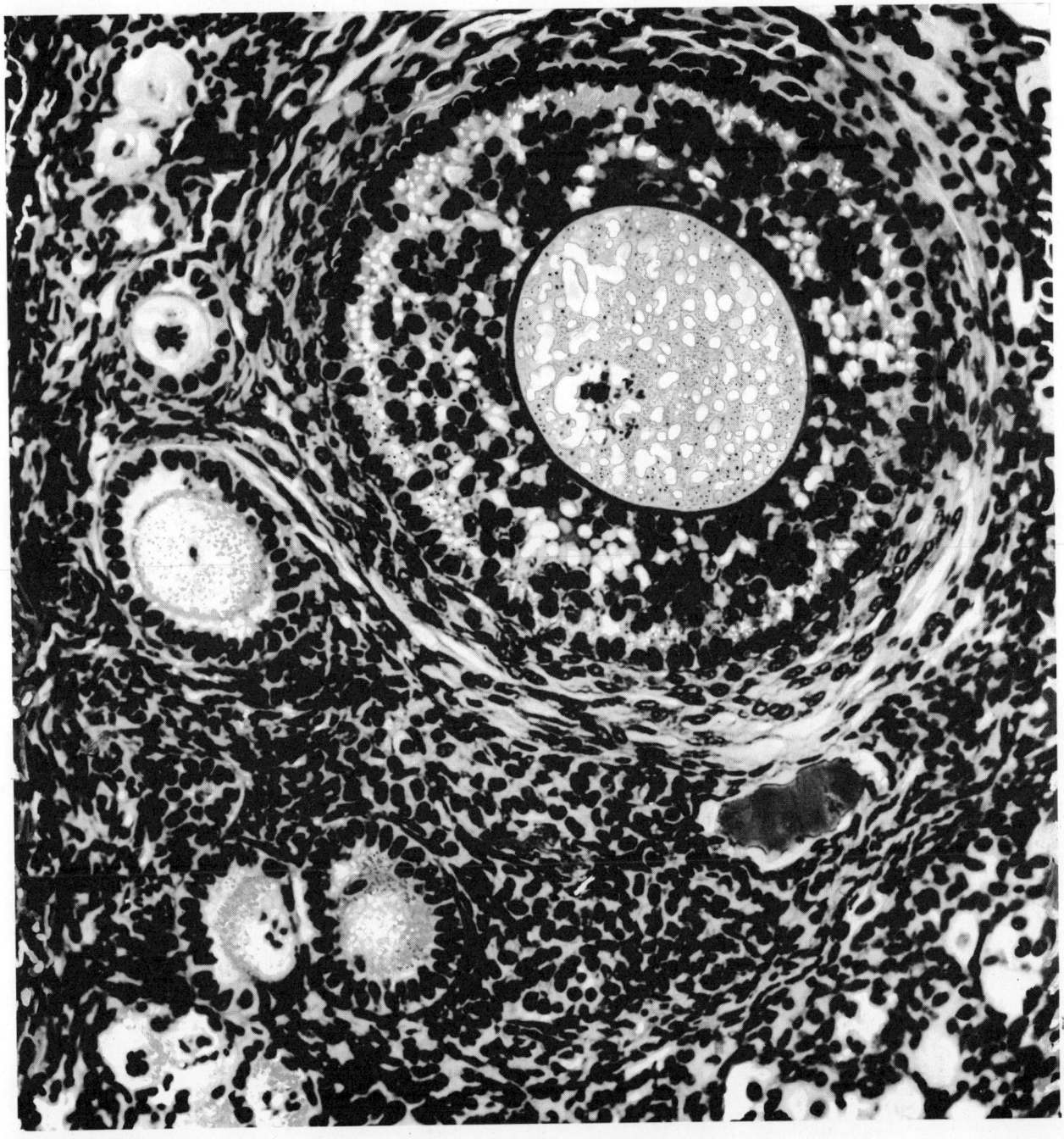

Opposite, 1. and 2. Ova of a mouse, removed from the dissected ovary.

3. Ova of a mouse. These ova were cultivated in the serum of an adult mouse. The nucleus has divided and the first polar body has been expelled.

4. Ova of a mouse after the second polar body has been expelled. The ova are now capable of being fertilized.

After Moricard, Palais de la Découverte collection.

Above, section of ovarian follicle in the ovary of a rabbit.

Poilpot.

emerge by pressing the animal's flanks. These same eggs, in the absence of pituitary treatment, would have remained in the ovaries until the following spring. If a male frog is similarly injected the sexual reflex will occur, together with the emission of semen.

It is not known exactly how the pituitary hormone acts on the ovarian gland. Certain biologists have claimed that the action can be observed *in vitro* on ova removed from the frog's body and placed in a suitable solution to which the hormone has been added. The reality of the effect is, however, far from having been proved. The chemical composition of the gonadotrophic hormones is still ill defined; they belong to the family of the polypeptides.

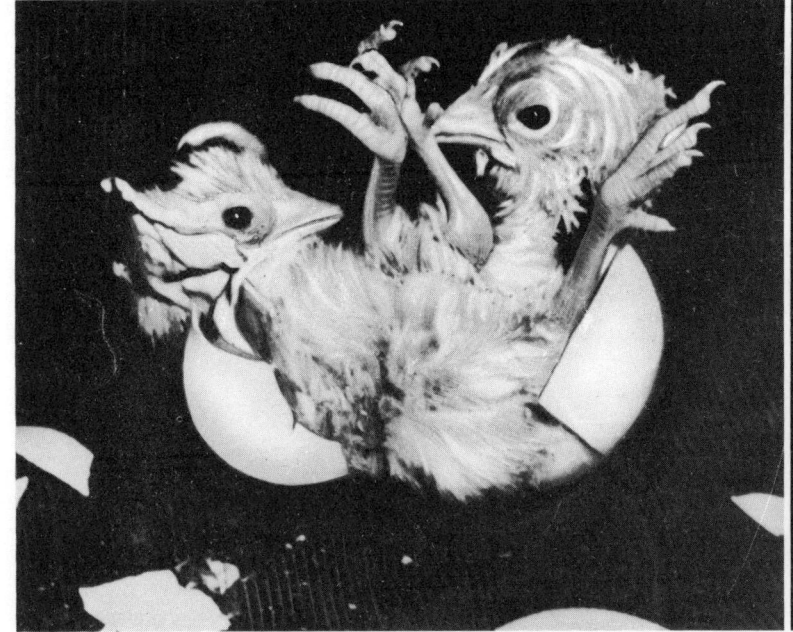

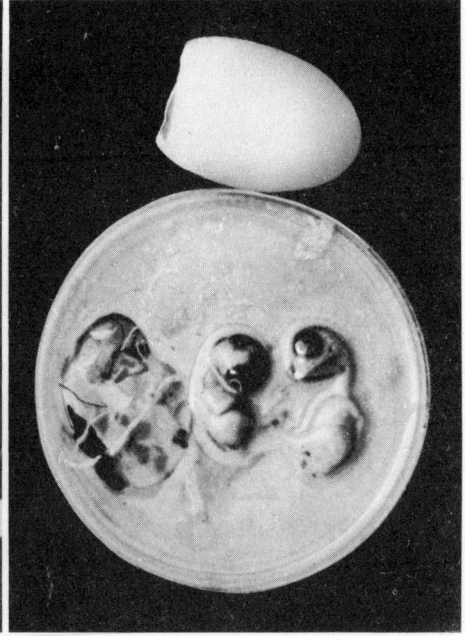

Rabbits are used in a wide variety of research programmes and they are shown here waiting for their daily medical.

B.B.C.

On the left, the egg has just hatched and two chicks appear, born of two yolks. *On the right,* the very rare case of an egg with three embryos.

Agence intercontinentale. X Photography.

The biological diagnosis of pregnancy

Women from the first weeks of pregnancy produce, and through their urine eliminate, certain hormones which have similar properties to the pituitary gonadotropic hormones. These, in consequence, are easily detected by the effect they have on the sex glands of certain animals. Thus it is possible to diagnose pregnancy very early and to be correct ninety-eight or ninety-nine times out of 100. The urine of a woman presumed to be pregnant is injected into an immature female rabbit, a young male mouse, a male toad or a male frog. In the case of the young rabbit typical changes in the ovary are observed at the end of a few days. In the case of the male toad or frog the passage of semen in the urine is observed within a few hours. By removing the semen with a fine glass pipette and examining it under the microscope the wriggling spermatozoa can easily be seen.

This diagnosis furnishes doctors with invaluable information when it is a case where urgent intervention is necessary, and in cases where one must differentiate between pregnancy and a fibroma, a cyst or any other tumour.

Superovulation

The gonadotrophic hormones have been employed in animal breeding to increase the number of ova simultaneously produced by the female, and hence the number of her offspring. In this way among sheep where the ewe usually bears only one lamb at a time, multiple births can be caused by the injection of such hormones. The urine of a pregnant mare, unusually rich in gonadotrophic hormones, is generally used for the purpose.

The work of the Russian biologist Zawadowsky with tens of thousands of animals of different breeds has shown the practical value of super-ovulation in cattle raising. Forty-seven to fifty per cent of sheep so treated produced twins, while six to eight per cent produced triplets, and one to two per cent had between four and six lambs at a time. With cows comparable results have been obtained.

Valuable as it may be, the method of super-ovulation has its drawbacks and must not be applied without certain precautions, for it may happen that females are sterilized by hormone treatment. Charles Thibault contributed interesting facts on this point by establishing with careful statistics that to avoid harmful effects the gonad-stimulating hormone must be administered only at a certain moment in the animal's reproductive cycle.

Mortality, moreover, is rather high among twin lambs and even higher among triplets and quadruplets, which reduces the benefit of the treatment except in the case of Bukhara lambs whose pelts are commercially valuable at birth.

The technique of super-ovulation has not yet been deliberately applied to the human race, although so-called 'fertility drugs' have had this effect. There is no *a priori* reason why it should not be effective, but as the exact results cannot be accurately determined the candidate for twins might risk having triplets or even more!

145

Influence of light on the development of the sex organs.

The experiments of Professor Benoit, Palais de la Découverte collection. Larousse.

1. The testes achieve their full development in the duck submitted to intense lighting.

2. The testes remain undeveloped when the animal is kept in the dark.

In natural circumstances the bearing of false or fraternal twins is almost certainly connected with a slight over-activity of the pituitary body.

Eggs with two yolks

Among birds super-ovulation can result in the laying of eggs with two yolks. This phenomenon occurs when the eggs are emitted so rapidly that the oviduct has not sufficient time to manufacture the albumin of which the white is composed, or the shell required by each egg. Certain breeds of hen are more subject than others to laying eggs with multiple yolks. In one well-known case a hen in old age began to lay enormous eggs with two yolks every three days. In another, ten eggs were laid by the same hen, from which nine sets of twin chicks were hatched.

The pituitary gland and light

It is by the intermediary effect of the pituitary, and the gonadotrophic hormones it secretes, that light exercises on the sexual glands of certain animals those remarkable effects which were demonstrated by the researches of Bissonnette and above all of P. Benoit, to whom we owe a series of most instructive experiments on the domestic duck.

Ducks kept in the dark from infancy develop their sex glands only after a long delay. The same delay occurs if, though given access to the light, their heads are completely protected by a hood of dark material. On the other hand, if the whole body is covered by a protective coat and only the head exposed to the light the sex glands develop. For this to happen it is enough that the eyes alone receive light. If the optic nerves are severed light no longer has its gonad-stimulating effect.

The light rays reach the retina of the eye and the pituitary gland then functions. This sensitivity of the

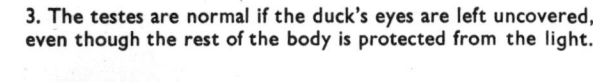

3. The testes are normal if the duck's eyes are left uncovered, even though the rest of the body is protected from the light.

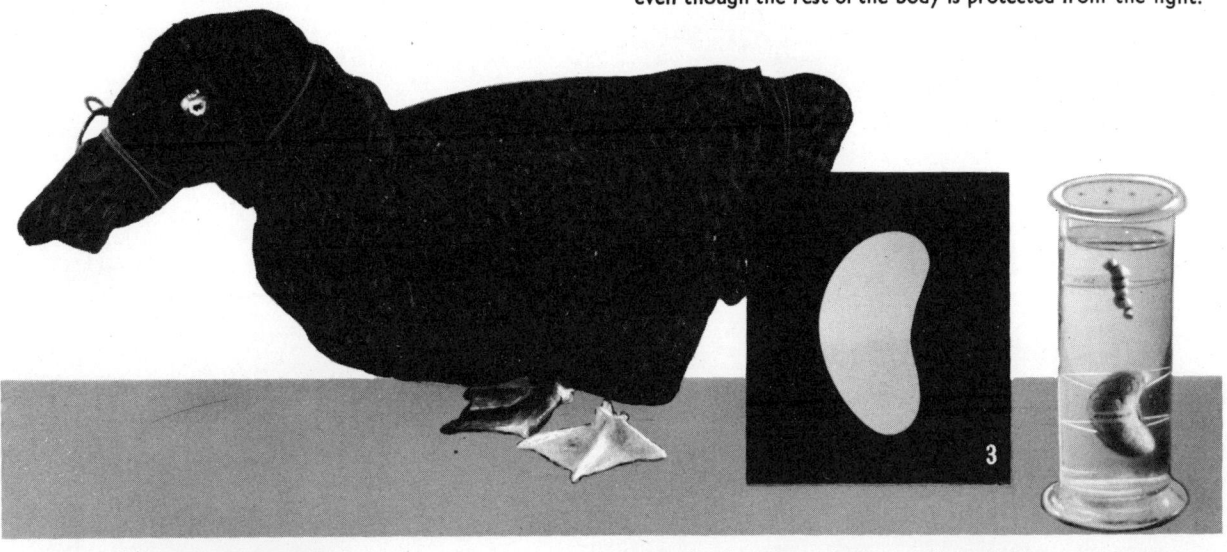

4. The testes remain undeveloped in ducks whose bodies are exposed to the light but whose eyes are masked. Thus it is the ocular region which is sensitive to the light: the light excites the pituitary gland, whose gonad-stimulating hormones bring about the development of the testes.

retina is not, however, the same as its visual sensitivity, for among ducks the retina is most sensitive to yellow and it is not the yellow light rays which stimulate the duck's sex glands.

It has been experimentally demonstrated that the pituitary gland of ducks which receive a proper amount of light is richer in gonadotrophic hormones than that of ducks kept in the dark. Analogous facts have been observed among other birds, and also among several mammals including ferrets and rabbits. In man it would seem that light has only a slight action on the functioning of the sex glands, but reliable data on the subject are lacking, although it is thought that the menstruation of Eskimo women becomes irregular during the long polar nights for similar reasons.

Other sensory stimulants, olfactory, auditory, and tactile, can affect the functioning of the pituitary and hence the functioning of the sex glands. But such effects, less striking than those of light, have been much less studied. Sound stimuli, in any event, have little effect on ducks; Benoit and Assenmacher have without result exposed immature ducks to a radio playing twelve hours a day for a month.

The presence of another member of the same species can sometimes have a gonad-stimulating effect. Among female pigeons the sight of another pigeon, or even the sight of herself in a looking-glass, will stimulate ovulation. Social relations, amorous display, and courtship hasten the moment of coupling as much among mammals as among birds. In the human species it is commonly agreed that psychological and social factors can accelerate puberty and stimulate sexual activity. It has even been suggested that the lowering of the age of puberty in recent generations can be partly attributed to the eroticism seen in films, books, and on television.

Artificial insemination

Artificial insemination may be defined as the putting of the ova and sperm of a given species into contact with each other with a view to bringing about fertilization. No difficulties are involved in the operation among oviparous animals like amphibians and fish

147

Filling pellets with diluted semen.

Centre d'enseignement zootechnique, Rambouillet. Larousse.

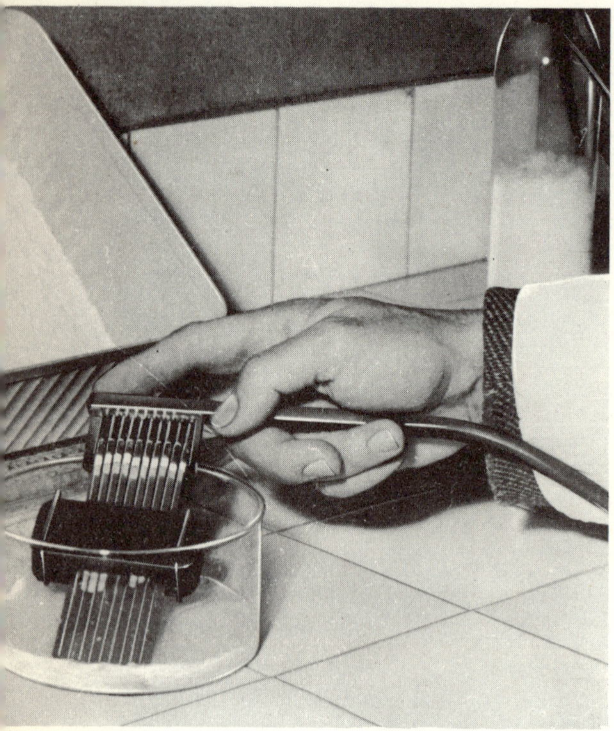

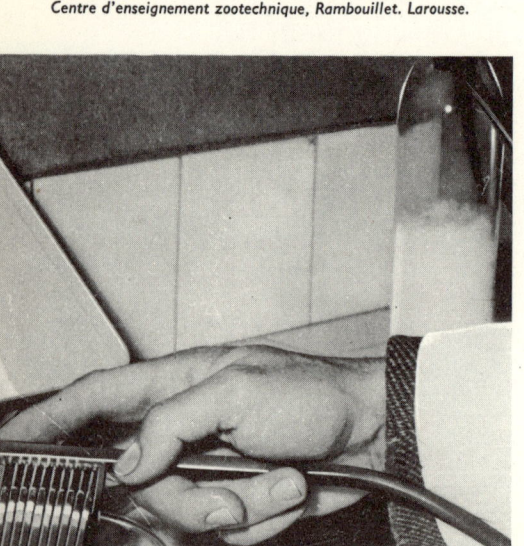

Artificial insemination by means of the 'vaginascope'.

"Vie à la campagne".

Collecting bulls' sperm for use in artificial insemination.

Despoisses. Atlas-Photo.

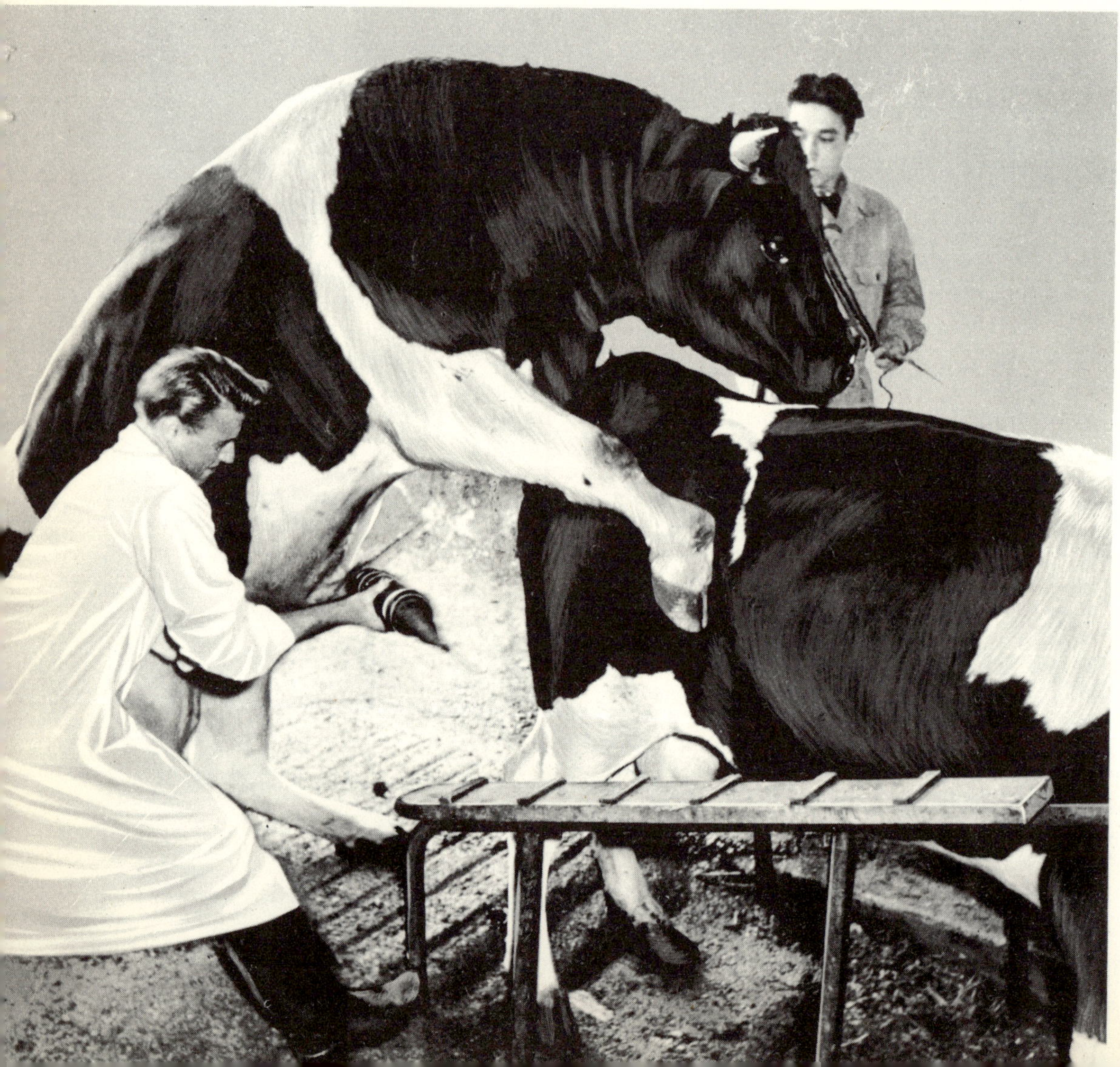

An artificial vagina for cattle.

Centre d'enseignement zootechnique, Rambouillet. Larousse.

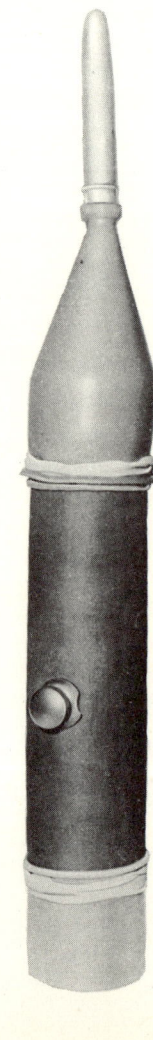

whose fertilization is external. With trout, for instance, sperm is collected during the spawning season by squeezing the male's flanks. It is then poured over the eggs, which are afterwards well sprinkled with water. Artificial insemination is currently practised by fish breeders, and a few drops of semen is all that is needed to fertilize hundreds of eggs.

Among viviparous animals like mammals whose fertilization is internal the operation is obviously more delicate, since, after having collected the semen, it must be introduced into the genital tract of the female. It is nevertheless practised with considerable success, providing that certain precautions are taken, especially that of keeping in mind that the female produces her ova only at definite intervals. The ova rapidly lose the capacity for receiving male sperm and insemination during the wrong season will of necessity be fruitless.

The first artificial insemination of a mammal was the work of the celebrated physiologist Lazare Spallanzani who in 1780 fertilized a spaniel bitch by means of a syringe, and obtained three puppies resembling the dog which had furnished the semen. Following this memorable experiment Charles Bonnet wrote the following lines to Spallanzani:
'I rather think that what you have just discovered may one day have undreamed of applications in the human species, the consequences of which will not be trivial...'
This prophecy was realized in 1781 when Hunter undertook the first artificial insemination of a human being. His example was soon to be followed by numerous doctors.

Among animals this method was, from the end of the last century, increasingly employed and owed much to the work of Russian biologists and stock breeders, as well as to the French veterinary surgeon Repiquet (1887). Today artificial insemination is widely practised in all civilized countries.

Its advantages are many. In the first place it assures a prompt genetic improvement of livestock by permitting the maximum exploitation of selected sires. For instance, 1,500 cows were fertilized in one year by the sperm of a single bull, and 15,000 ewes by the sperm of a single ram. In addition, artificial insemination allows us to overcome sterility in certain females and to prevent the spread of diseases which are transmitted during mating: trichomoniasis, foot and mouth disease, granular vaginitis, brucellosis, equine syphilis. In spite of a widely held belief to the contrary, artificial insemination results in an appreciably higher percentage of success than natural insemination.

The techniques employed obviously vary greatly according to the kind of animal inseminated. Semen can be gathered in different ways, either by fitting a 'rubber collector' to the animal's penis, or by causing the male to ejaculate into an artificial vagina at the moment he is about to mate. The male can often be taught to copulate with a 'dummy' simulating a female, or even with a simple trestle or a bench padded with horsehair. Laplaud reports the case of a ram which would arrive and paw at the door of the shed where the semen was collected.

With rams and boars electric stimulation is sometimes employed. One electrode is inserted into the rectum of the animal while the other is applied to the skin, at the level of the fourth lumbar vertebra. This technique has been simplified by using a single two-point electrode in the rectum.

Once gathered the semen is diluted in a liquid of appropriate composition — one part semen to from ten to fifteen parts liquid. The volume of semen gathered varies considerably from animal to animal. Approximate quantities are:

Bull	3— 4	cubic centimetres
Horse	75—150	cm³
Pig	200	cm³
Dog	7	cm³
Ram	0.8	cm³
Rabbit	0.7	cm³

The semen of the bull is much richer in spermatozoa (800,000 per cubic millimetre) than that of the horse (600,000) or of the pig (100,000).

Although a single spermatozoon is sufficient for each ovum a certain concentration of sperm is necessary to ensure fertilization. The sperm is injected into the vagina of the female by means of a glass tube attached to a syringe, or with an inseminating 'pistol'. Before being injected it is carefully inspected, not only for the quantity of spermatozoa it contains, but for the quality of the spermatozoa themselves. This is estimated by their mobility and by certain coloration tests.

Semen is generally used from between twelve to forty-eight hours after collection. It can be preserved in a sterile tube in a refrigerator kept at 5 °C. It is, of course, injected into the female at the moment when fertilization is most likely to take place. This depends on the species. Among cows, twelve to twenty-four hours after they are first on heat and among mares, three days afterwards.

A medium frequently employed for preserving semen is Watson's medium, composed basically of egg-yolk to which phosphates are added. But other substances will serve to dilute the semen, cow's milk, isotonic sea-water, and so forth. Sometimes the semen is 'gelatinized', which allows it to be placed in the vagina in solidified cartridges.

Semen can be kept alive for at least several days and even as long as three weeks. It can therefore be shipped, and males of superior pedigree can thus reproduce at a distance. For example, the semen of a ram has been collected in Cambridge, put into a thermos flask and sent to Poland by air. Two days and three hours later the semen was injected into a Polish ewe who in due course bore healthy lambs which resembled their distant English sire. Other 'long range' inseminations have taken place between the United States and the Argentine, between San Francisco and New York, Missouri and New Jersey, North America and Italy, and so on.

It is rarely realized that the first experiment in *telegenesis* was made towards 1880 by the anthropologist Vacher de Lapouge. In his book on *Social Selections* the following curious passage may be found: 'Semen can be transported: in one of those idiotic experiments which Darwin recommends I obtained in Montpellier the fertilization of a rabbit with semen sent from Beziers by post.'

Conservation of semen for longer periods

The semen of some animals can support very low temperatures, at least for short periods of time. Thus the fertilizing cells of the male rabbit will for a few minutes resist the temperature of solid carbon dioxide ('dry ice') —79 °C below zero.

Some years ago a method was discovered of preserving semen for long periods at low temperatures by the use of glycerine as a protective agent. The first experiments of this nature were due to J. Rostand who worked with the semen of the frog. The semen of the frog can resist freezing temperature only for a few hours, but if it is placed in a glycerine medium of one part glycerine to nine parts semen it will, without perishing, support a temperature of 6 °C below zero for some twenty days. Similar results have been obtained with the semen of other batrachians such as toads and water salamanders.

Parkes and his colleagues, English biologists, have largely and brilliantly developed this method and its applications are steadily being extended. They have shown, for instance, that the semen of the cockerel can resume its fertilizing activity after having been submitted in a glycerine medium to a temperature of minus 79 °C.

The effects observed in this case are truly spectacular. If the cock's semen is diluted with an equal part of pure Ringer's fluid and then chilled to a temperature of minus 79 °C for twenty minutes the spermatozoa, when thawed out, show no signs of revival. But if the dilution is made with Ringer's fluid containing fifteen per cent glycerine the spermatozoa recover all their original mobility. As far as movement is concerned there is no difference between semen thus treated and identical semen not treated: it still shows the same wave-like motion that is characteristic of a cock's seminal fluid. If the glycerine is then eliminated by slow dialysis the 'vitrified' semen can be successfully used for artificial insemination.

Chicks have already been obtained with semen kept for thirty-three days at a temperature of 79 °C below zero. In the same way cows have been successfully inseminated with semen frozen for eight days. In 1950 Bunge and Sherman achieved the artificial insemination of a woman with semen which had been preserved for six months. The baby which was born of one of these 'vitrified' spermatozoa was perfectly normal.

The creation of 'semen banks' on a large scale can henceforth be seriously envisaged. Some biologists have even gone so far as to state that it is the duty of the human race immediately to store seminal reserves which can be sheltered from atomic radiation in the grim event of such radiation irreparably damaging the genetic patrimony of mankind.

Artificial insemination of poultry and bees

Artificial insemination is also applied to poultry with good results. The semen is gathered either by massaging the part of the cock's abdomen through which the *vas deferens* passes, or by fitting a collector, usually a glass cupel, to the bird's cloaca. The semen of birds survives only for a brief time and must be used at once, unless it is preserved at 5 °C on a sterile coating of vaseline, when it will last for several hours. Alternatively, if it is stored at a temperature of minus 79 °C after the addition of glycerine it will last several months.

Even bees, in spite of their small size, can be artificially inseminated. Watson was the first to succeed in collecting the semen of the male (the drone) and using it to inject into the queen bee. His method was perfected by Laidlaw, and by Mackensen and Roberts. The queen, stupefied by a whiff of carbon dioxide, is injected with semen obtained by simply pressing it from the abdomens of five or six males which have been previously chloroformed. The operation is successful sixty-five to eighty times out of 100; but queens fertilized in this way use up their seminal reserve at the end of six months. After that time they lay only virgin eggs and, in consequence, can give birth only to males.

Artificial insemination of the human race

In the human species artificial insemination was first used in 1781 in order to enable a man suffering from a malformed penis, known as hypospadias, to procreate. But it was not until the opening of the twentieth century that it began to become current medical practice, having survived much opposition.

Many cases of sterility, due to malformation of the wife's or of the husband's sexual equipment, can be overcome by artificial insemination. Furthermore, by employing the semen of a donor (hetero-insemination) the woman whose husband is sterile can become a mother. Even when the semen used is that of the husband there are even today, objections of a religious nature to artifical insemination. When the semen of a donor is used these objections are much stronger, and the question becomes a moral or a social issue.

Transplantation or grafting of ova

The ova of one female have, for many years now, been successfully transplanted to another female. The technique was developed especially by Raymond E. Umbaugh of the University of Texas, who worked with cows. In the field of the production of milk and meat, he reported that genetic characters determined the quality and the economic value of cattle. Artificial insemination therefore enabled desirable male genes to be widely distributed by means of semen. A comparable genetic selection to raise the quality of females is much more difficult, since a cow produces only one calf a year and the average number of calves she produces during the course of her existence rarely exceeds five. Thus it was thought possible that the offspring of cows of high quality might be appreciably increased by causing them to 'super-ovulate', and by transferring the ova in this way produce less valuable cows to serve as 'nursing mothers' to the embryos. Super-ovulation can easily be stimulated by the action of gonadotrophic hormones administered as wax pellets or by intravenous injection. In this way an average of 23.4 ova per cow can be obtained.

When the ova have been fertilized (by artificial insemination) they are transferred into the uterine tubes of another female who is on heat. There they accomplish their development and in due course give birth to healthy calves. Naturally the hereditary characters of these calves are in no way changed by the fact that their embryonic development took place within the womb of a cow which was not their genetic mother. For the stock breeder, then, no disadvantage results from the fact that the 'ova-carrying' or 'nursing' cows are of ordinary, unselected quality.

The first attempt to 'graft eggs' was made by the French physiologist Paul Bert, a pioneer in animal grafting. 'To my learned master, M. Gratiolet,' he wrote in 1863, 'I owe the idea of an experiment which in execution presents the greatest difficulties. The problem is to take from the Fallopian tubes or the uterine cornua of mammals fertilized ova, but nothing else, and to transfer them into the peritoneum of another animal in order to see if the eggs will develop. If the two animals were of the same species and the operation well performed the experiment would very probably be successful. But what if the animals were of different species? Would the egg of one species develop in the abdominal cavity of another very closely related species — that of a wharf-rat, for instance, in the peritoneum of an ordinary rat? In my opinion it seems likely that it would. And, if such is indeed the case, would it not be possible to connect those strange apparitions which so puzzle naturalists with some modification of the egg at the moment of its formation? Unfortunately I have not succeeded in grafting eggs.'

This passage from Paul Bert is remarkable not only because in it he suggests the experiment of grafting eggs but also because he plainly raises the problem of the influence of a foreign maternal medium on the embryo's development. The experiment was, in fact, successfully made in 1890 by Heape with rabbits. Fertilized ova of pure Dutch breed were placed in the uterine tubes of a female rabbit of the variety known as the Belgian Hare, which had previously been mated with a male Belgian Hare. She gave birth to two kinds of young. Some, of pure Belgian Hare variety, were strictly her own; while others, of pure Dutch breed, had obviously developed from the transplanted ova.

Later, the technique of grafting eggs was improved by Hammond, Nicholas, Chang, etc., and employed with mastery by the Americans G. Pincus and V.

Enzmann who also chose rabbits as a subject for experiment.

Doubling the number of chromosomes artificially

We have already spoken of those individuals or of those races of animals as well as of plants which carry in their cells several sets of chromosomes instead of the two sets which characterize normal organisms. This phenomenon, known as polyploidy, is found fairly often in nature, and can be brought about at will by artificial means.

It was among plants that the doubling of chromosomes was first achieved, and by a very simple chemical process. It is sufficient to immerse plant seeds in a dilute solution of a certain poison — colchicine, an alkaloid extract of autumn crocus. Under the influence of colchicine the mechanism of cell-division is disturbed in such a way that the chromosomes divide within the nucleus, while the nucleus itself does not divide. Thus cells are formed which contain twice the ordinary number of chromosomes: instead of the normal two sets (diploid) they have four sets and are tetraploid. If the germinal tissue is thus affected it will produce diploid gametes instead of the normal haploid gametes (with a single set of chromosomes). Two diploid gametes, when they join, will of course form a tetraploid zygote (or fertilized ovum) which will grow into an individual with the same tetraploid chromosome formula. If two tetraploid individuals unite they can become the point of departure for an entire race of the same type.

By the repeated action of colchicine even octoploid — with eight sets of chromosomes — individuals and races have been obtained.

Instead of plunging seeds into the solution of colchicine the young plant can be sprayed with the solution or, again, the buds can be daubed with vaseline containing colchicine. There are many ways of inducing polyploidy and treatment varies in efficacy and convenience from species to species.

In France Marc Simonet applied these methods to flowers like the iris and to various vegetables such as the tomato and the aubergine or egg-plant. Interesting 'creations' have resulted, for polyploid plants are often bigger and more vigorous than normal plants.

Other chemical agents besides colchicine have proved capable of producing similar effects. Such are phenylurethane, acenaphthene, naphthalene, oil of lemon, and apiol (which is an extract of parsley seeds).

According to Häggqvist and Bane in 1950 colchicine may also bring about polyploidy in animals. They believe that they have obtained rabbits which are triploid — that is with three sets of chromosomes — and are larger than normal rabbits, as a result of insemination with semen treated with colchicine. The value of their results is, however, still controversial. On the other hand there is no doubt that animal polyploidy can be produced, and even easily produced, by chilling or heating the recently fertilized egg.

151

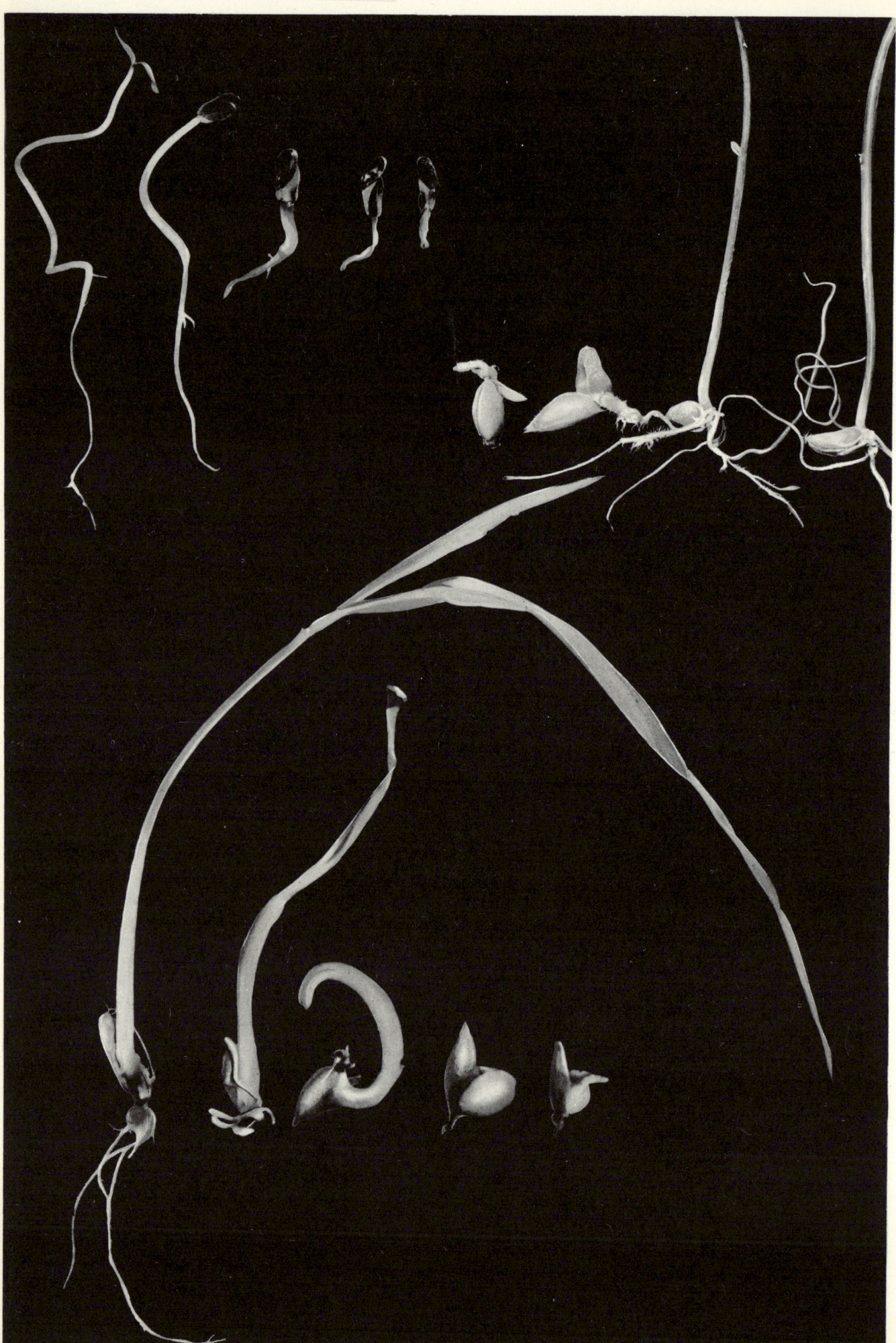

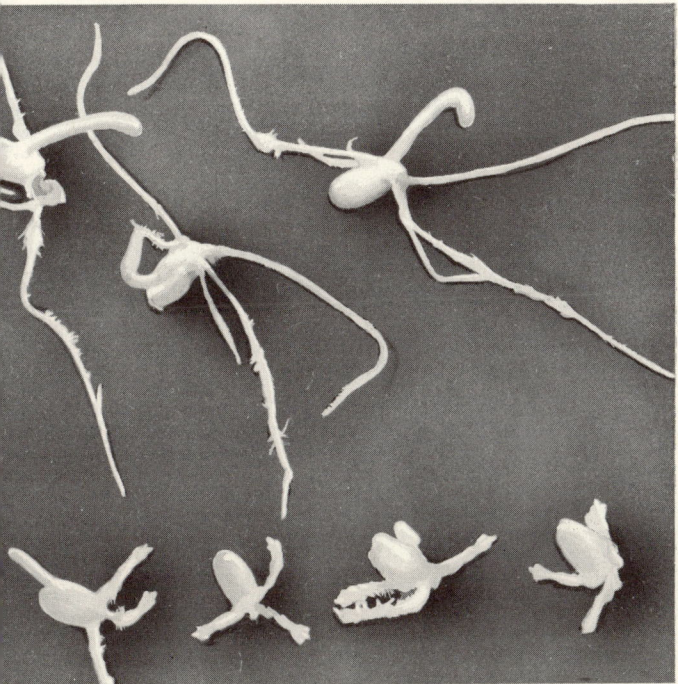

Opposite page from top to bottom
Flax seeds treated with colchicine. *From left to right,* normal control seed; colchicine strengths 0.01, 0.02, 0.05, 0.10. Germination is increasingly retarded, while the plantlet thickens.

From left to right, seeds treated with acenaphthene; with colchicine; control seeds.

Barley seeds treated with colchicine. *Left to right,* control; colchicine at 0.01, 0.02, 0.05, 0.10. Again the colchicine slows down germination and causes the plantlet to thicken.

Top left, the flowers of the *Arabis alpina,* tetraploids; *top right, Arabis alpina,* diploid (2N) *below,* and *above,* tetraploid (4N). The tetraploid flower is appreciably larger.
Above right, Gaillardia. *On the right,* normal gaillardia, double and diploid (2N = 36); *on the left,* tetraploid (4N = 72), flowers of the same variety after treatment with colchicine. The tetraploid flowers are larger.

J. Vincent.

Above left, the germination of wheat. Three control plantlets with, beneath them, four plantlets treated with colchicine.

After Simonet.

153

This process has since 1938 been current practice with amphibians, in for example the experiments of Fankhauser and Griffiths, inspired by those of J. Rostand. If shortly after being laid, that is to say shortly after being fertilized — for in this case the egg is fertilized at the same moment as it is laid — the eggs of the water salamander or triton are submitted to a bath of iced water they will produce as many as seventy-five to eighty per cent triploid larvae. This method, so simple and effective, has been extended to other salamanders — *Ambystoma* — and also to frogs and toads. The treatment is ineffective, however, unless it is applied within half an hour of the eggs' being fertilized. Very probably it acts by suppressing the ovum's reduction division which among amphibians occurs only after the ovum has been penetrated by the spermatozoon. Triploidy would thus be due to the doubling of the ovular chromosomes.

Heating the egg is no less effective than chilling it: eggs of the triton or of the frog, heated to 35-37 °C for about fifteen minutes will yield a high percentage of triploids — eighty-four per cent.

Fankhauser and Humphrey succeeded in rearing triploid salamanders to the age of sexual maturity. Males with the triple set of chromosomes are sterile, but the females can reproduce when coupled with normal males. In their offspring a few tetraploid subjects appear.

Both by chilling and by heating ova Beatty and Fischberg have obtained embryos of triploid mice. The embryos have reached the age of nine and a half days, but so far it has been impossible to keep them alive long enough for them to reach maturity.

Since the number of ovular chromosomes is doubled in the triploid egg it was obvious that efforts should then be made, by combining polyploid treatment with hybridization, to create individuals that carried two chromosome sets of the maternal species and only one set of the paternal species. Such individuals should presumably reveal an accentuation of maternal characters when compared with normal hybrids. This, in fact, was what Fischberg found when he crossed tritons, and G. Hertwig confirmed when he crossed toads. Triploid subjects were obtained by crossing a female *Bufo bufo,* or common toad, with a male *Bufo viridis,* or green toad. The hybrids carried a double set of maternal chromosomes and a single set of paternal chromosomes and were, indeed, more like the common (maternal) toad than are normal hybrids of the same two species. As the experiment demonstrates, the biologist can by this process create composite organisms which have a double dose of one species and a single dose of another.

Polyploidy and gigantism

An interesting problem is that of the connection between the size of an organism and its chromosome equipment. Broadly speaking polyploidy in animals as in plants brings about an increase in the size of the nuclei and of the cells, an increase roughly in proportion to the number of chromosomes the nucleus contains. The effect of polyploidy on the size of the

Above, on the left diploid and on the right tetraploid flowers of *Iberis.* The photograph on the right shows diploid (on the left) and tetraploid flowers of the same variety of daisy. In both cases the tetraploid are larger than the normal diploid flowers.

J. Vincent.

Larger polyploid ear of rye.

Diploid ear of rye.

actual organism varies greatly according to the species under consideration. Triploid and tetraploid moths *(Solenobia)* are distinctly larger than diploids of the same species. Triploid woodlice, *Trichoniscus,* tetraploid and octoploid *Artemia,* also show a certain degree of gigantism. On the other hand polyploidy in no way alters the general dimensions of silkworms and amphibians, for in these organisms cellular gigantism is compensated for by a corresponding reduction in the total number of cells. In a diploid silkworm, for instance, there are 680 cells in the silk-producing glands, while the same gland in a triploid silkworm has only 443. The reason for this singular phenomenon of regulation is still unknown.

Polyploidy does not alter the organism's speed of growth, or faculty of regeneration; it does not affect its sensitivity to the hormones of growth or of metamorphosis. In the triton it appears to cause only a difference in pigmentation, arising from the fact that the pigmentary cells (melanophores) are larger and less numerous in polyploid animals. Beyond a certain degree of polyploidy — pentaploidy or five sets of chromosomes — the animal's reactions and vitality are reduced. Even among triploids intelligence seems to be slightly diminished. According to Fankhauser and his colleagues triploid tritons learn to find their way through an experimental labyrinth a little less quickly than ordinary tritons — which is explained by the fact that their nerve-cells are less numerous. In practice the greater size of the cells does not, it would seem, compensate for their reduction in number.

Hybridization

Everyone knows that the sexual union of two unlike species is always sterile, whether it occurs in the course of natural coupling or whether it has been brought about by artificial insemination. Today no-one would try to cross a dog and a cat as did the illustrious Abbé Spallanzani towards 1780. Even less would we, like Réaumur, cause a male rabbit to cohabit with a hen in the hope of obtaining rabbits covered with feathers or chickens clad in fur!

While agreeing that a too heterogeneous union cannot bear fruit, the biologist cannot foresee with absolute certainty what will be the issue of a cross which has not yet been attempted. It has been found that fairly distantly related species will occasionally unite to bear offspring while two closely related species — sometimes even two breeds of the same species — will not.

There are many different events that can occur when two gametes — ovum and sperm — belonging to different species are brought into contact. Most frequently the spermatozoon will not penetrate the ovum, and the ovum will remain perfectly inert, receiving no influence from the fertilizing cell. This is what happens if, for instance, the sperm of a cock is put in contact with the ovum of a bitch or the ovum of a frog. It can happen, however, that the spermatozoon does not penetrate the ovum, but nonetheless has a slightly stimulating effect on it by simple external contact. The ovum is thus activated. It begins to divide, then almost immediately stops. The phenomenon is

155

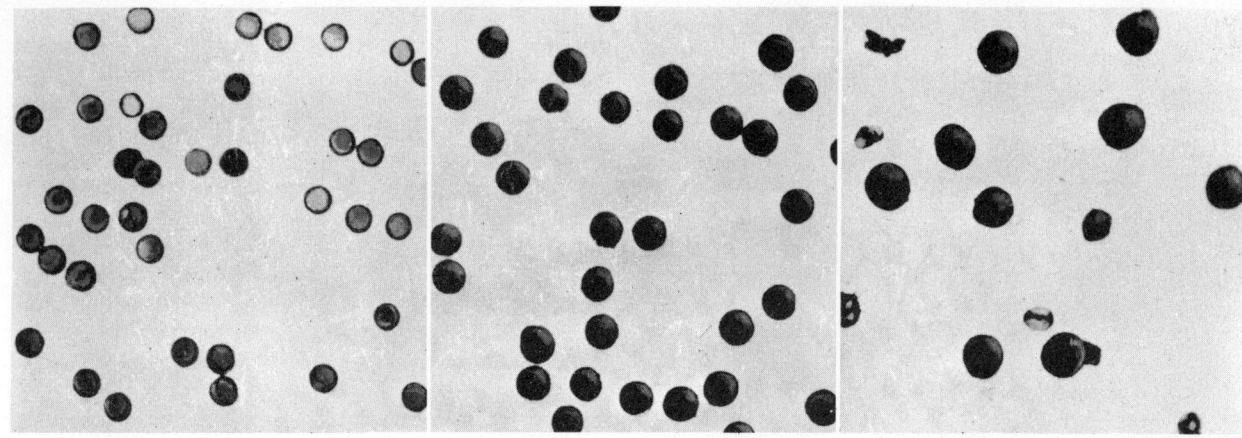

Pollen of *Datura stramonium*. *From left to right,* diploid (2N), tetraploid (4N), octoploid (8N).

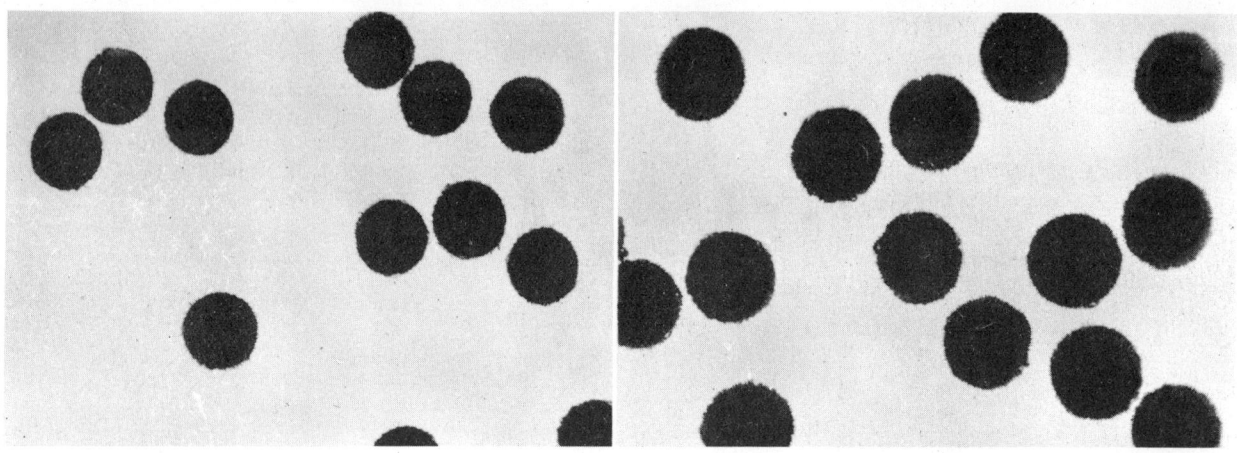

Pollen of diploid and tetraploid gourd. The size of the pollen grain is roughly proportional to the number of chromosome sets it contains.

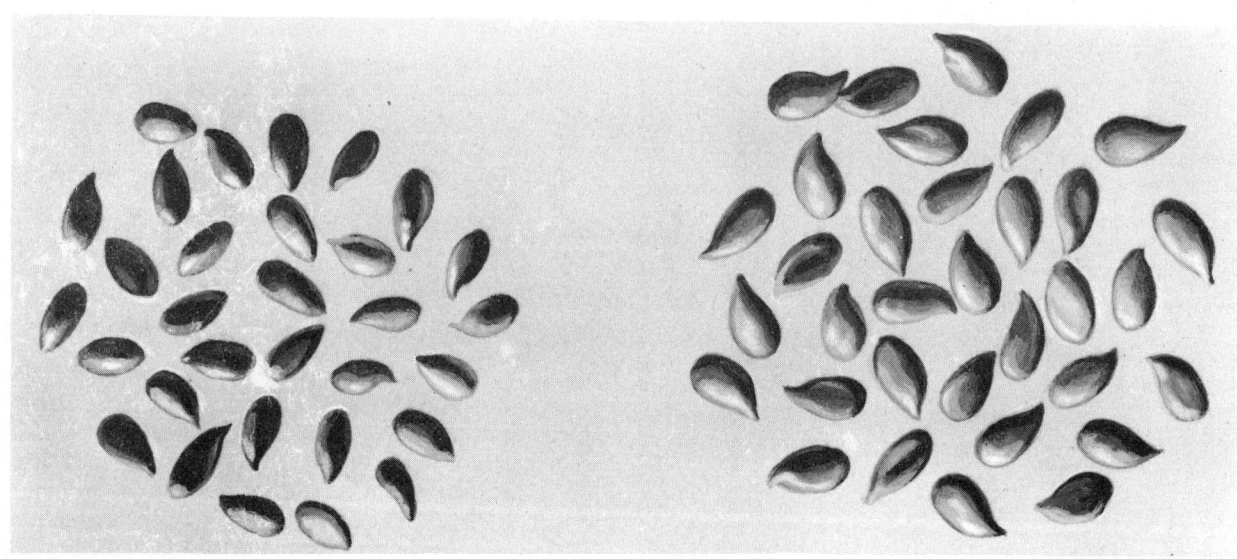

Left, diploid linseed with 30 chromosomes; *right,* tetraploid linseed with 60 chromosomes.

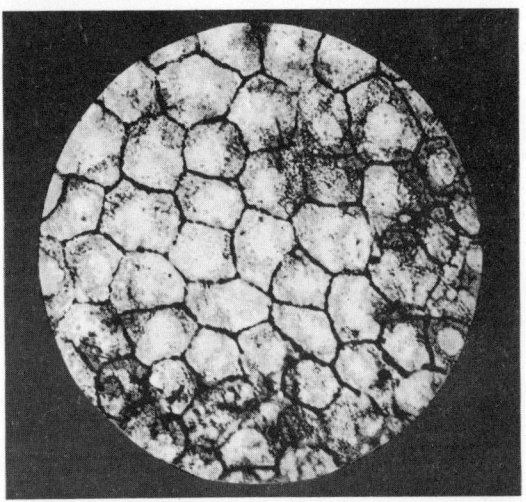

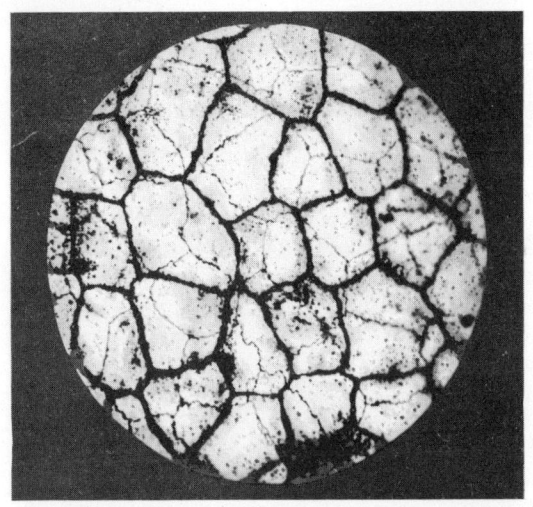

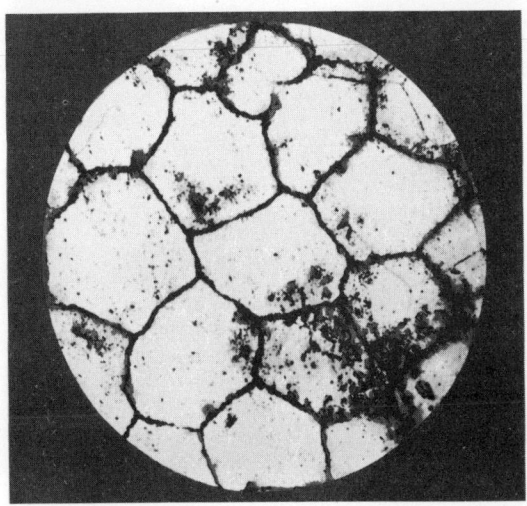

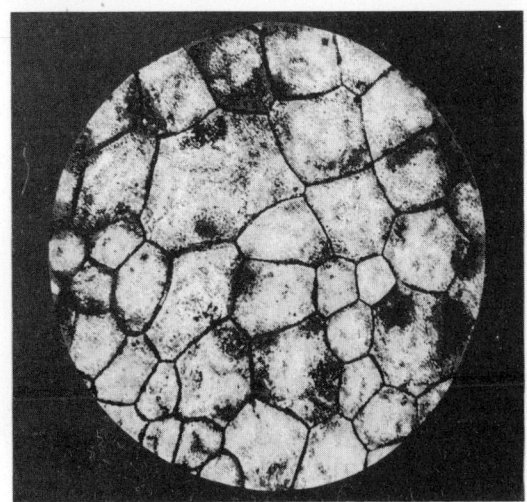

Top left, cells of the tail of a haploid larval *Pleurodeles* (N = 12). Cells with a single set of chromosomes are smaller than normal cells with two sets.

Top right, cells of the tail of a normal or diploid larval *Pleurodeles* (2N = 24). By a special technique the boundaries of the cells are made clearly visible.

Above left, tail cells of a triploid *Pleurodeles* larva (3N = 36). Cells with three sets of chromosomes are larger than cells with the normal double set.

Above right, tail cells of a *Pleurodeles* larva of heterogeneous constitution, (mosaic or chimaera). According to regions they differ in the number of chromosome sets they contain, and hence in size.

After Gallien and Mugard. Palais de la Découverte collection. Larousse.

Trichoniscus provisorius, an isopod crustacean. This little animal exists in two forms, one diploid and reproducing sexually and the other triploid and parthenogenetic. *Above*, the female diploid; *below*, the female triploid, which is noticeably larger.

After Vandel.

157

observed when, for example, the ovum of the toad is put into contact with the sperm of a triton. Only very attentive examination by a specialist reveals any difference between this and the first case. In some cases the spermatozoon enters the ovum and brings about its development without, however, contributing its paternal chromosomes. The result is that curious phenomenon of 'false hybridism', which has already been discussed in connection with gynogenesis. The spermatozoon penetrates the ovum and cooperates in its embryonic development. In this event there is a whole range of possibilities, from early arrest of development to development which continues until a viable and normal individual is produced.

When a common female toad is crossed with a common male frog the eggs develop normally until the so-called *blastula* stage of the embryo is reached. This, in normal conditions of temperature, takes thirty-six hours of evolution. From then on the embryo degenerates and soon dies.

Much the same thing will occur in various crosses between bony fishes, or when the domestic rabbit is crossed with the wild rabbit. The hybrids of minnows and goldfish last at most for twelve days. In the case of the common frog, Moore demonstrated that crosses between races which come from widely separated regions can result in such faulty embryonic development that the larvae are completely abnormal and incapable of achieving metamorphosis. Here racial difference alone is sufficient to create incompatibility between hereditary factors.

The conflict of hereditary factors is somewhat slow in making itself felt when a sheep is crossed with a he-goat: development can continue at a normal rate until the forty-fourth day of gestation, but no young will be born. It has, indeed, often been claimed that this cross can result in viable offspring; until now, however, no convincing proof seems to have been advanced that such hybrids actually exist.

Crossing the domestic cock and the turkey hen — or the hen and the turkey gobbler — gives rise to embryonic development which can approach the hatching stage, but never ends in the birth of live birds.

Sometimes cross-breeding is possible in one direction and not in the other: the egg of the common toad, fertilized by the sperm of the common frog, produces

The tigron, a hybrid of the tiger and the lioness. Its characters are intermediate between those of the two parent species.

Larousse.

158

an embryo; while the egg of the frog is not even penetrated by the sperm of the toad. In the same way the egg of the green frog will produce an embryo after being fertilized by the sperm of the common frog; but the egg of the common frog cannot be penetrated by the sperm of the green frog.

The eggs of the green toad, fertilized by the common toad, give normal or almost normal hybrids. The eggs of the common toad, fertilized by the green toad, give larvae which perish two weeks after birth.

Successful crosses

Among fruitful crosses between different species or genera the following may be listed: lion and tiger (the tigron); jaguar and leopard; wild ass and donkey; horse and she-ass (the hinny); mare and he-ass (the mule); stallion and female zebra (the zebrinny); mare and male zebra (the zebrula); horse and quagga; zebra and onager; bull and American bison (the catalo); bull and yak; yak and zebu (the ozo); sheep and moufflon; domestic goat and Himalayan ibex; wolf and dog; dog and jackal; dog and coyote *(Canis latrans);* brown bear and polar bear; marten and polecat; wild Brazilian guinea-pig and domestic guinea-pig; macaque and mandrill; macaque and *Cercopithecus;* cock and pheasant-hen; cock and peahen; cock and guinea-hen; peacock and guinea-hen; goldfinch and canary; muscovy duck and common duck, and many more. Some of these crosses have been achieved by means of artificial insemination, as sex attraction does not always exist between the species cross-bred.

It is still considered impossible successfully to cross the dog and the fox, the rat and the mouse, the mouse and the fieldmouse, the hare and the rabbit, although it is not known what causes sterility.

Controversy has long raged over the possibility of a fertile union between the hare and the rabbit. Buffon, who took much interest in problems of animal hybridization, was the first to attempt crossing the two species. He reared male rabbits with female hares, and male hares with female rabbits, but, he wrote, 'these attempts came to nothing and only taught me that these animals whose form is so similar are nevertheless of a nature so different that they do not even produce — like the horse and donkey — a hybrid.'

The hybrid of a Muscovy drake and a domestic duck.
Théret.

Left, the embryo of a hybrid (turkey hen and domestic cock) and, *on the right,* the normal embryo of a turkey cock. After 22 days of incubation the hybrid embryo is more advanced than the normal embryo is after 24 days.

159

Foal produced by a mule and a she-ass. Mules are very rarely fertile.

Théret.

Characteristics of hybrids

Buffon believed he could, however, affirm 'that the rabbit males sometimes really coupled with the female hares; at least it was often certain that the male, in spite of the female's resistance, was satisfied.'

The characteristics of hybrids are, in general, somewhere between those of the two parent species. Among them there is often a certain structural variability. Thus mules and hinnies sometimes have 'chestnuts' on all four legs like horses, and sometimes only on the forelegs. They sometimes have five lumbar vertebrae like the ass, and sometimes six like the horse.

The products of a given hybridization can differ according to how the cross is made: thus the he-mule, which is the offspring of jackass and mare, is a little different from the hinny, the offspring of she-ass and stallion. The reason for these differences is not yet fully understood: they may arise from the effect of maternal cytoplasm which is not the same in the two cases, or again from the effect of the uterine and humoral environment in which the foetus was developed.

Hybrids, though generally vigorous are often sterile, both among themselves or when bred to one or other of the parent species. The inability of he-mules and hinnies to produce offspring has been known from the days of Aristotle at least. And yet both kinds of animal show every sign of sexual ardour. The males produce semen but it does not contain spermatozoa. This inability to produce inseminating fluid is, at least in part, due to the dissimilarity of the parents' chromosome equipment, a dissimilarity which is a hindrance to the pairing off of chromosomes which — as we have already seen — precedes reduction division.

If he-mules are always and basically sterile, the same is not true of she-mules. Buffon long ago spoke of a fertile she-mule who was said to have given birth six times between 1763 and 1776. Phenomena of this sort are not especially rare, at least in Mediterranean countries like Spain, Italy, and North Africa. When a fertile mule is mated with a stallion she produces, it would seem, colts of pure or nearly pure type; when she is mated with an ass, however, she produces typical he-mules.

There is one famous recorded example of an Algerian mule which in 1873 gave birth to a female. Both

The hybrid of a pheasant and a domestic hen.

Théret.

The hybrid of a he-goat and the ewe. This cross is often held to be impossible.

Théret.

mother and daughter were examined by a veterinary surgeon, together with the sire, a Barbary stallion. All three were bought by the *Jardin d'acclimatation* in Paris and their full history there recorded. The mare mated with the same stallion again and in 1874 produced another female offspring. The characteristics of both 'fillies' were purely equine and in no way resembled those of an ass. The mule was next bred to Egyptian asses and gave birth to two males, one in 1875 and the other in 1878. Both these animals were typical he-mules, in every way like he-mules born of jackass and mare. In 1881 she produced her last offspring — a typical colt by her original partner, the Barbary stallion.

The two mares proved fertile and coupled with stallions produced equine colts, which, however, did not live long. The first he-mule proved to be sterile, exactly like a typical he-mule. The colt proved to be perfectly fertile and sired, with a Tarbes mare, a filly which was in all points normal.

The second-born mare was studied by the veterinary school of Lyons, where she was found to be a complete reversion to the equine type, the influence of the ass 'grandparent' being apparent in hardly more than the

disposition of the vocal apparatus.

To explain this rapid extinction of heredity deriving from the ass the following hypothesis has been advanced.

The cells of the she-mule contain a set of chromosomes contributed by the horse and a set contributed by the ass. When the reproductive cells — the gametes — mature they receive a single set of chromosomes. Now it appears that more or less only the equine chromosomes are retained to the detriment of the chromosomes of the ass. The mule, then, daughter of a mare, would behave from the genetic point of view as though she were herself a mare. The case has been reported of a female hinny who proved fertile after union with a he-ass; the product was a typical ass.

If the preceding hypothesis is admitted, it was, in this case, the chromosomes of the ass which were retained by the ova of the hinny to the detriment of those of the horse. Thus the hinny, daughter of a she-ass, would behave from the genetic point of view as though she were herself a she-ass.

Among hybrid animals the normal sex ratio is often characteristically unbalanced. Thus the cross between the domestic hen and the golden pheasant

161

cock results in a super-abundance of males. The same phenomenon occurs when goldfinches and canaries are crossed.

Even in cases where males and females are produced in equal numbers one of the sexes may be stricken with partial or total sterility. Haldane's law tells us that in the hybrids of mammals it is usually the male descendants which are sterile, while in hybrid birds and moths the reverse is true.

In the human species — in which all races can successfully interbreed — the cross between Whites and African Blacks produces a marked excess of females. The possibility of a hybridization between man and any other species of the animal kingdom is impossible because of the zoological distance between them. The closest resemblance is between man and the great ape, which is certainly very pronounced from the anatomical point of view, and extends to the form of the spermatozoa, the number of chromosomes, the way, in which ova divide, and so forth. But in spite of these similarities the distance is far too great to be compatible with fruitful cross-breeding. At the very most all that could possibly occur would be that a human spermatozoon might penetrate the ovum of an ape (or vice versa) and in that way give rise to gynogenesis. In this event the result would be a false hybrid, purely ape or purely human according to the case.

A theory was held among geneticists of the last century of 'telegony', in other words, that of a female, first fertilized by a male of a race or species other than her own, giving birth by another mate to offspring which exhibit features inherited from her first husband. This has now been almost universally dismissed as sheer illusion.

The hybrid of the turkey cock and the guinea-hen.
Barraud.

It has been shown, however, that when a mare is carrying a mule embryo antibodies, or agglutinins, form in the bloodstream as if humoral conflict had broken out between the two genetically dissimilar organisms.

The agglutinins pass into the mare's milk, especially into her first milk (colostrum) and if they become too concentrated they can give the young mule serious jaundice.

The way of avoiding this is both simple and efficacious: all that is needed is to prevent the newborn animal from suckling, for in the course of its first feed it will swallow an enormous quantity of the antibodies. It is therefore removed and supplied with a substitute mother.

Development

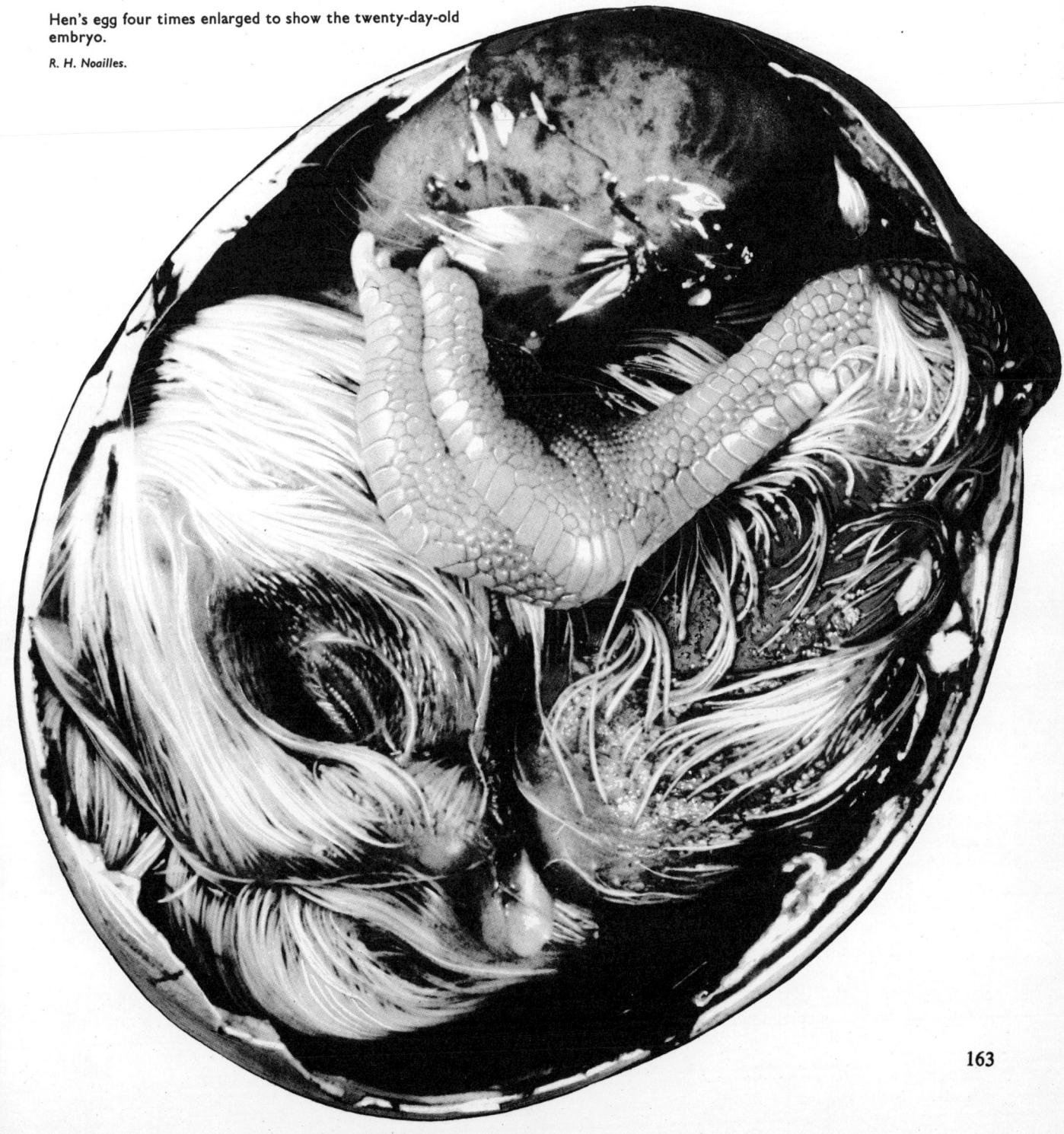

Hen's egg four times enlarged to show the twenty-day-old
embryo.

R. H. Noailles.

Embryonic development

'Of all the mysteries of nature and organic economy,' wrote Cuvier in 1817, 'the birth of organized beings is the greatest.' The mystery is still far from being solved, though today we have one considerable advantage over our predecessors: we can' at least definitely dismiss explanations which are too wild, crude, or simple.

After the seventeenth century many naturalists and philosophers, discouraged by the difficulty of the problem, decided purely and simply to ignore it by assuming that every organized being carried within itself ready-made germs which were actual miniatures of itself and contained the future being in all its structural complexity. Thus there was no necessity for the new being to form itself, to construct itself little by little. To become exactly like its parent all it needed to do was enlarge its initial dimensions. In other words, embryonic development was synonymous with growth.

The origin of these convenient germs went back to the remote ages of the creation. The first representatives of each species contained germs which themselves contained like germs and so on to infinity or thereabouts. Thus from the beginning the species disposed of a practically unlimited supply of almost infinitely minute individuals endowed with potential life. This extravagant theory of germs within germs was upheld by learned men as eminent as Swammderdam, Malebranche, Leibnitz, and Charles Bonnet. Cuvier himself hesitated to reject it, as it seemed hard to find an alternative doctrine which was more plausible.

Thanks to the work of the embryologists, we know that the germ in no way resembles the being it is called upon to produce, because the germ is nothing else but a cell or egg formed by the fusion of two reproductive cells detached from the bodies of the parent organisms.

It goes without saying that a cell contains no organs and no structure which will belong to the future being. It is not a miniature of this being; it is not even a confused blueprint of it. Whether the fertilized ovum is destined to produce a man or a whale or a toad it is always a small spheroid mass in which the magnifications of the most powerful microscopes succeed in detecting nothing which bears the faintest resemblance to an eye, a head or a foot.

The problem is therefore to understand how an organized being can arise from an object so different from itself. Admittedly the cell is also a very complex organization enclosing, as it does, an inner nucleus which contains particles or chromosomes themselves made up of an ordered assembly of multitudinous elementary particles called genes, each of which plays a definite part in development. But this complicated organization has nothing in common with the organization of the developed being. Moreover this nucleus with its chromosomes and genes is found duplicated in every cell of the body, and yet every cell is not capable of producing a new being. What is it then that distinguishes the egg-cell from ordinary cells, and confers upon it the singular power of reproduction and a capacity for development?

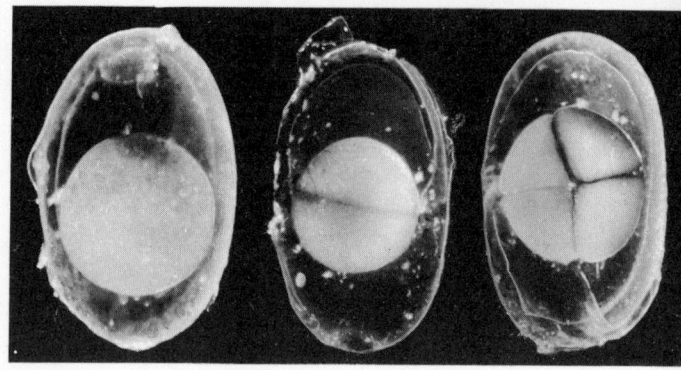

The triton's egg in various stages of its evolution.

R. H. Noailles.

Such are the problems which the embryologist attempts to solve. To clear the ground slightly he must, in the first place, attentively follow all the aspects of this transformation of egg into animal which constitutes development. One of the essential discoveries he will make is the extraordinary diversity of means which nature has at its disposal to produce an animal from a single cell.

Sometimes development is direct. The organism appears at birth in the form which persists throughout its life. Sometimes the development is indirect and before assuming its definitive form the organism undergoes transformation or *metamorphosis*. Sometimes it goes through a more or less complicated series of *larval stages*.

At times development is slow, at other times rapid; at times it affects the entire egg, at others only a limited part of the egg — for instance, the germinal disc of the hen's egg. But at all events and in whatever manner development takes place it always occurs within a given species, with strict regularity and perfect consistency down to the last detail. It is always in the same fashion that the egg of the frog gives birth to a frog, that the egg of a hen gives birth to a chicken, that the human egg gives birth to a baby. Deviations are extremely rare and when, exceptionally, they occur the result is the formation of abnormal organisms or monsters.

The study of embryonic development is not, of course, founded only on an external examination of the embryo. It demands the preparation of very fine incisions which permit internal changes to be studied. For this purpose a special technique is commonly employed. Certain regions of the embryo are impregnated with colouring matter such as Nile blue, which has the property of staining the living cells without damaging them. It is then easy to follow the changes and eventual displacements occurring in such 'tattooed' zones during the course of development.

Extremely diversified though they may be, three essential processes are inevitably encountered: namely, cell-division, cell-growth, and cell-differentiation.

There is division because from a single cell must arise the myriad cells which will constitute the completed animal. There is growth because the animal will

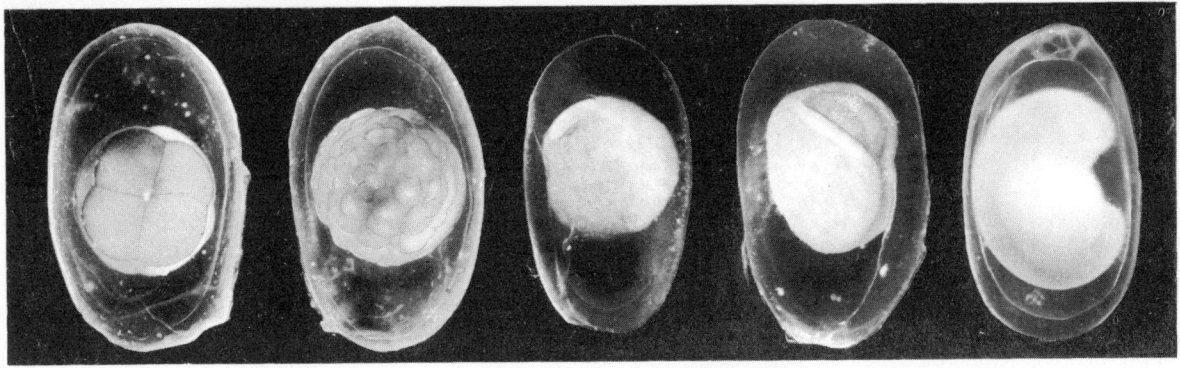

Frog's eggs expelled by pressing the abdomen.

Steiner.

Development of the egg of the web-footed triton; the eggs are laid on leaves of *Elodea* and are in various stages of development.

R. H. Noailles.

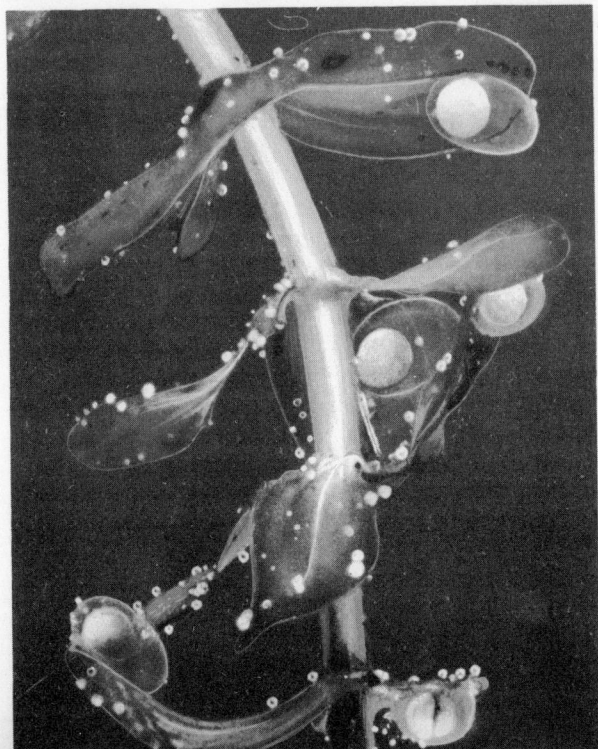

be much larger than the egg-cell. Finally there is differentiation because the cells which constitute the animal, far from being all alike, have different characteristics according to the tissues they will form.

To the process of growth we shall return a little later. As for that of division we may recall that when a cell divides the two daughter cells are, in theory, identical with each other and identical with the mother cell except in size. In particular, the mechanism of division is such that it does not in any way modify the constitution of the nucleus: the two daughter cells contain the same chromosomes and the same genes as the mother cell.

Thus the egg-cell — and this is the first act of the development — divides into two cells which in their turn divide again. This process is repeated again and again until all the cells which compose the animal are formed. At the beginning of development the embryonic cells are all alike, or nearly alike, both in form and in function. But sooner or later according to the species a certain stage is reached when the cells are no longer alike: they begin to differentiate, some — at least in organisms sufficiently high in the evolutionary scale — becoming skin-cells, others nerve-cells, muscle-cells, bone-cells, or blood-cells. The question of how this, differentiation comes about can today be at least partially answered, thanks to the brilliant researches of Hans Spemann on the egg of the triton for which he was awarded the Nobel Prize for biological sciences in 1935.

With the very young embryo of the triton the embryologist is in a position to foresee the future behaviour of a given cell in normal conditions of development. He can with great certainty say that cells in a given zone will give rise to skin-tissue and those in another to nerve-tissue. If a small group of cells were removed from the zone producing skin (the region of presumptive skin) and transplanted to a zone producing nerve-tissue (the region of presumptive nerve-tissue) and if this operation were made in reverse, embryonic development would in no way be disturbed. By the substitution of one group of cells for another the cells originally destined to produce skin would produce nerve-tissue instead and those originally destined to produce nerve-tissue would produce skin. This proves that all the embryonic cells begin by being alike and are all interchangeable. The cell begins by being a jack-of-all-trades, unspecialized and undifferentiated. Its future is not fixed or determined, but depends on the situation which it occupies in the embryonic whole.

If, however, the experiment were repeated by exchanging not cells but tissues with an embryo a few hours older than in the preceding case, the results would be very different. If a shred of presumptive skin were grafted on to the region of presumptive skin, although the skin tissue occupied a place where nerve-tissue should be it would nonetheless produce skin. The nerve-tissue, transplanted to the region of skin, would continue to produce nerve-tissue. In other words, the cell's future at a certain stage is fixed and its destiny inscribed within it. It can no longer be changed after that time by simply changing its position.

Thus we are led to distinguish two periods in embryonic development: a period of indetermination when cells can change their role in development, that is to say a period during which their potential future is greater than their actual future; and a period of determination, when cells can no longer change their role in development, that is to say a period in which their potential future and their actual future are the same.

What influences bring about this cellular determination? We might have expected cellular determination to take place simultaneously throughout the embryonic mass, but this is not so. It begins in a particular place, in a narrow region situated in the neighbourhood of the primitive 'mouth' or *blastopore*. The blastopore is the source of an 'inductive' effect which first of all causes the rudiments of a nervous system to appear; this *neural plate* will itself 'induce' the formation of other organs and in this way the embryo little by little takes shape by a complex of *inductions*.

By displacing the inductive zone one can cause the embryo to form in an unusual position. By removing the zone and grafting it on to another embryo already provided with its own inductive zone one can cause not only the normal organs to appear but also a second set of organs, in other words an extra embryo.

This zone which is so active and potent may be considered the real organizing centre of the embryo and has therefore been named the *organizer*.

When an organizer, thus grafted, induces the genesis of a supernumerary embryo, the extra embryo is not formed from the transplanted tissue — which only instigates its formation — but at the expense of the tissue of the embryo receiving the graft. This can be shown by grafting an organizer of one species on to the embryo of another, differently pigmented species. The supernumerary embryo will display the pigmentation of the species of the organism grafted, and not that of the species furnishing the graft.

A fact of capital importance is that the organizer does not possess zoological specificity: in other words, the organizer of one species can bring about the organization of an embryo of remote species. Waddington, for example, was able to organize an embryo chicken by means of the organizer of a duck and even with the organizer of a mammal. Watt organized an embryo triton with the organizer of a bird!

The effect of organization has been obtained with tissues of other origin: from the liver of a lizard or calf, mollusc liver, crushed *Daphnia,* the lymph of moths.

Organizing power is transmitted by simple and prolonged contact between active and inactive tissue. It behaves as though it were connected with some chemical substance — a kind of hormone or 'organisin' — which diffuses from one tissue to another.

The power does not depend on cellular vitality; it persists after rough treatment like being immersed in boiling water, in powerful solutions of alcohol for a week, or in weaker solutions for several months. Heating can even cause it to appear in tissues where it was previously lacking.

For the moment its exact constitution is unknown;

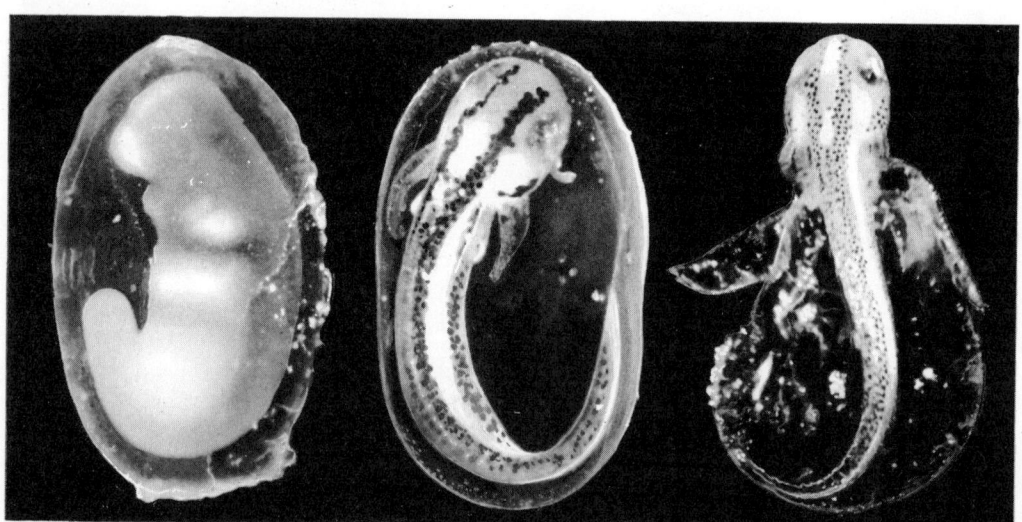

Larval form and hatching of the triton.

Older larvae.

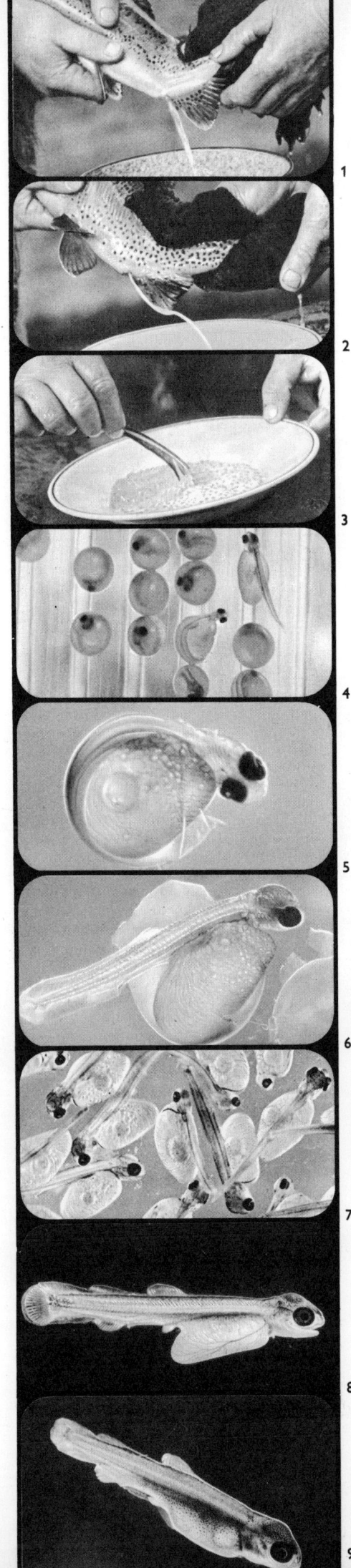

but it is a step of great importance to have detected the intervention of chemical substances at the beginning of development. It may be assumed that throughout development actions of this sort continue, each stage of the embryo's formation being marked by the elaboration of new substances which bring about changes in form which in their turn create new chemical activity.

To cite only a single example of partial 'inductions', we may recall that the crystalline lens of the eye is produced under the 'inductive' influence of a chemical substance emanating from the eye cup. If the eye cup is grafted into another region of the animal's body a lens will be formed there.

Differentiation by nucleus or by cytoplasm?

We have just seen, on the one hand, that cellular differentiation takes place only when a certain stage of development is reached and, on the other hand, that it takes place under the influence of a certain region of the embryo, the organizer. Although these are two fundamental conclusions of the utmost value to the biologist, they give rise to two questions. What does cellular differentiation consist of? Which parts of the cell are affected by it? And how does it come about that the cells of only a certain region of the embryo acquire that power to organize which we have just discussed?

Concerning the nature of cellular differentiation, it was supposed *a priori* that it did not affect the chromosomes of the nucleus, since cell-division in no way alters the cell's chromosome content. But, in recent years, a very elegant method has become available which enables us experimentally and directly to

Artificial fertilization and development of the trout's egg.

R. H. Noailles.

1. Obtaining milt from the male.
2. The female is made artificially to lay eggs.
3. Milt and eggs are mixed.
4. The eggs are placed on trays and the embryo develops.
5. Development of the embryo.
6. Birth of the alevins, or young fish. These have scarcely hatched.
7. and 8. Alevin in course of development.
9. Alevin two months old.
10. Alevin after 25 days.

167

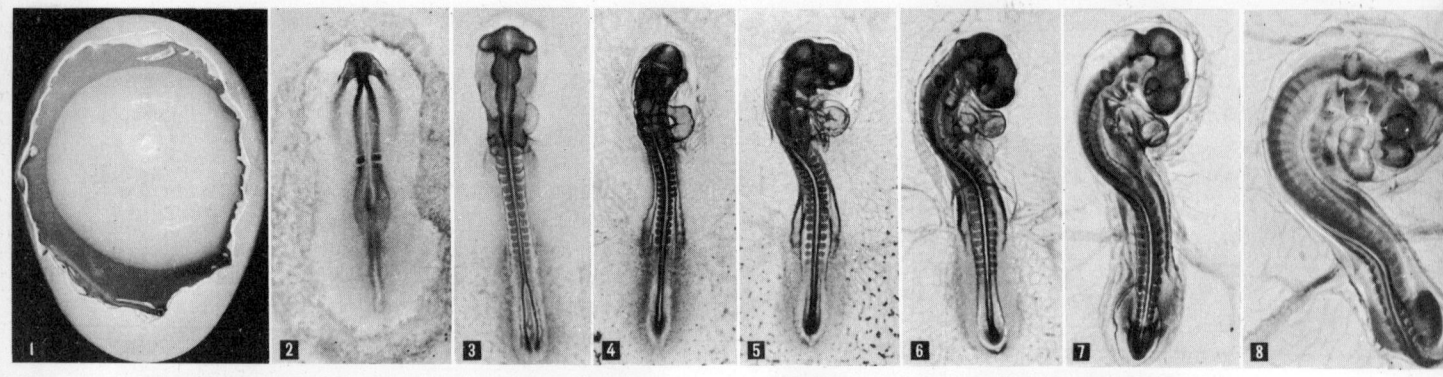

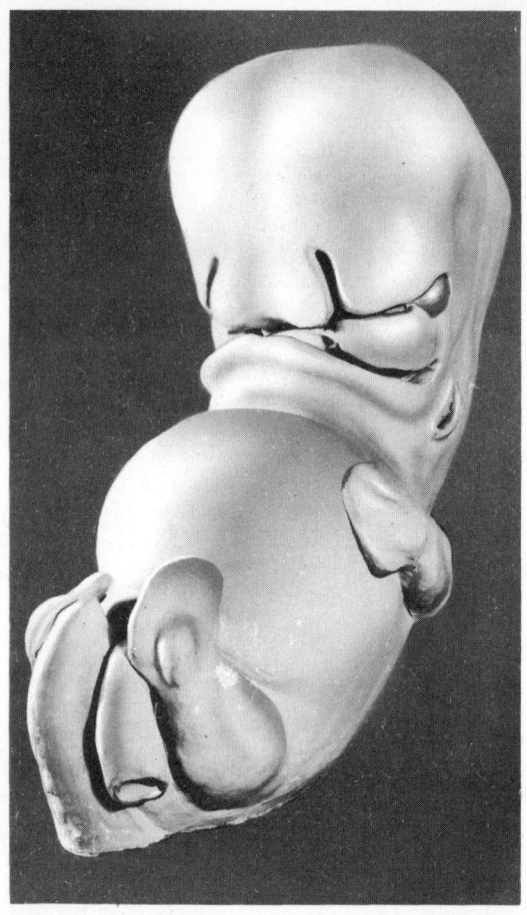

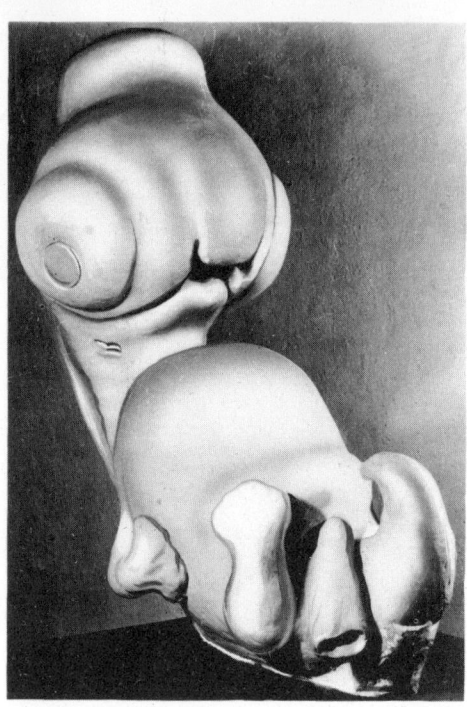

Embryo of a chicken after five days of incubation, fifteen days before hatching. *On the left,* a human embryo at the age of about two months. (Clay models from the Palais de la Découverte.)

Larousse.

approach the problem of the modification of nuclei during the course of embryonic development.

If nuclei undergo no change after the first cleavage of the egg they must of necessity remain identical to the nucleus of the egg itself and, in consequence, we should, without hindering normal development, be able to replace the nucleus of an egg by any nucleus taken from any cell in the embryo.

This very significant experiment was made by two American biologists, Robert Briggs and Thomas J. King. They used the leopard frog, a common American frog *(Rana pipiens),* in their researches and this is the way they proceeded.

The nucleus of a virgin egg was extracted by means of a fine glass needle. Then, under the microscope, they isolated a cell from an embryo which had reached the *blastula* stage, that is, the stage reached after twenty-four hours of evolution when the embryo

comprises several hundreds of cells. The blastula cell was extracted by suction with a micro-pipette a little smaller in calibre than the diameter of the cell so that the cell was jammed, crushed, and its nucleus freed. The nucleus was then delicately injected into an egg which had just had its own nucleus removed.

A large number of eggs which had thus been provided with a blastula nucleus developed regularly. The larvae which they produced were complete and formed normally. The results obtained were certainly not due to any error in the experiment, for all the control tests showed that the amputation of the nucleus was effective and that eggs thus deprived of a nucleus never developed.

We may then accept it as established that the blastula nucleus retains all the formative capacities which belong to the nucleus of the egg. It can serve as a substitute for the egg's nucleus and bring about the

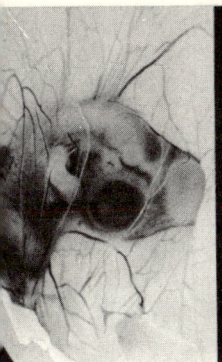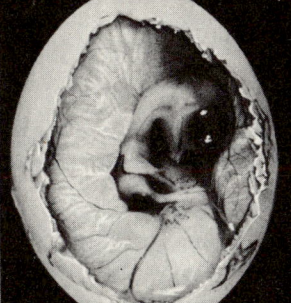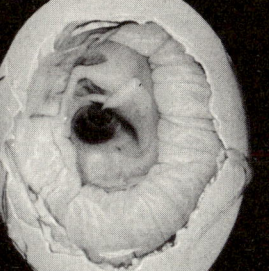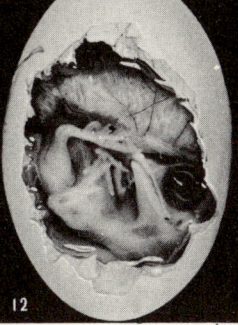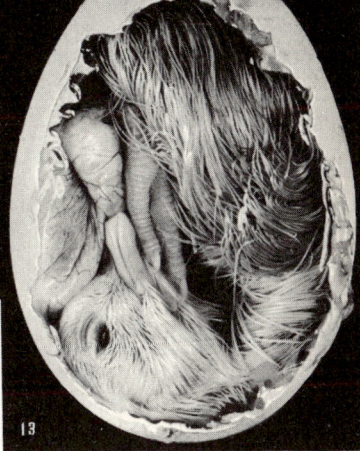

Development of the embryo chicken

After R. Hugh. R. H. Noailles.

1. Egg opened to show an embryo after 12 hours of incubation.
2. After 22 hours.
3. After 30 hours.
4. After 38 hours.
5. After 42 hours.

6. After 46 hours.
7. After 62 hours.
8. After 4 days.
9. After 6 days.
10. After 9 days.
11. After 10 days.
12. After 12 days.
13. After 16 days.

At the beginning of the present chapter the embryo is illustrated after 20 days of incubation, ready to hatch.

complete development of a larva. At this, the blastula, stage of embryonic development no signs of nuclear differentiation can be detected. Nor are they found at a later stage — that of advanced *gastrula* — when cellular differentiation has already begun. But if the experiment is repeated at a still later stage different results are obtained, and it would seem that after a certain critical stage in embryonic evolution is reached the nuclei undergo changes which deprive them of their initial 'totipotence'. For the moment we have no idea of the nature of these changes.

It is interesting to note that this method of substituted nuclei — recently extended to the triton and *Ambystoma* — enables the biologist to produce great numbers of genetically identical individuals. Since a blastula contains many hundreds of cells, all with identical nuclei, one can, by transplanting the nuclei to eggs from which the nucleus has been removed, produce hundreds of larvae having exactly the same chromosome heritage. In this way 'cuttings', as it were, can be implanted on the embryo 'stock'.

Preorganization of the egg

Experiment seems to prove that a certain differentiation affects the nucleus of the cell in later stages of development. The cytoplasm of the cell is also affected, though we have not the slightest idea in what way the two phenomena are connected. It may be that cytoplasmic differentiation depends, at least partially, on nuclear differentiation. But whether or not this is so, the possibility of a *primary* differentiation of the cytoplasm is conceivable, for the egg is far from being a rigorously homogeneous cell, uniform throughout. It almost always presents polarity, with an upper and lower pole, due to the unequal distribution of nutritive substances. It contains enclaves and granulations of various kinds which are more numerous in certain regions of the cell than in others. As well as this visible heterogeneity

other, more subtle, dissimilarities which escape our powers of analysis undoubtedly exist which would justify Albert Dalcq's remark that 'the egg is more than a cell, it is a germ.' Furthermore the act of fertilization introduces new causes of heterogeneity into the egg: in some ova the point at which the spermatozoon enters more or less strictly determines the new being's plane of bilateral symmetry. For all these reasons it is understandable that the egg, simply by the act of cleavage, can from the very first divide into cells in which the cytoplasm content is not uniform. As a result of interaction between cytoplasm and nucleus these dissimilarities can only become increasingly accentuated as development continues, and when the time is ripe such differentiation becomes apparent in the special properties that we have observed in the embryonic 'organizing' zone.

Body cells and germ-cells

While the vast majority of cells which make up the embryo become specialized in the tasks they are to perform a certain number of them remain undifferentiated, preserving their primitive character of young embryonic cells. It is these which later produce the reproductive cells or gametes.

In many cases we can follow, from the first divisions of the egg, this line of special cells which remain, as it were, apart from the general development of the organism. They are known as *germ-cells* in opposition to all other cells of the organism which are known as *somatic cells*.

Thus in the worm *Ascaris,* which is a parasite of the horse's intestines, one can discern as early as the fourth division of the egg, when the embryo is composed of only sixteen cells, the cell which will be the ancestor of the germinal line and therefore of all the reproductive cells. In the little crustacean *Pisha* the two cells from which the germinal line arises can be isolated towards the sixth division. Among certain vertebrates it is estimated that the isolation of the germinal line begins at the eighth cell division, while in man it apparently takes place when the embryo measures 2½ millimetres. Sometimes in the egg itself, before cleavage has begun, the region which will contribute to the formation on the germinal line can be recognized. In such cases, preparations for the next generation are already taking place in the still unsegmented egg. The egg of an insect, for instance, reveals at its posterior extremity protoplasm of particular appearance which confers germinal properties on those cells which in division receive it. In the egg of the frog there also exists a 'germinal determinant' represented at the inferior pole of the egg by a protoplasmic area which is distinguished from the ordinary surrounding protoplasm by its structure and the way it can be stained.

If germ-cells owe their special quality to some specific material we should, by destroying this material in the egg itself, be able in theory to prevent the formation of the germinal line and thus obtain an individual that is completely sterile. This theory has been tested by experiment and found to be true not only in the case of the fruit fly and the silkworm but in that of the frog. In all cases the experimenter has obtained products totally devoid of germinal tissue, thus achieving castration at the earliest conceivable moment — in the egg itself!

Rays of short wave-length are employed to reach and destroy the future germ-cell. Bounoure, in one of his remarkable experiments, extracted the frog's eggs from their gelatinous matrix and placed them in a quartz vessel, quartz being permeable to ultraviolet rays while ordinary glass is not. Then, for a moment or two, he submitted the lower pole of the eggs to ultraviolet radiation. In this way their germinal potentialities can be radically and permanently abolished. Many of the frogs which issue from eggs so treated are found to have atrophied genital glands which contain no trace of reproductive elements. A curious detail is that such frogs are larger than normal frogs, as though their bodies had also absorbed that portion of the embryo's nourishment which is usually taken by the germinal tissue.

Working with chickens Vera Dantchakov destroyed the mother cells of the germinal line with radium. Again castration was premature, and the results she obtained were embryos devoid of all sex-cells. The facts elicited from such experiments suggest an essential distinction between germ-cells and somatic cells, or, in the terms used by Weismann who was one of the first to insist on the importance of this distinction, between *germen* and *soma*.

According to the earlier conception — before the days of Weismann — the germen produced the soma which itself produced a new germen, capable in turn of producing a new soma, and so on. In brief, germen and soma alternated.

Under the influence of Weismann's theory of the continuity of the germ plasm the situation has been presented very differently. The germen produces a soma on the one hand and, on the other, a new germen. There is direct continuity between germen and germen: 'the germinal line is immortal, unlike somatic or body cells which are only its perishable bearers and administrators'. This conception, with its basic opposition between germ-cells and somatic cells, has been strongly challenged, and it does in fact seem a little sweeping; for in certain cases — among the *Ascidia,* worms, molluscs, etc. — observers believe they have noted the belated appearance of germ-cells arising from somatic tissues which are already more or less differentiated. The biologist Paul Brien wrote: 'The *soma-germen* duality of the organism is a myth.' On the whole, however, it is true to say that in embryonic development a very early segregation takes place between the mass of cells which are differentiated to form the body or soma, and the little group of undifferentiated cells which are the ancestors of the future germ-cells.

In any case it may be remarked in passing that the problem of the soma-germen duality is less important today than it seemed to be during the last century. We now know that there is no essential difference, as far as chromosomes are concerned, between somatic cells and germ-cells; and that, in their capacity to multiply, somatic cells too are potentially immortal — as the researches of Carrel and his successors have shown. In addition, the problem of soma-germen duality and the problem of the inheritance of acquired characters are now entirely dissociated.

Today the majority of biologists deny that acquired characters — characters, that is, acquired by the soma — can be inscribed in the germ-cells and thus transmitted to the offspring. They deny this not, as is too often said, because they make any theoretical distinction between soma and germen, but simply because it is almost impossible to conceive how such acquired characters could be inscribed in any cell at all. The problem of imprinting a somatic character in a germ-cell is merely the same problem as imprinting it in any somatic cell, epithelial, hepatic, or renal.

Regulation

In every egg there exists, as we have seen, a certain preorganization. But this preorganization is far from being so rigid that development cannot often be made to take a direction quite different from that which, left to itself, it would have taken.

For example, take a sea-urchin's egg which has just completed its first division. If the two resulting cells — or *blastomeres* — are now separated from each other by means of a fine glass needle, we shall find to our astonishment that from each cell a complete embryo will form and a complete larva be born. Now, had the cells been left in place each would evidently have produced only a half-embryo, a half-larva. By separating the first four blastomeres it is even possible to obtain four complete larvae.

This experiment, made by the embryologist Hans Driesch, is of extreme significance because it shows that a fragment of germ can produce in isolation something which it would not have produced in the course of normal development. The scientist has, as it were, forced each bit of the egg to reveal its potentialities and to surpass its normal performance.

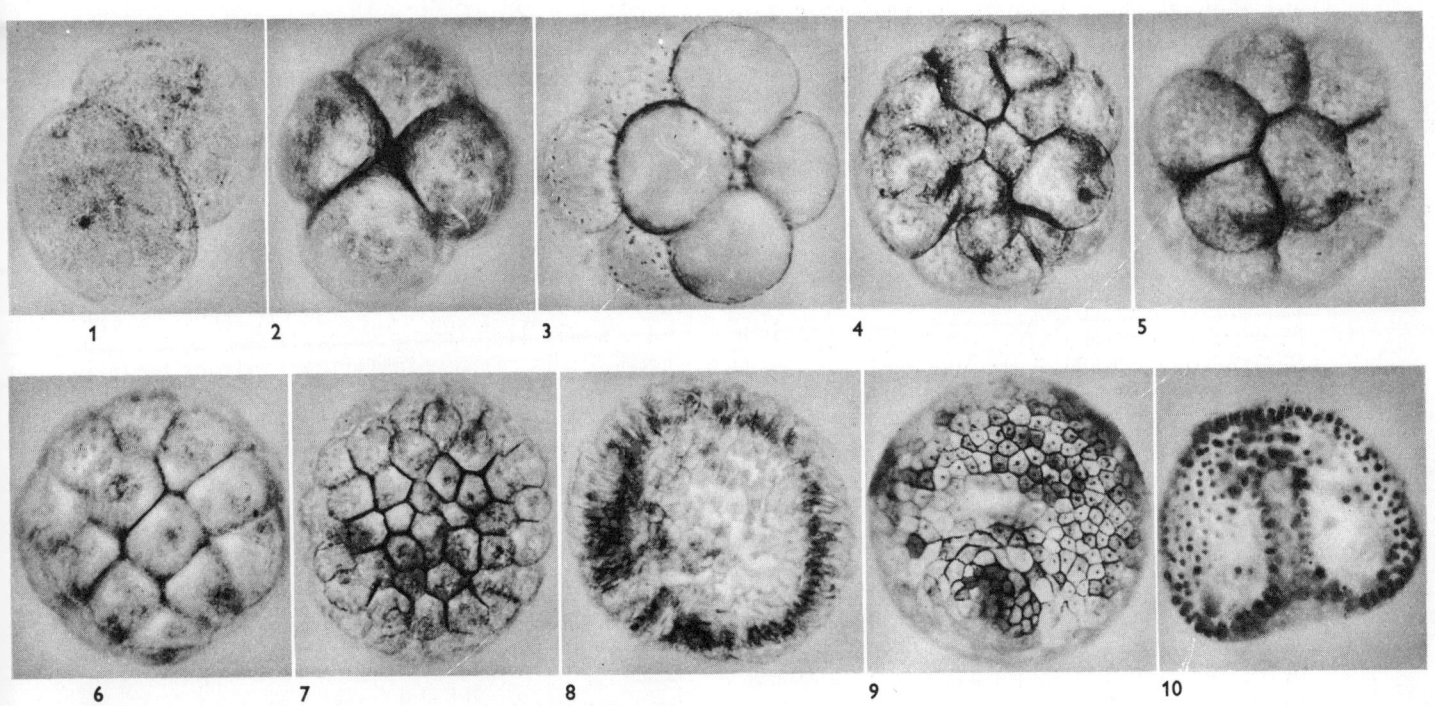

The development of the sea-urchin's egg.

H. Mugard.

1. Stage 2. Segmentation into two blastomeres.
2. Stage 4.
3. Stage 8.
4. Stage 32 (young morula) with micromeres.
5. Stage 32 (morula).

   (Stages 2, 4, 8 and 32 are stained with silver.)

6. Advanced morula, and 7, blastula (both stained with silver).
7. Section of blastula with blastocoele. (Stained with hemalum.)
8. Gastrula with blastopore. (The silver stain reveals a primitive mouth.)
9. Gastrula. (Hemalum reveals the primitive intestine.)
10. Gastrula. In this stage the larva is prepared.
11. Young larva or pluteus. The anus is already formed.
12. Pluteus with anus and mouth.
13. Pluteus or larva at a more advanced stage.

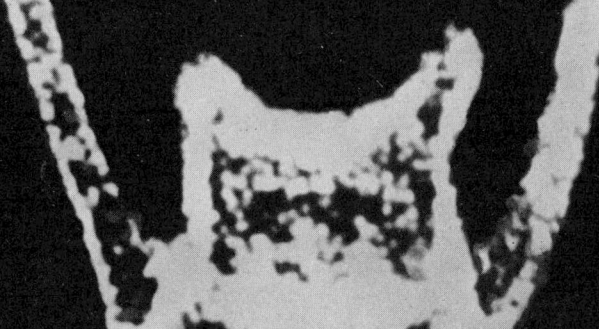

The great embryologist Hans Driesch, who demonstrated germinal regulation in the sea-urchin's egg.

Larousse.

171

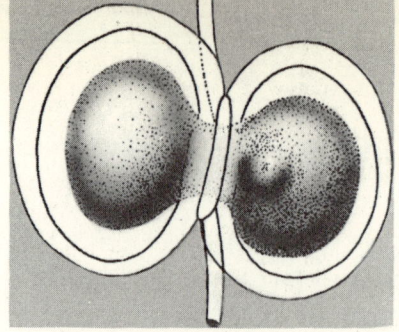

Spemann's experiment: the artificial production of twins in amphibians. *Above*, a triton's egg in the two-blastomere stage is separated by means of a fine looped and knotted hair. *Below*, the result — twin larvae in the same enveloping membrane.

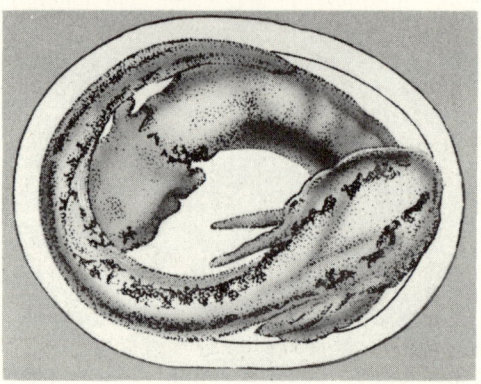

As Dalcq remarked, 'It isn't necessary to be a great student of biology to realize that obtaining a complete embryo from a fragment of egg is a striking phenomenon. The natural tendency of our minds is to think of an egg as a germ which potentially contains the future organism, and we try to imagine a special configuration which corresponds to this thought. If now we find that a part of the system is as capable as the whole system of forming a complete embryo we encounter enormous difficulty in picturing the plan which had begun to take shape in our minds. We come up against a problem which is all the more remarkable for being specifically biological.'

This striking and instructive experiment has been made not only with the sea-urchin but also with animals belonging to widely different species: *Amphioxus, Balanoglossus*, amphibians, birds, and mammals. It teaches us that among all these animals, as with the sea-urchin, the germ bears within itself the stuff of more than one organism.

It is not, however, true of all eggs. There are eggs in which separation of the blastomeres only results in splitting the embryo and bringing about the formation of incomplete organisms — half an organism, or a quarter. Such is the case with the eggs of *Ascidia*, of certain molluscs, and of certain insects.

For this reason it was formerly held that there were two distinct categories of germs: one, like the sea-urchin's egg, capable of *regulation;* the other, *mosaic eggs*, like that of *Ascidia*, incapable of regulation.

In practice this distinction has had to be almost abandoned as being too rigid, for even in so-called mosaic eggs it is rare that the biologist cannot, by means of certain artifices, cause some degree of regulation to appear. An aptitude for regulation is now held to be a fundamental property of germ-cells. In nature the necessity for regulation very seldom arises and it can therefore confer little advantage on the germ-cell possessing it. But its usefulness in the remote past may, as the eminent embryologist A. Dalcq suggested, have been great.

According to him the evolution of animal species entailed variations of a special kind — *onto-mutations* — which instantly and profoundly affected the constitution of the egg. Such sudden and convulsive mutations would inevitably have led to germinal catastrophe and to extinction, had not the egg been endowed with that power of regulation which the investigator today still finds it to possess.

It may have been by virtue of this power that evolution could afford to be truly revolutionary, and, as it were, take chances. Thus regulation may have been one of the potent auxiliaries of evolutionary progress, although we are unable to decide whether it is a primary property of living matter or whether it was afterwards acquired and strengthened by natural selection.

172

Double monster in the shape of a reversed Y, obtained by slitting the lower part of a germinating chicken with a micro-needle. The result is the development of fused heads and two diverging trunks.

After H. Lutz. Et. Wolff.

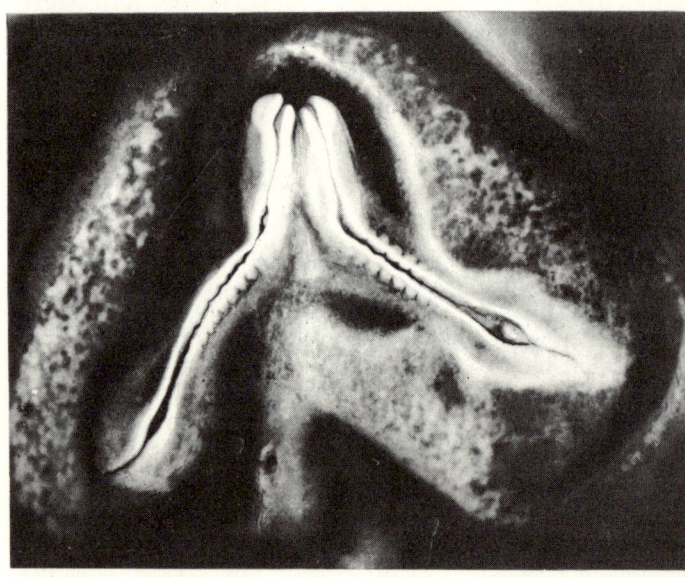

A Y-shaped double monster obtained by slitting upper part of germ with a micro-needle. Result: two heads and a single trunk.

After H. Lutz. Et. Wolff.

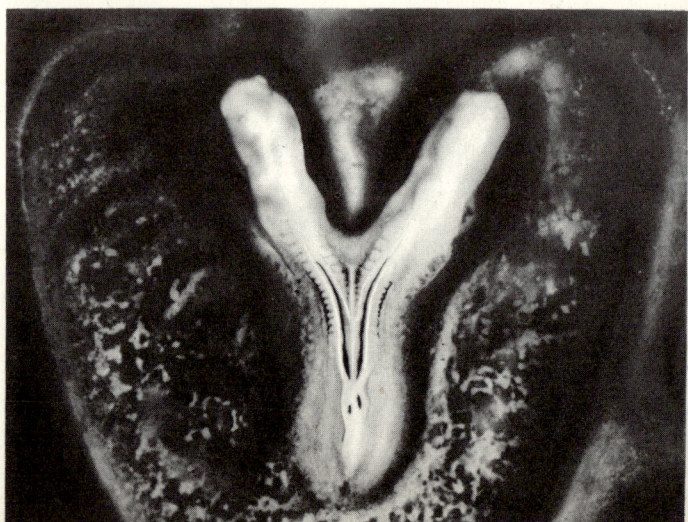

Polyembryony

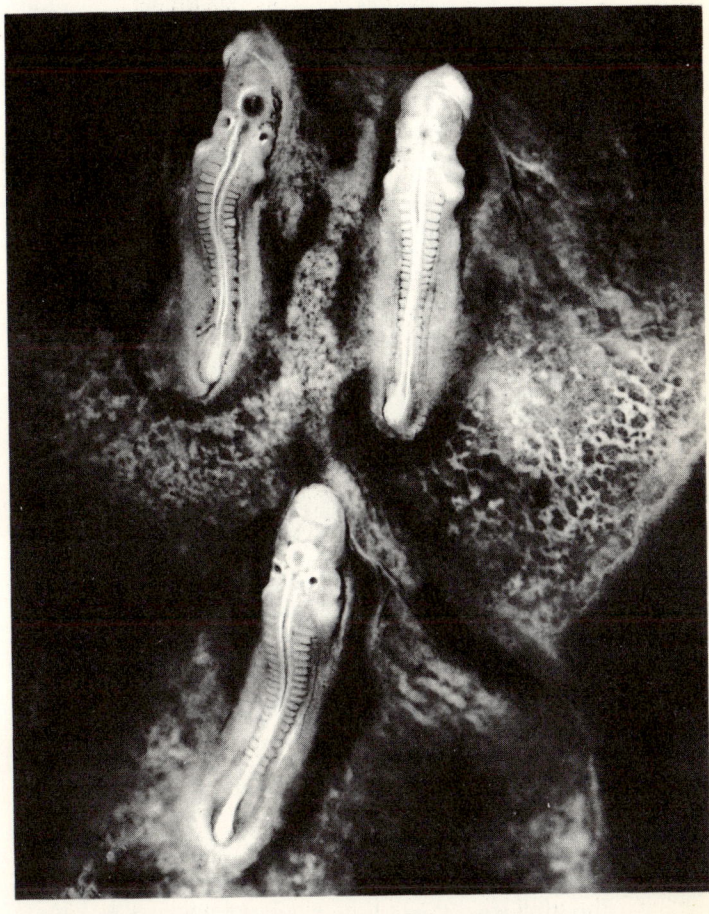

Artificial polyembryony

There are many methods by which the research biologist can induce polyembryony, in other words, cause a single egg to produce several embryos. If, however, he incompletely separates two embryos the egg will develop into a twin or double monster.

Artificial polyembryony can be brought about among sea-urchins by developing the eggs in sea-water from which calcium has been removed, for without calcium the cells do not remain united. The same effect can be obtained by adding a small quantity of tin salts to the sea-water. Among starfish carbonic gas will cause the phenomenon.

Bataillon induced polyembryony in lampreys by submitting the egg to strongly salted or sugared solutions. Stockard, with trout, obtained double monsters by means of low temperatures and with anaesthetics — ether and alcohol. To double the embryo of the frog it is sufficient to press the egg between two sheets of glass and turn it over: fifty per cent of the eggs thus treated produce double monsters. Among tritons one can, as Spemann did, separate the two first cells by means of a ligature, made with a fine hair. With birds artificial twins are obtained by mechanical fissuration of the embryonic disc or blastoderm, as the experiments of Wolff and Lutz demonstrated.

If, in the embryonic disc of a duck's egg, an incision is made parallel to the major axis of the egg — and thus perpendicular to the embryo's plane of symmetry — a large proportion of double monsters will be obtained, lying head to head and in opposite directions. If the cut is made parallel to the minor axis of the egg — hence along the plane of embryonic symmetry — double monsters will still be obtained, but this time lying side by side and in the same direction. Even quadruplet ducks — or rather 'quadruple monsters' — can be obtained by making a cross-shaped incision in the embryonic disc: each quarter will produce a complete embryo. As many as six embryos have been obtained from a single duck's egg. From a single rat's ovum J. Nicholas and H. V. Hall produced two embryos by separating the two first blastomeres.

Localized irradiation with X-rays, ultraviolet rays, etc., are also among the methods the biologist uses to induce polyembryony in eggs.

Natural polyembryony

The human ovum belongs to the category of eggs which is capable of polyembryonic development, and in our species nature itself fairly often performs the 'experiment' which consists of producing more than one individual from the same egg.

Among human twins, identical twins — which alone are due to polyembryony — are roughly half as common as fraternal or false twins which are due to super-ovulation. Much rarer than identical twins are

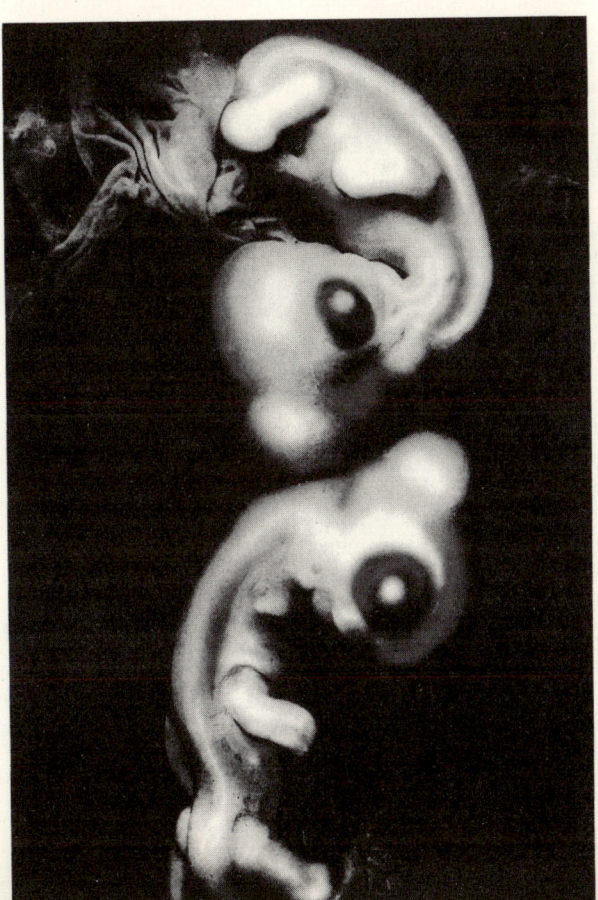

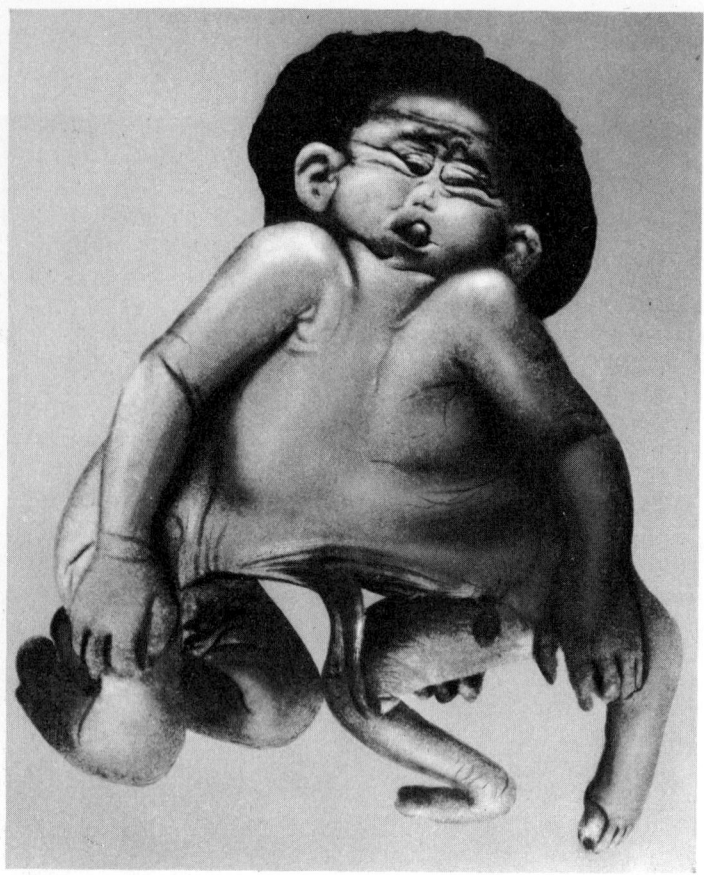

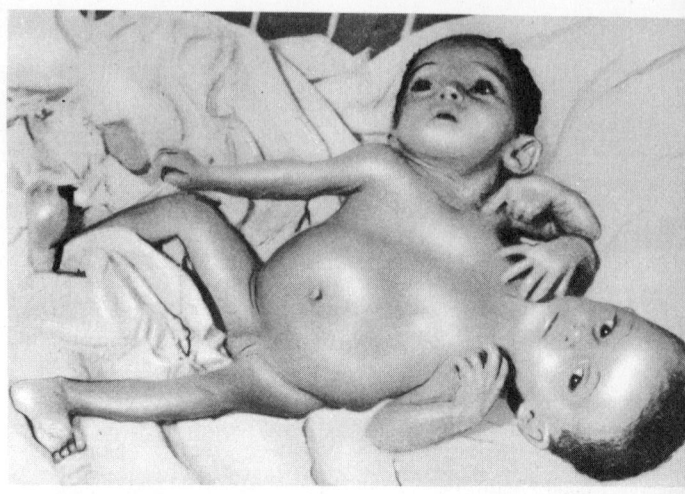

Female Siamese twins. Ira-Galya, Moscow, 1938.

Observation Anockine.

Human Siamese twins with two bodies, a single head, but two faces. The two components are joined belly to belly. In the photograph only one face can be seen, due to the union of the half-face of one component with the half-face of the other. Each face is laterally disposed and looks sideways. This is the parallel axis type of double monster.

Et. Wolff.

identical triplets, identical quadruplets, and so on. Quadruple pregnancies are very approximately one hundred times rarer than triple pregnancies which are themselves a hundred times less common than double. In the annals of human reproduction very few cases of identical quintuplets are mentioned.

Human polyembryony is not, as one might suppose — and as it was originally believed — due to the separation of the first cells derived from the egg. The polyembryonic ovum begins its evolution exactly like a normal ovum and it is only at a relatively advanced stage of its development that the anomaly appears — as a result of the formation of two or more centres of embryonic organization.

The causes of human polyembryony are still obscure. It is fairly likely, but by no means proved, that hereditary factors, perhaps of paternal origin, play a part in the fragmentation of the egg. Perhaps, also, the polyembryonic egg is slow in becoming implanted in the uterine wall, a delay attributable to faulty mechanism in the hormone secretion that controls nidation. Maternal syphilis has also been blamed, though without convincing evidence.

The only well supervised and thoroughly studied case of identical quintuplets is that of the Dionne sisters, born in Canada. Not unnaturally they aroused immense curiosity in scientific circles. From the moment of their birth they were the object of close and constant investigation, for the opportunity of studying no less than five individuals with an identical hereditary patrimony was far too rare to neglect. A team of

biologists and psychologists was attached to them to scrutinize in the minutest detail the progress of their physical growth, how their reflexes were acquired, the awakening, of their sensibilities, the formation of their language. Not one of their gestures, not one of their glances escaped being put on record; their babbling, the slightest nuance of their infant prattle became subject matter for learned exegesis. More often photographed than Hollywood stars, treated at the same time like princesses and laboratory animals, submitted to perpetual inspection and innumerable tests, observed at every minute of their daily life — asleep, at table, at play, in tantrums, when sulking, and even in the privacy of bathroom and lavatory! — the little Dionne girls, whose relief was doubtless great when they were freed from such impertinent surveillance, are biologically better known to us than any human being in recorded history. Already the details of their existence are set down in heavy volumes filled with photographs and diagrams, with tables, graphs, and figures.

In antiquity multiple births were considered as good omens if the number of children born was odd. According to Phlego of Tralles, the emperor Trajan ordered that quintuplets should be brought up at his own private expense. Dionysius of Halicarnassus reports that certain triplets were fed at their parents' board but at the state's expense.

On the other hand the simultaneous birth of an even number of children was considered a malevolent omen. Pliny speaks of quadruplets whose birth announced

a famine. The death of the emperor Claudius was, they said, foretold by the birth of a double monster of hideous aspect. The birth of Romulus and Remus was held to forebode evil, especially as they were born to a vestal virgin!

Siamese twins

In approximately every 100,000 births two are of Siamese twins, that is to say of individuals who are physically joined and are, in reality, identical twins imperfectly separated.

Siamese twins — who are, of necessity, always of the same sex — can be joined together in all possible manners: *pygopagus* twins are attached to each other by the buttocks, *metopagus* by the forehead, *cephalopagus* by the top of the head, *ischiopagus* by the hypogastric region, *xiphopagus* by the sternum, etc.

Sometimes the bodies are united below and independent above, in a Y-shaped formation: sometimes, united above and independent below, the Y-shaped malformation is reversed. Sometimes Siamese twins can be surgically separated, but the operation is always very delicate and often entails the death of one if not both of them. Triple monsters also exist, but they are extremely rare.

Among the most celebrated Siamese twins were Chang and Eng, who were xiphopagus. They were born in 1811 of Chinese parents who lived in the kingdom of Siam, and towards 1835 they were exhibited in the chief capitals of Europe. Between them they fathered

more than twenty children who were themselves perfectly normal.

Isidore Geoffroy Saint-Hilaire's report on Chang and Eng is of great interest.

'When,' he wrote, 'the Siamese twins were equally calm or equally animated they breathed and their hearts beat simultaneously. This, however, was not always the case: one day when one of the brothers bent down to examine the works of a watch his pulse at once quickened, while that of his twin underwent no noticeable change.

'In their other functions the Siamese twins also displayed a remarkable concordance, which was not, however, invariable as newspapers in the United States, London and Paris delighted in repeating and as, indeed, Chang and Eng themselves told people whose curiosity was satisfied by asking a few vague questions. Their hours of hunger, of sleep and of waking, their pleasures, their anger and their pains are shared in common. But it is not true that the two brothers always feel at the same moment and to the same degree the pangs of hunger, nor that the slightest indisposition of one is always felt by the other, nor, finally, that they always fall asleep and awaken at the same instant so that one brother has never been able to see the other asleep.

'Each of them — they speak English — individually follows a separate conversation with different interlocutors; but they almost never address each other, and when they do it is only to say a few words, apparently disconnected, and scarcely intelligible to others.' When Doctor Harris suggested an operation to separate the Siamese twins he only succeeded in putting them into a rage.

Mention should also be made of the pygopagus twins Helena and Judith, who were born in 1701 in Szony, a town in Hungary, and died at the age of twenty-two. They were described and depicted in Buffon's Natural History and celebrated by the poet Alexander Pope. They had neither the same temperament nor the same character. Their menstrual flow differed in time, duration, and amount, although they possessed only a single genital orifice. The need to evacuate they felt simultaneously, but that of urinating occurred to each separately. They could sleep at different times.

Female pygopagus twins can marry, become pregnant and bear children normally — as did one of the two sisters, Rosa-Josepha Blazek.

Rita-Cristina, who was born in Paris possessed two legs, four arms and two heads. She died in 1829 at the age of eight and a half months and was made famous in scientific literature by the anatomist Serres in a work entitled *Recherches d'anatomie Transcendante,* 1832.

'What a subject for meditation,' wrote Isidore Geoffroy Saint-Hilaire, 'The spectacle of this double being with two wills and double sensations! While one of the two heads was deep in slumber the other cried for and avidly seized the suckling breast; or else,

175

both awake, one would scream in protest while the other smiled peacefully at its mother. If the arm of one of the two sisters was tickled she alone perceived the sensation. When the right leg was touched only Rita felt it; when the left was touched Rita felt nothing — only Cristina was aware of it. Both were hungry at different times, but almost always felt need to expel faecal matter at the same time.'

Xiphodyme Siamese twins lived until the age of twenty-eight at the court of King James of Scotland, where they had been engaged as musicians. It was claimed that they sometimes quarrelled bitterly.

Among Siamese twins who have left their trace on history we may recall the case of one that was born in 1569 and died a few moments afterwards. A certain Jacques Roy, who dissected it, was thereupon inspired to draw — in verse — a somewhat astonishing moral parallel, bidding the reader to . .

'. . . . behold the twins thus intertwined,
With single body, double heart and mind:
One dead unbaptized, while the other,
His soul preserved, survives his brother'

He came to the conclusion that the first represented the false Huguenot and the latter the true Catholic religion!

Parasitic deformities

Related to the phenomenon of double monsters are those cases in which one twin, being much smaller than the other and less completely formed, is in fact a kind of parasite growth which may be represented by no more than a head or a pair of limbs. Montaigne described one of these parasitic monsters in his *Essays* (Book II, Chapter CCCV). Another was studied in the seventeenth century by the celebrated physician Bartholin: the subject was named Collerado and he exhibited himself in brilliant array with, attached to his sternum, a parasite twin consisting of a head, two arms and one leg.

When the parasite is reduced to one extra head, the extra head is usually sensitive but immobile — or only capable of very slight movement. If the parasite is reduced still further it becomes that 'monstrosity by inclusion' which is well known to the medical profession.

Displaced organs

With double monsters it frequently occurs that the internal organs of one of the twins display a more or less complete reversal of position. The lower pole of the axis of the heart is inclined towards the right so that the base of the heart is situated beneath the right nipple instead of beneath the left.

This inversion of the heart's position not unnaturally entails a functional inversion: it is the left side of the organ instead of the right which receives blood from the veins and, via the pulmonary artery, sends it up to the lungs; it is the right side instead of the left which receives arterial blood and, via the aorta, sends it through the entire body.

Other organs are also reversed in the thorax or in the abdomen: a right lung (recognizable by its division

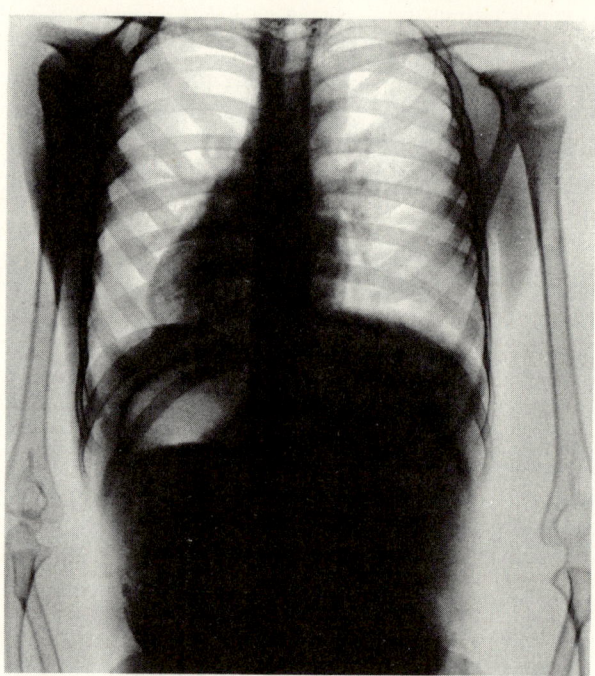

Situs inversus. X-ray photograph showing an abnormal orientation of the heart.

M. Lamy.

into three lobes) occupies the left part of the thoracic cavity, while a left lung (with two lobes) occupies the right. The liver is situated on the left, and also the ascending colon, the caecum and the appendix. The spleen and the descending colon are on the right.

Mirror symmetry in twins

As we have just seen, the disposition of the viscera in such cases more or less exactly resembles the normal disposition of the same viscera as they would appear in a looking-glass. Various names have been given to this anomaly: transposition, inversion, reversal, *situs viscerum inversus*, etc. Spectacular though such mirror symmetry may be, it need not entail any special disadvantages if it affects all the organs and leaves their relative positions and normal connections unchanged.

The less complete the inversion is the more serious are its consequences, and it can then lead to irregularities in the circulatory system. *Situs inversus* seems to be almost the rule among Siamese twins joined by the thorax. Generally speaking the anomaly affects the left-hand twin, that on the right having its organs arranged normally. Among identical human twins, *situs inversus,* though infrequent, sometimes affects one or the other. In any case if one twin has the anomaly the other's organs are situated normally — *situs solitus.*

Certain indications of mirror symmetry, often found among identical twins, may be attributed to a slight degree of organic inversion: for instance, the right-hand fingerprints and palmprints of the one may resemble the left-hand prints of the other, and this resemblance can be further accompanied by a functional similarity — hence the frequency of left-handedness among identical twins. Again, one may have a cowlick which sweeps to the right, while the other has one which sweeps to the left; one may have a freckle on the right cheek, while the other has a similar mark on the left cheek, and so forth.

The experiments of Spemann and Falkenburg with tritons have shown that among twins produced by doubling the embryo artificially a large proportion are obtained with the *situs inversus* anomaly, which above all affects larvae derived from the right half of the embryo. Cases of *situs inversus* have similarly been noticed in the course of artificial-twin experiments with ducks. When the germ-cell is separated in 'spring eggs' — that is, in eggs which have already attained a certain stage of development — almost all the embryos are found to reveal *situs inversus*. As to what brings about this curious phenomenon many points are still obscure; but there is no doubt that it is closely related to the doubling of the egg or of the embryo. If so, what of those cases of *situs inversus* which are found among human beings who have no twin? Such individuals, it is commonly agreed, had in fact a twin which perished in the embryonic stage. Thus every case of *situs inversus* would be the sole survivor of a pair of twins.

Situs inversus was already well known in the seventeenth century, when it could, of course, only be discovered accidentally during post mortem. Today it can be detected by radioscopy.

Concerning the frequency of *situs inversus* the estimates of radiologists are wildly contradictory. Some say that the anomaly is found in one out of 4,000 subjects examined; others say that the proportion is more like one to 40,000. In international medical literature some 650 cases have been recorded.

Without throwing doubts on the efficacy of the stethoscope and palpation, it would be interesting to know whether cases of *situs inversus* have been diagnosed by doctors without the aid of X-rays.

Polyembryony and doubled organs

In many animals polyembryony occurs accidentally and results in a more or less complete doubling of the body or of certain parts of the body. Since the days of antiquity double monsters have been reported among mammals and birds — calves, fawns, pigs, cats, lambs, hares, pigeons, ducks, chickens, and so on — and also among reptiles and fish. Such monsters, varying enormously, are relatively common and have been the subject matter of innumerable published descriptions.

In connection with double monstrosity we may consider those not infrequent cases of supplementary legs — polymelia — in mammals and birds, and also the case of dicephalism. Double-headed chickens (*Pullus dicephalos*) are described by Aldrovandi in the sixteenth century and dicephalous pigeons were presented to the Academy of Sciences in Paris by the Cardinal de Polignac in 1733.

Among birds the proportion of double monsters is much higher in ducks' eggs than in hens' eggs, being in

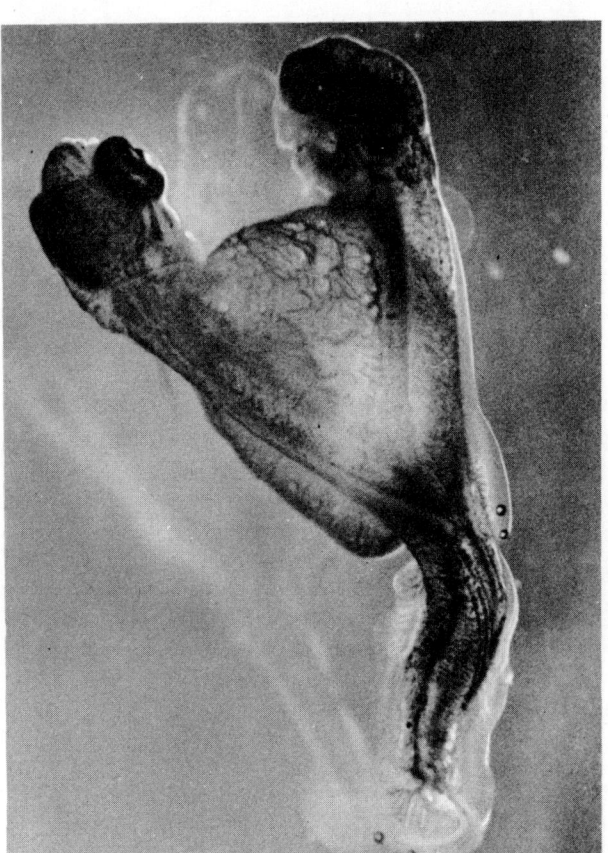

Twin alevins of a trout with a single vesicle and a single tail.
R. H. Noailles.

the former case some eighteen per thousand, and in the latter 1.3 per thousand. The reason for this difference would seem to be that the egg turns over within the uterine cavity of the hen less frequently than in that of the duck, where only forty-four per cent of the eggs retain their original position, in other words, the small and forward. By artificially turning over eggs in the uterus of pigeons Ancel and Vintemberger have induced the formation of two double monsters, an extremely rare anomaly in this species.

Two-headed and even three-headed snakes were known to Aristotle and according to Lacepede naturalists of his day believed in the existence of dicephalous species. A dicephalous snake was described at length by Francesco Redi. It lived for three weeks and its right head died seven hours before its left head. It had two hearts, two livers, two stomachs, etc. Snakes with a double tail are also known. In dicephalous lizards the two heads can be independent of each other: if a fly is offered to one the other will switch round and try to snap it up. The existence of a tortoise with two heads and two pairs of front limbs has been reported.

Barbour in 1888 described a tortoise of New Haven whose two heads led an entirely autonomous life; each on its own account looked and listened, ate and drank, slept, breathed, and moved. There seemed to be no cooperation between the right side and the left. If by accident the tortoise fell on its back, the two heads would work in opposite directions to right the shell; thus the animal could not turn over without help. The

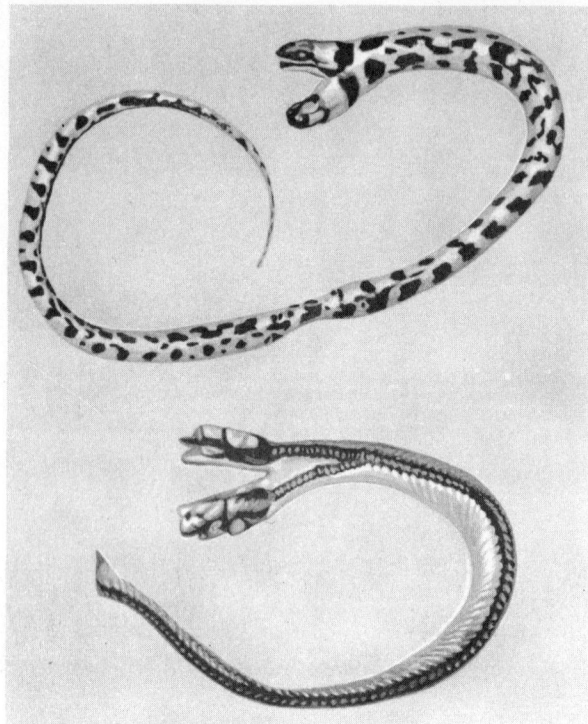

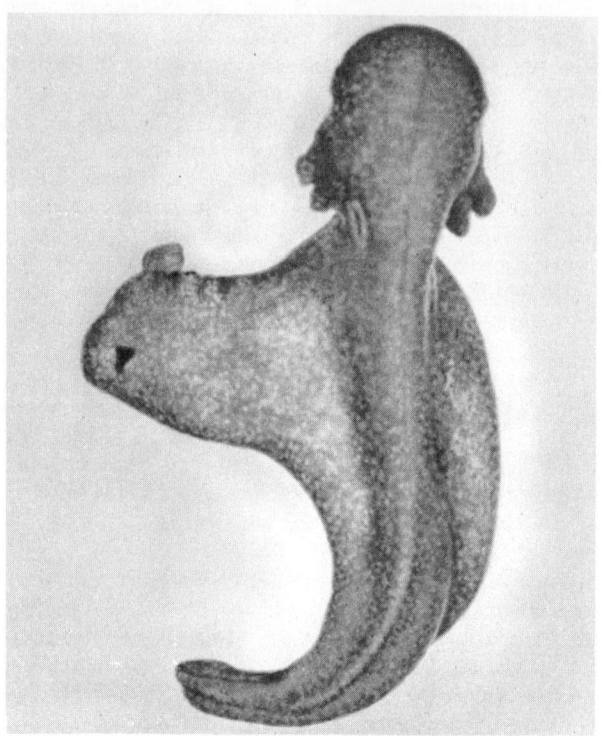

right head was more timid, irritable, and quicker to withdraw than the left head. In the case of tortoises and snakes double monsters can survive in a state of nature, unlike double-monster birds or mammals. Among invertebrates, especially king-crabs, double and even triple monsters are found.

Normal polyembryony

In connection with the determination of sex we have already mentioned the polyembryony of the armadillo and also that of the parasitic hymenopteran *Encyrtus fuscicollis*. The same phenomenon occurs in other insects of the same order. The egg of *Litomastix truncatellus*, the parasite of *Plusia gamma,* can give birth to one thousand individuals. Polyembryony also seems to exist in an insect of the order Strepsiptera. There is also the case of certain bryozoan cyclostomes in which a single egg produces some hundred larvae, and that of the worm *Lumbricus trapezoides* whose egg always, or almost always, gives birth to two individuals.

The factors which determine natural polyembryony are unknown. In the case of the armadillo the egg passes a long period in a free state before it becomes implanted in the uterine wall. The question naturally arises whether this delayed implantation, entailing insufficient nourishment and oxygenation, may not have something to do with polyembryonic development. The argument that it does is unconvincing, for among

many other mammals — roe deer, bears, martens, badgers — the implantation of the egg is also appreciably delayed and yet these animals show no tendency to polyembryony.

A single organism derived from two eggs

As we have seen, an egg may divide in such a way that it produces several individuals. The reverse phenomenon — that is the production of a single individual from two eggs — can also occur.

Among jellyfish (medusas) Metchnikov observed that two eggs sometimes fuse together to produce a single embryo of greater size, and the same phenomenon occurs with certain slugs and certain roundworms *(Ascaris)*. In proboscis worms *(Nemertea)* the phenomenon can be brought about in the laboratory by developing the eggs in sea-water deprived of calcium. With tritons two eggs pressed together will evolve into a single embryo. By the fusion of many sponge larvae a single individual can be obtained.

The fusion of two eggs, natural or experimental, can in certain conditions result in a double monster. Formerly it was widely believed that double monsters in the human species derived from fusion of this sort; but today we know that their origin is a single egg which has been doubled. On the other hand, among certain insects — *Carausius* — the source of double monsters is in fact the fusion of two eggs.

Single Monsters

In the last chapter it was stated that embryonic development in a given species always proceeds with perfect regularity down to the last detail. But occasionally the process is disturbed, a deviation occurs and the result is inevitably the formation of an abnormal organism. Every degree of anomaly exists, ranging from slight malformation or simple structural oddities to serious anomalies or monstrosity in the proper sense of the word. In the human race, according to certain statistics sixty-one genuine monstrosities and 454 slight anomalies occur in every 100,000 births.

Throughout history monstrosities have fired the human imagination, and in this field as in so many others the human imagination has been given free reign to create fantastic theories which have perished only slowly in the harsh light of judgments founded on reason and observation. One of the baked clay tablets which composed the library of Ashurbanipal, King of Nineveh, gives in cuneiform characters a catalogue of monsters observed in Nineveh in 2,000 B. C., together with the events they 'presaged'. Throughout the ancient world — except in the East where monsters were venerated because of some imagined resemblance to a sacred animal — the birth of a monster, single or double, was viewed as an evil omen, a mark of divine anger, the announcement of a scourge, a calamity, or a public catastrophe. Thus it was the custom to appease divine wrath by sacrificing the monster, and sometimes the mother to whom it was born! During the Middle Ages monstrous births were attributed to the intervention of the devil and this belief persisted until the eighteenth century, in spite of the protests of the humanists and the teachings of Pietro Pomponazzi who in his *Treatise on Destiny* (1529) remarked, 'that only fools attribute effects of which they do not understand the causes to God or to the Devil.'

Little by little explanations of a physical nature were advanced, such as contaminated semen, the effect of food or drink on the parents, sexual intercourse with

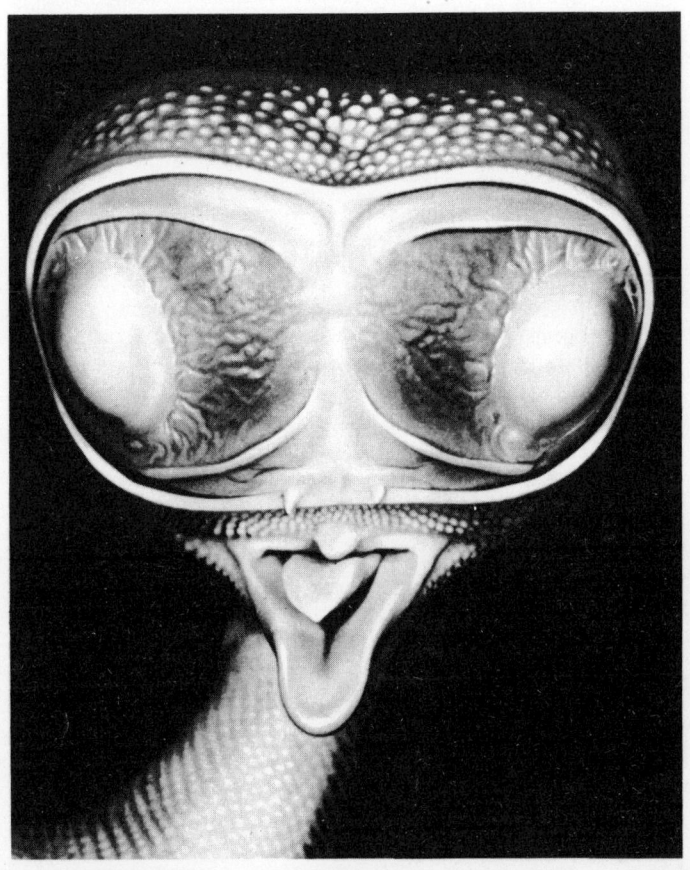

Head of an embryo chick with a cyclopian malformation — cyclocephalus — the result of accurately localized irradiation. No upper beak, two eyes in contact within a single orbit, sharing a common elliptical eyelid.

After Et. Wolff.

animals, the influence of the stars or planets, the imagination of the mother, fright suffered during the course of pregnancy, and so on.

In the eighteenth century, Winslow sustained that monstrosity pre-existed in the germ, while Lémery held that its origin was accidental. The arguments wielded by the two antagonists were as theological as they were anatomical: if the monster is in the germ then God from all eternity willed it — which is incompatible with the idea of divine wisdom and goodness. To which it was replied that to deny God the right to break the laws of nature was to limit His power. Happily the quarrel was not one of those which — in Fontenelle's phrase — possessed the privilege of being eternal. We are now in a position to know that there are two kinds of monstrosity, that which pre-exists in the germ-cell, proving Winslow's case, and those which result from an accident, proving Lémery's case.

Monsters which issue from a monstrous germ, in other words from a genetically abnormal ovum, are those which the geneticist studies under the name of mutations. Such are *acheiropodia* (or suppression of the feet and hands), various malformations of the bones (*polydactylism,* or the presence of supernumerary digits, *syndactylism,* or the union of digits), funnel-shaped thorax, certain forms of dwarfism, congenital absence of the iris, absence of sweat glands, albinism or total absence of pigmentation, and ichthyosis or fish-like skin.

The genetics of monsters, known as teratological genetics, is intimately connected with medical genetics since a great many complaints and diseases are due to anatomical or functional deficiencies transmitted by the parents.

Accidental monsters, on the other hand, are those which derive from a genetically normal ovum, but during the course of development suffer some *teratogenic* or disturbing influence.

A large number of abnormalities can be either genetic or accidental, and it is often impossible from the subject's appearance alone to be sure of their origin. Only in the subject's offspring can one be certain, as genetic anomalies are by definition heritable while other anomalies, being acquired by the individual, are not.

We shall not list here the extraordinarily varied forms of monstrosity which occur in the human race, but merely recall that they can comprise either an excess of development or a lack of development. The monster is often a badly finished product whose imperfect structure corresponds to a more or less early stage of embryonic development. For instance the 'harelip' corresponds to the foetal stage when the two lateral sections from which the lip is formed have not yet grown together.

Deformities are found at every level of the evolutionary scale, from Ciliophora to mammals.

Deformities in insects

Monstrosity in insects has been the subject of much research, and observers have reported cases of fused, lacking, and supernumerary members, of missing

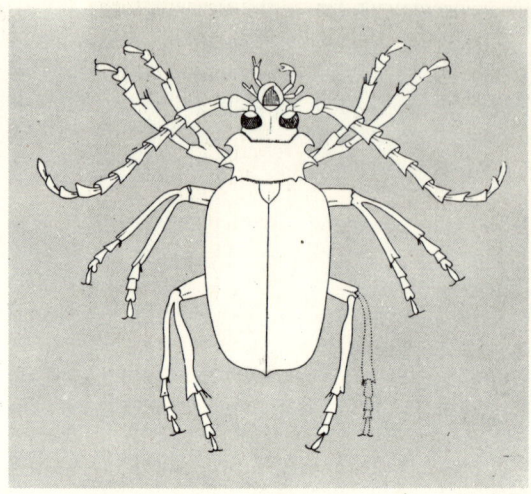

Prionus californicus, found in the state of Washington. Binary schistomelia of legs and feelers. A very rare natural anomaly.

After Jayne.

eyes, of cyclopia, microcephaly, etc.

One of the most curious of these insect monsters is a *Prionus californicus,* found by Morrison in the state of Washington during the last century. All the beetle's legs, its maxillary feelers and its left labial feeler, were doubled. We do not know if this monstrosity was hereditary or acquired.

Experimental production of monsters

As the biologist learns more and more about the mechanisms which control the organism's formation so he becomes increasingly able to manipulate the strings and to disturb development at will, thus bringing about the birth of monsters which are more or less comparable to those produced by nature itself. This experimental production of monsters — or *teratogenesis* — often throws light on the factors which lead to monstrosity in nature, as well as on the processes of normal embryology.

It seems unnecessary, perhaps, to point out that the creation of monstrosities in the laboratory is not undertaken to satisfy a taste for the aberrant or the sensational, but because these creations help us to penetrate the secrets of development. In the chapter on genetics we shall see the numerous ways in which artificial mutations can be brought about and how, by modifying the sex-cell of the parents, hereditary anomalies can be produced. As for non-genetic or acquired anomalies, these can be produced by interfering with the embryo itself during the course of its development. The first

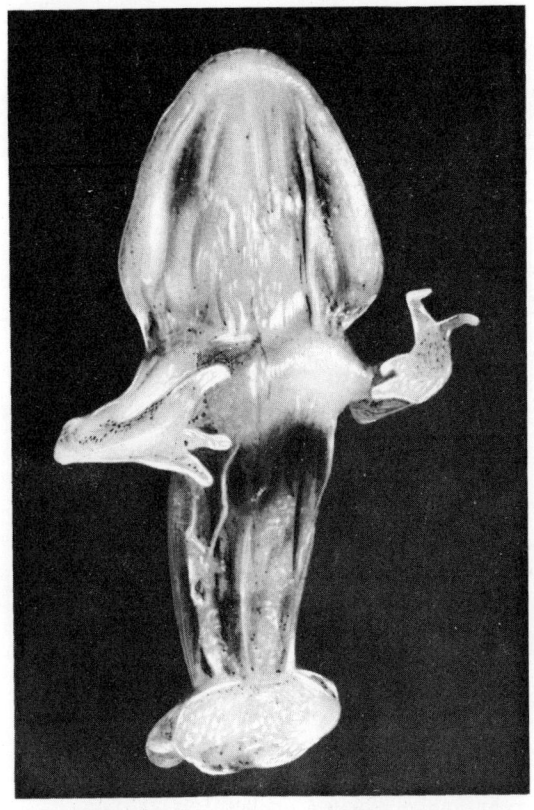

The leopard frog, *Rana pipiens*, with supernumary back legs.
A natural anomaly.

After C. G. Henry, Tunica, Mississippi. Larousse.

The E anomaly in *Rana temporaria*, the common frog. Complete
suppression of lower legs.

After J. Rostand. J. Vincent.

Midwife toad, *Alytes obstetricans*, with supernumary forelegs.
A natural anomaly.

After J. Rostand. J. Vincent.

attempts to alter embryonic development were made
in the beginning of the nineteenth century by the great
naturalist Etienne Geoffroy Saint-Hilaire, who upset
the embryonic development of chickens in various
ways. He incubated the eggs for two or three days in
the normal way. Then, with varying degrees of violence,
he shook them, or at selected points perforated the
shells. Other eggs he kept in a vertical position, either
on the broad end or upside down, or again he coated
part of the shell with wax or varnish to make it impermeable to air. Among the chicks hatched from this
disturbed incubation he found some that were abnormally formed: the anomalies which he had thus
induced resembled those which very occasionally
appear in chicks normally hatched. Etienne Geoffroy
Saint-Hilaire may thus be called the founder of experimental teratogenesis, one of his chief claims to fame.

With the way thus open, Dareste set to work in the
same field, also experimenting with the chicken's
egg. Between 1877 and 1891 he not only repeated
Saint-Hilaire's experiments but tried out many other
methods as well, such as retarded incubation — or
ageing the egg — overheating or chilling, unequal
degrees of heat applied to the shell, and so forth.
In the course of his experiments Dareste obtained
many thousands of artificial monsters, representing
almost every type of single monstrosity.

Today the experimental production of monsters has
become an important branch of biology, and there is
scarcely any physical or chemical agent which has not
been employed in the laboratory for the purpose of
trying to understand the causes of natural deviations
from the normal. One obvious technique is that of
surgery on an organism undergoing development.

Among the young larvae of frogs and toads the legs first appear in the form of small shoots or buds. If one of these shoots is surgically divided into two or three parts a double or a triple leg will be produced instead of the normal member. As early as 1905 Tornier in this way created *Pelobates* with eight legs — six behind and two in front. These results were confirmed and developed by Maurice Lecamp with *Alytes* or midwife toads, and by the same method Wolff has obtained birds with duplicated legs.

Another and particularly effective way of inducing teratogenesis relies on X-rays, rays which have a destructive action on growing tissues, that is to say on tissues in the course of cell-division. Since it is now possible with great precision to direct a beam of X-rays on a given part of the embryo and, moreover, accurately to control the desired dose, it is obvious that arrested development can be localized and therefore a wide choice of monstrosities produced at will. Etienne Wolff, in a series of now classic experiments, employed cylindrical beams of X-rays which he sent through long tubes of lead. The part of the chicken's embryo to be irradiated was placed where the beam emerged, almost in contact with the tube. In this manner Wolff reproduced most of the spontaneous monstrosities which are known to occur among chickens. The type of malformation produced depends not only on the situation of the zone irradiated but also on the stage of embryonic development and the period of irradiation. Wolff concluded that 'any embryo can be turned into the type of monster decided upon in advance if intervention takes place at the appropriate stage and under appropriate conditions of localization.'

This is a marked advance on the days when Dareste wrote: 'The causes of monstrosity are not, as medical practitioners say, specific causes, since they do not produce effects which can be determined.' Wolff, with his method of localized irradiation, even obtained monstrosities which have never appeared in nature, or at least which have never been reported, such as *anterior symelia* or *sympteria,* that is, fusion of the wings into a single central wing.

Among other physical agents which can produce anomalies in the laboratory are radium rays, ultra-

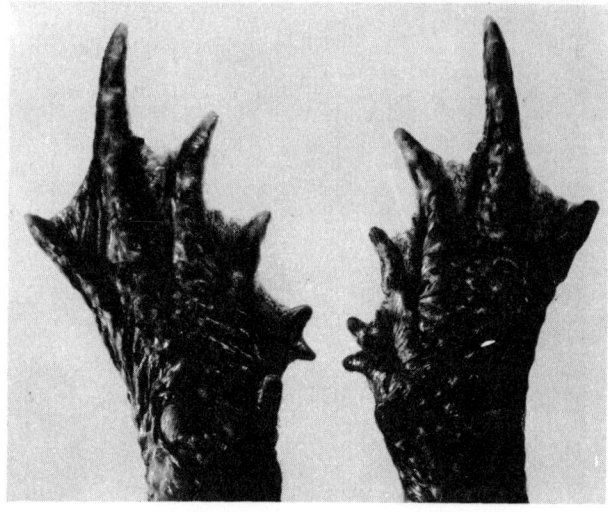

Bilateral polydactylia in the common toad *Bufo bufo.* Feet with six instead of five toes. A very rare anomaly. Polydactylia is hereditary in toads.

After J. Rostand. Feher.

Apparatus used to localize irradiations. The X-rays which issue from the generator above are canalized by a lead tube of narrow interior diameter. A shaft of brilliant light plays on the beam of X-rays which, emerging from the tube, strikes the embryo in the egg at the precise point desired by the experimenter.

After Et. Wolff.

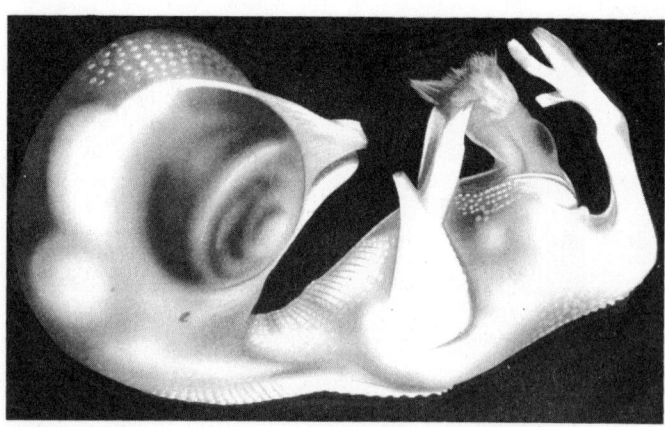

Monster embryo of chicken experimentally produced by accurately localized irradiation. The two budding members are fused into a single leg situated in the prolongation of the dorsal axis.

After Et. Wolff.

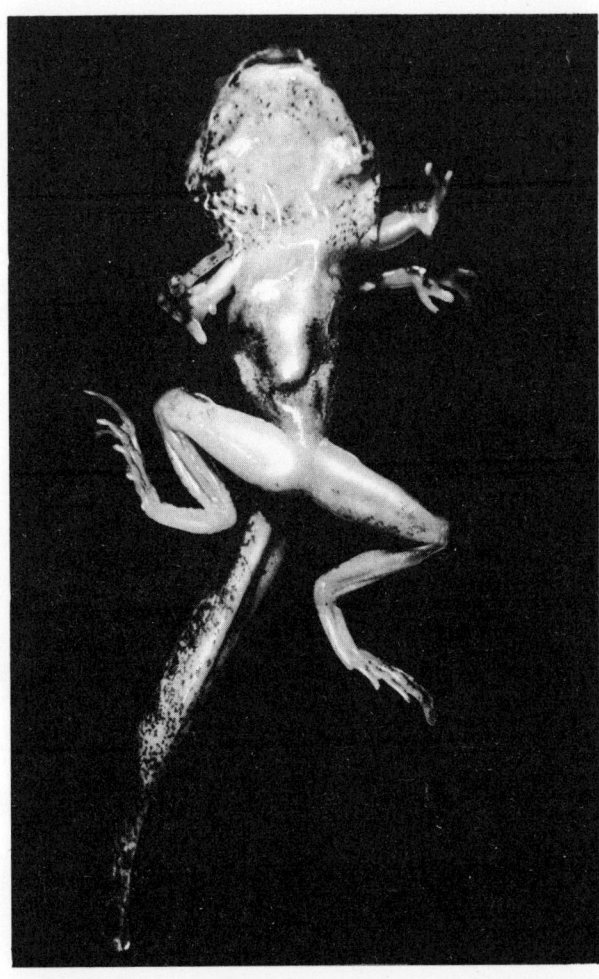

Above, doubling of the left foreleg and *below*, multiple anomalies of the hind legs. These malformations were produced in frogs by ultraviolet irradiation.

After J. Rostand. J. Vincent.

violet rays, and ultra-sounds. With the common frog. J. Rostand easily obtained both electromelia and polymelia (atrophy or duplication of limbs, anterior and posterior) by exposing newly hatched larvae to ultraviolet rays for a few minutes. Schinz, with *Xenopus,* obtained duplicate limbs by the action of X-rays on the embryo. Between 1893 and 1901 Charles Fere tried out a then new method of producing monsters: namely, the action of chemical substances introduced into the white of the egg during incubation. He used mineral salts, alcohols, sugars, peptone, alkaloids, creatine, hydrocyanic acid, cantharidin, antipyrine, poisons, microbe toxins, etc. Also, he exposed the egg to mercury vapour, the vapours of ammonia, phosphorus, musk, various essences, tobacco, ether, and chloroform. By these means he obtained a large number of malformations, but was unable to establish an exact relationship between the use of a given substance with the appearance of a specific anomaly.

Later the researches of Herbst, Gurwitsch, and others revealed the specific action of certain substances, notably that of lithium salts on the development of amphibian eggs, while Stockard in his turn showed that with the fish *Fundulus heteroclitus* it is possible in almost all cases to produce cyclopia by submitting young embryos to the action of sea-water which contains an excessive amount of magnesium chloride. The results of this treatment are so consistent that Stockard assures us that, if sea-water contained a greater amount of magnesium, cyclopia would be the normal characteristic of the species and monstrosity would consist of the presence of two eyes!

In 1933 Robinson reported that chickens fed on grain containing selenium frequently gave birth to chicks with malformed beaks and claws. In 1938 Suzanne Lallemand demonstrated the specific role of colchicine in causing *strophosomia* — or the turning over of the embryo — among chickens. Her findings were the point of departure for a long series of fruitful

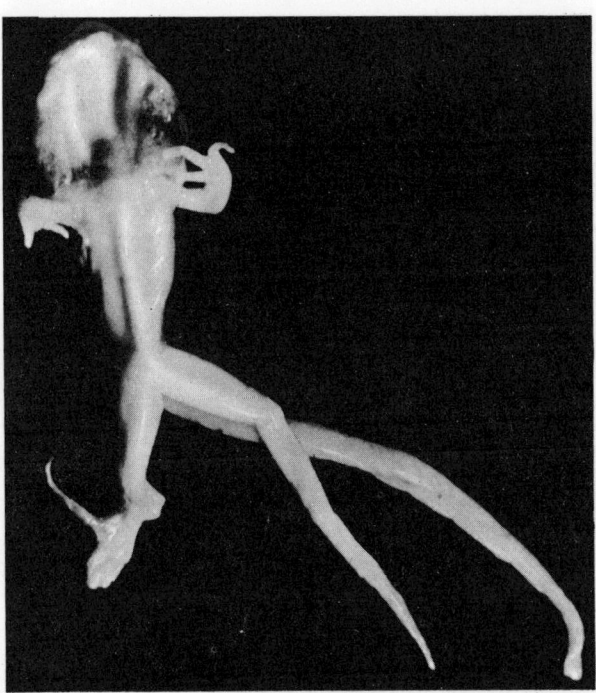

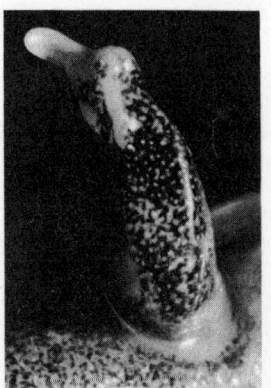

Chemical teratogenesis in the common frog. Malformation of the forearm after treatment with eserine-trypaflavin.

After J. Rostand. J. Vincent.

183

researches into the selective sensitivity of embryos towards particular chemical substances.

Also with embryo chickens, the great biologist P. Ancel produced the following abnormalities: *anuria* (taillessness), using arsenical salts; *celosomia* (hernia of all the viscera), by colchicine, saponine, and ricine; *ectromelia* (absence of limbs) and *brachymelia* (shortening of limbs) by sulphamides and eserine; *ectrodactylism* (absence of one or more digits), by arsenical salts, eserine, and colchicine; *syndactylism* (fusion of digits), by colchicine and trypaflavin; 'parrot-beak' by sulphamides and eserine. The sensitivity of the embryo to these chemical agents varies with its age, and treatment is only effective if made at the right time. The identical malformation can be brought about by substances of quite different chemical constitution, while substances of very similar constitution can produce monstrosities of quite different kinds.

Among the very numerous anomalies which can be chemically induced, mention should also be made of 'cross-beaked' chickens as a result of urethane and certain deformities of the legs among frogs by means of the extract of sweet-peas *(Lathyrus odoratus)*. There are many anomalies among mammals caused by pilocarpine hydrochlorate, thallium salts, or boric acid. Another well-known induced deformity is that with Coleoptera of *Tribolieum confusum* (doubled legs), using ethylouinone. J. Rostand, working with the common frog, has produced ectrodactylism and syndactylism by means of certain chemical compounds, such as trypaflavin and eserine.

From the theoretical point of view the result of these researches is of great significance, for it shows us that certain anomalies produced by chemical substances are faithful reproductions of anomalies which are of germinal origin and can thus be inherited. The taillessness, the shortening of limbs, the malformation of the beak which Ancel has produced among chickens are in every way comparable to the same anomalies which appear in chickens by mutation and constitute the characteristic traits of certain breeds. Thus the biologist is led to believe that even in the case of hereditary anomalies the appearance of the abnormal character is related to the production of chemical substances, synthesized by the organism itself in the course of its development.

Among the agents of teratogenesis the hormones occupy a place apart. As we have already seen, the sex hormones, by stimulating or inhibiting embryonic genital development, can create inter-sexed anomalies. As W. Landauer demonstrated, the pancreatic hormone, *insulin,* when introduced into the hen's egg during the first three days of incubation, brings about the partial or total absence of the tail. Introduced into the egg a little later it affects not the tail but the beak and eyes, and also the legs which it shortens — a condition known as achondroplastic micromelia.

It is thought that insulin causes such anomalies by lowering the sugar content of the embryo's blood, but for the moment the connection between lack of sugar and skeletal malformation is not fully understood.

Adrenalin, acting on the foetus of the rat or of the rabbit, causes haemorrhage at the extremity of the limbs, resulting in necrosis.

Cortisone, administered to pregnant female rats, leads to the appearance of harelip in from eighteen to 100 per cent of the offspring. Sensitivity to the treatment varies according to the breed of rat.

Also in rats, an excess of vitamin A results in foetal malformation of the head, the eyes, the palate and the spinal column. Insufficient vitamin A leads to a reduction of the eyes. A lack of vitamin B2 can result in various deformities — cleft palate, fusion of digits, absence of certain bones. A lack of vitamin D can produce curvature of bones or of limbs. The work of Warkany and Giroud in this field has been very instructive. A deficiency of calcium can also lead to malformation.

The causes of anomalies in human beings

The lessons learned from all these investigations are of immediate importance to the human race, for they show us that certain anomalies can be caused not only by agents commonly employed in therapeutics, such as X-rays, and medical products like sulphamides and alkaloids, but also by vitamin insufficiency, hormone disturbances and hyperthermia.

It has been directly established that radiotherapy of pregnant women is not without danger to the unborn child, and all doctors now agree that the greatest caution must be used in employing X-rays in such circumstances. The most dangerous period would seem to be between the first and the third months of pregnancy.

Investigations made at Hiroshima and Nagasaki in 1951 have tragically demonstrated the disastrous effects which atomic radiation has on the unborn. Within a radius of 2 kilometres from the place the bombs exploded twenty-eight per cent of the women who were pregnant (ninety-eight out of the 1774 survivors of child-bearing age) had premature, stillborn foetuses. Among those which were conceived less than three months before the date of the explosion the mortality rate was highest. Even among those children which were born at the normal time there was

The P anomaly in *Rana esculenta.* Very serious forms, characterized by extreme deformity of the members, especially the hind legs (1, 2, 3, 4, 5) and the appearance of tumoral excrescences (4). Serious forms (6 and 7) characterized by thickening of the thighs and the presence of protuberances. The feet of adult polydactyls (8, 9, 10); forefoot with bifurcated first digit (8); hind foot with eight instead of five toes (9); forefeet with six instead of four toes (10). 11. The lower left member has been amputated in a young larva and a normal member has grown in its place. The regenerated leg contrasts with the abnormal right leg.

After J. Rostand. Feher; Larousse; J. Vincent.

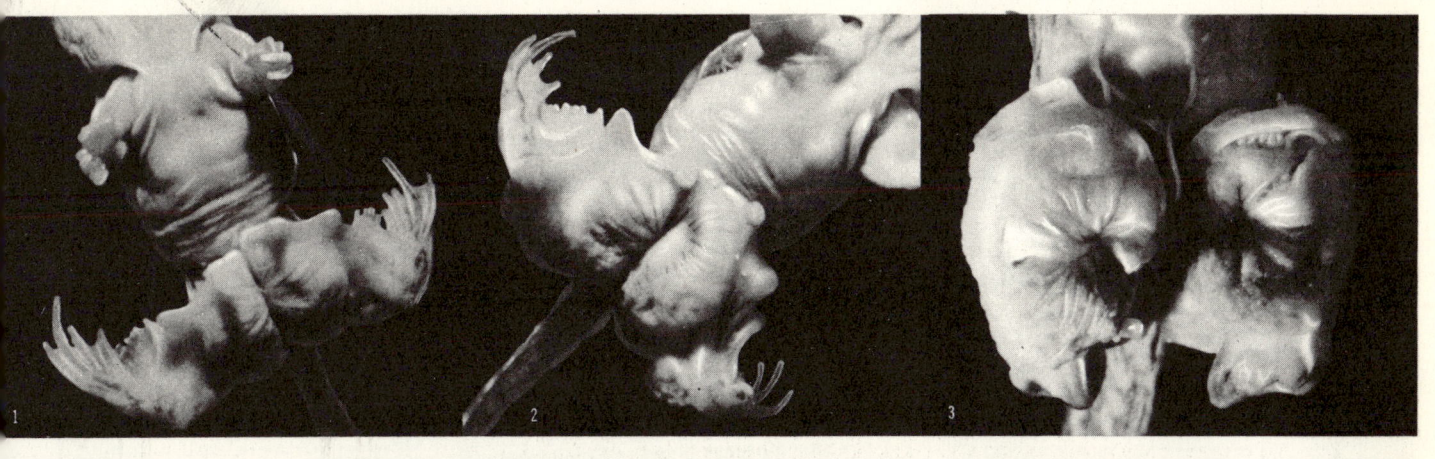

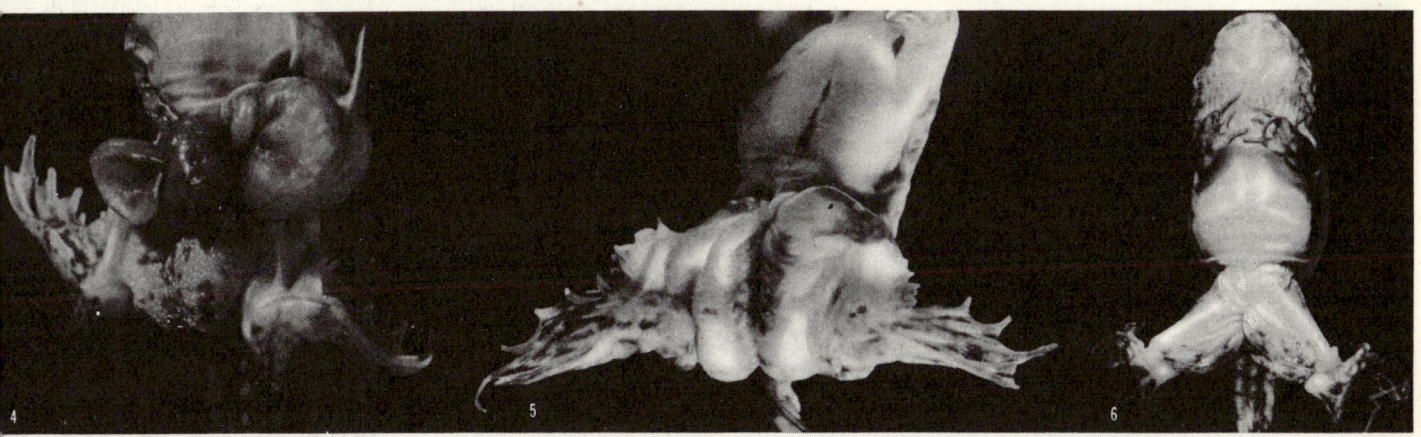

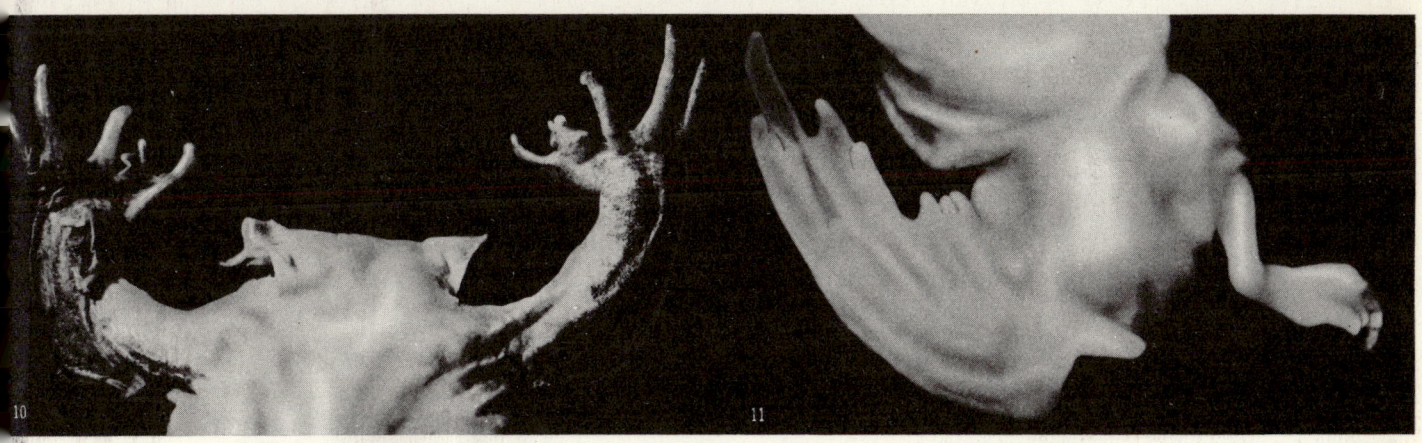

185

an abnormally high mortality rate, delayed growth, various physical or mental deficiencies and, notably, certain cranial deformities.

In so far as the secretion of hormones depends on the sympathetic nervous system, it must be admitted that violent maternal emotions can be the origin of certain foetal malformations — which forces us to revive that very ancient and tenacious popular belief which scientists once thought fallacious.

The responsibility of viruses in the genesis of human anomalies is very great. Doctor Gregg, an Australian ophthalmologist, was the first to demonstrate the teratogenic action of the German measles virus during the course of a serious epidemic which broke out in Sydney in 1940.

The earlier the expectant mother catches the disease the more pronounced its effects can be: according to recent statistics, if the virus attacks before the seventh week as many as ninety-five per cent of the babies born may be affected; before the eighth week, eighty-six per cent; between the ninth and the twentieth, twenty per cent.

The malformations may affect the heart, the eyes, the ears; after the third month only deafness results.

The teratogenic menace created by German measles — a disease so mild in ordinary circumstances — seemed to the Australian government sufficiently dangerous to warrant preventative measures, encouraging girls to have German measles so that they will be immune during those years when pregnancy is most common.

Medical abortion has also been considered in cases of maternal German measles, but the deliberate elimination of a human being — even if it is almost certainly doomed to deformity — always involves a moral dilemma for doctor, lawyer, and public conscience alike.

It is possible that other viruses, including measles, mumps, smallpox, chickenpox, influenza, poliomyelitis, shingles, scarlet-fever, may be the cause of embryonic malformations, perhaps even when the infection is not apparent in the expectant mother herself. Experimentally, the inoculation of embryo chickens with the virus of influenza, or of mumps, has produced malformations of the head and abnormal curvature of the body.

We cannot for certain say that the human embryo is more sensitive to the virus of German measles than it is to other viruses. On the contrary, it is possible that it is the mildness of the German measles virus which permits the embryo to survive. In recent years the connection has been discovered between the tranquilizer thalidomide and a new form of 'polyneuritis', an incurable nervous disorder causing various malformations of the foetus. A large proportion of pregnant women who had taken the drug gave birth to deformed children. The main effect appears to be abnormal development of the limbs, which in extreme cases, are non-existent.

Finally, among the causes of human anomalies, we must not forget those of a purely mechanical order, such as abnormal position of the foetus, abnormal tightening of the amnion, loops in the umbilical cord, and so on, giving rise to compressions, vascular disturbances, and amputations.

The prevention of monstrosity

In recent years embryologists have discovered that certain substances have the power to prevent the appearance of certain monstrosities, both genetic and acquired. In chickens Landauer has arrested or at least reduced the teratogenic effects of insulin by means of nicotinamide. The same substance also combats the monstrosity-producing effects of eserine, sulphamides, and arsenical salts, while pyruvic acid gives protection against certain teratogenic effects of insulin.

The teratogenic effects of X-rays can be lessened, if not neutralized, by cysteamine and methylamine which, broadly speaking, have the property of protecting living tissues from the deleterious action of radiations.

The majority of embryo chickens which were experimentally treated with cysteamine before being submitted to harmful radiations developed normally, while others which received the same amount of irradiation developed typical abnormalities. Similar results have been obtained with methylene blue, whose protective quality seems to be connected with its power as a reducing agent. Certain 'anti-teratogenic' treatments can be applied to hereditary anomalies as well as to acquired anomalies.

While normal chicken have only four claws, some breeds — Dorkings, Houdans, Brahmas, Faverolles — have five. As Sturckie and Warren proved, the appearance of this 'polydactyl' anomaly can be prevented if, between the second and third day of incubation, the eggs are submitted to a temperature $1\frac{1}{2}°$ below the usual temperature of incubation. Only four per cent of the embryos thus treated will have five claws, while without the treatment 100 per cent are polydactyl.

W. Landauer obtained results of a comparable nature by introducing insulin (the hormone of the pancreas) into the embryo. This hormone treatment does not completely suppress polydactylism, but it lessens its frequency and especially it reduces the hereditary five-claw anomaly to one foot only.

P. Ancel and J. Courtial demonstrated how a certain alkaloid of vegetable origin, colchicine, also exercises an inhibitory action on polydactylism when it is applied in very dilute solution to the two day old embryo. While the frequency of the anomaly in the untreated eggs was ninety-seven per cent (eighty-three per cent on both feet, fourteen per cent on one only) that of eggs treated with colchicine was sixty-four per cent (fifty-three per cent bilateral, eleven per cent unilateral).

Among guinea-pigs the advancing age of the mother exercises an anti-teratogenic effect: for instance, in one line of guinea-pigs twenty-nine per cent of the offspring of young females — less than six months old — had extra toes, while females more than fifteen months old gave birth to only six per cent with this anomaly. In other lines of guinea-pigs the percentages have been thirty-four and twelve, sixty-eight and twenty-two, eighty-one and thirty, for young and old mothers respectively. We do not yet know exactly how to interpret this curious relationship between ageing mothers and their normalizing effect on the foetus, but it is doubtless connected with some chemical change in the uterine environment.

The P anomaly

Among free-living green frogs a very curious anomaly is encountered, the cause of which still remains obscure. Jean Rostand found that of the tadpoles in certain ponds near Concarneau in Brittany as many as thirty-five to forty per cent were deformed. These animals not only have extra toes and fingers (poly-dactylism) but also supernumerary legs, bony excrescences of various types, and finally a characteristic malformation of the limbs known as the P anomaly. There can be as many as twenty toes on the back legs and as many as seven on the forelegs.

Apart from these deformities the animal is completely normal; it shows no signs of illness, swims actively, and eats voraciously, exactly like an ordinary tadpole. If, however, the anomaly is especially pronounced it infallibly leads to death at the moment of metamorphosis, that is to say at the moment when aquatic life gives way to life on land. This early death appears to be the consequence of purely mechanical defects

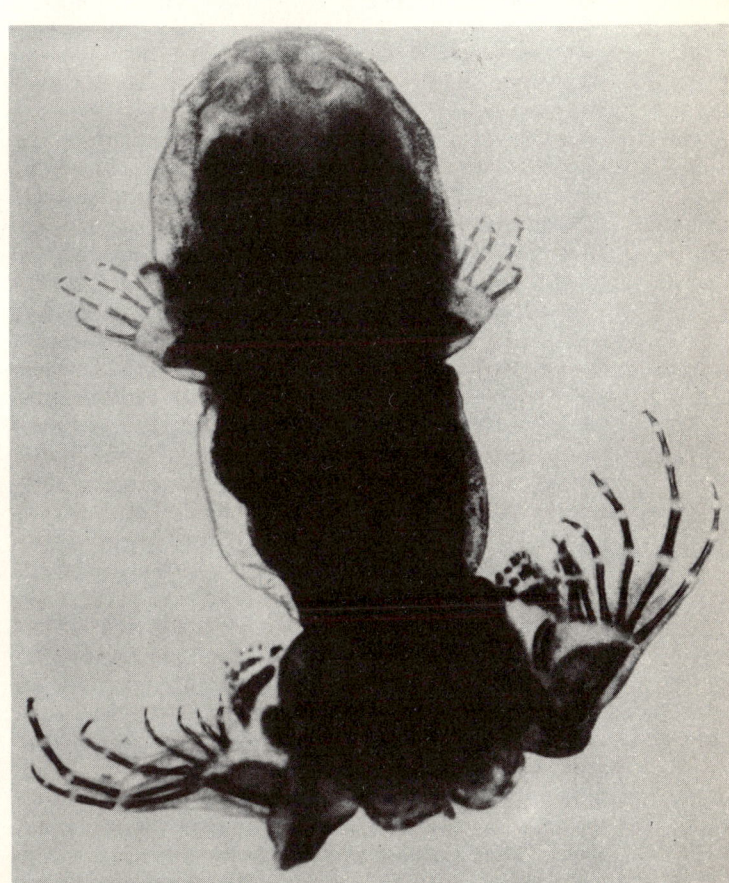

Abnormal young frog lightened with potassium and coloured with alizarin to show bone structure.

After J. Rostand. J. Vincent.

The P anomaly in *Rana esculenta*. Above, feet with seven toes. Below, skeleton of foot with ten toes.

After J. Rostand. Larousse.

187

linked to the deformity of the skeleton.

In a state of nature, then, we may assume that all animals seriously affected by the P anomaly die before leaving the water or at least soon after the beginning of their air-breathing existence. That is why, among adult frogs in these ponds, none are found with supernumerary legs and feet, or with bony excrescences. In its severe form the anomaly is lethal: only individuals attacked by *simple polydactylism* survive, though these can have as many as nine toes and six digits. It is, moreover, interesting to note that if, as sometimes happens among green frogs, the tadpole does not undergo metamorphosis during its first year of life, it can in spite of the gravity of its deformity, continue to live.

An examination of abnormal subjects, lightened by alkalis and coloured red by alizarin, reveals the varied and profound modification of the skeleton. The long bones are much thickened and shortened. Their deformity recalls that which is found in certain cases of foetal rickets. The P anomaly appears very early in the animal's development. It can be seen in the embryo as a characteristic flattening and broadening of the sprouting limb. The question arises as to whether it is a hereditary variation, in other words, a mutation comparable to hereditary polydactylism in toads. This is extremely doubtful. Numerous crosses have been made in both directions, that is, between normal frogs and polydactyl frogs, and between polydactyls themselves (seven-toed males and seven-toed females). In all cases the result is offspring with a completely normal number of toes. The anomaly is thus not transmissible, or at least not according to the classic laws of heredity.

In addition, if the hind leg of a very young tadpole suffering from the P anomaly is amputated it will always grow again as a normal leg — a proof that the anomaly is not constitutional in so far as it depends on the genetic composition of the tissues.

We are thus led to conclude that the cause of the anomaly is external, brought about in the very young tadpole by some environmental factor. The first thought that occurs is of a chemical action, exercised either by the water in which the animal lives or by the vegetation on which it feeds. But by rearing larvae from the egg itself in water taken from the ponds in question the P anomaly cannot be reproduced. Again, no factor is known, physical or chemical, including radiation, which is capable of bringing about the malformations characteristic of the P anomaly. It has been suggested that there is some monstrosity-producing virus, and indeed such a hypothesis would explain the aspect of the disorderly anarchic proliferation which the tadpoles display, and also the impression is given of a morbid process not unlike that which takes place in cancer. In extreme cases of the P anomaly genuine tumours do, in fact, develop in the inguino-ventral region of the animal. Several types of cancer such as Rous' sarcoma in poultry, Shope's papilloma in rabbits and Lucke's carcinoma in the Canadian leopard frog are caused by viruses. It is true that among animals no virus is known which is capable of making extra toes or feet appear but among plants viruses are found which stimulate the formation of outgrowths on leaves.

This is a phenomenon known as *enation*.

The P anomaly is not limited to the pond in Brittany where it was discovered to be so common and endemic. Many cases of simple polydactylism have been observed in many other countries of the world, all resembling the abnormalities found in the adult polydactyls of Brittany. It is reasonable to assume that among them all polydactylism is merely a certain benign form of the P anomaly.

If polydactylism is indeed caused by a virus, presumably the virus is either found exclusively in certain regions, or even in certain ponds, or else only certain racial strains are susceptible to contamination. The P anomaly has until now been found only in the edible green frog *(Rana esculenta)*. Other frogs, the common frog, the tree-frog; toads, and all tailless amphibians appear to be immune. Possibly the anomaly is related to other curious abnormalities such as the unilateral duplication of posterior legs, which was discovered by Dr Voitkevitch among Causasian frogs, perhaps also to the 'massive polydactylism' reported in 1947 by the American naturalist, Bishop, among an amphibian of the order Urodela.

In a lake situated at a great altitude near Boulder, Colorado, nineteen salamanders of the species *Ambystoma mexicanum* were collected and found to have extra feet and toes on their posterior limbs. Most of the animals were dead by the time they reached the laboratory and, as the few survivors did not reproduce, no data could be obtained as to the possibility of the abnormal characters being transmissible. Although Bishop himself was strongly inclined to believe in the genetic nature of the anomaly it seems more likely that here again we have a case of acquired polydactylism and polymelia, which is therefore not transmissible. The causes are probably the same as those of the P anomaly, and if the enigma of the deformed frogs of Brittany is solved the mystery of the deformed salamanders of Colorado will doubtless be cleared up at the same time.

Effects of atomic radiation

In the course of the summer of 1957 the P anomaly occurred again in dramatic fashion in Holland. Newspaper headlines announced that deformed frogs had been gathered from a small canal in Amsterdam into which radio-active waste from the Institute of Nuclear Research had been dumped.

Biologists of the Zoological Museum at first held that the anomalies could be due to radio-active contamination of the water. The physicists, on the other hand, insisted that the water showed not the slightest trace of radio-activity and that there was no proof that radio-active waste had in any way affected the frogs' eggs. An official enquiry revealed that the phenomenon was a typical case of the P anomaly. This meant elimination of the radio-activity factor, especially since there was no institute of nuclear research near the ponds in Brittany and the degree of natural radio-activity of the pond water, examined by competent technical authorities, was perfectly normal. In addition, neither the abnormal frogs nor the abnormal tadpoles registered the slightest degree of activity when tested with a Geiger counter, which shows that they contained no selective accumulation of radio-isotopes as has been known to occur.

Finally, there is no reason to believe that radio-activity, even in the hands of a skilled operator, can produce malformations comparable to those which characterize the P anomaly. Though the action of X-rays and of ultraviolet rays on the larvae of amphibians does result in important deformities of the limbs, the resemblance between these artificially induced deformities and the P anomaly is merely superficial.

It is interesting to note that among the frogs found in the canal in Amsterdam in 1957 the proportion of monsters was approximately the same as that found in the ponds of Brittany. Whatever the monstrosity-producing agent may be it seems to have operated everywhere in a comparable fashion and without attacking all the individuals in the same aquatic environment.

Infection and variation

With these acquired anomalies, caused perhaps by a virus, we may compare those structural modifications which certain plants reveal under the influence of intracellular parasites.

In sweet-peas, *Bacterium tumefaciens* causes the fasciation (coalescing and abnormal thickening) of stems and branches. In *Datura* a certain mosaic virus causes the glabrous condition of the capsule and the division of the tubular corolla into separate petals. This infection can even on occasion become hereditary and thus give rise to a new variation of the species. Such was the origin of certain mutations of the evening primrose *(Oenothera)*, earlier observed by the great Dutch botanist, Hugo de Vries.

One of the commonest of the mutant species which thrived in the famous garden of Hilversum was the dwarf evening primrose *(Oenothera nanella)*. This is no more than 20-30 centimetres high, in other words, scarcely a quarter as tall as the plant from which it was derived — *Oenothera lamarckiana*. As soon as it reaches some 10 centimetres it begins to bloom and the flowers are all the more striking because of the plant's small size. Its leaves are broad and short, especially at its base. This dwarf evening primrose is not, as we might suppose, due to the mutation of a gene. It is simply an *Oenothera lamarckiana* infested by a particular microbe which, contaminating the seeds, is regularly transmitted to subsequent generations. All the characters proper to the species *Oenothera nanella* are due to the presence of the micro-organism in the tissues of the plant.

A. Chevalier went so far as to wonder whether all the mutations, studied by de Vries, even those which seemed exempt from parasites, may not have been brought about by bacteria or viruses which at that time infested all Dutch evening primroses. 'Oenothera of the family Onagraceae,' he wrote,' have an especially

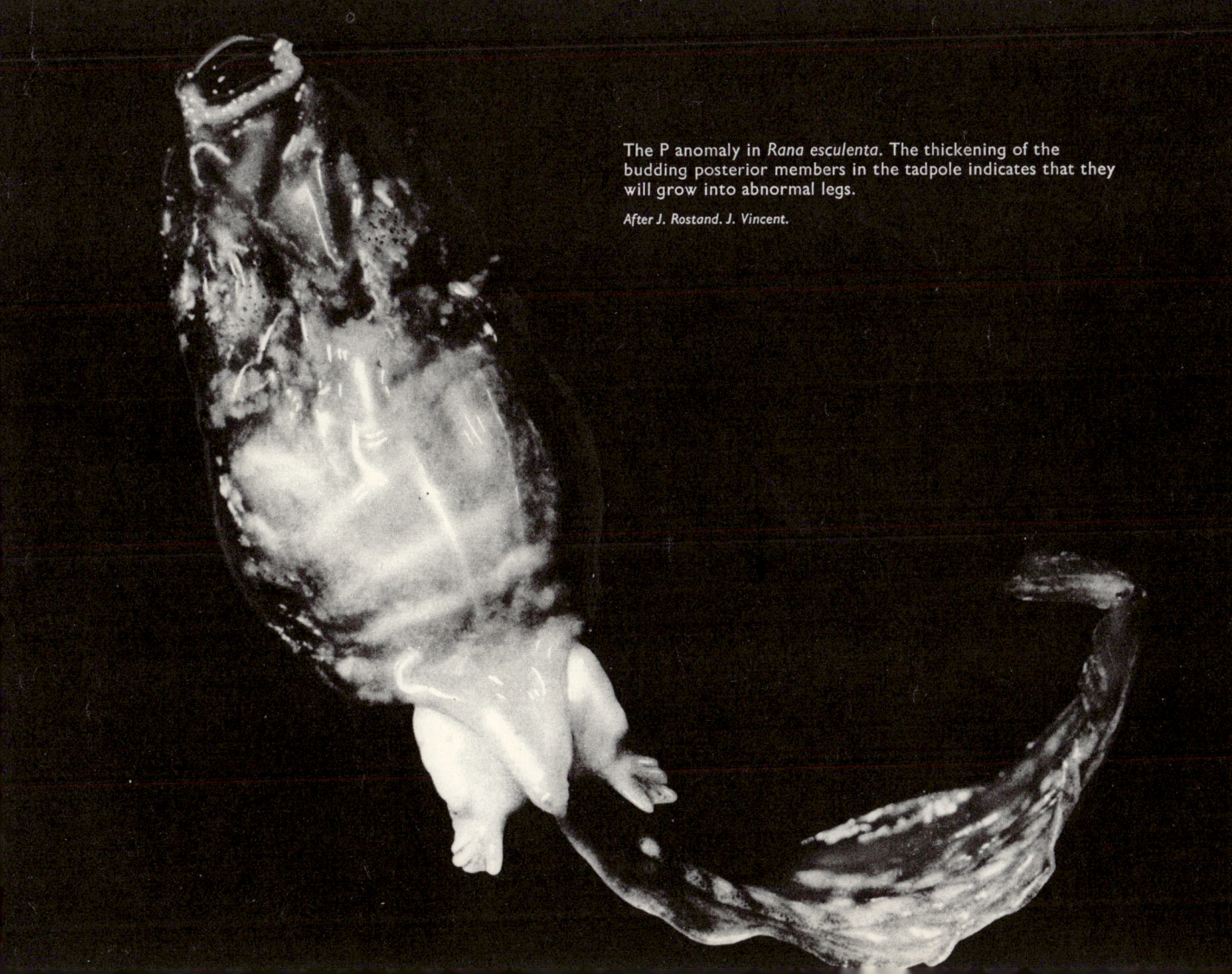

The P anomaly in *Rana esculenta*. The thickening of the budding posterior members in the tadpole indicates that they will grow into abnormal legs.

After J. Rostand. J. Vincent.

complex chromosome equipment, but this does not exclude the possibility that variation may have been brought about by viruses secreted by bacteriaceae, or by mutant viruses inducing hereditary changes... A plantation of *Oenothera biennis* which we have cultivated for several years also reveals anomalies of foliage which bear witness to the fact that most of the plants are attacked by a virus.'

An analogous question arises with regard to the curious ivy-leaved toadflax discovered by Touton.

The common ivy-leaved toadflax, or *Linaria cymbalaria,* is a small and pretty plant frequently found growing on old and damp walls. It belongs to the Scrophulariaceae family and is closely related to the speedwells and to cow-wheat. The corolla of its delicate mauve flowers is formed by two lips like those of a miniature snapdragon. Its stem straggles and its leaves are shaped somewhat like ivy leaves, although the lobes are blunt. It begins to flower in the spring.

In 1936 the botanist J.-B. Touton found, on an old wall in Laval, three tufts of an aberrant ivy-leaved toadflax which differed from the typical plant chiefly by its laciniate leaves, the majority of which were divided into three lobes. Having gathered and cultivated these odd specimens he obtained seeds from which he grew plants which were exactly like the parent plants. As the anomaly continued from generation to generation he was naturally led to conclude that it was a case of hereditary variation, or mutation. Botanists were in agreement in considering the new ivy-leaved toadflax as a true species and it was named Touton's ivy-leaved toadflax, or *Cymbalaria toutoni.* The structural characteristics of its flowers and of its fruits are, in fact, distinctly different from the normal flower. On the appearance of the two cotyledons during germination special characteristics are already apparent, and as the plant spreads out in spring it too displays unusual characters: small slightly fleshy leaves, deeply trilobate or even trifoliate, and at times reduced to a single lanceolate leaflet.

The old wall on which the first specimens of the 'mutation' grew has today been torn down, and nowhere else since then has Touton's ivy-leaved toadflax been found. But the species has not for that reason been lost, for it continues to be cultivated not only by M. Touton himself but by numerous gardening enthusiasts and in several botanical gardens, notably the alpine garden of the Museum of Natural History in Paris.

The plant has sown itself, been sown by gardeners, or been multiplied by cuttings. The characters proper to the new species have remained constant and unvaried. Touton's ivy-leaved toadflax is, in appearance, as vigorous and resistant as the original plant. Without showing the slightest signs of degeneration it continues to produce stems which bear fruit and flowers, and can attain a length of 1½ metres. When, however, it competes with the normal species its inferiority is marked. The same observation is true for a number of mutations which in natural conditions and without the protection of man are not long in becoming extinct. But is Touton's ivy-leaved toadflax really a mutation?

According to A. Chevalier the genesis of the plant arises from an exceptional and most interesting me-

chanism. It would seem that no change of genes or alteration in the number of chromosomes is involved, but instead a kind of infectious disease due to a virus which, by way of the seeds, is transmitted from generation to generation.

In defence of this hypothesis Chevalier points out that the leaves of *Cymbalaria toutoni* sometimes bear excrescences, adventitious little tongues resembling those by which the virus of enation reveals its presence on the leaves of tobacco and tomato plants. The excrescences tend to disappear from the leaves of *Cymbalaria toutoni* as the new species becomes longer established. It is as though 'the virus was less active and tended towards a state of equilibrium, although the new species retains the specific characters which differentiate it from the common ivy-leaved toadflax, characters which persist and establish its right to be classified as a Linnaean species, producing fertile seeds which breed true.'

If this mutation is really the work of a hereditary virus we must then agree that the virus is present not only in the seeds but also in every cell of the plant, since the aberrant type can be propagated just as well by cuttings as by seeding. Thus, in some cases, a virus infection could become the creator of species.

It is thus not inconceivable that evolutionary changes may be brought about by kinds of virus-producing alterations which cannot be considered as pathological. A. Dalcq did not exclude the idea that such phenomena of symbiosis may have had much to do in major changes of species. It was also the view of Noel Bernard and J. Magrou that symbiosis, or the association of dissimilar organisms, played a major role in the evolution of the vegetable kingdom.

Ivy-leaved toadflax *(Linaria cymbalaria).*
Larousse.

190

Growth

Growth is characteristic of living organisms, and life itself might almost be defined as the ability to grow. Every living being, in fact, tends not only 'to preserve in being' but also to enlarge itself at the expense of the material or food that its environment supplies.

Let us, for example, consider a ciliophoran, a single-celled organism, in a liquid — a broth of hay — where it finds suitable nourishment. Little by little its dimensions increase and when it has reached a certain size it divides into two smaller organisms which in their turn have only to grow to become exactly like their parent. And so it will continue, the ciliophorans dividing every day or every second day. The biologist Metalmikoff kept ciliophoran (Paramecium) cultures for more than twenty-two years, that is to say for more than 8,000 cellular generations. Had all the individuals which could have descended from the first ciliophoran to appear on earth been preserved they would form a mass of protoplasm larger than the sun! Danysz calculated that the cholera bacillus could in one day supply 10 tons of bacillary matter.

In the formation of an organism, a human being for instance, growth and division alternate. It is of course true that in the first stages of development the egg divides without increasing its mass: it is simply split up into smaller cells by *segmentation,* but growth soon ensues and, from one division to another, the cellular mass is augmented.

Some sixty cell-divisions suffice to produce, in nine months, a new-born baby from the initial cell which was the egg. The rate of cell-division in this case is not, then, very rapid. But growth will continue after birth for eighteen to twenty years. Post-natal growth is due both to the acquisition of new cells and to the enlargement in certain tissues of the cells themselves.

The rate of human growth is illustrated in the following figures: the human egg measures from one to two tenths of a millimetre and weighs about one thousandth of a milligram: the new-born baby mea-sures some 19½ inches and weighs some 6½ pounds. Thus the human organism has increased roughly 5,000 times in length and several thousand million times in weight. From birth to maturity it will become about twenty-five times heavier and a little more than three times taller.

The phenomenon of growth, by the very fact that it is so widespread, raises a fundamental biological problem: namely, how can a substance as complex as protoplasm increase at the expense of the relatively simple and ordinary materials which it finds in the outer world?

It is obvious that the growth of a living organism is in no way comparable to the growth of a crystal which, placed in liquid of the same constitution as itself, little by little grows larger. In the case of the crystal, like matter is simply deposited on like matter, while in the case of the living organism, matter, or food, is trans-formed, first broken down, then built up or synthesized. In a word, the living being must convert nourishment into its own substance or, as we say, *assimilate* it.

A full understanding of the phenomenon of growth — or of assimilation, which comes to much the same thing — would be an important step towards the solution of the problem of life. Whether or not it is legitimate to speak of growth and assimilation in connection with intra-cellular genes and viruses has been much debated. In the case of these elements it does not seem that division in the true sense of the term takes place. When a gene (or a virus) is replaced by two daughter genes (or two daughter viruses) it is thought to be the entire cell which produces a 'copy' of the element instigated by gene or virus. In this hypothesis the gene (or virus) does not reproduce itself: it causes itself to be reproduced by the cell.

By and large growth is much more active during the early period of the individual's life. Embryonic tissues divide more than those of the young animal. There are periods in which growth is accelerated: in the human race two spurts of growth are evident, one towards the age of five and a half and the other at puberty. The rate of growth varies greatly in different animal species. It is, for example, rapid in the pigeon whose weight doubles within forty-eight hours of its birth and in twenty days increases from 25 grams to 435 grams.

To double its weight at birth a rabbit requires six days, a dog nine, a pig fourteen, a sheep fifteen, a calf fifty-seven, a horse sixty, and a man 180. Thus the human being's rate of growth when compared with that of other animals is remarkably slow. Similarly the human egg takes 280 days to attain a weight of 6½ pounds, while that of the cow in 300 days attains a weight of 77 pounds, the weight of the new-born calf. Later we shall examine what consequences this singular delay in human development may have entailed.

All parts of the young individual's body do not grow at a uniform rate. Thus the relative size of various organs and parts of the body alters appreciably with age. As everyone knows the new-born baby's head is disproportionately big and its limbs disproportionately short when compared to those of an adult. From birth to maturity the length of the head is only doubled while the length of the trunk is trebled and that of the arms quadrupled.

Some organs grow more quickly than the rest of the body while others grow more slowly. The brain of the new-born baby accounts for twelve per cent of its total weight, while that of the adult is only two per cent. On the other hand, the growth of the sexual apparatus is belated.

Among animals such as toads and snakes growth is accompanied by periodic changes of skin — sloughing. Among those that are imprisoned in a hard shell, the crustaceans for example, growth can take place only in the interval between shedding one shell and growing another. Why in the majority of organisms growth is arrested when the body has reached a given size, known as the limiting size, is a problem which has not been solved. Nevertheless, even in an animal which has achieved its full stature, a certain power of growth persists in certain tissues. These include the skin, glandular epithelium, blood tissues, etc., which constantly renew themselves. These temporary tissues differ in this respect from permanent tissues like striped muscle and nerve tissues which, once formed in the young organism, are never again renewed.

Due to influences which are not well understood a capacity for exaggerated growth is sometimes awakened in certain tissues which then begin to proliferate in a disorderly fashion to the detriment of normal tissues, thus producing a malignant tumour or cancer. It may be hoped that the biological study of normal growth will help us to find effective means to combat these abnormal growths, which are so costly in human lives.

Long-tailed mandrill with young.

L. Frédéric-Rapho.

Arab filly and her colt.

Ylla-Rapho.

The factors of growth

The growth of an organism is only possible when various conditions are favourable, conditions both external and internal.

The external factors of growth pertain to the realm of general physiology, and we shall limit ourselves to recalling in brief the roles of temperature and food. Among so-called 'cold-blooded' animals whose temperature varies — the invertebrates and the lower vertebrates — growth can be deliberately stimulated, moderated, or almost arrested, by altering the external temperature. As Jacques Loeb demonstrated, the speed of growth roughly obeys the law which Van t'Hoff enunciated for the speed of chemical reactions: in other words, for every 10°C the temperature is raised the rate of growth is slightly more than doubled.

If two groups of young frog larvae, that is, tadpoles, are reared at temperatures of 12°C and 22°C respectively, the first group will take twice as long, and even a little more, to reach the same size as the second.

As for diet, it is common knowledge that certain vitamins such as A, B, C, and D, encourage or are necessary to growth. The same is true of certain amino acids: no matter how abundantly mice are fed on proteins which lack *lysine* they soon cease to grow and remain as though arrested at that state of their evolution. If the diet has not been too prolonged they will simply resume their growth when given food which contains lysine.

Certain hormones are equally indispensable. Among vertebrates the hormones which chiefly affect the phenomena of growth are those of the thyroid gland, of the thymus, and above all of the anterior lobe of the pituitary.

When this lobe is removed from a young mammal — a delicate operation but perfectly compatible with life — the animal loses all capacity to grow. The capacity can be restored by grafting another similar gland or by injecting extracts of the gland under the skin. The pituitary growth hormone *(somatotrophin)* increases the sugar content in the blood. Its dosage is measured by the action it exerts on an animal deprived of its pituitary gland. The 'rat-unit' of somatotrophin is that quantity necessary to make a twenty-eight day old pituitary-less rat gain 10 grams in ten days, receiving one injection of the hormone daily. By injecting extracts of pituitary Evans has been able to bring about experimental gigantism in rats which thus doubled their normal weight.

In men it sometimes happens that the pituitary gland secretes an excessive amount of the growth hormone and, if the individual is still in process of growth,

The Russian giant Machnov who measured 2.85 m (9 ft 4 ins) weighed 182 kg (nearly 380 lbs.) His hands measured 32 cms (13 ins) and his feet nearly 51 cms (20 ins).

Chusseau — Flaviesn.

A dwarf — possibly Don Sebastiano de Morra — by Velasquez, in the Prado museum, Madrid.

Anderson — Giraudon.

'The Dwarf' by Zuloaga in the Musée du Jeu de Paume, Paris.

Anderson — Giraudon.

193

exaggerated tallness or gigantism results. If the individual has already reached maturity the result is a morbid development of the extremities known as acromegaly.

Growth in different breeds

Attempts, more or less successful, have been made to connect structural differences in human individuals with differences in hormone activity. It has even been proposed to classify human 'types' according to the preponderance of this or that endocrine gland.

Dealing with differences between breeds of the domestic dog Stockard brilliantly developed similar ideas. He attributed the shortened legs of the Basset hound to insufficient secretion of the pituitary body at the time the legs were formed; the Saint Bernard and the Borzoi, or Russian Wolfhound, are giants or acromegalic because of pituitary over-activity; the legs and cranium of the Bulldog are achondroplastic because of thyroid insufficiency.

There is evidence that some specific characters are also connected with variations in hormone conditions, variations either in the quantity of hormones produced or in the sensitivity to them of the affected tissues. The eyes of the mole *Talpa europaea* seem to be a case in point.

A hormone which opens the eyes

Since the days of antiquity it has been known that moles possess rudimentary eyes, more or less covered by skin, and incapable of sight. In his *History of Animals* Aristotle wrote: 'Moles are blind; externally no eyes are apparent, but if the rather thick skin which covers the animal's head is lifted we discover, in the place where the eyes of other animals are usually found, eyes which are useless to the mole but which nonetheless lack no part proper to the organ of sight. We distinguish the white of the eye, the iris and, in the middle of the iris, the pupil. But these parts are smaller than in animals whose eyes are exposed, and none of them appears on the surface because of the thickness of the skin. It is as though the mole was blinded from the moment of its formation.'

But why is the mole thus provided with diminished and useless eyes?

The question may be asked not only in the case of the mole's eyes, but also in that of all other animals, vertebrate and invertebrate, whose visual apparatus is defective when compared with that of related species.

Lamarck suggested that they derived from an ancestor of the mole which had lived above ground and, like other mammals, possessed well-formed eyes. Then, little by little, in its subterranean environment where the mole made no use of sight, its visual organs atrophied, in conformity with the general law that organic structure is modified by the organism's habits and that an organ withers away from lack of usage.

The hypothesis proposed by Charles Darwin was quite different. According to Darwin the regression of the mole's eyes was brought about by natural selection, since it constituted an advantageous factor for the animal by sparing it inflammation of the cornea.

Today both Lamarck's theory that the mole's eyes atrophied under the influence of a lightless environment, and Darwin's theory that it was reduced by natural selection, have been replaced by fairly general agreement with the view that the phenomenon was caused by one or several accidental variations, in other words, by mutations. Mutant 'proto-moles' with eyes thus degenerated would have been able to survive in an underground environment where blindness was no

The giant South African boxer Ewart Potgieter.

Agence Intercontinentale.

194

drawback, while above ground they would have been wiped out by species provided with normal sight. None of these hypotheses supplies the reason why the eyes of present-day moles do not attain structural normality. Whatever the original cause of the animal's ocular regression, the phenomenon must have a present embryological cause.

As Lucien Cuénot pertinently remarked: 'Very often it can be observed that rudimentary organs in adults — the eyes of the mole, etc. — appear in the embryo more or less as highly developed as the corresponding organs of ordinary animals. But at a certain stage the organ's growth slows down and stops. There is, then, in the embryo itself some present cause which motivates this slowing down.'

Jean Tusques, having observed that in young rats the opening of the eyelids depends on the hormone thyroxin, secreted by the thyroid gland, began to wonder if in those animals like the mole which suffer from a more or less complete occlusion of the eyelids even in adult life the thyroid mechanism might not in some way or other be defective. In spite of the difficulties involved in rearing moles and their particular vulnerability in the hands of the experimenter, Tusques succeeded in causing the young animals to absorb quantities of thyroxin. The hormone in solution, administered in the proportion of one part in a thousand, was given through the mouth with a medicine-dropper, three to ten drops daily for twenty days. Under the influence of this treatment noticeable changes in the ocular apparatus became evident. The opening of the palpebral fissure, or incipient eyelids, enlarged from less than ¾ millimetre (as in untreated moles) to 1 millimetre. This wider aperture was sufficient to permit the eyeball to protrude.

Furthermore the eyeball itself had more than doubled in diameter, measuring 2 millimetres instead of 0.9 millimetres as in ordinary moles.

'The effect produced resembled that of an exophthalmia, but not of the type known as Basedow's disease in which the eyelids will not cover the protruding eyeball. Rather, the opening of the palpebral fissure and the development of the eyes gave the animal the appearance of a normal mammal instead of the more or less blind aspect habitual to moles.'

It is then reasonable to suppose that in moles either the thyroid hormone is secreted in lesser quantity than in mammals with well-developed eyes, or else that the visual organs of the mole are less sensitive to thyroxin. The second hypothesis, which seems more plausible, is that which J. Tusques adopts. 'We can conclude that one of the important characters of the species *Talpa europaea* is its high threshold of sensitivity to the thyroid hormone which controls visual structures.'

If this is indeed the case, such experiments demonstrate 'the mechanism of a specific variation'. Their great interest lies in the fact that they throw light, if not on the actual genesis of hereditary variations, at least on the manner in which variations affect the organism through the agency of hormone processes.

It should be added that within the species of *Talpa europaea* some individuals have wider eye apertures than others. Such differences must arise from individual variations, perhaps genetic, in local sensitivity to the thyroid hormone. The differences can be so marked that it was debatable whether or not two distinct species were involved: *Talpa europaea* and the blind mole *Talpa caeca*. When, however, all is taken into consideration a single species seems to be indicated — *Talpa europaea* — which may in some cases have its eyes open and in other cases closed, unlike the true blind mole whose eyes are invariably covered by skin and which is exclusively native to Switzerland and Italy.

The example of the moles treated with thyroxin shows us that the biologist can include characteristics in the appearance of an animal which it does not normally possess but which belonged to its ancestors. Moles given extra thyroxin resemble moles of former times — proto-moles. Thus it might be said that a sort of 'experimental atavism' has been achieved.

The case of the mole is certainly not unique and it may be presumed that the development of rudimentary organs in other animals will be realized by the use of thyroxin or other hormones.

Etienne Wolff has suggested that the atrophy of the horse's lateral toes may be caused by some hormone. Possibly 'a special secretion intervenes in every case where an embryonic organ atrophies in the normal individual'.

The facts reported by A. and J. Raynaud about the development of mammary glands in mice may be considered in this connection. In the normal embryo of the mouse the primitive buds of these glands are so arranged that they give rise to a single row which, towards the twelfth day of development, bifurcates at the distal extremity. Now, under the influence of strong doses of female hormone — *folliculine* — with which either the pregnant female or the embryo itself is injected — important alterations in the evolution of the mammary glands can sometimes be observed: either the vague beginnings of multiple buds appear, or the mammary row is duplicated, as it is in other species of mammals.

Thus it is possible, by means of a simple hormone treatment, to change the specific plan of the animal's mammary structure to a different structural plan which more nearly resembles that of other species. Once again, by the use of a chemical substance, the biologist has brought about a deviation in development which more or less faithfully mimics evolutionary variation. Indeed variations in the way the endocrine glands function must have played an important part in the evolution of species.

Metamorphosis

Among moths, flies, and frogs development involves a profound transformation of the organism, while among the majority of animals the transition from youth to maturity is marked by little more than an enlargement of the body and a few changes in its initial proportions. The transformation of organisms like insects is called metamorphosis, the juvenile phase being known as the larval stage, while the sexually mature adult is known as the imago.

These changes of form also imply changes of diet, habits, and habitat. The winged butterfly that sucks nectar from flowers begins its existence as a crawling creature that devours leaves and stems. The frog emerges from its egg provided with a tail with the appearance of a small fish, to acquire, on maturity, limbs and air-breathing apparatus. In trying to explain the possible reason for such transformations it is hardly necessary to point out that such questions touch on the immense problem of the origin of living beings, which has not yet been fully understood. But if biologists find themselves hard put to it to explain why in the course of evolution such strange complications should have been introduced into the development of certain animals, they have at least the satisfaction of being able to analyse with relative accuracy the present causes of metamorphosis. The physiological mechanism of metamorphosis is so well understood that they are in a position to bring about the phenomenon, delay it, or even suppress it at will.

We shall examine in turn the two chief types of metamorphosis: that found in amphibians like the frog, and that among insects like the moth. When the tadpole emerges from the egg of the frog it is furnished with small tufts, or external gills, which enable it to breathe the oxygen dissolved in the water, and with a long tail for swimming. It has no trace of legs, either front or hind.

Soon the external gills dwindle and disappear, replaced by internal gills. Then adumbrations of the limbs appear, only the posterior pair being at first visible. These lengthen and become legs with toes. When the hind legs are well developed the front legs appear one after the other. Finally the tail atrophies and vanishes. Meanwhile the tadpole's entire organism has undergone various alterations: the horny beak and teeth disappear, the mouth is enlarged, eyelids and a muscular tongue appear, the lungs grow bigger and the intestine shorter.

The tadpole, on becoming a frog, has changed its diet. Previously herbivorous and a great eater of vegetable detritus, it now accepts only living prey—worms or insects. It is prepared for life on land, both by its breathing apparatus and by the structure of its legs which are now even better adapted to walking and jumping than to swimming.

This entire transformation or metamorphosis has taken about two and a half months. The transition from life in the water to life in the air entails a veritable physiological 'crisis', which is not infrequently fatal.

Like many other biological phenomena metamorphosis depends on a hormone mechanism, as the Austrian Gudernatsch demonstrated many years ago. In an aquarium he placed a group of young tadpoles which in the ordinary way would have undergone metamorphosis at the end of a month or two. That is, they were in the stage when their hind legs were beginning to form. A little fresh or dried thyroid gland was added to their diet. The thyroid gland, as we know, secretes a hormone. Situated in the region of the neck it plays, in all vertebrates, a considerable part in controlling the rate of metabolism.

The effect of the thyroid treatment on the tadpoles was as prompt as it was violent. After only a few days the characteristic signs of metamorphosis were apparent: the animal's body, previously globular, thinned down and assumed the shape of a violin, the hind legs began to lengthen and then the front legs, while the tail shortened and at last was entirely resorbed. Within

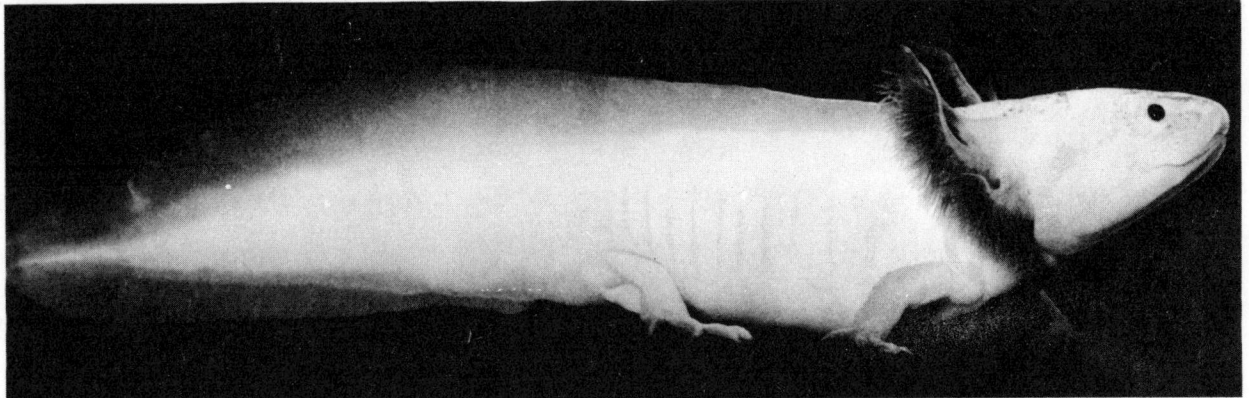

Mutant albino of the Mexican *Ambystoma*; an example of
neoteny or prolongation of the larval stage. External gills
preserved.

Gallien. Larousse.

roughly the space of a week the tadpole had become
a frog, a minute frog scarcely bigger than a fly. It was
normally formed in every way, but excessive precocity
had made it a dwarf.

This experiment is highly instructive because it
demonstrates that the thyroid gland contains a substance capable of hastening the hour of metamorphosis.
This substance can only be the particular hormone
which the gland secretes, a hormone which has been
isolated and prepared in a pure and crystalline state,
chemically identified as thyroxin. Indeed, by the use of
synthetic thyroxin the same 'metamorphogenic' effects
are obtained as with thyroid extracts.

Thyroxin, an iodine-containing amino acid, can be
replaced by other substances containing iodine in the
organic state, and notably by *thio-iodotyrosine,* which
is still more powerful since it will bring about metamorphosis in concentrations as small as one part in ten
thousand million, while the concentration of thyroxin
must not be less than one part in twenty million.

When concentrations fall below this strength partial
metamorphosis results, affecting only the organs which
are most sensitive to the hormone. Thus the growth of
the limbs and the reorganization of the intestine can be
brought about with doses which are incapable of
causing the resorption of the tail.

If very strong doses are given, in the region perhaps
of thyroxin concentrations of one in a million, the tail
is resorbed too promptly, before the front legs have
emerged or the hind legs reached their full length. Such
violent and disorderly metamorphoses might be called
'caricatures of metamorphosis'.

In natural conditions of development it is obviously
the thyroid hormone of the tadpole itself which brings
about metamorphosis. If the thyroid glands of young
tadpoles are removed it is found that the animals are no
longer capable of metamorphosis, unless they are
subsequently grafted with a fragment of thyroid or
given thyroxin. By the suppression of the thyroid gland
the tadpole is condemned to remain indefinitely in the
larval stage.

Metamorphosis can also be retarded or prevented by
submitting the tadpole to a diet lacking in iodine.

In the normal tadpole the thyroid gland does not
function of its own accord. To secrete or rather to
liberate its hormone it must be prompted by another
hormone supplied by the anterior lobe of the pituitary
gland, lodged beneath the brain. Tadpoles deprived
of this lobe by cauterization or by surgical operation
are incapable of metamorphosis, exactly like those
deprived of thyroid. Furthermore they acquire a silvery
colour which is caused by the lack of a 'pigmentation
hormone', also secreted by the pituitary.

The removal of the pituitary gland results in the
atrophy of the thyroid gland. If an animal from which
the pituitary has been removed is injected with extract
of pituitary of grafted with a fragment of pituitary its
capacity for metamorphosis is restored.

Exactly how the thyroid hormone acts on the tissues
is not yet fully understood, but it is observed that it
stimulates cellular multiplication.

Neoteny

Neoteny is a phenomenon that can be described as the
prolongation of the larval stage. In certain species of
Mexican salamanders *(Ambystoma)* the tadpole, or
larva, is fairly often incapable of achieving metamorphosis. It continues to grow and, still in its larval
form, reaches the stage of sexual maturity. These giant
larvae have long been known as *axolotls* and are able
to mate and reproduce. Although spontaneous metamorphosis is rare among axolotls artificial metamorphosis can be induced by thyroid or pituitary
treatment.

An aptitude for spontaneous metamorphosis varies
according to the environment. *Ambystoma tigrinum*
remains in the larval stage indefinitely in the high, cold
Rocky Mountains of Colorado, while in the warm
plains of the east metamorphosis takes place.

In the axolotl it seems that the 'metamorphogenic'
hormone is present in the thyroid gland, but is not
discharged into the bloodstream in sufficient quantities. Possibly its pituitary gland is deficient. Or possibly
the animal's entire organism is less sensitive to the
effects of the hormone, and demands greater quantities
of it in order to achieve metamorphosis. In any case the
axolotl has long fascinated zoologists.

As early as 1600 Hernandez, in his *History of the
Animals of New Spain,* spoke of a fish which inhabited
'Lake Mexico' and was called the *Atolacalt.* In 1767

197

Pleurodeles whose pituitary gland was removed during embryonic development. Because the hormone which controls pigmentation was lacking, the animal, pictured at the age of one year, is colourless. It also retains its larval state, revealed by the external gills. In the absence of the pituitary body the thyroid gland is not stimulated and the animal does not undergo metamorphosis. *On the right,* a control animal of the same age is shown; it is pigmented and has undergone metamorphosis.

Gallien. Larousse.

Silkworms of various ages on mulberry leaves.

R. H. Noailles.

Johnston affirmed that the axolotl was only a 'watery illusion'; but Cuvier, who had read Humboldt's notes on the animal, was aware that the axolotl must be the larva of some large 'unknown salamander'.

Napoleon III's ill-fated expedition to Mexico resulted at least in the acquisition of six axolotls (five males and one female) which were presented to the French Museum of Natural History in 1864.

On the fourth of January 1865 the female began to lay eggs and it was thought that the great Cuvier had made a mistake — since a larva does not lay eggs. Shortly afterwards, however, the axolotls underwent metamorphosis and Cuvier's opinion that it was a salamander was justified.

Today these fertile larvae are laboratory animals and currently employed by biologists. They provide an excellent example of the phenomenon of neoteny.

Total neoteny occurs in animals like the axolotl which reproduce in their larval form; partial neoteny occurs in others which, though attaining unusual size as larvae, do not achieve sexual maturity.

There are other tailed amphibians which can exhibit neoteny under the influence of cold and altitude: in the lakes of Lombardy alpine tritons are found which reproduce in the larval stage. In other species of tritons total neoteny is limited to the female sex. In still other amphibians neoteny is only partial. One example of this is the midwife toad and another the edible green frog whose tadpoles can reach extraordinary lengths — as much as 12 centimetres instead of the usual 4 — and never acquire the faculty to reproduce.

Artificial neoteny

Attempts have been made to bring about neoteny in amphibians by artificial means. As we have already seen, metamorphosis can be prevented and, as a result, permanent and more or less giant larvae obtained, either by removing the animal's pituitary or thyroid, or by submitting the animal to chemicals with anti-thyroid or antipituitary effects — thio-uracil, basolan, etc. But in these ways total neoteny is not obtained, for the sex glands do not mature in laboratory specimens. Michel Delsol did, however, obtain giant larvae with relatively large sex glands of a *Discoglossus* toad.

Permanent neoteny

Among some amphibians belonging to the family of the Perennibranchiata the specific adult type displays all the characteristics which we are accustomed to consider larval characters. For example, the olm, *Proteus anguinus,* which lives in the subterranean lakes of Carniola in Austria, has external gills and two little forelegs; its eyes are rudimentary and in colour its skin is very pale pink.

It is tempting to assume that *Proteus* is a case of permanent and definitive neoteny. *Proteus* would, accordingly, be the descendant of an ancestor which had a different adult form, but which in the course of its evolution lost the faculty of metamorphosis. In any case metamorphosis, even partial metamorphosis,

A silkworm emerging from its egg.

R. H. Noailles.

The silkworm when fully grown.

Le Charles.

Female of the silkworm moth, *Bombyx mori*, and the eggs it has laid.

R. H. Noailles.

cannot today be induced in *Proteus*.

No matter what metamorphogenic treatment the creature is submitted to, whether it be thyroid hormone or pituitary hormone, or in what amounts the hormones are administered or how they are introduced — by ingestion, by sub-cutaneous or intra-coelomic injection — *Proteus* remains unshakably *Proteus*. Its gills refuse to regress, its legs show no signs of growth. Hence the hypothesis of definitive neoteny remains no more than an unverified hypothesis.

As long ago as 1817 Spix suspected that *Proteus* might be the larva of a terrestial salamander. He took a few live animals with him to Brazil, hoping that the change of climate would induce their metamorphosis.

In the *Last Days of a Philosopher* by Sir Humphry Davy (1828) we find the following imaginary dialogue on the subject of *Proteus* of the Grotto of Madalena at Adelsberg.

'*The Stranger:* At first sight it might be supposed that this animal is a lizard, but its movements are similar to those of a fish . . . . There where the eyes should be there are only two small points, as though to preserve the analogy to nature. It is here that the Baron Zoïs discovered it, but since then they have been found, though rarely, at Littich, some hours distant from here, thrown up by the waters of a subterranean cavity.

Eubathes: Do you not think it possible that this creature may be the larva of some large unknown animal which inhabits these subterranean caverns?

The Stranger: I cannot suppose that they are larvae. I do not believe that there is in nature a single example of a transformation analogous to this kind of metamorphosis of a mature animal into an immature animal.'

The hypothesis of permanent neoteny has been extended to include other amphibians, such as *Typhlomolge, Necturus, Siren,* and *Pseudobranchus*.

In the majority of them, as in *Proteus*, the thyroid gland is capable of functioning and contains the hormones which bring about metamorphosis. In addition their pituitary gland is capable of stimulating thyroid activity. Therefore their neoteny — if indeed it is neoteny — must be due to an incapacity of the tissues to undergo metamorphosis, having presumably lost the faculty to respond to hormone stimuli.

Neoteny may be one of the basic processes which evolution employs to bring about important changes in organic species. According to Bolk, G. R. de Beer, Cuénot, and many others, neoteny has played its part in the genesis of our own species. Man, it is true, resembles the giant foetus of the great ape far more than the adult ape; and it is thus tempting to imagine that the passage from ape to man involved some process of 'foetalization', connected with variations of hormonal equilibrium and a slowing down in the rate of growth.

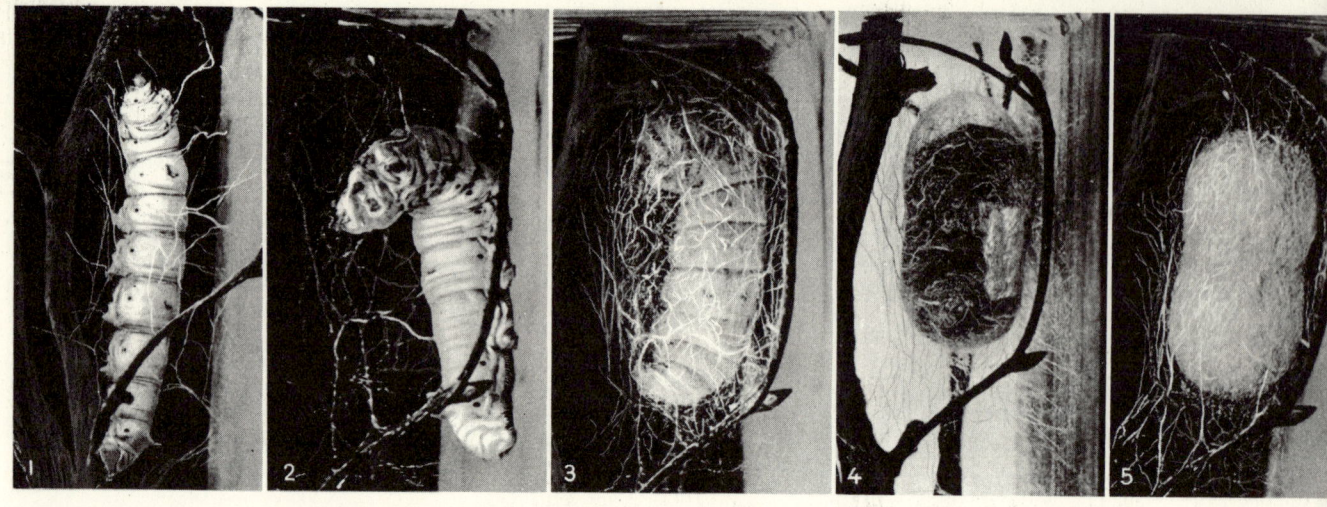

The metamorphosis of insects

The metamorphosis of insects is a considerably more complicated event than the transformation of tadpole into frog, for the butterfly is totally different from a caterpillar, and the fly is completely unlike a maggot.

Between the two stages, characterized not only by a difference in form but by differences in life, habits, and diet, there is in addition an intermediate stage of repose, represented by the nymph — chrysalis or pupa. It is during this stage of repose and organic torpor that those profound transformations take place which characterize insect metamorphosis. It is then that the entire organism is recast, remodelled, and renewed; and a detailed analysis of the process would be beyond the scope of the present volume.

For many centuries the transformation of caterpillar into butterfly has fascinated naturalists and philosophers. It was once thought that the moth was hidden or 'masked' by the caterpillar. Such was Swammerdam's idea, a theory which was on a par with that of germinal pre-formation. For this reason the name larva was given to the moth's first stage, the word being borrowed from the Latin *larva* which means 'mask'.

Swammerdam believed he detected the features of the future moth in the caterpillar itself. He obviously deceived himself, for although it is true that the moth, in a certain measure, pre-exists in the caterpillar, it pre-exists in a very rudimentary fashion. This is merely in the form of minute islets, tiny enclaves in the tissue, which are known as *imaginal discs*. During larval life these discs remain in repose. Then at a certain moment, which is the moment of metamorphosis, they begin to enlarge and develop, giving rise to wings, legs, and compound eyes. At the same time other parts of the organism, including the silk glands, degenerate while still others are more or less reorganized: the digestive tube, the nervous system, the excretory system.

It is known that the metamorphosis of insects, like that of amphibians, is under the control of specific hormones. J. J. Bounhiol demonstrated in a delicate and ingenious series of experiments, that there exist certain restraining hormones which prevent metamorphosis. These inhibitory hormones are secreted by little glandular organs situated in the head close to the brain and are known as *corpora allata*.

During its larval evolution the silkworm moults or changes its skin four times: then, after the cocoon is spun, comes the nymphal moult from which the chrysalis emerges. The chrysalis in turn undergoes the imaginal moult, and the finished insect — the *imago* or moth — appears.

If after the third larval moult the *corpora allata* are surgically removed the worm will undergo a premature metamorphosis: it spins its cocoon and turns into a chrysalis. If, however, the operation is performed after the second larval moult the same metamorphosis will take place and larval life will have been thus reduced to one half of its ordinary duration. Such moths, obtained by premature metamorphosis, are distinctly dwarf, but it is not inconceivable that even earlier metamorphosis could be obtained in this way, giving rise to even smaller moths. The surgical operation is impracticable, but it might prove possible to bring about the functional suppression of the *corpora allata* by indirect means, such as irradiation of the caterpillar's head with X-rays.

The 'anti-metamorphosis' effect of the *corpora allata* can be demonstrated by performing the opposite experiment: that is by grafting these glands on to worms about to undergo their nymphal moult. In this way their metamorphosis is prevented and their larval life prolonged, though for a very short space of time because they soon succumb from inability to nourish themselves.

Pflugfelder succeeded in grafting *corpora allata* on to an orthopteran, *Carausius morosus,* and in this way induced the insects to undergo five supplementary moults (sixteen instead of eleven) and in the process become twice their average size, which is 7½ centimetres.

The anti-metamorphosis hormone secreted by the *corpora allata* has been called the 'juvenile hormone' and it is its disappearance that determines the change of the larva into the mature insect.

It is not, however, the only hormone which plays a part in insect development. A second hormone, connected with moulting, is also metamorphosis-inducing, and has its source in certain glands of the thorax or the

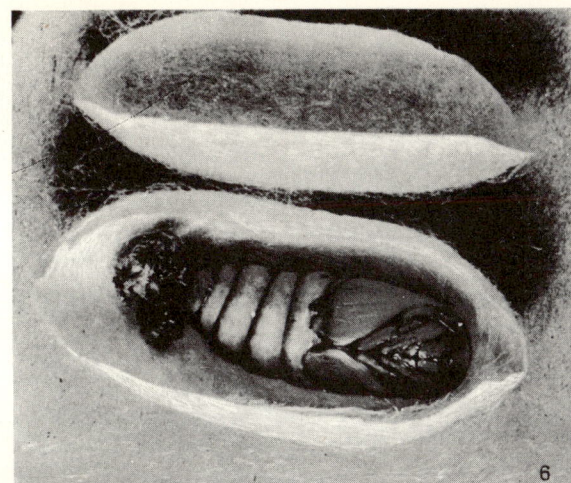
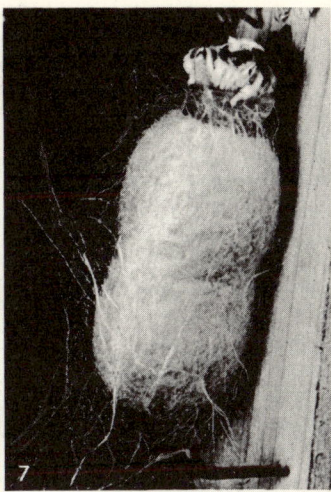
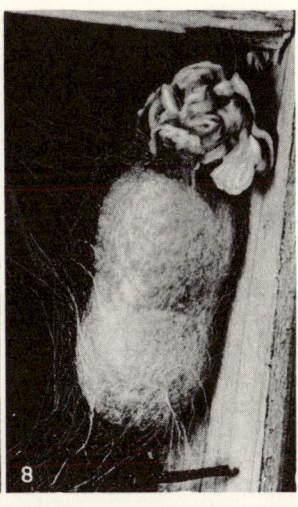

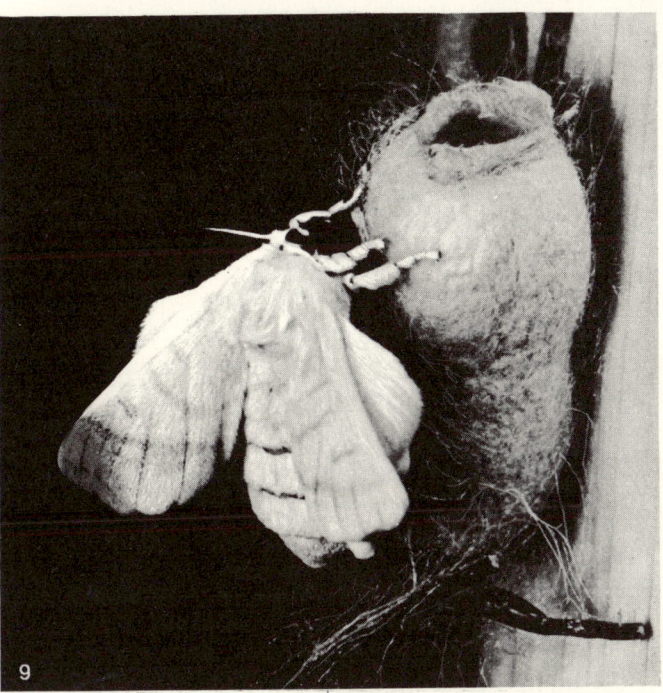

1, 2, 3, 4, 5. The worm spins its cocoon. 6. Cocoon of the silkworm cut in two to show the chrysalis, beneath which the remains of the caterpillar may be seen. 7. The moth's head emerges. 8. Its body follows. 9. The wings are dried. 10. Female moth after hatching.

R. H. Noailles.

Larvae of mosquitos. The lower extremity of their bodies brushes the surface of the water and in this manner the larvae breathe.

Bayard.

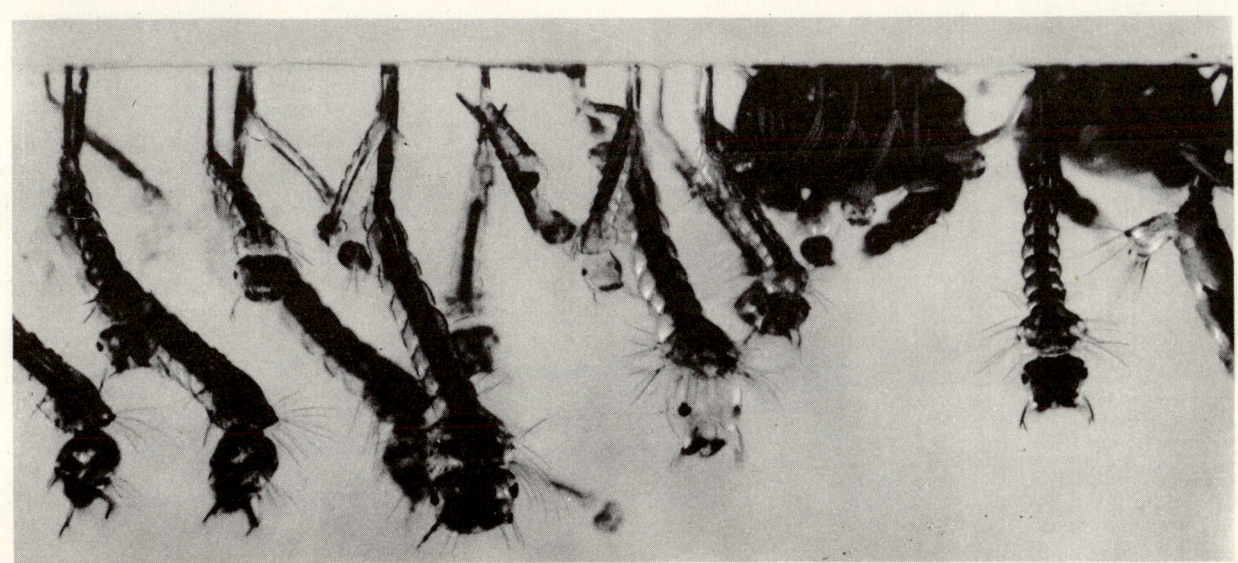

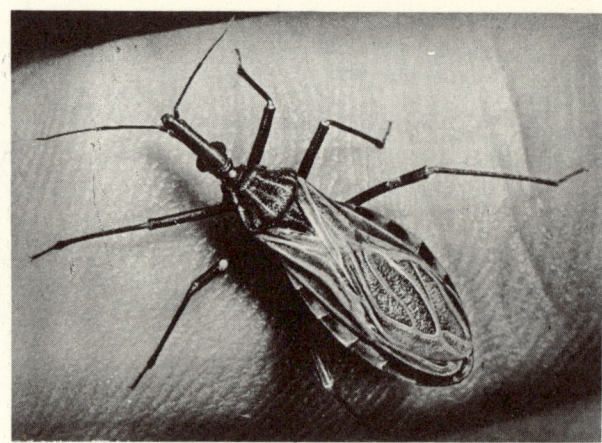

The head of the phasmid *Carausius morosus*. It is in this region that the *corpora allata* are situated.

R. H. Noailles.

Rhodnius prolixus, the South American hemipteran which Wigglesworth used in his fundamental researches into the hormones which control moulting.

J. Carayon.

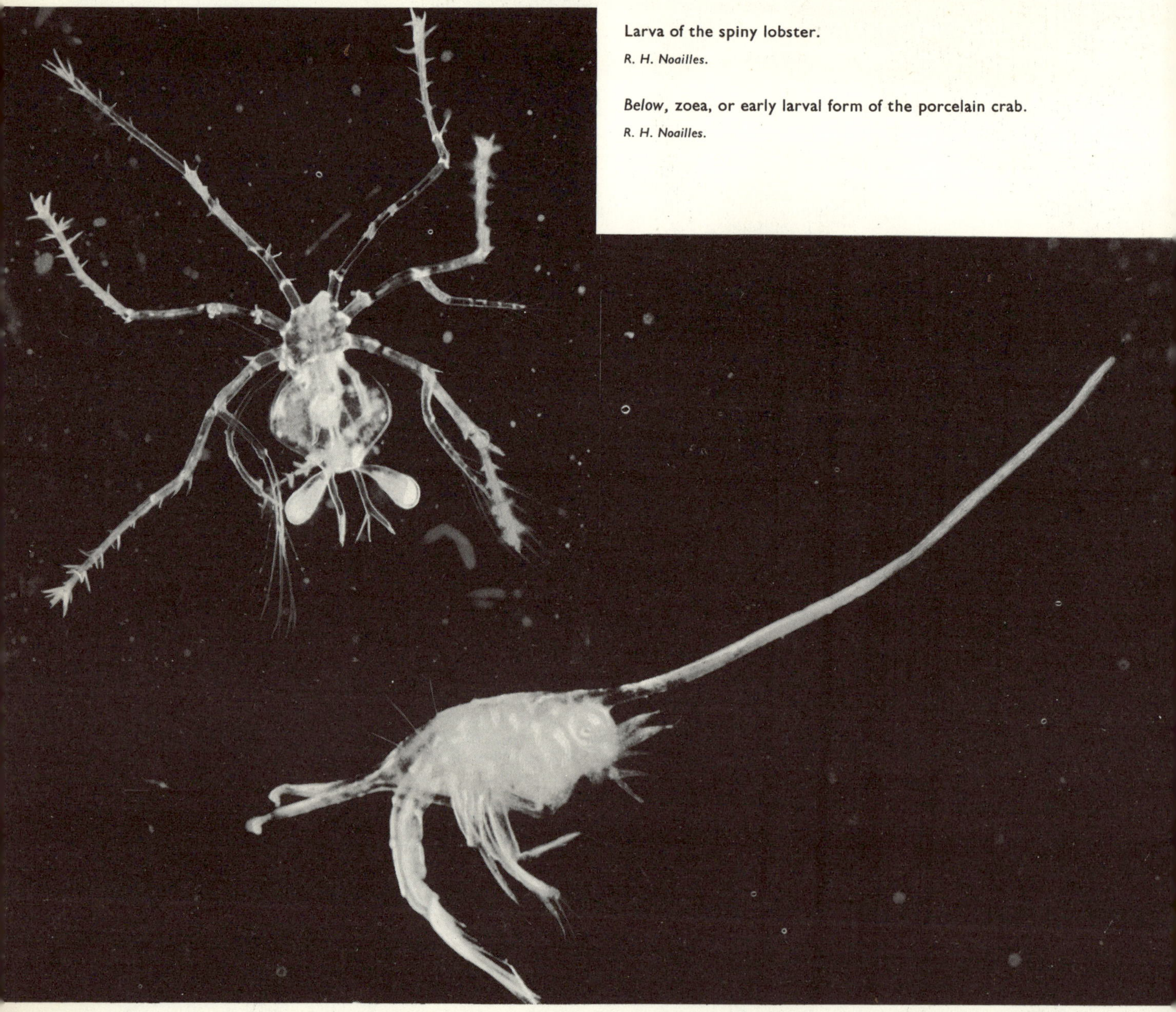

Larva of the spiny lobster.

R. H. Noailles.

Below, zoea, or early larval form of the porcelain crab.

R. H. Noailles.

Caterpillar and moth of the Indian moth (*Actias selene*).

Caterpillar and butterfly of the peacock (*Nymphalis io*).

Caterpillar and moth of the garden tiger (*Arctia caja*).

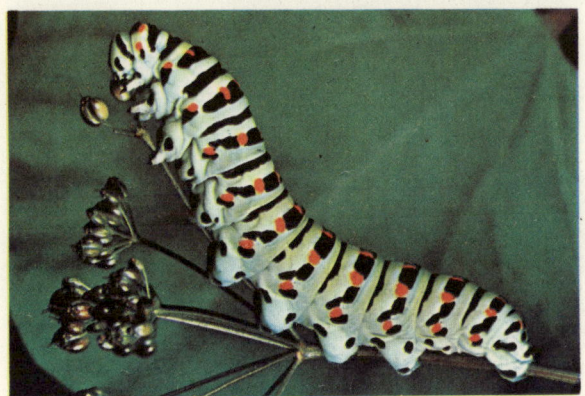

Caterpillar and butterfly of the swallow-tail (*Papillio machaon*).

prothorax. The secretion from these glands is moreover stimulated by a third hormone, which originates in the brain.

Thus we may roughly say that metamorphosis is controlled among the majority of insects in this way: as long as the 'moulting hormone' coexists with the 'juvenile hormone' the changes which take place will be larval moults but as soon as the juvenile hormone disappears the change becomes a nymphal or imaginal moult, and leads to metamorphosis.

The part played by these two antagonistic hormones is clearly demonstrated by the work of the English biologist Wigglesworth who in 1936 began to study the development of *Rhodnius prolixus,* a large South American bug. The metamorphoses of *Rhodnius* are incomplete, but the last of its five moults corresponds to an imaginal moult, that is to the moult of metamorphosis. The bug is nourished on blood and a single meal is enough to stimulate it to moult. This occurrence in the case of a larval moult takes place in twelve to fifteen days after the meal and in the case of the imaginal moult twenty-six to twenty-eight days afterwards.

If the insect is decapitated immediately after its meal of blood the moult will not take place, although the insect itself can survive for more than a year. If, however, the decapitated larva is grafted to another larva (this is done by surgical joining or 'parabiosis' which creates artificial 'Siamese twins') and the second larva is not decapitated until two or three days afterwards and is thus still able to moult, the first larva will accomplish its larval moult, thanks to the hormonal stimulus it receives from the second.

The imaginal moult is also determined by a hormone, a fact which can be demonstrated by parabiotically joining a young decapitated larva to an old larva, also decapitated. Under the influence of the hormone it receives from the older larva the young larva will accomplish not only a larval moult but also an imaginal moult — and become a dwarf imago!

The explanation of why *Rhodnius* will moult after its meal of blood is that the meal distends its abdomen; this sets up a nervous stimulus which releases the secretion of the cerebral hormone, which in its turn excites the glands producing the 'moulting hormone'.

Recently the 'juvenile hormone' has been successfully extracted from the abdomens of the male moth *Platysamia cecropia,* a large lepidopteran. Though the hormone is insoluble in water it can be dissolved in ether. After evaporation in a double-boiler a golden-yellow oil is obtained which, even in minute amounts, is very active and capable of exercising its specific effects on insects of various orders: Coleoptera, Hemiptera, Orthoptera.

The reason why the tissues of the abdomen are so rich in the juvenile hormone is doubtless because they solidify it at the moment which precedes metamorphosis. Why the hormone cannot be extracted from the abdomens of female moths is unknown. If the chrysalis is daubed with the juvenile hormone the insect obtained will have characters halfway between those of a chrysalis and an adult. Such insects are not viable and the possibility of using the hormone as an insecticide has been discussed.

Around the oesophagus of larval Diptera (flies), among whom metamorphosis is even more profound than among moths, the existence of a secretory organ has been revealed. The central part of this organ, known as Weismann's ring, produces the juvenile hormone, while the lateral parts produce the moulting or pupation hormone. If the Weismann's ring of young maggots (the larvae of flies) is amputated the larval state is prolonged and permanent larvae are obtained which can survive as long as forty-five days after the operation, but are incapable of metamorphosis. If, on the other hand, young maggots are grafted with a Weismann's ring taken from an old maggot, premature nymphosis takes place.

These experiments are interesting not only because they help to solve the problem of insect metamorphosis but also because they throw light upon the way the hormones act in general, that is, among all living organisms. Metamorphosis is by no means limited solely to insects and amphibians. A great number of living organisms undergo various degrees of transformation during the course of their development — sponges, worms, bryozoans, echinoderms, ascidians, molluscs, crustaceans, fishes, etc. We are, however, still very poorly informed on the physiological conditions of such changes. It would seem that copper salts have some 'metamorphogenic' effect on the bryozoans and the ascidians.

The larvae of the great lamprey are known as ammocoetes. They live in fresh water and are very different from the adult lamprey, which lives in salt water. The transition from larva to adult lamprey involves not only a complete reorganization of the mouth and the gill apparatus, but also the development of the eyes and the atrophy of the gill ducts.

No hormone treatment, thyroid or pituitary, has proved capable of inducing metamorphosis in lampreys. The physiological mechanism of the eel's metamorphosis is equally unknown.

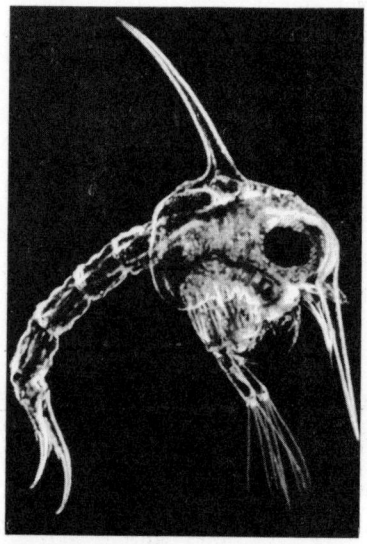

Larva of the common shore crab, *Carcinus maenas.*
D. P. Wilson.

Old Age and Death

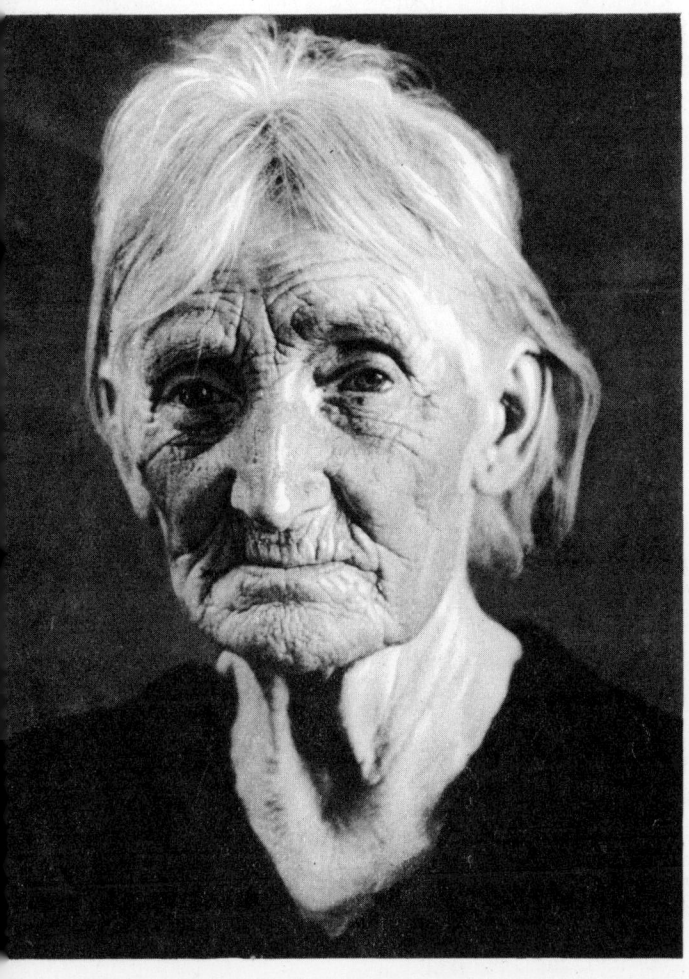

S Weiss — Rapho.

Two things seem to all of us to be naturally and inevitably connected: life and death. We tend to think that every creature born is, for that reason, condemned to die. All that begins must end.

And, in fact, among the majority of living organisms — in any case among those known to the layman — the duration of the individual's life is limited. When the organism has finished its growth, then passed through a longer or shorter period of apparent stability, it begins to run down, to grow feebler. Its organs function less efficiently; slight deteriorations set in, which little by little are accentuated until a state of decadence is reached, the state called senility which leads inevitably to death — to what we call 'natural death'.

But this view of life and death is by no means valid for all organic beings. In spite of the general opinion, senility and death are not the lot of all living things. It is not true that every birth is but a promise of death. Death, in fact, is not the inevitable consequence of life.

Immortality

In the first place all unicellular organisms such as the ciliophorans and microbes, are potentially immortal, in that they can, if the conditions of their environment are favourable, multiply indefinitely without the slightest decline in their vital activities or faculty to propagate.

Even in 1838 Ehrenberg remarked, 'The reproduction of the Ciliophora by division abolishes the possibility of their individual destruction and renders them immortal.'

The great naturalist Maupas concluded from his patient research that cultures of Ciliophora suffered from a kind of ageing at the end of a few hundred divisions and that this cellular decrepitude could only be repaired by conjugation, that is, by the union

205

of individuals two by two. In reality, however, these animalcules can be cultivated indefinitely without resort to conjugation, providing only that they are given appropriate nourishment. In this way Calkins obtained 700 generations of *Glaucoma* in his laboratory; Hartmann 2,500 generations of *Paramecium aurelia;* and Metalnikoff 8,704 of the same organism.

To follow and to count accurately the successive generations of his microscopic subjects Metalnikoff employed shallow plates into which were placed from three to eight drops of a hay infusion or other nutritive medium. Then, with the aid of a capillary pipette a single *Paramecium aurelia* was introduced. On the following day the paramecia newly produced by division were counted. In this space of time the paramecium ordinarily divided once or twice. All superfluous paramecia were put into a culture specially reserved for them so that only one remained in the plate. The nutritive medium was changed either every day or every two days. The experiment, begun in 1910, continued for twenty-two years and five months, and there were 386 generations a year.

Potential immortality is also possessed by microbes and, as Carrel demonstrated with his tissue cultures — described later — by the cells which compose higher organisms.

There even exist multi-cellular organisms which are immortal; such are certain planarians, small ciliated worms which live in fresh water, and above all the hydras whose perennial existence was studied so brilliantly by the Belgian biologist Paul Brien.

Brien took a hydra, placed it in a small glass of water and, maintaining a constant temperature, fed it lavishly. One of the methods by which the hydra reproduces is by budding, and as buds become detached from the parent Brien isolated them. The water is renewed daily by transferring the hydra to another glass. In this manner Brien kept the same animal alive for nearly five years without observing the slightest lessening of its vitality or the slightest slowing down in its reproduction by budding.

Under strictly controlled laboratory conditions a single hydra produced 702 buds between 29th February 1949 and 6th October 1952. At the later date it was as young and active as the liveliest of its descendants.

This perennial vigour of body, this immortality of the individual hydra is assured by the continual renewal of the cells which constitute it. The forepart of the animal grows while the hind part withers away. This can be observed by staining the forepart blue with colouring matter, Nile blue, and grafting it on to the hind part of another hydra. Little by little the blue tint replaces the natural tint until, at the end of four weeks, the hydra so grafted will be entirely bluish, a demonstration that the hind part has been renewed at the expense of the forepart.

Longevity

Since some organisms are immortal why is death natural to those which constitute the great majority of the animal kingdom, and to all the higher animals?

No satisfactory answer has yet been made to this question. Before examining some of the theories which attempt to explain ageing we may note the extreme diversity in the length of life of various animals.

Extent of life		
Man		115 years
Indian elephant		70
Whale		37
Horse		62
Mandrill		28
Chimpanzee		26
Lion		18
Bat *Pteropis giganteus*		18
Rhinolophus hipposideros	7 to	16
Miniopterus schreiberi		9
Vampire bat *(Desmodus rotundus)*		12
Dog		34
Beaver	2 to	7
Mouse	2 to	3
Shrew		10 months
Vulture	68 to	118 years
Bateleur eagle		55
Goose		80 years
Parrot		82
Swan		70
Carrion crow		100
Dove		30
Duck		29
Chicken		20
Blackbird		8
Giant tortoise of Mauritius		140
Common or Greek *(Testudo graeca)* tortoise		125
Box tortoise		123
Mississippi alligator		56
Giant salamander		60
Natterjack toad		15
Frog		6
European eel		14
Goldfish		10
European pike		10
Seahorse	4 to	5
Crab		20
Cicada	13 to	17
Black-beetle		10
Tridacna (giant clam)	100 to	300
Sea-anemone		66
Leech		20
Earthworm		10

These figures, some rather dubious, are based on isolated examples and in any case represent maximum limits of longevity.

From a comparative examination of animal longevity it is almost impossible to draw any general conclusions. Furthermore the processes of senility may not, perhaps, be the same among all living creatures, and the natural death of a butterfly and of a mammal may bear only an apparent resemblance.

Even within the same group, mammals, for instance, it is hard to explain the great differences in length of life-span. Broadly speaking there would seem to be some correlation between longevity and size, and between longevity and slow growth. It is probably

Hydra (*Hydra viridis*) contracted (*on the left*), and fully extended.

H. Mugard.

not a pure coincidence that man is the mammal which is both slowest to mature and longest to live.

Human senility

No human being is certainly known to have lived for more than 115 years. The case of Thomas Parr, said to have died at the age of 152, is often cited but, although an autopsy was performed by the great Doctor Harvey, there is no way of verifying his actual age.

Among men the picture of senility is varied in the extreme. Some individuals retain their teeth, others their hair, others their muscular vigour, and still others the suppleness of their joints. In one case the circulatory system wears out, in a second the nervous system, in a third the digestive system. There can apparently be 'local senescence' and often enough a person's physiological age seems to have little to do with his or her legal age.

Among the most tell-tale signs of approaching old age may be listed: fragility of the skeleton by loss of calcium; hardening of the ligaments and arteries by sclerosis; hypertrophy of the connective tissues; myasthenia or alteration of the muscular fibrilla attended by muscular weakness and fatigue; increasing deafness; presbyopia or long-sightedness; atrophy of the skin causing wrinkles to appear; wavering of the voice; accumulation in the blood of glutathione and cholesterol; decrease of the eosinophilic leucocytes; disturbance of nitrogen balance; changes in sulphur metabolism; accumulation of yellow pigment in the motor neurons; decrease in the auto-antibodies of the blood; alterations in the composition of the plasma which make it a poorer medium for the growth of tissues; slowing down of the speed with which the skin heals; functional decline in liver, kidneys, thyroid and sex glands.

But in these many and varying symptoms what is cause and what effect? Are all these phenomena of loss and decline perhaps only a multiple expression of a single fundamental process? General theories of the cause of senility are unsatisfactory and purely hypothetical.

According to Child, Mühlmann, Harms, and others, old age is brought on by the slow determination of the nerve-cells which become incapable of renewing themselves by division.

Other authorities agree with Minot in holding that cellular differentiation is itself incompatible with immortality. Thus decline could be said to start at the moment when the cells begin to differentiate, in other words during the first stages of embryonic development. Accordingly, decline would be most rapid during these very stages, since the embryo ages more quickly than the child, the child more quickly than the adult, and the adult more quickly than the aged. In this view senility and death could be described as the cost of organic superiority, 'the price,' as Minot puts it, 'we must pay for the privilege of acquiring specialized structures and functions.'

Still other authorities lay emphasis on the changes which take place in the substances which form the connective tissue, tissue which was once believed to be merely 'packing' tissue, but to which contemporary physiologists tend to attach great importance. Thus it would not be the cells themselves that age but their connecting links. 'The immortal cells,' in the words of Albert Delaunay, 'are the stones; the connective tissue is the cement. It is because the cement cracks and becomes incapable of holding the stones together that the house falls into ruin.'

In support of this viewpoint the experiments of Verzar may be mentioned. Verzar demonstrated that under the influence of hot water the contracting force of the rat's tendons — almost entirely composed of connective fibres — increases with age, due to the alteration of the substances which compose them, namely collagen and mucopolysaccharides.

Other investigators have considered the possiblity of somatic mutations; when we reach the age of sixty, one out of four of our body cells will, they suggest, have undergone genetic change, a mutation which is irreversible and disadvantageous.

Still others have laid the blame on changes in the intracellular ferments, or on progressive flocculation of the protoplasmic colloids, or again on the auto-intoxication of the organism, either by the toxic action of microbes in the lining of the intestines or by the humoral liquids themselves in which certain substances, necessary to certain tissues, could in the long run have noxious effects on the assembly of the body's cells.

Also suggested have been the effects of cosmic rays, the accumulation of 'heavy-water' in the tissues, and so forth. It is evident that none of these many theories is based on totally persuasive arguments.

In other words, we are still in the dark concerning the factors which cause old age and lead to natural death. Our uncertainty is such that some biologists have even questioned whether old age really exists and suggested that the downfall of the organism may simply be the result of a host of small accidental mishaps. In this hypothesis, which is by no means absurd, the body does not age; it simply wears out like an old chair, motor-car or typewriter, and collapses.

Influence of heredity

It is certainly a fact that an aptitude for living for longer or shorter periods is hereditary. Like many other organic characters it is intimately associated with the quality of the 'genes' transmitted by the parents.

It would be tedious here to list the figures on which this affirmation is based, but the careful statistical enquiries which we owe to many biologists, and especially to the American Raymon Pearl, author of the important work *The Biology of Death,* leave no doubt as to its accuracy. There are families in which longevity is the rule.

It is hardly nesessary to add that one inherits, as with other qualities, only a potential aptitude for long life. The actual result will, of course, depend in good measure on the life that one leads, and the individual who has inherited excellent genetic capital can squander it by excess and reach a less advanced age than another individual, less well provided for, but of more temperate habits.

According to a report on octogenarians to the French

'The Old Women', by Goya, from the Musée de Lille.
Giraudon.

ninety-eight and 100 in 1901, ninety-nine were men and 214 were women. In the United States in 1890, among 3,981 persons who were almost centenarians there were 1,398 men to 2,583 women. In Germany in 1885, among 5,648 nonagenarians 2,081 were men and 3,567 women. Among 947 people aged between ninety-five and 100 there were 306 men and 641 women.

French statistics in 1930 registered the deaths of ninety-one centenarians of whom twenty-six were men and sixty-five women. In 1931 the deaths of 119 centenarians were registered, thirty-five being men and eighty-four women. The census made in Moscow in 1939 gave the figures for centenarians as six men and forty-nine women; for nonagenarians as ninety-one men and 520 women.

It is then evident that from the point of view of maximum longevity women have the great advantage over men. Furthermore the average life-span of women is significantly longer than that of men. (The average life-span or 'life expectancy at birth' is, of course, the average duration of individual lives in any given population). In France the life expectancy of a man at birth is 63.6 years; for a woman it is 67.4 years. In the United States the figures are 69.3 for men and 73.6 for women.

To account for this clear superiority of women both in maximum longevity and average life-span it has been pointed out that women escape a number of social hazards, such as military service war risks, heavy physical labour, and so forth and also that women are less apt to indulge in such toxic products as tobacco, alcohol, morphine, and cocaine. It would seem, however, that such things only partly account for the matter, and in the last analysis fail to explain the deeper causes of earlier male mortality on which only biology can throw some light.

If we compare the mortality rate of the two sexes at different stages of development and age we discover a fact of primary significance: namely that the organic superiority of the female sex is plainly evident from the earliest stages of existence, in embryonic life and in childhood, before any difference in working conditions are possible or habits like smoking and drinking acquired. As we have seen in an earlier chapter, this female superiority must be derived from the female constitution itself, and depend on hormone or genetic factors. Some authorities even hold that the female hormones have a beneficent effect on the blood vessels.

The struggle against old age

During this century the average length of life has in all civilized countries increased by nearly twenty years, rising from some forty-five to sixty-five, and this notwithstanding deaths caused by technical progress, world wars, motor-car accidents, plane crashes, and so forth. The child who was born, say, in 1957 had a distinctly greater life expectancy than the child who was born in 1900, for his chances of escaping the perils of extreme youth and reaching adult maturity were very much better. This increase in the average length of life is due partly to the advance of medical science and partly to better social legislation which allows a larger

Academy of Medicine forty-nine per cent had at least two grandparents either paternal or maternal who had lived for more than eighty years. Seven per cent had an octogenarian grandparent on both sides; twenty-six per cent had one octogenarian grandparent, while only eighteen per cent had none.

The occurrence among identical twins of identical or almost identical life-spans is also a striking indication of the genetic character of longevity. Statistically the life-span differs among people of different nationalities, being for instance appreciably longer in Norway than in France. In the fruit fly very precise experiments have shown that racial longevity is regularly transmitted like a Mendelian character.

Influence of sex

Length of life also differs greatly according to sex. At all times and in every country the number of centenarians, and of the aged in general, is made up of a great many more women than men. In 1850, for example, there were 334 people in France older than ninety-nine, 101 being men and 233 women. Of 313 aged between number of people to benefit from applied scientific knowledge in the fields of medicine, surgery, hygiene. Surgeons now operate with success on vital organs — the heart, the lungs, the brain — which not long ago they would not have dared to touch. Antibiotics cure many diseases formerly considered extremely grave and often deadly, such as streptococcus infections and tubercular meningitis. Today pneumonia is not very much more dangerous than a cold in the head. Thanks to vaccination and antitoxins, diphtheria and tetanus have practically vanished from the medical scene. Nor should we forget the therapeutic hormones such as insulin in the treatment of diabetes, cortisone and the others; or again the vast improvements in anaesthetics, grafting, prosthesis or artificial limbs, blood transfusions; and finally the innovations of psychiatry such as 'shock treatment', prolonged sleep, lobotomy, and so forth.

The discoveries of modern medicine are by no means the final word in scientific progress. Tomorrow, no doubt, a method will be found to cure afflictions caused by viruses, or arteriosclerosis or malignant tumours, and suddenly a new advance will be registered in the average length of human life. But that is not the end of the matter. If medicine aims at the almost total elimination of death caused by disease, biology in the broad sense of the word envisages a struggle against death itself by attacking the natural cause of human decease — that is to say, old age.

Working from data established by Carrel concerning the alteration of the blood with the onset of old age, attempts have been made to rejuvenate senescent organisms by transfusions of young blood. This treatment sometimes results in a stimulation which could be taken for rejuvenation, but the effect lasts for a very brief time; in a few days the new blood takes on the characteristics of senile blood.

All kinds of processes have been tried with a view to warding off old age or reducing its symptoms: the grafting of glandular tissues; the injection or implantation of sex hormones (testerone for males, folliculine for females); washing in the urine of the new-born; ingestion or injection of embryonic extracts; inoculation with living cells (Niechans' method); vaccination against senile toxins, and so on. Attempts have been made to stimulate the connective tissue (Bogomoletz's serum); the royal jelly of bees has been tried and the 'biogen stimulines' which tissues held in suspense are believed to produce (Filatoff's method), and so forth.

Of all these methods it would seem that it is still the implantation of sex hormones which has so far given the best results. Huet, L. Binet, and Bourlière are well known for their encouraging observations concerning the very old of the masculine sex. Following the implantation of the male hormone a general improvement has sometimes been noted, a renewal of strength and a return of memory, while on occasion troubles of the sphincter muscles have been cured.

Embryonic extracts, prepared by grinding up the foetus of cattle in sterile vessels refrigerated to a temperature of under minus 8 °C, can be injected hyperdermically and are said to improve the psychological state of the aged, to increase their muscular strength and hasten the healing of wounds. At the same time a general physical improvement is said to occur with modifications of the blood, including the decrease of glutathione and increase in the number of eosinophil leucocytes, and also of basal metabolism.

These efforts of 'embryotherapy' are directly inspired by experiments made long ago by P. Carnot who, at the beginning of the century, detected in embryos and in tissues in process of regeneration the presence of substances which stimulated cellular multiplication, and suggested that they might have a therapeutical application.

In 1957 a derivative of novocaine called H 3, was presented by the Rumanian woman doctor, A. Aslan, to the medical congress at Karlsruhe. Claims were made that it could perform miracles, even for centenarians! However, we must recognize that cures for old age have proved to be of small value and, at best, can lay claim to very modest results. Slight and ephemeral stimulation, amplified perhaps by auto-suggestion, is all that can be observed, and no serious-minded doctor or biologist would at present dare to speak of rejuvenation or even of the preservation of old age. Nevertheless it remains probable that more effective treatments will be invented. Gerontology, the science of old age, is only in its infancy and there is no *a priori* reason why the processes of ageing should permanently defy the efforts of science, when science has grappled so successfully with many other vital phenomena: growth, development, and sex.

In the first place some hope can be founded on the general technique of grafting. Although this process has long been considered impractical when the graft is transferred from one person to another, the method is being ceaselessly improved and already its progress allows us to foresee a day when all the organs of those who accidentally die in the fullness of youth and health can, for others, be used as replacements, spare parts. Reservoirs or banks of living organs can then be formed, like the eye, bone and blood banks of today. Reserve supplies will be available of hearts, pancreases, kidneys, testes, ovaries, pieces of skin, and so on.

Another source of hope lies in methods of refrigeration. As early as the eighteenth century the mathematician Maupertius believed that cold might slow down 'the vegetation of our bodies'. The great American biologist Loeb estimated that the lowering of our body temperature by only 1° would result in an appreciable lengthening of human life. Experiments of varied inspiration suggest that refrigeration may contribute to the solution of the problem, and indeed, there are many who have faith in the idea that their bodies, doomed to die by some incurable disease, can be put into 'cold storage' until such a disease is curable.

The life-span of the small moth *Galleria mellonella* is, at the temperature of 37° C, invariably seven days. If every second day the insect is put into a refrigerator and kept in a state of suspended animation for twenty-four hours at a temperature of 1°C, these physiological pauses will much more than double its life-span. Moths so treated live from thirty to thirty-five days and lay twenty-five to thirty-five eggs instead of the usual nine to fifteen. This interesting experiment was made by L. Destouches and would seem to indicate that enforced periods of repose can lengthen an organism's life.

Denise Marcoux also observed the rejuvenating effect of chilling, and more generally, of a slowing down of vital processes, in the case of certain mushrooms.

Even among mammals a lowering of internal temperature seems to have a restorative effect, as the remarkable experiments of the Yugoslav physiologist Giaja demonstrated.

A rat, chilled to an internal temperature of 15° C, can live for some twenty hours in a state of complete inertia and insensibility. After being warmed up again the animal promptly revives, and the artificial 'suspension of life' has inflicted no damage on its organism, quite the contrary, in fact.

Giaja and his colleagues showed that the organs of the refrigerated rat were actually more resistant than those of similar animals which did not undergo the treatment. This unexpected result was demonstrated as follows: after being chilled to 15° C and kept at this temperature for from one to three hours the rats were slowly warmed up again by being exposed to a temperature high enough for them to recover their normal temperature by their own efforts. When this was done breathing was arrested by puncturing the bulb of the aorta and the thorax was opened to expose the heart. The same operation was performed on other unchilled rats in order to compare the resistance of the heart in both cases. The heart of the unchilled rat never beat longer than six minutes, while that of the animal which had been chilled continued to beat for as long as twenty-five minutes. As the physiologist concluded:

'It could be that lowering the temperature of a warm-blooded animal for a limited time imposes on its organism a cessation of metabolism and expenditure of energy which it never knows in normal life. The brief repose could also be physiological, that is, a restorative repose after the manner of hibernating animals, of most 'cold-blooded' animals and plants — an obligatory repose in their life cycle.'

In addition, Giaja discovered in the blood of chilled animals tonic substances which had some power to stimulate the heart.

Finally, another Yugoslav biologist, Doctor Andjus, has demonstrated that by using a special technique of 'super-congelation' rats can be chilled to and kept without damage at an internal temperature of minus 3° C. Apparently dead, they come to life again when warmed up again and are physically unaffected by the experience.

Thus the prospect of 'therapeutic refrigeration' seems to face us, in accordance with Carrel's striking prediction. 'Perhaps,' he wrote, 'we can prolong the duration of human life, cure some disease, and make fuller use of exceptionally endowed individuals, by making them hibernate from time to time.'

The beneficial effects of slowing down life in this fashion may be compared with those of undernourishment. As McKay has shown, rats whose growth when young has been greatly retarded by insufficient food live longer than rats fed in the ordinary way. One of these undernourished animals attained an age of 1,421 days, which is a record for the species.

Finally, by giving a little rein to the imagination, the prospects of anti-senile therapeutics become still more cheerful. If the ageing of nerve-tissue is in truth the chief cause of general senescence, then grafts of nerve-tissue or injections of extracted embryonic nerve-tissue, is a possibility to be envisaged.

Since longevity is hereditary and hence linked with the quality of the genes, we may assume that it will one day become possible to borrow substances from the cells of the very aged — individuals who have proved their capacity for endurance — substances extracted from cultures of their tissues.

This assumption would seem to find some support in recent experiments in chemical hybridization by means of D N A. Great hopes may perhaps be founded on the use of D N A extracted from cells of long-lived organisms.

Lastly, in gerontology, as in every other science, the unexpected can occur, the startling and unforeseen discovery. Who, a generation ago, could have suspected that the treatment of infectious diseases would be totally revolutionized by a little green mould commonly found on decaying cheese?

The Suspension of Life

Can life be interrupted? Under the influence of conditions like cold and drought can all signs of an organism's vital activity cease, only to reappear when normal conditions are restored?

This question, much debated in the past, first arose in connection wtih microscopic organisms such as rotifers, Tardigrada, and Anguillula, which live in the damp moss on walls or in the moist sand of gutters. These animalcules, essentially aquatic, are adapted to life in a liquid environment, and are very active as long as they find a little humidity. But, if the necessity arises, they are perfectly capable of resisting extreme conditions of dryness. In this case they change form, contract, curl up and become completely inert. They have every appearance of being dead. If, however, they are placed in as little as a droplet of water they quickly recover their original appearance and mobility.

In the eighteenth century the celebrated Abbé Spallanzani maintained, on the evidence of his own experiments, that the desiccated animalcules, rotifers and Tardigrada, had really succumbed and that their reanimation was nothing less than resurrection! Other investigators insisted that in the inert and dried up creatures life persisted, though in a form so attenuated as to be inobservable.

The quarrel was continued into the nineteenth century between Doyère, who believed in the 'resurrection' theory, and Pouchet, who condemned it. It was towards this time, 1861, that Edmond About wrote his tale of the *Man with the Broken Ear* in which a French colonel is put in a state of suspended animation by being dried up, just like a common rotifer.

The problem of suspended life also arises in the case of the eggs of silkworm moths, in the case of plant seeds, and in that of organic ferments or yeasts. All these organisms are or can be put into a state of 'latent' life, as Claude Bernard called it; in other words a state of total inertia, of 'chemico-vital indifference', a mode of existence widely found in nature, which will doubt-

less some day explain 'a great number of phenomena which today seem mysterious'.

In plant seeds, notably, Claude Bernard denied the slightest persistence of 'manifest life': between the seed and its environment the rupture of relations was complete; no exchanges were made; the life of the seed was purely potential. The seed was as inert as dead matter: all vital processes, those of destruction as well as of construction, were provisionally suspended — which was precisely why it could, without perishing, await for such long periods the return of conditions which would again allow it to display its vitality.

Biologists of that epoch did not share Claude Bernard's opinion. In particular his pupil, Albert Dastre, denied the possibility of such a suspension of life and refused to believed that after indefinitely prolonged slumber the seed or the Anguillula, emerging from its torpor, could, like the Sleeping Beauty of the fairy tale, resume the tenor of its existence at the point where it had been interrupted and, as it were, leap across the centuries.'

As far as seeds are concerned, it is today well established that the partisans of a total interruption of life were wrong. Though life in these organisms is immensely slowed down and enfeebled it has by no means stopped. Between the seed and its environment one can, contrary to what Claude Bernard thought, detect faint gaseous exchanges. Furthermore the seed ages, and wears out; for the processes of protoplasmic destruction continue within it so that in time (the period varies from species to species) it loses the power to germinate.

Among short-lived seeds the capacity to germinate lasts only for a few days or months. Among long-lived or macrobiotic seeds, which are characterized by the impermeability of their integuments, longevity can exceed a century. In 1807 Desfontaines saw a hundred-year-old bean from the herbarium of Tournefort germinate. Paul Becquerel succeeded in growing seeds

212

Paul Becquerel, whose researches into the resistance of living organisms to low temperatures are of fundamental importance.

X Photography.

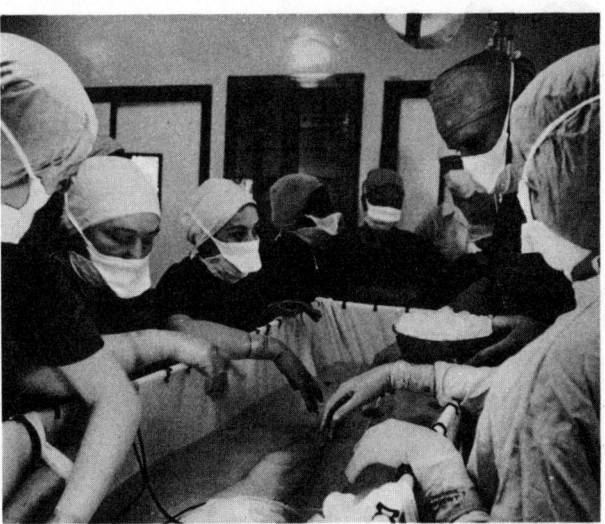

The 'deep freeze' is sometimes used together with drugs in treating cancer patients, who are cooled in a bath to which ice is added.

Camera Press:

of *Cassia multijuga* from the herbarium of the botanist Adanson which were 158 years old. This, however, approaches the limit, and claims that older seeds have been made to germinate are without scientific basis. There is, incidentally, no serious foundation for the tales we have heard of the survival of seeds removed from Egyptian tombs. Maspero, in a letter of 15th July 1901 addressed to Griffon, summed up the matter succinctly by stating that tomb seeds bought from the fellahs almost always sprouted, while those collected from the tombs by the investigator himself never did. The Arab guides, in fact, made a profitable industry of selling seeds to tourists which, in the tourists' presence, they 'discovered' in ancient tombs, seeds which they had hidden there the night before. Occasionally even more startling discoveries were made in some antique burial site: grains of maize, a plant of American origin, unknown in the Old World before the expedition of Christopher Columbus! The eminent botanist A. de Candolle held that the germination of 'mummy wheat' was not an impossibility, but attempts — such as those of F. Gain, with seed from the authentic collections of Boulaq which are some thousands of years old — have produced no results. Much younger seeds taken from the Merovingian tombs have also failed to germinate.

Plant seeds, then, do not in normal conditions fall into a state of totally suspended life. The question, however, arises: could they be made to do so by submitting them to exceptional conditions?

This was the problem which Paul Becquerel set himself, and to which he devoted an important series of experiments. Becquerel at first discovered that by dehydrating seeds and removing their gaseous content the duration of their life could be considerably prolonged. They pass into 'a more advanced state of protoplasmic solidification', which brings about a further decrease in their exchanges with the environment and allows them to retain the power of germinating for

a longer time. But this is still not a total suspension of life. Later, however, it would seem that Becquerel did indeed achieve the actual interruption of life he aimed at. He reported the result of his researches to the French Academy of Sciences in December 1950 and January 1951.

Using advanced cryogenic, or freezing, techniques Becquerel succeeded in submitting seeds of alfalfa, clover, tobacco, petunia, as well as a variety of organisms (mould spores, *Bacillus subtilis, Bacillus mesentericus,* rotifers and Tardigrada of various species) to a temperature very closely approaching absolute zero — or minus 273° C. The experiment took place in the cryogenic laboratory of the University of Leyden and lasted two hours. The organisms, during this intense cold, were also submitted to the most perfect vacuum conditions possible. After rewarming and rehydration, all the organisms came to life again. The seeds germinated quite as well as the 'control' seeds, the mould spore and microbes produced abundant cultures and the animalcules all revived.

It is obviously very hard to believe that in such conditions protoplasm can preserve its colloidal state or, more generally, retain a structure compatible with the exercise of its vital processes. Thus in this case the hypothesis of a discontinuity of life would appear to be plausible.

As Becquerel said, 'Under the combined action of dehydration, vacuum, and a temperature within a few thousandths of a degree from absolute zero, the protoplasm of the cells and their contents — vacuoles, nuclei, chromosomes, nucleoli, plastids, mitochondria, diastases, hormones — without traces of liquid water or air, were completely solidified.'

When we recall Van t'Hoff's law, which states that the speed of chemical reactions is roughly halved for every 10° C the temperature is lowered, and, in addition, refer to Guye's theories on the rate of molecular dissociation, we arrive, according to Becquerel, at

213

the following conclusions: at a temperature of minus 100° C, the vital chemical reactions of the seeds are 85,300 times slower than at a temperature of plus 20° C. At minus 200° C they are 4,840 million times slower. At minus 250° C they are seventy-one billion (10^{12}) three hundred thousand million times slower. And finally at minus 273° C (absolute zero) they stop altogether. If these figures are translated into 'duration of suspended life' we arrive at 'the stupefying result that a seed which lives only for a year between 10 and 20° C could in theory germinate at the end of seventy-one billion three hundred thousand million years if it were kept at a temperature of minus 270 degrees.'

This indeed would be immortality by refrigeration! But of course it is still only a theory, and it is by no means certain that protoplasm could support such prolonged periods of inactivity.

The only way of finding out would be by 'long-term experiments in sealed tubes containing seeds sheltered from abiotic radiations and plunged into liquefied gases. Every ten or fifty years one could test whether or not their power to germinate had varied.'

According to Becquerel it would be thanks to the cold that life on Earth could be preserved in the event of some cosmic disaster — such as the withdrawal or extinction of the sun — when our atmosphere would liquefy as the atmospheres of Jupiter, Uranus, Neptune, and Pluto have done.

'The vegetable kingdom and the animal kingdom would disappear. But spores, egg-cells, seeds and animalcules like rotifers would long survive in a state of suspended animation, ready for a new evolution if favourable cosmic conditions returned.'

The picture recalls Anatole France's moving description in the *Jardin d'Epicure* of a dead and frozen world where all that remained were marble statues and the material vestiges of human thought. After every man on Earth had disappeared, and all the animals and all the higher plants, minute organisms might still remain, preserved in the immense deep-freeze into which our globe had been transformed.

Cell preservation

We have seen in a previous chapter that the seminal cells of higher animals can at very low temperatures and with an admixture of glycerine be kept alive for months, and even for years. The same method has been applied in the preservation of blood-cells, glandular cells, and others. Human blood can in this way be kept for two years.

The ovaries of mice after being submitted to a temperature of minus 196° C in the presence of glycerine have been grafted with such success that two weeks after the operation they functioned quite normally.

Survival of an embryo chicken's heart after deep freezing. When the eggs have been incubated for nine days an opening is made near the lungs, every possible precaution to maintain aseptic conditions being taken. When the shell is completely opened the embryo is carefully withdrawn and detached from the mass of egg yolk.

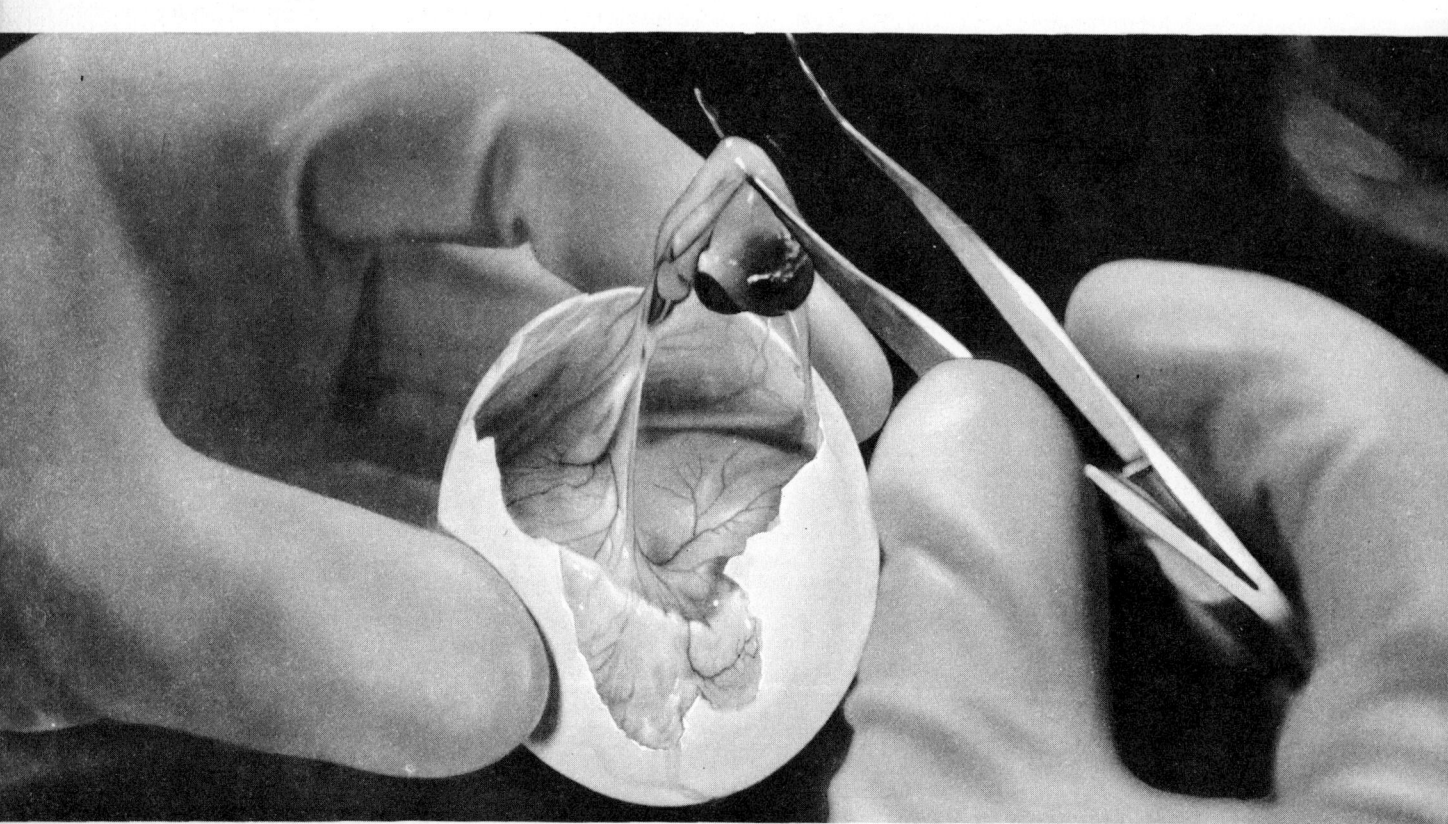

Entire hearts of embryo chicks, treated in a similar manner, have when rewarmed begun to beat again.

Efforts have been made to analyse the way in which glycerine protects the cells by direct observation under the microscope of what occurs in the frozen cells. It can be seen that ice crystals are not formed in the interior of the cellular bodies. Furthermore, the ice crystals which do form in the glycerine medium outside cause less damage to the cells than they would in an ordinary medium, because they are much smaller and less closely packed together.

Actually the protective action of the glycerine relies on a very complex mechanism which is only beginning to be understood, thanks to the research of Lovelock, Louis Rey, and others. A substance closely related to glycerine in chemical composition, namely glycol, is in some cases capable of producing the same effect. In 1951 Federico Gonzales and Basile Luyet, scientists who have long specialized in the biological effects of cold, observed that fragments of heart removed from an embryo chick could without perishing support a temperature of minus 195° C, which is the temperature of liquid nitrogen, for several weeks, providing that they had been previously immersed in a thirty per cent solution of glycol.

The persistence of cellular vitality was shown by the fact that the heart fragments could still produce tissue cultures, characterized by the displacement and proliferation of cells. Fragments submitted to glycol for one minute and to the intense cold for one, two, three, or four weeks could not be distinguished from control fragments chilled for one minute; all were identical in density and in the extent of cellular growth and spread. In fact, being kept for four weeks in liquid nitrogen does not in any discernible way reduce the power of embryonic tissues to survive.

In the light of these results we can now envisage the possibility of successfully putting the somatic and germinal cells of higher animals in a state of suspended life as Becquerel did with plant seeds and rotifers. In the words of Louis Rey:

'There are very good reasons to believe that in future research a satisfactory balance will be found between preliminary freezing, the choice of protective liquids, the temperature of desiccation, the desired degree of dehydration, and the process of reconstitution. The conservation of life will then be possible, and the present gap between higher organisms and the Tardigrada and rotifers bridged. When these problems are solved it will be possible to suspend vital activity for perhaps indefinite periods of time.'

The embryos are placed in a bath of tepid physiological solution to remove the last traces of yolk. They are then placed on a bed of black paraffin wax, firmly affixed with pins, and dissected under microscopic control in order to remove the hearts.

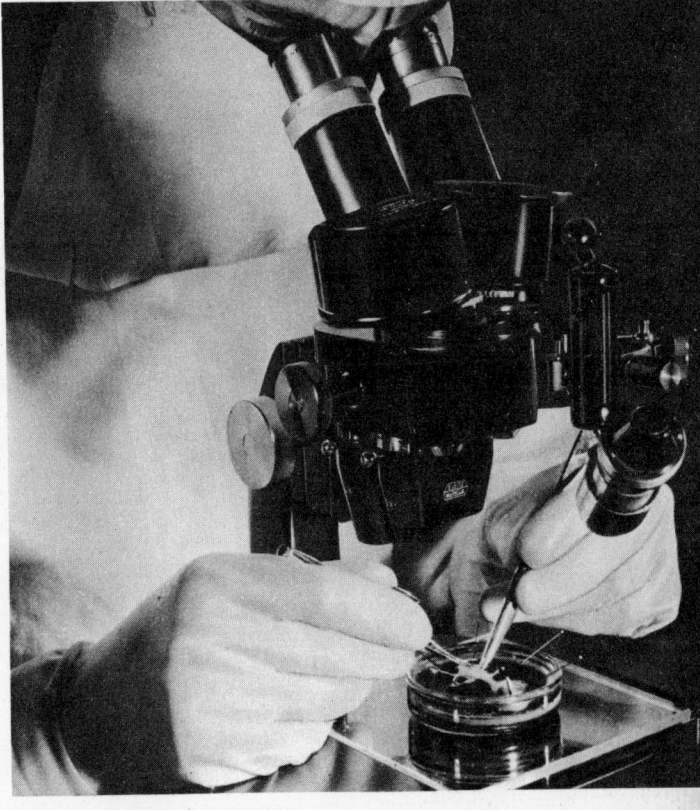

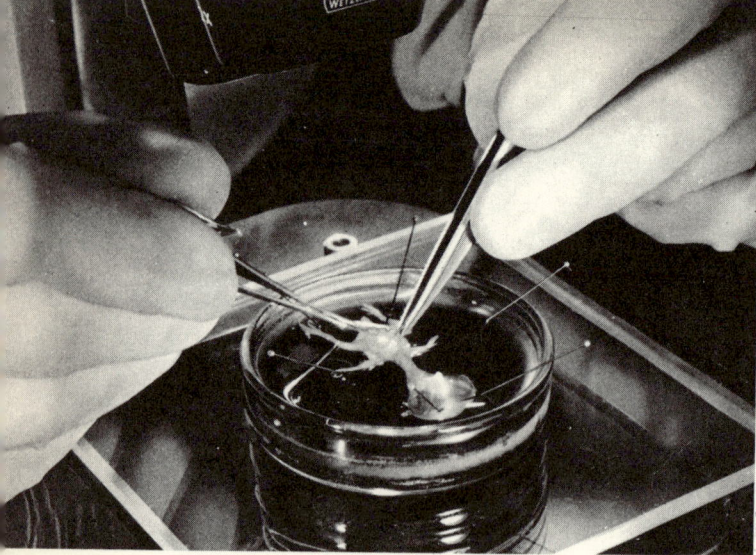

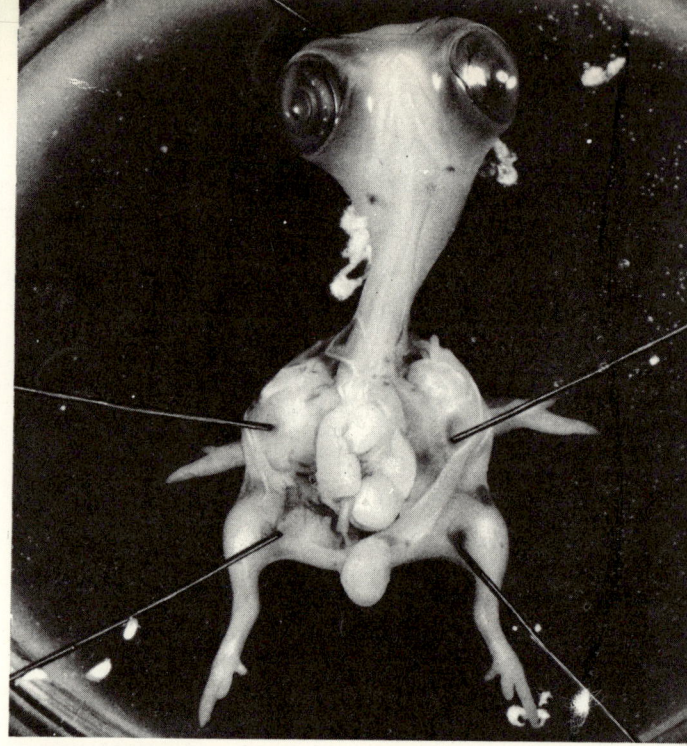

With the aid of special scissors — known as Patcheef-Wolf scissors — the abdominal wall is opened by longitudinal incision.

The heart is thus laid bare and, with great care, it is possible to remove it.

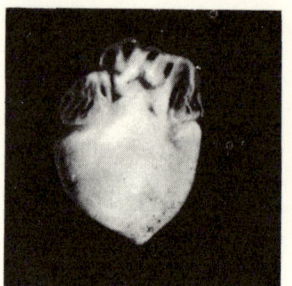 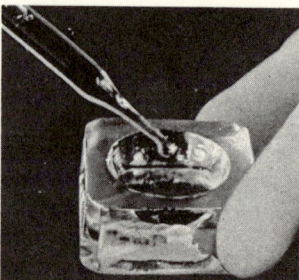

The isolated heart is then put into a physiological solution containing glycerine.

Afterwards it is delicately arranged on a thin strip of aluminium which, being an excellent conductor of heat, facilitates the rapid freezing of the organ.

Preparing the physiological solution into which the embryo chicken's heart will be placed.

L'Air liquide.

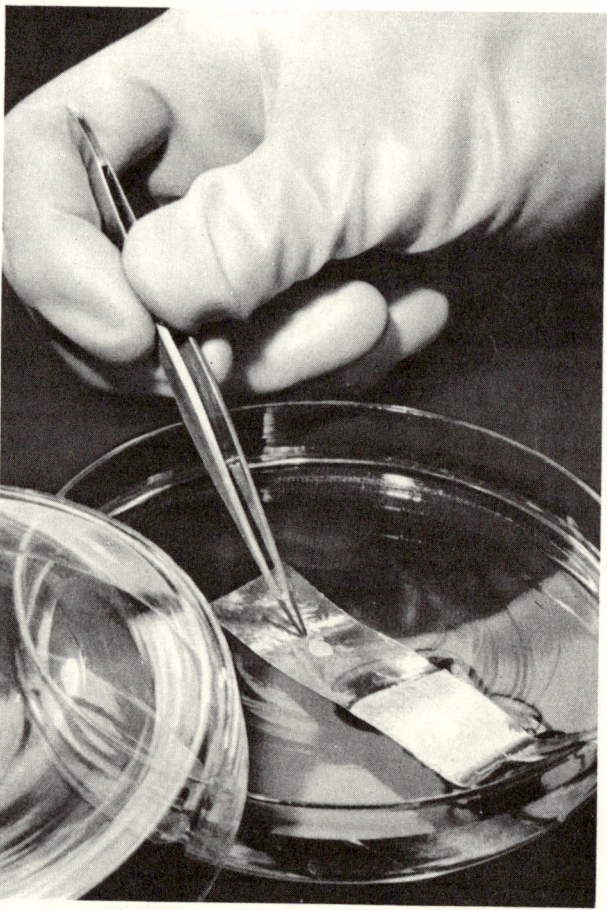

Finally, the aluminium strip carrying the heart of the embryo is plunged into a Dewar flask containing liquid nitrogen, which boils at minus 96° C. Freezing is therefore almost instantaneous.

Regeneration

All living beings possess in varying degrees the faculty of regrowing or regenerating certain parts of their anatomies that have been removed. This regenerative faculty plays an important part in nature and has been the subject of innumerable experiments since naturalists in the eighteenth century first remarked on it.

There is perhaps no creature which shows greater powers of regeneration than the little freshwater animal, belonging to the phylum Coelenterata, which is known as the hydra. Two species are commonly found, the brown hydra and the green, differing not only as their names suggest in colour but also in that the brown hydra is slightly larger than the green, though both measure less than one centimetre. They inhabit stagnant waters where they are generally found affixed to the lower surface of aquatic leaves.

The hydra's structure is relatively simple. It consists of a kind of elongated tube which at the upper extremity terminates in a mouth surrounded by a crown of filaments or tentacles. It is with the aid of these mobile and flexible appendages that the hydra captures the minute prey it feeds on.

Its power of regeneration was discovered in about 1740 by Abraham Trembley and it was under the name of the 'fresh-water polyp' that it achieved biological celebrity in the volume which Trembley himself published in 1744 to record his observations and experiments.

Trembley, operating on the little animal with astonishing patience, dexterity and ingenuity, discovered not only that by cutting a hydra in two he could produce two hydras but also that a single fragment of the hydra could give rise to an entire new animal.

For example, he cut a hydra in four, taking care to feed each of the four parts well. When they had reached sufficient size he divided them again into two or three parts, and again encouraged them to grow. Once more he subdivided them, and in this manner he cut the polyp into fifty parts. All fifty became complete animals, capable of all normal functions including that of reproduction.

It would not, perhaps, be inappropriate to date the beginning of experimental biology from these researches. For the first time in the history of the science of living things an investigator had actively and 'creatively' intervened in the basic processes of development; he had worked with nature and changed it.

The history of the 'freshwater polyp' enjoyed enormous success. The animal which could 'sprout' like a plant was discussed at court and in drawing rooms, for it seemed to be a link between the animal and the vegetable kingdoms, and to fulfil the definition of those ambiguous beings which were said by Leibnitz, the philosopher of continuity, to exist!

Then again the fragmentation of the polyp aroused difficulties of a metaphysical and even theological nature. What became of the soul of an animal which, without dying, was subdivided by the experimenter? Since it is the soul which gives life, and the soul is by definition an indivisible essence, how was it that each portion of the polyp could retain life? Could the soul, after all, be cut into pieces, or had it taken provisional refuge in one of the fragments of the creature? Such burning questions offered subject matter for endless controversy.

After the basic researches of Trembley the power of regeneration was sought out and studied in all levels of the animal kingdom where it is found unequally distributed, but never entirely lacking. Even unicellular organisms possess it. If a ciliophoran — paramecium — is cut in two, the two fragments pursue a very different course: that which contains the nucleus survives and regenerates itself, while that which contains no nucleus promptly degenerates and dies.

In large ciliophorans like *Stentor* each section will become a new individual because each section contains part of the nucleus.

The arms of the starfish will grow again when cut off. In this animal a singular phenomenon known as *autotomy* is often observed. If an arm is wounded or merely brutally handled it will of its own accord become detached at the base in a kind of spontaneous self-amputation. In the case of other echinoderms, the Ophiuroidea, it is enough to seize the animal by an arm for the arm to fall off.

The phenomenon of autotomy is particularly striking in Holothuroidea or sea cucumbers. When this animal is handled or merely disturbed it will eject all its internal organs, including its digestive tube. This spontaneous 'self-evisceration' will then be followed by complete regeneration.

An even more curious case of regeneration is that of the sponge which will survive being ground up and strained through a sieve so that the animal is split up into little groups of cells. The little groups will then, at least if they contain certain types of cells, gather together again and finally produce an entire sponge.

Among hydroids like *Tubularia*, the faculty of regeneration is very stronge and comparable to that of the hydra. It is almost equally strong among the little flat and ciliated worms known as planarians. These tiny animals live in fresh water among rotting vegetation. Several genera exist, each with several species. Some of them can be cut into small pieces, and each piece will grow into a complete animal. In fact the worm *Lineus* can be divided into 100 such sections and at the end of a few weeks 100 complete worms will result.

If an earthworm *(Lumbricus)* is cut in two by a spade or other tool the back part can regrow the front part, together with a head, while the anterior section can regrow the posterior section together with a tail, on condition that the animal has been bisected at one of the appropriate rings.

Among molluscs, the snail and the slug offer inte-

Woodcuts from Abraham Trembley's *Memoires* on the 'history of a genus of freshwater polyps with arms in the form of horns, 1744'. Fig. 8 shows a budding Hydra.

Regeneration of freshwater worms.

After Charles Bonnet, in his 'Observations on certain species of freshwater worms which, cut in pieces, become as many complete animals'.

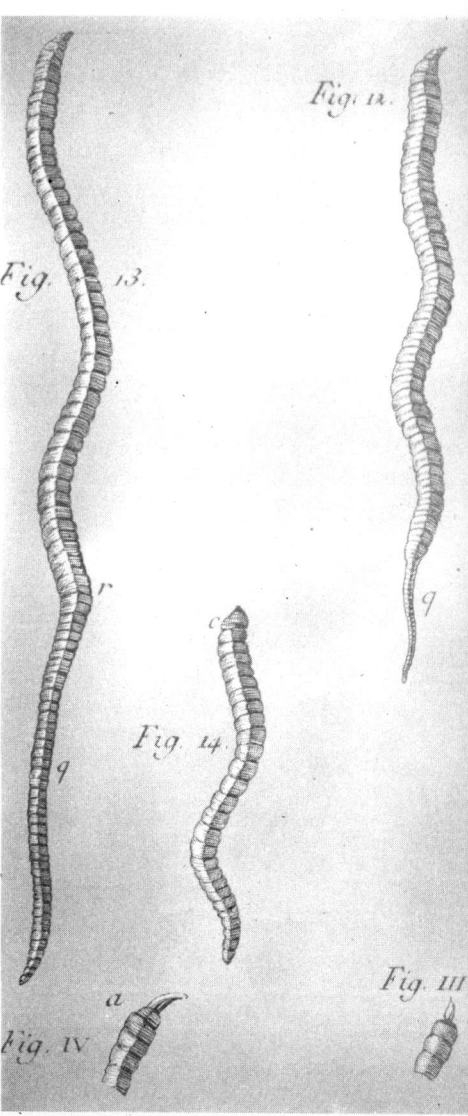

resting examples of regeneration. If the snail's head, or at least the greater part of its head, is cut off the missing elements will regrow: the face, the mouth, the radula or rasping tongue, the two pairs of tentacles one of which carries the eyes. In view of the structural complexity of these organs the snail's regeneration is particularly striking. A similar phenomenon also occurs among slugs, but their regeneration is less perfect.

It was in the eighteenth century, towards 1768, that the regrowth of the snail's head was first observed by the Abbé Spallanzani. It was one of his first discoveries and caused a sensation. At that time the phenomenon of regeneration was known only in the hydra, an animal of much lower category. The snail was several stages higher in the animal kingdom, and at first many naturalists denied the reality of Spallanzani's discovery. Cutting off snails' heads became the rage, and two camps were formed, pro- and anti-Spallanzani. Among

the snail beheaders were men of no less distinction than the great chemist Lavoisier and Voltaire himself, who related his personal experiments in two small pamphlets entitled *The Singularities of Nature* and the *Snails of the Reverend Father Escarbotier*. The regeneration of the snail's head delighted Voltaire who held that natural history had never produced anything more astonishing, marvellous, and wonderful. The miracle, he insisted, was worthy of Saint Denis, who when decapitated walked six miles with his head in his hands. That a head should be restored to 'an animal so large and visibly alive, an animal whose genus is unequivocal, is an unheard of prodigy, but a prodigy which cannot be denied. In this case no suppositions have to be resorted to, no microscopes employed, no errors of calculation feared. Human reason, and above all scholastic philosophy, is confounded by the testimony of our eyes. We consider the head in living beings as the guiding principle and cause of movements,

Attempts to graft the polyps.

After Trembley, 1744.

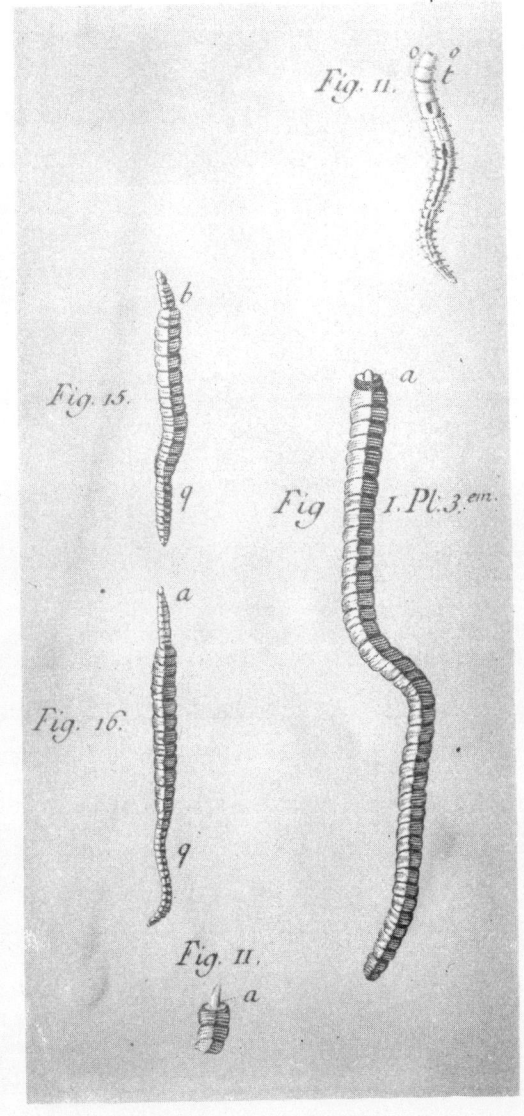

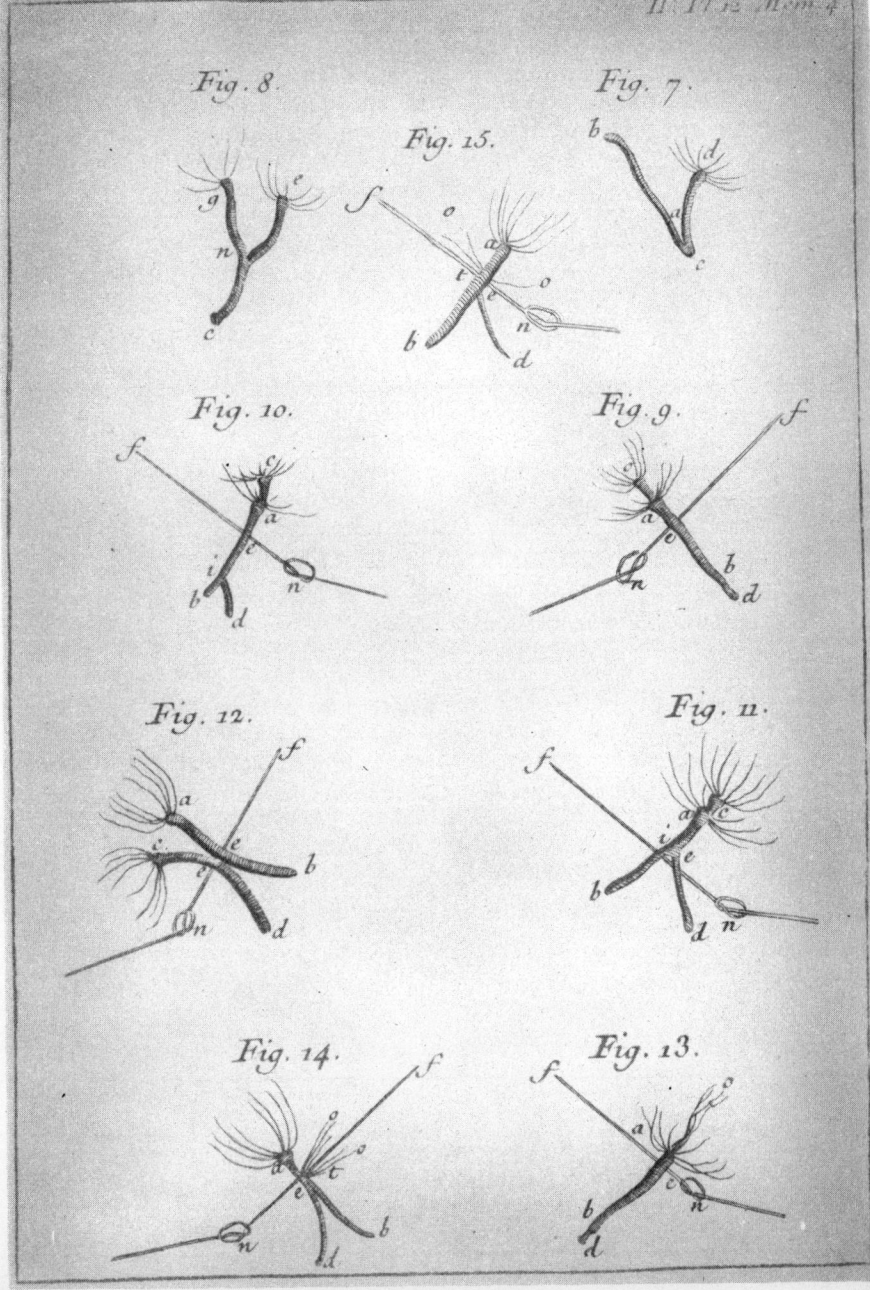

sensations and perceptions. Here it is just the opposite. In the course of a fortnight or three weeks the head which is reborn receives, from the body, fibres, nerves, a circulating liquid which serves for blood, telescopic tentacles, eyes, a brain, sensations, ideas. I say ideas because feelings are impossible without at least a confused idea of what is felt. What now becomes of the primary cause of the living animal? Will we be forced to return to the Greek principle of harmony, and thus discard ten thousand volumes of metaphysics as absolutely useless? If, at least, the re-formation of these heads forces certain gentlemen to doubt their own convictions the snails will have rendered great service to the human race.'

The regeneration of the snail's head was carefully studied by biologists. It was at first believed that for the head to grow again the decapitation had to be made in a way which spared the cerebral ganglions which in fact constitute the animal's brain. Actually the regrowth of the head does not depend directly on the presence of nerve centres, but on the integrity of certain tissue zones which alone have the power to regenerate the head.

Among arthropods such as insects, crustaceans, centipedes, spiders, and scorpions the faculty of regeneration is much reduced, although as long as the animal's growth is incomplete it can regenerate feet and antennae.

The phenomenon of autotomy is frequently observed in crustaceans. For instance, when the leg of a crab is wounded it will fall off, and the spontaneous amputation of the leg always occurs at the same level, namely, at a joint at the base of the leg which seems prepared in advance for the purpose.

The walking legs of the crayfish and spiny lobster break off less easily than those of crabs, except for the first pair which are provided with large claws and also have a joint ready for autotomy. When a leg is detached by autotomy very little blood is shed. Autotomy is controlled by a nervous mechanism; it is a reflex action in which the brain plays no part. If, however, the ventral nervous chain is destroyed, the phenomenon will not occur.

Free-living spiny lobsters often lose their claws. Rathburn reported that among hundreds of spiny lobsters (or sea crayfish) freshly caught in Narragansett Bay in 1880 a quarter had lost at least one of their two claws, while according to Herrick's statistics, quoted by Morgan, seven per cent of 725 spiny lobsters captured at Woods Hole had lost one or two claws.

The origin of the phenomenon of autotomy has been much debated. To some extent it may be considered a means of defence for the animal. If, for instance, a crab is seized by a leg it can escape by sacrificing the leg, a sacrifice which, if the animal is young, is not too painful since young crabs have the power to grow new legs. Some authorities see in autotomy an arrangement to facilitate regeneration, and there are in fact cases, notably that of the crab, in which regeneration takes place more successfully in a zone of autotomy than in other regions of the animal's body. The problem is particularly difficult because it raises the larger problem of natural selection. One can only say that it is far from certain that the advantages bestowed by

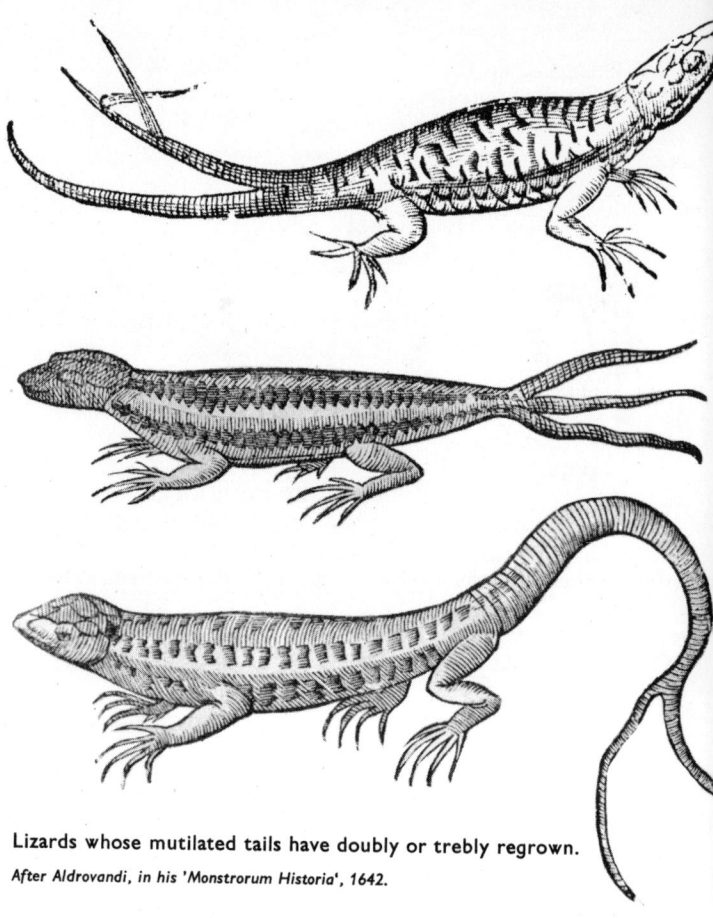

Lizards whose mutilated tails have doubly or trebly regrown.

After Aldrovandi, in his 'Monstrorum Historia', 1642.

autotomy are of a kind which favour the survival of the animal in competitive existence.

The crab *Alpheus* has claws of unequal size, one very large and one small. The large claw can be either the right claw or the left. If it is amputated a small claw will grow to replace it, while at the same time the small claw opposite will begin to grow until it becomes the size of the originally amputated large claw. At the end of the process there will in other words be an inversion of the normal asymmetry of the crab's claws. It is as though the large claw had, until amputated, exercised a restraining or inhibitory influence over the growth of the small claw. The process would seem to be controlled by a nervous rather than a humoral mechanism.

In other crustaceans, which also have claws of unequal size, regeneration does not result in a reversal of asymmetry. In these cases the physiological equilibrium between the two claws is, at least if the animal is an adult, definitely fixed.

Regeneration in vertebrates

Among higher animals, that is among vertebrates, we also find remarkable examples of regenerative power. The animals which furnish these examples all belong to one and the same class, namely the amphibians: and among them it is the tailed amphibians, or Urodela,

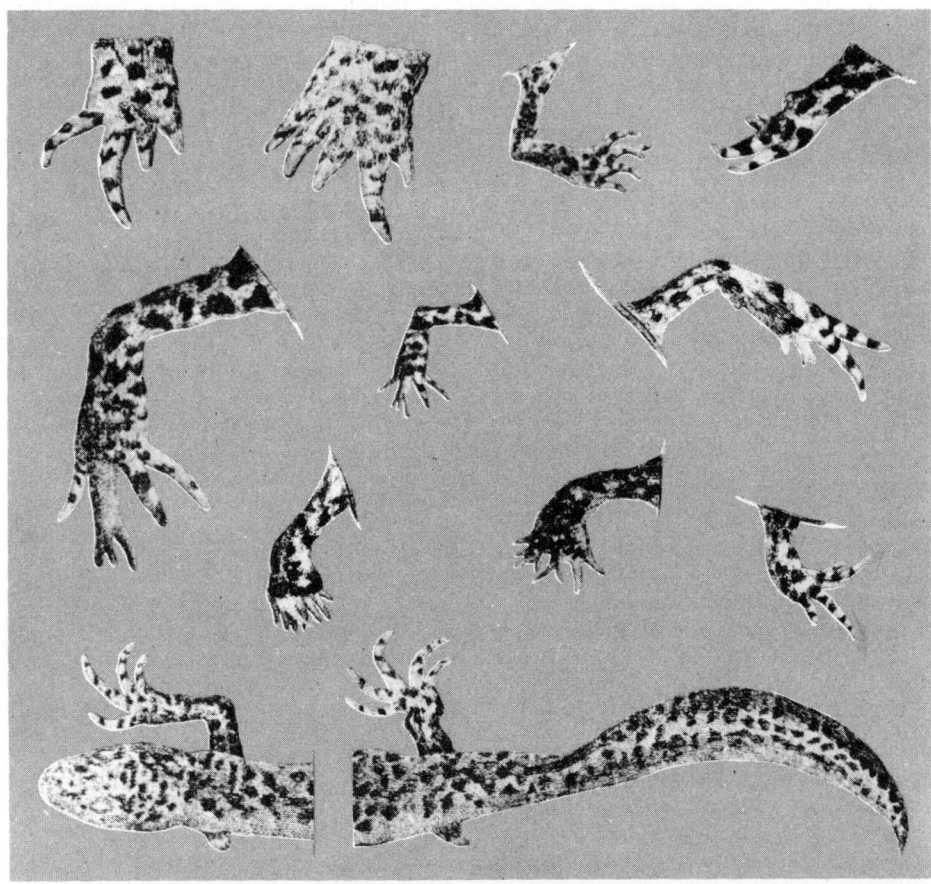

Regeneration, normal or abnormal, of the triton's legs.

After Charles Bonnet, 'Memoires on the reproduction of the members of the aquatic Salamander'.

which exhibit the most highly developed faculty of regeneration.

The triton is in spring commonly found in pools and ditches filled with water. The triton can, after amputation, regrow all four legs, its tail, its jaws, and even its eyes.

The structure of the regenerated organ is in every way comparable to the original organ and it is extraordinary to see how, within a few weeks, a structure as complicated as a vertebrate's leg, with all its bones, muscles, blood vessels and nerves, will grow again. The stump of a triton's leg will almost always restore the missing parts in their entirety, no matter how or at what level of the leg they have been removed.

Several successive regenerations of the same organ can take place: within a period of three months Spallanzani obtained the regeneration of the four legs and of the tail six times running. He calculated that in the course of these restorations no fewer than 647 bones had been newly formed.

All successive regenerations are equally rapid. The rapidity of regeneration depends on various factors, themselves influenced by the animal's vitality. It is increased if the animal is well fed, if the surrounding temperature is raised, and so on, although even complete inanition does not prevent regeneration.

Among tailless amphibians like frogs and toads the faculty of regeneration is not great, at least if the animal has reached maturity. If the leg of a frog is cut off, a stump and nothing more than that will form where the amputation took place. If the same operation is performed on the larva of the frog, the tadpole, the leg will grow again. The tadpole can also grow its tail again, its external gills, and so on, but not its eyes.

The only tailless amphibian which can when mature regrow its limbs, or at least its toes, is *Xenopus,* the South African amphibian which is so commonly used in laboratories because of the ease with which it is reared. In fishes the regeneration of fins and part of the jaw has been observed. Among reptiles, the slow-worm (or blind-worm) and the lizard can regenerate their tails, or at least an appendage which resembles a tail without being altogether structurally the same. Some birds can partially regenerate their beaks. Among mammals the power to regenerate the external parts of their bodies is entirely lacking.

It used to be said that the extra finger of 'men with six fingers' was capable of growing again. 'It is a curious thing,' wrote Doctor Apert, 'that the supplementary digits of polydactyls, when they remain rudimentary and devoid of calcareous bone, enjoy the same properties of regeneration as the digits of lower animals and embryos. They can grow again not only when cut off in the middle — which could be explained as an exaggerated growth of the remaining stump — but also when cut off at the base.' White of Manchester is reported to have seen one such finger regrow three times, each time re-forming the fingernail. Darwin spoke of a man who was polydactyl, as were two of his brothers and one sister. 'He had fourteen children of whom three were polydactyl. The removal of his own supplementary fingers, and those of one of his children, was on two different occasions followed by their regeneration. In the case of the child the extra finger contained a bony axis and was amputated through the joint. Nonetheless it grew again.'

In spite of the authority of the names cited by Doctor Apert, it would seem that the observations are totally without value. Mammals, like birds and other vertebrates, do on the other hand possess some ability to regenerate their internal organs.

Ponfick removed as much as a half and even three-quarters of the liver from dogs and from rabbits. At the end of a few weeks he found that the remaining part of the liver had greatly increased in size. In some cases it had trebled. The regeneration of the hepatic tissue begins as soon as thirty hours after the operation.

The salivary gland also has great capacity for regeneration, as the experiments of Podwyssoski and Ribbert demonstrated. When five-sixths of a rabbit's salivary gland was removed the gland recovered its normal size by the end of three weeks. The regenerated tissue functioned perfectly and was exactly like the original amputated tissue.

Regeneration has also been observed in other organs: the spleen, the pancreas, the thyroid gland, the testis, the kidney, etc.

The growth of new bone is well known. When a bone is broken the two parts almost always knit together again; for between the two fragments the periosteum — that is to say the membrane surrounding the bone — forms a more or less solid callus which contains newly formed bony tissue.

The muscles have some capacity for regeneration; the nerve fibres have even more. When a dog's sciatic nerve is cut it will grow again, daily lengthening by about 1 millimetre.

The sciatic nerve of a man grows at the rate of 1.2 millimetres a day, and can in this way reach a length of 1 metre.

The peripheral nerve end, as it degenerates, plays an important part in the new growth, drawing the budding nerve towards it. This attraction is due to the emission of chemical substances which might be called 'necro-hormones'. The same effect can be induced by ground up nerve-tissue.

The brain, the optic nerve and the spinal cord have scarcely any regenerative-capacity. R. M. May, however, observed a slight regeneration of cerebral nerve fibres when cerebral tissue of a newborn mouse and a sciatic nerve end, or muscle end, of the same species is implanted in the anterior cavity of an adult mouse's eye.

The existence of regenerative power in internal organs and tissues clearly demonstrates, as Morgan emphasizes and various others have denied, that there is scarcely any correlation between the power to regenerate an organ and the vulnerability of the organ itself. Thus we may doubt if the regenerative power has during the course of the ages been acquired and developed by natural selection.

Related to the phenomena of regeneration are those of compensatory hypertrophy — or the enlargement of organs or tissues. If a kidney of an animal is removed the other kidney will increase in size. The same thing occurs with most of the other paired organs the adrenal glands, the sex glands, the mamma. Hypertrophy is due to hormones which circulate through the blood stream.

Physiological regeneration

Physiological regeneration is the periodic renewal of tissues such as the blood tissue, glandular tissues, intestinal lining, uterine lining, skin, nails, hair, and so forth.

In man the red blood-cells live on average one month, some 200,000 per cubic centimetre of blood being destroyed every minute. The white blood-cells live for only a few days.

Human skin consists of two layers: one the protective outer layer called the *stratum corneum,* and the other, deeper, called the Malpighian layer in which active cell-division takes place. The outer layer continually wears away (desquamates) and is replaced by cells from the inner Malpighian layer.

The antlers of the stag fall every year and are replaced by still larger antlers.

A few general principles emerge from the study of regeneration. The faculty is, as we have seen, more marked among the lower members of the animal kingdom than among the higher, although the rule is not hard and fast. Amphibians, which are vertebrates, have a greater capacity for regeneration than insects. Again, within the same group we find very unequal powers of regeneration, the frog and *Xenopus* for instance. Why these inequalities exist we do not know.

Regeneration is, broadly speaking, more active in young animals than in adult animals. We have seen that the larvae of many amphibians will grow new legs, which the adult is totally incapable of doing.

In embryos regeneration is even more active than in young animals and larvae. The phenomena of embryonic regeneration are associated with those of germinal regulation, which we have already discussed in connection with development. It may be remembered that germinal regulation is particularly striking in the case of eggs like those of the sea-urchin and birds.

The regenerative power of a given part of the body relies on the intrinsic properties of the tissues themselves and not on the bloodstream. If an individual is incapable of regenerating an organ, and the organ is removed and grafted on to an individual of another species which is capable of regenerating it, no regeneration occurs. In other words, the tissues of the transplanted organ lack regenerative properties which the blood stream of the new environment is unable to supply.

The most important step in our study of regeneration will consist, as the great biologist Jacques Loeb once said, of increasing our control over it; in other words, of finding ways to produce it experimentally in cases where it does not normally occur. If we could do that, we could, he pointed out, also analyse the mechanism

Three and a half months after the tail of the slow-worm is severed the stump is covered by scales and the exposed vertebra is no longer visible. The stump will grow again into a more or less typical tail.

R. H. Noailles.

Xenopus, a native of South Africa, is the only amphibian which is able to regenerate its toes after amputation.

After J. Rostand. A. Steiner.

of the phenomenon more thoroughly.

Such a step was, in fact, made several years ago, beginning with the study among frogs of the regeneration of the legs. We have previously mentioned that the frog, once it has reached the adult stage, can no longer regenerate its legs. It is, however, possible to make a frog's leg grow again by submitting the amputated stump to certain chemical treatments: immersion in strong saline solutions (sodium chloride), alkaline solutions (sodium bicarbonate) sugar solutions (lactose); or to certain physical treatments: heating to 55°C for one minute, dilaceration of the muscles, ablation of the skin, ligature of the stump, and other methods. Exactly how these 'hyper-regenerating' treatments work is not yet understood, but their common effect seems to be to prevent or retard the healing of the wound.

Possibly they bring about a necrosis of the tissues with emission of stimulants such as peptones, ribonucleic acid, etc.

When from the eye of a chicken or a rat the crystalline lens is surgically removed the iris never regenerates that organ. If, however, a piece of the iris is first put into an artificial culture medium and then implanted in the eye of an animal of the same species it will, within a few weeks, induce the formation of a true crystalline lens. It is as though the iris during culture had acquired regenerative power, or rather, as Guyenot explained, 'it had lost something which was preventing it from exercising its latent power of regeneration.'

The value of this type of experiment, which may lead us to discover ways and means of artificially increasing the regenerative faculties of human beings, is too obvious to need emphasis.

The process of regeneration

All regeneration ordinarily begins by the formation, at the surface of the wound itself, of a kind of small bud or *blastema,* a mass of undifferentiated cells which develop into the regenerated member or organ.

In some ways regeneration is a renewal of the phenomena of development and growth. 'If,' as Abeloos stated, 'one thinks of the arrest of growth in the adult as an expression of a state of dynamic equilibrium between the parts of the body, then an amputation, disturbing this equilibrium, may be said to unleash a new spurt of growth tending to restore the balance.' It is important to remember, however, that the regenerated organ is not necessarily formed in the way it is formed during the course of development.

Regeneration has an entirely new genesis, and the cells which produce the new organ can derive from a very different part of the body from that which, in the embryo, originally produced it. New muscle does not necessarily arise from old muscle, nor bone produce new bone, as we can see in the case of the triton. If all the bone of a triton's leg is first removed, the leg, when cut off, will nonetheless regenerate another leg with normal bones.

Another example: if the crystalline lens is removed from a triton's eye a new lens will be formed, but at the expense of the cells on the edge of the iris, while in the embryo the lens is formed from cells of the superficial ectoderm, that is to say, from a skin-tissue whose origin is not at all the same as that of the iris tissue.

This difference in the process by which organs are normally developed and the process by which they are regenerated is of great importance from a theoretical point of view. It shows that the organism can by different means and different processes attain the same final result.

It is usually found that the bigger the amputation is the more rapid and vigorous will be the regeneration of the missing part. If we cut off a small piece of the tail of one earthworm and a large piece of the tail of a second earthworm, regeneration will take place much more quickly in the second case. In fact the worm which suffered the more serious amputation may be whole again almost as soon as the one which underwent the lesser operation.

Among higher animals analogous phenomena are encountered in the healing of wounds. Carrel, and more especially Lecomte du Nouÿ, demonstrated cases in which the greater the surface of the wound, the faster it healed. They even expressed the phenomenon in mathematical terms.

It sometimes happens that regeneration begins by the reconstruction of the extremity of the section amputated. Among earthworms, which have been the subjects of many experiments of this kind, the amputation of the entire rear half is first of all followed by a total regeneration of the terminal rings, the intervening regions being completed later.

In vertebrates, hormones secreted by the thyroid and pituitary glands play their part in regeneration as they do in growth. As for the nervous system, its role is not yet entirely clear, but it must be of considerable importance.

At first it was even thought that the nervous system determined the structural type of the regenerated organ for on deviating the course of a triton's sciatic nerve (the leg nerve) it was observed that a supplementary leg was forming at the new point where the nerve ended. Subsequent research, however, showed that the sciatic nerve has no power to cause a leg to appear: like every other nerve it can merely excite the proliferation of tissues which, in accordance with their own formative potentialities themselves produce whatever they are capable of producing. If, in the earlier experiment, the sciatic nerve appeared to cause the growth of a supplementary leg it was simply because it had been deviated into a tissue-zone capable of producing a leg. When the nerve is made to end in another zone, say the zone which can produce a tail, then it will cause the formation of a supplementary tail.

Guyénot designated these zones or regions of different potential regeneration by the name of 'territories'. According to him the organism of the triton is a 'mosaic' of invisibly boundaried territories which differ in their formative powers, physiological properties, etc. If one of these territories is removed no other territory can supply those regenerative potentialities it possessed, and the animal thus mutilated will for ever be deprived of the organ or part of the body which it has lost. In this way tritons have been obtained which can never grow new jaws, tails, legs, and so forth. If one of these territories is moved to a different region of the organism it will retain its properties: the territory of a leg grafted on to the tail will there give rise to a little leg, and vice versa. If, returning to the early experiment, we deviate the sciatic nerve so that it ends at the dividing line between the leg territory and the tail territory a composite appendage will form, a mixture of leg and tail.

It would seem that a given territory's specific qualities are chiefly due to the connective cells. From a lizard Guyénot removed the skin of the foreleg and replaced it with a band of skin from the tail which is exclusively made up of connective tissue. After healing, he amputated the foreleg at the level of the graft: the member which was then regenerated was not a foreleg but a small tail.

Territories evolve or 'ripen' independently of each other. Thus in the frog's tadpole the 'leg territory' loses its regenerative capacity at a stage of development when the 'tail territory' still retains its powers of regeneration.

Guyénot emphasized the importance and the general applicability of this concept of territory. It helps to explain a great number of physiological findings, such as the fact that different territories of the same organism are unequally sensitive to hormones, pathogenic agents like microbes and viruses, or cancers, etc.

In all studies of localized physiology and pathology the concept of territory is obviously of the greatest value.

Travelling cells

In some animals, and especially in flatworms (planarians) the regenerative cells — or *neoblasts* — can originate in all parts of the body. This fact was demonstrated by Francoise Dubois with much skill and ingenuity.

If the head of a flatworm is severed behind its eyes regeneration will begin at once, and a new head will be formed within a few days. But if before decapitation the forepart of the animal has been powerfully irradiated with X-rays, regeneration will be much delayed and it will be more than a month before the new head is formed.

The reason for this delay is that the cells in the forepart of the animal, damaged by the X-rays, are unable to accomplish the task of regeneration. The work has been taken over by cells coming from the rear part of the worm, and of necessity some time is required for them to arrive at the spot where the new head must be formed.

It is of great interest to discover that the head of the animal can be thus renewed by travelling cells, endowed with migratory properties: a 'mobile repair service' as it might be called. Francoise Dubois measured the speed at which the cells travel. She calculated that it required some ten days for them to cover a tenth of the journey, and a hundred days to travel seven-tenths of the length of the worm's body. The cells are immediately mobilized when the animal is decapitated, and even a simple incision is enough to raise the alarm. It is as though the wounded surface had sent out a kind of S. O. S. It is still a debatable point whether the message is conveyed by the nerves or by body fluids — a 'wound-hormone' — emanating from the injury.

Lender, also studying flatworms, performed fascinating experiments which throw light on the regeneration of the eyes. He discovered that the species *Polycelis nigra* has some ninety eyes on an average, arranged in a small fringe around the head. If a part of the eye-fringe is cut away the eyes will grow again, but they will not be regenerated when the animal's brain has been removed. The brain, then, possesses power to induce the regeneration of the eyes, and this power is not exercised by means of nerve threads but by body fluids. This can be shown by grafting experiments. If a fragment of eye-fringe is transplanted to a position near the brain or even in the front third of the worm's body the eyes, when removed, will grow again. This action at a distance from the brain does not reach as far as the tail. A brain badly damaged by X-rays will still induce the regeneration of the eyes, a phenomenon which can also be produced by 'head-juice', that is, ten or fifteen worms' heads crushed in 10 cubic centimetres of water.

'Head-juices' obtained from other species of planarians — *Dugesia lugubris, Dugesia gonocephala* — are found to be almost as effective as that of the same species. The eye-inducing substance resists alcohol, desiccation and heating to 60 °C. In a centrifuge it is present in the juice which remains floating on top. From two to four days of contact are necessary before it produces its effect. When it is very concentrated — twenty heads in 10 cm³ — regeneration is accelerated, being from twenty-four to forty-eight hours quicker than normal regeneration. This eye-inducing substance seems to be present in the central section of the worm's body, though in very small amounts. But the power to regenerate eyes in this part of the body can be aroused by treating it with alcohol or heat.

Lender also studied the regeneration of the brain in *Polycelis nigra* and observed that 'head-juices' prevent this regeneration, while 'tail-juices' have no such inhibitory effect.

The task of explaining the mechanism of regeneration still remains. It is simpler to say what it is not than to say exactly what it is, easier to dismiss unsatisfactory explanations than to propose one that can stand up to critical examination. It is not after all very surprising that the phenomena of regeneration have not yet been fully explained, since they are closely linked with the phenomena of development which are themselves far from being entirely understood.

Obscure though certain aspects of regeneration may be, the same obscurity veils many other facts of life, and to solve the problems it poses we have at our disposal only the experimental approach, the sole method which in biology, as in every other science, is fruitful.

Some investigators have compared organic regeneration with the regeneration of crystals. It is known that a crystal, when broken, can rebuild itself and recover its own specific form, provided that it is placed in a solution of its own chemical composition from which it can borrow the elements of its reconstitution. But this analogy, which was strikingly advanced by Herbert Spencer, between crystalline or chemical regeneration and vital regeneration is manifestly untenable. One cannot legitimately compare the ability of a crystal to recover its geometrical shape with the faculty which a living being possesses of regenerating a limb or an organ with its complex structure and its thousands or millions of cells harmoniously grouped into tissues.

In total contrast to this materialistic viewpoint, we find an explanation which might be called philosophic and attempts to describe regeneration as an essentially vital faculty of mysterious nature. This is the concept of entelechy, a 'regulating force' or 'directing force', more or less intelligent, more or less transcendent. This, however, would seem to be just as enigmatic as the mystery it attempts to solve.

The illustrious biologist Thomas Hunt Morgan, in his *Embryology and Genetics,* put the matter well when he said that it was perhaps superfluous to criticize such a view since it is admittedly purely metaphysical. 'But,' he continued, 'one must at least point out that in the first place the explanation is as obscure as the facts it pretends to clarify; in the second that it tends, in view of its air of finality, to hamper further search for a natural explanation; and finally that certain facts concerning regeneration flatly contradict the theory that entelechy is a beneficent directing force.'

It is these facts, to which Morgan refers, that we must now examine.

Faulty regeneration

The truth of the matter is that regeneration does not always restore the missing limb or organ in its original state. Errors occur — *heteromorphoses* — in other words, the part regrown is not always the part needed to complete the organism.

In some crustaceans, such as crayfish and prawns, when the ocular peduncle or stalk is cut off near the base, it is not a new eye which the basilar part of the peduncle regenerates but an antenna. This phenomenon was first discovered by Herbst. The animal operated on in this way has, when regeneration is complete, an eye on one side and on the other side a supplementary antenna. For a new eye to be grown (instead of the antenna) the amputation must leave the optic ganglion intact.

Among the insects of order Phasmida — creatures which resemble little sticks of wood — the amputation of the antenna may be followed by the regeneration of a minute leg, comprising a tibia and a tarsus with four joints. This heteromorphosis, as Lucien Cuénot has observed, takes place when the antenna is cut off at a specific level, more precisely when it is cut off at the second joint. If it is cut off higher up regeneration is normal and the severed antenna is replaced by a normal antenna.

It can also happen that the amputated organ is regenerated with abnormal characters. This can be observed among various insects such as cockroaches. When the end of a leg is cut off it will grow again, but instead of a tarsus with five joints which is normal in cockroaches it will have a four-jointed tarsus. Some naturalists, arguing from the fact that insects of closely related genera have only four joints in the tarsus, have interpreted this error in regeneration as a kind of atavism: regeneration, they say, reproduces ancestral characteristics of the race. The hypothesis is

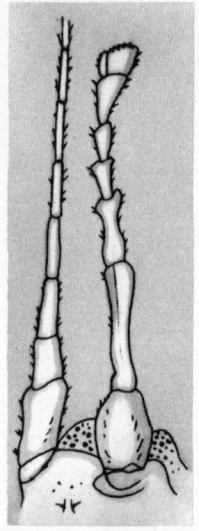

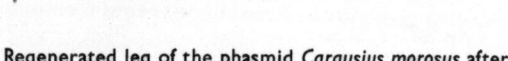

Regenerated leg of the phasmid *Carausius morosus* after amputation of the right antenna.

After Cuénot.

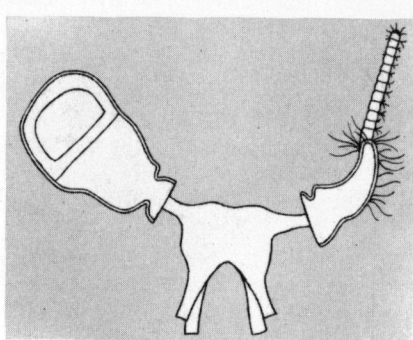

Heteromorphosis in the prawn *Palaemon*: regeneration of an antenna in the place of the right eye, removed with its optic ganglion.

After Herbst.

ingenious, but quite without foundation.

In tritons, on the contrary, the legs regrown by regeneration are often furnished with a greater number of digits than normal legs: six, seven, or eight.

Sometimes it even occurs that the amputation of a member will be followed by the growth of supplementary members. When, for instance, a lizard's tail is cut off two or even three tails may grow to replace it.

Broadly speaking when an amphibian with supernumerary feet or toes is encountered it can be assumed that the animal was wounded when young and that the trauma brought about excessive or hyper-regeneration; but supernumerary formations arising from other causes are also found among amphibians.

If larger and larger sections are cut off the earthworm a point is reached when the subsequent regeneration is no longer normal. Instead of the large missing part being reproduced, the small remaining part doubles — resulting in the formation of a composite monster which is not viable.

Let us, for instance, detach a small anterior part of the worm, consisting of the head and a few rings only: this small section of worm, instead of regrowing the rest of its body with the tail, will regenerate a second head pointing in the opposite direction.

In the same way, if a piece of the worm's posterior comprising the end of the tail is removed, it will not regenerate the remainder of the body with the head, but will form a second tail instead.

The forepart of the planarian *Procerodes lobata,* if cut in two just behind the eyes, will always produce a new head. As the operation is repeated farther and farther from the head, the percentage of these heteromorphoses falls, and at a certain point vanishes altogether. Then, as the tail is approached, a point is reached behind the base of the pharynx, when the posterior section will always reproduce another tail.

Numerous cases of heteromorphosis are known among hydras, hydroids *(Tubularia),* etc.

The results of these experiments are of twofold interest. In the first place they show that all levels of the body do not possess identical powers of regeneration; for the extremities reproduce a head or a tail as the case may be, as though at a certain point there occurs an 'inversion of polarity'.

In the second place the experiments reveal that regeneration does not always take place in a manner calculated to restore the organism's original integrity. Regeneration simply does what it can, and is not directed by some mysterious and infallible force acting in the interest of the organism.

Rosine Chandebois studied the anomalies of regeneration in the planarian *Procerodes lobata:* these included double heads, cyclopean heads, pointed and asymmetrical heads. He experimented with these abnormalities, inducing them or increasing their frequency by various chemical or physical influences.

When worms are bisected at a level which in normal conditions always results in normal regeneration, heteromorphosis can nevertheless be produced by means of salts, strychnine or low temperatures. Faulty or teratogenic regeneration, experimentally produced, is of course closely linked to the experimental production of monsters, described in an earlier chapter.

The Culture of Tissues

The culture of animal tissues

With the exception of certain organisms which are so small that the naked eye cannot distinguish them, all living creatures are made up of an assembly of minute elements or cells whose number, usually immense, varies from species to species.

In man, for example, though the figure can only be approximate, it is estimated that there are several billions (10^{12}) of cells, and that each cubic millimetre of human blood alone contains many millions.

Each cell is itself a small complete organism with an individual life of its own. It is, however, dependent on all the other cells, since all contribute to the harmonious existence of the whole organism which, in return, assures those working and living conditions which each cell requires. These conditions are met with in the internal environment — blood and lymph in higher organisms — which supplies oxygen, nourishment, salts, growth factors, and so forth. If, from this harmonious organic community, we detach a small group of cells, such as, for example, a fragment of liver removed from a mouse, what will become of it? What will be the fate of the cells when they have been withdrawn from the whole of which they once formed a part?

The answer will depend on the new conditions to which they are submitted. If they are placed in a small flask containing a physiological solution, that is to say water with the same saline concentration as blood, and if the solution is carefully maintained at 37 °C, the temperature of the animal's body, and if during the course of these operations all bacterial contamination is avoided, the cells thus placed in warm and sterile physiological saline can survive for a few hours or even for a few days. But they will inevitably perish, if only from lack of nourishment. If, however, instead of putting the fragment of tissue into the physiological saline, we put it into blood plasma, which is the liquid which remains when the corpuscles are removed from the blood, and if we take the plasma from the same species of animal as that which furnished the fragment of liver, the cells survive for a much longer period, perhaps several weeks or more. They will, however, neither divide nor multiply as they would have done had they remained in the body of the mouse. Prolonged survival has been attained, but not growth and reproduction. Something is still missing which is in our power to supply.

If a little juice, obtained by grinding up an embryo such as that of a chicken, were added to the blood plasma, the nourishment thus supplied would allow the cells to grow and multiply. By the end of a day or two the fragment of tissue would have increased to twice its original size. If a portion of the enlarged fragment were removed and placed in another flask containing fresh plasma with the juice of an embryo added and the process repeated every day or every second day, new tissue could be grown indefinitely. The process is, by analogy with the culture of microbes, known as the culture of tissues. The analogy is close, for it is essentially the same technique which is employed in both cases: i. e. the successive planting out of cells in new and rigorously aseptic media.

Two names dominate the history of tissue culture: that of the American scientist Harrison who in 1907 first kept fragments of amphibian tissues alive in blood plasma and observed, *in vitro,* how the nerve fibres lengthened; and that of the Frenchman Alexis Carrel who in 1910 showed that to obtain cell-division in culture it was essential to add a little embryonic juice to the medium. This improvement in technique may appear small, but its consequences were great, for it was only after the work of Carrel that real progress in tissue culture could be achieved.

Carrel worked with tissues called fibroblasts, which were removed from the heart of embryo chicks. Some

Alexis Carrel (1873—1944).

Manuel.

1   2

The culture of animal tissues. 1. Fragment of a colony of fibroblasts of an embryo chicken derived from a stock isolated by Carrel in 1912. When this photograph was taken the stock was already 26 years old. 2. Fragment of a culture of the heart of a nine-day-old embryo chicken. On the periphery, fibroblastic growth. 3. Detail — more highly magnified — of the fibroblasts in figure 2. In the centre, mitosis. 4. Fibroblastic growth. In the centre, metaphase. 5. Epithelial layer formed at the expense of a culture of the kidney of a new-born rat.

After Ebeling. Parker (1). Lab., prof. Verne. Hébert (2—5).

of his cultures, begun in 1912, were still alive and thriving in 1939, in other words twenty-seven years later! For purely accidental reasons they have since perished; otherwise they could perfectly well be alive today. Thus cells removed from an organism can live in a glass jar much longer than cells which continue to form part of the organism. In twenty-seven years thousands of cellular generations succeeded each other *in vitro*. If all the cells which were born of the initial fragment of tissue could have been preserved, the resultant mass would have weighed thousands of tons.

We have already seen that single-celled organisms are potentially immortal, as are the germ-cells of multicellular organisms. From the culture of tissues we further learn that the somatic cells — the cells of the body — are also endowed with potential immortality.

Among those whose work prepared the way for Carrel's important discoveries Justin Jolly must be mentioned. Jolly in 1903 observed that the red blood-cells of the frog would survive in blood maintained in sterile conditions at normal temperature: the cells

would continue to divide *in vitro* for some two weeks.

The method of tissue culture described above has been applied to a great number of tissues and has proved successful in numerous cases: with blood-tissues, glandular epithelia, epithelia of the retina and of the iris, the crystalline lens, dermal fibrocytes, and so on. On the other hand, nerve-tissue and striped muscle tissue cannot be grown in culture and are incapable of cell-division, even in the organism itself.

Human tissues have been cultivated: Thomas, for several years, maintained pure cultures of fibrocytes taken from the skin of a young man. Thus it might be fair to say that science can already bestow immortality on a human being — or at least cellular immortality!

'To say that such and such a person is dead,' as Dastre put it, 'is not so much a statement of fact as a prognosis. In the corpse before our eyes how many elements are still alive and remain potentially capable of rebirth!'

From every corpse it would, in fact, be possible to remove living cells and preserve them in culture; and inasmuch as they would continue to live it would not, biologicaly speaking, be strictly accurate to say that the human being who supplied them had totally disappeared. The culture of tissues derived from exceptional individuals, such as geniuses and centenarians, may in the future be systematically undertaken, in order to extract from them substances with desirable properties.

The question arises of why the juice of an embryo is indispensable to the successful culture of tissue. The answer is that the juice contains substances which are necessary to tissue growth and proliferation. Carrel

228

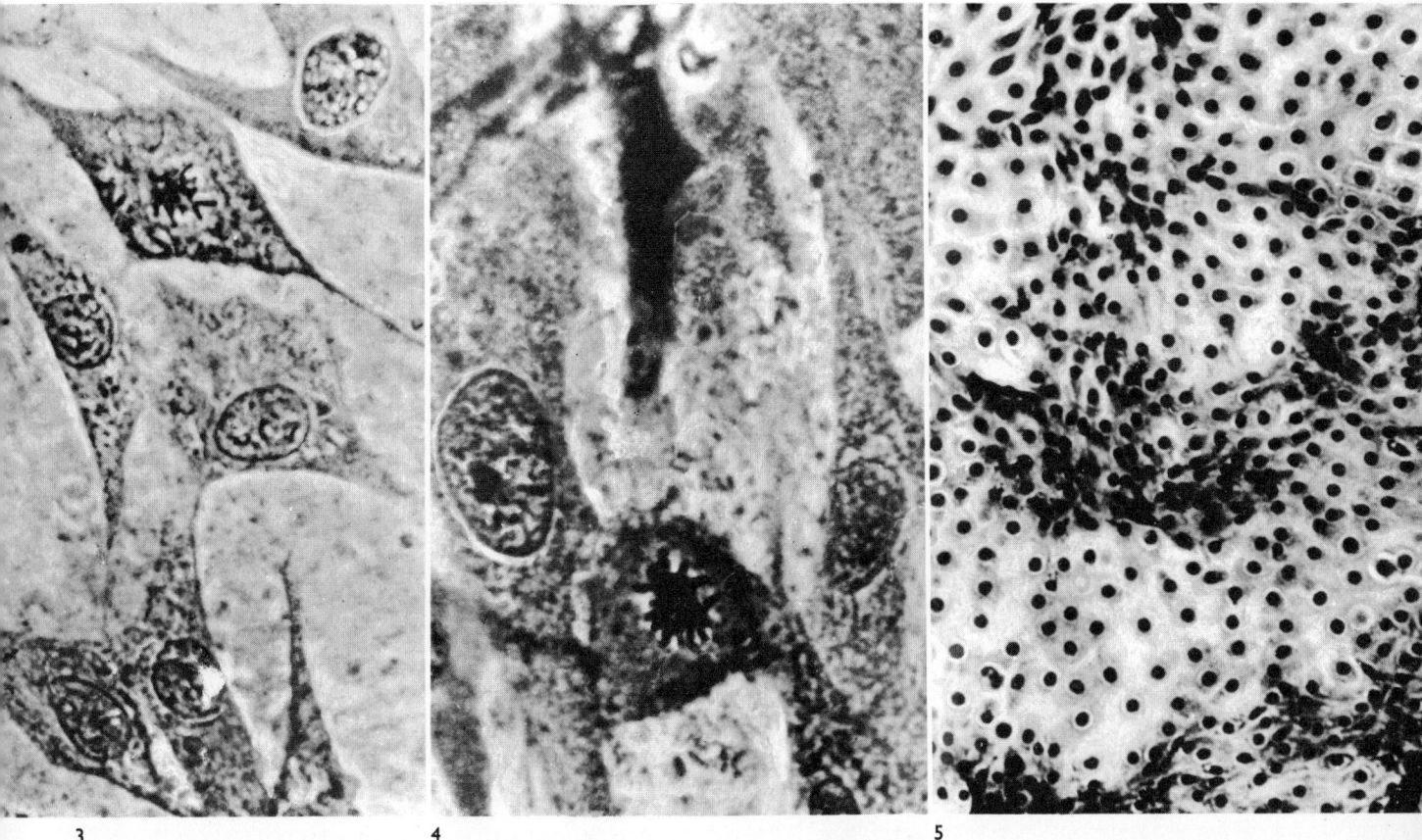

3   4   5

called these substances 'trephones', and their chemical composition is still uncertain. They are fragile, but support desiccation in a vacuum at low temperatures. They can be precipitated by alcohol and appear to be related to nucleo-proteins. These 'trephones' are found not only in embryos but also in tissues in the process of regeneration, where Carnot in 1907, and before Carrel, detected and studied them. He called them 'poietines'.

Indefinitely prolonged culture of animal tissues has not yet been achieved in a synthetic medium; but in 1946 White succeeded in keeping tissues alive for several weeks in certain very complex culture media containing salts, sugars, amino acids, vitamins, and hormones.

For a long time it seemed impossible to form cultures from one cell only; but in 1948 Sanford, Earle, and Likely succeeded in doing so, employing special media, previously modified by abundant cultural growths.

Thanks to the culture of tissues, which technical progress constantly makes more available to investigators, we can now follow the manifestations of cellular life in glass jars. Cells which were previously so veiled in secrecy can be watched, as it were, in full daylight as they live, function, and work. Tissue cultures are not only observed under the microscope, but they are photographed and filmed so that their movements can be seen and studied in detail.

It is of immense fascination and value to find that cells in culture can continue to function and accomplish their specific tasks: the muscular cells contract, the secretory cells produce their secretions, the pigmentary cells manufacture their pigments, the liver cells elaborate their glycogen, and so forth.

It appears that there is a kind of antagonism between the phenomena of growth and those of function. When conditions of temperature and nutrition are so arranged that the cells' multiplication is accelerated their normal functioning is slowed down. If, on the contrary, the rapidity with which the culture grows is moderated the cells' normal activity is increased.

In the early days of tissue culture it was thought that cells thus cultivated lost their specific characters and reverted to a more primitive state of non-differentiation. Today we know that this was a misapprehension. No matter how prolonged the culture may be or in what conditions it is made the cells remain what they were when it began: they retain their specialization or differentiation. Even though they sometimes change a little in appearance they undergo no actual regression, and remain permanently incapable of producing daughter cells of a different type from their own.

From the culture of tissues we further learn that the age of cells has an influence on their capacity for growth, a fact which is not, of course, very surprising. Tissues taken from an adult organism grow less well in culture than tissues taken from a young organism which, in turn, grow less well than tissues taken from an embryo.

The exact cause of these differences in growth capacity is still unknown; if the cause could be discovered much light would undoubtedly be shed on the processes of ageing. In any case the differences seem to be connected less with changes in the cells themselves than with an accumulation around the cells of substances which hamper their growth. The growth of adult tissues is, in fact, improved when they are previously

229

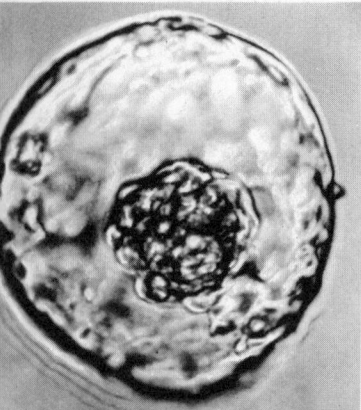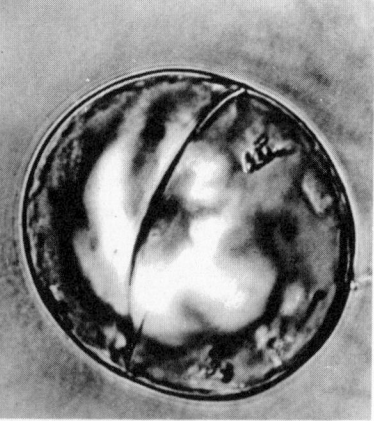

Proliferation of isolated cells of the pea. *Left*, a cell which has been in culture for twenty-four hours. In its centre the nucleus, surrounded by grains of starch, can be seen. *Right*, at the end of five days in culture the isolated cell has enclosed itself in new walls.

Torrey.

An illustration of the luxuriance of plant tissues in culture. *Left*, a colony of blackberry tissues after two weeks in culture. *Right*, a similar colony after six months of culture. It can be seen that the colony has practically usurped the place of the nutritive medium.

Gautheret.

submitted to the action of certain digestive ferments such as trypsin. Trypsin is an enzyme which attacks albuminoid substances and acts as though it 'scoured out' the adult tissue by digesting noxious matter.

The growth of cells in culture can be used as a most convenient method to test the action of various physical and chemical agents. In this way Carrel studied alterations in power of cellular growth by submitting cells in culture to the action of blood serum taken from animals of different ages. He was thus able to observe that cellular growth was more successful in the serum of a young animal than in the serum of an old animal. There is a twofold reason why this is so: not only does the serum with age become poorer in stimulating substances but at the same time it grows rich in substances which impede cellular growth.

Tissue culture has helped to solve the problem of how white blood-cells are formed, for *in vitro* they can be seen to change: from monocytes into macrophages, and from macrophages into fibroblasts, while lymphocytes turn into myelocytes and leucocytes.

The same technique allows us to observe the direct action of radiations, of poisons, of hormones and, more generally, of all external agents on the living cell. It has supplied a method of studying the way lethal genes work, genes, that is, which prevent the development of the organism when doubled, in other words, when inherited from both male and female parent.

It was extremely interesting to find out how tissues taken from an organism suffering from this genetic fatality would behave in culture. The experiment, successfully undertaken by Boris Ephrussi, showed that they could multiply indefinitely, much like normal tissue. Thus it is evident that the lethal gene kills not the actual cells but the organism as a whole. When the embryo dies prematurely under the influence of lethal genes it is not because the cells individually lack any factor essential to life but because conditions have been created in the organic assembly which oppose the continuation of development.

Many branches of medical science have been benefited by tissue culture. By introducing bacilli into the culture of normal cells we can observe the formation of giant cells and tubercles. Unlike visible microbes, viruses cannot be cultivated in any nutritive medium; but they can be cultivated on tissue cultures. In this manner it is possible to maintain stocks of the poliomyelitis virus and of the viruses of yellow fever, influenza, sheep-pox, rabies, cow-pox, exanthematic typhus, etc. In 1954 the American scientists F. C. Robbins and T. H. Waller received the Nobel prize for having successfully cultivated the virus of poliomyelitis on animal tissues — a discovery which would lead to the preparation of effective vaccines.

Finally, tissue culture has enabled us to study cancerous cells with great care, and to specify certain differences between these and normal cells. From the purely structural point of view there is scarcely any difference between a normal cell and a cancerous cell of the same histological (or tissue) type. Nevertheless physiological and biochemical differences between the two are apparent in tissue culture. The malignant cell demands less nutrition than the normal cell, and it has rather peculiar fermentary properties. We still do not know if the peculiarities are due to cellular mutation or to the presence of a foreign virus. Efforts have been made to induce cancer artificially in tissue cultures. The cells are submitted to the action of cancer-producing substances which in the living organism almost always leads to the formation of experimental tumours. In the presence of methylcholanthrene it has sometimes been observed that normal cells change into malignant cells; but it is by no means certain that the

230

Colony of carrot tissues derived from a mutation of the stock illustrated above. The effect of this mutation has been to stimulate the cells to elaborate a large amount of carotene. This gives the colony a reddish-orange colour.

Colony of Jerusalem artichoke tissues, two months old, derived from a stock isolated in 1940 and pricked out 114 times. Here the tissues are normal and can be cultivated only in the presence of indolacetic acid or other stimulating substances of the same type.

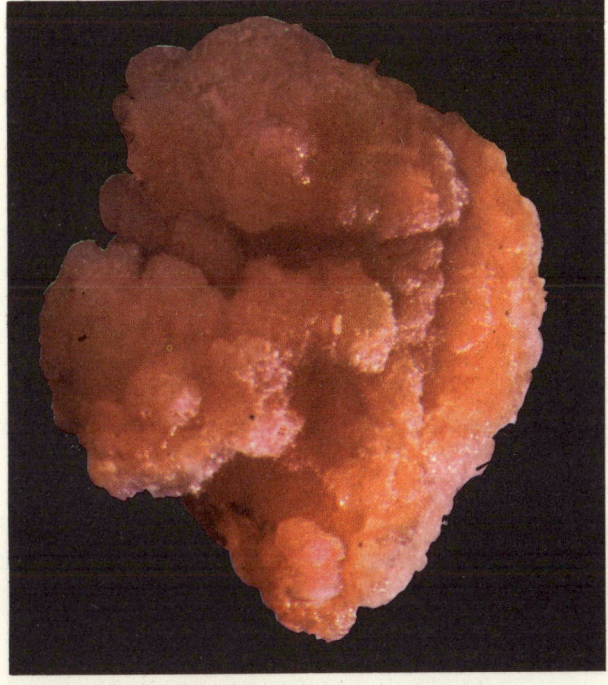

Colony of carrot tissues at the age of two months, derived from a stock isolated in 1937 which has been pricked out 132 times. The tissues show traces of chlorophyll.

Colony of black salsify crown-gall tissues, aged a month and a half, derived from a stock isolated in 1946, and thinned out 78 times. This tissue is of a cancerous nature and can be cultivated without a medium, for by itself it elaborates the factors required for proliferation.

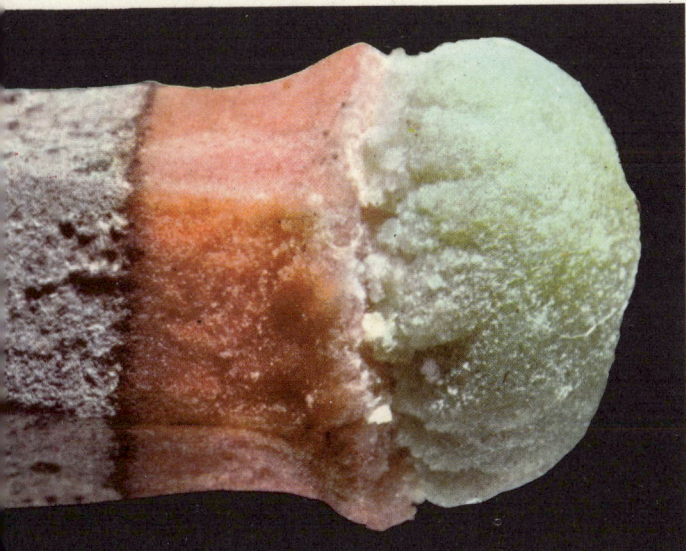

Piece of a carrot root after two months in culture. It has produced a large protuberance of parenchyma which has elaborated chlorophyll. The greyish-violet region is the part which was immersed in the culture medium.

Professor Gautheret. Larousse.

Giant colony of carrot tissues, aged ten months. The initial fragment, which weighed about half a thousandth of a gram, has grown into a mass of tissue shaped like an irregular disc some 12 cms (5 ins) in diameter, 2 cms (¾ in.) thick, and weighing more than 150 grams (5 oz).

Gautheret.

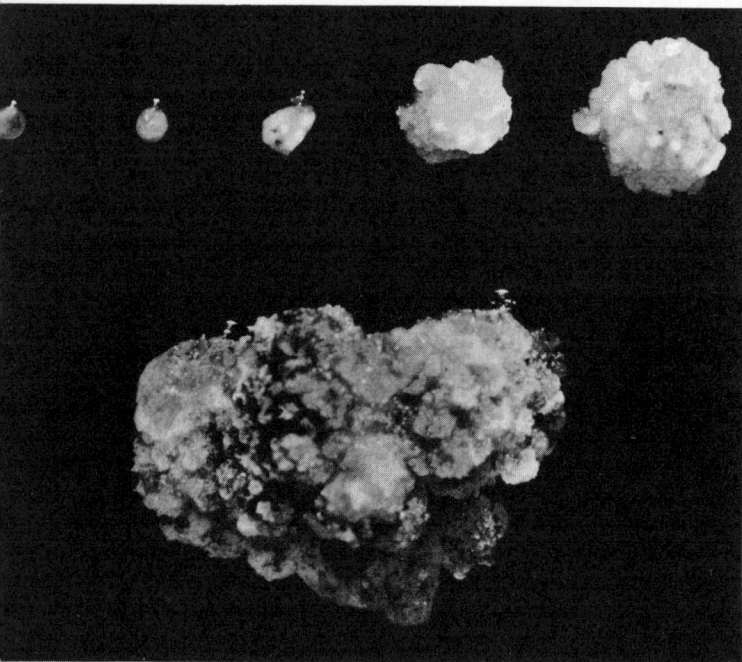

Cultured fragments of sunflower crown-gall. In the upper left corner the original explant can be seen, followed by fragments which show the monthly proliferation of the cultured tissues.

White.

action is specific, for some investigators believe that they have observed similar changes in the control cultures.

The technique of tissue culture has also been valuable in the diagnosis of various ailments. For instance, the nasal epithelium of a person suffering from hay fever can, *in vitro,* react allergically to the irritant which causes it.

The culture of plant tissues

While the culture of animal tissues was achieved as early as 1910 it was not until 1937 that plant tissues were successfully cultivated; and yet it was with plants that experiments of this kind were first attempted — by Haberlandt in 1902. They failed completely. In 1922 Robbins succeeded in keeping young tissue (meristem) from the root-tips alive *in vitro*. It proliferated at first but soon began to differentiate, producing a root which ceased to grow. It then seemed impossible to obtain true plant tissue cultures. Some ten years later Gautheret tried to cultivate cambium, the meristem layer of actively dividing cells which, forming cylindrical layers, brings about the process of secondary thickening of stems and roots. This time unorganized cellular masses were obtained, but their growth stopped at the end of a few months, and further research, due to Nobécourt and to Gautheret himself, was still necessary before, in 1937, the successful culture of plant tissue for indefinite periods was achieved.

Today all kinds of plant tissues are cultivated: carrot, salsify, endive, Jerusalem artichoke, cabbage, turnip, tobacco, sunflower, vines, Virginia creeper, brambles, hawthorn, as well as many root-tissues and those of herbaceous stems, creepers, and bark. The tissues of all groups of dicotyledonous plants are cultivated; for the moment the monocotyledons appear more refractory.

The maintenance of vegetable-tissue cultures, like that of animal-tissue cultures, requires periodic planting out — in other words, removal to a new medium. During the operation the greatest care must of course be taken to assure strictly aseptic conditions. Transplanting is less frequent than in the case of animal tissues. It is often sufficient to plant out the tissues once a month or even once in two months, instead of every second day which is the rule with animal tissues.

Some plant tissues have already lived much longer than the actual plants from which they were taken.

'While a carrot,' Gautheret wrote, 'dies within two years, the tissue cultures which we have made from fragments of carrot tissue have already lived more than eight years, and there is no reason to expect their development to cease in the future.'

The notion of potential immortality, already applicable to animal tissues, can therefore safely be enlarged to embrace plant tissues.

The speed of growth depends on the origin of the tissue cultivated. In the case of the carrot the weight of the initial fragment, which must be at least half a milligram, is increased tenfold in nine or ten weeks. Vine tissues grow more rapidly, while those of the

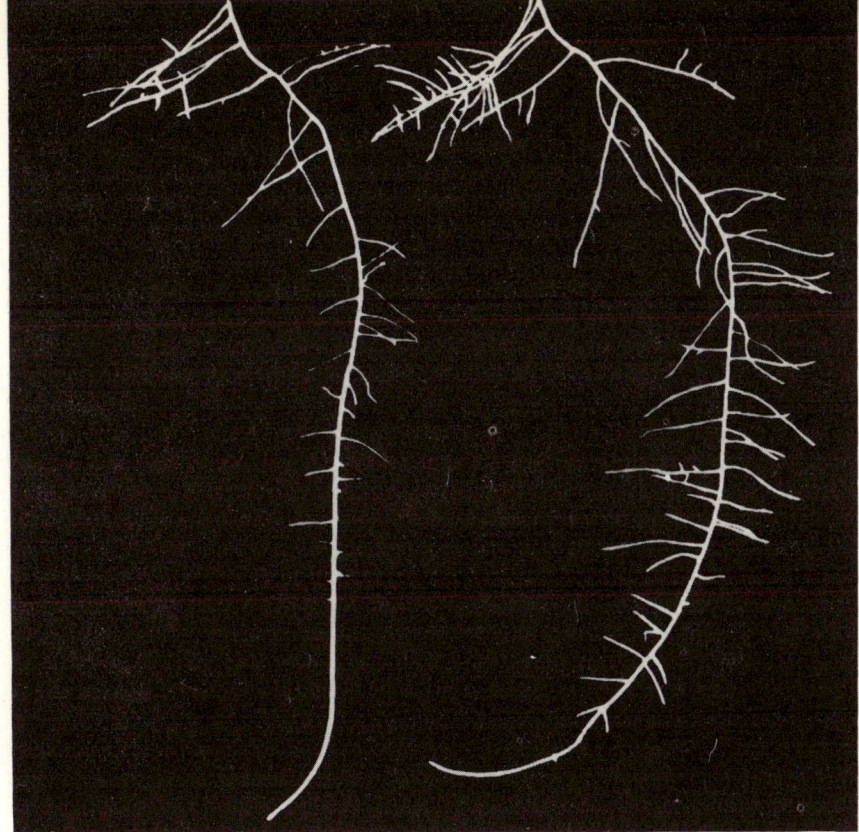

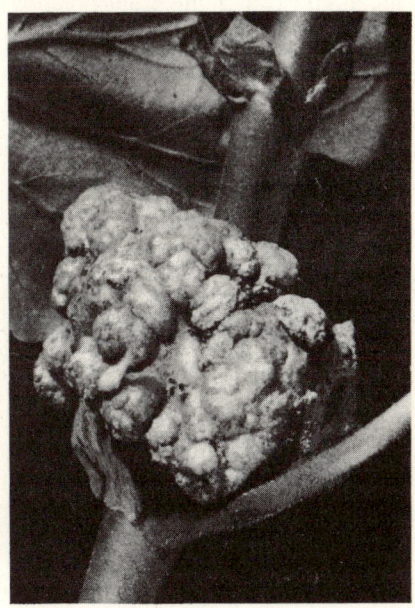

Tomato root from White's stock — 23rd pricking out. *Left,* the end of the root when pricked out. *Centre,* the root twelve days later. *Right,* the root at the age of seventeen days.

White.

Tumour produced by *Agrobacterium tumefaciens* on the stem of a pelargonium.

J. Vincent.

Jerusalem artichoke grow a little more slowly.

If the jar used is of sufficient size, and the growing tissues are not planted out, giant cultures can be obtained. An initial mass of half a milligram will, in ten months, yield a mass weighing 150 grams, 12 centimetres in diameter and 2 centimetres thick.

The culture medium is of a definite and relatively simple composition, which is one of the advantages which the culture of plant tissue has over that of animal tissue. With the exception of gelose or agar which is used to solidify the medium, the other ingredients are all well-known substances, obtainable in pure and crystalline state; mineral salts and sugars, growth elements, vitamins, etc.

The growth factors needed by plant tissues vary greatly. They are required by the tissues of Jerusalem artichokes, whereas those of the carrot and the endive, less demanding, are satisfied with a purely mineral and sugar medium. Three growth factors are necessary for the culture of hawthorn tissues.

A number of problems concerned with cellular physiology have been studied by means of the culture of plant tissues, problems of growth, respiration, the rising of sap, polarity, the mechanism by which roots, buds, leaf organs, vessels and generative zones are formed, the differentiation and de-differentiation of tissues, and so forth. By the same means we can compare the nutritive requirements of various tissues of the same species or of the same tissue in different species.

As a rule the cultures are comprised of several types of cells; but, as in animal tissues, pure cultures have been obtained, composed of one type only.

Grafts of tissues have been made *in vitro* and of special interest is the fact that it has been possible in this way to study cancerous tumours in plants (tumours more or less comparable to the malignant tumours of animals), not only those spontaneous cancers which appear in certain hybrids of tobacco but also bacterial cancers like crown gall, produced by *Agrobacterium tumefaciens.* The tumours caused by this microbe can be transmitted by inoculation and sometimes attain a size of several centimetres in diameter, giving rise to secondary tumours, or metastases.

In culture the malignant cells of plant cancers do not need the growth factors which normal tissues of the same plant require. This fact is interesting in view of what has previously been said about the lesser nutritive demands of malignant animal cells.

Cancer can be produced experimentally in plant-tissue cultures by the addition of certain growth factors. For this reason it was not unnaturally supposed that factors of this sort were secreted by the cancer-producing microbe, *Agrobacterium tumefaciens.* But this leaves the secondary crown gall tumours to be explained; for they do not contain the microbe in question and can, nevertheless, be cultivated for months, even years, without losing their malignant character.

Must we then assume that the tissue, once it is made cancerous by the microbe, remains cancerous even in the microbe's absence? Or that the cancer is due not to the microbe but to an invisible virus which is usually associated with the microbe but can, in the microbe's

Carrel and Lindbergh's 'Perfusion Apparatus' by means of which organs can survive in a nutritive medium provided with oxygen.

Palais de la Découverte. Larousse.

absence, remain within the cell?

The culture of plant tissues has enabled us to cultivate mushrooms in ordinary media like vine-mildew in which they were previously ungrowable and, finally, to cultivate disease-producing viruses like the tobacco mosaic virus. White was the first to cultivate this virus on the roots of tomato plants. Segrétain reported that the virus would retain all its virulence after eleven months of culture.

The culture of plant organs

Not only plant tissues but entire organs of plants are grown in culture: roots, embryonic seeds and even ovaries.

Full-blown tomato flowers can be placed in a nutritive medium, solid or liquid, to which a little fresh tomato juice, previously sterilized, has been added and at the end of a week the ovaries will begin to enlarge and little by little turn into tiny tomatoes. Three weeks later the little tomatoes will redden and ripen, exactly as though, still attached to the parent plant, they had received from it the nourishing sap. These tomatoes, ripened for instance in a flask, would contain no seeds because the flowers would not be fertilized; but there is no doubt that complete and normal fruit, provided with fertile seeds, could be obtained by the same method.

These experiments of fructification *in vitro* were made by Nitsch but have not yet found any practical application, and for the moment their chief interest lies in the fact that they furnish means of accurately observing how fruit is formed. It is always of value to

biology when, in the simplified conditions of the laboratory, phenomena can be produced which have not previously occurred, except in the complex interior of the living organism.

The culture of animal organs

Among plants the culture of organs preceded the culture of tissues. In the case of animals the opposite occurred, and it was not until 1936, twenty-six years after Carrel's famous experiments, that animal organs were cultured. Again it was Alexis Carrel, with the collaboration of the aviator Lindbergh, who first succeeded in making an animal organ live and function outside the animal's body.

During their first experiments Carrel and Lindbergh employed a very ingenious and relatively simple apparatus, a series of glass vessels with a 'pump' by means of which a nutritive and oxygenated liquid could be made to circulate through the isolated organ. The composition of the liquid was carefully prepared and it was constantly renewed. The apparatus was placed in an incubator and maintained at the desired temperature and, finally, the strictest precautions were taken to avoid any bacterial contamination of the culture.

Since that time other more elaborate apparatuses have been designed and built. In France J.-André Thomas has perfected a robust and accurate apparatus which is furnished with an artificial heart and an artificial lung. It has many advantages over the original apparatus: in particular it allows blood to be used as the nutritive fluid and regulates the physiological conditions of circulation — blood pressure, rate

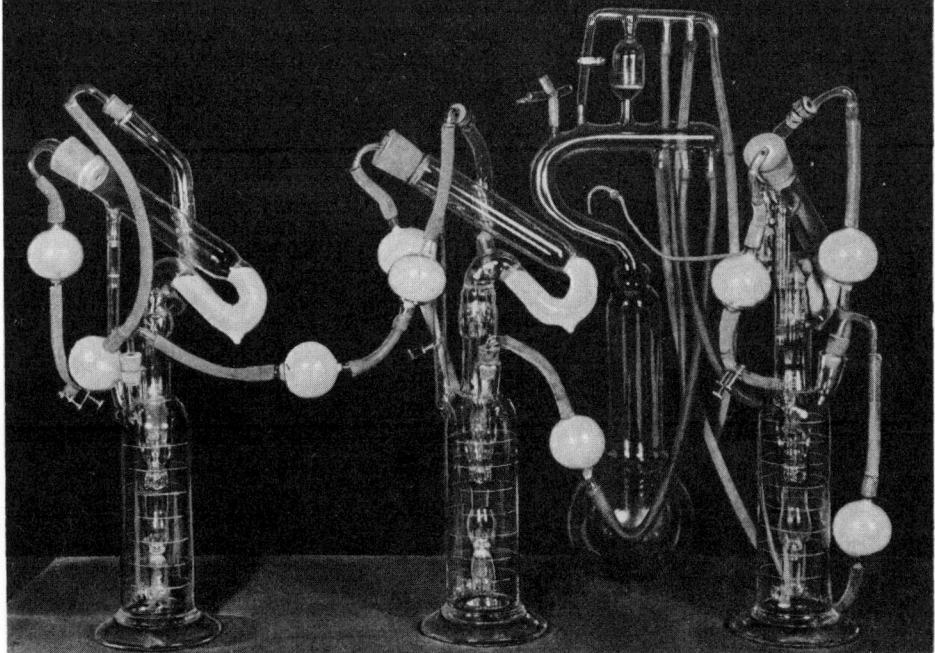

The more highly perfected apparatus of J.-André Thomas, designed for large organs.

Palais de la Découverte. Larousse.

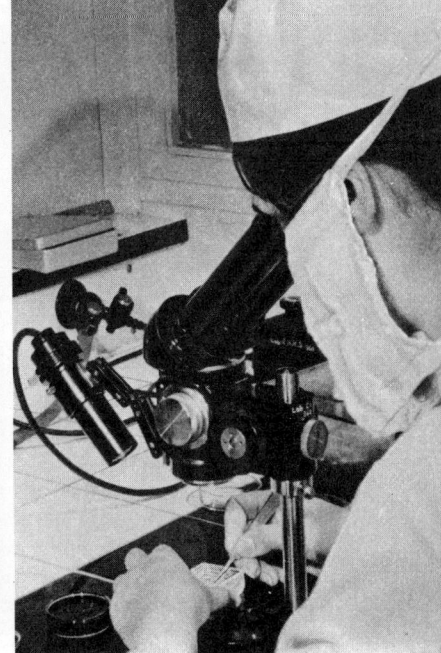

The aseptic culture of organs. In this phase of the technique an organ of a chicken's embryo is placed on an appropriate medium in a glass 'salt cellar'.

Et. Wolff. Larousse.

of pulse, and amount of oxygen contained in the blood. It is suitable for the culture of organs weighing as much as 20 pounds.

In his apparatus Carrel was able to make a number of organs live and function, organs usually taken from cats.

'In more or less diluted blood serum,' he reported, 'the thyroid gland, the parathyroid, the ovary, the Fallopian tube, the adrenal capsules, the arteries, the veins, the pancreas, the spleen, the lymphatic ganglions, the upper cervical ganglion, can remain alive for several days, for several weeks, and perhaps much longer still.'

During all this time each organ continued to function, very much in the way it would had it remained in position in the organism. An ovary produced its ova and a thyroid gland secreted its colloidal hormone. At the end of eighteen days the thyroid gland, still much alive, had undergone no reduction in size nor alteration in appearance. Microscopic examination revealed that its structure was unchanged and that its capacity to function persisted. An English biologist kept the mammary gland of a cow alive, and the glands continued for several hours to secrete milk.

The fascination and value of thus studying the life and behaviour of organs isolated and separated from their normal connections can easily be imagined. Among other things it supplies a means of finding out which organs can function by themselves and which depend on the total organic system from which they have been cut off. Also, study can be made of how the structure and functioning of the organ varies with the composition of the fluids used to nourish it.

It is obvious that such methods open out a vast field for experimental research and allow us, as Carrel

said, to study the human body while alive. Human organs, in fact, can be removed during the course of an operation, or even immediately after death, and be put into culture. How they function can then be observed and, if the organ removed is an endocrine gland (pancreas, thyroid, adrenal, or sex gland) the hormone it normally secretes into the bloodstream can be collected. 'Human organs themselves manufacture *in vitro* the therapeutic substances which today we collect from horses and rabbits.'

It is agreed that hormones, generally speaking, are without zoological specificity and that, in consequence, animal hormones have the same effect on human beings as human hormones. Nevertheless it might well be that human hormones may prove superior in practice to the animal hormones now medically employed.

Carrel goes so far as to envisage removing a diseased organ from the patient's body and putting it into a culture medium, much as the patient himself is put into hospital. *In vitro* the organ could be more effectively attended to than in the body of its owner. The organ when cured would then be reinstalled by means of grafting.

Aseptic survival

Among cold-blooded animals like amphibians the survival of parts separated from the body can be assured by a method much simpler than organ culture. This is known as aseptic survival. The separated part is merely put into a nutritive medium which has been made sufficiently sterile for the purpose by the addition of antibiotic substances.

235

In this manner Thomas has kept fragments of tritons' limbs, toes, regeneration buds, and so forth, alive for nearly a month. He has been able to observe the repeated moulting of some of the fragments, the healing of the amputated surface, and even a certain evolution, marked by the growth of connective tissue. It would seem that the nutritive substances are utilized by the surviving fragments, though this is not absolutely certain as the fragments can also survive in pure or slightly salted water.

It may be mentioned that as early as 1858 Vulpian demonstrated that pieces of the tadpole's tail could be kept alive for from eight to ten days in pure water, where they would heal and even bud slightly at the amputated surface. For that period, over a century ago, the experiment was remarkable and pointed the way to all that was afterwards done in the field of survival *in vitro*.

Occasionally the fragments of an organism will survive for some time, even in conditions which are not particularly favourable to their survival. Among insects, for instance, the head when separated from the body will remain alive for a while. Patijaud reports that the severed head of the clover bombyx will survive for two and a half hours, that of the emperor moth for six hours, and that of the mole-cricket for as long as thirty-eight hours.

The head of the male *Lucanus,* a large coleopteran better known as the stag beetle and famous for the horns which adorn the male's head, will, when cut off, live for a week, while that of the female stag beetle, when placed in a damp container at ordinary temperature, will survive as long as nineteen days. To the end its antennae move in a lively manner and contract at the slightest contact.

The stag beetle's severed abdomen can live for a month; its thorax two or three days only. But the abdomen and thorax together (i.e. the remainder of the decapitated insect) can live for nearly three months. It would seem, indeed, that among stag beetles, and also among certain moths, the result of decapitation is to prolong rather than shorten their lives.

The experiments of J. Rostand have shown that the survival of the silkworm moth's abdomen, and even fragments of its abdomen, is also remarkable. Male abdomens, sewn up after amputation, have lived for thirty-five days, whereas no male of the species in similar conditions of temperature will live for more than eighteen days. The abdomens of females survive for a somewhat shorter period, their maximum life span being twenty-two days. A seven-ringed section of a male abdomen lived for nineteen days; a fragment with five rings for twelve days; one with four rings for twenty-two days.

The long survival of the abdomen must be attributed to its state of semi-inertia; that, no doubt, is why decapitation increases the length of the animal's life. It is important to note that most of the detached abdomens when they finally perish no longer contain a trace of their fatty reserves, while insects dying in the normal way are far from having consumed their reserves — at least in the case of the males who have not used them in the production of eggs. It would seem that the severed abdomen makes fuller use of its reserves than the complete insect, and in truth dies of hunger. The same cannot be said of the complete insect which, according to Metchnikoff, succumbs to a kind of uraemic intoxication.

Isolated abdomens of *Bombyx mori* (amputated, after ligature) are capable of fertilizing eggs.

If, immediately after the male *Bombyx* has mated with the female, its abdomen is isolated by a ligature followed by amputation the abdomen will remain affixed to the female for from seven to twenty-four hours. It then falls away, after which the female lays fertilized eggs exactly as though she had coupled with a complete male insect. Fertilization of the female can also be obtained with morsels of the male abdomen, the abdomen having been amputated at a suitable level instantly after mating has begun. As long as the operation spares the five posterior rings not only does the severed abdomen remain in place and quiver slightly as it does in normal acts of fertilization but it almost always succeeds in fertilizing the female. Of ten females thus mated with masculine abdominal fragments comprising five rings, eight laid fertilized eggs, only two laying virgin eggs. The fragments had remained in place for from six to thirty-four hours.

When only the four posterior rings of the male's abdomen are cut off they will remain attached for from nine to twenty-four hours and sometimes fertilize the female, though in a majority of cases they will not: of twenty-three females eight laid fertilized eggs and fifteen unfertilized eggs.

When only the three posterior rings of the abdomen are amputated the fragment will not remain in position: it almost immediately drops from the abdominal extremity of the female.

The tail of the adult triton can live after amputation; it will continue to move for several hours and sometimes, if the surrounding temperature is relatively low, for as long as a day and a half. When the head of a frog or of a toad is cut off the eyes and jaws continue to move for a few hours.

Among the same animals it has long been known that the heart can be kept alive for long periods. If the frog's or the toad's heart is separated from the large blood vessels, delicately extracted and, to avoid drying, placed in a humid container, it will continue to beat for hours. If it is first washed in a physiological saline solution to which glucose is added, the heart will survive and beat for several days at ordinary temperature.

In tortoises the resistance of the organism's fragments is especially marked, as the illustrious physiologist Francesco Redi had observed as early as the seventeenth century.

According to G. Fano, the heart of the turtle *Emys europea* will survive in a humid vessel for at least two weeks, even when it has the additional work to do of moving a lever to inscribe its rhythm on a recording cylinder.

Boris Rybak undertook a series of interesting experiments on the survival of the isolated hearts of fish. He has found that when the detached auricle of the spotted dogfish was periodically shaken its contractile activity was considerably prolonged — in fact for as long as ten hours instead of the customary one.

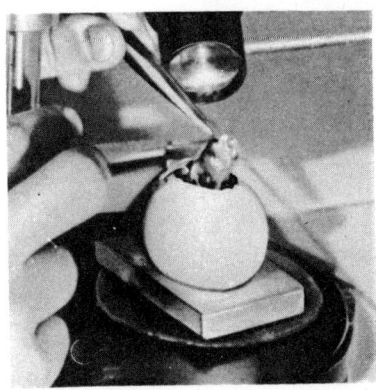

Extracting the body of the embryo whose detached organs will be put in culture.

Et. Wolff. Larousse.

Rybak also discovered that a cardiac element as small as a single sino-auricular valve could supply impulses powerful enough to be recorded by electrocardiograph.

Among birds and mammals the heart in isolation soon stops beating, at least if it comes from an adult or from a completetely formed organism. If, however, it is taken from an embryo it will survive for a fairly long time. While the heart removed from a newly hatched chick will stop beating at the end of three minutes, the heart of the five day embryo can beat for several hours. The embryonic heart of the higher animal thus behaves to a certain extent like that of a lower animal. It is claimed that the heart of hibernating animals, including the marmot, the dormouse, the bat, and the hedgehog, is endowed with unusual powers of survival.

The culture of embryos

By using apparatuses of the type perfected by J.-André Thomas it is possible to cultivate large mammal embryos. The gravid cow or ewe is brought to the slaughter-house where, after the umbilical cord is cut, the foetus is extracted under sterile conditions. Its circulatory system is washed out with pure blood serum and connected up with the chief artery of the apparatus. The liquid of perfusion, that is to say the liquid which is circulated through the body of the foetus, is composed of defibrinated blood of the same species to which the foetus itself belongs. In the foetal organism the blood circulates in the opposite direction from that of the animal after birth. It arrives by one of the umbilical arteries and returns to the apparatus by one of the umbilical veins. Blood pressure is kept fairly low and the heart-beat maintained at eighty-five pulsations per minute. The respiratory mixture contains eighty per cent oxygen, seventeen per cent nitrogen, and three per cent carbon dioxide. The apparatus is placed in an incubator and kept at 38 °C.

In these conditions Thomas kept embryos alive for about two days. As the technique is perfected there is little doubt that it will be possible to increase this period very considerably. It is impossible not to recall the famous passage in Aldous Huxley's *Brave New World* which described ectogenesis or babies born in bottles. Although total ectogenesis (in other words, the complete development of the human embryo from the egg) seems unlikely to become a reality it is not at all inconceivable that partial ectogenesis may be achieved, and embryos aged from four to five months be placed in culture. A Doctor Greenberg constructed some years ago an apparatus which he called a 'spare uterus' and which, he claimed, could permit the foetus to develop outside the maternal womb.

While these ambitious experiments in the culture of foetuses have progressed, research has also developed along other lines, research which employs the methods of tissue culture in the culture of very young embryos and detached embryonic organs.

It may be remembered that the segmentation of the egg of a mammal, kept in blood plasma, can be observed under the microscope. *In vitro* the opening phase of development promptly stops, but at a later stage of embryonic evolution culture again becomes possible. A. Brachet has kept the blastoderm vesicles of a rabbit alive in plasma, and similar experiments have been successfully made by other investigators.

Among the most instructive of these are the researches of Jolly and Lieure. They removed from the female rat a fertilized ovum which had developed for about nine days and placed it in a vessel filled with plasma of the same species. The recipient was then covered with a thin sheet of glass, affixed with a little sterile vaseline, put into an incubator and kept at 30 °C. In these conditions the egg continues to develop. When first put into the culture all that it revealed was an amniotic vesicle and a germinal disc with only an indication of the primitive streak. It soon formed an embryonic axis, with a brain, a rudimentary medulla, optic and auditory vesicles (the rudiments of eyes and ears), a notochord, ten or twelve somites or lateral muscular blocks, an allantois, a rhythmically moving heart, a suggestion of blood vessels, and a little blood containing red blood-cells. All of this formed in culture within forty-eight hours.

Results which are not dissimilar can be obtained with the egg of the guinea-pig, the ovum being removed and put in culture at the age of thirteen days, which roughly corresponds to an age of ten days in rats.

Jolly and Lieure reported that the nutritive medium was absorbed with extraordinary energy and that the egg swelled so rapidly that its growth could be visually followed under the binocular microscope. To be a spectator of these phenomena of development which normally take place in the privacy of the maternal womb is indeed a fascinating experience. Unfortunately at the end of forty-eight hours the enterprise failed: the heartbeats became feebler, and the embryo perished. Its food requirements had by this time become so great that the mere absorption of plasma could no longer supply them. Clearly new experimental techniques must be devised which are better adapted to the growing demands of this stage of embryonic development.

Other investigators have specialized in the culture of organs detached from the embryonic body. Waddington cultivated parts of the embryo chicken on hen's

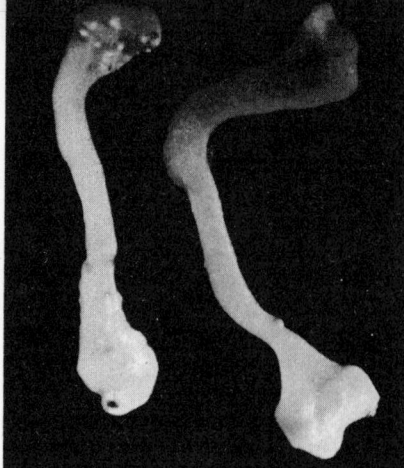

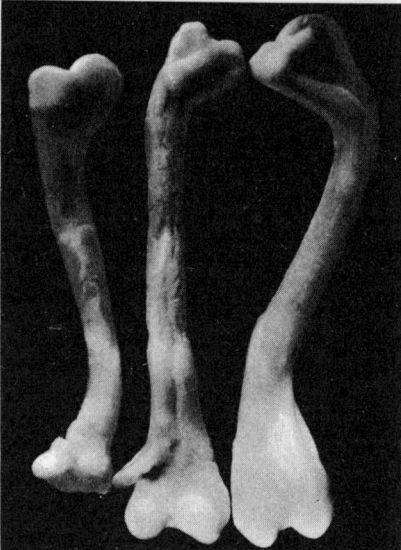

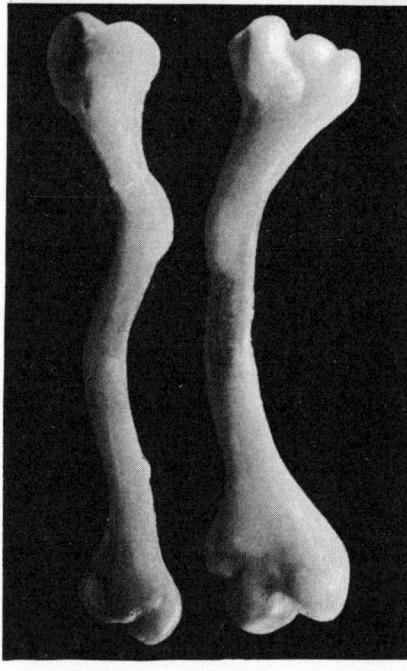

plasma and observed that, in some cases, the differentiation of the part takes place *in vitro* exactly as it would have done *in vivo* during the course of ordinary development. When, for example, a budding member is put in culture when it is still no more than a formless mass without structure it soon gives rise under the eyes of the observer to a complete member, provided with bone, joints, and so forth.

Etienne Wolff greatly improved the technique of the culture of embryonic organs by applying methods suggested by the fact, previously not fully understood, that the most favourable conditions for the culture of tissues are the most unfavourable for the culture of organs. In tissue culture, media which are very rich in nutritive matter are employed; they are generally composed of blood plasma and embryonic juice, the latter supplying stimulating substances, or 'trephones', which all cultures, embryo or tissue, require. Such media bring about cellular proliferation which is rapid, disorderly, and uncoordinated. Now such tumultuous cell-division is precisely what must be avoided if the structural unity of an organ in culture is to be safeguarded. Wolff therefore used relatively poor media, containing embryonic juice but not blood plasma. He also solidified his media by the addition of agar.

Once detached from the embryo – as a rule the embryo of a duck or chicken — the organ destined for culture is delicately disposed on the surface of the nutritive medium, into which it is gently and lightly pressed. In this way close contact is assured between one side of the organ and the agar from which its nourishment is drawn. The culture takes place inside a recipient which, after the organ has been placed there, is covered with a sheet of glass which is sealed with paraffin wax. Finally the culture is put into the incubator. In this way the organ remains dry, though in surroundings which are saturated with humidity. It lives and breathes, in fact, like a small terrestrial organism.

The first experiments of Wolff and Haffen were made with the sex glands of the duck. The glands were removed from eggs which had been incubated for from six to nine days. At this stage of embryonic development sexual differentiation is not apparent. After a few days in culture the glands not only increased in dimension but very nearly acquired the structure they would have acquired had they remained in place in the living embryo. Sexual differentiation was perfectly accomplished *in vitro*. The sex of the gland could be diagnosed simply from its appearance, while microscopic examination of its tissue confirmed the diagnosis. The male gland had become a small, roughly cylindrical structure, the female gland a flattened little strip surrounded by a transparent fringe.

The result of this experiment was in itself of the greatest significance, for to have brought about the differentiation of the sex glands in culture meant that the 'explants', that is, the organs transplanted, themselves contain the factors which cause them to evolve into male or female glands as the case may be. In other words, the sexual differentiation of the gland is the work of the organ itself and not of the total organism: it is 'autodifferentiation'.

The next question was, what would occur if the embryonic gland was submitted to extra-ordinary

Development in culture *(in vitro)* of the tibias of chicken embryos on nutritive mediums of varying suitability. The torsion of the tibias is an obvious indication of growth. This twisting, which does not occur in the normal embryo, is due to the fact that the ends of the bones are anchored in a thick jelly which hampers their development in length.

After Et. Wolff and Kieny.

conditions: if, for example, instead of cultivation in a medium of normal composition it was cultivated in a medium containing certain hormones.

Wolff's earlier experiments with living embryos have been described in a previous chapter. It may be recalled how they showed that the evolution of the male sex could be profoundly modified by the influence of female hormones.

To attempt to reproduce the 'feminizing' effect *in vitro* was obviously indicated. This could be done either by adding a soluble hormone preparation to the culture medium or by sprinkling its surface with drops of hormone in an oily solution. Under the influence of female hormone applied in these ways it was seen that the budding male glands developed female characters and finally acquired the structure of an ovary, or at least an intermediate structure between that of an ovary and that of a testis.

By coupling two embryonic glands of different sex, analogous effects are obtained. In this case the feminization of the male gland is even more marked than in the preceding experiment. Between a testis thus feminized and a true ovary the difference is so slight that 'it would doubtless escape the observer if he did not already know the genetic sex of the organs'.

With good reason Wolff insisted on the fact that glands thus treated did not change sex, but simply underwent their first sexual differentiation in a false direction, that is, in the direction opposed to their proper genetic determinism. He also pointed out that in these experiments of coupled glands, two embryonic organs having been grafted together, we have a demonstration that the female gland elaborates its active hormone even before it has reached the stage of sexual differentiation. In addition it can be seen that this hormone acts directly and immediately on the male gland, without any intermediary agency, hormonal or nervous, as one might otherwise have supposed.

The culture of embryonic organs has supplied a method of studying the development and differentiation of those organs which evolve in different ways according to sex, organs like the syrinx, the Müllerian ducts, etc., discussed in an earlier chapter.

The method is now widely applied to the culture of numerous embryonic organs, notably in the case of the eye and the skin of embryo chickens. In culture the budding optical apparatus grows into pigmented eyes, with crystalline lens, choroid and iris; the skin, at first smooth, bristles with long feather-papillae, pigmented or colourless according to the breed of chicken which supplied the explant.

The possibilities which the culture of embryonic organs offers to the biologist are innumerable. Not only can an organ be withdrawn from the complex assembly of the total organism and its development be followed, but the conditions of its development can be varied at will. The composition of the nutritive medium can be modified or an organ associated at choice with any other organ. In other words, the precise functions, capacities, reactions to stimuli, and so forth, of a given embryonic part can be studied in isolation. The value to biology of such a simplification of vital processes can well be imagined.

'To advance experimental analysis,' Claude Bernard

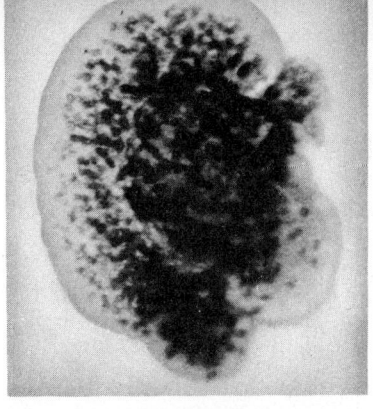

Association, *in vitro*, of two genital glands removed from embryo ducks. The gland at the top is female and the one below it male. After seven days in culture the female genital gland has feminized the male gland.

After Et. Wolff and K. Haffen.

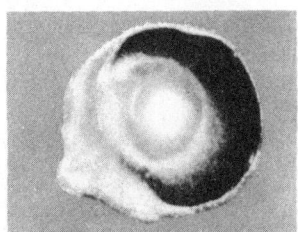

Eye of an embryo chicken developed *in vitro. On the right*, the pigmented layer of the retina, shaped like a crescent, envelops the ocular sphere. *In the centre*, the developing crystalline lens is surrounded by a light circle which corresponds to the iris.

After Et. Wolff.

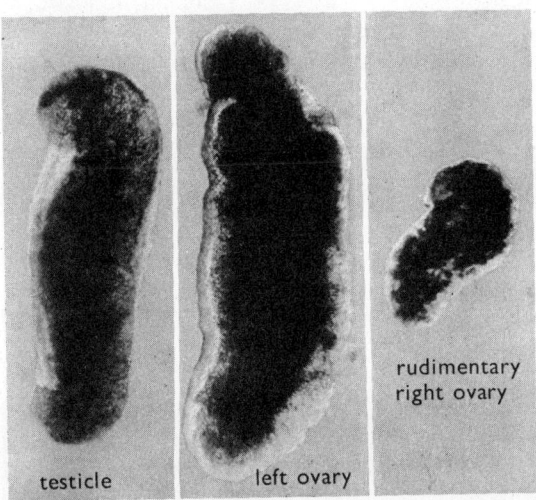

rudimentary right ovary

testicle   left ovary

The genital organs of embryo ducks. Beginning as buds, which were in appearance identical, the normal sexual differentiation of these organs took place *in vitro*. Ducks, like most birds, have only one ovary, situated on the left side. The right ovary remains rudimentary.

After Et. Wolff and K. Haffen.

wrote, 'we must, as far as we can, move physiological acts outside the organism. In isolation we can observe and more fully understand the hidden conditions under which phenomena occur, so that, later, we may interpret their vital role in the living organism.'

The work of Etienne Wolff has done much to fulfil the programme Claude Bernard advocated, and the attempt to 'move physiological — or embryological — acts outside the organism' was as successful as it was daring.

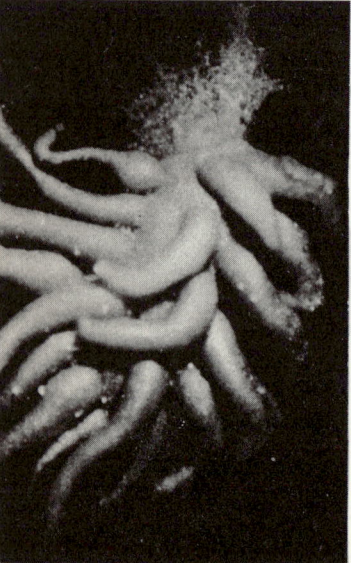
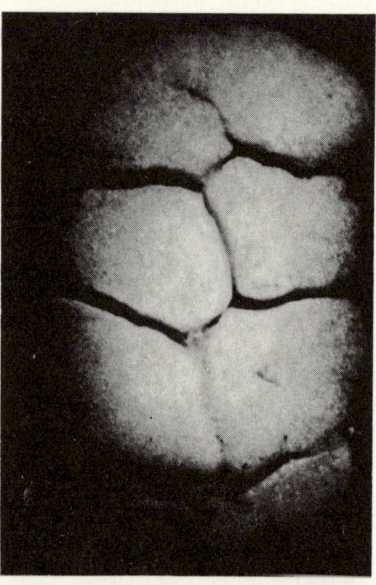

Skin of an embryo chicken grown in culture. Beginning with a fragment of smooth and homogeneous skin the feathers *(on the left)* are differentiated or, *(on the right)* scales develop, depending on whether the skin was taken from the back or from the legs. *Below,* the exuberant growth of feathers in an explant of skin removed from the embryo before any differentiation had taken place and afterwards cultivated on a medium rich in an extract of embryonic chicken brain.

After Sengel.

Wolff and his fellow workers also set themselves the task of achieving the culture of embryonic organs in synthetic media, and succeeded in replacing embryonic juice — previously held to be strictly indispensable for every culture — with a complex mixture of amino acids, hormones, and vitamins. This achievement was all the more remarkable in that the culture of tissues in a medium of definite chemical composition had never been successful. Why an entire organ can find what it needs for growth and differentiation in a nutritive medium in which tissues prove incapable of proliferating is unknown. But the fact remains that tibias removed from embryo chickens, as well as sex glands and syrinxes from ducks, have greatly increased in size in a synthetic medium. In particular tibias, explanted at the embryonic stage when the cartilage is forming, regularly develop in culture. Their epiphyses form and their diaphyses lengthen; they gain thirty per cent in length and thirty-five per cent in weight. In culture some explants, notably those of the liver, show signs of intense movement. The action *in vitro* of embryonic extracts of abnormal breeds of poultry — for example, *Creeper* — on the embryonic tibias of normal breeds has been studied, and it can be shown that their influence causes abnormalities.

Embryonic tissues have been grown in culture in association with adult tissues, as have tissues from different organs (thyroid-pituitary, liver-kidney) and homologous — or fundamentally similar — tissues from unrelated species (chicken-mouse, duck-mouse).

In the last case mixed organs *(chimaeras)* can be seen to form — bronchia, seminiferous tubules, etc. — and to the biologist it is an astonishing spectacle 'to behold, beyond the frontiers of species, genus, and even of zoological class, a manifestation of such affinities between tissues we consider as homologous, but which in nature have never had occasion to confront each other.' The affinity would seem to suggest some chemical analogy between the tissues and, therefore, the possibility that such 'temporary associations' could be made use of to facilitate grafts.

Another application of Wolff's technique is concerned with the culture of tumours. If cancers, like sarcoma in mice, are associated *in vitro* with the embryonic organs of chickens it is seen that the malignant cells invade the organ and destroy it, often sparing nothing but the surrounding membrane. The association is of profit only to the cancer which uses the organ as a culture medium. Thus *in vitro* we can see how the primary cancer produces the secondary cancer in the organ which nourishes the tumour. In this method of investigation the two adversaries, tumour and organ, meet as it were face to face.

Grafting

A part detached from a living organism continues to live and often to function for a certain length of time. Can this part be re-attached to the body from which it came — or to that of a different individual — and again become an integral working part of the organic whole?

Such is the problem posed by transplantation or grafting.

Plant grafting

Let us first examine the grafting of plants, an art which has been known since antiquity, and is today a current practice in horticulture.

The part grafted — the *scion* — is usually a shoot or a twig but can also be a bud. It is inserted into the root-system — or *stock* — of another plant by various means, such as vertical slits in the bark of the stock into which the tapered scion is fitted, cleft grafting, etc. The insertion of the scion is so made that the young cells, more precisely the cells of the cambium layer, come into close contact with those of the stock. At the same time the exposed tissues are protected from drying out by being tightly bandaged together, and the wound covered with 'grafting wax', a mixture of fresh cow dung, finely chopped hay, and clay. Strictly aseptic conditions are not, as in animal grafting, essential.

'Approach grafting' — or in-arching — is also practised, and consists of binding together the stems or branches of two neighbouring plants, in each of which a deep incision has been made so that their cambium layers are snugly fitted together.

Grafting is usually done in spring when the sap is rising or in autumn when its circulation is slowing down. Whatever technique of grafting is employed the end result is that on a given plant branches and buds from another plant are made to live and grow. The scion may belong to a different variety, to a different species, or even to a different genus. Thus grafts can be exchanged between the various species of rose bushes, the white lime can be grafted on to the common lime, the privet on the lilac, the sunflower on the Jerusalem artichoke, the chrysanthemum or the wormwood on the sunflower, the tansy on the wormwood, the fennel or the parsnip on the carrot, garlic mustard *(Alliaria)* or the mustard on the cabbage, the capsicum, the potato or the aubergine (egg-plant) on the tomato, the tomato on the black nightshade, the woody nightshade on the deadly nightshade, the French bean on the broad bean, the salsify on the *Scorzonera,* the service tree on the hawthorn, the wild cherry on the cherry laurel, etc.

Grafting is frequently employed in the cultivation of fruit trees, the apple tree being grafted on the pear tree; the pear on the quince, the medlar or hawthorn; and the peach on the plum or almond tree.

Generally speaking attempts at grafting between plants belonging to distantly related families have not proved successful.

The French botanist L. Daniel, who specialized in the study of experimental plant grafting, claimed that it was possible to graft the cabbage which is a crucifer on the tomato which is solanaceous, the haricot bean which is leguminous on the castor-oil plant which is euphorbiaceous, and the Jerusalem artichoke which is a composite on the black nightshade which is solanaceous. It is, however, almost certain that these were false grafts.

The reasons why one plant cannot be grafted on another plant are not well understood. The limitations of grafting are not, however, of necessity the same as the limitations of crossing, for to graft *Alliaria* on the cabbage, the capsicum on the tomato, and the sunflower on the wormwood is possible, while hybridization between these pairs is not.

Grafting is sometimes possible in one direction and impossible or extremely difficult in the other: for instance, it is easy to graft the pear tree on the quince

241

while grafting the quince on the pear almost always results in failure.

It is unnecessary to dwell on the practical advantages which agriculture and horticulture derive from grafting, but to mention a few: it permits us to cultivate delicate plants by giving them a hardy supporting stock; it permits us to cultivate a plant in soil to which its own roots are not adapted; it permits us to propagate plants which reproduce with difficulty or even, in the ordinary way, not at all; it permits us to save time by grafting shoots of young trees on older stock. Finally, as Van Tieghem says, grafting constitutes 'a valuable means of fixing and preserving those variations which occur only once in the seed, because grafting is itself incapable of producing the slightest further variation.'

Some botanists, however, have claimed that grafting can in fact bring about variations, transmissible or not, either in the scion or in the stock. Violent controversy has raged over this question, for it touches on the very mechanism of heredity and variation.

Experiments made by Daniel once seemed to indicate the possibility of hybridization by grafting, in other words, asexual hybridization, but his findings have remained unconfirmed.

It is important to be clear about what we mean by our terms. After grafting we can of course observe modifications in the plant's size, or date of flowering, or time of forming its reserves, or in its fertility. For example, Rivière and Bailhache reported that the average weight of fruit grown by a pear scion on a quince stock is unmistakably greater than that of fruit grown by a pear grafted on another pear. The sugar content is also said to be higher. Similarly the peach tree grafted on the plum tree is said to be more precocious though less hardy; the White Soissons French bean grafted on the Black Belgian French bean and the deadly nightshade grafted on the tobacco plant become smaller, while white lime on the common lime and the service tree on the hawthorn usually become larger. When French vines are grafted on American vine-stocks they become less resistant to cryptogamic diseases while the grapes lose something in quality.

When all is said and done, however, these are only superficial modifications, the result of ordinary nutritional influences — modifications comparable to those which can be caused by changes in methods of cultivation or composition of soil. It is obvious that a plant grafted on another plant receives its nourishment from roots not its own, and thus grows under conditions which are more or less abnormal — whence the possibility of variations of a quantitative order such as those we have mentioned. But they have nothing whatever to do with true hybridization by grafting in the sense that Daniel and his followers understood the term.

Daniel, for instance, claimed that after a Tours cabbage (which has tender green leaves and a conical head) was grafted on a Saint-Brieuc cabbage (which has dark green leaves and a round head) some of the grafted scions acquired deeper coloured leaves and a more rounded head. Again, when a tomato plant with round yellow fruit was grafted on a tomato plant with red, flattened fruit some of the scions formed flattened tomatoes which were reddish in colour. Finally, after grafting capsicum on a tomato plant some

Flowers and fruit of the graft hybrid *Pyrocydonia winkleri* (top) with its two parent plants the pear, *Pyrus communis* (centre) and the quince, *Cydonia oblonga* (below). This quince is the usual stock to which cultivated pears are grafted.

Muséum national d'histoire naturelle. Larousse.

of the scions formed fruit which resembled tomatoes.

Had it been possible to verify Daniel's observations they would have provided examples of genuine 'transference of racial characters' brought about by grafting. But these alleged facts of asexual hybridization have been unable to withstand the light of critical examination; the result of the experiments and the conditions under which they were made were scientifically unsatisfactory. A racial or specific character of a plant has never, in fact, been transmitted to another plant by grafting. Henri Colin stated the matter well by saying, 'it is possible to deprive a plant of its own resources and oblige it to model its demands on those of the plant on which it has been grafted. It is true that its normal diet is more or less modified after grafting, but its specific chemical qualities are not thereby basically altered. The fruit of the pear grafted on a quince stock does not take on the flavour or the shape of a quince; the sweet cherry grafted on the mahaleb cherry tree does not produce bitter fruit like its host; the cabbage grafted on the turnip does not acquire the taste of a turnip. When, as in Chablis, a white Pinot Chardonnay is grafted on American vine-stocks the grapes grown do not resemble American grapes.'

A good example of the mutual chemical independence of the scion and the stock has been provided by a telling experiment made by Henri Colin. The sunflower and the Jerusalem artichoke, which belong to the same family, the Compositae, each manufactures its own particular carbohydrate: the sunflower makes starch, the Jerusalem artichoke makes inulin. If the sunflower is grafted on the Jerusalem artichoke it is found that the graft in no way alters the chemical nature of the two plants. In spite of their intimate association not a trace of inulin will be found in the sunflower or a trace of starch in the Jerusalem artichoke. The carbohydrate proper to each of the two plants does not cross the boundary of the graft or, if it does, it is rapidly transformed by the foreign cells.

It is, however, not impossible that certain special substances, passively conveyed by the sap, are transmitted after grafting from one plant to the other. When the deadly nightshade is grafted on the tomato plant the alkaloid of the deadly nightshade, atropine, probably passes from scion to stock. In the reverse graft — tomato on deadly nightshade — atropine does not pass from the nightshade stock to tomato scion.

But this passive diffusion of substances — which, incidentally, has been questioned by some authorities — must not be confused with hybridization by graft which implies a change in the actual constitution of the cells and their biochemical properties.

Soviet biologists — of the school of Michurin or Lysenko — have in these last years maintained that grafting can bring about important modifications — especially when it is effected at the embryonic stage — and, in particular, that it can alter the plant's capacity for hybridization. They claim that embryo wheat grafted on rye albumen — or endosperm — produces plants which can be more easily crossed with rye. A similar result is said to be obtained if the same embryo wheat is grafted on the albumen of a certain variety of Chinese wheat which is itself easily crossed with rye.

These phenomena are still of dubious authenticity and even if they are verified they leave a wide margin for interpretation. Rather than cases of true asexual hybridization they may, it would seem, be more nearly related to those phenomena of 'immunological tolerance' which can — as we shall later see — be produced among animals by special treatments of the embryo.

Chimaeras

We must now examine certain odd cases of so-called graft-hybrids which have long disturbed and puzzled botanists.

Adam's laburnum is the result of grafting *Cytisus purpureus*, a purple laburnum, on *Cytisus laburnum*, which has clusters of yellow flowers. Adam's laburnum not only reveals the characters of both scion and stock, bearing clusters of both purple and yellow flowers, but also produces clusters with every possible combination of the two colours.

The flowers of pure colour are fertile and will reproduce either pure *Cytisus purpureus* or *Cytisus laburnum* as the case may be. The flowers of intermediate colour are sterile as those of hybrids generally are — sexual hybridization between the two species, *purpureus* and *laburnum*, being impossible.

According to Yves Delage the origin of this extraordinary shrub was as follows: in 1830 a gardener named Adam grafted a piece of bark (with a bud) of *Cytisus purpureus*, a small and delicate species, on a *Cytisus laburnum*, which is a hardy species of good size. The bud remained dormant for an entire year; it then produced several shoots of which one, stronger than the others, was sold by Adam as a variety of *purpureus* even before it flowered. It was only later that the singular off-shoots of this branch were observed and became the subject matter of a report addressed to Adam by Poiteau.

In the same way the famous medlar of Bronvaux, near Metz, which was the result of grafting a medlar on a hawthorn, revealed characters which were intermediate between those of scion and stock: like the hawthorn the branches were thorny and the inflorescence was a corymb, but the flowers had the shape and colour of those of the medlar. The fruit, though small and flattened, was that of the true medlar. The first samples of this equivocal tree were presented by Louis Simon to the Horticultural Congress at Paris in 1898. Since then the Bronvaux medlar has been propagated by cuttings and been named *Cratoegomespilus* to recall that it is a fusion of two species: the hawthorn or *Cratoegus* and the medlar or *Mespilus*.

Among oddities produced by grafting there is also the orange tree named the *Bizarria* because it produces at the same time bitter oranges *(Citrus aurantium)*, Florentine citrons *(Citrus medica)* and fruit of mixed type. It was created by a gardener in Florence in 1644 by grafting. The scion has died and the ambiguous shoots are produced by the stock. A rigorous analysis of all these aberrant cases, however, proves that the hypothesis of hybridization by graft is quite without foundation.

Histological study of plants with mixed characters

has shown that they are in fact made up of a collection of tissues derived from the two grafted species. In other words, the plants are chimaeras — organisms in which tissues of different genetic origin are associated. These tissues retain their own characters and autonomy but, in association, of necessity lead to the appearance of mingled characters in the plant.

Thus, in the case of Adam's laburnum, Buder has been able to show that the epidermal or superficial tissues are of *purpureus* origin while the deeper tissues derive from *laburnum*. The two species, as Caullery says, are thus encased one within the other, and each retains its individuality intact. The over-all form of the plant is the resulting compromise between the growth-tendencies normal to each species. Grafting does not then create a new species by the fusion of scion and stock.

Winkler, with tomatoes, artificially reproduced the phenomena which gave rise to the creation of plant chimaeras. He grafted a branch of black nightshade *(Solanum nigrum)* on to the stock of the tomato, *Solanum nigropersicum*. When the graft had 'taken' he cut through the stem at the point where the graft was made and observed that the cross-section thus revealed was composed of two kinds of tissue: *nigrum* in the centre and *nigropersicum* on the circumference. Had adventitious buds arisen from the junction of the two kinds of tissue they would of necessity have produced branches of mixed genetic origin, in other words, chimaeras.

The fact that the 'variegated' or 'mosaic' character can pass either from scion to stock or from stock to scion was formerly advanced as an argument in support of hybridization by graft. This transmission had already been noted by the great Darwin, following the graft of a bud from a plant with variegated leaves on a plant of the same species but with leaves of uniform colour.

Today we know that variegated leaves are often symptomatic of an infectious disease produced by an invisible virus. The virus can of course be communicated by scion to stock, or vice versa, and such cases are simply cases of contamination.

To sum up, all the facts advanced in favour of hybridization by graft can be explained by mechanisms already known to biology. We must therefore conclude that when two plants of different species are united by grafting each retains its individuality and independence, and neither influences the specific nature of the other.

Influence of grafting on subsequent generations

Although two plants united by grafting retain their specific qualities, what of their offspring? It is at least conceivable that the reproductive cells of the stock are influenced by those of the scion or that the reproductive cells of the scion are influenced by those of the stock.

L. Daniel believed in this kind of indirect hybridization. When, for example, he grafted *Alliaria* — sometimes known, because of its smell, as garlic mustard — on a cabbage stock he gathered the seeds formed by the scion and planted them. Now, according to him, these seeds grew into *Alliaria* which differed appreciably from normal *Alliaria*. Their roots were more developed and thicker; their stems were more tender, numerous and not so tall; their leaves, closer together, were broader and greener and smelled like a combination of cabbage and garlic mustard. By grafting Daniel believed, in brief, that he had created a new variety with hereditary characters. He also believed he had created a new variety of cabbage by grafting a kohl-rabi on a cabbage.

The results obtained by Daniel and his followers can, however, be explained by such every-day influences as soil and weather, by unperceived sexual hybridization, by impurities in the material employed, by throwbacks or atavism, by the formation of chimaeras, by mutation, and so forth.

Nor must we forget that most of these experiments were made during the first years of the century before the rigorous methods of Mendelian genetics had shed light on and introduced precision to the study of hereditary variations.

Since that time all research of this kind is preceded by a careful study and selection of the material utilized, so that the botanist knows very accurately what characters can and what characters cannot appear in the offspring obtained. Under these stricter conditions grafting has never been found to result in the formation of a new variety with modified hereditary characters.

Animal grafting

The scientific study of animal grafting was inaugurated towards 1860 by the celebrated French physiologist Paul Bert. Well before then grafting had, of course, been practised on human beings in a somewhat hit or miss fashion. Some important applications had been discovered, for the grafting or knitting of bones was understood and also skin grafting. Paul Bert, however, was the first to undertake direct and experimental study of the question, to grasp its full importance from the biological point of view, and to define clearly the many problems posed by the seemingly simple operation of displacing an organ or of transplanting it to another organism.

In his own words transplantation or animal grafting occurs in the first place 'when a part is detached from one animal's body and transplanted to another's body where it continues to live,' and in the second place, 'when, in the same animal, a part completely separated in a single or in several operations from its connections recovers those connections or acquires new ones.'

In the vocabulary of today we say, in the first case, that a homograft or a *homoplastic* graft has taken place if the donor and the host are of the same species, a *heteroplastic* graft or heterograft if donor and host are of different species. In the first case — when both donor and host are the same individual — we say that *autoplastic* grafting or autografting has occurred.

In an earlier chapter we have already seen how deeply indebted embryology is to the technique of pre-natal grafting. Embryonic grafting will therefore be omitted from our rapid review of grafts achieved among various animal groups.

Among Protozoa, or animals composed of a single cell, two pieces derived from different individuals can be joined or grafted together, and the nucleus of one protozoan can be removed and replaced by the nucleus of another.

Homografting is not difficult with sponges, hydras, planarians, earthworms, starfish, etc. Among moths, Crampton grafted the two halves of different chrysalises by uniting them with a ring of melted paraffin wax. In this way he obtained composite insects, true chimaeras, animals whose anterior section may have belonged to one species while the posterior section belonged to another. The fusion, however, was superficial and did not involve the internal organs.

In a series of widely heralded experiments made in 1923 W. Finkler claimed that it was possible to graft the heads of various coleopterans and orthopterans. According to him the exchange of heads produced very curious results of a psychological nature when the head grafted came from an insect of the opposite sex. Females of the family Hydrophilidae (water beetles) bearing the head of a male tried to couple with other females, while males grafted with female heads displayed the sexual passivity characteristic of the female. Thus in the determination of amorous instincts it was the head, the seat of the cerebral ganglions, which ruled the rest of the body!

Finkler furthermore claimed that this cephalic graft was possible between insects belonging to different species and even different genera. He stated that he had grafted the heads of water beetles of the family Dytiscidae on the bodies of hydrophilids, and vice versa; and in addition that he had observed the subsequent influence of the grafted head on the coloration of the host's body. A dytiscid with the head of a hydrophilid was reported to have lost the yellow bands which adorn its elytra (or modified front wings) and assumed the uniformly sombre array of the hydrophilid.

Plaviltsthikov, a Russian, affirmed in 1927 that he too had grafted cockchafers' heads on the bodies of dung beetles, and vice versa, whereupon the dung beetles began to browse off leaves while the cockchafers burrowed in the dung!

These extravagant reports have not been confirmed by other investigators. If, however, we are less ambitious and confine ourselves to *autografting,* success may be achieved. J. Rostand has cut off the head of an insect — the phasmid *Timarcha* — and immediately grafted it on again. In such cases the regrafted head will survive for a considerable time and occasionally a regeneration of the intermediate tissues is observed.

Among amphibians G. Born as long ago as 1896 performed remarkable experiments in grafting the larvae of frogs. When a tadpole is cut in two it is possible to re-unite the two parts and even — by operating on two tadpoles — to unite the front part of one with the back part of the other. In this way composite larvae are obtained, larvae which are perfectly viable, develop normally, and can achieve metamorphosis. The graft is successful even between tadpoles of different species, and in this way Harrison

The physiologist Paul Bert (1833—1886), one of the great pioneers in scientific grafting.

has created chimaera frogs, the front half belonging to the species *Rana virescens,* and the rear half to the species *Rana palustris.*

Tadpoles and young frogs constitute ideal material for this kind of experimental research, and it is possible to graft a tail or a budding member — front or hind leg — on the animal's head or back, and to graft a tympanic membrane on its back, or an eye where the ear is located, and so forth.

Among amphibians, homoplastic grafting (that is between animals of the same species) of the eye and, indeed, of many other organs is successful even when the animals have reached the adult stage. Not only does the grafted eye survive, but it resumes its sensory activities and vision is restored. When the operation is performed on the triton the retina of the grafted eye quickly degenerates, leaving its debris in the region of the iris. This debris, however, will in a few days entirely regenerate a new retina, the cells of which then reconstitute the optic nerve by which they are connected with the brain.

A few weeks after the operation the fact that vision has been restored can be shown by appropriate tests. If, for instance, a piece of red rubber is attached to a string and dangled before the triton with the grafted eye the animal will watch, then attempt to leap forward and snap it up, as all normal tritons do in the presence of a coloured moving object. A blind triton, on the contrary, does not react to the movement of the bait. Among green tritons the same eye can be grafted on different hosts as many as three or four times in succession.

It is even possible to graft the eye of a salamander — *Ambystoma* — on a triton, though the reciprocal graft is a failure. The experiment is all the more remarkable in that there are, between salamander and triton, differences of visual power, the triton having the more acute eyesight. It is found that a triton with a salamander's eye reacts more sluggishly than a normal triton, but in a more lively fashion than a normal salamander. The triton, in this case, not only sees with the eye of another individual but of an individual that belongs to another amphibian genus. In other words, a composite animal or chimaera has been created by grafting; the triton provided with a salamander's eye is no longer altogether a triton. In human terms such a creation would correspond to a man who saw with the eyes of an ape!

245

Grafting in higher vertebrates

When we reach the higher vertebrates, birds and more especially mammals, the distinction between auto-plastic grafting and homoplastic grafting assumes its full importance. In general grafting is extremely difficult, between two different individuals, though we shall later see that this statement requires a few reservations.

It seems that the individual refuses to adopt an organ or fragment of an organ derived from another individual, from an 'organic native land' which is not his own, as Paul Bert put it.

Between one triton and another, between toad and toad, the tissues are, so to speak, interchangeable. An organ from Toad A can live and function in the organism of Toad B. The case is quite different among rats. An organ from Rat A cannot live and function in the organism of Rat B. Among higher animals it is as though each individual is more individualized, more definitely itself, more set apart from its fellow creatures. On the other hand, autoplastic grafting among higher animals is constantly and successfully practised with many organs, and in particular with almost all glands.

Most people have heard of grafting a cock's spur on its crest, the crest appearing to offer a particularly favourable field for grafting because of its many blood vessels. When a cock's spur is thus grafted on its crest it grows to a much larger size than a spur left in the normal position on the animal's leg. An experiment made by the physiologist Caridroit is of interest in this connection. In hens the spur is always rudimentary because its development is inhibited by the presence of the hormone secreted by the animal's ovary. Now if this rudimentary spur is grafted on the hen's own crest it will begin to grow. Simply by displacing the spur from its usual position, the inhibitory action of the hormone has been nullified.

According to Mantegazza the spur of a cock which was grafted on the ear of an ox attained a length of 10 inches but no-one else has ever been able to repeat the phenomenon which contradicts everything that is known about the problem of heteroplastic grafting among higher animals.

Eye grafting — even autografting of the eye — would seem to be impossible in birds as well as in mammals — although Koppanyi in 1923 claimed that he had performed the operation with a rat.

The reason for the failure is that in warm-blooded animals the retina or sensitive membrane of the eye degenerates with extreme rapidity when it is no longer irrigated by blood. Now, before vision is recovered, it is necessary that the nerve-cells of the retina rebuild, or regenerate, the fibres of the optic nerve and thus re-establish the connection with the brain.

Organic individuality

What is the reason for this individuality of tissue which so often causes the failure of homoplastic grafting in higher animals? Basically it arises from the genetic

246

Frog with three eyes. The animal's left ear was replaced by an eye bud at the embryo stage.

May, Palais de la Découverte collection. Larousse.

A frog with five legs. The animal, while an embryo, was grafted with the bud of a supplementary leg.

May, Palais de la Découverte collection. Larousse.

diversity of individuals. We know that every organism is unique in the collection of genes that it receives from its two parents; and, if it refuses to accept a graft from a donor, that is because its genetic individuality is not the same as the donor's.

Naturally, between individuals born of the same egg — in other words identical twins — grafting is always possible. Since the two subjects are identical from the genetic point of view the homograft in this case is equivalent to an autograft.

Among various animals, such as mice and guinea-pigs, one can create pure strains, that is to say, strains in which the individuals are practically identical from the genetic point of view. In these cases it is found that organs can in fact be freely transferred with almost constant success from one individual to another. They are, so to speak, interchangeable.

In a pure strain of Japanese mice — the variety known as Japanese waltzing mice because of their faulty locomotor equipment — the grafting of the spleen from one animal to another is constantly successful. If these waltzing mice are crossed with mice of a pure albino strain hybrid mice are obtained, these being waltzing albinos. On these hybrids it is possible to graft a spleen donated by mice of either the waltzing variety or of the albino variety. The reciprocal operation is not possible. If a waltzing mouse or an albino mouse is grafted with a spleen taken from a waltzing albino hybrid the operation will fail, because the host does not carry all the hereditary genes carried by the donor.

Even within the confines of a pure strain it is sometimes found that females will not accept grafts derived from males. The reason for this intolerance is that the male cells carry genes on the Y-chromosome which are not found in the female cells. Males, on the other hand, accept grafts derived from females, because the female cells carry no genes which are not also carried by male cells.

It is not known why biological individuality is an obstacle to grafting among higher animals when it presents no difficulty among fish and amphibians. But the physiological consequence of this individuality, inasmuch as it affects resistance to grafting, has been carefully studied. In the case of homografts, and even more so in that of heterografts, the grafted animal, the host, reacts and defends itself against the donated organ which is destroyed by necrosis and soon eliminated.

'There is today no doubt,' wrote Raoul-Michel May, 'that the reactions of an individual to a graft from another individual of the same or of a different species come within the framework of the general laws of immunity.'

If a foreign substance is introduced into the bloodstream of an organism the organism normally protects itself. It sets up a defence against the toxic action of the foreign substance by producing a new substance which acts, as it were, as a counter-poison. The defensive substance is called an *antibody* while the offensive substance is called an *antigen*. Antibodies belong to the protein family, or more precisely to the group of proteins known as globulins. During the course of the organism's life antibodies frequently play their part in

neutralizing microbes or poisons. Antigens are usually proteins, though they can also be carbohydrates (sugars) and lipids (fats). Their presence evokes the formation of the antibodies which combine with them chemically. Thus the cells of the donated graft bring about the production of antibodies in the host organism, antibodies which attack and destroy the invader and, in consequence, prevent the success of the graft.

Antibodies are formed only after the graft is made, which is why the graft seems to be tolerated for the first few days. Homografts begin by healing just as autografts do, and the difference between the two is only revealed later when the homograft is finally destroyed and resorbed while the autograft persists.

The time which elapses before the homograft shows signs of regression is exactly the same period of time which the organism needs to manufacture its antibodies.

Once formed the antibodies persist, and the organism's immunity to further grafting of the same kind is therefore increased. If skin is grafted on an organism which has already received and resorbed a graft of the same genetic type the new graft will survive for a distinctly shorter period of time than the first graft. It is apparent that the first graft in some way 'vaccinated' the organism so that it was better able to resist the second graft. The antibodies responsible for the anti-graft reaction seem to possess special properties: unlike antibodies responsible for defence against bacteria they do not circulate through the bloodstream but remain associated with the lymphatic cells. However, the intolerance of the organism towards grafts can be appreciably reduced by the action of X-rays or of cortisone, and recently use has been made of this fact for transplantation surgery both in mammals such as dogs and in Man himself. The number of human hearts which have been transplanted is now (July 1968) over two dozen and the grafting of kidneys is now commonplace. Grafting of such organs as the lungs and liver is also receiving extensive investigation at the present time.

Many heart transplants have so far failed and the patient has died. The reason for this is very interesting. The actual surgery involved in heart transplantation is laborious but not overwhelmingly difficult. The production of antibody-carrying lymphocytes can also be suppressed, and so the failure of the patient to survive is often not due to rejection of the graft. It is simply due to the fact that the body's ability to resist infection by foreign organisms has been deliberately lowered and so the patient is much more susceptible to any stray infection, (unrelated to the operation itself) which he may catch. The medical world has now become well aware of this danger and transplant patients are carefully isolated from possible sources of infection.

There is no doubt that human transplantation surgery is in its infancy and there are many technical problems which it will have to overcome. Nevertheless the rewards to be gained from using such a method — the replacement of diseased organs, and indeed the continuance of life — are sufficiently valuable to stimulate the very greatest efforts to overcome these short-term difficulties and establish transplantation techniques as routine medical practice.

Brephoplasty

The difficulties which we have just discussed apply chiefly to the grafting of adult organs on adult organisms. The biological personality which, as we have seen, imposes such narrow limits on the possibility of grafting is often, as Raoul-Michel May emphasized, less pronounced when we are dealing with young organs and young organisms, and even less so with embryonic organs and organisms. It is as though biological individuality is reinforced and accentuated with age. A young bird or a young mammal may tolerate a graft from an equally young donor when the same graft, between adult and adult, would fail. It is often found that an organ removed from an embryo can be successfully transplanted to an adult organism.

Because the tissues of the sex-glands are in many ways comparable to embryonic tissues they, too, can sometimes be transplanted from one individual to another, even among the higher vertebrates. Finally, because the cancerous cell is, at least in its power to proliferate, rather similar to the embryonic cell, homoplastic grafting of cancerous tissues is sometimes possible in animals, even when grafting of the corresponding healthy tissues is not.

The technique of embryonic grafting or *brephoplasty* (from the Greek *brephos* or embryo) was first employed by Paul Bert, the great pioneer in animal grafting. It was later improved, extended, and perfected by numerous workers in the same field. Voronoff and Didry, for example, successfully grafted the embryo pancreas. Having surgically removed the pancreas of a dog they grafted on the animal a pancreas taken from a canine foetus. With this substitute pancreas the dog was able to live for a very long time without displaying the symptoms of serious diabetes which, in control animals, inevitably followed the removal of the pancreas.

R. M. May removed the thyroid gland of a one-day old rat and then grafted it on a rat aged four to five weeks, the organ being inserted into the anterior chamber of the eye, a small natural cavity where the graft found conditions particularly favourable for survival. At the end of a week he removed the rat's own thyroid gland. After this operation it was observed that the rat grew and thrived in a completely normal fashion, while the development of other rats, which had had their thyroid glands removed but not been previously grafted with an embryonic thyroid, suffered severely from lack of the hormone secreted by the thyroid. It can therefore be concluded that the grafted embryo gland fulfilled its purpose and replaced the missing thyroid — something which a grafted adult gland could not have done.

Other experiments of a similar nature reinforce the same conclusion: experiments with the parathyroid, the adrenal, the pituitary, etc. — all glands which produce hormones.

In particular the intra-ocular graft of an embryo anterior pituitary lobe will result in the performance of the many hormonal functions which normally pertain to this important gland: the animal host, in spite of the absence of its own pituitary, will grow in the customary manner, accomplish its sexual cycle, reproduce, and so forth. The graft can be placed not only in the eye, but in the kidney or in the renal parenchyma or specific tissue of the kidney.

Apart from the technique of embryonic or brephoplastic grafting other methods exist for overcoming the intolerance of host to donor. Aron and his fellow workers have, for instance, explored the possibilities of the intra-testicular graft, and found that prolonged survivals of homografts can be obtained when they are implanted in the testis.

These experiments were made with guinea-pigs and the technique is simple. It consists of introducing fragments of the organs into the testis by means of a hollow needle, or trocar. The organ most frequently transplanted was the anterior pituitary, a gland which secretes hormones that have a direct and rapid action on the male gland. This permits the functional survival of the pituitary to be easily supervised.

The graft remained healthy in fifty-four experiments, eighteen of which lasted for four weeks and some as long as sixteen weeks. After a phase of de-differentiation which lasted about two weeks the graft recovered its typical structure and left no doubt about its secretory activity. After a month of functional survival a graft may be considered as having succeeded.

Aron and his colleagues obtained comparable results by grafting fragments of thyroid gland in the same fashion. With fragments of kidney, spleen, pancreas, liver and muscles the results were less satisfactory, but survival was still appreciably prolonged.

If the host reacts to the homograft by forming an immunity to it, thus leading to the failure of the graft, we must, Max Aron concluded, 'assume that in the testis the transplant does not give rise to this reaction or else that it evades it'.

Continuing his researches Aron attempted heterografts with human cancerous tissue implanted in the testis of guinea-pigs. The grafted tissue generally survived for six days though, in the case of a fragment of renal cancer, cells were still found living in the grafted tissue at the end of three weeks.

Although for the moment one cannot pass judgement on these investigations it may be hoped that the technique of intra-testicular grafting might open a fresh approach to the study of homografts and their ultimate employment in medicine. If the results already obtained with guinea-pigs are verified with human subjects their practical value will obviously be great.

The work of Medawar

Tackling the problem from another angle, the English biologist Medawar, in a very important body of research, established that among mice a state of tolerance to homoplastic grafting can be created by submitting the individual to foreign attack while it is still in the embryonic stage.

This is Medawar's now classic experiment: while still in the uterus the embryos of Mouse A were injected with cells taken from Mouse B. When the little mice were born to Mouse A they were grafted with the skin from B. The graft persisted instead of being resorbed

Pleurodeles, which at the stage when the embryo measured 2 millimetres, received on its right flank a 'territorial' graft of the foreleg of another embryo. The bud has developed into a leg.

Larousse.

Pleurodeles, whose right flank was grafted with a foreleg at the beginning of embryonic development.

Larousse.

as it would certainly have been had not the embryo been 'prepared' for it in this fashion.

Medawar's results have also been obtained with other animals, both mammals and birds like turkeys, chickens, pheasants, etc. A state of tolerance can sometimes be created even after the birth or hatching of the animal: in new-born rats by subcutaneous or intra-muscular injections; in new-born mice by intra-venous or intra-peritoneal injections. The state of tolerance created by the injection of cells of one type — blood-cells, for example — is brought about by any type of cell which has the same genetic constitution. The substances responsible for the 'Medawar effect' are localized in the nucleus of the cell but do not merge with the DNA.

Embryonic tolerance to grafting is sometimes so great that the foreign cells with which the embryo is injected become an integral part of its organism with the result that it becomes a cellular 'mosaic' or 'chimaera' containing cells of dissimilar genetic types.

This is the case with twin cattle which, though not identical twins, share the same placenta and are often found to have 'mosaic blood', each of the two animals having received and adopted from the other the cellular material from which the blood is formed. Each, more-over, will tolerate a skin-graft donated by the other — which, Medawar points out, could have been expected.

Such 'blood chimaeras' have been identified among other mammals such as sheep and also among chickens born of an egg with two yolks.

The relationship between graft and host

In plant grafting we have seen that when two plants of different species or varieties are associated by grafting both scion and stock retain their individuality. Each keeps its specific or racial characters and the basic constitution of its cells undergoes no modification at all. In brief, we concluded that hybridization by graft, that is, asexual hybridization, does not exist.

We shall now find that in the animal kingdom the same conclusion is valid. Doubtless, influences of a trophic or hormonal nature are sometimes apparent. For instance, when the eye of a large-eyed salamander is grafted on a small-eyed salamander it will remain smaller than an ungrafted eye; but this is simply because the grafted eye, in its new environment, finds less favourable conditions for attaining its full growth. It can also occur that some chemical substance with a special action passes from graft to host or vice versa. Effects may then be observed which resemble a kind of hybridization by graft, though the mechanism is altogether different.

Phenomena of this kind are responsible for the effects observed among insects and reported by Kühn and Caspari, and by Beadle and Ephrussi. One strain of the fruit fly *Drosophila* has vermilion coloured eyes and another has brick-red eyes. Now if an eye-bud of the former is grafted on the larval abdomen of the latter

249

it will produce a brick-red eye. The host (brick-red) has therefore incontestably influenced the graft (vermilion). But the influence also works in the opposite direction, that is to say the graft can affect the host, for when an eye-bud of brick-red strain is implanted in the abdomen of a larva of vermilion strain the fly which the larva produces will have brick-red eyes instead of the vermilion eyes it ought to have according to its genetic constitution. A careful analysis of these phenomena has revealed that the influences are exerted through the intermediary of a well-defined chemical substance — *kynurenine* — which, diffused through the insect's body fluids, intervenes at the moment when the eye-pigment is formed. So again in this case, though the colour of the eye was changed by the graft, the organ underwent no basic modification: in other words, its cells were not altered.

Nor are the cells of frogs affected by grafts: in the chimaeras obtained by Harrison, the front halves of which were *Rana virescens* and the rear halves *Rana palustris,* each half developed in conformity with the type of its species.

Danforth — experimenting with very young chicks — grafted skin from the rump, choosing donors and hosts of different breeds. He transplanted skin removed from a White Leghorn chick, a breed which of course has white plumage, to a Plymouth Rock chick, a breed with greyish-white feathers barred with bluish black. When the bird reached the adult stage the graft could be recognized by the white feathers which cut across the rest of the plumage, a proof that the graft had not been influenced by the host as far as its pigmentation was concerned.

A decisive demonstration that grafting does not bring about hybridization is given by the graft of the sex glands. If the host could exercise even the slightest specific influence on the cells of the graft and the graft is a gland which forms germinal cells, then we should of necessity note in the offspring at least some modification of characters. Now the experiment has been made many times and always produces negative results.

Let us, for instance, remove the ovary from the young larva of the fruit fly *Drosophila* belonging to the strain with vermilion eyes and graft it on a larva of the red-eyed strain. When the host-larva becomes a fly and is capable of reproducing it will be simple for us to verify — by means of appropriate cross-breeding — that the grafted (vermilion) ovary produces functional ova of rigorously pure vermilion type. The fact that these ova have been matured and nourished in an organism of the red-eyed strain has in no degree altered their character: the host's influence on their eye-colour is nil.

The same experiment can be repeated with *Lymantria dispar,* the gypsy moth. Klatt removed ovaries from caterpillars of the normal variety — which is not black — and then grafted them on caterpillars of the black variety. When the caterpillars underwent metamorphosis he united the black moths with normal male moths and obtained offspring which in colouring were perfectly normal, the germinal cells of the grafted ovary having received no influence from the organism which housed and nourished it.

Finally there is the celebrated experiment made by

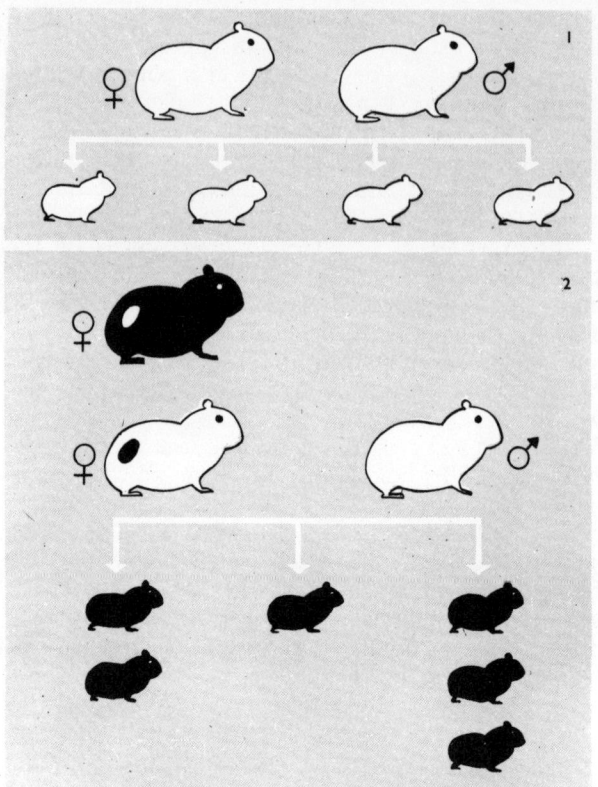

The experiment of Castle and Philipps (1901).
1. Reproduction of a pair of albino guinea-pigs.
2. Coupling of an albino male with an albino female which had been grafted with the ovary of a black female. The reproductive cells of the donated 'black' ovary were in no way affected by the receiver's albinism.

After L. Bounoure.

Castle and Philipps at the beginning of the century. They removed the two ovaries of a young guinea-pig of pure white breed and replaced them with ovaries taken from a young female of pure black breed. In guinea-pigs the white or black coat is a strictly hereditary character. Now, a few months later when the grafted white female had reached the age of reproduction they mated her with a male of pure white breed. After a normal gestation she gave birth to two young guinea-pigs whose coats were completely black. Again she produced a black offspring and afterwards three more — all black. In all there were six perfectly black offspring — black like the female which had furnished the grafted ovaries.

In the course of certain experiments with insects singular phenomena have been observed which mimic graft hybridization, but must be carefully distinguished from it.

Thus in the case of the small flour moth *(Ephestia kuhniella)* when the female larva of the albino strain is grafted with an organ removed from the larva of the common pigmented strain, the organ secretes a chemical substance which diffuses through the albino's body fluids. This substance not only modifies the pigmentation of the insect host but also accumulates in the eggs it lays. As a result even the offspring are affected and will, at least when young, display the characteristic pigmentation of the normal flour moth. The pigmentation fades little by little and disappears completely during the course of development. It is not transmitted to the following generation. This phenomenon of 'pseudo-heredity' through maternal chemical influence is extremely remarkable, and shows how much caution is required before announcing that a modification of hereditary patrimony has been obtained by grafting.

The experiments of L'Héritier and Scoeux on the transmission of 'sensitivity to carbon dioxide' in fruit flies illustrate the point from a slightly different angle. This sensitivity is peculiar to certain stocks of fruit fly and can be transmitted by grafting. When the ovaries of the resistant strain are grafted on sensitive females whose own ovaries have been removed the sensitive females will still procreate a more or less large number of sensitive offspring. Is this not, then, an example of hybridization by graft? Not at all: the 'sensitivity to carbon dioxide' depends on a kind of virus which passes from the body of the sensitive host into the resistant grafted ovaries. In other words, the ovaries are merely contaminated by the host, and the case is not unlike that of the transmission of virus-induced 'variegations' in plants.

In brief, among animals as among plants, no true case of hybridization by graft has yet been observed. Between host and graft or reciprocally, a virus may pass or a substance of specific effect be exchanged; but the basic constitution of the cells is never modified and no variation, however slight, occurs in their hereditary patrimony.

Grafting as a method of research

Many problems of a biological nature can be rewardingly studied by means of animal grafting. Apart from the special problems pertaining to grafting itself, such as those of the genetic individuality of tissues, of the conflict between host and graft, of the influence of host on graft and vice versa, the method offers us opportunities to investigate a wide variety of phenomena. As Raoul-Michel May says in his excellent book on *Animal Transplantation,* 'the method of grafting permits tissues and organs to be freed from given influences while submitted to others. Thus it continues and will continue to play an important part in the study of those anatomical and physiological problems in which, to observe the behaviour of cells more clearly, we wish to isolate them but at the same time do not wish to withdraw them completely from all the organism's influences.'

Much earlier Paul Bert had said that 'animal grafting is not so much a given problem as a kind of crossroads where many problems which seem to have no connection with it will meet.' And, in fact, animal grafting has been widely employed in the study of growth, of organic correlations, of metamorphosis, of cell differentiation, of the relationship between the nerves and nerveless organs, and so on.

We have already suggested the pre-eminent role it has played in physiology: the study of the internal

'Chimaera' embryo, produced by grafting the front half of *Rana virescens* to the rear half of *Rana palustris.*
1. Shortly after the operation.
2. A more advanced stage.
3. Adult frog with the pigmentation of *Rana virescens* in front and of *Rana palustris* behind.
4. Normal *Rana palustris.*

secretory glands is founded on grafting experiments, since it is by displacing or transplanting an organ that it is possible to discover what part of its activity arises exclusively from an endocrine function.

Another application of grafting is illustrated in the excellent work of Beadle and Ephrussi on the humoral determinism of eye-pigmentation among fruit flies.

Very suggestive research among amphibians has revealed the structural modifications which the spinal cord undergoes after being grafted with supernumerary members. When such new conditions are created by grafting, the spinal cord has more work to do; to meet the situation the nerve-cells on the side of the supernumerary member increase and the cord thus loses its symmetry. If less work is required the number of nerve-cells is on the other hand reduced. These facts throw light on the flexibility of the nervous system.

In the evolution of an organ, grafting provides a means of distinguishing between the part played by its own intrinsic potentialities and the part played by the humoral environment in which it develops. For instance, a salamander of species A can be grafted with an eye taken from species B, the two species being chosen because the size of their eyes is noticeably different. If this is done, the eyes of the large-eyed species, even when grafted on a salamander of the small-eyed species, will become larger than those of the

251

host. They will not, however, become as large as they would have if they had remained in place on the animal of the large-eyed species. Thus it can be shown that two kinds of factors influence the growth of the eye, factors intrinsic in the grafted organ itself and factors present in the humoral environment of the host.

Interesting experiments have led to the discovery of the conditions a grafted organ can be submitted to without thereby compromising the success of the graft. Before being grafted an organ can be kept for a long time, provided that it is kept at a low temperature. It can also undergo a fair amount of dehydration: the ovaries of some mammals have been almost completely dried out and afterwards grafted with success. The same is true of tumour fragments. Experiments of this sort contribute to our knowledge of the resistance and vitality of cells.

One of the most curious applications of grafting technique is seen in the work of Hadorn, who produces mutations by the treatment *in vitro* of the ovaries of fruit flies. The embryonic ovaries of four-day old larvae are immersed in a very weak solution of phenol — one part in 10,000 — and then replaced in the insect's body. When the insect reaches sexual maturity it lays eggs which, in consequence of this chemical treatment, give birth to a very high percentage of mutations.

Finally there is the not inconsiderable part which grafting has played and still plays in the study of cancer. In mice, especially, it is possible to maintain particular stocks of cancerous tissue by grafting them in series from one animal to another. The cancerous tissue, though grafted, conserves not only its own malignant properties but also the characters of the particular stock employed. In this way numerous and rewarding experiments have been made to throw light on the genetic and physiological conditions involved.

Parabiosis or Siamese grafts

The term parabiosis or Siamese graft is applied to a special kind of graft in which two entire animals are united. The alternative name is derived from the famous Siamese twins who were, as we know, physically attached to each other from birth.

Paul Bert was again the first to conceive the idea of thus artificially coupling two organisms and of joining them together by organic bonds so that between them a kind of vital solidarity would be established. In his own words he wished to know if between two animals it was possible to obtain the exchange of nutritive material and, in a word, to create 'double-monsters' after the animals' birth.

The experiment was made in 1862 on two-week-old albino rats. A superficial incision was made along the length of one rat's left flank and along the right flank of the other. Contact between the two bleeding surfaces was assured by a suture and a bandage treated with collodion.

The results of this operation were very simple. In a few days the wound closed without the slightest suppuration. At the end of a week it was possible to remove the bandages by which the two rats were united. They walked side by side, joined by a band of skin about 4 centimetres wide. Soon, however, they began to pull against the band which united them and to gnaw at it with their teeth. Paul Bert therefore decided to sacrifice them, having first shown them to many people including the great physiologist Claude Bernard.

Paul Bert repeated this curious experiment several times. He succeeded in keeping one pair of surgically joined rats for more than two months. They were, he wrote, 'the most patient couple I possessed.' He submitted them to various experiments designed to reveal the communications between their circulatory systems. In particular he injected into the veins of one 'Siamese' rat a solution of atropine, an alkaloid which promptly causes the dilation of the pupils. At the end of half an hour he observed this characteristic reaction in the pupils of the second rat, a clear demonstration that the atropine had passed from the bloodstream of one into the bloodstream of the other. Since the historic experiments of Paul Bert — made over a century ago — the technique of parabiosis has become of considerable importance in certain branches of biology.

Though Paul Bert applied the term 'Siamese graft' to the new method of grafting it should be pointed out that he was himself fully aware that the double-monsters he created and those created by nature were not produced by the same process. 'When I first showed my two rats united by a cutaneous suture,' he wrote, 'many people said to me, 'you have created Siamese twins.' I insisted then, and I still insist, on denying such an interpretation of my experiments. In my opinion animal grafting is a totally different thing from double-monstrosity. The processes of the former are utterly unlike the causes of the latter.'

Paul Bert was, of course, correct in drawing the distinction, for, contrary to the accepted opinion of his day, Siamese twins — and double-monsters in general — are not two beings, originally distinct, which have been joined in the course of their development, but two beings arising from the same egg-cell, that is, identical twins which are imperfectly separated.

The work of Paul Bert was continued and extended by various physiologists, notably by Sauerbruch and Heyde who operated on rabbits, and by Morpurgo who worked with rats.

The first fact which emerges from all such research is that parabiosis, like every other graft, is only possible between very closely related animals. The almost certain success of the operation can be assured simply by choosing animals which carry the same hereditary factors, the same genes: in other words, animals belonging to the same 'pure strain'.

It is possible to keep subjects of different sex alive in parabiosis for some time, and this without damaging the fertility of either. A female rat after being grafted to a male can be fertilized by another male and, after a normal period of gestation, give birth to normal rats.

Morpurgo very clearly demonstrated the solidarity, indeed the functional unity, of animals parabiotically joined: if the two kidneys of one such twin are removed the animal will not succumb to the operation; instead, the elimination of its waste matter is assured by the other twin whose kidneys will enlarge sufficiently to accomplish the extra work.

Parabiosis and transplantation of gonads in *Xenopus laevis*.
1. Parasite fused to the ventral wall of the autosite (or more nearly complete component of the 'Siamese' twins.)
2. Parasite fused to the left flank of the autosite.

Chih-Ye-Chang.

Even if a third kidney is removed the parabiotic couple will be able to survive for a few days with the single kidney which remains.

Similar experiments have been made on glands of internal secretion like the pancreas which supplies the bloodstream with the hormone insulin, without which the animal will promptly die of diabetes. The pancreas of one parabiotic rat can be removed without subsequent appearance in the other of any insulin insufficiency.

It is, as a matter of fact, in the study of hormone mechanisms that parabiosis has been chiefly utilized, for it supplies a means of detecting certain hormones which classic methods of investigation are unable to observe.

Parabiosis can also be employed in the study of senescence, for example by associating an aged animal with a younger animal. Again it is used to gain information about abnormal subjects which are unable to live independently, but can be made to live by the research biologist if he grafts them on to a normal subject. In this way Gallien, Beetschen, and Signoret have grafted the monstrous larvae of *Pleurodeles* on normal larvae. Couples thus formed may live to achieve metamorphosis, which allows the evolution of the parasitical monster to be followed. Its abnormality may be the result of a genetic mutation brought about in the egg by a thermal shock or else the result of the serious disturbance of its embryonic development brought about by the chemical action of lithium chloride.

The value of parabiosis in the study of sex differentiation, metamorphosis, etc., has been indicated in earlier chapters.

Opposite top, parabiotic or artificial Siamese twins. Pair of young diploid *Pleurodeles* before metamorphosis; *above,* pair of diploids at the age of 420 days, having undergone metamorphosis.

Gallien.

Opposite, parabiotic *Pleurodeles*.
1. *On the left,* a malformed component (hypomorph), and *on the right,* a diploid, 193 days old.
2. *On the right,* the diploid autosite, and *on the left,* the triploid hypomorph.
3. Diploid autosite with, *on the left,* very reduced hypomorph at the age of 158 days.
4. Haploid, aged 5 months, carried by a diploid.

Human grafting

Having examined in some detail the possibilities of grafting in animals we must now glance at its applications to the human species. Human grafting dates from very remote periods if we are to believe Wilford who claims that Indian priests long ago knew how to replace a lost nose with skin taken from the forehead or buttocks — a jealously guarded secret by which they retained power and influence.

Paul Bert, in his treatise on *Animal Grafting,* stated that Celsus and Galen also restored noses by means of grafting. But it was above all in Italy and Sicily where, on one pretext or another, numerous noses were cut off, that the art of restoring noses — or *rhinoplasty* — was born. Among the earliest of the 'nose-makers' were the Brancas of Sicily and Vianco, a Calabrian.

In the family of the Boianis, who lived in the Calabrian village of Tropea, the art of nose-grafting was handed down from father to son. It was doubtless a Boiani who taught it to Gasparo Taliocotius (1546—1599), an Italian physician who towards the end of the sixteenth century wrote a treatise on the subject *(de cavitarium chirurgia per insitionem,* Venice, 1597), and to whom the grateful citizens of Bologna erected in their amphitheatre a statue showing him with a nose in his hand.

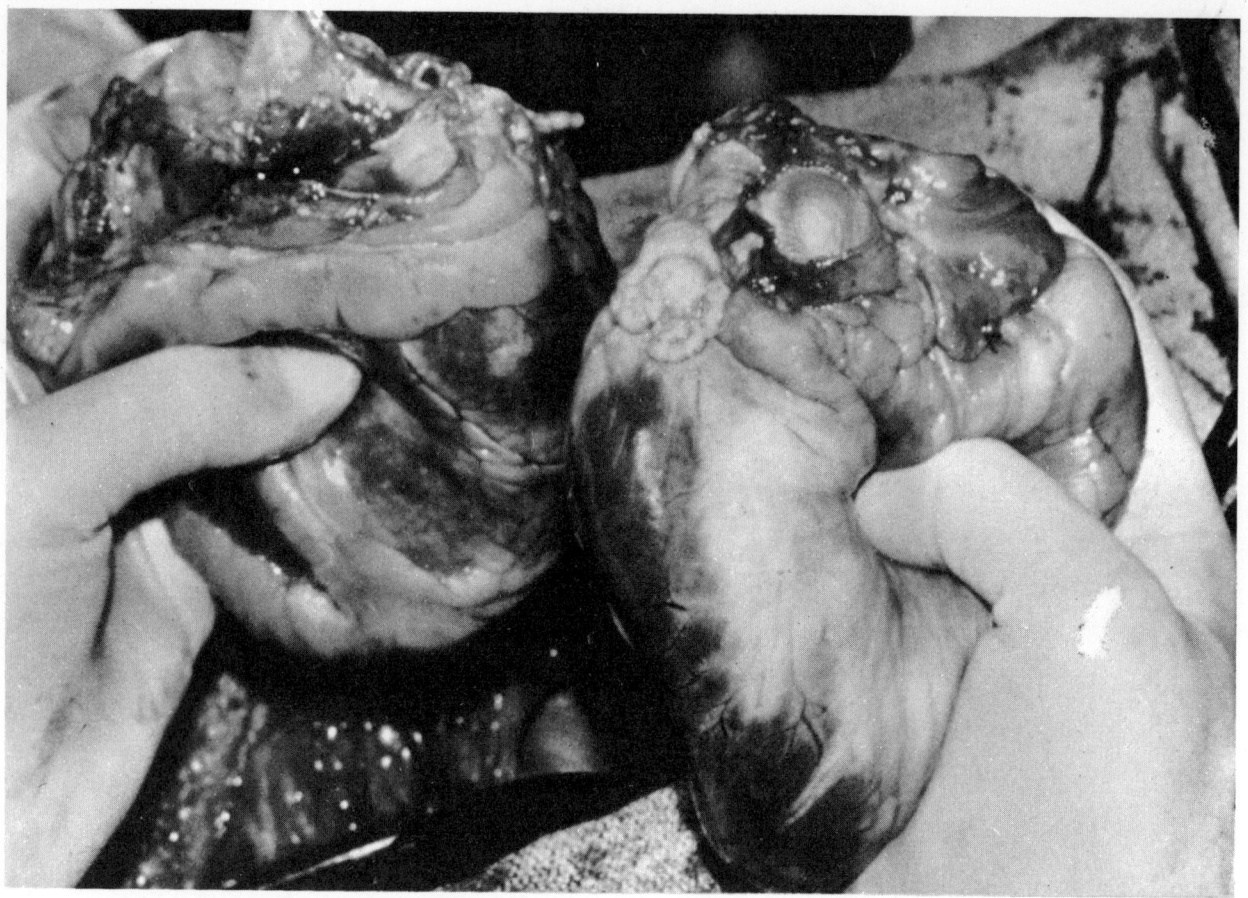

A photograph taken during a heart transplant operation. *On the left,* the ailing heart; *on the right,* the heart from the donor.

P. Popper.

The operation of grafting as it was practised by Taliocotius — also called Tagliacozzi — was, according to Fontenelle, so bold as to be scarcely credible; but skilled practitioners like Ficinus and Fabricius Hildanus were, it seems, witnesses to its success.

This is how Fontenelle described the operation: from the body of the man whose nose was to be restored a piece of skin was taken. The piece was removed from that part of the arm — the biceps — which is nearest to the nose when the hand rests on the top of the head. But, when the flap of skin was lifted each end of it was at first left attached to the arm. The end which was to form the upper part of the mutilated nose was then detached and sewn in place. Meanwhile the other end was left clinging to the arm and only cut off when the flesh of the first end was perfectly united with the flesh of the mutilated nose. Finally, the skin was fashioned into the shape of a nose, two nostrils were pierced, and it was joined by a second suture to the top of the upper lip.

The following reference was made by Francis Bacon (1561—1626) in one of his texts:

'One also speaks of certain individuals who, having a nose of enormous size or bizarre shape and wearied by the pleasantries which their deformity attracted, took the course of cutting off these excrescences, and afterwards having an incision made in the arm where they could be inserted and kept sewn up for a time, an expedient by the aid of which they procured for themselves an endurable nose.'

In the seventeenth and eighteenth centuries tales of noses being cut off and restored to position were a popular subject of mirth. Most educated men believed such stories were untrue and absurd. The 'nose-makers' did not escape the raillery of Voltaire ever eager to doubt and to mock. Taliocotius he discribed as the Etrurian Aesculapius who would make you a new nose by borrowing flesh from a poor man's buttocks. But, alas, on the death of the lender, your nose falls off and, Voltaire concludes, in the coffin is often justly restored to its proper place — the dead man's rump. In Voltaire's day it was naively believed that the graft followed the fate of the donor, so that when the donor died the graft dropped off the host.

It was only during the course of the nineteenth century that rhinoplasty began to be taken seriously and with it other forms of grafting. Today, as everyone knows, human grafting is current medical practice and its success is only limited by the obstacle created by biological individuality.

Among men, in fact, as among all mammals, the one graft which generally succeeds, at least with adults, is the autograft. A strip of skin cannot be transplanted from one person to another; but is is on the contrary easy to transplant skin from one part of the patient's body to another part: for instance, from the arm or thigh to the forehead or face. Such autografts are widely and successfully employed by specialists in facial plastic surgery.

Some glandular homografts — that of the thyroid, for example — appear to give hopeful results; but these must be attributed to the very slow resorption of the graft which thus sets free active elements. The glandular homograft is, in fact, not a true graft, but rather a kind of living organotherapy, or treatment by the administration of animal organs or their extracts.

Heterografts, such as the 'rejuvenating' operations performed by Serge Voronoff, are even less practicable. Voronoff, it may be remembered, grafted his patients with organs removed from great apes — chimpanzees. In view of all that is known today of the biological factors involved in grafting it is obvious that the organ of an ape cannot continue to function normally in the human organism, close though the blood and tissue relationship between man and great ape may be. It is, of course, possible that Voronoff's operations sometimes produced temporary stimulation, for the reason — organotherapy — which was mentioned in the preceding paragraph.

In order to avoid a not infrequent misunderstanding, it is as well to point out that blood transfusion, which is common practice between people who belong to the same 'compatible' blood groups and is discussed in a later chapter, is not in any way comparable to a 'blood graft': the foreign elements introduced by transfusion survive in the blood stream of the receiver only for a brief time.

This confusion between blood transfusion and grafting dates from Paul Bert himself who saw in transfusion 'a true graft, a liquid graft . . . in which the blood corpuscles play the chief role.'

Grafting between identical twins

The exceptional case in which homoplastic grafting is found to be consistently possible in man is that between identical twins, who are formed from the same egg and have the same hereditary factors. This exception actually confirms, the rule, for identical twins are, so to speak, the 'same individual of whom two copies have been printed'.

The graft of the kidney had often been attempted and, in spite of perfect surgical technique, always with complete failure until, in 1954, the operation was successfully performed in Boston between two identical twins, Richard and Ronald Herrick, aged twenty-four.

Richard Herrick had, as a consequence of scarlet fever, suffered from chronic nephritis, and towards the end of 1953 he was threatened with death by uraemia. It was then decided to attempt grafting him with a kidney removed from his twin brother.

The operation took place on 23rd December 1954. The two brothers were anaesthetized and one team of surgeons removed one of Ronald's kidneys while a second team prepared a place for the graft in Richard's abdomen. The vascular sutures presented difficulties. The ureter was joined up with the bladder. The double operation lasted in all for five and a half hours. Even during the following day the transplanted kidney showed signs of activity and excreted more than half a litre of urine. A few weeks later Richard showed no trace of uraemia; he was in perfect health and had gained more than 12 pounds. In March 1956 one of his diseased kidneys was removed and in June the other. The two brothers are at present in perfect health, and since this epoch-making operation nine renal grafts between identical twins have been successfully made in the United States.

Thus it might be said that the human being who possesses an identical twin is unusually fortunate, for, in case of accident or sickness, he has a handy reserve of spare organs — always assuming that his twin is generous enough to make the necessary sacrifice.

The case may be mentioned of Charles Madeira whose legs and abdomen were so seriously burned that only a skin graft could prevent his death. Happily he had an identical twin who flew half across the world to bring his brother the 300 square centimetres of skin which could save his life.

This tolerance to grafting between identical twins has sometimes been of legal value. In maternity wards it has on occasion occurred that two new-born identical twins have been attributed to different families. Thanks to a 'micro-graft' of the skin, however, their genetic identity can be recognized: if a bit of skin removed from one survives when grafted on the other it means that they are identical twins.

Intolerance towards a graft can sometimes be overcome by means of drugs and by irradiation, and in this way homografts have become possible between individuals who are not identical twins. As we have already seen, the study of homograft rejection and how it can be avoided is currently one of the most rapidly expanding fields of medical science.

Human embryonic grafting

The technique of brephoplasty — or embryonic grafting — is applicable to the human species, although it encounters the practical difficulty of obtaining the organs to be grafted, which must be removed from a human foetus or embryo immediately after death. It is, as May says, obviously impossible to foresee where or when these embryos will be available. Even in great cities where the probabilities are high that the various hospital services can supply them, their employment as graft donors is only possible if death was not caused by specific or infectious disease.

Nonetheless, remarkable successes may already be expected from this method of grafting, especially in the field of the endocrine glands.

The adrenal glands of a still-born child have been grafted on a woman suffering from Addison's disease, a very serious condition which results from adrenal insufficiency. Improvement was marked and persisted at least for several months, which suggested that a true graft had taken place and not simply the temporary survival of the grafted glands.

Embryonic parathyroids have been grafted with success on patients suffering from post-operative tetany — after the removal of a goitrous thyroid gland. The pituitary of a still-born baby has been grafted on a girl suffering from Simmonds' disease and foetal thyroids and pituitary glands on cretins and individuals with arrested development of the mongoloid type. May and Huignard successfully grafted thyroids from a still-born baby on a myxoedematic cretinoid child. A year after the operation the child had grown 6 centimetres taller and become considerably more intelligent, rising to tenth place in a class of thirty pupils. Foetal human skin has been successfully grafted on large burns of the arm, and such grafts have survived for several months.

Thus it is undeniable that a graft of still-born or foetal origin can be tolerated by an adult subject, be integrated into the adult's organic system and there fulfil its proper physiological function. Admittedly brephoplasty is by no means always successful. Its occasional failure, according to R.-M. May, is due to the fact that certain individuals display a more marked antagonism to the graft, though this intolerance can neither be foreseen nor explained.

Efforts have been made to improve the results of brephoplasty by cultivating the foetal graft for two or three weeks in the blood plasma of the individual destined to be its host. Foetal parathyroid glands thus treated have been grafted on patients suffering from tetany, and several total cures have been effected, especially among younger patients, that is, between sixteen and twenty-six.

Brephoplastic grafting of the kidney is not practicable, for the kidney of the new-born has no functional value to the adult organism. The method seems, however, to be full of promise for the grafting of the spinal cord, a graft which sometimes enables an organism to repair serious damage caused by exposure to intense irradiation.

The researches of Medawar on the acquired tolerance of embryos have not yet found medical application; although the advantages of creating such a state of tolerance in human babies — even with regard to a single future donor — have been seriously considered.

In any case it is reasonable to apply Medawar's conclusions as to the possible formation of 'cellular chimaeras' to mankind. At least one human being has been known to be a 'blood chimaera', a certain Mrs. McK., who at the age of twenty-five possessed blood which contained a mixture of two different types of blood-cell: A and O. This singularity was due to the fact that during her embryonic life she had shared the maternal uterus with a twin of male sex, who died at the age of three months. She was unaware that such a fraternal twin had existed until laboratory examination revealed the 'chimaeral' state of her blood. The fraternal twin, of blood group A, had supplied her with a certain number of blood-cells which became grafted on her bloodstream and continued, even when she became an adult, to manufacture A cells. Had this twin survived it is evident that a graft between him and Mrs. Mc K. would have been possible. From this case it is apparent that the graft-test for identical twins is not absolutely decisive. When an exchange of skin proves possible between two individuals the two are almost always identical twins; but there remains the remote possibility that they are only fraternal twins.

According to Medawar the blood of fraternal twin cows is almost always 'chimaeral', while in human beings the phenomenon is exceptional, occurring once in about one thousand cases.

The dead graft

The dead graft is so called because in such grafts the grafted tissue does not survive, or at least does not long survive, in the host organism. All the cells it contains eventually perish and disappear, but are slowly replaced by new cells, cells contributed by the organism itself. In other words the foreign graft serves as a framework or support: it is colonized, reinhabited, and brought to life again by native elements. Many grafts which were formerly supposed to be true living grafts are, in reality, grafts of this kind. Such, generally speaking, are grafts of tendons, arteries, and nerves.

In practically all these cases it is moreover possible to use, instead of a living graft, a graft in which all the cells have been previously destroyed by heat or cold, by alcohol, formaldehyde, etc. As Raoul-Michel May said, 'one grafts a piece of dead tissue and, after a while, one finds a living graft.'

In the case of dead grafts, heterografts are often found to be as effective as homografts. Thus the tendons of a dog can be grafted on a man. These grafts of animal donation are not, of course incorporated and adopted in their original state, but are 'revitalized' by the cells of the human host.

The practical applications of the dead graft have long been known, but it was the physiologist Nageotte who with much sagacity interpreted the processes which lead to the revivification of the graft. Most, if not all, bone grafting is a form of dead grafting. The grafted bone induces the process of osteogenesis, which results in the formation of new bone. Bones to be used as grafts are cleaned and sterilized, and then kept at a low temperature. 'Bone banks' — not to mention banks of blood vessels and tendons — are not, as it is sometimes mistakenly believed, reserves of living organs, but reserves of dead organs.

The graft of the cornea

Most people have heard of grafting the cornea — or keratoplasty. It is an operation that can be of benefit not only to the blind but to those who have lost their sight when, as the result of a burn, infection, or injury, the cornea loses its transparency and becomes opaque. It would seem that keratoplasty also belongs to the type of graft known as a dead graft: the cells of the grafted cornea do not remain alive in the eye of the host, but are little by little replaced by live cells which the host supplies.

Keratoplasty is, however, peculiar in that the cornea grafted must as a rule be of human origin. There is no need for the graft to be newly removed from the donor's body; on the contrary the operation appears

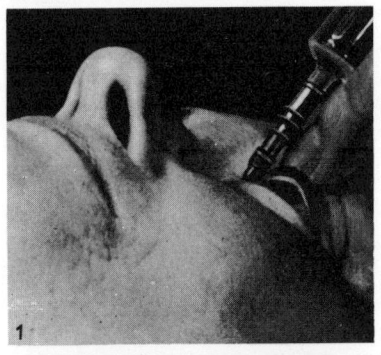 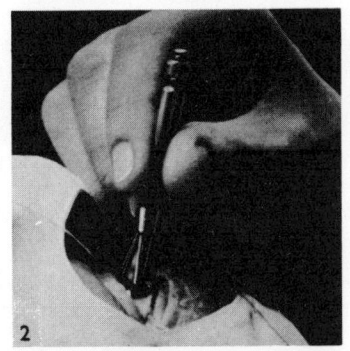

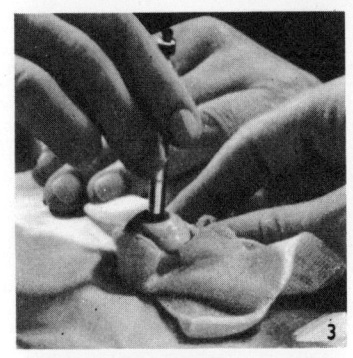 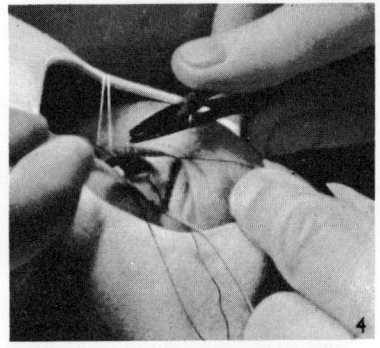 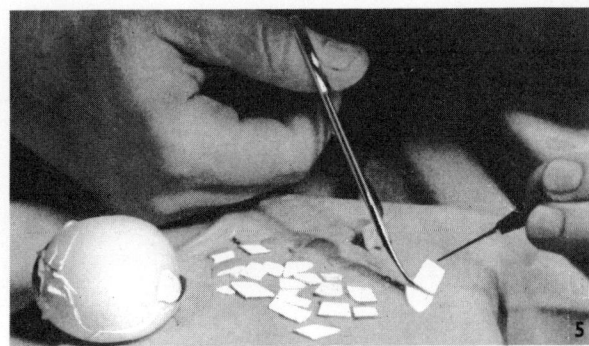

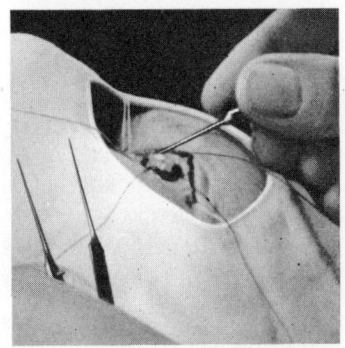 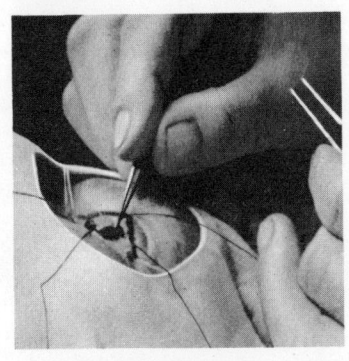

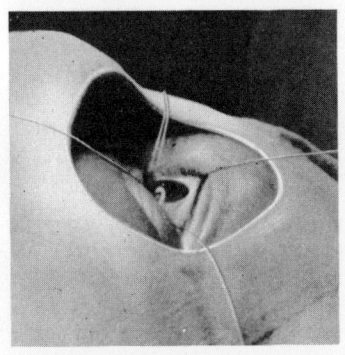 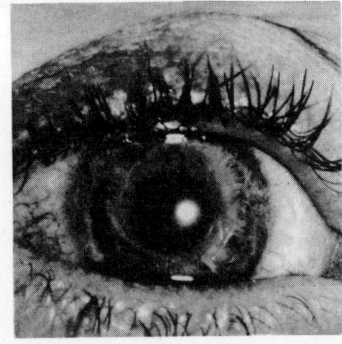

Keratoplasty, or graft of the cornea.
1. The eye is anaesthetized.
2. With the aid of a special punch a disc of the diseased cornea
is then removed.
3. A corresponding disc is removed from the healthy cornea
of a corpse.
4. Replacing the disc from the diseased cornea with the disc
removed from the healthy eye of the corpse.
5. Meanwhile, a small fragment of hard-boiled egg to be used
as a dressing is prepared.
Top left, a network of threads maintains the graft in position;
top right, the graft is covered with the prepared egg membrane;
above left, the eye is kept wide open by weighted scissors;
above right, the cornea is grafted by transfixion.

Renard, Hôtel-Dieu, Paris.

'Bone bank'. The bones, having been pulverized, are stored until required for grafting operations.

to be more successful when the graft has lost some of its vitality by being kept for a few hours at low temperature. Indeed, good results have been obtained with corneas frozen and then dried.

Apparently the idea of grafting the cornea first occurred to the French oculist Pellier de Quengzy. The technique of the operation has been vastly improved by modern practitioners, especially those of the Soviet school.

The operation takes place under local anaesthesia and lasts about half an hour. A disc of cornea from 3—6 millimetres wide has first been removed from the deceased 'donor's' eye with the aid of a special trepanning tool with a fine cutting point. A similar disc is now removed from the opaque cornea of the receiving 'host'; the graft is fitted into the space left vacant and maintained in position by silk threads, From fifty-five to sixty per cent of such operations are more or less completely successful.

It is legally possible in many countries for a person to 'bequeath' his eyes for medical purposes in the event of his death. Although it is sometimes said that a living being, after the graft of a cornea, sees with the eyes of a dead man, the phrase is, in fact, inaccurate, since it would appear that the graft does not entail cellular survival. Certain authorities, it is true, hold that the cells of the graft do survive in the hose, their survival being due to the protection against the defensive reaction of the organism which the corneal membrane affords.

The status of the individual

In the case of embryonic grafting we have seen that a human individual can adopt a foreign gland, benefit from its secretions, and with its aid survive, grow, and become more intelligent. Thus the individual in a sense lives with the collaboration of another; he exists because of an organic contribution another has made. In these circumstances it might seem legitimate to question whether he is still altogether himself. At the present time this violation of individuality, this assault on personal uniqueness by grafting, is unobtrusive; but it is easy to imagine that with the progress of science the fabrication of human chimaeras, of mosaic organs, or of tissues of varying origin, will become increasingly simple. It is not inconceivable that some day it will be possible to graft nervous tissue; and then what will be the position of psychic individuality?

The jurist Aurel David sought to reassure us on this point by declaring that the human person — on whom law and morality are founded — cannot be this carnal composite of spare parts science strives to create. Such a creature is, in his opinion, a mere 'protoplasmic robot' and as distinct from the true person as his furniture, jewellery, or motor car . . .

But where exactly does this true person reside, this person who cannot be menaced by grafts or by any other intervention of a material nature? The distinguished jurist does not tell us, or at least not in a manner which is satisfactory to the biologist.

Genetics

Heredity, or the resemblance of offspring to parents, is a concept universally accepted and a phenomenon daily observed. From the moment a child is born family and friends search for resemblances between the baby and its father, its mother, its grandparents, uncles and aunts. As the child grows up so its own characteristics become more marked, those which are like and those which are unlike the family's. The child is rarely an exact reproduction of either of its parents. Usually it reveals a mingling of the characteristics found in its direct progenitors and those observed in collateral branches of the family.

The transmission of hereditary features often seems capricious. This is especially so in pathological cases: for example, a half-forgotten anomaly or a disease from which a remote ancestor suffered will suddenly reappear. These resurgences reveal that the abnormality has not, as one had supposed, been eliminated. Ever present, such flaws remain hidden, but always form part of the family's hereditary patrimony. For this reason heredity was long considered to be a blind and mysterious force, obeying no law and scattering likenesses and unlikenesses to no apparent plan.

The livestock breeder and the gardener have a more accurate conception of heredity. They work with breeds of animals or varieties of plants whose characteristic qualities of form and physique remain unvaried and, as long as breed is crossed with breed and variety with variety, are regularly transmitted. To preserve a breed admired for its beauty or valued for its productivity — the quantity or quality of its meat or milk, for instance — cross-breeding with a different race must be avoided.

On the other hand, to acquire new and improved characters it is necessary to cross different breeds with that of the animal which possesses the character desired. But when the character is obtained every livestock breeder knows how difficult in may be to maintain it in subsequent generations.

Johann Mendel (1822—1884), the founder of genetics or science of heredity.

"Science et Vie", *Library of the Muséum national d'histoire naturelle. Larousse.*

The historical background

As early as 1683 Leeuwenhoek pointed out that when the wild grey male rabbits caught in the dunes bred with white, black, or blue females the little rabbits produced were always grey.

In the eighteenth century Maupertuis took great interest in heredity. He himself recounts the efforts he made to maintain a breed of Icelandic dog with slate-grey coat and yellow head, observing in the process phenomena of atavism or the reappearance of a former character which for several generations had disappeared.

261

In 1764 the botanist Koelreuter methodically crossed various species of tobacco, pinks, and henbane. He noted that such crossing resulted in hybrids which varied in many ways from the 'parental' types.

In 1826 Sageret crossed cantaloupes with another variety of melon and obtained hybrids which possessed the characters of the parental species in various combinations. For instance, one hybrid had yellow flesh and furrowed rind, two features of the cantaloupe, and white seeds and an acid flavour, two features of the other variety. These experiments showed evidence of the segregation of characters, that is, of characters acting independently of each other.

Another botanist known for his experiments in hybridization was Charles Naudin who worked in the *Jardin des Plantes* in Paris from 1854 to 1861. He crossed numerous and varied species — *Linaria, Datura,* tobacco — in an unsuccessful effort to obtain new and permanent species. He observed that while the hybrids of the first generation were uniform those of the second generation showed great diversity and sometimes a reversion to the original parental types.

Naudin grasped the fact that there must be a separation of characters in the gametes of hybrids, but he thought of it as a separation of two hereditary patrimonies, of two 'specific essences' temporarily united by hybridization. He considered the hereditary material as an indivisible block which was transmitted in its entirety. To envisage the transmission of specific characters, not as a block but separately, awaited the genius of Mendel. Mendel thought in terms of characters and not in terms of race or of the individual.

In Switzerland, Colladon, crossing grey mice with white mice, had already in 1824 obtained the fundamental data of heredity, but his results were not published.

In an empirical way the phenomena of heredity had long been made use of; it was much later that they were studied scientifically.

Mendel, the founder of genetics

Johann Mendel was born in 1822 in Heinzendorf bei Odrau, a little village in the former province of Moravia, on the farm which his parents worked. His father was interested in the cultivation of fruit trees. Johann was a brilliant student and the village schoolmaster encouraged him to pursue his studies. At the age of eleven he was sent by his parents to a secondary school in Leipnik, and after that to the school of philosophy in Olmütz. When he was twenty-one he decided on the monastic life for which his modest, peaceful and studious tastes seemed suited. He entered the Augustinian house of Saints Thoma at Brünn (now Brno) as a novice, and four years later as Father Gregory he was ordained priest. After two years further study at the University of Vienna he was appointed deputy professor at the monastery school at Brünn where for fourteen years he taught boys physics and natural history.

The researches which have immortalized his name took place in the monastery garden at Brünn. These researches date from 1856 when he began to cross different varieties of peas. At first a simple pastime, his experiments soon filled him with passionate interest and for eight years he persisted with them. Alone and totally unaided he artificially pollinated hundreds, of plants, cultivated and, with the greatest care, examined as many as twelve thousand specimens.

Monohybridism

Mendel's first experiments were concerned with monohybrids. That is to say he crossed two varieties of peas which differed in a single character: for example, a variety with a round seed and a variety with a wrinkled seed. He then observed the behaviour of these two opposite characters, round and wrinkled, in the hybrids.

In this manner he undertook the examination of seven pairs of contrasting characters: the form of the seed, *round* or *wrinkled;* the colour of the cotyledons, *yellow* or *green;* coloration of the flower, *white* or *red;* position of the flower, *axillary* or *terminal;* shape of the ripe pod, *straight* or *moniliform;* colour of pod, *green* or *yellow;* length of stem, *short* or *tall.*

The pea, it should be remembered, is a self-fertilizing annual in which the flower, while still a bud, is fertilized by its own pollen. Thus to bring about a cross between two varieties of peas, A and B, it is essential to prevent this natural self-fertilization in one flower and to replace it by artificial cross-fertilization with pollen from the other. To do this the stamens of variety A flower are removed before they reach maturity and, at a suitable time, its pistil, or more accurately its stigma, is brushed with pollen from a flower of variety B. Inversely the flower of variety B, deprived of stamens, receives pollen from a flower of variety A.

Mendel's historic cross was made between a variety of pea with round seeds and a variety with wrinkled seeds. Before the experiment Mendel had carefully verified the fact that each of the varieties always produced peas of the same form; in other words, that the strains were perfectly pure and stable — a state incidentally which the pea's self-fertilization makes easier to obtain.

The cross produced hybrid peas, and all of them were round. In appearance they were all alike and identical to the seeds of one — the round — parent plant. The character *round* was therefore called *dominant.* Such were the peas of the first generation — usually designated by the abbreviation F_1, meaning first filial generation. When sown they gave rise to hybrid plants whose flowers were self-fertilized in the normal manner, and produced 7,324 peas of which 5,474 were round and 1,850 wrinkled. This proportion of 5,474 round to 1,850 wrinkled peas reduces to 2.95 to 1, or very nearly three to one.

The peas of this second generation, F_2, three-quarters of which were round and one-quarter wrinkled, were thus no longer all alike in appearance. The wrinkled peas possessed a character which had disappeared from F_1, and resembled one of the grandparents. They were a throwback and the character *wrinkled,* hidden for one generation, was therefore

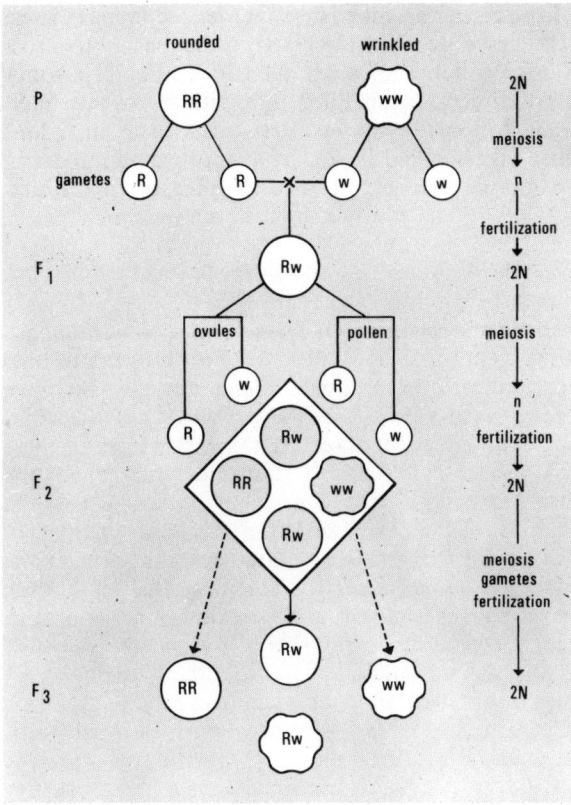

Mendel's classic experiment: crossing round and wrinkled peas.

After Darlington and Mathers.

characters, round and wrinkled, were in fact present, but only one — masking the other — was visible. This illustrates the law of dominance.

To explain the second generation it was necessary to assume that in the hybrids of F_1 the gametes or germ-cells have been inherited in equal numbers from both parents, that half are therefore pure round and half are pure wrinkled. This is Mendel's second law, the law of the segregation or divorce of characters.

These gametes will then combine according to the laws of chance. In the course of the hybrids' self-fertilization four combinations are possible:

(a) male gamete (round)   X female gamete
(b)   ,,   ,,   (round)   X   ,,   ,,
(c)   ,,   ,,   (wrinkled)   X   ,,   ,,
(d)   ,,   ,,   (wrinkled)   X   ,,   ,,

(round)   → egg (round/round)
(wrinkled)   → ,, (round/wrinkled)
(round)   → ,, (wrinkled/round)
(wrinkled)   → ,, (wrinkled/wrinkled)

These hybrids give rise to offspring composed as follows. (a) ¼ plants bearing round peas of pure strain which will always produce plants with round seeds. (b and c) ½ plants bearing round (dominant) peas which however possess the (recessive) character wrinkled. They always produce in the same proportion three plants with round peas to one plant with wrinkled peas. ¼ plus ½ or ¾ of the plants have round peas which are in reality different but not visibly identifiable. (d) ¼ plants bearing wrinkled peas of pure strain which will always produce plants with wrinkled seeds.

It has been found convenient to represent dominant characters by a capital letter and recessive characters by a small letter. The initials of the dominant and the recessive characters may be used: for example, in the cross just described R would represent the character round and w the character wrinkled. Alternatively, the initial, in lower case, of the recessive character may be used and the same initial, in upper case, used for the dominant character: thus the above example would be w again for wrinkled but W for round.

Mendel pointed out that each gamete possessed one or other of the characters R or w. Thus the fertilized egg formed by the union of two gametes would be represented by two letters. The pure and true-breeding peas of the parental generation would be symbolized as RR or ww; their gametes would be either R or w and, when the gametes united to form hybrids, the formula would be Rw. The Rw hybrids would produce two kinds of gametes in equal numbers, namely R and w. The F_2 would then be composed of the possible combinations between these two kinds of gametes. Thus:

female gametes		*male gametes*	
R		R	w
w		RR	Rw
		Rw	ww

The diagram shows how one plant bears round RR peas of pure and stable strain, how two hybrid plants

called *recessive*. These wrinkled peas, when sown, always grew into plants which produced wrinkled peas.

On the other hand, the round peas of F_2, when sown, did not all behave alike. One-third of the round seeds — which, as we have seen, comprised in all three-quarters of the peas obtained in F_2 — grew into plants which always produced round peas, while the remaining two-thirds grew into plants which produced round and wrinkled peas in the same proportion of three round to one wrinkled — though this numerical proportion could only be observed when the number of peas was sufficiently great.

Repeated crossings with the seven pairs of contrasting characters all gave analogous results. In F_1 the following characters appeared: yellow cotyledon, white flower, axillary flower, straight pod, green pod, tall stem. These characters, then, were dominant and their contrasting characters were recessive. The plants of F_2 also were numerically composed in the same way, the proportion being three plants which displayed the dominant character and one plant which displayed the recessive character.

The consistency of these figures deeply impressed Mendel and, analysing them, he became the first biological statistician. He grasped the fact that the characters behaved like discrete units which were transmitted unchanged from generation to generation. In the peas produced by the parental cross the two

bear round *Rw* peas whose offspring will be heterogeneous, and how one plant bears wrinkled *ww* peas of pure and stable strain. These symbols or genetic formulae tell us the genetic constitution of the individual; they represent the individual's hereditary constitution or *genotype*.

The two contrasting characteristics, round/wrinkled are known as *alleles,* and are said to be *allelomorphs* of each other.

The true-breeding *RR* or *ww* plants, which are represented by two capital or two small letters, are called *homozygotes. RR* is the dominant homozygote and *ww* is the recessive homozygote. Homozygotes of course produce only one type of gamete, *R* or *w*.

The *Rw* hybrid plants, represented by a capital and a small letter, are known as *heterozygotes*. Heterozygotes produce two kinds of gamete, *R* and *w*.

The dominant character always makes its appearance in a dominant homozygote *RR* or a heterozygote *Rw*. The recessive character only appears in the recessive homozygote *ww*.

The peas *RR* and *Rw* are round and it is impossible to tell them apart from their appearance. They are said to be of the same *phenotype*, the round phenotype. Round *RR* peas and *Rw* peas are thus of the same phenotype but of different genotype.

Crossing two *RR* dominant homozygotes will produce descendants of pure strain composed exclusively of *RR* individuals. The same is true when two recessive *ww* homozygotes are crossed. All subsequent generations will be alike and of pure strain.

Crossing two *Rw* heterozygotes will, however, give rise to heterogeneous descendants: the generations which follow will not be alike.

Crossing two varieties which differ only in a single pair of contrasting characters or alleles is known as *monohybridism*.

Dihybridism

After his experiments in monohybridism Mendel undertook further experiments in dihybridism, that is, he crossed varieties of peas which differed in two pairs of contrasting characters.

For instance, he crossed a variety of pea with round seeds and yellow cotyledons with a variety which had wrinkled seeds and green cotyledons. The first generation, F_1, was exclusively composed of plants with round peas and yellow cotyledons. The characters round and yellow were dominant. F_2 was composed of plants with peas resembling the parental types, but in addition there appeared new combinations — plants with round seeds and green cotyledons and plants with wrinkled seeds and yellow cotyledons.

To interpret this result Mendel assumed that each pair of characters (alleles) behaved as though it were independent. The characters segregated and assorted freely. This is the third law of heredity, and it means that gametes corresponding to the various possible combinations exist in equal numbers and combine at random.

Using the symbols described above, the parental seeds could be represented as *RRYY* and *wwgg*. These plants form *RY* and *wg* gametes which when pollinated produce *RwYg* hybrids with round peas and yellow cotyledons.

Each of these hybrids will form gametes of four kinds and each gamete will contain one character of seed-shape and one character of cotyledon-colour. The four possible gametes are *RY, Rg, wY, wg*, and they exist in equal number. By self-fertilization the four types of female gametes can be pollinated by the four types of male gametes. Thus sixteen combinations are possible and the following plants will be produced:

9 plants with round peas and yellow cotyledons,
3 plants with round peas and green cotyledons,
3 plants with wrinkled peas and yellow cotyledons,
1 plant with wrinkled peas and green cotyledons.

If a sufficient number of peas is grown the above numerical proportion — 9, 3, 3, 1 — appears. It will be noted that the proportion 3 to 1 previously observed in monohybridism is contained in this new numerical relationship: there are 9 plus *3* or 12 round peas, and 3 plus *1* or 4 wrinkled peas; there are 9 plus *3* or 12 yellow cotyledons and 3 plus *1* or 4 green cotyledons.

Of the sixteen peas only two — the doubly dominant homozygote *RRYY* and the doubly recessive homozygote *wwgg* — will produce the pure-breeding parental types. The other fourteen peas are intermediates of varied genotypes.

Such, much simplified, were the researches of Mendel, whose intuition had led him to the discovery of hereditary discontinuity, and whose mind, unhampered by the concepts of race and individual, grasped the essentials of hereditary factors. He was well served by his mathematical skill which enabled him to analyse

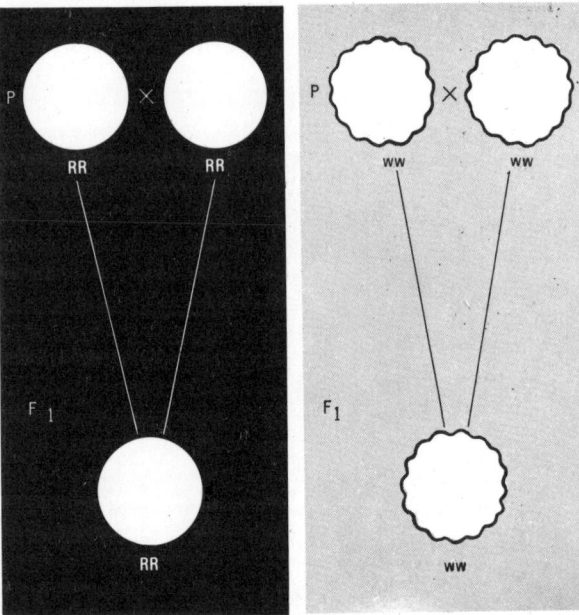

Mendel's experiment of crossing peas.

not only the qualitative but also the quantitative aspect of his results, by his inexhaustible patience, and finally by his choice of the pea which lent itself perfectly to his experiments.

Mendel published his findings in 1865 in the journal of a small local society. They aroused no interest and passed totally unnoticed even by men of science like Nägeli, who was then professor of plant physiology in Munich. His work was too original to be understood or appreciated by minds unprepared to accept such startling conclusions. Thirty-five years were to elapse before full justice was rendered to the founder of modern genetics.

For Mendel was not a precursor of the science of heredity: he was its true creator. His experiments were the point from which the science developed some three decades later. Whether Colladon and his experiments with mice some thirty years earlier had any influence on Mendel is not known.

The rediscovery of Mendel's laws

In 1900, working quite independently, three botanists, the Dutchman Hugo de Vries, the German C. Correns and the Austrian E. von Tschermak, rediscovered the laws of plant hybridization. At the same time two zoologists, the Englishman Bateson and the Frenchman Cuénot, applied the laws to animals.

Mendel's name was rescued from oblivion, and in honour of his genius the laws of hybridization became Mendel's laws, the character-unit became the Mendelian characteristic, and the new theories were called Mendelism. By the turn of the century the progress of biology, especially newly acquired knowledge of the mechanism of fertilization, had prepared minds to receive the idea that heredity could be controlled by discrete factors.

The new science was immensely successful and at once made considerable strides. Mendel's laws were found to apply to a large number of plants and animals and to govern numerous transmissible characters — sweet peas, antirrhinums, marvels-of-Peru, Chinese primulas, maize, various rodents including rats, mice, and guinea-pigs, poultry, insects, crustaceans, molluscs . . . even the human race furnished significant examples, especially in the field of abnormal characters and certain hereditary morbid conditions. The interest of doctors and psychologists was awakened by these discoveries, which clearly demonstrated the immense importance of heredity for a fuller understanding of man's moral and physical nature.

In 1906, during the course of the third international conference of the Royal Horticultural Society, Bateson coined the word *genetics* as a name for the new science which seeks to elucidate the phenomena of heredity.

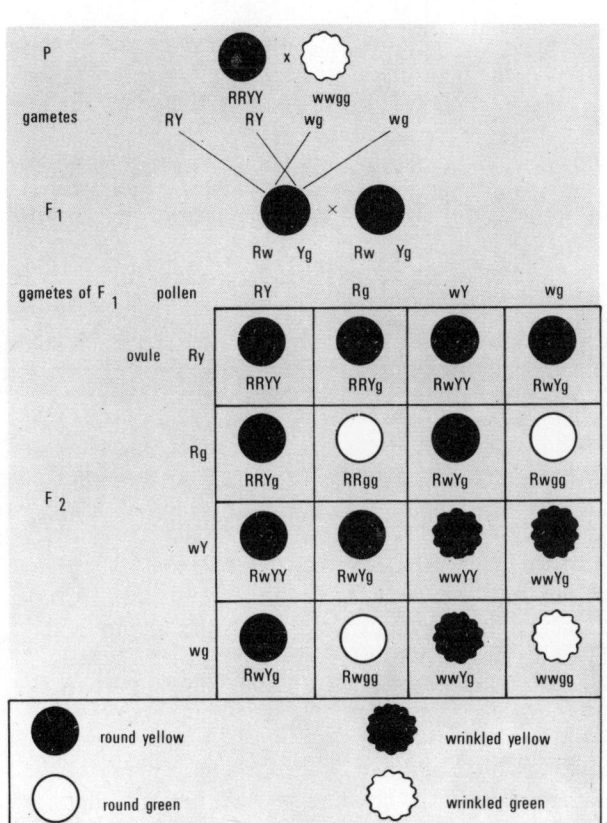

Mendel's experiment of crossing round, yellow peas with wrinkled, green peas.

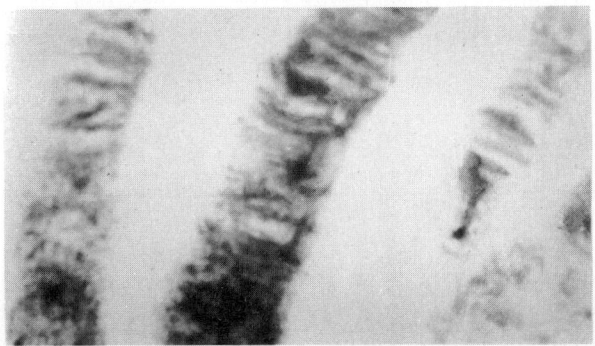

The 'giant' chromosomes of the salivary glands of Diptera.
H. Mugard.

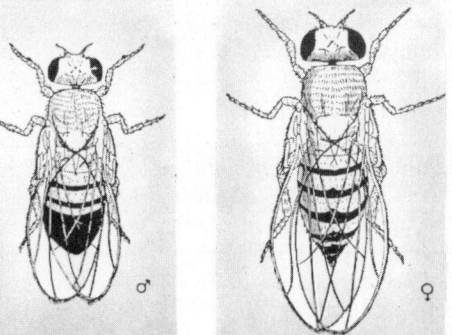

Drosophila melanogaster. The male and, *on the right,* the female.
After T. H. Morgan, 1932.

265

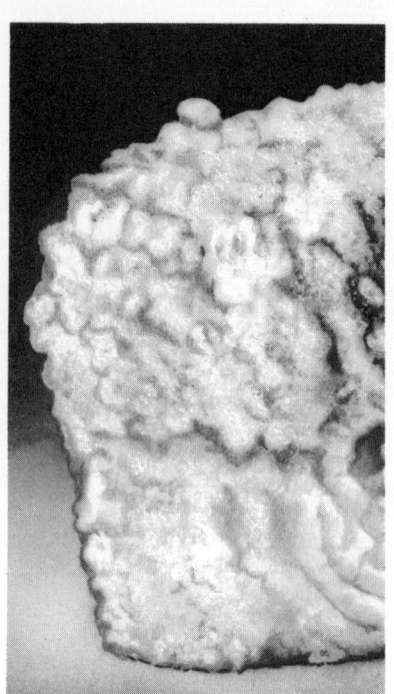

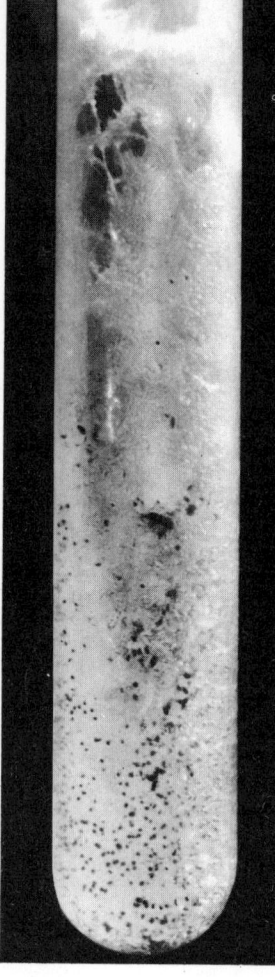

Left, a fragment of bread, putrified by a colony of *Neurospora sitophila.*

Larousse.

Right, culture of *Neurospora tetrasperma* on an agar-agar medium. The perithecia is at the base of the test-tube and the conidia at the top.

Muséum national d'histoire naturelle, Paris.

Modern genetics and the fruit fly

From 1900 genetics developed rapidly and many new facts were acquired. Then, after 1910, the face of the new science was changed by the entrance on the scene of a new animal — a very small fly, the fruit fly, or, more accurately, *Drosophila melanogaster.*

It was Thomas Hunt Morgan, in New York, who hit upon the idea of choosing this unexpected 'guinea-pig'. The advantages it offered to genetical research were many: it is an insect easily procured; swarms of the little flies can be found on slightly over-ripe fruit. Its cycle of development is short; in the laboratory forty generations a year are obtained. The species is prolific; a single couple produces 200—300 offspring. Rearing is extremely simple and takes place in special jars of 100 cubic centimetres capacity in which a standard nutritive medium is enclosed. The morphological

differences between male and female are clear cut, which enables the two sexes to be quickly and easily identified. Microscopic examination presents no difficulties. There are only four pairs of chromosomes, that is, eight in all. The salivary glands — as among Diptera — have giant chromosomes which can be studied with particular ease. Some 2,000 generations of *Drosophila,* comprising millions and millions of individuals, have been reared. In comparison, since the days of the Egyptian pharaohs, there have been fewer than 200 human generations.

The unchallenged ascendancy of *Drosophila,* however, may be in decline. Various investigators are experimenting on the ascomycete fungus, *Neurospora.* Broadly speaking, present-day research is turning towards micro-organisms, new material which can perhaps offer even greater opportunities to the research geneticist.

The application of Mendel's laws to animals

In the cases of cross-breeding discussed below it should be understood that the parental generation is always composed of individuals of pure strain, that is, of individuals which breed true. In addition the animal's sex will not be given, as the sex of the individual has no bearing on this type of cross.

Cross between two animals which differ by one pair of allelomorphic characters

Crossing a wild grey mouse GG with an albino, or pigmentless, mouse aa

This cross produces a first generation composed of grey mice *Ga,* the character grey being dominant. When these grey hybrids are crossed among themselves the offspring are both grey and albino, in the proportion of three grey mice to one albino mouse.

The albino mice, crossed among themselves, always produce albino mice: the strain breeds true.

The grey F_1 mice are all of the same phenotype but not all of the same genotype, though the difference cannot be visibly recognized. In order to distinguish between them all that is needed is a *backcross,* that is to say, a cross with one of their parents, in the present case with the albino parent. When this is done some of the grey mice will, with the albino parent, produce grey mice only. They are, then, genetically *GG,* for when crossed with the *aa* albino parents the offspring were grey heterozygotes *Ga.* Other grey mice, with the albino parent, will engender grey and albino mice in equal numbers. These, then, are grey heterozygotes *Ga* which, coupled with *aa* albinos, therefore produce grey heterozygotes *Ga* and albinos *aa.*

The first cross is analogous to Mendel's historic crossing of peas, and is an illustration of complete

The fruit fly *Drosophila melanogaster*. This small insect is very easy to rear and was used by the American biologist T.H. Morgan to demonstrate the chromosome theory of heredity. Hereditary variations, or mutations, are numerous in *Drosophila* and affect, among other characters, the colour of the eyes which in the normal or wild type are brick-red.

The mutation 'white eyes' is characterized by the absence of pigmentation of the eyes. In crosses it is dominated by the normal brick-red colour and is transmitted according to the laws of sex-linked characters.

In the so-called 'cinnabar' mutation the eyes are of a special red which is brighter than the normal brick-red eye. In crosses the cinnabar mutation is also dominated by the normal colour.

R. H. Noailles.

267

dominance. It can be shown thus:

Parents	Grey Mice GG X Albino mice aa

Parents	Grey Mice GG	X Albino mice aa
Gametes	G G	a a
F_1 (all grey)	Ga Ga X	Ga
Gametes of F_1	G a	G a
F_2	G X $G \rightarrow GG$	
	G X $a \rightarrow Ga$	
	a X $G \rightarrow Ga$	(all three grey mice)
	a X $a \rightarrow aa$	(albino mice)

Crossing a hen with normal plumage nn, and a cock with curly plumage CC

This cross produces a first generation composed of fowl whose plumage is semi-curly. Thus the curly plumage character is not completely dominant. It is as though the intensity of the character has been diminished by the presence of the normal character. The curly plumage character illustrates incomplete or semi-dominance. The F_1 hybrids, crossed among themselves, produce fowl with very curly feathers, moderately curly feathers, and normal feathers in the proportion 1, 2, 1.
This can be seen thus:

Parents	Normal hen nn X	Curly cock CC
Gametes	n n	C C
F_1	Cn Cn X	Cn
		(all semi-curly)
Gametes of F_1	C n	C n
F_2	C x $C \rightarrow CC$ (curly)	
	C x $n \rightarrow Cn$ (semi-curly)	
	n x $C \rightarrow Cn$ (semi-curly)	
	n x $n \rightarrow nn$ (normal)	

In cases of incomplete dominance, which are fairly infrequent, the difference between the dominant homozygote CC (with very curly feathers) and the heterozygote Cn (with semi-curly feathers) is, of course, apparent to the eye.
Another example of semi-dominance is supplied by crossing the red and white varieties of *Mirabilis jalapa* or marvel-of-Peru.

Cross between two animals which differ by two pairs of allelomorphic characters

Crossing a grey mouse with normal gait GGNN, with an albino waltzing mouse aaww

When these two kinds of mice are crossed the first generation is composed exclusively of grey mice of normal gait, $GaNw$. These dihybrids when crossed among themselves engender polymorphic offspring comprising grey mice of normal gait, grey waltzing mice, albino mice of normal gait and albino waltzing mice in the proportion 9, 3, 3, 1.
The dihybrids form in equal numbers four types of gametes: GN, Gw, aN, aw. Thus in fertilization sixteen combinations are possible as indicated in the following table:

		Male Gametes			
		GN	Gw	aN	aw
Female	GN	$GGNN$	$GGNw$	$GaNN$	$GaNw$
Gametes	Gw	$GGNw$	$GGww$	$GaNw$	$Gaww$
	aN	$GaNN$	$GaNw$	$aaNN$	$aaNw$
	aw	$GaNw$	$Gaww$	$aaNw$	$aaww$

Among these sixteen combinations there is only one doubly dominant homozygote $GGNN$ and only one doubly recessive homozygote $aaww$ which is identical to the two grandparents. The other fourteen combinations are homozygous for one character or heterozygous for both characters. Grey mice with normal gait are of the genotypes $GGNN$, $GGNw$, $GaNN$, $GaNw$. Grey waltzing mice are of the genotypes $GGww$, $Gaww$. Albino mice of normal gait are of genotypes $aaNN$, $aaNw$. The albino waltzing mice are of genotype $aaww$.

Cross between two animals which differ by three pairs of allelomorphic characters

Crossing a guinea-pig with black, short, curly hair BBSSCC and a guinea-pig with albino, long, uncurly hair bbsscc

The first generation is composed exclusively of guinea-pigs with black, short, curly hair $BbSsCc$, these three characters being dominant, while the contrasting characters white, long, and uncurly are recessive. These trihybrids crossed among themselves produce a polymorphic second, F_2, generation composed of:

27	guinea-pigs with	black, short, curly hair
9	,,	black, short, uncurly
9	,,	black, long, curly
9	,,	albino, short, curly
3	,,	black, long, uncurly
3	,,	white, short, uncurly
3	,,	white, long, curly
1	,,	white, long, uncurly

Such trihybrids form in equal numbers eight kinds of gametes BSC, BSc, BsC, bSC, Bsc, bSc, bsC, bsc. Fertilization is thus capable of producing eight times eight or sixty-four combinations. Among these sixty-four combinations there is only one tri-dominant homozygote $BBSSCC$ and only one tri-recessive homozygote $bbsscc$ which is identical to the two grandparents. Six new combinations have appeared: black, long, uncurly — black, long, curly — white, long, curly — black, short, uncurly — white, short, uncurly — white, short, curly. These diverse phenotypes correspond to diverse bi-dominant or mono-dominant genotypes.

Cross between two animals which differ by four, five, and more pairs of allelomorphic characters

As the number of allelomorphic characters increases so the number of different gametes produced increases and, therefore, the greater will be the number of combinations possible when gamete meets gamete in fertilization.

Actually the number of gametes and the number of combinations which result from their fusion can be calculated by a simple mathematical formula. The number of pairs of characters (n) will yield 2^n types of gametes and $(2^n)^2$ possible combinations. For example, with seven pairs of allelomorphic characters there will be 2^7 (or 128) kinds of gametes, and 128^2 (or 16,384) possible combinations. But crosses involving large numbers of characters are of little experimental value, for far too many generations are necessary in order to obtain all the possible combinations.

The various crosses we have examined, as well as countless others which have been made among both plants and animals, are replicas of Mendel's original experiments with peas, and further illustrate his three fundamental laws of heredity which may thus be summarized:

1. The law of dominance.
2. The law of segregation or separation of characters.
3. The law of independent assortment, or chance distribution of alleles to the gametes (ova and spermatozoa).

The statistical nature of the Mendelian laws

The Mendelian laws of heredity are universal and apply to all living things, plants, animals, and man. It is important to remember, however, that they are statistical laws, that is to say, the numerical proportions they predict are accurate only when we are dealing with large numbers of individuals. In this connection it may be helpful to recall the simple principles on which such laws are based.

If a coin is tossed into the air it will fall heads or tails, and the probability that it will fall heads is of course equal to the probability that it will fall tails, namely one half. But if the coin is tossed four times it will not necessarily fall heads twice and tails twice. It can fall: heads four times and tails not at all, heads three times and tails once, heads twice and tails twice, heads once and tails three times, and finally heads not at all and tails four times. There are thus five possibilities, although the probability of their occurrence differs. Now if we toss the coin not four times but 100 times the probability that the results more nearly approach the theoretical average of fifty heads and fifty tails will increase. Heads may, for instance, turn up

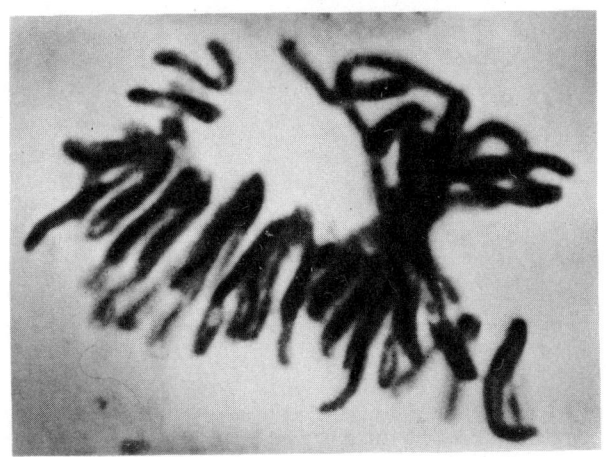

Chromosomes of a triton.

H. Mugard.

forty-seven or forty-eight times and tails fifty-three or fifty-two times. If the experiment is repeated 1,000 times the average of fifty per cent heads and fifty per cent tails will be still more nearly approached, and in general the greater the number of times the coin is tossed the nearer the results obtained will be to the calculated probability of a half.

Mendelian proportions behave in exactly the same way. Crossing a grey mouse GG with an albino mouse aa will, in F_2, produce $\frac{3}{4}$ grey and $\frac{1}{4}$ albino mice. This proportion will be accurate enough when as a minimum 100 mice are involved — roughly 75 (say, 73, 74, 76, 77) will be grey and 25 (27, 26, 24, 23) will be albinos.

If the F_2 offspring consists of only twelve mice these proportions (three to one) will not necessarily be found. The twelve may, by chance, consist of the theoretical nine grey mice and three albino. But the following numerical combinations are also possible.

12 grey mice to	0	albino mice
11	,,	1	,,
10	,,	2	,,
9	,,	3	,,
8	,,	4	,,
7	,,	5	,,
6	,,	6	,,
5	,,	7	,,
4	,,	8	,,
3	,,	9	,,
2	,,	10	,,
1	,,	11	,,
0	,,	12	,,

The probability of any one of these combinations appearing naturally varies: the probability of the combination of 0 grey mice and 12 albino mice appearing is obviously slight.

In the case of our cross between guinea-pigs differing by three pairs of allelomorphic characters the same mathematical principles are of course involved. The tri-recessive homozygote *aalluu* has a one in sixty-four probability of occurring. This is small when compared

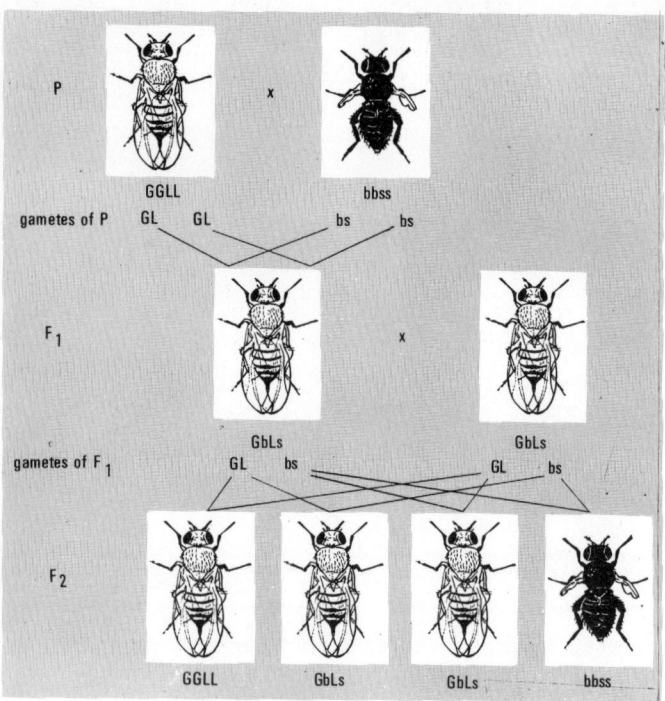

Diagram illustrating the cross between grey, long-winged *Drosophila melanogaster (GGLL)* and black, short-winged *Drosophila melanogaster (bbss)*.

Diagram of the cross between grey, short-winged *Drosophila melanogaster (GGss)* and black, long-winged *Drosophila melanogaster (bbLL)*.

with the twenty-seven in sixty-four probability which the black, short, curly tri-dominant homozygote has of occurring. A very large number of descendants is essential before the theoretical proportions — 27, 9, 9, 9, 3, 3, 3, 1 — can be closely approached.

The chromosome theory of heredity

The second law of heredity — that allelomorphic characters segregate or separate into different gametes and thence into different offspring — presupposes the existence of discrete discontinuous hereditary material, in other words, of hereditary units or factors. The concept of hereditary units is not purely intellectual but is imposed by experimental evidence.

What, then, are these hereditary units? They are called *genes,* or hereditary factors, and are carried on the chromosomes of the cell nucleus. They are material particles capable of segregation, and the totality of inherited genes constitutes the organism's genotype.

In all the preceding crosses it was a matter of indifference whether a given character was contributed by the male or by the female: the results were identical. This general observation implies a strict equality between sperm and ovum. Ova and spermatozoa, however, differ profoundly in size and shape: they possess only one element in common — the nucleus. In the nuclei only the chromosomes are identical in form and number. During cell-division or mitosis the chromosomes are, it will be remembered, first duplicated and then separated, so that one goes to each daughter

nucleus, with the result that both daughter nuclei have the identical chromosome, and hence gene, complement to the cell from which they derive. There is no intermingling of chromosomes and they preserve their unity. In this way hereditary factors remain constant.

With the aid of this chromosome theory of heredity the crosses we have examined are easy to interpret.

Let us re-examine the first cross, that of grey and albino mice, and assume that the grey gene G and the albino gene a are situated in a pair of homologous chromosomes, homologous chromosomes being pairs of chromosomes which contain identical sets of *loci,* or gene-positions. The mice of pure grey strain have their two G genes in one pair of homologous chromosomes; the albinos possess the same pair but the corresponding gene-position is occupied by two a genes.

Following meiosis (or reduction division, which results in the production of germ-cells) the male gametes have only a single chromosome with a G gene, while the female gametes have a single chromosome with an a gene. (Or, of course, vice versa.) When the two gametes fuse in fertilization an egg or zygote will be formed having two homologous chromosomes, one of which will bear the G gene and the other the a gene. The mice born of such eggs will be Ga grey hybrids. These hybrids will form two types of gametes in equal number, one type bearing the G gene and the other the a gene. During fertilization the two types will unite in the four combinations possible, namely, GG, Ga, aG, aa. If the number of individuals crossed is sufficiently high the numerical proportion of the dominant G to the recessive aa will be three to one.

By localizing hereditary factors or genes in the chromosomes we thus obtain an exact picture of how

1

2

Animal mutations.
1 and 2. The Long Evans breed of piebald rat — normal and two 'Hairless' spontaneous mutants.
3 and 4. Mice of the Strain C57 Bl/6, one with a black coat, the other piebald.
5 and 6. Albino rat (6, *right*) and hairless spontaneous mutants.
7 and 8. Normal mouse (weighing 20 to 25 grams) and two obese mutants (weighing 45 to 50 grams).
9. Flame-coloured guinea-pig with wavy fur.

3

4

9

5

6

7

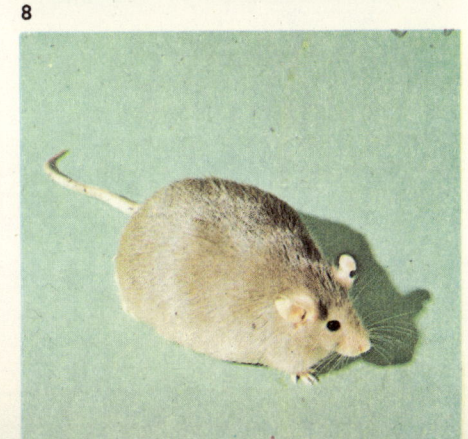

8

271

characters segregate. The more complex crosses of dihybridism, trihybridism, etc., can be explained in the same way by placing the various pairs of allelomorphic characters (alleles) in different pairs of chromosomes. Two pairs of alleles are thus localized in two pairs of chromosomes, or three, in three pairs.

Linkage — the association of two or more genes on the same chromosome

The independent segregation of genes which is illustrated in all the preceding crosses is only possible if the genes are situated in different pairs of chromosomes. Thus it would appear that the number of hereditary factors or genes could not be greater than the number of chromosome pairs. This, however, cannot be so, for genes are vastly more numerous than the number of chromosome pairs. In *Drosophila*, for instance, more than 1,000 gene alleles have been found, while the insect has only four pairs of chromosomes. Among the animals and plants which have been studied, including guinea-pigs, mice, rabbits, rats, chickens, maize, peas, etc., the same fact emerges: they all possess more pairs of alleles than pairs of chromosomes.

How is the apparent difficulty to be explained? It is obvious that if several genes are situated in the same chromosome they must participate in all moves that the chromosome makes and will thus act as a unit, or like a single pair of alleles having more than one effect. Put in another way, genes localized in the same chromosome must be totally associated or *linked* and passed from generation to generation as a group.

If we cross a *Drosophila* with a grey body *GG* and long wings *LL* with a *Drosophila* with black body *bb* and short wings *ss* the first filial generation will be composed of hybrid flies with grey body and long wings. Thus the genes for grey body and long wings are both dominant. These F₁ hybrids when crossed among themselves give rise to a second generation composed of grey flies with long wings and black flies with short wings in the three to one numerical proportion which is characteristic of a cross between two individuals which differ by only one character. In this cross between the two *Drosophila*, then, it is as though only a single pair of characters were involved. This is only possible if the two characters, grey body and long wings, are situated on the same chromosome, and the characters black body and short wings are both situated on the corresponding homologous chromosome.

The cross can be represented in this way:

Parents   Grey *Drosophila*   X   Black *Drosophila*
             with long wings   with short wings
             GGLL   *bbss*
Gametes: *GL*   *GL*   *bs*   *bs*
F₁   *GbLs* *GbLs* X *GbLs*
             (grey *Drosophila* with long wings)
Gametes of F₁ *GL*   *bs*   *GL*   *bs*
F₂ *GL* X *GL* → *GGLL*
    GL X *bs* → *GbLs*

272

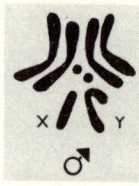 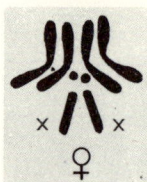

The four pairs of chromosomes of the male and of the female *Drosophila melanogaster*. The Y-chromosome of the male is easily distinguished.

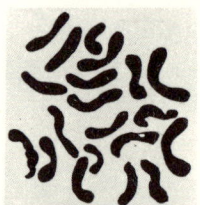

The ten pairs of chromosomes of maize.

Crossing-over.

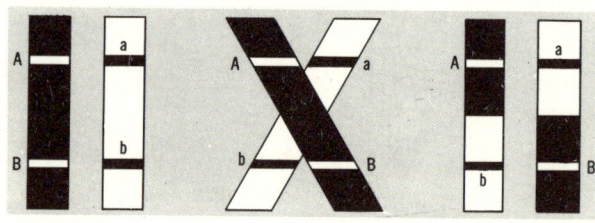

bs X *GL* → *GbLs* (all three grey *Drosophila* with long wings)
bs X *bs* → *bbss* (black *Drosophila* with short wings)

The characters grey body/long wings and black body/short wings are linked and behave like an allele which has two different effects.

As we shall see the inverse cross gives different results. Linkage is independent of the nature of the genes; it depends only on the manner in which the genes are associated in the parental generation. No chemical or physiological attractions influence the phenomenon of linkage.

Innumerable crosses made with *Drosophila* enabled Morgan and his fellow workers to show that the number of *linkage-groups* is exactly the same as the number of chromosome pairs. *Drosophila melanogaster* has four pairs of chromosomes, which can be identified by shape and size; it also has four groups of associated genes. The pair of dot-like chromosomes, which are particularly small, carry few genes while the long chromosomes contain a great number.

Other species of *Drosophila* have three, four, and five pairs of chromosomes and, as expected, a corresponding number of gene linkage-groups.

Maize has ten pairs of chromosomes and ten linkage-groups have been observed. The antirrhinum has seven pairs of chromosomes and seven linkage-groups.

The linkage-groups of numerous species are not yet known in their entirety, but the number of those which have been detected never exceeds the number of chromosome pairs. The phenomenon of linkage and linkage-groups is an additional proof of the existence of genes and of the fact that they are arranged in single file or 'linear order' along the length of the chromosome. Each gene occupies its own fixed and well defined position known as its *locus*.

Crossing-over: failure of linkage and recombination of genes

The experiment of the grey *Drosophila* with long wings crossed with the black *Drosophila* with short wings illustrated the linkage, in this case, between 'grey-long'

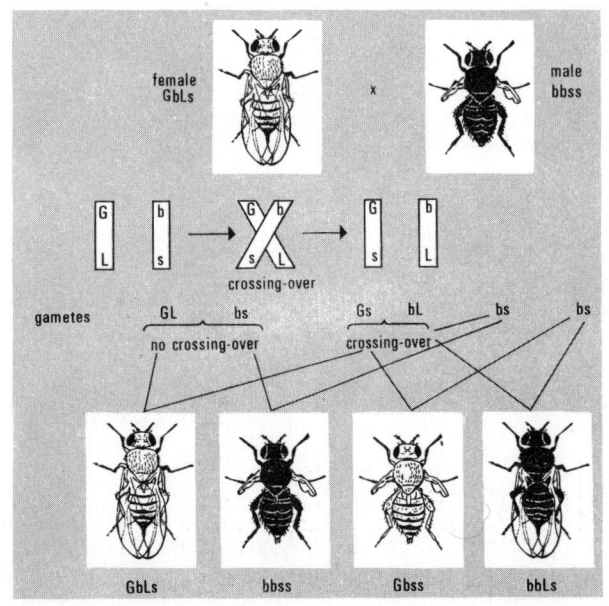

Diagram showing the part played by crossing-over, a phenomenon which affects certain chromosomes of the grey, long-winged female (*GbLs*).

Diagram of the frequency of crossing-over, that is, the cross-over value. (c. o. v.)

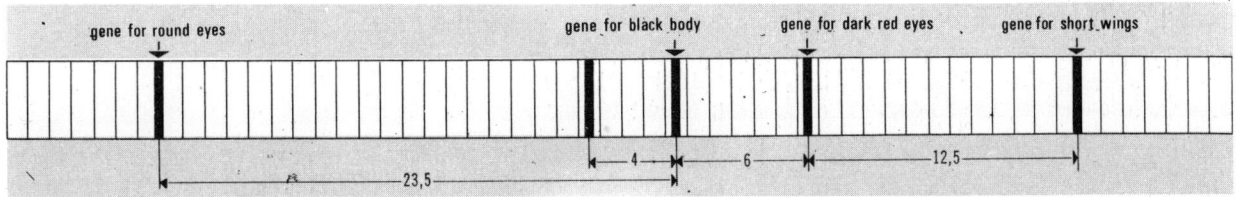

genes and 'black-short' genes. The question of whether these linkages always exist is answered by further experiments in cross-breeding.

The cross between the grey *Drosophila* with long wings and the black *Drosophila* with short wings produced, as we remember, grey hybrids with long wings having the genetic formula *GbLs*.

1. Let us cross one of these F_1 male hybrids *GbLs* with a black female with short wings *bbss*. The offspring is composed of grey flies with long wings and black flies with short wings in equal numbers — thus:

Parents ♂ (Male) Grey, long wings X ♀ (female) black, short wings

	GbLs		*bbss*
Gametes	*GL bs* (linked)		*bs bs* (linked)
F_1	*GL* x *bs*	*GbLs*	
	GL x *bs*	*GbLs* (both grey, with long wings)	
	bs x *bs*	*bbss*	
	bs x *bs*	*bbss* (both black, with short wings)	

The composition of the offspring is thus in conformity with absolute linkage between the genes *G* and *L* and between the genes *b* and *s*.

II. Now let us cross not the male hybrid but the grey, long-winged female hybrid with a black, short-winged male *Drosophila bbss*. The offspring is more complicated than in the preceding cross. It comprises not fifty

per cent but only 40.75 per cent grey flies with long wings, 40.75 per cent black flies with short wings and, in addition, 9.25 per cent grey flies with short wings and 9.25 per cent black flies with long wings.

This result proves that the linkages between *G* and *L* and between *b* and *s* hold firm in most of the chromosomes — in 81.5 per cent — but not in all. In 18.5 per cent of the chromosomes the linkage has been broken. This failure of linkage only takes place among female hybrids and never among males. This difference in the behaviour of the sexes is not general: among other species the failure of linkage is observed also in the male, or in both sexes. The reason for different behaviour of the sexes in different species is obscure.

To explain the appearance of the two unexpected combinations Morgan proposed the hypothesis of crossing-over.

The female hybrid *GLbs* has two homologous chromosomes carrying *GL* and *bs* respectively. Morgan assumes that the two chromosomes break at an equal distance between *G* and *L* and between *b* and *s*. Then the extremity of one becomes affixed to the extremity of the other, and reciprocally. Hence the pair of homologous chromosomes is now *Gs* and *bL*. Female hybrids with this formula, *GsbL*, crossed with the doubly recessive male *bbss* produce grey flies with short wings and black flies with long wings in equal proportions.

The unexpected offspring of the cross can thus be explained as a result of partial linkage; while linkage occurred as usual in the majority of the chromosomes, in two homologous chromosomes crossing-over, followed by breaking and rejoining of the exchanged parts, occurred.

The percentage of crossing-over is 18.5. This figure would seem to be a property of the genes in question, and independent of the dominant or recessive combinations to which the two genes may belong.

Crossing-over and chromosome maps

A simplified diagram of crossing-over is illustrated on page 272. The two chromosomes (on left) are homologous; one, black, carries the dominant genes *A* and *B*, inherited from one parent. The other, white, carries the recessive genes *a* and *b*, derived from the other parent. The two homologues then cross (in centre), forming what is known as a *chiasma*. Entwining causes the two chromosomes to break, and the rupture is followed by a joining up of the two segments exchanged (on right). Instead of the two original homologous chromosomes there are now two mixed chromosomes, one carrying genes *aB* and the other *Ab*.

In the diagram the point of rupture is shown in the middle of the chromosomes. This position is not privileged and there is no reason why a chromosome should break at that point rather than at any other. But if genes occupy fixed positions or *loci* the break is more likely to occur between two genes which are widely separated than between two which are close together.

To make this clearer let us imagine a chromosome divided into 100 equally spaced points, numbered from one to 100. If one gene is situated between points 20 and 21 and a second gene between points 50 and 51, there are thirty (that is, 50 minus 20) chances in 100 that the break will take place between these two genes. On the other hand, if one gene is situated between points 20 and 21 and the second between points 22 and 23, there are only two (22 minus 20) chances in 100 that the break will take place between them. In the second case the probability of a rupture is, then, relatively slight.

If the number of ruptures that is, crossings-over, is known, the above reasoning can be used to deduce the distance between the genes concerned. For example, if cross-breeding results in thirty per cent crossing-over between genes *A* and *B*, and two per cent crossing-over between genes *C* and *D*, then we can assume that the distance between *A* and *B* is fifteen times as great as the distance between *C* and *D*.

Thus by the frequency of crossing-over the distance which separates two genes along a chromosome may be measured. The measure is, of course, purely relative, and is expressed by the percentage of gametes in which one of the two genes has been exchanged for its allele from the homologous chromosome: this is the *cross-over value* (or c. o. v.)

In the cross just discussed the loci of the genes *b* (black body) and *s* (short wings) are 18.5 units apart, since the cross-over value is 18.5.

By crossings in which several linked genes are involved it becomes possible to draw up chromosome maps. In *Drosophila* the linked genes for black body and dark red eyes have a cross-over value of six. The linked genes short wings and dark red eyes have a cross-over value of 12.5. The dark red eye gene is, therefore, placed between the genes for black body and short wings at six units from the first and 12.5 from the second. A fourth gene modifies the shape of the eye so that it resembles a French bean instead of being round as is normal. This gene *H* is dominant to the normal eye gene *h*, and cross-over values indicate that *H* is 23.5 units from the black body gene and 17.5 from the dark red eye gene. A fifth gene has a cross-over value of four with the black gene and a cross-over value of ten with the dark red gene, and must therefore be located four units to the left of the black gene.

And thus, progressively, the map of each chromosome is built up. Maps of the four pairs of chromosomes of *Drosophila* have been fully analysed. Maize has ten pairs of chromosomes, maps of which are well advanced. For mammals chromosome maps are much more difficult to establish and progress has been correspondingly delayed.

Mutations

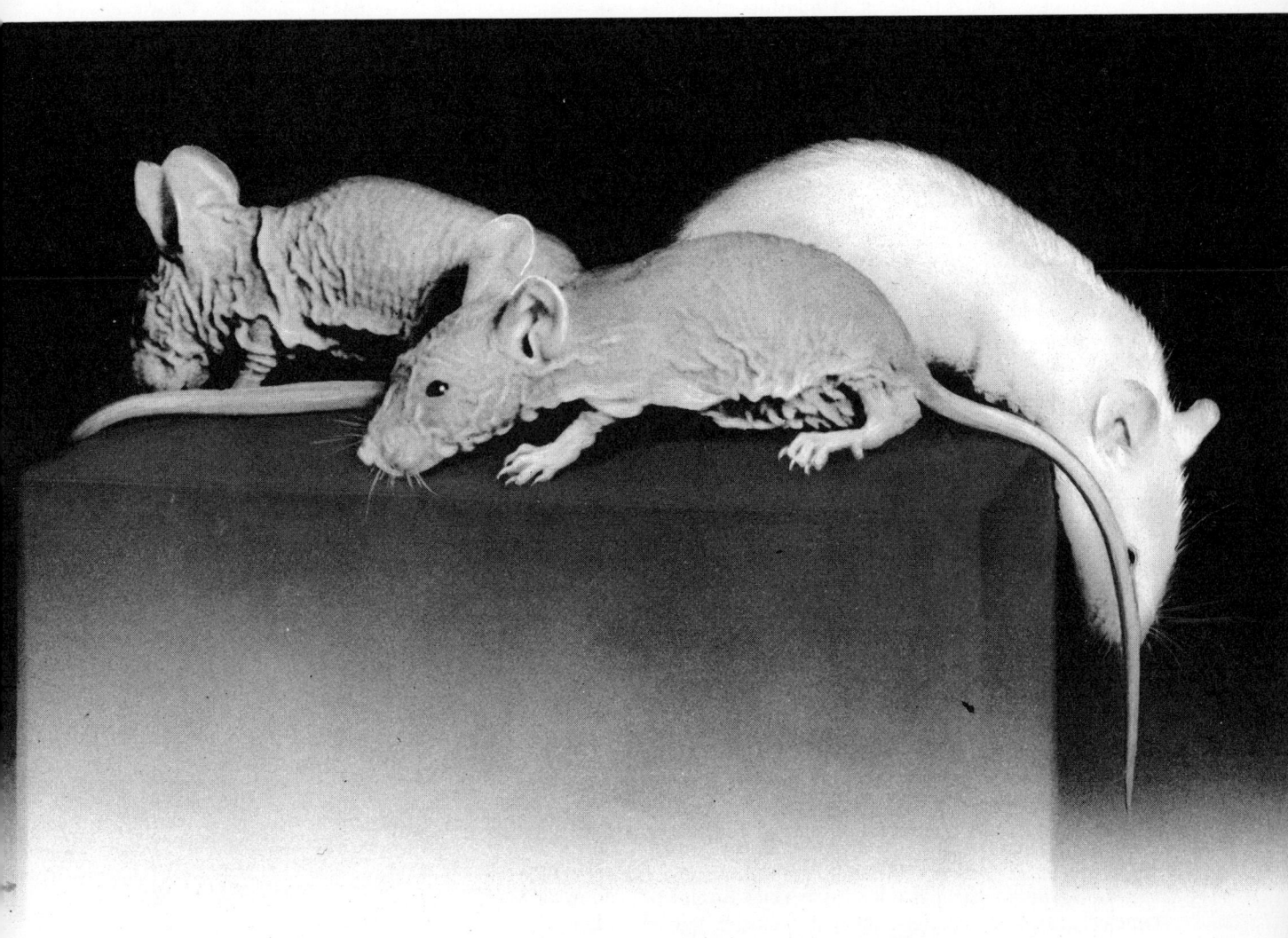

Two hairless mice and a normal mouse.
Dr Kobozieff.

Although the regular transmission of hereditary factors assures the continuity of living species it is nonetheless evident that a certain amount of continuous variation occurs.

Continuous variation

As an example of continuous variation we may consider the height of adult men. A given population always contains adult men of different heights, tall, short and with every intermediate height between the two extremes. This uninterrupted gradation can be shown graphically by a curve expressing for each height the number of corresponding individuals. To construct the figure, the various heights are plotted along the horizontal axis and the corresponding number of individuals on the vertical axis. The peak of the resulting curve indicates the most frequent height. There is only one peak in this case, but the same operation involving populations of different races would, though producing similar curves, have different peaks.

With a heterogeneous population, composed for example of English and Chinese, the curve would have two peaks. A classic example of such a double-peaked curve was given by de Vries for the number of florets on the flower-heads of *Chrysanthemum segetum* counted in one thousand plants gathered from various gardens. The frequency curve has two peaks, one at thirteen florets and another at twenty-one florets. These two peaks reveal that two strains of chrysanthemums had been mixed, one most typically possessing thirteen florets, the other twenty-one.

Continuous variation is observed in all quantitative characters, dimensions and weight. We find such variation, for example, when analysing the individual weight of a large number of French beans of the same variety. The weight of the beans varies between 20 and 90 centigrams and is distributed as follows:

Weight	Number of beans
20 to 25 centigrams	4
26 to 30	27
31 to 35	65
36 to 40	179
41 to 45	364
46 to 50	587
51 to 55	533
56 to 60	418
61 to 65	260
66 to 70	132
71 to 75	52
76 to 80	34
81 to 85	19
86 to 90	2

There are 587 beans weighing between 46 and 50 centigrams and they form the most frequent category. Above and below this category frequencies diminish. A curve drawn to represent these frequencies would show the variability of a collection of seeds having no precise genetic relationship. What, it may be asked, would happen if we planted the extreme types of these

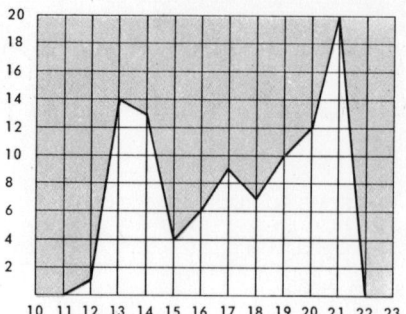

Variation in the number of ligulated florets on the flower-heads of *Chrysanthemum segetum* gathered in several localities. The curve of frequency has two peaks which indicates that there is a mixture of at least two populations of different genetic constitution. On the X-axis (numbered 10 to 23) the number of ligulated florets; on the Y-axis, their frequency.

After Hugo de Vries.

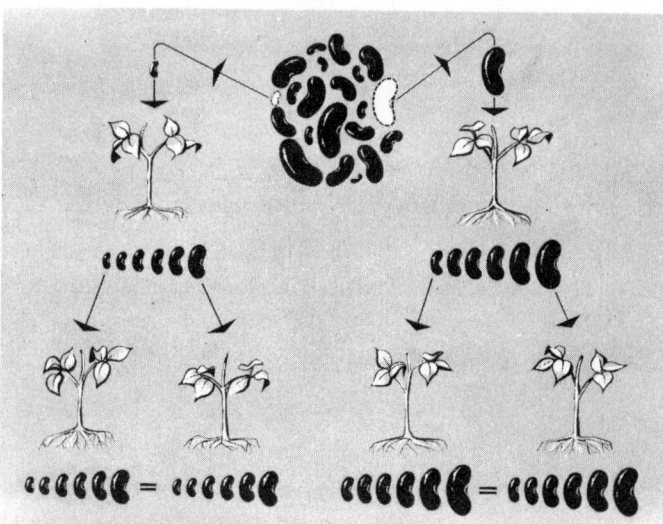

Johannsen's experiment with beans of different weights.

seeds, the largest and the smallest? Would the resulting beans resemble the general average or would they tend to resemble their large — or small — parent?

Johannsen has studied this problem. When he planted the heaviest of the seeds they grew into plants that, when self-fertilized, produced beans for which a curve of frequencies was drawn. The weight of the beans varied between 36 and 90 centigrams. Thus:

Weight	Number of beans
36 to 40 centigrams	2
41 to 45	5
46 to 50	9
51 to 55	14
56 to 60	21
61 to 65	22
66 to 70	24
71 to 75	23
76 to 80	17
81 to 85	6
86 to 90	2

Head of a ñato bull, a native of Chile.

After Camille Dareste, 1888.

Head of a hornless Hereford steer from Kansas.

X Photography.

Angora rabbit.

Larousse.

The category of highest frequency is thus of beans weighing 66 to 70 centigrams. The curve cannot be superimposed on the preceding curve; it has moved towards the right, that is, towards the heavier weights. When the same experiment is repeated by sowing the lightest beans the result is analogous, but the curve will move to the left, towards the lighter weights: weights will vary between 20 and 65 centigrams. The two experiments have brought about a kind of selection in the simplest meaning of the word, enabling us to obtain beans which are increasingly large or increasingly small.

What now happens when we again plant the heaviest of the beans derived from the grandparent heavy beans, namely those weighing from 86 to 90 centigrams? The resulting plants will produce beans having exactly the same frequency curve as their parents. The same is true of the lightest beans, and indeed of all seeds of a given category. In a pure strain selection no longer intervenes.

Hence it is possible in a population to isolate a series of strains which are kept pure by self-fertilization. All samples of a pure strain possess the same hereditary material; they are of the same genotype. The slight differences in weight between seeds of a pure strain arise from conditions in the environment at the time of growth, in particular the nourishment they receive, the amount of sunlight, and the number of beans in the pod. All such conditions of environment affect the outward appearance of the bean, that is to say its phenotype. That selection no longer affects a pure strain could have been expected, since all samples are of the same genotype. The genotype — the inherited factors — is independent of the environment.

The continuous variation observable in species is the result, then, of hereditary differences on the one hand and, on the other, of phenotypical differences which are individual and due to the environment.

Thus continuous variation is a common or garden phenomenon which is observed among all plant and animal species. From the evolutionary point of view it has no special importance.

Discontinuous variation

There is, however, another kind of variation, namely discontinuous variation which, unlike continuous variation, is rare. Now and again in nature — usually among domestic animals — an individual appears which differs from its fellows by some character or, at times, by several characters. It falls well outside the normal curve of probability. An abrupt change occurs, a step is taken without intermediate stages, and at once the new character is hereditary. This phenomenon is known as spontaneous mutation.

Darwin observed mutations which he called 'sports' and to which he attached little importance. Throughout history naturalists have noticed such exceptional occurrences and given them various names. Entire volumes have been dedicated to their enumeration.

Originally considered as curiosities, tricks played by nature, such sports have assumed their true importance since the work of Morgan and his colleagues on the mutations occurring in breeding *Drosophila*.

277

Fragaria monophylla (monophyllous strawberry plant) gathered by J. Paillot on the 19th of May 1862 in the Doubs.

Jardin botanique de l'Ecole de pharmacie, Paris. Larousse.

Gene mutation

A given gene, occupying an exact position on the chromosome, determines a given character — for instance, the formation of normal wings in *Drosophila*. A change in this gene *L* (for long) can transform it into another gene *s* (for shortened or dumpy) which determines the formation of short wings in *Drosophila*. Gene *L* has mutated into gene *s*, and the change is a genetic or gene mutation. The two genes *L* and *s* are alleles. A repetition of the same phenomenon can take place for any gene present in a chromosome, and the mutation of a gene will automatically give rise to its own allelomorphic gene or allele. The existence of a gene is thus revealed by its mutation: the gene *s* for shortened wings betrays the presence of a gene *(L)* which produces the normal long wings of *Drosophila*.

The mutant — that is to say, the individual with the new character — when crossed with the original or wild type will have offspring in accordance with the Mendelian laws of heredity. The very fact that this cross obeys Mendel's laws proves that the newly appeared character is indeed a mutation. The 1,000 mutants, dominant or recessive, which have been employed in drawing up the chromosome maps of *Drosophila* were all born by mutation. Mutation, then, produces new hereditary types.

In every organism two homologous chromosomes are always paired. Each of the homologous chromosomes carries the same gene. When mutation occurs in the gene of a single chromosome the result is a heterozygote; if both genes are mutant the result is a homozygote. Mutation being rare, heterozygotes are much commoner than homozygotes.

In nature mutations happen spontaneously, abruptly and without apparent cause. In large populations more than one or two individuals are rarely affected. When the mutation is recessive the date of its first appearance cannot be known because it may have already existed in heterozygotes and remained masked and unobserved. But if the mutation is dominant — a much rarer occurrence — the date of its appearance marks the birth of a new kind of organism.

Spontaneous mutations

Towards 1791, on a farm in Massachusetts, twin lambs, a male and a female, were born to one of the sheep. They had the characteristic aspect of a basset hound with its long back and short, crooked legs and were the origin of an ephemeral breed known as otter-sheep or ancons. In 1919 an ancon lamb was born in Norway.

Snub-nosed oxen or ñatos, (the South American form of the Spanish word *chato* meaning snub-nosed) with short head and neck and an undershot jaw revealing the lower teeth, were in 1760 regarded as a curiosity in Buenos Aires. They probably came from the herds of natives south of the river Plate. The mutation has reappeared in the Argentine and on several occasions in France. Ñatos, which feed with more difficulty than other breeds, are no longer reared.

The mutation *lack of horns* has frequently appeared among cattle. Hornless oxen were already known in the Bronze Age. The mutation was reported in 1861 in France, in 1874 in Sicily, and in 1889 in Atchinson, Kansas, and was the origin of the hornless Hereford breed. The different hornless breeds of Europe, Africa and Asia are of independent origins.

The so-called 'castorrex' rabbit, covered with downy fur and kemp, appeared on a French farm in 1919. Two of these animals were born to the same pair of rabbits on different occasions. The same mutation was reported to have been observed in the neighbourhood of Bordeaux fifty years before.

Among rodents the mutation of black eyes to creamy white is frequent. Albino mice, of Chinese origin, have been known since 1654.

The silver fox 'platinum' appeared in 1933 on a fox farm in Norway. Since then it has recurred on numerous fox farms.

Among mink there are various mutations of colour. The first silver mink was reported in 1931 by a mink breeder in Wisconsin. The same mutation occurred four years later on another mink farm in Wisconsin and in 1942 it occurred again in Oregon. 'Royal pastel' mink made its appearance in Canada in 1936-37. The albino mink has been known only since 1940 in Oregon.

In America the cotton rat (*Sigmodon hispidus*) was reared for ten years, and among roughly two million animals born all had the form of the agouti which is characteristic of the wild cotton rat. Then in 1947, in two separate stocks, the same mutation appeared: a white individual with small agouti spots and black eyes. This example shows how rare spontaneous mutations are.

Mutations are also observed among birds. The 'hen of Padua' with cephalic hernia was known in ancient Rome as we can see by statuettes made nineteen centuries ago. The tumbler pigeon dates from 1850. The yellow canary appeared between 1677 and 1713 in birds from the Canary Islands and Madeira. In the Australian parakeet *(Melopsittacus undulatus)* mutations of colour have often been observed and the dates when they occurred are known: yellow, towards 1872; blue, towards 1878; olive in 1919; mauve and grey in 1921; whitish in 1927.

Among *Drosophila* stocks reared in laboratories

Chelidonium majus.

Chelidonium laciniatum.

Laciniate beech.

Jardin du Luxembourg. Larousse.

Mercurialis altera foliis invarias et inæquales lacinias quasi dilaceratis.

Mercurialis with laciniate leaves. One of the first known mutations.

After Marchant, 1719. Larousse.

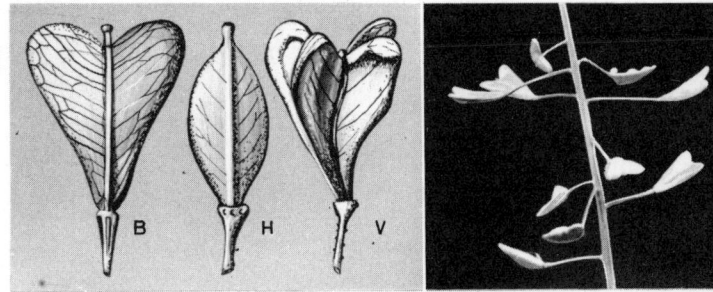

Mutation of the crucifer *Capsella bursa-pastoris* (shepherd's purse) affecting among other characters the constitution of the fruit or silicula.

After Massart.

B. Silicula with the two indented carpels of normal form.
H. Silicula with the rounded carpels of the mutation *heegeri*.
V. Silicula with the four indented carpels of the mutation *viguieri. On the right, Capsella bursa-pastoris.*

Larousse.

Mouse with recessive markings.

X Photography.

279

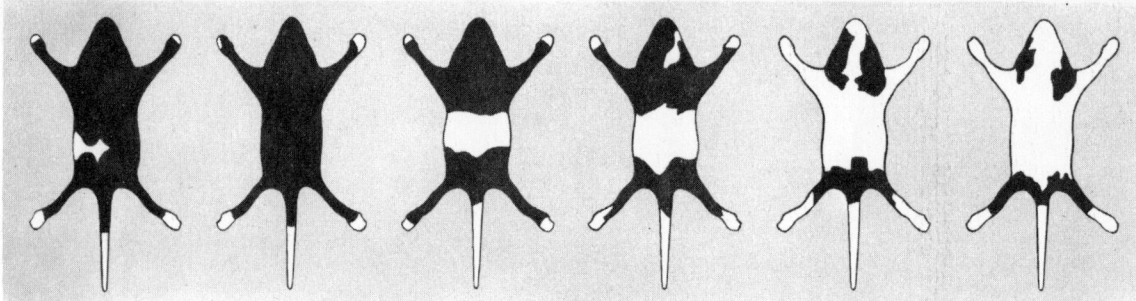

many hundreds of mutations have been observed, mutations of general colour, of colour of the eyes, shape of wings, number of bristles, sex-ratio, and so forth. These mutations are not necessarily the result of captivity, for among seventy-eight female *Drosophila* collected in the neighbourhood of Berlin, Timofeeff-Ressovsky counted eighteen with mutated characters. The proportion of mutants varies according to the season and the insect population.

Examples of mutation are also frequently found in plants and some such examples have become classic. The greater celandine with deeply incised leaves and petals *Chelidonium laciniatum* appeared in 1590 in the garden of an apothecary named Sprenger in Heidelberg. Since that time it has been cultivated in botanical gardens where it grows beside the lobe-leafed celandine *Chelidonium majus*. It has often been found growing wild.

There are at least two mutations of the Crucifera shepherd's purse *(Capsella bursa pastoris)*. The *heegeri* mutation with rounded carpels was found in 1897 by Heeger in the market place at Landau and the *viguieri* mutation in which the fruit is composed of four indented carpels instead of the normal two was discovered in 1908 by Viguier at Izeste in the Basses-Pyrénées. The new character bred true and has never been found again in Izeste.

In a nursery garden in 1855 the false acacia *(Robinia pseudacacia)* produced a monophyllous mutant with a large single foliole which breeds true. The monophyllous Duchesne strawberry plant appeared in a garden in the Faubourg Saint-Honoré in 1763 and still grows in botanical gardens.

Impatiens sultani, a plant from Zanzibar which was introduced into Europe towards 1880, had carmine-red flowers. Propagated by cuttings and seeding the plant bred true until 1890. Nine different plants appeared. They arose from one sowing and there was a single example of each, with flowers of varied shapes and colours ranging from pale pink to purplish-blue. The mutations were as stable as the original plant.

Shirley poppies display a wide variety of colours and are derived from a stool of *Papaver rhoeas* reported in 1882 by Wilks.

The copper beech has made its appearance on at least three occasions: in the fifteenth century in the South Tyrol; before the seventeenth century in the canton of Zurich, and in the eighteenth century in Thuringia. All the copper beeches scattered over the earth's surface have derived from these three mutations. The dwarf beeches of the forest of Verzy near Rheims, and those of the forest of Dain-en-Saulnois in the Moselle, have also made more than one original appearance.

Examples of various dorsal markings which occur on pied mice.

After Charles.

Rhinoceros mouse at the age of ten months, showing the folds in the skin.

After Howard.

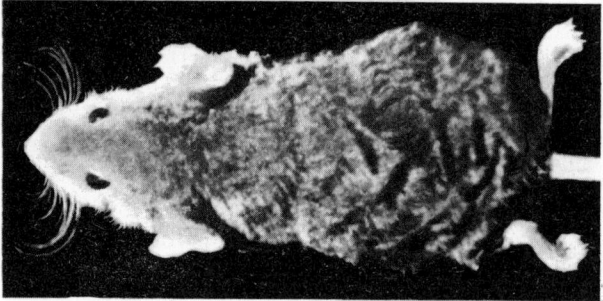

'Waved' mouse.

After Keeler.

Offspring of 'crinkled' and 'fuzzy' mice at the age of 19 days. The crinkled mouse is leaden grey in colour, the fuzzy is black.

X Photography

The 'trembler' mutation. Twenty-six-day old mouse in a typical posture during a convulsion.

After Falconer.

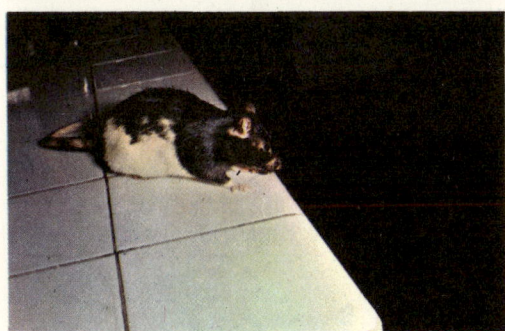

Top left, a pair of mice of the strain C 57 Bl, which are black. In this strain very few spontaneous tumours occur. Top centre, a pair of chocolate-brown mice of the strain C 57 Bl. Above left, adult male guinea-pig of the strain W, with white coat and pigmented eyes and skin. Above centre, rat of 'Long Evans' strain. Black with markings which are determined by known genes. Right, rat of 'August' strain. Markings and beige colour determined by known genes.

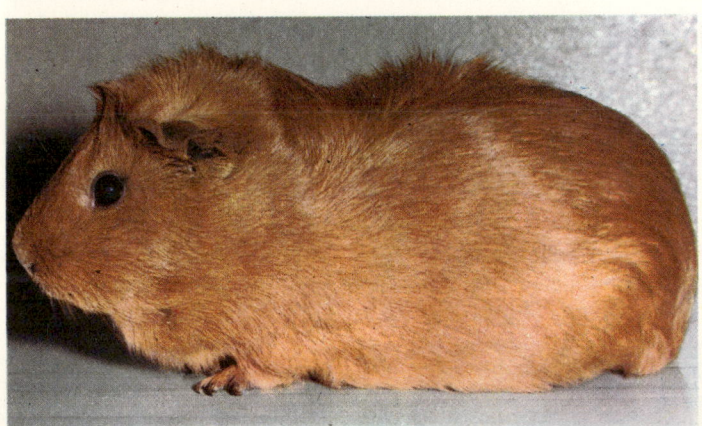
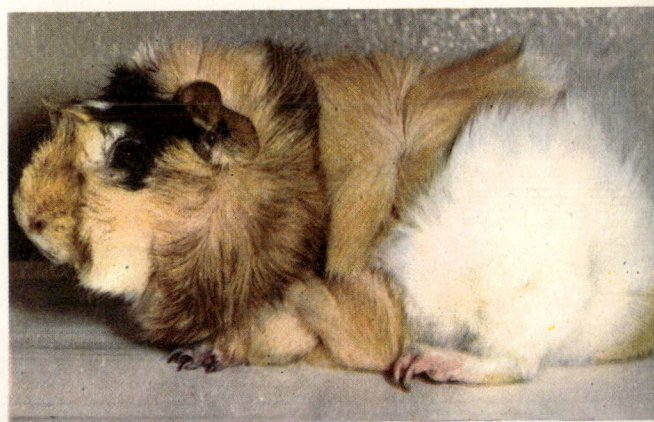

Above, adult male guinea-pig of the strain F with tawny coat and pigmented eyes and skin. Above right, guinea-pig of 'Abyssinian' type. Fur grows in rosettes over the entire body. Right, castorrex rabbit. Its coat is composed of beaver-like down with short hairs of kemp.

Centre d'élevage d'animaux de laboratoire at Gif-sur-Yvette. Larousse.

Mutations in vertebrates

Owing to the ease with which they are reared, mice are among the most thoroughly studied vertebrates. The wild mouse, with its yellow, black or brown pigmentation, appears grey. Its coat is smooth and its tail long.

Mutations which involve the *coat* include colour of fur (chinchilla, albino, agouti, yellow, black, black and tan); variegation formed by more or less extensive patches of white hair; appearance of hair (wavy, caracul, 'rex' with short kemp, wavy and irregularly distributed); and lack of hair (bare-skinned mice often appear.) In the 'rhinoceros' mouse the skin forms great transverse wrinkles. The coat then remains normal until the age of fourteen days, when the hair begins to fall out around the eyes, the head, the shoulders, and the belly. A few hairs remain on the feet, the tail and the ears. From the third week wrinkles in the skin appear across the nose, back and belly, and along the four legs. In another hairless mutation the skin is very thin and the hairs break off at the surface of the epidermis as a result of incomplete keratinization.

Mutations of the *tail* result in short-tailed mice, tailless mice, and mice with tails bent in one or two places or twisted up like a corkscrew.

Mutations of the skeleton include duplication of the posterior extremity entailing the formation of four kidneys, four gonads, two bladders, two rectums, four back legs, two tails. Other mutations lead to: absence of tibia, oligodactylism or reduction in the number of toes, polydactylism or extra toes, a tail stump, shortening of the trunk and anomalies of the spinal column, harelip, abnormalities of head and jaws.

Eye mutations include a total or partial anophthalmia in which the eyeball and muscles are lacking, though the socket, eyelids and the conjunctiva are present; reduction in size of the eyes; retina lacking rods with attendant blindness; eyelids open at birth; cataract.

Ear mutations include short 'dog-ears'.

In the nervous system there can be the absence of the *corpus callosum*, that is, the great commissure which unites the hemispheres of the brain; anencephalia or absence of the cranial vault and cerebral hemispheres, leading in almost all cases to embryonic death; hydrocephalus or dome-shaped cranium following dilation of the lateral ventricles due to the accumulation of cerebrospinal fluid. There are deaf waltzing mice which gyrate and move in circles or figures of eight; overactive, excitable mice with chorea or St Vitus dance; short tailed mice with disturbances of equilibrium; and mice with myelencephalic bullae or blisters with attendant abnormalities of the eyes, the extremities, the fur on the flanks and sometimes of the kidneys. The formation of these bullae in the embryo results from an excessive production of cerebrospinal fluid or an inability to resorb it. The fluid lodges in blisters in the skin and these compress the underlying tissues, impairing development.

All these mutations among mice have been the object of detailed investigations. We know when they first appear, whether they are dominant or recessive, and whether they are compatible or incompatible with life. In some cases the mechanism of the gene's behaviour is understood. The importance of the mutations varies considerably. Many lead only to anomalies or blemishes; some result in true monstrosities which gravely handicap the animals afflicted. They are not without resemblances to the flaws and monstrosities found in the hereditary anomalies of the human race.

Among other mammals mutations are as varied as they are in mice, but they have been much less thoroughly studied. They can, for instance, affect the coat and its colour. Such mutations are frequent in all mammals, those of fur-bearing animals being particularly well known because of their commercial value — mutation fox and mink, for example. There can be mutations in the texture of coat: wavy and curly hair is found in rats, hares, pigs, cows, horses, sheep, guinea-pigs. Cases of hairlessness are known in rats, rabbits, cows, sheep, goats, horses, pigs, dogs, cats, shrews, moles, orang-utans. Absence of tail or short tails are frequently seen in dogs and cats (Manx cats) though more rarely in horses, cows, pigs, foxes, and sheep.

In the *skeleton* achondroplasia or general disturbance of the cartilage and bone formation can lead to forms of dwarfism, or else the cranium and the vertebrae can be affected while the legs are not and — as in the case of ñatos and English bulldogs — dwarfism does not result. Again, the cranium and vertebrae may be normal and the legs short and twisted, as in basset hounds, otter-sheep or ancons and Dexter cattle. In Paraguay 'basset' jaguars are known, and in Guinea basset goats; there are also basset horses and zebras. Various other anomalies of the cranium and face are known: refusal of the parietal bones to knit, shortening of the lower jaw and harelip have been reported in dogs, pigs and sheep.

Ectromelia or the absence of front or back legs has been observed among dogs, horses, donkeys, cows, goats, sheep, rabbits, guinea-pigs, pigs and cats; polydactylism, or supernumerary digits, among cows, horses and guinea-pigs; syndactylism, or fusion of digits, and the fusion of two phalanges or the absence of phalanges have been reported in pigs and rabbits.

In addition there are breeds of cattle, goats and sheep without horns; rabbits, dogs, horses and cows with cataract; various abnormalities of the eye in rats and guinea-pigs; anomalies of the ear in rabbits, bulls, and sheep; hydrocephalus in pigs and dogs; paralysis in dogs and sheep; epilepsy in rabbits; various visceral abnormalities in pigs and horses; hereditary anaemia in rats; haemophilia in pigs and dogs; and jaundice in rats.

As with mice, many of the mutations of these mammals are pathological or lead to monstrosity. They are found chiefly among domestic animals and are, of course, best known among laboratory animals. Mutations which are useful, ornamental or, if viable and able to reproduce, merely curious have long been encouraged by animal breeders and made the point of departure for new breeds.

Mutations in birds are not rare, particularly in domestic fowl such as chickens and pigeons. Fantail pigeons, pouters, jacobins and tumbler pigeons have been known since 1600.

White blackbird with black patches (pied blackbird).

Binet, doyen of the Faculty of Medicine, Paris.

J. P. Wütenburg.

Porcupine pigeon. The character 'porcupine' is recessive to the normal character.

After Cole and Hawkins.

Cock and hen with naked necks.

Larousse.

Above, the normal four-toed right foot of a cock and, below, the right foot of a Houdan cock with five functional toes. The fifth digit derives from a doubling of the first (or thumb).

Larousse.

Peacock pigeon.

Larousse.

Karacul sheep.

Larousse.

Imperfect albinism is found in chickens such as White Leghorns, Silky Whites, White Dorkings, and also parakeets, canaries and sparrows. Some breeds of poultry derive their characteristic appearance from silky feathers due to the absence of the hooks or *hamuli* which secure the barbules, from feathers with reduced barbs and barbules, or from very fragile feathers. There are chickens which are unable to fly, chickens covered with wool, 'porcupine' chickens, and chickens which are either totally devoid of feathers or have naked necks which recall those of Egyptian vultures.

Achondroplastic dwarfism is also found in chickens of the 'creeper' type which are 'basset' chickens. In these the dimensions of the body are normal but the legs and wings are shortened. When standing they give the impression of normal hens sitting. Finally there are chickens without rumps, others with multiple spurs (Black Sumatras), or no spurs, polydactyls (Houdan, Crèvecoeur) and brachydactyls, while cross-beaked hens often appear on chicken farms.

Reptiles and amphibians are difficult and slow to rear, and their mutations have therefore been much less studied. A case of melanism in an American snake has been reported and one of dwarfism in an axolotl which developed a bulldog-like appearance. Albinism has occurred in frogs and toads, and polydactylism in toads. It is still unknown whether or not numerous malformations in amphibians are hereditary.

Some mutations are known among aquarium fishes: the goldfish of the Far East, *Platypoecilus maculatus* of Mexico, *Lebistes reticulatus* of Brazil, *Aplocheilus latipes* of Japan, the swordfish *Xiphophorus helleri* of Mexico. The mutations especially affect colouring, the markings made by pigmentation, shape of the body, increase in size or doubling of fins, and telescopic eyes.

Mutations in invertebrates

The great fertility of insects and the speed with which they develop means that large numbers can be reared so that the probability of mutations occurring is much increased.

More than a thousand mutations have taken place among fruit flies and have been the object of detailed and precise analysis. Whole volumes have been devoted to them. The mutation is described and the place on the chromosome (locus) of the gene responsible is known. Its manner of transmission is made clear by various crosses. The mutations affect the form of the wings, the veining of the wings, the shape and texture of the eyes, the colour of the eyes, the number of bristles, the shape of the abdomen, the legs, the general colour of the body, the insect's fertility and vitality. The cheese fly, a variety of mosquito, and other Diptera have displayed mutations of colour and of wing shape. Tower studied in detail the mutations of the Colorado beetle or potato bug *Leptinotarsa decemlineata*. Among 207,891 specimens collected in different parts of the United States he discovered 118 mutants. These mutations which had occurred in nature afterwards reappeared in the laboratory. They affected chiefly the colour and design of the pigmentation. Various mutations have been observed in other insects, crustaceans, molluscs, and even single-celled animals.

Mutations in plants

Plant mutations are as numerous and varied as animal mutations. Certain plants have, like the fruit fly *Drosophila,* been studied with particular care. These include in particular the antirrhinum, the Gramineae and the Solanaceae.

The snapdragon, or antirrhinum, is, like the pea, reproduced by self-fertilization which makes its mutations easier to preserve. Numerous varieties of antirrhinum are in origin mutations and new mutations continue to appear. They affect the shape of the flower (smaller petals, partially united, absence of anthers, radiate flowers), the form of the leaf (acicular, curled, concave), the colour of the flowers, leaves and cotyledons, and the size of the plant (dwarfism).

Mutations are especially common in maize, or American corn, and barley, affecting the shape and colour of the leaves, the form of the ear and the grain it bears, the height of the plant, and the stigmas. In wheat and oats spontaneous mutations are rare. Among the Solanaceae it is above all the *Datura*, tobaccos, and tomatoes whose mutations have been observed.

Leguminous plants like the pea, bean, and lupin, crucifers like Shepherd's purse, primulas, the cottonplant, marvel-of-Peru, pineapples, and sunflowers all mutate spontaneously. Spontaneous mutations have also been reported among ferns, mosses, and mushrooms.

Mutations of bacteria and bacteriophages

Mutation also exists among bacteria. Biochemical mutants which are capable of fermenting various sugars are known, as well as mutants which are able to resist the attack of bacteriophages and certain antibiotics.

Among the bacteriophages, or viruses which destroy bacteria, one — that which destroys the colibacillus — is fairly well known. It contains from thirty to fifty genes, each of which is capable of mutating before contact with the bacteria host. This relatively small number of genes may make it possible to isolate them — an extremely complicated process, but one that is much simpler than that of isolating one of the roughly 10,000 genes on the chromosome of multicelled organisms.

From the above long and incomplete inventory of spontaneous mutations a certain pattern seems to emerge: among allied species it would appear that many mutations are repetitive.

In all classes of vertebrates, for instance, perfect albinos are found. In the same way melanic mutations occur in men, rats, mice, hamsters, squirrels, cats, panthers, cattle, etc., and xanthic mutations in mice and canaries. Angora coats are seen among cats, dogs, goats, sheep, guinea-pigs and rabbits. In the two snails *Helix nemoralis* and *Helix hortensis* exactly the same mutation — coloured bands — takes place. The bulldog aspect of the dog of that breed is found again in the South American ox, the ñato; the basset hound's appearance is repeated in the ancon sheep.

Mutant rabbits, dogs and pigs have pendent ears. We find hairless dogs and horses, 'rhinoceros' mice and cats, pigeons and hens without feathers. The same species may include both giants and dwarfs.

In the vegetable kingdom similar repetition is observed: weeping willows, birches, ashes, beeches, elms, *Sophora* and cedars; dwarf beeches, oaks, elms and walnut trees. Trees with grain twisted to right or left are common to many species.

Frequency of mutations and radiation

To calculate the exact percentage of mutations which occur spontaneously is difficult, more especially as it varies according to species and variety, age of cells, composition of the genome — the set of all the different chromosomes found in the nucleus — the genes themselves, and the physiological state of the cell.

That one species is more apt to mutate than another is shown by statistics. Some genes are more stable than others. Stable genes, indeed, may mutate over a million times less frequently than genes which are known for their instability. The latter are presumably responsible for most cases of variegation, particularly among plants.

Experiments demonstrate that the mutation rate rises 'with the age of the germ-cells or gametes. It is also affected by temperature.

The experimental production of mutations

From the earliest days of genetics attempts have been made to bring about mutations experimentally. In 1927—1928 Muller succeeded in inducing mutations in *Drosophila* by means of X-rays and gamma rays of radium. During the Second World War Auerbach and Robson demonstrated that certain chemical compounds also cause mutations and their results were published after hostilities ceased.

The effect of electromagnetic and ionizing radiations

Electromagnetic waves which induce mutations range from the extremely short and penetrating gamma rays to those of a wave-length of 313 mµ in the ultraviolet range. All ionizing radiations, alpha rays, beta rays, cosmic rays, are effective.

These rays produce mutations by losing energy, that is to say, by being absorbed by living matter. Whatever their nature their effects are qualitative and quantitative. The majority of artificial mutations are similar or identical to known spontaneous mutations. It is impossible at will to obtain one mutation rather than another, and the type obtained is always unpredictable. Lethal mutations, incompatible with life, and semi-lethal genes which reduce vitality, are ten to fifteen times more frequent than non-lethal mutations.

The rays operate not by delayed action but directly and at once, and to be efficacious they must strike the gonads. For example, the irradiation of the head and thorax of a *Drosophila* whose abdomen is protected by a screen will have no effect.

Among plants a mutation of the bud can be brought about by irradiation while the rest of the plant remains normal. Such mutations, affecting only the somatic cells, are not hereditary and can be maintained only by asexual reproduction, by cuttings. Similar somatic mutations also appear spontaneously.

Irradiation can also produce mutations in reverse, that is, the mutant gene may re-acquire its original structure and function. It can also mutate again. The phenomenon of reversibility proves that the gene is not destroyed but simply altered.

Quantitative analysis shows that the frequency of radiation-induced mutations depends on the wavelength and total amount of exposure, not on intensity of exposure or on factors inherent in the cells.

The number of mutations is directly proportional to the amount of radiation. The manner of administrating the given amount is immaterial: a dose of 3,000 Roentgen units given continuously for a short time or given discontinuously over a longer period — in several exposures more or less spaced out — produces exactly the same effect. Mutation does not result from a cumulative effect; it is as though, in the course of bombardment, a projectile charged with a quantum of energy found the correct range and hit the bull's-eye.

It now seems that all radiations are not equally effective. Maximum success appears to be obtained with gamma rays; then as the wave-length increases — with neutrons and alpha particles — effectiveness diminishes.

As in the case of spontaneous mutations the physiological state of the cell and its age also play their parts.

Temperature acts in a particular manner: at ordinary temperatures radiation has no effect, while cold augments its action. Air pressure also alters the frequency of mutations.

In our atomic age the effect of radiation on gene mutation is of redoubled interest: the question is, in brief, are radiations dangerous? The effects of irradiation have been studied chiefly in the fruit fly; but more recently similar experiments have been made with mice. From these experiments it appears that the probability of induced mutation is ten times greater in mice than in fruit flies. How would human beings react to the treatment? Do those who work with atomic energy, and are exposed to the accumulative effect of radioactivity, endanger their descendants? Now, normally, 200 undesirable mutations appear in each generation for every ten million individuals. If throughout the population this mutation rate were doubled the rate in ten generations would be stabilized at 400 mutations annually. The unfortunate human race would then be burdened with 400 injurious mutations.

In the last fifty years exposure to radiations has been greatly increased through the abuse of X-ray diagnosis and radiotherapy. A certain amount of radiation occurs in nature, but the British Medical Council estimates that the gonads of the average Englishman artificially receive twenty-two per cent more than this amount while in America the figure is 100 per cent.

An American report has been published on the 'bomb children' born in Japan to mothers and fathers who were exposed to varying degrees of radiation. There was a change in the normal sex-ratio: of the

1

2

3

4

5

6

7

8

9

Plant mutations.

1. Comfrey *(Symphytum officinale)* with normal inflorescence. 2. Comfrey with fasciated inflorescence: the concrescent peduncles are flattened.

Montegut.

3. Violet and mauve irises and the mutant 'Tangerine'.

Lauros - Atlas - Photo.

Animal mutations: Japanese quails.

Lauros - Atlas - Photo.

4. Wild quails, male and female, aged three months. 5. Black-coloured mutants. 6. Dove-coloured mutants. 7. White mutants. 8. Quail chicks aged eight days. The wild ones are striped and the mutants black and white. 9. Wild quails' eggs of different colours. The white eggs are mutations.

1   2   3

4   5   6

7   8   9

Mutations of *Drosophila melanogaster*.

Élevage Bösiger, at Gif-sur-Yvette, Service of Professor Teissier.
R. H. Noailles.

1. Dark coral eye. 2. Sepia eye and ebony body. 3. Indented wings. 4. Fringed wings. 5. 'Bar' eye. 6. Widely separated wings. 7. Vestigial wings. 8. Crumpled wings. 9. Minute wings.

Mutants of albino type appear among the descendants of a lettuce treated with ethyl oxide at a concentration of 1.5 per thousand. The two control plantlets are green, the two mutants are discoloured.

Somatic mutation of a scion (William's striped pear) which, before grafting, was irradiated with the gamma rays of cobalt 60 — the dose being 6,000 Roentgens. A branch has appeared with white leaves due to a total absence of chlorophyll.

P. Dommergues of the Institut national de la Recherche agronomique. Larousse.

1,445 children born in Nagasaki to women submitted to about 300 Roentgens, 746 were girls and only 699 boys. In addition numerous diseases affected a large part of the exposed population.

Atomic industry does not yet seem to have altered the radio-activity of the atmosphere to any appreciable extent. The bomb explosions are, however, another matter. Atomic fall-out, in the form of dust, is disseminated in the upper air and takes months, even years, to reach the ground. Among the by-products of fission is strontium 90 which combines with the calcium used in bone formation. Experiments undertaken in England show that among rabbits the young in particular fix this radio-active element. An analysis of human bones, from recent European and American corpses, reveals a radio-activity of 0.12 micro-curies per gram of calcium. Every child born in America accumulates one to two or even five per cent of the maximum concentration which the body will tolerate. In England, where the figures are smaller, it is felt that the situation of American children is not without danger. Though isolated doses may be harmless the question arises whether the organism tends to retain and concentrate them. Unfortunately the answer seems to be yes.

It is already known that mineral elements contained in a cubic metre of sea-water can be concentrated in a single gram of living matter. The by-products of fission are liable to similar concentration. At Bikini herbivorous fish absorbed ten times as much radio-active matter as carnivorous fish, due to the accumulation of radio-active matter in the algae which herbivorous

fish consume. Since 1946, in the region of Hiroshima, it is established that the oysters absorb radio-activity through the shell, then the viscera and finally the muscles. In Japanese waters the scales of tunny fish betray the radio-activity of barium 140, strontium 89 and uranium 237. The accumulation of the by-products of fission in algae, plankton, and migratory fish is a source of genuine danger.

Migratory tunny fish, eels and salmon can carry concentrations of radio-active elements in their bones and viscera for vast distances. Then again the action of these products of fission on micro-organisms and on bacteria could be considerable, endowing them perhaps with new and accrued pathogenic powers.

To be dangerous in itself sea-water would have to contain a hundred times more radio-active matter than it contains at present; but owing to the concentration of this matter in various food products radio-active elements could ultimately reach the human organism.

For all these reasons the future, even the immediate future, is disquieting, and it has been felt necessary to work out the admissible doses of radio-activity in terms of their biological effects. The measure used is the *rem* or Roentgen-Equivalent-Man. The maximum dose during the course of human life is calculated to be 200 *rems* distributed as follows: until the age of about thirty — the average age of reproduction — 50 *rems* at most; then in the course of each following three decades 50 more — in other words 5 *rems* per year.

Numerical increase of chromosomes

In certain circumstances the entire stock of chromosomes can be doubled: the cells, instead of two sets of chromosomes, then have four sets. Instead of being diploid they are tetraploid. The phenomenon, rare in animals, is commonly seen in plants. Such mutations give rise to giant plants which attract attention.

The most celebrated example is that of the giant variety *(gigas)* of the evening primrose *(Oenothera lamarckiana)* a single example of which appeared in 1895 among the plants cultivated by de Vries. Compared with the normal evening primrose the new variety was taller, had broad leaves, larger flowers and thicker stems with shorter internodes. The cells of *Oenothera gigas* are larger than those of normal diploid plants. The new variety is tetraploid and its cells contain four sets of chromosomes; that is, each chromosome is present in the cell nucleus four times instead of only twice as in diploids.

The multiplication of the single, or haploid, number of chromosomes (as found in germ-cells) by 3, 4, 5, etc., is known as polyploidy. Triploids have three sets of chromosomes, tetraploids have four, pentaploids have five, and hexaploids have six sets of chromosomes.

Generally speaking, polyploidy brings about a series of modifications which tend to produce gigantism. Increase in the size of the plant cells entails increase in the height of the plant. Other parts of the plant also reveal gigantism: it is found in the diameter of the pollen grains, the thickness of stems, the total surface and thickness of leaves, the dimensions of the flowers, the weight of the seeds. Polyploid plants are often slower to bloom and flower over a longer period.

Among the very numerous polyploids many plants belonging to a diversity of families may be listed: roses, dahlias, chrysanthemums, poppies, campanulas, periwinkles, plum trees, tobaccos, hawthorns, petunias, irises, wheat. There are kinds of wheat, for example, which are diploid with fourteen chromosomes, tetraploid with twenty-eight and hexaploid with forty-two. Chrysanthemums exist in even wider degrees of polyploidy, diploids with eighteen chromosomes, tetraploids with thirty-six, hexaploids with fifty-four, octoploids with seventy-two, and even decaploids with ninety chromosomes. Many large varieties of strawberries, pears and tomatoes are polyploid.

Apart from gigantism polyploidy often confers qualities of resistance on plants and permits them to be grown in climates which are fatal to diploids. The octoploid rose with fifty-six chromosomes is subarctic and grows in polar regions. The diploid tomato is killed in two hours when exposed to one degree of frost while the tetraploid will support a temperature of —5 °C for ten hours. Among polar and alpine flowers there are many polyploids: in Spitzbergen eighty per cent are such. In the Pamirs, eighty-five per cent of the Gramineae are polyploid, and in the Altai mountains sixty-five per cent. The more vigorous polyploids spread with great success. While diploid woody nightshade, fumitory, and shepherd's purse

Somatic mutation of a lettuce after treatment of the seed with a 2 per thousand concentration of neutral ethyl sulphate. One section of the young plant's leaves is yellow, being deprived of chlorophyll.

Chemically induced mutations

Ionizing radiations are not alone in causing gene mutations. The same effect is produced by chemical compounds of various kinds, the most important being yperite or mustard gas.

Chemically induced mutations have the same characteristics as spontaneous mutations or mutations caused by irradiation. No chemical substance brings about a specific mutation. As in the case of radiations the physiological state of the cells influences the frequency with which mutations occur. The detailed mechanism by which these substances bring about mutations remains unknown.

Artificially induced mutations

Mutations have been brought about in the laboratory by Griffith, Avery, Boivin, and others among pneumococci, colibacilli and dysenteric bacteria. Some virulent pneumococci are surrounded by a capsule which protects them from the phagocytes which defend the host. This capsule has a characteristic chemical structure and is hereditary.

Chromosome mutations

In addition to gene mutations there are mutations which involve the entire chromosome. Some modify the number of chromosomes while others alter the genetic structure of the chromosome.

Diploid plant *(on the left)* compared with triploid plant *(on the right)*. The triploid is larger.

X Photography.

remain localized in certain areas their corresponding polyploids are found everywhere. The tetraploid herb paris *(Paris quadrifolia)* has spread from Asia into Europe, while diploid varieties of the same plant are confined to Japan, Taiwan, and the Himalayas. It seems that polyploids of the Sahara are more resistant than their diploid relations living under the same conditions.

Polyploidy is not, however, always an advantage: some polyploid plants are no more — and even less — hardy than the corresponding diploids. Frequently they are sterile, especially triploids: they must multiply by means of vegetative propagation — for example the bulbils of tiger lilies, and the bulbs of tulips.

Polyploidy results from an anomaly in cell-division. After the anaphase (the stage of mitosis in which the duplicated chromosomes separate and move apart) the partition which normally forms between the chromosomes does not appear, and the cell, undivided, thus possesses two daughter nuclei which fuse. In this way the number of chromosomes is doubled.

Physical injury can promote polyploidy. For instance, if a young tomato plant (diploid with 24 chromosomes) is decapitated so that a callus develops where the stem was cut, some of the cells will double their chromosome complement. Often triploid branches with thirty-six chromosomes will then develop, and tetraploid branches with forty-eight chromosomes. Grafting, which is in fact a form of injury, will also at times produce polyploidy.

Since the experiments of Blakeslee in 1937 we know a sure method of inducing polyploidy, namely by using colchicine, the alkaloid extracted from the bulbs of the autumn crocus. Solutions of colchicine are applied to the growing tissues: the extremities of the stem or roots, and to seeds when germinating. During cell-division colchicine brings about specific anomalies, anomalies which also result from the employment of narcotics. The action of colchicine, however, has the great advantage of being specific; it does not entail other reactions and can be used in various concentrations.

By the use of colchicine, then, polyploid plants are easily obtained. After treatment, the meristem of the plant is usually composed of tetraploid cells mingled with cells which have remained diploid. But branches which are completely tetraploid may be formed together with diploid gametes (gametes being, of course, normally haploid), which will engender a new generation of tetraploids.

Animal polyploidy is much rarer than plant polyploidy. Triploid and tetraploid frogs and tritons are found in nature; as well as triploid and tetraploid silkworms, and triploid *Drosophila*. There are small crustacean phyllopods which are tetraploid and triploid woodlice. Some moths, some molluscs, some earthworms, and some roundworms are polyploid.

In animals colchicine does not exert its inhibitory action during cell-division as effectively as it does in plants. Recently Rostand has experimentally brought about polyploidy in a vertebrate by a simple device: thermal shock. The recently fertilized egg of a toad was submitted to a long period of chilling which resulted in a doubling of the maternal chromosomes. Fankhauser, systematically applying this method to tritons and salamanders, obtained a great number of triploid individuals. If chilling had taken place within half an hour of fertilization as many as seventy and eighty per cent of the larvae produced were triploid. Rapidly heating the eggs produces similar results.

Reasonable progress is being made in experimental animal polyploidy but plant polyploidy, which is facilitated by asexual or parthenogenetic reproduction, remains much easier to produce.

There is another means by which the number of chromosomes can be multiplied, namely hybridization. This form of modifying hereditary material takes place only among plants, many of which are created in this

The flowers of *Coreopsis tinctoria. On the left,* tetraploids. Tetraploidy was induced by colchicine.

J. Vincent.

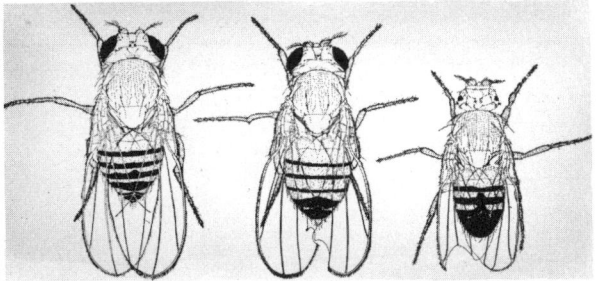

Drosophila melanogaster. Left, female triploid; *centre,* intersex resembling a female; *right,* intersex resembling a male. The latter is also affected by the 'bar' eye mutation.

After E. Wallace.

manner. Crossing two plant species produces hybrids which are generally sterile and can only be maintained by vegetative reproduction. This is because the paternal and maternal chromosomes, being different, cannot couple during cell-division and therefore pairs of homologous chromosomes are not formed. By chance, however, it can happen that certain gametes are formed which contain both the paternal and the maternal set of chromosomes. The fusion of two such gametes will result in an ovule which has two sets of paternal chromosomes plus two sets of maternal chromosomes. It is therefore a double diploid (or tetraploid) and also fertile, since each chromosome has a partner with which it pairs during cell-division. Many species arise in this manner, either spontaneously or artificially.

The salt-marsh grass known as rice grass *(Spartina townsendii)* may be the only polyploid plant which arose spontaneously by hybridization during the course of recent history. It was discovered in England on the coast near Southampton in 1870. By 1907 it had spread so prolifically along the south coast of England that on occasion it choked up harbours. Then it crossed the Channel and reached the French coast, thriving in salt water where little else would grow. Morphologically speaking *Spartina townsendii* reveals the mixed characters of *Spartina stricta,* a European species, and *Spartina alterniflora,* a species of American origin which arrived in Europe with ships' cargo. In nature the two species, *stricta* and *alterniflora,* grow beside *townsendii.* Cytological examination shows that *S. townsendii* has sixty-three pairs of chromosomes, and that *S. stricta* and *S. alterniflora* have twenty-eight and thirty-five pairs respectively. The spontaneous cross between these two species produced a sterile hybrid with twenty-eight or thirty-five or sixty-three chromosomes, comprising one set from each parent. It has been suggested that, by chance, gametes were formed with sixty-three chromosomes. The fusion of two such gametes produced an ovule with 126 chromosomes;

comprising two sets of chromosomes from each parent. Thus each chromosome was able to pair with its homologue, and the new species, *Spartina townsendii,* with sixty-three pairs of chromosomes became an independent species, true-breeding, hardy and with a tendency to eliminate its parent species. However, it must be said that this view of its origin is not accepted by all botanists.

Artificially, many new species have been created in a similar way.

Transformation of the chromosome's genetic structure

We have already seen that when chromosomes 'cross over' and break the detached segments can be exchanged, thus altering the chromosome's gene content. Other spontaneous breaks can be followed by rearrangements which profoundly modify the structure of the chromosome.

The following are some of the modifications which may occur. *Deficiency:* a section of the chromosome carrying one or more genes is lacking. In the case of homozygotes such deficiency brings about a modification which is normally incompatible with life. In the case of a heterozygote the modification can be dangerous or not, depending on the genes which are missing.
Duplication: a part of the chromosome becomes fused with the same part of its homologous chromosome.
Inversion: a segment of the chromosome is reversed, at the same time retaining its linear position on the chromosome. This can happen when a loop, followed by rupture and re-fusion, occurs.
Translocation: two non-homologous chromosomes

291

break and then mutually exchange segments. This is called *reciprocal translocation* and the homologues of the chromosomes involved remain normal. The translocation can also consist of a transfer of part of a chromosome into a different part of a homologous chromosome.

These various modifications of structure take place during mitosis when the pairing of homologous chromosomes leads to complicated patterns. In pairing it is as though each gene on one chromosome exerts an attraction for its homologue on the other.

The giant (that is, microscopically visible) chromosomes which occur in the salivary glands of dipterous insects — notably *Drosophila* — permit these modifications to be detected and allow genes to be located so that chromosome maps can be verified. Changes in chromosome structure give rise to phenotypes which reveal new characteristics. The eye of *Drosophila* provides a good example. The normal eye is, like that of all insects, a compound eye with a certain number of visual facets or ommatidia. The eye of wild *Drosophila* has about 780. A series of mutants have eyes with fewer ommatidia. The heterozygote of the 'Bar' mutation has 358; the homozygote of the same mutation sixty-eight. The 'double-Bar' heterozygote has forty-five, and the double-Bar homozygote has twenty-five. Examination of the chromosome map shows that the Bar mutation results from the repetition, in the same order, of six bands. It thus corresponds to a duplication of the X-chromosome in the region 16 A (see diagram below). The double-Bar mutation has the same bands three times and hence there is a trebling of the X-chromosome.

It would seem, then, that the progressive intensity of the Bar character is related to the number of times the region 16 A is repeated.

Section of an abnormal chromosome in a cell of the salivary gland of a larval *Drosophila melanogaster* which has formed a loop.

After Dobzansky.

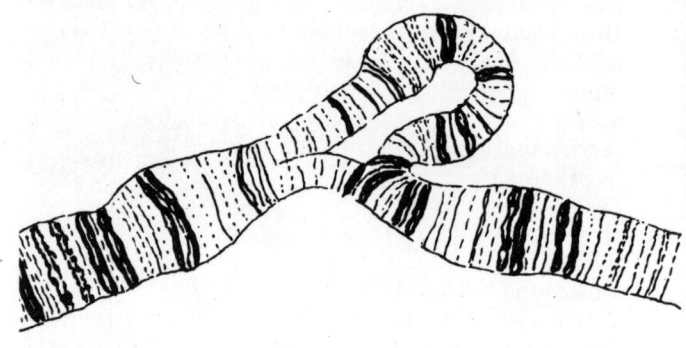

Diagrams of chromosome translocations.

| A B C D E | A B C |
| A B C D E | A B C D E D E |

Two homologous chromosomes showing a deletion and a duplication. *Below*, two non-homologous chromosomes undergoing mutual translocation.

| A B C D E | A B C P Q |
| M N O P Q | M N O D E |

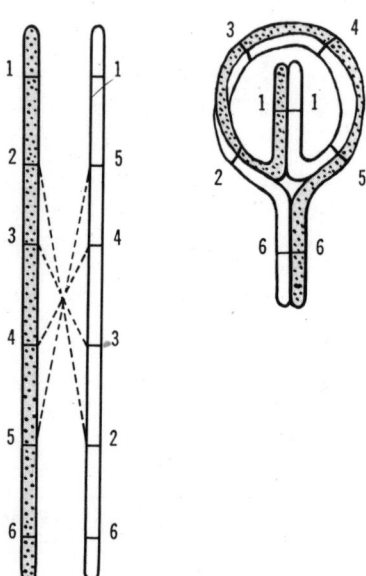

Left, diagram of a pair of homologous chromosomes, one of which has undergone inversion at loci 2, 3, 4, and 5. *Right*, position taken by the two chromosomes so that corresponding loci shall face each other.

After Herberer.

Diagram of the probable mechanism of inversion in a single chromosome.

After Shull.

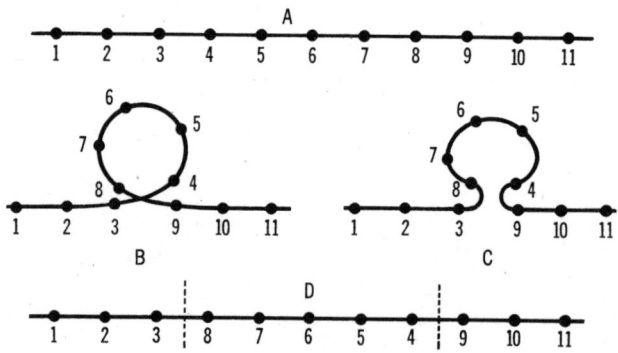

A, Chromosome before the inversion; B, formation of the loop; C, double break of the loop between points 3 and 4 and between points 8 and 9; D, the chromosome after inversion. Points 1 to 11 are the loci of the genes.

If we agree that each region composed of six bands corresponds to a gene which conditions the number of the ommatidia, then the X-chromosome of the wild *Drosophila* carries a single Bar gene, the mutant Bar *Drosophila* two such genes, and the ultra-Bar *Drosophila* three. But these characters depend not only on the number of genes, but also on the position of the genes. Sturtevant observed that when two Bar genes are carried on two chromosomes the effect is not the same as when they are carried side by side on the same chromosome. In the former case the insect's eye has sixty-eight facets; in the latter only forty-five. This discovery led him to the study of the *position effect* which was found again in numerous other mutations.

It is evident that such chromosome rearrangements can lead to immense complexity, a fact which would explain the racial diversity of populations inhabiting different geographical areas.

The 'bar' eye mutation of *Drosophila melanogaster*.

After E. Wallace.

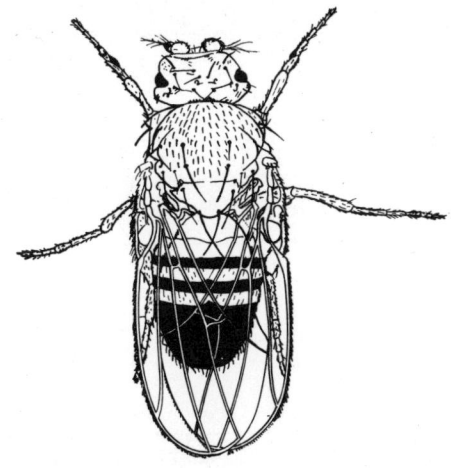

The eyes of *Drosophila melanogaster*.

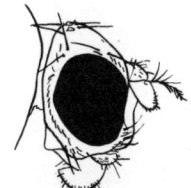

Female wild type.

Female bar homozygote.

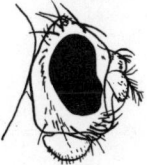

Female bar heterozygote.

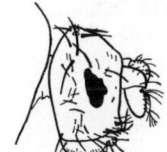

Male double-bar.

Female infrabar homozygote.

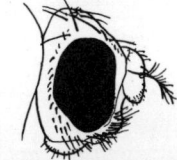

Female infrabar heterozygote.

The 'bar' region of the X-chomosomes of the salivary glands.

After Bridges.

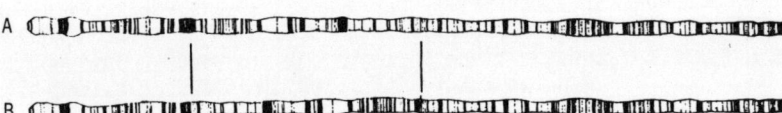

Male double infrabar.

Inversion in *Drosophila melanogaster*. A, normal position of the genes on the third chromosome; B, inversion situated between the two vertical lines.

After Moore.

Special Cases of Heredity

Some characters, though they obey Mendel's basic laws of heredity, are transmitted in certain special ways. Such is the case with sex-linked characters, with the collaboration of genes which have different effects, with multiple genes, multiple alleles and lethal genes.

Heredity linked to the sex chromosome

In the crosses that we have so far examined the sex of the organisms involved was immaterial. The results were the same whether a character was contributed by the male or by the female. This, however, is not always the case. Sometimes the results of crossing differ according to whether a given character is present in the male or in the female.

This slightly disconcerting phenomenon arises from the fact that the male and the female chromosome equipment are not quite the same. There is one pair of chromosomes which differs in the two sexes. These two chromosomes are called the *sex chromosomes* to distinguish them from the remaining chromosomes which are known as *autosomes*.

In the female the pair of sex chromosomes is homologous while in the male the pair is dissimilar. In *Drosophila* — which has eight chromosomes — the female has six autosomes and two identical sex chromosomes called X-chromosomes. The male *Drosophila* also has six autosomes, but two different sex chromosomes, called the X-chromosome and the Y-chromosome respectively.

During the formation of gametes the female will form ova all of the same type, containing three autosomes and one X-chromosome. But the male will form two types of spermatozoa in equal numbers: one with three autosomes and an X-chromosome, the other with three autosomes and a Y-chromosome. Thus the female is homogametic (all gametes alike), while the male is heterogametic (unalike gametes, of two kinds). Some insects and fish, frogs and all mammals, form gametes in the same manner as *Drosophila*. Among *Drosophila* fertilization presents two possibilities:

1. ovum with three autosomes plus X-chromosome unites with spermatozoon with three autosomes plus X-chromosome and produces zygote with six autosomes and two X-chromosomes. The zygote (XX) is female.

2. ovum with three autosomes plus X-chromosome unites with spermatozoon with three autosomes plus Y-chromosome and produces zygote with six autosomes, one X-chromosome and one Y-chromosome. The zygote (XY) is male.

In other words, the zygote's sex is, as we have seen in an earlier chapter, determined by which sex chromosome (X or Y) the heterogametic male parent contributes.

A variant of this type of fertilization is found among many insects and among roundworms. The male, having only one sex chromosome X, has one chromosome less than the female. For example, the male of a certain nematode worm has eleven chromosomes, one of which is an X-chromosome, while the female has twelve, two being X-chromosomes. The spermatozoa are of two types, one type with five chromosomes and the other with six (five autosomes plus one X-chromosome). The ova are all alike with six chromosomes (five autosomes plus one X-chromosome).

Finally tritons, birds and some insects including moths, belong to another group which is the inverse of the one described above. In this second group the male is homogametic and has two similar sex chromosomes, which may be represented by the symbol Z. The female is heterogametic and has two different sex chromosomes, Z and W.

In this case the possibilities of fertilization are also two, but are as follows:

Sex-linked heredity.

Marmalade cat.

Black cat.

Tri-coloured cat.

295

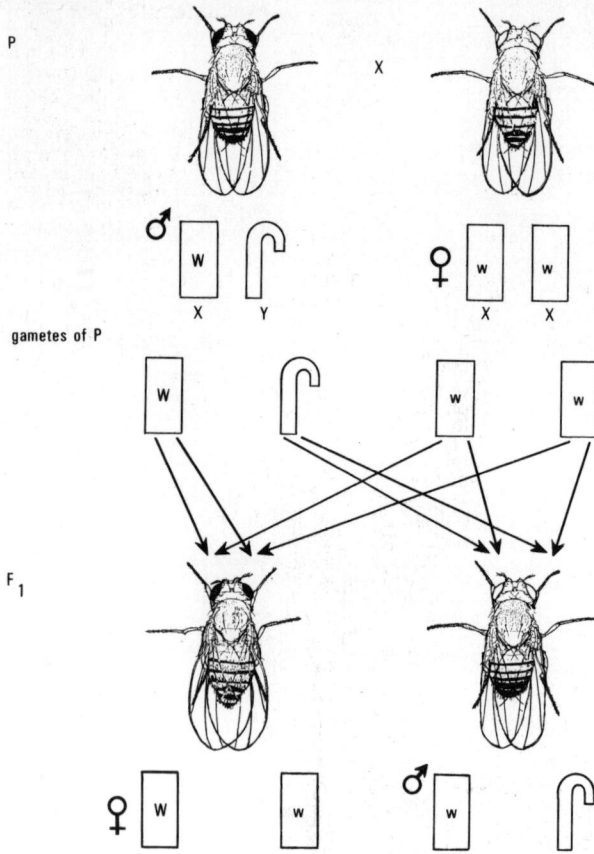

P

X

♂ | W | ⌐ |
   | X | Y |

♀ | w | w |
   | X | X |

gametes of P

| W | | ⌐ | | w | | w |

F₁

♀ | W | | w | ♂ | w | ⌐ |

Sex-linked heredity in *Drosophila melanogaster*. Cross between male with normal eyes and a female with white eyes. *Below, the reverse cross.*

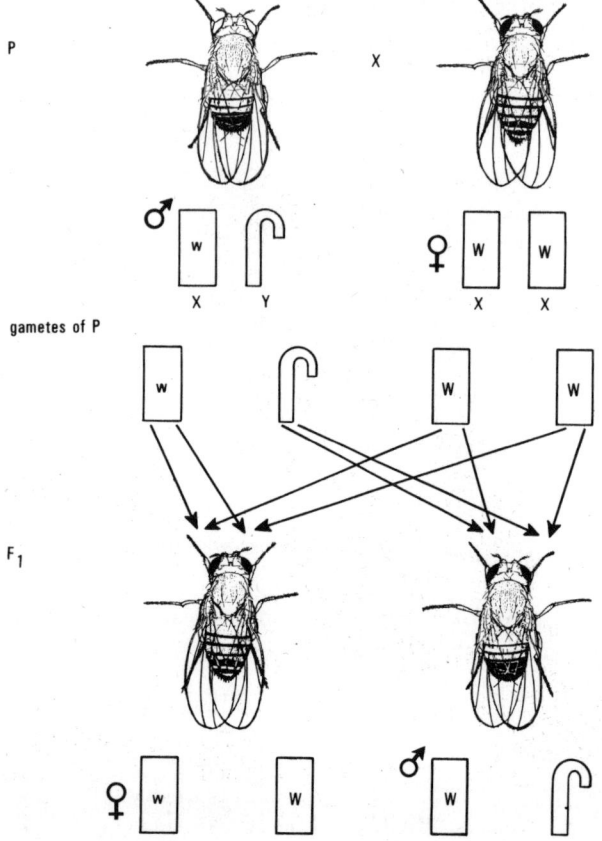

P

X

♂ | w | ⌐ |
  | X | Y |

♀ | W | W |
  | X | X |

gametes of P

| w | | ⌐ | | W | | W |

F₁

♀ | w | | W | ♂ | W | ⌐ |

1. an ovum with *n* autosomes plus a Z-chromosome unites with a spermatozoon with *n* autosomes plus a Z-chromosome to produce a zygote with 2*n* autosomes and two Z-chromosomes. The zygote (ZZ) is male.

2. an ovum with *n* autosomes plus a W-chromosome unites with a spermatozoon with *n* autosomes plus a Z-chromosome to produce a zygote with 2*n* autosomes, one W-chromosome and one Z-chromosome. The zygote (WZ) is female.

Hereditary characters linked to a sex chromosome are determined by the genes carried on the sex chromosomes and follow the destiny of these chromosomes.

Crosses between animals of the first group

The cross between a male *Drosophila* with red eyes and a female *Drosophila* with white eyes produces males with white eyes like the mother and females with red eyes like the father. The opposite cross — between a male with white eyes and a female with red eyes — produces males and females all with red eyes.

The two crosses give different results according to whether the character red eyes is carried by the male or by the female parent. How can this be explained?

The gene which determines the colour of the eyes is situated on the X-chromosome. The gene for white eyes is recessive and is symbolized as *b*. The gene for red eyes is dominant and is symbolized as *B*. In the first cross the female forms ova which all carry the white-eye gene *b*. The male forms sperm of two kinds, one with an X-chromosome carrying the red-eye gene *B*, the other with a Y-chromosome that carries no gene for eye colour. All female offspring are *Bb* heterozygotes and have the dominant red eyes. All male offspring have the recessive gene *b* but no corresponding allele; their eyes are therefore white. Thus:

Parents　　Red-eyed Males　X　White-eyed Females
　　　　　　　(X$_B$Y)　　　　　　　　　X$_b$X$_b$
　Gametes　　X$_B$　　Y　　　　X$_b$　　X$_b$
F₁　X$_B$ x X$_b$ → X$_B$X$_b$
　　X$_B$ x X　→ X$_B$X$_b$　　　(red-eyed females)
　　Y　x X$_b$ → X$_b$Y
　　Y　x X$_b$ → X$_b$Y　　　(white-eyed males)

Hence the father transmits his character to the daughters and the mother her character to the sons.

The reverse cross can be represented thus:

Parents　　White-eyed Males　X　Red-eyed Females
　　　　　　　(X$_b$Y)　　　　　　　　　X$_B$X$_B$
　Gametes　X$_b$　　Y　　　X$_B$　　X$_B$
F₁　X$_b$ x X$_B$ → X$_b$X$_B$
　　X$_b$ x X$_B$ → X$_b$X$_B$　　(red-eyed females)
　　Y　x X$_B$ → X$_B$Y
　　Y　x X$_B$ → X$_B$Y　　　(red-eyed males)

In this kind of sex-linked transmission the dominant or the recessive character always appears in the male since it is not masked by another gene allele.

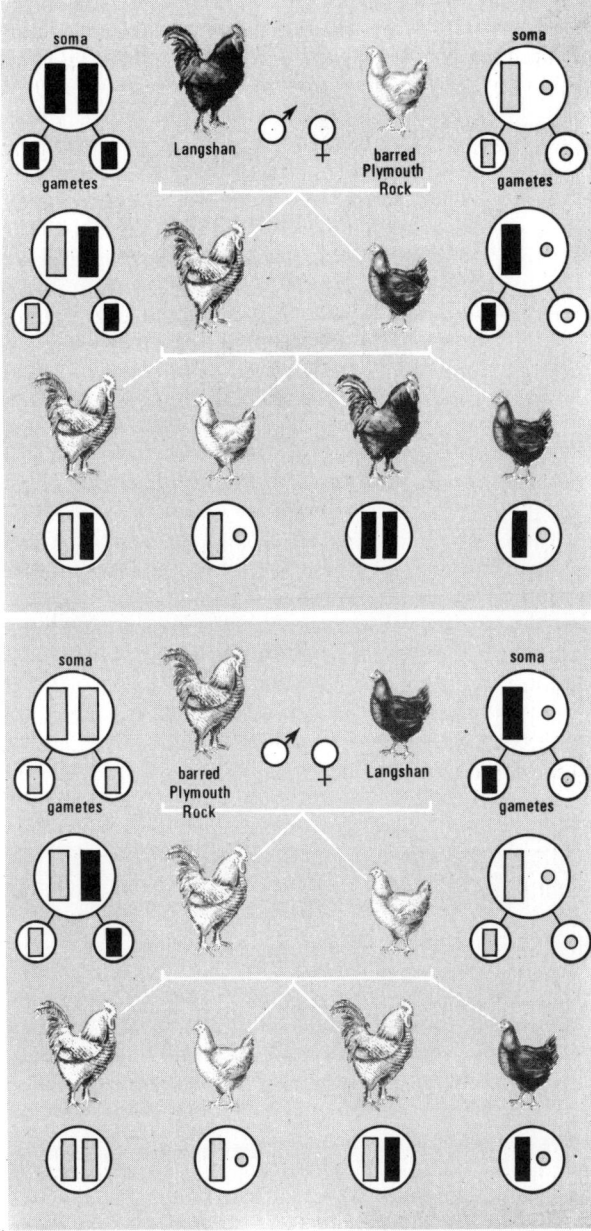

Diagram of the cross between two breeds of poultry, Barred Plymouth Rock and Langshan, illustrating sex-linked heredity.

After Morgan and Goodale.

The upper couples represent parents of pure race; those in the middle rows are F_1 hybrids; F_2, or the lower rows, are the result of crossing the hybrids.

The large circles drawn beside the individuals show the constitution of their somatic cells, the small circles depict the constitution of the gametes in so far as their sex chromosomes are concerned. The chromosome which carries the barred factor is shaded; that which carries the black factor is black and the small Z-chromosome peculiar to the female is indicated by a dot.

Cats of two colours

The cross between a black and an orange cat gives rise to male kittens of the mother's colour and to bi-coloured female kittens, the two colours appearing as a mosaic, that is to say, appearing side by side, neither being dominant.

The black male carries the gene b on its X-chromosome and the orange female carries the gene o on its two X-chromosomes. The cross is represented as follows:

Parents   Black Male (X_bY) x Orange Female (X_oX_o)
Gametes X_b Y X_o X_o
F_1 X_b x $X_o \rightarrow X_bX_o$
         X_b x $X_o \rightarrow X_bX_o$ (females, bi-coloured black
                                and orange)
         Y x $X_o \rightarrow X_oY$
         Y x $X_o \rightarrow X_oY$ (orange males)

Thus it is possible to tell the sex of a bi-coloured or — if white variegations are added — of a tri-coloured cat at a distance: for they are always females. Rare two-coloured males do exist, but they are sterile and their behaviour is that of a castrated animal. Often they are intersexes.

In *Drosophila* many characters are sex-linked: colour of the eyes (white, vermilion, cerise, eosine or rose-coloured), rudimentary wings, abnormal shape of the abdomen, colour of body (yellow, black), etc.

Genes situated on the small Y-chromosome

Genes on the Y-chromosome exist only in males, of course, and are transmitted from father to son.

There is a variety, *maculatus,* of the small Trinidad fish *Lebistes reticulatus,* which is characterized by the presence of a black spot on the dorsal fin of the male. All the males transmit this black spot to their sons, but never to their daughters. The character is linked with the Y sex chromosome.

The dominant allele of the gene 'bobbed' which modifies the bristles and abdominal bands of *Drosophila* is situated on the Y-chromosome.

Crosses between animals of the second group

The cross between a Langshan cock which is uniformly black and a Plymouth Barred Rock hen with dark plumage barred with whitish markings produces barred cocks, like the hen, and uniform hens like the cock.

The reverse cross — Plymouth Barred Rock cock and uniformly black Langshan hen — produces cocks and hens which are all barred.

As in the crosses we have just examined the results differ according to which plumage character is carried by which sex.

The dominant gene B for barred feathers and its allele b for uniform feathers are carried on the Z-chromosomes. The two crosses are represented thus:

Left, head of a black Bresse cock: single comb, regularly indented, back lobe detached from the neck.

Larousse.

Centre, pea comb *(above)* and strawberry comb.

Right, head of a Wyandotte cock. Rose comb.

Larousse.

First Cross
Parents Langshan Cock (Z_bZ_b) x Plymouth Barred Rock
                                 Hen (Z_BW)

Gametes	Z_b Z_b		Z_B W	
F_1	$Z_b \times Z_B \rightarrow Z_bZ_B$	(Plymouth Barred Rock cock)
	$Z_b \times W \rightarrow Z_bW$	(Langshan hen)
	$Z_b \times Z_B \rightarrow Z_bZ_B$	(Plymouth Barred Rock cock)
	$Z_b \times W \rightarrow Z_bW$	(Langshan hen)

Second Cross
Parents Plymouth Barred Rock Cock X Langshan Hen
      (Z_BZ_B)   (Z_bW)

Gametes	Z_B Z_B		Z_b W	
F_1	$Z_B \times Z_b \rightarrow Z_BZ_b$	(Plymouth Barred Rock cock)
	$Z_B \times W \rightarrow Z_BW$	(Plymouth Barred Rock hen)
	$Z_B \times Z_b \rightarrow Z_BZ_b$	(Plymouth Barred Rock cock)
	$Z_B \times W \rightarrow Z_BW$	(Plymouth Barred Rock hen)

Cocks of two colours

Some crosses produce mosaic colour schemes analogous to those observed in cats. Thus a cross between a golden Leghorn cock and a silver grey Dorking hen, or between a Rhode Island Red cock and a white Light Sussex hen, will produce cocks with mosaic feathers, a mixture of the plumage of the two breeds crossed, and hens with plumage identical to that of the parent cock.

Sex-linked characters enable the sex of newly hatched chicks to be identified so that poultry breeders can if they choose eliminate the cocks. In the cross

between a Rhode Island Red cock and a white Light Sussex hen the chicks are easily distinguished, the males being white and the females red.

The collaboration of genes

It is rare that one gene alone determines a character; more often several genes collaborate to produce the visible feature or phenotype.

The cock's comb

In different breeds of poultry various types of cocks' combs are found, four of which we shall consider. The commonest type is the 'single' saw-toothed comb which is seen in Leghorns, Minorcas, and *Gallus bankiva* from which all domesticated breeds of poultry are thought to be derived. The 'rose' comb is characteristic of Bantams, Wyandottes, and Hamburgs; it is composed of small 'fingershaped' tubercles arranged in parallel series. The 'pea' comb, formed of three rows of tubercles, is found in the Brahma and Indian Game breeds. The Malayan breed has a 'strawberry' comb, more or less indented and recalling a strawberry or perhaps a walnut.

The cross between a 'rose' comb and a 'single' comb produces in the first generation — F_1 — rose combs, and in F_2 three rose combs to one single comb. Inheritance is of the standard Mendelian type, rose being dominant and single recessive.

The cross between 'pea' comb and single comb gives in F_1 pea combs, and in F_2 three pea combs to one single comb. As in the preceding cross the result is Mendelian and pea comb dominates single comb.

The cross between rose comb and pea comb gives rise in F_1 to a new type of comb, namely a 'strawberry' comb. The composition of F_2 is also unexpected: of sixteen chicks nine have the strawberry comb, three the rose comb, three the pea comb, and one the single comb. This last type of comb did not appear in either the parents or the grandparents of the chick.

It will be remembered that this proportion — 9, 3, 3, 1 — was characteristic of the F_2 generation in crosses involving two pairs of alleles carried on independent chromosomes. We also know that the characters rose and pea are dominant. If R represents the gene for rose comb, and P the gene for pea comb, the genetic formula of rose comb would be $RRpp$ and that of the pea comb $PPrr$. The cross between them is as follows:

Parents Rose Comb *(RRpp)* X Pea Comb *(rrPP)*
  Gametes   *Rp*   *Rp*   *rP*   *rP*
F_1 *RrPp*   *RrPp*   *RrPp*   *RrPp*

The genes R and P are dominant over the single comb, but not over each other. In the chicks of F_1 each of the genes R and P exerts its influence simultaneously, which results in a kind of mosaic or compromise — which is the strawberry comb.

The double-hybrids with strawberry comb now form four kinds of gametes, namely RP, Rp, rP, rp. Fertilization between these four kinds of gametes will give rise to sixteen possible combinations:

1 *RRPP*	1 *RRpp*	1 *rrPP*	1 *rrpp*
2 *RrPP*	2 *Rrpp*	2 *rrPp*	
2 *RRPp*			
4 *RrPp*			
9 strawberry	3 rose	3 pea	1 single

It can be seen that the strawberry comb always has the two dominants R and P whether homozygous or heterozygous, that the rose comb has the dominant gene R, the pea comb the dominant gene P, while the single comb has no dominant gene and is therefore a double recessive.

The cross is a good example of pairs of alleles which are transmitted independently but act together. The action of the pair of genes Rr is a function of the other pair Pp. In other words, if P is also present in the animal the final result will be a strawberry comb, since the two genes R and P affect the phenotype in collaboration. If, instead of P, p is present the final result is a rose comb, as only the dominant R reveals its presence. In brief, the two pairs of alleles affect the same character, namely the form of the comb.

Atavism

Atavism is the unexpected appearance of an individual that wholly or in part resembles an ancestor which for many generations has disappeared. Darwin described a case of atavism in pigeons. By crossing black pigeons with white pigeons, or with white pigeons with red tails, he obtained dark pigeons which, when crossed among themselves, produced pigeons with blue necks, white rumps, double bars on the wings, tails barred with black at the extremity and marked with white on the edges. The description exactly fits that of the rock pigeon which is believed to be the chief ancestor of present-day domestic pigeons.

It is now known that this phenomenon, apparently so extraordinary, is simply the result of a chance encounter of genes present in one or other of the parents, genes which combine in such a way as to produce some feature or combination of features which is identical to an ancestral form.

Colour of the coat in mammals

Characters which appear to be simple, such as the colour white or the shape of a cock's comb, are on analysis found to arise from the complex interaction of several genes. This is particularly clear when we examine the factors responsible for the colour of mammals' coats.

We find mice which are grey, black, brown, yellow, white, silver, chocolate, cinnamon, and cream; and each colour is determined by at least ten different genes. By means of suitable crosses a number of these genes can be identified and designated by conventional letters. Thus the genetic formula of the wild or agouti mouse can be written $CCAABBDDPPLLnnMMPPaa$ $SSiiGGll$. Variations of colour are brought about by the mutation of one particular gene while the others remain unchanged. For instance if C is replaced by its allele c all development of pigment is inhibited; there will then appear an albino with a pure white coat and pink eyes (absence of pigment in the iris allows the blood vessels in the depth of the eye to be seen). The allele c is said to be *epistatic* for all the other genes of coloration; it prevents their action. The other genes, being suppressed, are *hypostatic*.

The mutation of gene A into a produces a black mouse; the mutation of AB into ab produces a chocolate tint. It is easy to see that a vast number of genetic combinations are possible and hence that mice exist in a wide variety of colours.

These variations in the colour of the coat are of special interest to breeders of fur-bearing animals. The blackish-brown fur of the standard mink is provided by animals with the following genetic formula: $P\ G\ Ip\ Al$ $B\ Bg\ Bi\ Bs\ Ba\ Bm\ Bp\ C\ H\ O\ dfs\ cm\ eb$. Each of the genes has at one time mutated into its allele and produced a new colour mutation. When the gene P mutated into p a platinum or silver-blue mink resulted. When Al mutated into al the mink was Aleutian blue. When eb mutated into Eb the mink was ebony blue.

The interaction of genes explains how animals can be of identical phenotype and yet of quite different genotype, a difference impossible to identify by outward appearance. Minks with the genetic formula $SScc$, $Sscc$, and $sscc$ are all albinos because they carry the gene c twice, that is, they are homozygotes for that gene. In albino mice the same homozygous state exists. The white phenotype in poultry results from three different genetic formulae: the white Leghorn has a inhibitory gene which prevents the formation of pigment; a certain Japanese breed has the pigmentation gene but lacks a gene for the enzyme required in pigment-elaboration; while in the white varieties of Plymouth Rock and Wyandotte this enzyme gene exists, but the pigmentation gene itself is missing.

Multiple genes or polygenes

In the above examples a given character was the result of genes with different effects constrained, as it were, to collaborate. The result was a compromise. Other examples show that a character can also arise from the cooperation of several genes distributed among several pairs of alleles; in these cases each gene has the same though partial effect and the final result is the sum total of what each individually contributes. Their action is cumulative. They are known as multiple or polygenes and in association they determine the transmission of numerous characters of a quantitative nature: stature, weight, etc.

Let us, for instance, cross two breeds of mammals which are of different height. One, say, measures 100 centimetres and the other 40. We shall assume that the greater height is determined by three pairs of polygenes, $AABBCC$ — although in reality there would be many more. The various genes which condition the minumum height of 40 centimetres in individuals of the two breeds are not taken into consideration. The three pairs of genes $AABBCC$ control growth from 40 centimetres to 100 centimetres — in other words, 60 centimetres of growth — and each of the six genes is responsible for 10 centimetres of height. The height of a given individual will thus be the basic 40 centimetres plus n times 10, n being the number of growth genes carried. The shorter breed carries the three pairs of recessive alleles, and its formula is $aabbcc$. Crossing the breed $AABBCC$ and the breed $aabbcc$ gives rise in the first generation to trihybrids of the formula $AaBbCc$ which measure 40 plus 3 times 10 or 70 centimetres, a height halfway between the heights of the parents. Each of these hybrids will form eight kinds of gametes, $ABC, ABc, AbC, Abc, aBC, aBc, abC, abc$. The second generation will thus comprise 64 individuals whose heights will be determined by the number of growth-genes, that is dominant genes, present. The number of such dominant genes is indicated in figures:

ABC	ABC	ABC	ABC	ABC	ABC	ABC	ABC
ABC	ABc	AbC	Abc	aBC	aBc	abC	abc
6	5	5	4	5	4	4	3

ABc	ABc	ABc	ABc	ABc	ABc	ABc	ABc
ABC	ABc	AbC	Abc	aBC	aBc	abC	abc
5	4	4	3	4	3	3	2

AbC	AbC	AbC	AbC	AbC	AbC	AbC	AbC
ABC	ABc	AbC	Abc	aBC	aBc	abC	abc
5	4	4	3	4	3	3	2

Abc	Abc	Abc	Abc	Abc	Abc	Abc	Abc
ABC	ABc	AbC	Abc	aBC	aBc	abC	abc
4	3	3	2	3	2	2	1

aBC	aBC	aBC	aBC	aBC	aBC	aBC	aBC
ABC	ABc	AbC	Abc	aBC	aBc	abC	abc
5	4	4	3	4	3	3	2

aBc	aBc	aBc	aBc	aBc	aBc	aBc	aBc
ABC	ABc	AbC	Abc	aBC	aBc	abC	abc
4	3	3	2	3	2	2	1

abC	abC	abC	abC	abC	abC	abC	abC
ABC	ABc	AbC	Abc	aBC	aBc	abC	abc
4	3	3	2	3	2	2	1

abc	abc	abc	abc	abc	abc	abc	abc
ABC	ABc	AbC	Abc	aBC	aBc	abC	abc
3	2	2	1	2	1	1	0

Summing up, we have:

1 individual with 6 growth genes measuring 100 cm
6	,,	5	,,	90	,,
15	,,	4	,,	80	,,
20	,,	3	,,	70	,,
15	,,	2	,,	60	,,
6	,,	1	,,	50	,,
1	,,	0	,,	40	,,

Or arranged in another fashion:

Number of dominant genes:	6	5	4	3	2	1	0
Height in centimetres:	100	90	80	70	60	50	40
Number of individuals:	1	6	15	20	15	6	1

Thus F_2 is composed of animals of seven different sizes. The greatest number of individuals, twenty, are of the average size, 70 centimetres, and sizes above and below the average are represented by equal numbers of individuals. A curve drawn to depict this situation would be perfectly symmetrical, with the maximum number of individuals (those of average height) at the summit, and a single individual (the shortest and the tallest) at each extremity.

In practice the seven classes would not be so neatly arranged — because we have for convenience limited the number of growth genes to three — and there would be individuals of intermediate heights. Also chance would play a part and, though among sixty-four individuals *about* fifty would measure between 60 and 80 centimetres as indicated, a smaller sample might easily give an F_2 which closely resembled the F_1 — in other words individuals halfway between the parents in height.

Unless correctly interpreted such phenomena could give the false impression that characters transmitted by the additive effect of polygenes do not obey the Mendelian laws of heredity.

Numerous quantitative characters are controlled by polygenes. By careful selection animal breeders can create pedigrees in which desirable genes accumulate. Such selection finally leads to a strain of homozygotes.

Multiple alleles

Geneticists long believed that genes could exist only in two forms, dominant and recessive. In 1904 Cuénot showed that the colour of the mouse's coat depended on several forms of the same gene. These differing forms or states arise from mutation and constitute multiple alleles or allelomorphic series.

The normal colour of the eye of the wild *Drosophila* is red; its recessive allele, white, has long been known. But between these two extreme colours there is a whole series of multiple alleles which determine the following colours: coral, eosine, cerise, apricot, honey, buff, tinted, ivory. All these alleles occupy the same locus on the chromosome.

Series of alleles, both in the vegetable and the animal kingdoms are far from rare; some series are known which contain as many as twenty alleles. Multiple alleles are often responsible for the position and diversity of colour in flowers, fishes, birds, and the pelts of mammals. In *Drosophila* a series of alleles determines the size of normal wings and also the size of wings as they grow shorter and shorter.

Other allelomorphic series determine regular gradations of colour which can be arranged in order of decreasing pigmentation from the darkest to the lightest, or of dimension, from the largest to the smallest.

Generally speaking the gene which is highest in the series is slightly dominant. Thus the red-eye gene of *Drosophila* dominates the entire range of colours, while the white-eye gene is dominated by all the others. Each gene is, in fact, dominated by those which precede it, and dominates those which follow it.

Lethal genes

A lethal gene is, as its name suggests, a gene which is deadly to the individual who carries it, unless its lethal influence is corrected or counter-balanced by the presence of its normal allele. Its power to kill was demonstrated by Cuénot in crosses between yellow mice. When yellow mice were crossed they always produced a mixture of yellow mice and non-yellow mice (grey, black or brown) in the proportion of two to one. These results seemed to contradict the ratio laid down in Mendel's laws. But another fact was noted which gave a clue to the enigma: the litters contained fewer mice than was customary, usually twenty-five per cent fewer. The explanation was thus clear: the yellow mouse can only live when it is heterozygous. The female then produces in equal numbers ova or gametes with the gene A_y (yellow) and gametes with the normal allele A or a. The cross between two yellow mice A_yA will produce one A_yA_y, two $A_yA,$ and one $AA;$ but, as the gene A_y is lethal when homozygous the A_yA_y embryos die in the uterus, where their dead bodies were indeed found. The cross thus engenders two yellow mice which are always heterozygotes for one grey or black *(AA)* mouse. The gene A counterbalances the lethal effect of its allele A_y which is dominant as regards colour but recessive in its lethal effect.

Creeper fowls with the thickening and shortening of their long bones also exist only as heterozygotes. The homozygote creeper is not viable and embryos with two genes for this condition die in the egg before it is hatched. They are even more malformed than the heterozygote creeper.

Albinism in maize is a lethal character. The plant dies rapidly because, through lack of chlorophyll, it is incapable of manufacturing carbohydrates. But if nourished artificially with a solution of sugar introduced into its leaves, albino maize will survive for several months, grow in size, produce further leaves and even inflorescences like a normal plant. The lethal action of the gene for albinism is corrected, but the gene itself remains unchanged and will be transmitted to the following generations.

A gene may be lethal in one environment and not in another. For example, the gene which inhibits the synthesis of adenine in the mould fungus *Neurospora* is lethal when *Neurospora* is cultivated on a medium which does not contain adenine. On a medium which does contain adenine the mutant fungus grows as well as the normal *Neurospora*.

If a recessive lethal gene is situated on an X-chromosome (or sex chromosome) of *Drosophila* its presence will be revealed by the proportion of sexes obtained in a cross between a female with the lethal gene and a normal male. The cross will give rise to two females per male, whereas normally there would be an equal number of the two sexes. But in this case the males which received an X-chromosome carrying the lethal gene have died. This happens because the gene's lethal action was not countered by the presence of its normal allele.

Pleiotropic genes

A single gene can exert an influence on several and indeed dissimilar characters which at first seem to have no connection with each other. Such genes are called pleiotropic (from *pleion*, 'more', and *tropos*, 'that which turns'). It is increasingly believed that the majority of genes are pleiotropic and even influence the organism's chemical and psychic reactions, its vitality, longevity, and fertility.

For instance, the *c* gene of the mouse inhibits pigmentation and produces albino mice, but at the same time it suppresses wildness and diminishes resistance to infection. The yellow mouse grows fat as it ages; it has less intelligence than albinos, and shows a tendency to contract cancer of the lungs.

Among *Drosophila* the *club* gene produces crumpled wings, prevents the development of two thoracic bristles, reduces the size of the eyes, flattens the head, and twists the thorax and abdomen.

The *vestigial* gene of *Drosophila* produces wings which are reduced to stumps, alters the position of the bristles, changes the insect's poisers, decreases the number of ovarian tubes, lessens vitality, and restricts fertility. It is not always easy to detect this type of genetic action, but among animals a few cases of it have been analysed.

A breed of curly poultry, studied by Landauer, is

characterized by its shrivelled plumage. In addition, its basic metabolism is abnormally high and its blood over-abundant; its heart beats with excessive rapidity. All these abnormal characters are closely bound up with each other. The unusual form of the feathers provides less protection than normal plumage and the result is an important loss in body heat. In compensation the basic metabolism increases and the flow of blood increases, entailing more rapid circulation with attendant over-strain on the heart — tachycardia. These conditions all derive from the morphological anomaly of the feathers for which faulty keratinization is responsible. The fault in production of horny tissue — or keratin — is no doubt determined by a single gene. Since this gene is responsible for several characters it is pleiotropic.

Another case of the same phenomenon among rats has been thoroughly investigated by Grüneberg. Some rats after birth develop abnormally and die before reaching maturity. They show the following symptoms: a few days after birth trouble in breathing appears; the shape of the thorax is malformed and the ribs have thickened. In consequence respiration is possible only by means of the diaphragm. Death follows more or less rapidly. Animals which succeed in living for two months suffer from a defective occlusion of the incisors and, being unable to eat, die of starvation. Dissection of the dead animals reveals anatomical and physiological anomalies. The lungs, which were normal at birth, are abnormal in structure because insufficient pulmonary alveoli were formed. The small number of these air sacs brought about the abnormal shape of the thorax which led to the impossibility of costal respiration. The lack of respiratory surface was compensated for by an increase in the number of red blood-cells and quantity of haemoglobin. The lack of pulmonary respiration led to cardiac enlargement (of the right ventricle), to obstruction of the pulmonary blood vessels and, in consequence, to frequent haemorrhage of the lungs.

The rigidity of the thorax has a further consequence: in the nest the nostrils of young rats are often obstructed by dust, which they get rid of by sneezing. But with a rigid thorax it is impossible to sneeze; the nose remains blocked and the baby rat, unable to suckle, dies of starvation.

In addition to all these troubles there is an anomaly of the cartilage which appears in the course of development. Whether this structural modification is produced by a gene or is the result of hormone deficiency is a problem which can be solved by grafting experiments. Abnormal cartilage grafted on a normal rat remains abnormal. Inversely, normal cartilage grafted on an abnormal rat remains normal. In other words, the structure of the cartilage is determined by its genetic origin and not by the action of hormones. The structural abnormality is, then, the result of genetic influence. By a mechanism which is still unknown the gene affects the growth of cartilaginous tissue by provoking a series of anomalies.

Further examples of such chain reactions set off by a single gene have been observed among mammals and birds. A gene, for instance, will affect a certain tissue, producing an anomaly which will itself bring about a whole series of secondary phenomena, each being at first effect and afterwards cause. Such series of linked phenomena can be drawn up as a 'pedigree of cause' as Grüneberg calls it.

How genes act

Precisely how a gene exercises its influence, by what mechanism it determines the appearance of a character, is a question to which the geneticist can often give no answer. Since the gene is a chemical unit its activity may be of a chemical nature, and there are in fact cases in which genetic action can be observed on the molecular scale.

The colour of flowers

The close relationship between genes and biochemistry is particularly striking in the case of the coloured phenotypes of flowers. Flower colours are the result of pigments which, with the exception of carotene and xanthophyll, belong to two groups, one of which produces yellow and ivory tints, the other — the anthocyanins — red and blue and all the intermediate shades of violet and purple. A molecule of anthocyanin is composed of one or two sugar molecules combined with a pigment, red, reddish-pink, violet, or blue-black.

Chemical reactions of great complexity, themselves under the control of genes, determine the colouring of flowers.

The colour of the eyes of Drosophila

The wild type of *Drosophila melanogaster* has red eyes. Numerous mutations affecting the colour of the eye occur in laboratory-bred insects, mutations yielding eyes of varied colours, from sepia (an eye rich in pigment) to white (an eye devoid of pigment). In this wide range of mutant colours two mutations are of particular interest: cinnabar (gene *cn*) and vermilion (gene *v*).

Ephrussi and his colleagues grafted larvae of the wild type of *Drosophila* with eye-buds removed from cinnabar and vermilion mutants. In their new environment these buds developed and acquired the red colour of the wild type of host. Thus the colour is not yet definitely fixed in the ocular bud, but is influenced by external substances which are present in the larvae to which it has been transplanted.

A vermilion eye-bud transplanted into a cinnabar larva will also develop into the red eye of the wild type; but a cinnabar eye grafted on to a host with vermilion eyes will retain its cinnabar colour. From these experiments the existence of two substances is deduced: a cinnabar substance and a vermilion substance, both of which are present in the wild type. The vermilion substance exists only in the cinnabar mutant, while neither substance is found in the vermilion mutant. The action of both substances is required to bring about the red eyes of the wild type.

Further experiments have shown that the vermilion substance will, except among cinnabar mutants, change into the cinnabar substance. The two substances are thus two consecutive links in the same chain of reactions; the vermilion mutant lacks the vermilion substance and, of necessity, the cinnabar substance.

What is the nature of these substances? They exist not only in *Drosophila* but also in the larval blood of other flies, *Calliphora,* and in the caterpillars of moths, *Galleria*. They act in minute quantities: a gram of dried pupae contains a sufficient amount of the substance to alter the eye colour of 20,000 flies. a gram of the pure substance would be capable of changing the eyes of 4,200,000 flies. They also act when swallowed. Mutant vermilion and cinnabar larvae fed with wild type pupae, crushed and mixed with agar, acquire red eyes — a demonstration that the agent responsible was introduced by ingestion. The substance is kynurenin, which the insect derives from tryptophan. Daneel has obtained insects with red eyes simply by plunging the eyes, at a suitable stage of development, into a solution of kynurenin and water.

Thus the colouring of the eyes in the wild type and in the vermilion and cinnabar mutants would seem to follow this course:

Tryptophan
derived from food

        gene v^+

gene v substance v^+ = kynurenin

        gene cn^+

   gene cn substance cn^+ = chromogen

      gene br gene br^+

Eye vermilion cinnabar   brown   red
     Mutants   Wild type

gene v equals vermilion gene
gene v^+ ,,   normal allele of vermilion
gene cn ,,   cinnabar gene
gene cn^+ ,,   normal allele of cinnabar
gene br ,,   brown gene
gene br^+ ,,   normal allele of brown

Tryptophan derived from nourishment causes vermilion eyes if the mutant gene v is present.

In the wild type, tryptophan, in the presence of gene v, produces kynurenin which is not localized in the eye but diffused throughout the organism of the larvae and pupae. Kynurenin, in the presence of gene cn^+ in the wild type, is transformed into a chromogen or substance cn^+ which, like kynurenin, is diffused through the organism of larvae and pupae. Finally, the gene br^+ of the wild type transforms the chromogen into the red substance of the wild type's eye.

When the normal gene cn^+ has mutated into its cinnabar allele the cinnabar mutant contains kynurenin in its organism. Similarly, when the normal gene br mutates into its brown allele the brown mutant, like the red-eyed wild type, contains the two substances — kynurenin and the chromogen — in its organism. Only the vermilion-eyed mutant has neither substance.

Ephestia kuhniella.

Thus the various colours of the eye of *Drosophila* depend on the presence of a definite chemical substance which acts like a hormone. But the presence of this substance, or its absence, depends on genes. Not only are the genes essential but equally important is the position at which they intervene in the chain of reactions.

The colour of the eyes and testes in the flour moth, Ephestia kuhniella

Among the larvae and pupae of the flour moth *Ephestia* a phenomenon which is in every way analogous may be observed. One strain of the moth which has black eyes and purplish-blue testes gives rise to mutants with red eyes and colourless testes. Transplantation shows that within the organism of the dark strain there is a substance required for the melanic pigmentation of eyes and testes. The mutants lack this indispensable substance, but if by transplantation, their eyes and testes are supplied with it then both organs will become coloured.

Genes and enzymes

The hypothesis that genes act by bringing about the formation of specific enzymes which facilitate biochemical reactions has received experimental support in work with micro-organisms.

The ascomycetous fungus *Neurospora* grows in a normal fashion on a simple synthetic culture medium which contains certain mineral salts, some carbohydrates, glucose and biotin. From this 'minimum medium' *Neurospora* is able to elaborate all the materials necessary for its growth: the amino acids, vitamins, purines, and pyrimidines. All these substances are found in the mycelian filaments of the culture, made in strictly aseptic conditions — evidence that the substances were synthesized. That they were synthesized under the influence of genes was demonstrated by Beadle and Tatum.

When irradiated by X-rays or ultraviolet rays such cultures produce spores which, in germinating, give rise to a certain percentage of mutant *Neurospora*, characterized by an incapacity to grow on the medium of the original culture. If, however, a substance which the mutant is unable to synthesize is added to the culture medium the mutant will develop. For example, the addition of leucine to the minimal medium enables a mutant which has lost the ability to synthesize leucine to grow. Thus the mutation appears as an incapacity of the gene which controls the production of the enzyme necessary for the synthesis of leucine. Such a gene mutation has two consequences: the substance normally produced is no longer produced; and the intermediate substances which take part in the synthesis accumulate and can, as we shall see, have secondary effects.

In the seven chromosomes of *Neurospora* Beadle has localized several genes which take part in biochemical reactions. The mutation of a gene is always revealed by an inability to synthesize the enzyme required for a definite reaction. The synthesis of arginine demands the cooperation of seven genes: four of them control the synthesis of ornithine; two others in two stages transform ornithine into citrulline, and the last finishes the process by changing citrulline into arginine. The chemical formulae of these amino acids show how one is derived from another. It is interesting to note that this method of synthesizing arginine has already been observed in the liver of mammals.

When the chain of reactions takes place normally the intermediate products are not easy to detect. But when a gene mutation breaks the chain the intermediate product accumulates. Horowitz has demonstrated this skilfully with *Neurospora* mutants which are incapable of synthesizing the amino acid methionine.

He succeeded in establishing the fact that the chain of reactions was: cysteine → homocysteine → methionine. The two first links, cysteine and homocysteine, differ by the radical CH_2; chemically, it is difficult to conceive how this radical is incorporated. Now there are two strains of mutants that are unable to change cysteine into homocysteine, which suggests that the transformation is done in two stages each under the control of a different gene. Experiment confirms this supposition: one strain accumulates a substance on which the other strain can live. The accumulated substance is cystathionine, an intermediate product between cysteine and homocysteine. The chain is as follows:

gene 1   gene 2   gene 3
cysteine → cystathionine → homocysteine → methionine

The mutation of gene 2 gives mutants which accumulate cystathionine, the chain of reactions being broken at this point. Cystathionine added to the minimum medium enables mutants in which gene 1 has mutated to grow. The mechanism which introduced the radical CH_2 can now be understood.

Thus genes control these biochemical reactions through the agency of enzymes which act as catalysts. As the enzymes are characterized by the fact that one specific enzyme is required for one specific reaction we may wonder whether each enzyme corresponds to a single gene, or whether one gene controls the action of several enzymes. Recent investigations tend to confirm the latter view.

Genes and enzymes in mammals

The direct connection between genes and enzymes is also observed among mammals. The pigmentation of the rabbit's or of the mouse's fur depends on the action of an oxidase (or oxidizing enzyme) called tyrosinase and on a chromogen *(Dopa)* which is turned into melanin. This is shown by the fact that an extract of crushed coloured hairs contains tyrosinase, while a similar extract of white hairs — or of albino hairs — does not. Now the gene C conditions the coloration of mammals, and when the C is replaced by its allele c the formation of pigment is inhibited. Thus the gene C affects cellular chemistry in such a way that tyrosinase develops in the hair bulbs at the base of the roots. Albinos which carry the c allele are deprived of tyrosinase and their hair remains white.

In this way the gene controls the appearance of a phenotype through a chain of connected reactions interposed between gene and character. Breakages in the chain at various points alter the final result, in other words, the character. In cases of deficiency the chain may be repaired by supplying the animal with the substance which is not being formed. But it is also possible to alter the chain of reactions by, for example, modifying metabolism or external conditions. The effect of the environment on the visible expression of the genes is the subject matter of phenogenetics. Here are a few examples:

Russian or Himalayan rabbits have white fur: only the extremities — ears, paws, nose, tail — are black. When they live in warm conditions the rabbits become altogether white and the dark areas completely disappear. In winter, when their burrows are cold, the fur of the young rabbits is often slightly dark all over. And if a small patch of fur is shaved from the back, and the animal is then exposed to the cold, the hair that regrows will be intensely black. On the other hand, if the paws are carefully shaved, but wrapped in a bandage to preserve local warmth, the hair which regrows will be white. The two experiments lead to the same conclusion: the black pigmentation is controlled by temperature and only appears when the temperature is sufficiently low. A skin temperature above 33 °C inhibits the formation of the oxidizing enzyme (oxidase) whose action on the chromogen produces the black pigment melanin. Low temperatures, on the contrary, encourage the manufacture of melanin. Temperature thus influences the action of the Himalayan character gene c^h, which controls the formation of the oxidase. The quantity of melanin produced is greater in the homozygote Himalayan rabbit, $c^h c^h$, with its two c^h genes, than in the heterozygote $c^h c$ which has only one. The homozygote also takes less time to elaborate the pigment than the heterozygote. The appearance of the character determined by the gene depends, then, on an external condition, namely the temperature of the environment in which the rabbit is reared.

Genes and enzymes in insects

There is a strain of *Drosophila* with 'abnormal abdomen', characterized by incomplete abdominal segmentation. Morgan observed that the strain was perfectly stable when reared in a sufficiently damp medium. If, however, the nutritive medium was dry and poor, the character would disappear. From this we need not deduce that the character 'abnormal abdomen' is not hereditary. Although it is possible in a dry medium to obtain several generations of flies with normal abdomens, the moment they are put back into a sufficiently damp environment the anomaly will reappear in their offspring. The regular transmission of the gene determining the anomaly continues, even if external conditions — dryness — prevent its visual appearance in the individual phenotype.

Genes and enzymes in plants

The phenotype of many plants depends on climate and sunshine, while their genotype remains unchanged. Under the control of three genes sun-red maize develops a reddish colour which only appears with the aid of sunshine: the outer zones of the plant are red while the inner zones remain green.

Hydrangeas have white, pink, and blue flowers. The change from pink to blue depends on the amount of aluminium in the soil. It occurs naturally in Brittany, for instance, due to the presence of slate, but the change in colour can easily be brought about by adding an aluminium salt to the soil. Hydrangeas contain delphinidine which in acidic sap is pink, but turns blue in the presence of aluminium.

Hormone phenomena

The nervous system also influences the manifestation of genetic characters. Caridroit removed the nerves from a black hen's wing; he then plucked its feathers. The feathers which grew again were white; but, if the nerves were given time to regenerate, the new feathers were black. In other words the black pigmentation only appears when the wing's nervous system is normal.

A modification in the organism's internal conditions can also alter the phenotype. The presence of thyroxin, the hormone secreted by the thyroid gland, determines the dominance of the character white in the cross between White Leghorn and Golden Leghorn poultry. If the thyroid of a heterozygous cock is removed, red feathers will grow on its breast instead of the usual white plumage.

If Andalusians — a breed of poultry with blue plumage — are crossed with Rhode Island Reds the first generation will be composed of purely blue hens and of cocks in which blue and red are mingled. In the female the character blue is completely dominant, while in the male the dominance is only partial. That this difference in hereditary behaviour is dependent on a difference between the two sexes' hormone content can

be shown by removing the ovaries of one of the blue hybrid hens. Red will then appear in her plumage. If, on the contrary, one of the hybrid blue and red cocks is castrated the red will disappear from his plumage.

A cross between a silver Ardennais cock and a golden Ardennais hen will produce golden cocks and black hens. After removing the hens' ovaries, silver plumage will appear; after castrating the cocks, black plumage will appear.

The female hormone is by no means alone in upsetting the customary interplay of dominant and recessive characters. A silver Sebright cock crossed with a golden Sebright hen will produce offspring which are all silver. Now if the hybrid cocks are castrated the golden pigment will appear among them — a proof that in this case it is the male hormone which is responsible for the dominance of silver over golden.

These phenomena of sex-controlled heredity — which must not be confused with sex-linked heredity — are of special interest to the physiologist, for they throw light on the actual mechanism of the genes, by demonstrating the chemical conditions in which genes function. In addition the investigator can take advantage of such experiments to produce hereditary characters which were previously unexpressed or latent.

Finally, we may reasonably hope that the hormones — or chemical substances in general — will supply a means of controlling the appearance of certain morbid hereditary characters.

Physiological genetics gives us a glimpse of one way in which the genes act, namely through the agency of diffusible substances which are produced under the control of the genes themselves. We are now in a position to know at least some of the processes which constitute the chain of reactions connecting the gene or primary cause with the phenotype or final result.

The action of certain genes also depends on their position on the chromosome. A change of position may modify a character, a phenomenon known as the *position effect*. The position effect seems to be found as frequently in plants as it is in animals.

The inheritance of acquired characters

The inheritance of acquired characters is a problem as difficult as it is controversial. It has given rise to countless misunderstandings not only among laymen but between specialists. That acquired characters are transmitted is more or less implicitly held by many educated people. It is in a sense an ingrained manner of thinking and to many persons transmission appears to be a reasonable and obvious fact of life. Sainte-Beuve found it natural to write that the 'literary wealth which my father amassed during his life had time to become firmly inscribed in his organism. He transmitted it to me when he engendered me, and from infancy I loved books ... in a word I loved what he loved. The point at which my father arrived was embedded, organically and instinctively, in a corner of my brain, and that has been my point of departure'. Bernard Shaw held that the organs are the creation of the animal's desire and will-power, 'the spirit becoming flesh.' Each generation departs from a point which no human means can distinguish from a starting point, a point which is not, however, zero. Even among scientists the question is still disputed and an authority on poultry breeding recently wrote: 'The theory of the transmission of acquired characters is still valid, assuming of course that it is conceived and expressed in a relative sense.'

On the opposing side of the debate there are numerous biologists who flatly deny the inheritance of acquired characters and hold that the concept has no significance whatever. But apart from the convinced, believers and unbelievers, there remain the hesitant, the undecided. Such people feel reluctance and even regret that they cannot accept a theory of heredity which would so easily solve so many and varied problems. They would be very relieved if facts allowed them to defend it, and they live in hope that such facts may be discovered.

The ancient Greeks believed with Aristotle and Hippocrates in the transmission of acquired characters. 'Mutilated children can be born to mutilated parents,' wrote Hippocrates.

In the eighteenth century the partisans of this theory were challenged by those who denied it. Buffon thought that dogs whose ears and tails had been docked would give birth to earless and tailless puppies. The results of domestication and acclimatization were also held to be hereditary. Charles Bonnet, on the other hand, advanced shrewd and powerful arguments to refute the fallacy. 'For two centuries,' he wrote, 'the English have cut off the tails of their horses and yet their horses are constantly born with tails. For an even longer period the Hottentots have removed one testicle from their children, but all male Hottentots continue to be born with two. The blind man's children have two eyes, and those of the one-armed man have two hands.'

Such bluntly contradictory opinions reflect the general disagreement which the subject of heredity then aroused. Buffon, who held the epigenetic theory that the germ-cell is always newly created, argued, that the new being was derived from formless semen and that its development was the result of successive additions. Thus if the body was modified the semen would be correspondingly modified, and the inheritance of acquired characters obviously followed. The opponents of epigenesis, the 'preformationists' such as Bonnet, held that the new being already exists in miniature in the organized germ — so that the inheritance of acquired characters did not follow.

With Lamarck the inheritance of acquired characters assumed capital importance, for on it he based his entire theory of evolution.

In the beginning Darwin also attributed a certain value to the transmission of variations acquired by usage or from the environment. He suggested the existence of 'gemmules' or hereditary units which might be produced by the cells of the body; modifications of these body-cells would then entail the production of modified gemmules, and in this way the transmission of the modifications would be accomplished.

There are, of course, hundreds of unsubstantiated tales which people delight in repeating to prove that parents hand on acquired characteristics to their offspring: the cat with a crushed tail that produced tailless kittens; the bitch who had puppies with malformed hindquarters because she had been violently struck across the spinal cord; the man with convulsions who acquired fallen eyelids which he transmitted to two of his three children. Then it is said that the udders of cows grow larger from generation to generation as a result of milking; that colts born of well-trained parents are themselves much more easily trained. Such venerable stories, and many more like them, are based on no experimental evidence and are totally lacking in scientific value.

The first experimental approach to the problem was made in 1847 by a Dutch veterinary surgeon. He operated on six calves — three of each sex — so that as adults they were hornless. The adults were then coupled and gave birth to calves which, like their grandparents, had horns. Flourens mutilated dogs by cutting off their tails or removing their spleens. The puppies they engendered always had tails and spleens. On the other hand, the experiments of Brown-Séquard seemed to suggest that acquired characters were transmissible: guinea-pigs which had suffered nervous lesions gave birth to little guinea-pigs with various abnormalities, structural (exophthalmia and malformation of the limbs) or functional (epilepsy).

Finally in 1887 Weismann attacked this complex and controversial problem. Reports were meanwhile so contradictory that it was impossible to know whether mutilations were transmissible or not. He decided on the simplest of experiments. He cut off the tails of male and female mice and allowed them to reproduce. He then cut off the tails of their offspring who again reproduced. This he repeated for five successive generations. All the mice were born with perfectly normal tails. The conclusion was inescapable: mutilations were not hereditary. In addition mutilations practised by racial groups such as circumcision among the Jews, the bound-up feet of the Chinese, the drawing of incisor teeth, the holes in the lip, the nose, the ears — none of these age-old mutilations had ever been transmitted. The problem of inherited mutilations was thus definitely settled.

Weismann's conception of a hereditary material led him to deny the transmission of acquired characters. For him the germ plasm had its own architecture, its 'specific molecular structure' which had no resemblance to the structure of the body. He professed himself unable to grasp how a somatic traumatism could produce the same injury in the germ plasm. 'Why this direct modification . . . even supposing that the injury could modify the plasm of the germ-cells why should it bring about a corresponding modification of the molecular structure? On the contrary why should it not bring about any of the thousands of other possible modifications? A supposition of this sort hardly merits being called a scientific hypothesis.' Weismann held that the soma and the germen were independent; the germ plasm was handed on unchanged from generation to generation. It was immortal, while the soma or the body-cells were not. It was, he said, extremely difficult to imagine any 'secret sympathetic mechanism which

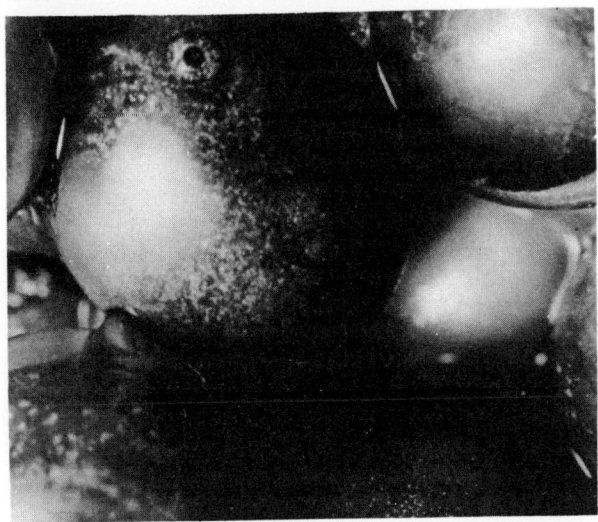

The midwife toad, *Alytes obstetricans. Below,* an enlargement of the eggs showing the rolled-up tadpole, its spotted skin, and one eye.

R. H. Noailles.

could enable each and every modification of the body to be photographed on the germ-cells.'

In Weismann's day a certain confusion in the meaning of words added to the general misunderstanding. He was well aware of this difficulty and defined exactly what was meant by 'acquired modifications'. An expression was required which should 'rigorously separate the two chief categories of modifications, namely the primary modifications of the body, and the secondary modifications which are the consequence of a germinal variation, whatever its origin may be. It is only the primary modifications which we have called *acquired;* they could also be called 'somatogenes' since they depend on the reaction of the soma to external influences. All other modifications of the body could, in contrast, be called 'blastogenes', that is to say the results of a germinal modification. In this way mistakes could be avoided.' To avoid all ambiguity it is sufficient to speak of characters acquired *by the soma* — that is,

by the body tissue as distinguished from the germ-cells.

After Weismann, and with the rediscovery of Mendel's laws, experiments became so numerous that it would be impossible to describe them all. All that need be said is that they unanimously revealed the weakness of the theory of the inheritance of acquired characters.

In 1923 the celebrated physiologist Pavlov had reported the inheritance of an acquired reflex in mice which had been trained to come for food at the sound of a bell. With each generation the descendants of these mice were more easily and rapidly trained. The hundred trials required to train the first generation became thirty in the second generation, twenty in the third, and only four in the fourth. But after Koltzoff's work in 1929 Pavlov would have recognized his error: he had, in fact, carried out a selection among strains of mice of different genotypes. Marxist biologists nonetheless lay great emphasis on this experiment of Pavlov's and attribute to him the honour of having 'formulated the idea that conditioned reflexes can be transformed in the course of evolution — that is, in the process of phylogenesis — into absolute conditioned reflexes.' Pupils of Pavlov are also said to have produced the hereditary transmission of conditioned reflexes. No one else, however, has been able to supply any serious confirmation of these alleged results, and they remain part of the Michurian mythology.

What are we to think of the school of Michurin and Lysenko? Much has been written about the war of the geneticists; here we shall recall only Michurin's technique of arboriculture, known as the 'mentor' process. Shoots or branches of fruit-tree A, which is a highly desirable species, are grafted on fruit-tree B when, it is claimed, the qualities of species A will be transmitted to species B, that is from scion to stock. Species A acts as 'mentor' or educator to species B, which thus acquires new and original qualities. In other experiments the mentor can also be the stock; cuttings made from such grafts are always modified and yet there is no exchange between the chromosomes in the cell-nuclei of the two plants. Nevertheless — so the argument goes — the hereditary characters of the scion A pass to the stock B. Hereditary characters are 'shaken up' by certain factors: grafting, environment, etc. Thus by means of grafting it is possible to transmit the special qualities of one species to another quite as easily as it is by sexual reproduction. Innumerable experiments made with potatoes, tomatoes, and fruit-trees demonstrated that 'vegetative hybrids are indistinguishable from natural hybrids' or hybrids which arise from a fertilized egg.

Michurin propounds the theory that plants can be developed, for he believes it possible 'to modify and to create species whose heredity has been directed by controlling the environment and the living conditions of the organisms. Heredity is the result of a concentration of those external influences which the organisms of preceding generations have assimilated.'

Lysenko adopted Michurin's theories and developed them. With his colleagues he carried out 'hybrid-grafting' on the grand scale in order to show that 'plastic matter', circulating from scion to stock and vice

307

versa, could hereditarily transform scion or stock into a more or less intermediate type of plant. He claimed to have demonstrated that the cells of the scion influenced the germinal cells of the stock. This influence included the transmission of such characters as form and colour to the descendants of the stock. Hence characters acquired by grafting were inherited.

It would be wearisome to relate in detail the criticisms of these astonishing claims, and we shall confine ourselves to reiterating that the experiments of the Marxist biologists have never, in spite of countless attempts, been successfully repeated in laboratories where due precautions have been taken to avoid sources of error. We may add, moreover, that the aggressive, partisan, provocative and political fashion in which the results of the experiments have been presented is enough to rob them of credibility.

It should, in passing, be remarked that if some experiments of the Michurian type appear to have a certain validity it does not follow that they prove the inheritance of acquired characters in the sense that Weismann clearly defined them: namely characters acquired by the soma, strictly somatic characters.

The crux of the problem of the inheritance or non-inheritance of acquired characters lies in deciding whether an acquired somatic character can be inscribed in a germ-cell. Various phenomena of pseudo-heredity, such as transmission of characters by the cytoplasm of the ovum, or of 'parallel induction', have nothing to do with the inheritance of acquired characters.

Partisans of the theory still use the argument of time: experiments, they say, cover too short a period of time to settle the matter. Or again, they argue, terrestrial conditions have changed so that today the phenomenon no longer occurs. On this level discussion is no longer possible.

A further argument is frequently advanced: certain structural characters whose genesis remains incomprehensible could be easily explained as the hereditary effects of usage or non-usage. Examples of such characters are the callosities of various animals like the camel, the ostrich and the wart-hog, or the abdominal curvature of the hermit crab which, in advance, takes on the exact shape of the empty shell which it will adopt. Further instances are the notch in the cat's lower lip which seems to be produced by the pressure of the teeth, and the notch in the right valve of the scallop *(Pecten)* which seems made to provide a passage for the byssus or filament secreted for purposes of attachment.

There are also some biologists who hold that we 'have no right to deny with absolute certainty the inheritance of acquired characters, even though it remains incomprehensible in our present state of knowledge.' Teilhard de Chardin was of the opinion that 'in spite of the assurance of our modern neo-Darwinians when it comes to denying everything which smacks of Lamarckism, it is rather hard to see how in animals, notably insects, numerous instincts which are surely hereditary today can have been established without fixation in the chromosomes of certain acquired habits — methods of building nests, hunting, etc. — which became gradually germinal by force of education repeated, with or without social

pressure, over a sufficiently large number of generations. Anatomically, it is true, man does not seem to have changed appreciably for some thirty thousand years. But, psychologically, is it so certain that we are the same? That is to say, are we quite sure that we are not born today with the faculty of perceiving and accepting as immediately obvious and natural certain dimensions, certain relationships, certain evident facts that would have escaped our distant ancestors? This alone would surely constitute sufficient proof that we are, biologically speaking, still on the move.'

The conclusion that we can, however, reach is that characters acquired by the soma as defined by Weismann are not hereditary. This conclusion, forced on us by experimental evidence, agrees well moreover with the present conception of what constitutes the hereditary substance, namely desoxyribonucleic acid, D N A. It is impossible to conceive how a somatic change could entail a change in the disposition of the atoms of a certain molecule of D N A, a modification which would be manifest in the following generation by the identical somatic modification. If this were so, we should be forced to assume that the constant kneeling of the camel brings about changes in the animal's system leading to the rearrangement of a molecule of D N A in its gametes, simply in order to produce a callosity on the knee of the camel's offspring!

Telegony

Almost as widely held as the belief in the inheritance of acquired characters is the conviction that heredity is affected by telegony — or impregnation.

Etymologically the word telegony (*tele,* distant, and *gonos,* semen) signifies the distant influence of the semen. It has been defined as 'the persistent influence exercised by the male over the hereditary potential of the female fertilized for the first time, so that this female will, in the course of future gestations, produce offspring which resemble the initial male, even though such offspring are engendered by other males.'

The supposition that the first male has an influence on the female's subsequent progeny is totally false. It has, however, been entertained by novelists as distinguished as Remy de Gourmont and Zola and, what is more serious, by the majority of animal breeders. Dog breeders have been particularly convinced of its reality.

Even in the human race telegony has its convinced defenders who will solemnly relate tales of white women who first have a child by a black man and afterwards give birth to children with negroid features although their fathers are white.

Countless experiments made under controlled conditions leave no doubt that the first male has no subsequent influence. So-called cases of telegony can always be explained when the genotypes of the progenitors are known. Although the myth should have long disappeared, it remains stubbornly rooted in the minds of many people who refuse to accept the indisputable evidence against it.

Extra-nuclear heredity

In the preceding chapters we have seen the capital importance of the chromosomes and the genes in determining and transmitting hereditary characters. Now, every cell is composed not only of a nucleus, containing chromosomes, but also of cytoplasm. The cytoplasm is the seat of reactions whose point of departure is genetic, and it plays an active part in the metabolism of the cell. But does it play an active part in heredity? In other words, is there such a thing as cytoplasmic heredity? Does the cytoplasm contain self-catalysing particles endowed with genetic continuity? Does it possess (as Wettstein suggests) a 'plasmon', or (as Darlington suggests) 'plasmagenes'?

Simple experiments in reciprocal crosses clearly indicate that the cytoplasm plays some kind of role; its influence is not negligible.

A cross between the female *Viola arvensis* and the male *Viola rothomagensis* produces good seed which germinates well. The reciprocal cross between the male *Viola arvensis* and the female *Viola rothomagensis* produces wrinkled seeds which do not germinate. Female *Epilobium hirsutum* fertilized by male *Epilobium parviflorum* gives rise to a small hybrid with wrinkled leaves which is almost totally sterile. The reciprocal cross produces larger hybrids with normal leaves and numerous flowers.

No reason which can be attributed to the chromosomes explains why these crosses should give different results, and the phenomena suggest an intervention of the cytoplasm. The essential difference introduced by the two cytoplasms arises from the fact that the female gamete is well provided with cytoplasm while the male gamete has a relatively minute amount. It is conceivable that the nucleus of the hybrid is incompatible with the cytoplasm of *Epilobium hirsutum* but compatible with that of *parviflorum*. But is the supposed character of the cytoplasm of *hirsutum*, as opposed to that of the cytoplasm of *parviflorum*, stable and independent? If it is, it must persist and be transmitted in all subsequent crosses. Further experiments with *Epilobium* suggest that this hypothesis is correct. The incompatibility of the cytoplasm with a foreign nucleus is transmitted independently of nuclear control.

The case of *Epilobium* is not unique; analogous cases have been observed in yeasts which transmit cytoplasmic characters by means of asexual reproduction.

Among animals similar phenomena are known. Kühn crossed two strains of the insect *Habrobracon* (of the order Hymenoptera), one of which is more highly pigmented than the other — a genetic character. The female hybrids were intermediary, but the offspring of the lighter hybrid females were appreciably less pigmented than the offspring of the darker females, although the genotype of all the hybrids was identical. This difference was also apparent among the haploid males parthenogenetically produced by the female hybrids.

A phenomenon of the same type is found among certain Gramineae in which the male is sterile; the sterility is transmitted by the cytoplasm. When all the chromosomes of a sterile male strain are replaced by chromosomes of the normal strain, the male strain remains sterile, although in succeeding generations the degree of sterility is less pronounced.

Variegated plants

Variegated plants, so often cultivated in public parks, have parti-coloured leaves: green and white, green and yellow, or green and red. The variegation may be uniform and symmetrical or irregular, some branches being totally green while others are totally white or yellow. In the cytoplasm of their cells the green leaves contain plastids which manufacture chlorophyll; the yellow and white leaves have plastids which do not form green pigment.

The ability to manufacture this green pigment depends on a pair of alleles: the normal gene controlling the manufacture of chlorophyll has a mutated allele, sometimes dominant, sometimes recessive, which inhibits the formation of pigment. The aspect of the plant thus depends on the plastids which the gametes contribute. Three eventualities are possible: firstly, the fertilized egg may contain the normal chloroplasts and will then engender a normal green plant; secondly, the fertilized egg may possess cytoplasm which contains normal and mutated chloroplasts, in which case the plant will be variegated; and finally, the fertilized egg may have cytoplasm containing only mutated plastids, and the resulting plant will be white and, being unable to accomplish photosynthesis for lack of chlorophyll, cannot live.

Flowers growing on the three types of branches — green, white, and variegated, — of the same plant have been fertilized by the pollen of flowers from a white, green, or variegated branch. The results are as follows:

Pollen from a Branch	*Fertilized Flower from a Branch*	*Plant grown from seeds of fertilized flowers*
White	White	White
	Green	Green
	Variegated	Pale, green, variegated
Green	White	White
	Green	Green
	Variegated	Pale, green, variegated
Variegated	White	White
	Green	Green
	Variegated	Pale, green, variegated

It will be seen that the three types behave in the same way; the type of pollen is immaterial. The essential factor is supplied by the female plant and it would seem that the maternal gametes transmit the white elements which produce white branches, the green elements which produce green branches, and the white and green elements which produce variegated branches. These elements are green chloroplasts and colourless plastids or leucoplasts. Thus variegation is a clear case of extra-nuclear heredity — although the action of certain genes on the green colour of the plants is not thereby excluded.

The killer Paramecium

Paramecium is a genus of ciliophoran which abounds in ponds containing decaying matter. Among these unicellular animals Sonneborn has reported a typical case of cytoplasmic heredity.

One strain of *Paramecium* — the killer *Paramecium* — secretes into its culture medium a toxic substance, paramecin, which kills almost all other strains. The killer *Paramecium* is immune to its own poison. It differs from non-resistant strains by the possession of a character which requires the co-existence of two factors: a dominant gene *K* and a cytoplasmic factor *kappa*. The kappa particles are responsible for the production of the toxic substance, but the self-reproduction of the kappa particles is controlled by the gene *K*. When this gene is absent the kappa particles disappear, and cannot be made to reappear unless reintroduced by other killer *Paramecium*. In other words, although the gene *K* is indispensable to the maintenance of the kappa particles it cannot by itself bring about their formation.

The kappa particles contain D N A; they measure from 0.2—0.8 microns and there are from 200 to 1,000 of them in a killer *Paramecium*. They are destroyed by high temperature and by X-rays. In this event the killer *Paramecium* is no longer immune. The particles are capable of mutation and then give rise to other varieties of *Paramecium*.

The kappa particles can be introduced into the cytoplasm of non-resistant *Paramecium*, either by conjugation when the cytoplasms mingle, or by placing the animals in contact with a killer extract rich in kappa particles.

The homozygote *KK* and the heterozygote *Kk*, with cytoplasm containing kappa particles, are killers. The homozygote *kk*, with cytoplasm which contains kappa particles derived from a killer *Paramecium*, is not sensitive to the poison, but becomes sensitive to it after a few cell-divisions, when the kappa particles have disappeared. Both the homozygote *KK* and the heterozygote *Kk* in cytoplasm which is devoid of kappa particles are sensitive.

Thus the kappa particle seems to be a cytoplasmic element which is able to reproduce itself and be transmitted from generation to generation. Its presence, however, depends on a gene in a nuclear chromosome.

Cancerous mice

Cancer of the breast in mice and other rodents is, determined primarily by genetic factors. By selection it has been possible to obtain highly cancerous strains — from seventy-five to 100 per cent — and other strains which are resistant to cancer.

Female hormones also play their part: if the ovaries of the females belonging to a cancerous stock are removed before they begin to function the females will no longer develop cancer. Males born of cancerous stock normally escape cancer, but if they are given continuous doses of oestrogen, the female hormone, a high proportion of them (as many as 100 per cent) will develop the disease.

Finally, reciprocal crosses between cancerous and non-cancerous strains and back-crosses with the parents show that females of F$_1$ have a high percentage of tumours (thirty-six to ninety per cent) when their mothers belong to a cancerous strain, and a low percentage (seven to zero per cent) when their mothers belong to a more or less resistant strain. This marked difference in the results of cross-breeding indicates the existence of an extra-nuclear maternal factor which is transmitted by the mother to her descendants.

This factor, which is found in the blood, the spleen, and the thymus, is transmitted by the milk. Baby mice from a highly cancerous strain when suckled by a mother belonging to a resistant strain very rarely develop tumours of the breast.

The formation of tumours in the females is determined by the interaction of genes, oestrogenic hormones, and the milk or Bittner factor. This factor multiplies in normal and cancerous cells. It contains R N A and may be a virus.

The maternal influence

Certain other crosses also show that there is a cytoplasmic maternal influence which is apparent in the offspring. The flour moth *Ephestia kuhniella* has black eyes and dark-purplish testes; its larvae are reddish or yellowish in colour. A mutant strain with the recessive gene *a* is characterized by absence of pigment, has red eyes, testes either colourless or pale yellow, and larvae which are white or very pale green.

Transplantation experiments show that in the homozygote *AA* the gene *A* is responsible for the production of kynurenin which is required for the formation of pigment. The recessive mutant *aa*, in the absence of gene *A*, is incapable of producing kynurenin.

The cross between a coloured female heterozygote *Aa* and a colourless male *aa* gives rise to offspring composed of heterozygotes *Aa* and homozygotes *aa*. The latter, which have no gene *A*, do not form pigment. Nevertheless the skin and eyes of the larvae are lightly pigmented like their mothers — pigmentation which disappears in the following generation.

How can this pigmentation be explained? The gene *A* conditions the production of kynurenin; but kynurenin is produced by the testes, the ovaries and the brain. In the ovaries of the mother *Aa* there is enough kynurenin to colour the skin and the eyes of the larvae. Larvae *aa*, having no gene *A*, do not, however, produce further kynurenin. Hence the quantity of it transmitted by their mother diminishes, so that their pigmentation is only a passing phase. The phenomenon is thus a case of the transmission of a substance by the cytoplasm of the maternal ovum, a substance which rapidly disappears in the absence of the gene responsible for its fabrication.

In conclusion it may be said that cytoplasmic or extra-nuclear heredity is the exception, while heredity conferred by the nucleus and its chromosomes is the almost universal rule.

Human Heredity

Less is known about human heredity than about heredity in plants and animals. The study of the latter progresses rapidly, and is a prerequisite to real advance in our knowledge of human genetics. For this reason most books on genetics are, to the disappointment of the average reader, filled with long discussions about animal and plant heredity. Most people demand facts about the heredity of the human race and are eager to know why their children, for instance, have blue eyes or curly hair. Why are there twins of the same sex who do not resemble each other? Is it wise for first cousins to marry? In what way does the Rhesus blood factor endanger marriages? It is natural to ask such questions and hundreds more about one's own species but, as we shall find, man constitutes poor material for genetic analysis. Fortunately the laws of heredity are the same for all organisms, and it is therefore legitimate to apply results laboriously acquired among plants and animals to the human species. We must not forget that the laws of heredity were first discovered by experiments with the common pea, that precise details of important mechanisms have been furnished by the fruit fly, that the processes of cell-division, of meiosis and of fertilization were first fully analysed in the sea-urchin, in frogs and in *Ascaris*, a parasitic worm. It is, indeed, impossible to speak of man alone; he forms part of a whole from which he cannot be isolated.

The data of human genetics accumulate at random and can hardly be based on carefully controlled experiments in cross-breeding! In the case of animals and plants we have seen the overwhelming importance of such experiments. Consanguinity and selection have enabled us to obtain homozygote stocks, strains which breed true. Successive generations, F_1, F_2, F_3, etc., and back-crosses reveal the behaviour of hereditary characters and establish the composition of the genotype.

The laws of heredity are, as we have seen, statistical laws; that is to say, their full significance only appears when dealing with large numbers of offspring. For various reasons, voluntary and involuntary, the human couple produces relatively few offspring. Defective or abnormal individuals frequently die in infancy before careful and detailed observations have been made. Many people remain unmarried and information which their descendants might have supplied is lacking. For such reasons the Mendelian ratios which apply to dominant and recessive characters and to normal and abnormal subjects alike are not clearly evident when we study human families.

Apart from the smallness of human families there are other difficulties. The very duration of the geneticist's life allows him at most personally to observe four generations. To complete a genealogical table he must have recourse to official records, parish registers, and so forth; to written material or, still less satisfactory, to oral accounts which can be twisted or falsified. One important source of error arises from illegitimate births and from adultery. The dispersal of families also adds another serious difficulty. Even worse is the refusal of many families of great genetic interest to allow themselves to be investigated.

The family enquiry usually begins by the chance discovery of an individual who carries some striking hereditary character, normal or pathological. Beginning with this person efforts are then made to examine as many of the members of his family as possible so that his pedigree or genealogical tree can be drawn. From this graphic representation, with its conventional signs for various characters, it is often possible to observe the manner in which a given character is transmitted.

More general enquiries can also be undertaken. For instance, a search may be made for all examples of some hereditary character in a given region, especially an isolated region like a remote mountain valley or an island. Such broad investigations have often been made in Scandinavian and Anglo-Saxon countries.

311

Examples are the studies made by Sjögren of Huntington's chorea or St Vitus dance in two parishes in the north of Sweden, and enquiries made by Hanhart in the Grisons valleys of Switzerland. Work of this nature, of course, requires a considerable expenditure of time and money.

In view of the present-day increase in atmospheric radioactivity it is obviously of importance to detect any modifications affecting hereditary factors at the earliest possible moment. It was for this purpose that H. B. Newcombe proposed certain methods suggested by the techniques used to gather and collate vital statistics. In Canada mechanical processes were devised which enabled perforated cards to be grouped and sorted out in a way that provides an immediately available card-index of families. A card could be devoted to every marriage celebrated after a given date. At the birth of each child a card would be added and classified by relationship. Cards would also be kept and similarly classified for foetal deaths and the decease of all descendants. All marriages between cousins could easily be identified and the births and deaths of their offspring recorded. In some such manner it would then be possible to obtain vital information concerning a great number of individuals and their inter-relationship.

Dominant and recessive heredity

The Mendelian laws of heredity are of universal validity and apply as much to the human species as to animal and plant species. Human genetics obeys the same laws and employs the same vocabulary as animal and plant genetics.

In both homozygotes and heterozygotes a dominant character is always apparent. A recessive character, on the other hand, only appears when the character is homozygous. In heterozygotes it is masked by its dominant allele. The consequences are: 1. In dominant heredity the inherited character must of course be transmitted by an individual who carries this character. When only one parent possesses the gene of a dominant character in a heterozygous state — that is, one dominant gene only — one child in two will on the average inherit the character. The dominant character is present in each generation, but it may at times appear to skip a generation or two. In this case the dominance is said to be irregular. Since it reappears the gene is always present in the hereditary endowment, but for some reason it does not express its presence in every generation. Each gene is characterized by its percentage of *penetrance* or frequency with which it expresses itself. A character which is regularly dominant has a penetrance of 100 per cent: a gene may have a penetrance of only one per cent, in which case, though dominant, its appearance will be rare. The genes which determine many illnesses, including asthma, sugar diabetes, leukaemia, polydactylism, harelip, and congenital torticollis, are of irregular penetrance. Weak penetrance accounts for the phenomenon of a skipped generation or irregular dominance. It is difficult to be certain whether a character with a penetrance of less than ten per cent which appears irregularly is the result of an incompletely dominant gene or of a recessive gene.

The degree with which a character is manifested may vary among members of the same family in which case the *expressivity* of the gene is then said to be variable. An example is a family of polydactyls in which individuals have a different number of supernumerary fingers or toes.

2. In recessive heredity the character is rarely manifested. It is often transmitted by individuals in whom it does not appear. It may, however, be found in the individual's pedigree. The character can skip many generations, after which it often reappears in the offspring of marriages between blood relations.

If a recessive gene determines a disease, two parents who are themselves apparently healthy but are both heterozygotes for this gene can engender one diseased child for three healthy children, two of whom will be heterozygotes, that is to say, carriers of the recessive gene. Two diseased parents (homozygous for the gene) will produce offspring composed entirely of children affected by the disease.

It would thus be eugenically valuable to be able to detect those who are heterozygous for a recessive defect, individuals apparently healthy but carriers of the gene responsible for the defect. Certain observations seem to indicate that recessive characters can cause slight anomalies even in heterozygotes. Thus drepanocytanaemia or sickle cell anaemia (red blood-cells are normally biconcave) is common among African negroes, but only fully manifest among homozygotes for the recessive gene of this affliction. Heterozygotes for the gene have numerous sickle-shaped red cells in their blood without, however, any other symptoms of the disease. The same thing is true of Mediterranean or Cooley's anaemia which seriously attacks only homozygotes, while heterozygotes suffer from very slight anaemia and reveal a few signs of erythrocytosis, or increase in the number of red blood-cells. The heterozygote for the gene which causes pigmentary retinitis shows a slight but unmistakable modification in the depth of the eye. The heterozygote for the gene of gout will have a higher than normal percentage of uric acid in his blood. It would therefore seem theoretically possible to identify heterozygotes who carry recessive genes, but to do so would require elaborate clinical, morphological, physiological, and biochemical analyses.

The following is a list of a few normal or sub-normal characters paired as dominant and recessive.

Dominant	Recessive
Dark iris	Light iris
Black or red hair	Fair hair
Curly hair	Straight hair
Grey hair at 25 (canities)	Hair greying normally
White frontal lock	Absence of white frontal lock
Hapsburg lip	Normal lip
Ear-lobe free	Ear-lobe attached
Normal pigmentation	Albinism

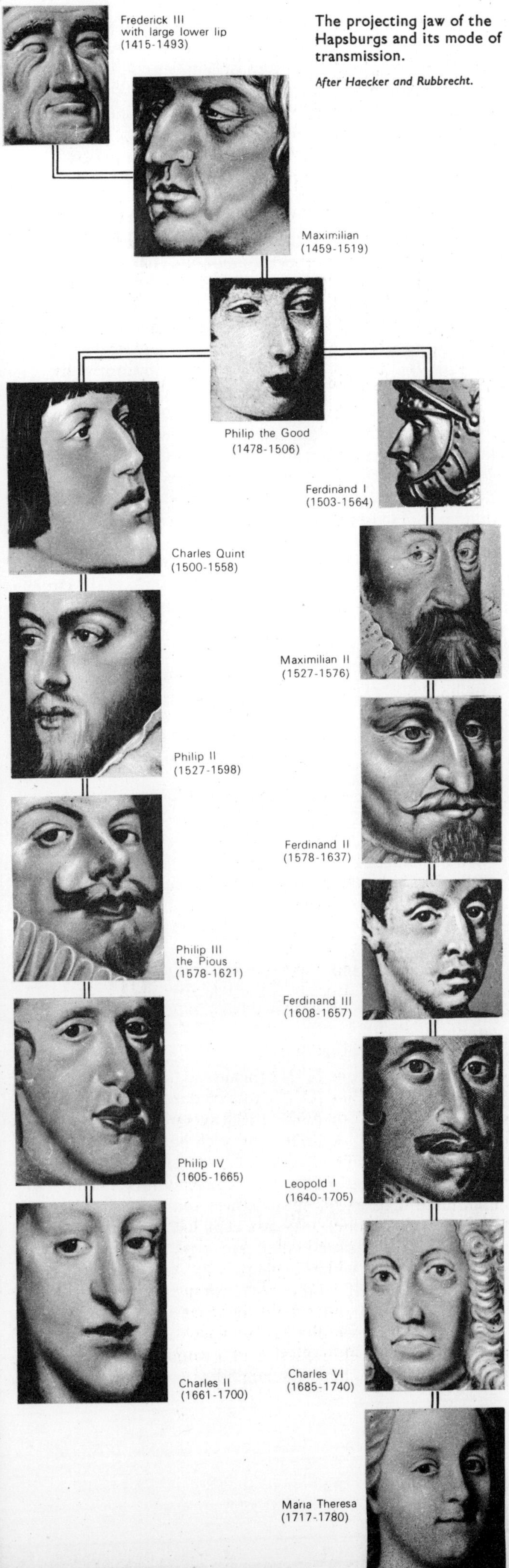

The projecting jaw of the Hapsburgs and its mode of transmission.

After Haecker and Rubbrecht.

Frederick III
with large lower lip
(1415-1493)

Maximilian
(1459-1519)

Philip the Good
(1478-1506)

Ferdinand I
(1503-1564)

Charles Quint
(1500-1558)

Maximilian II
(1527-1576)

Philip II
(1527-1598)

Ferdinand II
(1578-1637)

Philip III
the Pious
(1578-1621)

Ferdinand III
(1608-1657)

Philip IV
(1605-1665)

Leopold I
(1640-1705)

Charles II
(1661-1700)

Charles VI
(1685-1740)

Maria Theresa
(1717-1780)

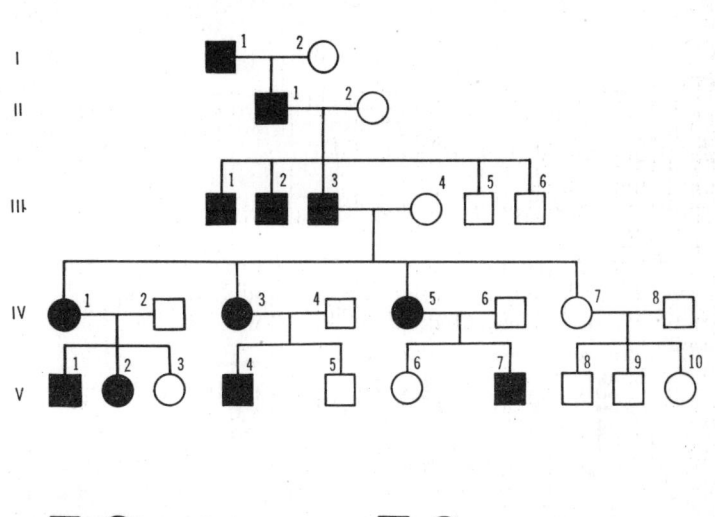

■ ● black frontal lock   □ ○ white frontal lock

Pedigree showing the transmission of the character 'white frontal lock' in a certain family.

After L. Pitch.

For 400 years the children of the Haanapel family of Doesburg in Holland have been born with black hands. The mother, *on the left,* has normal palms. *Below, right,* Jaap Haanapel holds his hand over a candle. The black skin of his palm is so thick that he does not feel the heat. The palms of his son reveal the same characteristics.

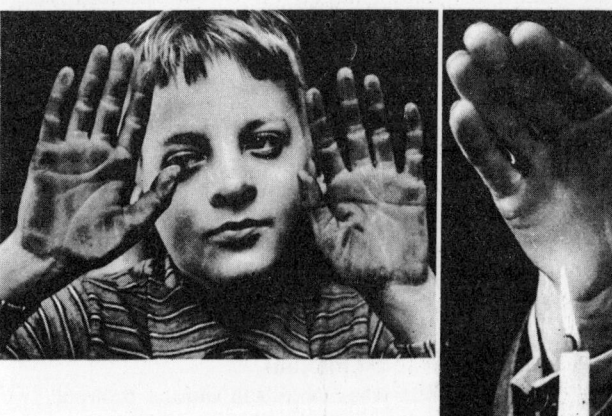

313

Freckles (Ephelides)	Absence of freckles
Hair on the middle phalanx of second, third, fourth and fifth finger	Absence of hair on these phalanges
Obesity	Thinness
Space between incisor teeth (diastema)	Absence of diastema
Polydactylism	Normal number of digits
Ability to taste phenylthiocarbamide	Inability to taste phenyl-thiocarbamide

Sex-linked heredity

All the genes situated on the autosomes of either men or women act in the same fashion; as they belong to the paternal or to the maternal genome (or set of all chromosomes found in each nucleus of a given species) their transmission is identical. This is not however true of the genes which are carried by the sex chromosomes, X and Y. Women have two X-chromosomes, while men have one X-chromosome, which is identical to those of women, plus a Y-chromosome, which is much smaller. In length the X-chromosome measures from 4—5 microns (thousandths of a millimetre) and the Y-chromosome 1.5 microns. Because of this difference in size the genes situated on the X and Y chromosomes are divided into three categories.

The first category includes those genes found in that portion of the X-chromosome which has no corresponding portion, or homologue, in the Y-chromosome. The action of such genes is always apparent in men because they have no alleles on the Y-chromosome to counterbalance it. They are known as the sex-linked genes and are responsible for such things as haemophilia, Daltonism, dipsomania, and optic atrophy. A father who carries one of these genes transmits it to all his daughters but to none of his sons. Women who are homozygous for these genes transmit them to all their offspring. In a woman who is heterozygous for a recessive gene the character does not appear because the gene is masked by its normal allele carried on her other X-chromosome. She is, however, a conductor — in other words, a carrier of the recessive gene, and can transmit it to half of her sons and to half of her daughters.

The second category of genes are those found in the homologous parts of the X and Y chromosomes. They behave like genes situated on the autosomes and are known as 'partially sex-linked'. Such, for example, are the genes determining pigmentary retinitis, day blindness or hemeralopia, total colour blindness or achromatopsia, and so on.

The third category of genes occupies that portion of the Y-chromosome which has no corresponding section on the X-chromosome. These genes are responsible for characters which appear only in men, a phenomenon known as holandric heredity. They are linked to the Y-chromosome, and examples of them are the genes of webbed-feet, keratodermia palmaris, hypertrichosis of the ears, ichthyosis, etc. The father transmits such genes to all his sons.

Let us examine what occurs in unions between two people who carry a sex-linked gene, the gene, for instance, of haemophilia. This disease is an affliction characterized by the delayed coagulation of the blood. Haemophiliacs do not bleed spontaneously but the slightest wound brings on a haemorrhage which no haemostat or bandage succeeds in stopping. An extracted tooth or a bleeding nose may bring on a slow and continuous haemorrhage which can last for several days. The case is cited of a haemophiliac who, after a dental operation, bled for twenty-two days and finally died.

This deficiency in coagulation arises from the absence of a globulin, the anti-haemophilic globulin, which is present in normal plasma and helps to convert prothrombin into thrombin. The disease is hereditary — it is transmitted by women and attacks men.

The gene of haemophilia, h, which is dominated by the gene H in those whose blood coagulates normally, is situated in the X-chromosome. Obviously only a woman who is homozygous for the gene X^hX^h will be a haemophiliac. The heterozygous woman X^HX^h will not suffer from haemophilia, but she will transmit the recessive gene h to half her progeny. On the other hand, however, every man who inherits a single h gene will be a haemophiliac, X^hY, since the gene is not masked by its dominant allele.

Let us imagine a marriage between a normal woman X^HX^H and a man with haemophilia X^hY. All the children will receive from their mother the normal gene H, while the daughters will receive from the father the gene h of haemophilia.

Parents X^HX^H X^hY
Gametes X^H X^H X^h Y
F_1 X^H xX^h → X^HX^h (conductor daughters)
       X^H x Y → X^HY (healthy sons)

The sons will be healthy. The daughters too will be healthy but conductors, for they carry the gene of haemophilia which they can transmit to their offspring.

That these apparently healthy daughters are dangerous is shown by their marriage to a normal man. They transmit the gene H of normal coagulation to half their children and the gene h of haemophilia to the other half. In this second half the boys, having the h gene, will have haemophilia, while the girls, although unaffected by the disease, will be carriers.

Parents X^HX^h X^HY
Gametes X^H X^h X^H Y
F_1 X^H x X^H → X^HX^H (healthy daughter)
       X^h x X^H → X^hX^H (carrier daughter)
       X^H x Y   → X^HY (healthy son)
       X^h x Y   → X^hY (son with haemophilia)

In the offspring of a woman carrier and a haemophilic man a haemophilic daughter can appear. Such a union will produce fifty per cent haemophiliacs of whom half are girls and half are boys. This type of marriage is rare, which explains why women haemophiliacs are rare. If both parents are haemophiliacs all their children will suffer from the disease.

In the case of heredity just analysed a haemophilic son was the result of the union of a woman carrier and a normal man. A male haemophiliac can engender a

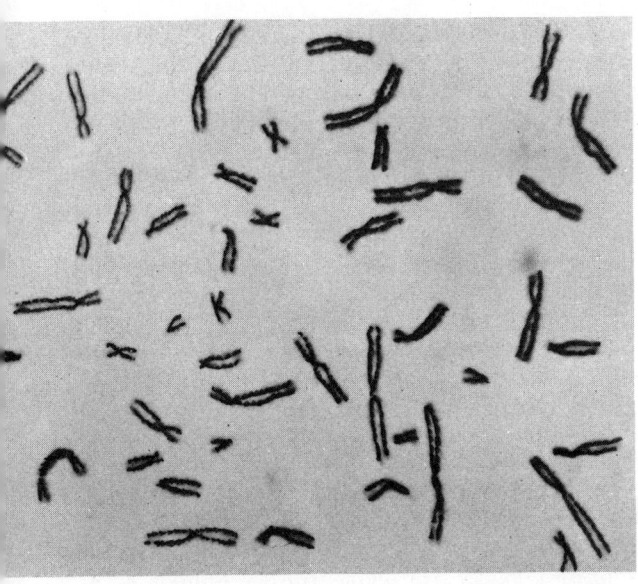

The 46 chromosomes of a man as they are dispersed in a normal cell.

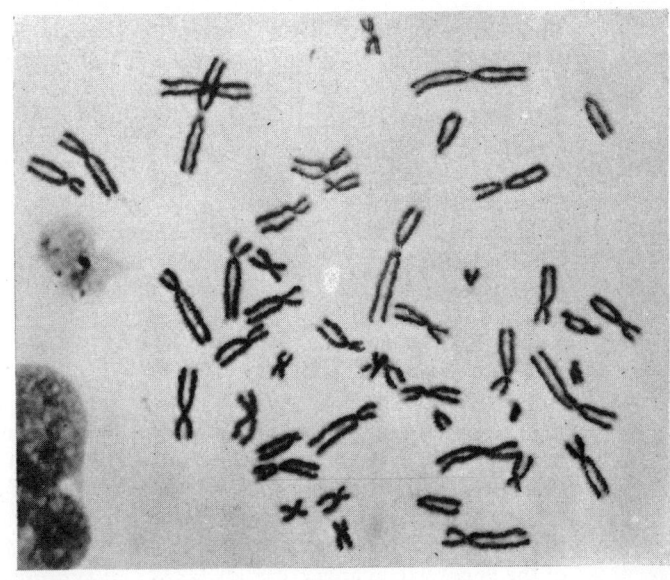

The 46 chromosomes of a woman as they are dispersed in a normal cell.

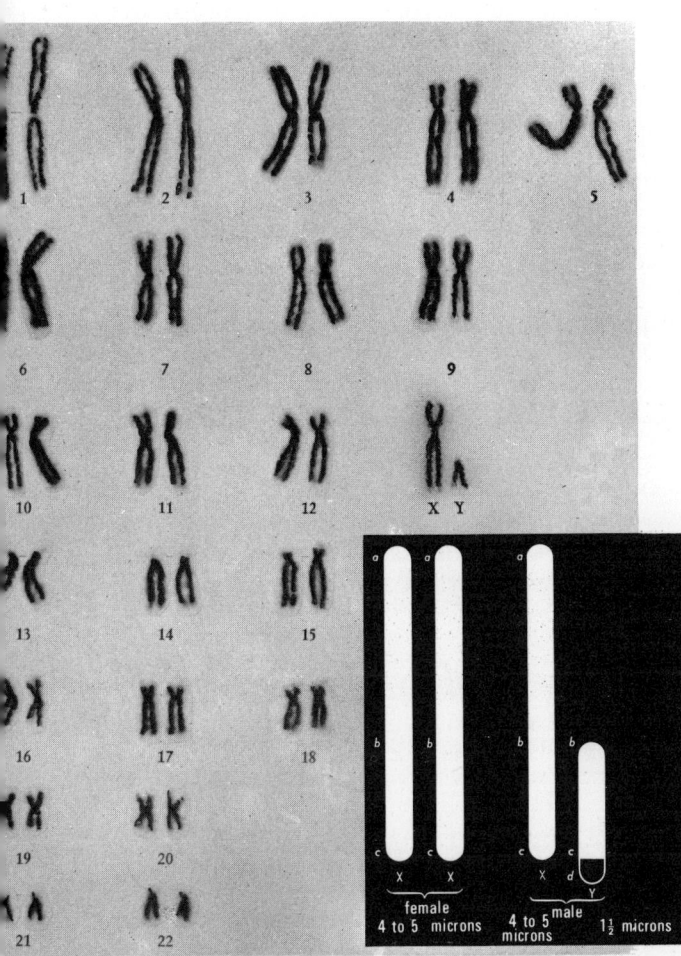

Karyotype showing a man's 46 chromosomes arranged in homologous pairs: 22 pairs of autosomes numbered from 1 to 22, and the two sex chromosomes X and Y.

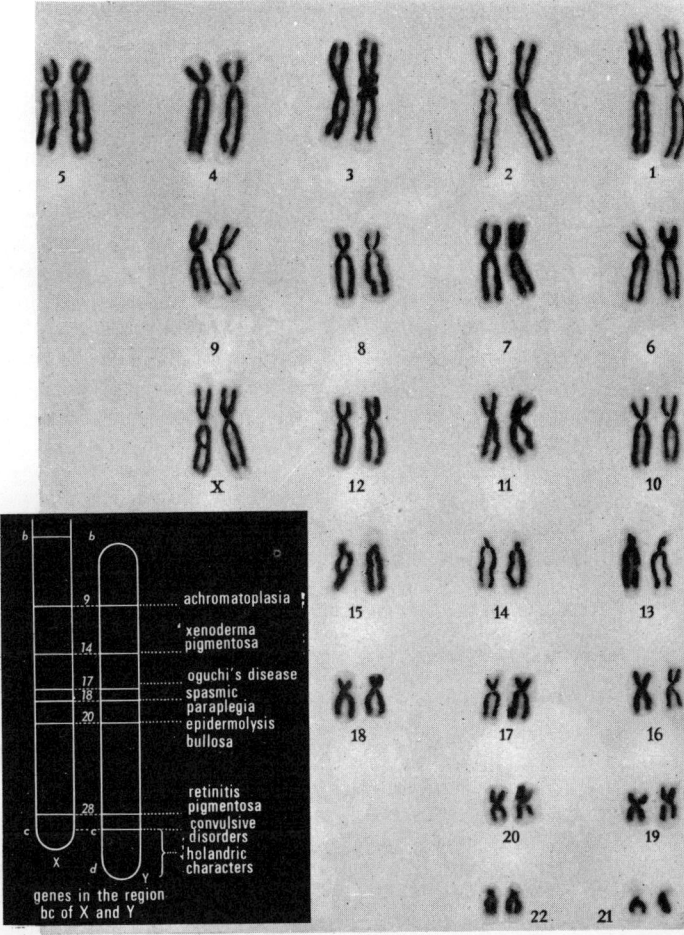

Karyotype of a woman's 46 chromosomes arranged in homologous pairs: 22 pairs of autosomes numbered from 1 to 22, and two X or sex chromosomes.

Lejeune, Faculty of Medicine, Paris.

315

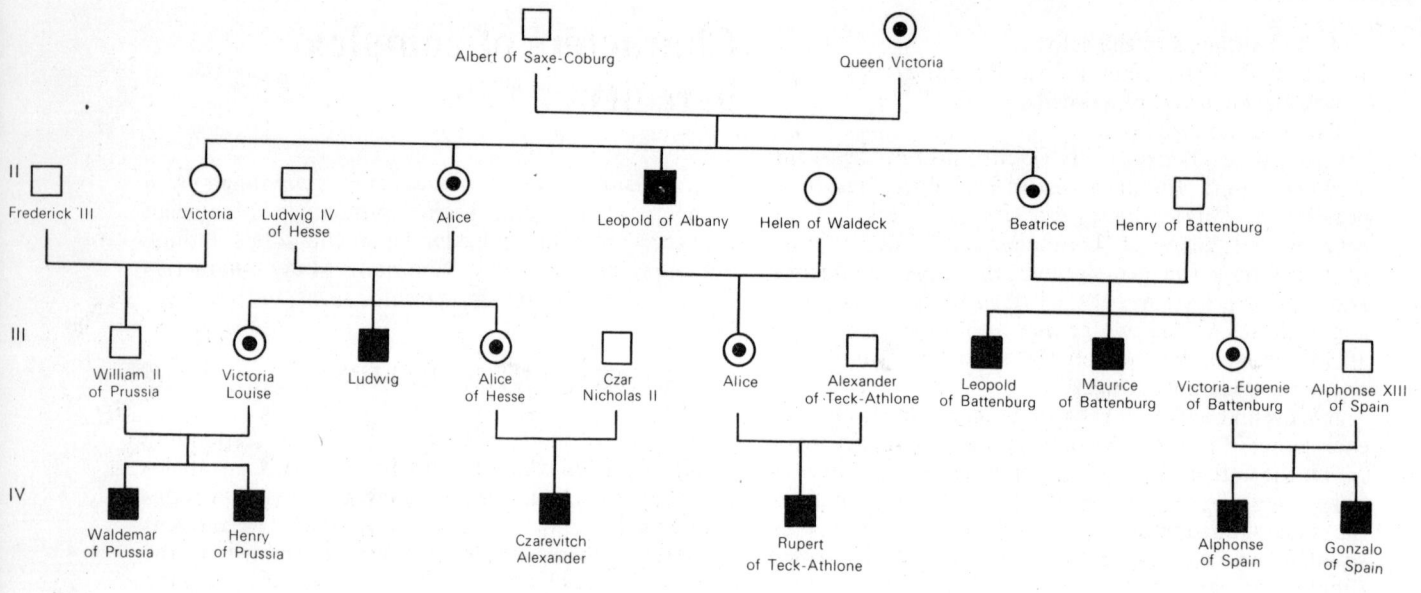

Frederick III · Victoria · Ludwig IV of Hesse · Alice · Leopold of Albany · Helen of Waldeck · Beatrice · Henry of Battenburg

III

William II of Prussia · Victoria Louise · Ludwig · Alice of Hesse · Czar Nicholas II · Alice · Alexander of Teck-Athlone · Leopold of Battenburg · Maurice of Battenburg · Victoria-Eugenie of Battenburg · Alphonse XIII of Spain

IV

Waldemar of Prussia · Henry of Prussia · Czarevitch Alexander · Rupert of Teck-Athlone · Alphonse of Spain · Gonzalo of Spain

Albert of Saxe-Coburg · Queen Victoria

haemophiliac

Genealogical diagram showing the transmission of haemophilia in the descendants of Queen Victoria. Haemophilia rarely affects females but it is often transmitted by them. They are then called 'carriers'.

carrier

carrier daughter, but not a son with the disease.

The most celebrated example of inherited haemophilia is, of course, found in the descendants of Queen Victoria. From the marriage of Victoria, who was a carrier, and her cousin Prince Albert of Saxe-Coburg, nine children were born, five daughters and four sons. Three daughters and three sons were normal. One son, Leopold, Duke of Albany, was haemophiliac; his daughter, Alice, was a carrier and transmitted the *h* gene to her son Rupprecht of Teck-Athlone.

The descendants of Victoria's two daughters Alice and Beatrice reveal that both were carriers of the gene *h*.

Alice married Grand Duke Louis of Hesse and seven children were born, of whom one son, Frederick, was haemophilic, and two daughters, Irene and Alix, were carriers. Irene married Henry of Prussia and their two sons, Waldemar and Henry, were haemophiliacs. Alix married the Tsar Nicholas II and their only son, Alexis the Tsarevitch, suffered from haemophilia.

The descendants of Victoria's daughter Beatrice were also deeply tainted with haemophilia. By her marriage with Henry of Battenberg she had two sons, Leopold and Maurice, both haemophiliacs and one daughter Victoria Eugénie, a carrier. Victoria Eugénie married the King of Spain, Alfonso XIII, and of their seven children two sons, Alfonso and Gonzalo, suffered from the disease.

Haemophilia is a clinical and genetic entity, but recent work has shown that there are two forms of the disease: haemophilia A, which results from a deficiency of the anti-haemophilic globulin; and haemophilia B, which arises from the deficiency of a new factor, called the Christmas factor after the name of the patient in whom its absence was first noted. As the two forms can only be distinguished by coagulation tests it still seems valid to treat haemophilia as a single disease.

To take another example, the gene of Daltonism, or the inability to distinguish red from green, is also situated on that part of the X-chromosome that has no Y-chromosome homologue. This gene, being dominated by its normal allele, is transmitted in exactly the same way as the gene of haemophilia. As in the case of haemophilia, Daltonism is much commoner in men than in women, four per cent of the former suffering from it as against about one per cent of the latter.

Characters variably transmitted

There are certain normal and pathological hereditary characters which are not always transmitted in the same way. They are sometimes dominant, sometimes recessive, and sometimes sex-linked. Such are the genes responsible for pigmentary retinitis, Charcot-Marie's amyotrophia, and hemeralopia or day blindness.

Linkage and crossing-over

Among plants and animals we have seen that certain genes situated on the same chromosome do not separate during cell-division and remain linked with each other. Sometimes total linkage is replaced by partial linkage, the genes being separated by crossing-over.

Researches in gene-linkage are extremely difficult and few human linkages have been detected with certainty. It would seem that there is a linkage between the following genes:
1. The blood-group A B O and Friedreich's disease.
2. The Rhesus blood system and ovalocytosis — or oval red blood-cells.
3. The Lewis blood factor and the non-secretion of

A B O antigens in the saliva.
4. The M N blood factor and drepanocytosis — or sickle-shaped red blood-cells.
5. Pale eyes and myopia.

Genes which belong to the X-chromosome or to the Y-chromosome are of necessity linked. It has been possible to obtain direct verification of the linkage between the genes of Daltonism and haemophilia. In one family the great-grandfather, two grandsons and a great grandson suffered from both haemophilia and Daltonism. The two genes responsible for the two afflictions were located on the same X-chromosome and were transmitted together.

It is even known that the two genes are fairly close to each other, for the cross-over value (frequency with which they are exchanged) is small. In the case of sex-linked genes crossing-over only takes place between the two X-chromosomes carried by women. Genes situated on the Y-chromosome cannot cross-over. Finally, for genes situated in the homologous regions of the X and Y chromosomes crossing-over takes place in the two sexes between the woman's two X-chromosomes and between the man's X and Y chromosomes.

Difficult problems are posed by the transmission of genes situated in the X and Y chromosomes, genes which are known as partially sex-linked. We shall only examine the case of genes which determine dominant anomalies.

A man who carries such a gene of abnormality on his X-chromosome will transmit it to all his daughters but not to his sons. If, however, crossing-over takes place he can transmit it to some of his sons while some of his daughters will not receive it. These daughters will then be normal and the sons abnormal, but they are very much less common than normal sons and abnormal daughters.

If the gene in question is situated on the Y-chromosome it will be transmitted to all his sons. Should crossing-over occur it may, however, be transmitted to some of his daughters while some of his sons do not receive it. In this case such abnormal daughters and normal sons will be born much less frequently than abnormal sons and normal daughters.

As we have seen, the sex chromosomes are already marked by independent genes which determine definite characters; but if research into linkage is to make further progress it is essential that the remaining twenty-two pairs of autosomes be similarly marked. The genes determining blood-groups make excellent markers, for they are invariable in their action and free from other influences. It is believed that nine pairs of autosomes are at present marked by the genes of the blood-groups ABO, MN, P, Lewis, Rhesus, Lutheran, Kell, Duffy, and by the gene for the ability to taste phenylthiocarbamide. It remains to discover characters, normal or pathological, which are more or less regularly transmitted with these marker characters.

When such linkages are better understood it will be possible to calculate in advance the chances which an individual who carries a given normal character has of possessing a given pathological character which is more or less strictly linked with it.

Characters of complex heredity

In the simplest case a character is determined by a single gene; the gene then represents a Mendelian character-unit. But such Mendelian characters, though not rare, are not very common. Many characters depend on the combined action of several genes.

Allelomorphic series

Sometimes a dominant gene has several corresponding recessive alleles which form what is called an allelomorphic series. The simplest example of such a series in man is that of the blood-groups A,B,O. The antigens present in the red blood-cells are determined by the genes A, B and O, which are three alleles occupying the same position on the chromosome.

Another example is furnished by Daltonism. Daltonism can be broken down into kinds of blindness: an inability to perceive red and an inability to perceive green. The gene R which is responsible for normal red vision has two alleles: r_1, the gene of protanomaly or slight blindness to red, and r_2, the gene for protanopia or total blindness to red. R is dominant to r_1 which in turn is dominant to r_2. Similarly the gene G, responsible for normal green vision, has two alleles: g_1, the gene of deuteranomaly or slight blindness to green, and g_2, the gene of deuteranopia or total blindness to green. Again G is dominant to g_1, and g_1 to g_2.

Multiple genes and blending inheritance

The appearance of a hereditary character sometimes requires the participation of several genes distributed in several pairs of alleles whose partial or special action adds or subtracts from the over-all effect, somewhat in the manner of an algebraic equation. The result is a blended inheritance.

This kind of heredity is of great importance in the human species and especially in the transmission of normal characters such as weight, stature, the general shape of the body, width of shoulders, pelvis, thorax, colour of eyes and skin, colour and type of hair, and so forth. The number of genes which take part in bringing about such characters remains in most cases obscure, and various hypotheses attempt to explain the results. The transmission of a unit-character is naturally much easier to study than that of a character affected by many genes. This fact would explain the abundance of literature which has been devoted to pathological characters which are most frequently determined by single genes, and the relatively small amount which has been written about normal characters for which blending inheritance is usually responsible.

Racial pigmentation

In racial pigmentation at least three pairs of genes play a part: *AA, BB,* and *CC* in negroes; and *aa, bb,* and *cc* in whites. As everyone knows the children of parents one of whom is black and the other white are mulattoes. Two mulattoes, in their turn, will engender offspring composed of further mulattoes of varying shades of colour. The genes for blackness do not completely dominate the genes for whiteness. In association they produce an intermediate colour which is more or less dark or fair according to which of the two kinds of genes, black or white, is more abundant.

The first generation mulatto born of, say, a white woman and a black man is of genotype *AaBbCc.* These three pairs of genes will segregate independently so that eight kinds of gametes can be formed: *ABC, ABc, AbC, Abc, aBC, aBc, abC,* and *abc.* Thus between the genes of two mulattoes 8 x 8 or sixty-four combinations are possible. Only one of the sixty-four will give birth to a pure white *aabbcc* like the grandmother, and only one will give birth to a pure black *AABBCC* like the grandfather. The remaining sixty-two combinations will be mulattoes of various shades depending on the number of black and white genes they have inherited. In other words, it is not impossible that from the union of two mulattoes a pure white or a pure black will be born, but the probability of this occurring is only one in sixty-four. The relatively small number of children born to the human couple merely creates an illusion of the stability of the intermediate character (mulatto) which appears in sixty-two cases out of sixty-four.

If racial pigmentation were determined by a single gene the union of two mulattoes would result in offspring composed of one quarter pure black, one quarter pure white and one half mulatto. The ratio, would be totally different from that observed in life.

The union of a mulatto and a white person can only give rise to children who are fairer than their mulatto parent; the mulatto's genes for blackness being numerically diluted by the addition of white genes, the resulting intermediate colour will be lighter.

Intelligence, aptitude for mathematics, music, drawing, etc., very certainly arise from the action of multiple genes, but our knowledge of the inheritance of non-physical qualities is still slight, and in this field it is difficult to estimate the part played by hereditary factors and by factors which are in the broadest sense of the word environmental. Identical twins, as we shall later see, furnish excellent material for studying and discriminating between the action of hereditary factors and the influence of the environment.

The great families of mathematicians, musicians, and painters certainly suggest that aptitudes are inherited and determined by numerous genes, comprising major genes, minor genes, and genes which act as modifying agents. Combinations of these different genes would explain, for instance, why musical families can vary so considerably, from the composer of genius to the member who has no musical sense at all.

Longevity is almost undoubtedly a hereditary character, as various observations have indicated. R. Pearl in particular has studied the question and drawn

Mulatto surrounded by blacks from Martinique.

Atlas-Photo.

Blackfoot Indian, Maria Sabina, born on 12th October, 1736, at Matuna, a plantation belonging to the Jesuits of Cartagena in South America.

Supplement from Buffon's 'Histoire naturelle'. Larousse.

318

up numerous statistics from which he concludes that there is a connection between the individual's life span and that of his six direct ancestors, i. e., his two parents and four grandparents. He estimates that the *TIAL* (Total Immediate Ancestor Longevity) is equal to the sum of the ages at which the six ancestors died. The higher a person's *TIAL*, the longer his life expectancy. Pearl cites the case of a man who was one hundred when the observation was made, whose parents lived to the ages of 97 and 101, and whose four grandparents lived to the ages of 98, 104, 106, and 93. The man's *TIAL* was thus 599, a figure about six times his age. The manner in which longevity is transmitted is quite unknown. Obviously the character depends on the general conditions of the organism, and must therefore involve numerous genes.

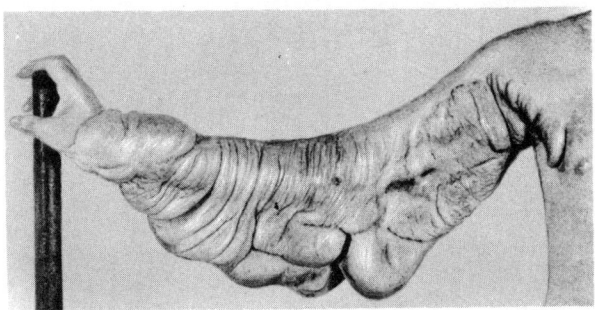

Arm of a patient suffering from Recklinghausen's disease which is hereditary and dominant.

Auzinger.

Pleiotropia

Pleiotropia is the tendency of an inherited character to influence or affect several parts or functions of the organism which, superficially, seem to have no connection. Most genes are probably pleiotropic, and influence different aspects of life — longevity, fertility and the organism's chemical and psychic activities.

An analysis of human pleiotropia proves difficult. The gene of albinism determines not only the absence of pigment in the skin, the iris and the hair, but also causes finer skin, sparser hair, and a certain timidity. Sufferers from mongolism have, in addition to the characteristic nervous symptoms, brown eyes and brown hair. Phenylpyruvic idiocy, the result of faulty metabolism, is accompanied by fair pigmentation of the skin, hair, and eyes. Genes connected with mental deficiencies seem to be pleiotropic and many symptoms of mental deficiency are attended by various morphological or clinical anomalies. Patients suffering from neurofibromatosis, or Recklinghausen's disease, a dominant affliction, have a tendency to tumours of the peripheral and central nervous systems. They are also abnormal in appearance, being small of stature, with large heads, elongated extremities and diffuse melanism. Bourneville's tuberous sclerosis, or epiloia, is a dominant and sub-lethal syndrome, characterized by sclerotic nodules the size of a cherry-stone or hazel nut in the superficial layers of the brain, by sebaceous adenomas around the nose and mouth, by tumours in various organs, by epilepsy and mental deficiency.

The various syndromes, with their complex of symptoms and different clinical manifestations, are probably determined by pleiotropic genes.

Lethal genes

A lethal gene is a gene which kills the individual who carries it unless he also carries its normal allele to counterbalance the lethal effect. Put in another way, the presence of paired lethal genes, as in a homozygote, is deadly.

In the human race there are numerous lethal genes which kill the embryo or the infant, as well as sub-lethal genes which cause the death of the child or the adolescent, usually before the age of puberty. Lethal genes are probably responsible for the not infrequent defects and monstrosities of still-born and new-born babies, and for many miscarriages which were formerly attributed to other causes. But, unless accompanied by abnormalities, lethal genes are hard to detect; their existence is only manifest in the sterility which often afflicts certain couples and cannot be explained by any other cause. An analysis of more than 50,000 births has shown that those couples who have already had one still-born baby are twice as apt as the average couple to have another.

Certain afflictions are lethal in the homozygous state: brachydactylia, telangiectasia, infantile amaurotic idiocy. Brachydactylia is characterized by the shortness of the middle phalanx of the fingers which sometimes fuses with the terminal phalanx; the anomaly, though not serious, prevents certain normal movements of the hand. The marriage of two people afflicted in this way produced a daughter whose toes and fingers were completely missing; she died at the age of one year. Doubtless she was homozygous for the gene of brachydactylia, having received one such gene from her father and the other from her mother. Thus a minor anomaly when heterozygous can become extremely serious and indeed mortal if homozygous.

Hereditary telangiectasia consists of an increase in the number and size of the capillary vessels so that angiomas form on the skin and mucous membranes, principally of the face. A married couple afflicted with this malformation had a daughter with naevi and angiomas of exaggerated size which led to her death by haemorrhage at the age of three months. The gene of telangiectasia in the homozygous state is also incompatible with life.

The recessive gene of *Xeroderma pigmentosum* may be qualified as semi-lethal. In homozygotes it brings about an eruption of pigmentary blotches, especially on the face. Sunlight can then cause ulcers of the skin which become malignant tumours and lead to death.

The recessive gene of congenital ichthyosis in the homozygous state brings about a serious abnormality of the skin which causes more or less early death *in utero,* in the new-born, or in the nursing infant. The recessive gene of congenital amyotonia, when homozygous, causes respiratory paralysis which leads to death at the end of a few days or a few months.

The blood-groups

In blood transfusions knowledge of blood-groups is of crucial importance and their heredity has been studied with particular thoroughness.

It has long been known that to mingle the blood of two individuals of the same species without due precautions can lead to serious consequences and even to death. That two kinds of blood are incompatible is revealed by characteristic symptoms: shivering, a rise in temperature, and retrosternal pains. Within a few hours jaundice sometimes develops; the kidneys are blocked, uraemia reaches a crisis and the patient dies. The cause of these post-transfusion fatalities was unknown until 1900 when Landsteiner discovered the blood-groups.

The blood-groups A, B, O

The red corpuscles or blood-cells carry an antigen, an *agglutinogen,* while the serum contains an antibody, an *agglutinin.* The agglutinogen is agglutinated (or clumped) by its corresponding agglutinin. Human beings have been divided into four classic blood-groups according to the types of agglutinogen and agglutinin their blood contains. These groups are designated by the letter by which the corresponding agglutinogen is known:

Group A: red blood-cells with agglutinogen A, serum with agglutinin anti-B.
Group B: red blood-cells with agglutinogen B, serum with agglutinin anti-A.
Group AB: red blood-cells with agglutinogens A and B, serum whithout agglutinins anti-A and anti-B.
Group O: red blood-cells without agglutinogens A and B; serum with agglutinins anti-A and anti-B.

The agglutinin anti-A clumps A corpuscles and the agglutinin anti-B clumps B corpuscles.

Transfusion will lead to trouble when the red blood-cells of the donor are agglutinated by the agglutinins of the receiver. For this reason people in group AB (universal receivers) can receive blood from all groups; their serum contains no antibodies and is therefore incapable of clumping any kind of red blood-cell. People in group O (universal donors) can give blood to all other groups; their blood-cells are without agglutinogens A and B and therefore cannot be clumped by the agglutinins in the other serums. Members of Group O, however, can only receive O blood, because they have the two agglutinins anti-A and anti-B.

The distribution of blood-groups varies according to peoples. Some percentages are given below:

	A per cent	B per cent	AB per cent	O per cent
W. European (Average)	41.3	12.5	3.9	42.2
Americans (black)	34.1	17.2	1.2	46.9
Americans (white)	46.8	9.8	1.9	41.5
Australians	38.5	3	1.5	57
Blackfoot Indians	76.5	0	0	23.5
Eskimos	56.2	11.2	8.7	23.9
Japanese	41	18	14	27
Senegalese	22.4	29.2	5	43.2

Congolese albino mother.

Mansell Collection.

The blood-group problem is further complicated by the existence of sub-groups A_1, A_2, A_3.

The individual's blood-group is a fixed character which is manifest from the earliest days of life and persists until death. No external influence can alter it, neither infectious diseases, vaccination, nor even blood transfusions.

The classic blood-group to which an individual belongs is a Mendelian character and depends on a single gene which occurs in three allelomorphic states. These three alleles are designated by the symbols A, B, and O, and correspond respectively to the presence of the agglutinogen A, of the agglutinogen B, and to the absence of the agglutinogens A and B. The three alleles occupy the same locus on a pair of homologous chromosomes. As everyone always has two genes with the same function — one from his father and one from his mother — the two genes determine his blood-group. The two genes may be AA, AB, AO, BB, BO, OO, forming six possible genotypes of which three (AA, BB, OO) are homozygous, and three — (AO, BO, AB) are heterozygous. Genes A and B are dominant while gene O is recessive.

Individuals of group A are of genotype AA or AO,

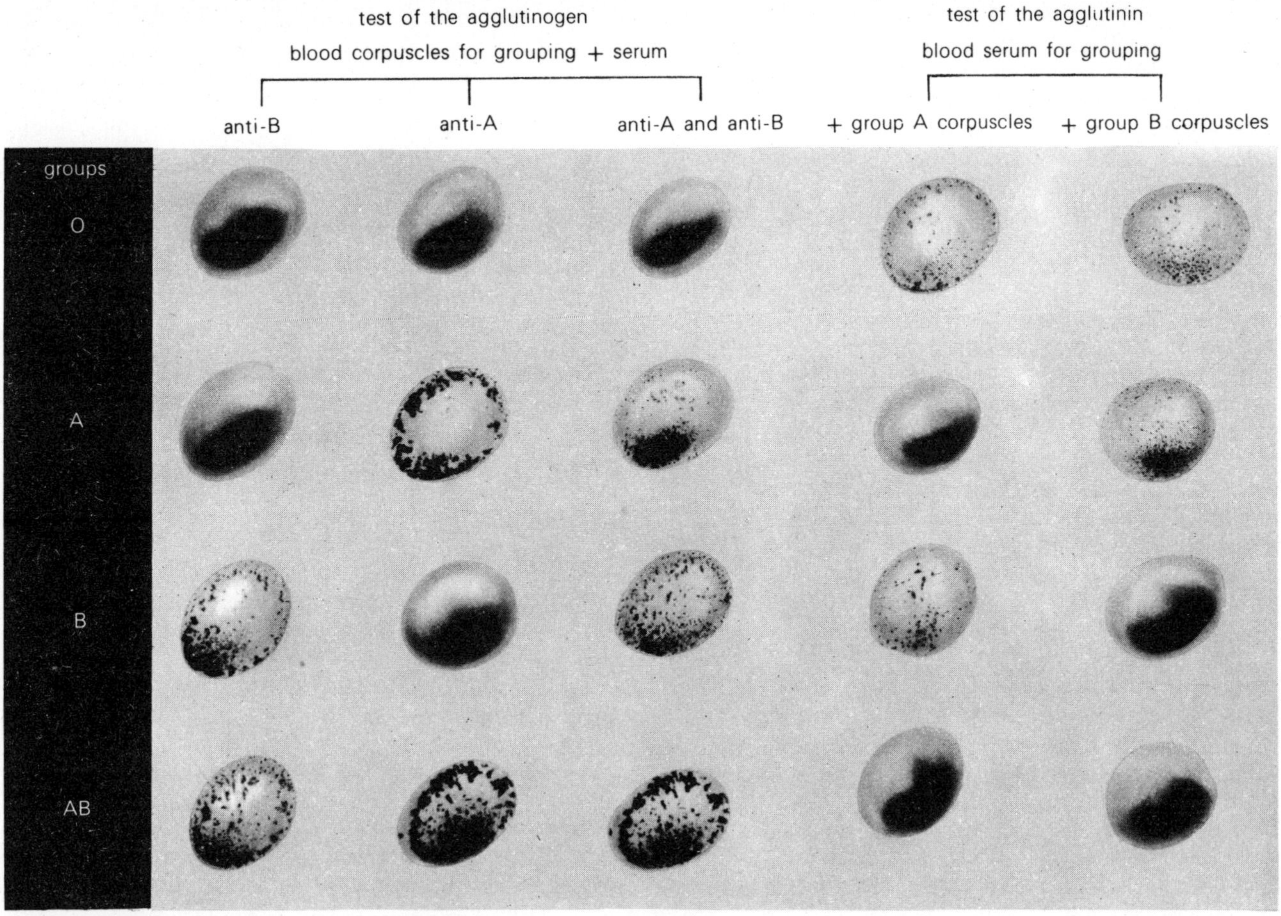

test of the agglutinogen
blood corpuscles for grouping + serum

test of the agglutinin
blood serum for grouping

| | anti-B | anti-A | anti-A and anti-B | + group A corpuscles | + group B corpuscles |

groups
O
A
B
AB

A diagram showing the reaction of the different ABO blood-groups to the different agglutinogens and agglutinins.

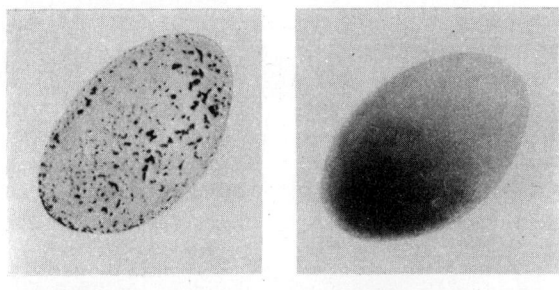

agglutination   no agglutination

Reaction of red blood-cells to a serum.

homozygotes or heterozygotes. Individuals of group B are of genotype BB or BO. Individuals of group AB are of genotype AB, the two genes expressing themselves equally and without dominance. Individuals of group O are of necessity recessive homozygotes OO.

Individuals of group A, if they are homozygotes AA, form only gametes which carry the gene A; if they are heterozygotes AO they form in equal numbers gametes which carry the A gene and gametes which carry the O gene.

In the same way an individual of group B will produce gametes which carry gene B only, or gametes half of which carry gene B and half gene O, according to whether he is a homozygote or a heterozygote.

An individual of group AB will produce gametes half of which carry gene A and the other half gene B.

Finally, individuals of Group O produce only gametes which carry gene O.

These facts allow us to foretell the possible blood-groups of children born to parents of various types. For instance, two parents of group O can only have children of group O. Two parents of group A can engender children of group A and of group O. Two parents of group B can have children of group B and of group O. Two parents of group AB can have children of groups A, B, and AB, but never of group O. If one parent belongs to group A and the other to group B their children may be of groups A, B, AB, and O.

Blood-groups M and N

In addition to the agglutinogens A and B the red blood-cells carry two other agglutinogens, known as M and N. The serum does not normally contain the corresponding agglutinins anti-M and anti-N. Hence blood transfusions are unaffected by the M and N factors.

The production of the M and N agglutinogens is controlled by a pair of genes M and N. Those who possess the two genes MM belong to group M; those who possess the two genes NN to group N; finally, those who have one of each gene belong to group MN. They are heterozygotes, while the MM and NN individuals are homozygotes. As in the case of genes A and B, there is no dominance between genes M and N; they act side by side.

About thirty per cent of the population belongs to group M, twenty to group N, while the remaining half belongs to group MN.

It is possible to foretell to what group (M or N) the children of a given union will belong by the same Mendelian laws which predict ABO inheritance. Two parents of group M will have only children belonging to group M. All children born to parents of group N will be of group N. If both parents belong to group MN half their offspring will belong to group MN, a quarter to group M and a quarter to group N.

The Rh or rhesus blood system

The Rhesus blood system was discovered in 1940 by Landsteiner and Wiener and is of great importance. It was given the name Rhesus from the name of the monkey, *Macacus rhesus,* in which its existence was first demonstrated. Its behaviour has explained the failures which occur during multiple transfusions between bloods which are compatible as far as the classic groups are concerned, and of blood transfusions during the course of certain pregnancies.

Demonstration of the Rh factor

When rabbits or guinea-pigs are injected with the red blood-cells of *Macacus rhesus* the cells are agglutinated by the serum of the rabbit or the guinea-pig. This anti-rhesus serum will also agglutinate the red blood-cells of eighty-five per cent of the white human race. Human blood thus contains the same agglutinogen as the blood of the macaque monkey. This substance is an antigen known as the Rh or Rhesus factor. Human blood which possesses this factor is called Rh-positive. Human blood which is not agglutinated by the anti-rhesus serum does not contain the factor, and is called Rh-negative. Some fifteen per cent of the white race is Rh-negative.

The Rh factor occurs independently of sex and the classic blood-group. In Paris, for instance, a blood survey of 380 men, women, and children showed that 85.26 per cent were Rh-positive, and 14.74 per cent Rh-negative. By sexes the percentages were: men, 84.55 Rh-positive and 15.45 Rh-negative; women, 85.77 Rh-positive and 14.23 Rh-negative.

Thus the difference in distribution by sex was negligible. But by races the distribution of the Rh factor varies widely, as the following table shows:

	Rh-positive (per cent)	Rh-negative (per cent)
American Indians (pure)	99.2	0.8
American Indians (half-breed)	93 to 98	7 to 2
Black race	92 to 95	8 to 5
White race	85	15
Yellow race	98.5 to 99.5	1.5 to 0.5

How the Rh factor is transmitted

The two blood types Rh-positive and Rh-negative are,

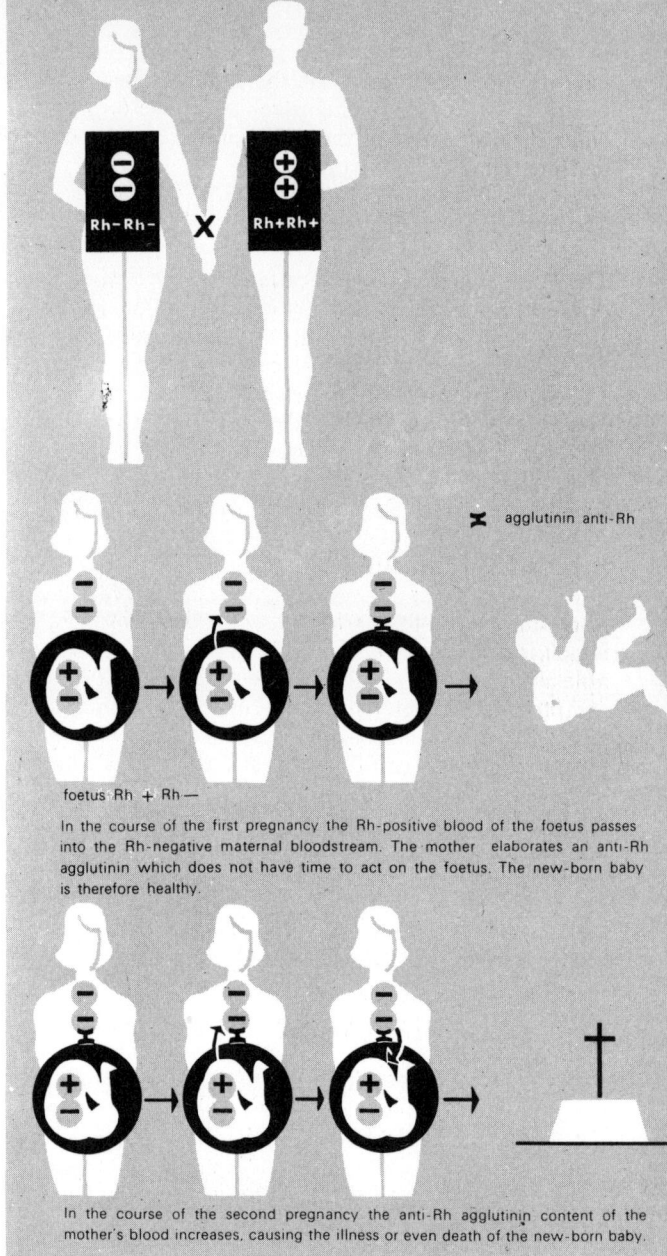

In the course of the first pregnancy the Rh-positive blood of the foetus passes into the Rh-negative maternal bloodstream. The mother elaborates an anti-Rh agglutinin which does not have time to act on the foetus. The new-born baby is therefore healthy.

In the course of the second pregnancy the anti-Rh agglutinin content of the mother's blood increases, causing the illness or even death of the new-born baby.

Marriage of an Rh-negative woman to an Rh-positive man (homozygote).

like the other blood-groups, hereditary characters. Hence the transmission of the Rhesus factor obeys the laws of Mendel, the Rh-positive factor being dominant and the Rh-negative factor recessive. Three genotypes for the factor are possible: two homozygotes, Rh-positive/Rh-positive and Rh-negative/Rh-negative, and the heterozygote Rh-positive/Rh-negative. There are two phenotypes: homozygotes or heterozygotes Rh-positive, and Rh-negative. Rh-negative individuals are of necessity homozygotes.

Two Rh-negative parents will, of course, engender only Rh-negative children. If one of the parents is Rh-negative and the other Rh-positive the offspring will vary according to whether the Rh-positive parent is a homozygote or a heterozygote. In the former case all the children will be Rh-positive; in the latter case half will be Rh-positive and half Rh-negative. If both parents are Rh-positive all the children will be Rh-positive, unless both the parents are heterozygotes, when three-quarters of the children will be Rh-positive and one quarter Rh-negative.

The importance of the Rh factor

The Rh factor is in itself harmless, for the blood serum does not normally contain its corresponding antibody. But anti-Rhesus agglutinins can be formed by Rh-negative organisms if they receive Rh-positive blood. The agglutinins remain in the serum and when a second injection is made they will then clump the red blood-cells received from the Rh-positive donor. In practice these conditions can arise when several blood transfusions are made, or during certain pregnancies.

Repeated blood transfusions

Sometimes the first transfusions made with blood which is apparently compatible produce no unfavourable reactions; then, after a third or fourth transfusion, symptoms characteristic of blood incompatibility suddenly appear. Analysis of the receiver's blood reveals the presence of an anti-Rh agglutinin which is responsible for the unfavourable reactions. What has occurred is this: the patient, Rh-negative, has received Rh-positive blood. After the first transfusion he has become sensitized and elaborated an anti-Rh agglutinin which, after later transfusions with the same Rh-positive blood, clumps the donor's red corpuscles.

To avoid such reactions it is essential that repeated transfusions are made with blood which is not only of the compatible ABO group but is also Rh-negative. Only fifteen per cent (of white races) are Rh-negative. Forty per cent belong to group O — who are universal donors. Hence there are only six per cent who are at the same time both O and Rh-negative.

Dangerous marriages: the haemolytic disease of the new-born

For many years medical practitioners were gravely concerned by the tragic obstetrical history of certain seemingly healthy married couples. Each pregnancy, with the exception of the first and sometimes the second, ended in the death of the foetus or of the new-born infant. In such families miscarriages were frequent until the discovery of the Rhesus factor — which enabled the facts to be understood.

Among the various possible unions between individuals who are Rh-positive homozygotes or heterozygotes and Rh-negative there are two combinations which are dangerous: the marriage of an Rh-negative woman and an Rh-positive man, homozygote or heterozygote. Either combination can produce a baby stricken with the haemolytic disease of the new-born.

In this grave affliction the three classic symptoms are present: erythroblastic anaemia, acute jaundice of the new-born, foeto-placental anasarca (dropsy). In the course of erythroblastic anaemia the red blood-cells are greatly reduced in number while the erythroblasts or nucleated cells, normally absent, are fairly abundant. Acute jaundice is characterized by its intensity: the

Macacus rhesus.
X Photography.

skin is dark brown in colour with a greenish tinge. Foeto-placental anasarca is observed in the prematurely still-born or in babies dying a few hours after birth. Their tissues are swelled by oedema and they are mongoloid in appearance with bull-necks and huge abdomens.

What brings about these symptoms? Haemolysis of the new-born results from an iso-immunization of the mother by the foetus. In the marriage of an Rh-negative woman and an Rh-positive homozygote man all the offspring will be Rh-positive heterozygotes. During the first pregnancy the Rh-positive blood of the foetus enters the maternal Rh-negative bloodstream. In response to this contribution of Rh-positive blood the mother forms an anti-Rh agglutinin. Because the anti-Rh agglutinin has not had time to attack and destroy the baby's Rh-positive red corpuscles the birth is without complications and the infant healthy. The first child is thus normal, but the mother's blood now contains a certain amount of the anti-Rh agglutinin.

During the second pregnancy Rh-positive foetal cells may again enter the maternal circulation, and the amount of anti-Rh agglutinin which the mother's blood already contains will be thereby increased. If the increase is not too great the second child may be normally born. It is more likely, however, that the anti-Rh agglutinin, having entered the bloodstream of the new-born baby, will attack its red corpuscles and result in haemolytic disease. In subsequent pregnancies the quantity of the antibody in the mother's blood has of course still further increased and will act still sooner, destroying the red blood-cells of the foetus, so that all further Rh-positive babies will die.

In the second type of dangerous marriage, that of an Rh-negative woman and an Rh-positive heterozygote man, half the children will be Rh-negative and the

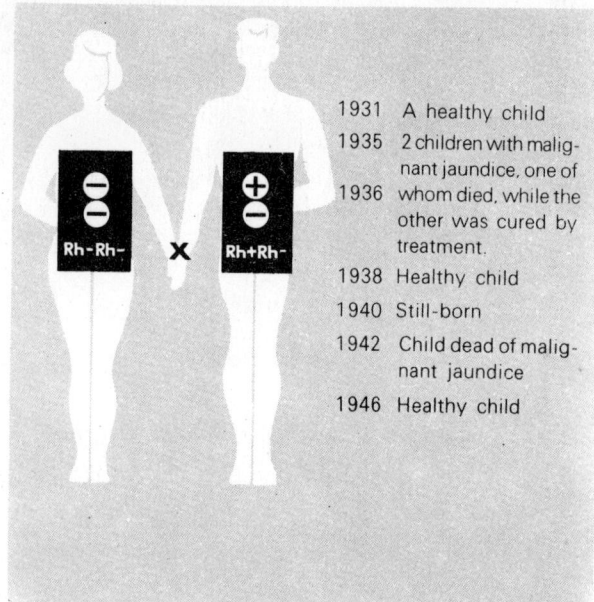

1931 A healthy child
1935 2 children with malig-
       nant jaundice, one of
1936 whom died, while the
       other was cured by
       treatment.
1938 Healthy child
1940 Still-born
1942 Child dead of malig-
       nant jaundice
1946 Healthy child

An actual example of the offspring of a Rh-negative woman and a heterozygous Rh-positive man.

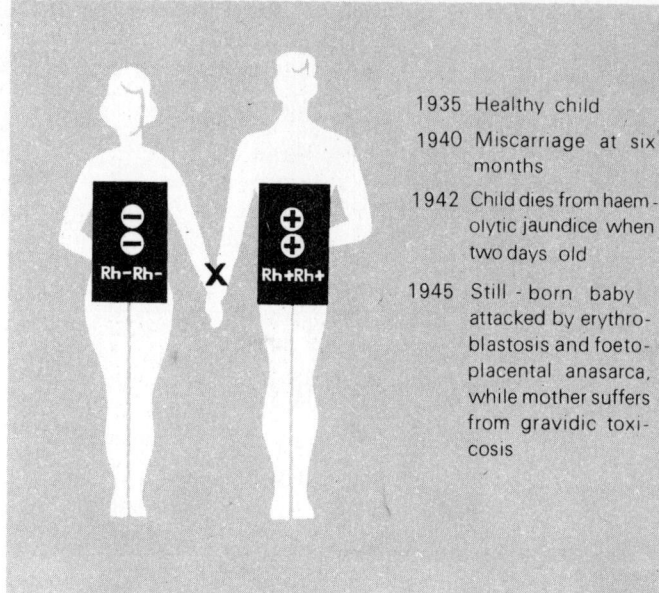

1935 Healthy child
1940 Miscarriage at six
       months
1942 Child dies from haem-
       olytic jaundice when
       two days old
1945 Still-born baby
       attacked by erythro-
       blastosis and foeto-
       placental anasarca,
       while mother suffers
       from gravidic toxi-
       cosis

An actual example of the offspring of an Rh-negative woman and a homozygous Rh-positive man.

other half Rh-positive. Foetal blood-cells may enter the mother's circulation in exactly the same way as before, but if the foetus is Rh-negative it will not immunize the mother and the maternal agglutinin, even if present, will have no effect on it. All Rh-nega-tive babies will therefore be healthy. If the foetus is Rh-positive, however, it will immunize the mother and the situation will be the same as in the first type of marriage: that is, Rh-positive babies will suffer from haemolytic disease of the new-born. The anti-Rh agglutinin has no effect on Rh-negative corpuscles and so the Rh-negative baby escapes the disease.

The frequency with which these two dangerous types of marriage occurs — Rh-negative women and Rh-positive men — is about one in ten. Happily the frequency with which they are followed by fatal results is much less, only one in about 250 births. There are several reasons for the discrepancy in these figures: many couples have only one child which is generally spared. Then, in the second type of marriage with a heterozygous man, half the children are normal and by chance such normal births may predominate — a maternal factor of placental deficiency probably con-tributes to this. Finally, all women do not inherit an aptitude to form antibodies; only five per cent are able so to immunize themselves.

Even the first child of an Rh-negative woman and an Rh-positive man may in rare cases suffer from a serious form of the haemolytic disease of the new-born. In such cases the woman, in childhood or as a girl, has received a blood-transfusion with Rh-positive blood, and in consequence formed an immunity to the Rhesus factor. This immunity can even fifteen years later be respon-sible for the haemolytic disease of the new-born in her first child.

Great care must therefore be taken never to admin-ister Rh-positive blood to a little girl or to a young woman who is Rh-negative for fear that her first child

may be afflicted with the disease. All women whose Rhesus group has not been identified should, unless they have passed the age of childbearing, receive transfusions of Rh-negative blood only.

It would be over-scrupulous to advise against a marriage between an Rh-negative woman and an Rh-positive man, especially in view of the fact that medicine is no longer incapable of saving the child. Two efficient methods of doing this are available: premature childbirth and an exsanguino-transfusion of the new-born baby. Treatment of the mother with specific antibodies is also being investigated, and appears very promising.

The Rhesus factor which has been discussed in the preceding pages is the 'standard Rhesus factor'. In reality the situation is much more complicated. There is not a single Rhesus antigen but a complex of anti-gens which are designated by the term 'Rhesus system'. The standard Rhesus factor corresponds to the antigen D. An Rh-positive individual has the antigen D which is determined by the gene D. The Rh-negative indi-vidual has the antigen d, determined by the recessive allele d. The antigen D is responsible for ninety-two to ninety-five per cent of the cases of Rhesus incompa-tability, so that the Rh standard factor is of the greatest clinical importance.

The other blood-groups

In addition to the ABO blood-groups, the M and N groups, the P group, and the Rhesus system, there exist still other, rarer blood-groups some of which have been discovered in recent years. Such are the Lutheran, Kell, Lewis and Duffy groups.

Basing our calculations on all the blood-groups now known it would be theoretically possible to identify nearly 30,000 different genotypes.

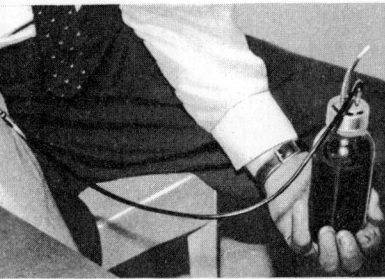
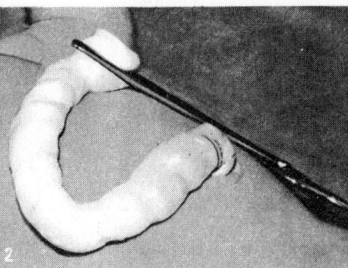
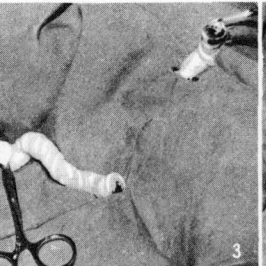
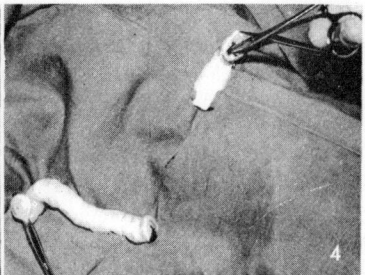
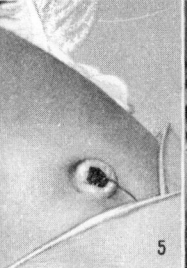
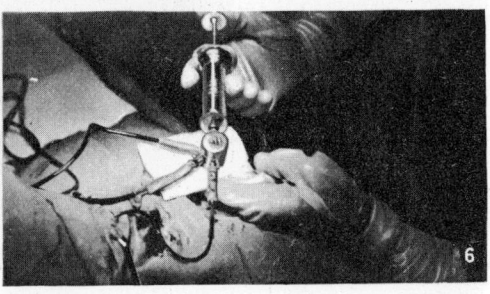

2. The baby's umbilical cord is still attached.

3. The operative field is arranged and the cord severed.

4. The wall of the cord is pinched by forceps while the tip of the scissors is inserted into the umbilical vein so that the catheter may be more easily put in position.

5. The baby is prepared for the catheterization, or withdrawal of blood, from the umbilical vein.

6. Exsanguino-transfusion by Tzanck's three-way syringe. At the end of the syringe the drum with its three outlets can be seen.

Larousse.

Exsanguino-transfusion.

1. Taking blood from a donor who belongs to a blood-group which is compatible with that of the new-born baby.

Secretors and non-secretors

The antigens which correspond to the ABO blood-groups are found not only in the red blood corpuscles but also in the saliva, the gastric juices, the urine, the bile and the amniotic fluid. This is true of about eighty per cent of the human race, and such people are called secretors. In the remaining twenty per cent the agglutinogens are found in the blood-cells alone: these people are non-secretors. The special gene of secretion is the gene *S* which is dominant over its allele *s* for non-secretion.

The non-secretor's genotype must therefore be *ss:* he is a recessive homozygote. The secretor may be the homozygote *SS* or the heterozygote *Ss*.

The blood-groups and forensic medicine

The blood-groups are of importance not only in blood transfusions but also in the legal identification of children and the establishment of paternity. If a man wishes to prove that he is not the father of a child attributed to him, or desires to recognize a natural child as his own, the blood-groups are of invaluable aid to justice.

Between the blood of a child and that of the man alleged to be its father a genetic incompatibility may be apparent, so that it is sometimes possible to prove that a given child cannot be the child of a given man. This is proof by exclusion. The opposite proof, that the child is the child of a given man, is still impossible to establish.

Let us suppose the child to be of group A. If its mother belongs to group O then its father cannot belong to group O. Similarly, if the child belongs to group B and its mother to group O, its father is necessarily of group B. A child of group O, with a group O mother, may have a father belonging to groups O or A (heterozygote) or B (heterozygote): but an AB father is impossible. Possible solutions obey the law: no blood character can appear in the child which does not exist in at least one of its parents.

It is worth dwelling on the fact that, although in certain cases the exclusion of paternity can be proved beyond shadow of doubt, it is still impossible to establish positive paternity. A more complete blood analysis involving a greater number of blood-groups naturally increases the chances of a man falsely accused of paternity proving his innocence. The following figures show the percentage of such chances:

Blood-groups	Percentages of exclusion
ABO	17.60
MN	27.41
ABO and MN	40.19
Rhesus	25.20
ABO & MN & Rhesus	55.26
Kell	3.79
ABO & MN & Rhesus & Kell	56.96
Lutheran	3.33
ABO & MN & Rhesus & Kell & Lutheran	58.39
Duffy	4.87
ABO & MN & Rhesus & Kell & Lutheran & Duffy	60.42
Kidd	3.16
ABO & MN & Rhesus & Kell & Lutheran & Duffy & Kidd	61.67

The chances of excluding paternity with certainty will continue to rise with the discovery of every new blood-group. Courts utilize the evidence of blood-group experts in the establishment of non-paternity. This must inevitably lead to a systematic demand for experts in blood analysis, although to obtain the necessary blood may be a more delicate matter. It cannot be obtained by force, for this would constitute an attack on the liberty of the subject.

The necessity for legal identification arises chiefly when babies have been exchanged in maternity hospitals. Wiener reports such a case from Chicago. As they left the maternity hospital, a couple named Bamberger received a baby with a tag on which was written the name Watkins. The Watkins, on the other hand, had been given a baby with a tag labelled Bamberger. The question was whether the babies had been exchanged or whether the tags had been exchanged. Blood analysis gave the following results:

Mr Bamberger:	group AB
Mrs Bamberger:	group O
Baby with Bamberger tag:	group A
Mr Watkins:	group O
Mrs Watkins:	group O
Baby with Watkins tag:	group O

The answer to the question was obvious: the babies had been exchanged but correctly labelled. In the Watkins family an A baby was impossible, and an O baby was equally impossible in the Bamberger family. The babies were thus restored to their real parents without the possibility of error.

Diseases to which the human chromosomes are subject

The pathology of human chromosomes has only recently been studied. In 1959 the researches of Turpin, Lejeune and Gautier led to the discovery of a supernumerary chromosome in a patient suffering from mongolism. At that time the oetiology of this curious disease, described as early as 1866, was still obscure. Mongolism is always characterized by certain morphological anomalies: shortness of stature giving a squat appearance, a round head, flattened face, oblique eye-slits with an internal fold of the skin. There are various malformations: cardiac abnormalities, cleft tongue, modified fingerprints and notably a single transverse crease across the palm as a result of the fusion of head and heart lines. But the basic symptom is feeble-mindedness: the mongol's mental age does not exceed that of a child of six or seven.

In man the normal diploid number of chromosomes is forty-six while the mongol possesses forty-seven. The supplementary autosome is V-shaped and identical to those of a pair of chromosomes normally present. In other words, the mongol has three examples of this chromosome instead of two. This chromosomal anomaly has been found in all the numerous cases of mongolism examined. It is evidently the result of a non-disjunction of two homologous chromosomes in a gamete of one of the parents, probably in the maternal ovum, which would thus contain not

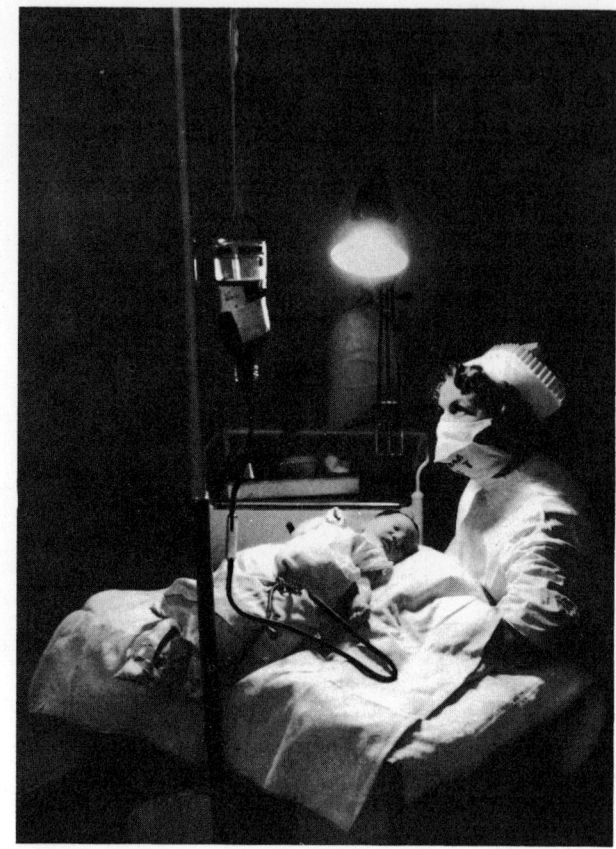

A baby being given a blood transfusion.

Press Association.

the normal single specimen of each chromosome but two which are exactly alike — twenty-four chromosomes instead of the normal haploid number, twenty-three. When fertilized by a sperm with the normal twenty-three chromosomes such an ovum would form a zygote with 24 plus 23 or 47 chromosomes.

Since the discovery of the first case of chromosomal pathology other examples of abnormal chromosome constitution have been reported. Persons afflicted with polydysspondylism show various symptoms: dwarfism, multiple malformations of the spinal column and cranium, feeble intelligence. They have only forty-five chromosomes. By translocation a small chromosome has become attached to another chromosome of medium size.

Sexual abnormalities can also arise from an incorrect distribution of the sex chromosomes. Normally women have two sex chromosomes, both of which are X-chromosomes, while men have one X-chromosome, and one Y-chromosome. Individuals suffering from Klinefelter's syndrome, characterized by defective development of the testes, have forty-seven chromosomes, including two X-chromosomes and one Y-chromosome. In Turner's syndrome, or defective development of the ovaries, the number of chromosomes is reduced to forty-five for only one X-chromosome exists. The syndrome of super-femininity, entailing secondary amenorrhea or cessation of menstruation and sex infantilism, is marked by the presence of three X-chromosomes, making forty-seven in all. The karyotype —

326

MOTHERS IN BABIES MIX UP

Daily Telegraph Reporter

MATERNITY hospitals in Wales were ordered yesterday to strengthen safeguards for identifying babies. The move follows a mistake at Morriston Hospital, Swansea, in which two mothers were given the wrong babies.

Mrs Barbara Taylor, 24, of Balaclava Road, Glais, Swansea, gave birth to a daughter at the hospital on Dec. 23. Mrs Ruth Painter, also 24, of Deepslade Close, Pennard, Swansea, who was in the same ward, gave birth to a daughter the next day.

The child handed to Mrs Taylor wore a plastic name band on her arm with the name Painter. Mrs Taylor pointed this out to a nurse and the next time the baby was brought to her it was without the name band.

Later, Mrs Taylor had doubts about whether the child she was given was in fact hers. She persuaded the hospital authorities to make blood tests.

She was told there had been a mistake only last week. The child Mrs Painter was given was taken to Mrs Taylor.

Yesterday a spokesman for the Welsh Hospitals Board said a full investigation was being carried out

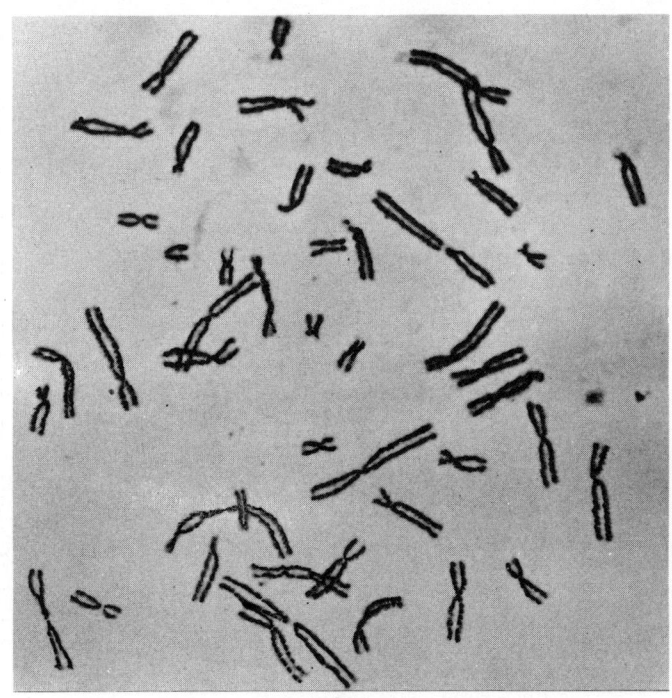

The 47 chromosomes of a mongoloid girl, dispersed in a cell.

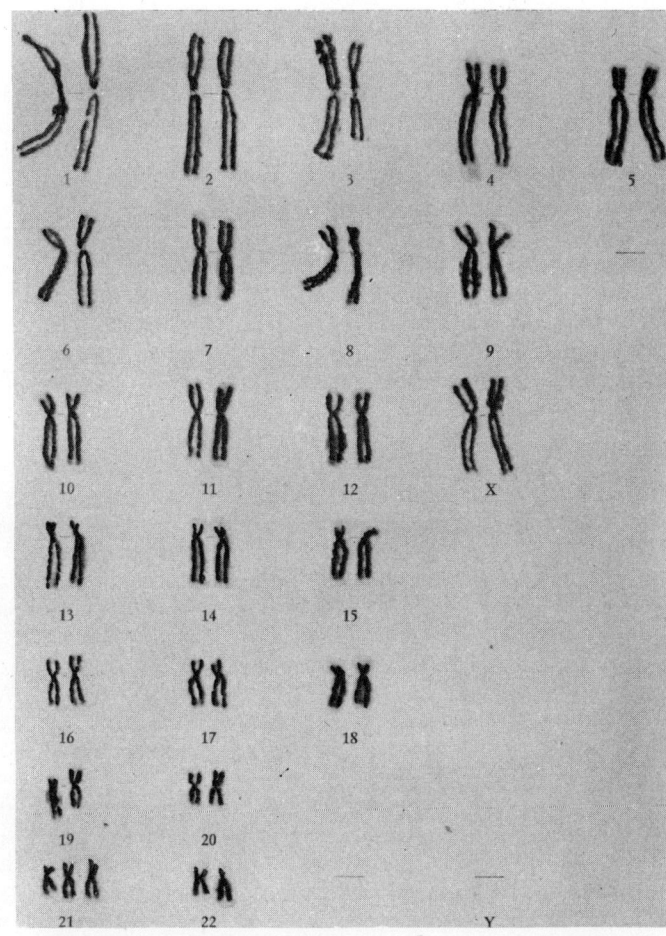

Karyotype showing the 47 cells arranged in homologous pairs. The supernumerary autosome is found in pair number 21 which thus comprises three chromosomes.

Dr Lejeune, Faculty of Medecine, Paris.

327

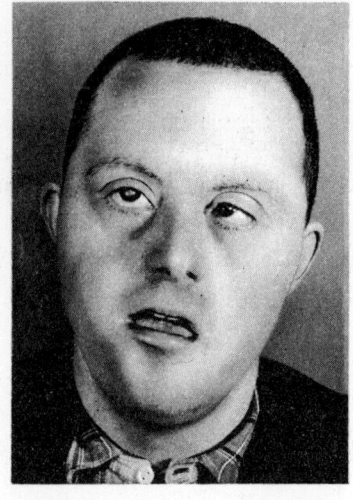

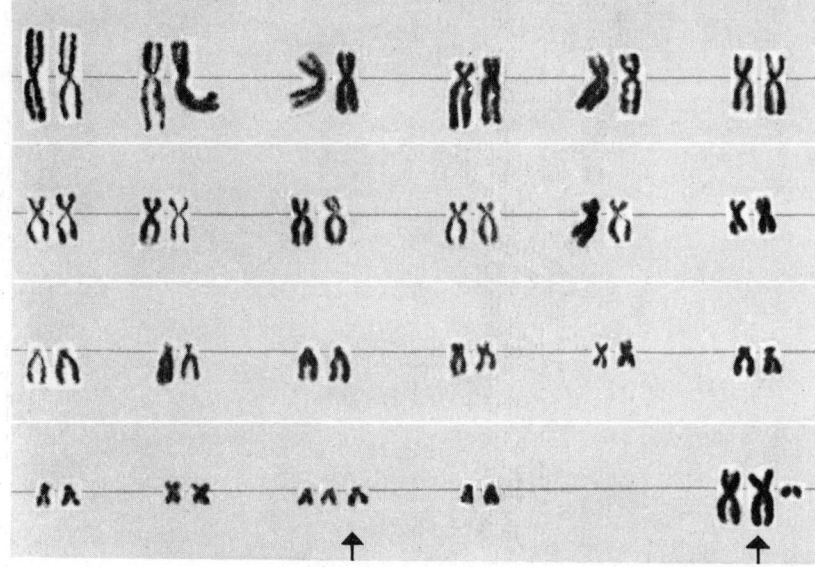

A Mongol who in 1961 was 35 years old. The mongoloid aspect is striking. In addition he suffers from a keratoconus and a bilateral cataract which is noticeable in the right eye.

Coll. Franceschetti. Cantonal Hospital of Geneva.

The 48 chromosomes found in a mongol aged ten months and suffering from Klinefelter's syndrome (testicular dysgenesis). The two supplementary chromosomes — a small autosome and a sex or X-chromosome — are indicated by arrows.

After Lauman, Sklarin, Cooper, Hirschorn.

or chromosomal constitution — of a six months old mongol who in addition suffered from Klinefelter's syndrome, comprised forty-eight chromosomes distributed as follows: the forty-six normal chromosomes, a small supplementary autosome characteristic of mongolism, and an extra X-chromosome, characteristic of Klinefelter's syndrome.

Mongolism and the syndromes mentioned above arise then from a mechanical fault in the division of the parental chromosomes. By this error gametes are formed which carry an abnormal chromosome complement consisting of either more or fewer chromosomes than the normal haploid number.

How the genes function

We have already seen something of how biochemical reactions in animals and plants are determined by genes. Further examples are found in the study of human heredity.

A rupture in the chain of reactions connecting the gene with the phenotype results in the appearance of an unforeseen character which, as inherited faults or deficiencies of metabolism demonstrate, is often pathological.

There are three diseases — alkaptonuria, tyrosinosis, and phenylketonuria — which arise from disturbances in the metabolism of phenylalanine and tyrosine. In the organism these two amino acids, which are derived from food, undergo a series of linked transformations. Phenylalanine, in the presence of an enzyme, is oxidized into tyrosine; tyrosine in its turn is transformed into p-hydroxyphenylpyruvic acid which then gives rise to 2,5-dihydroxyphenylpyruvic acid. This produces ho-

mogentisic acid, which is finally transformed into acetylacetic acid. Another series of reactions also begins with phenylalanine but progresses through phenylpyruvic acid to p-hydroxyphenylpyruvic acid, where the series of transformations again pursues the same course as in the first case. These reactions can be shown diagrammatically as follows:

$$
\begin{array}{c}
\text{enzyme} \\
\text{Phenylalanine} \rightarrow \text{Tyrosine} \\
\rightarrow \qquad \downarrow \quad \uparrow \\
\text{Phenylpyruvic acid} \rightarrow \textit{p}\text{-hydroxyphenylpyruvic acid} \\
\text{PHENYLKETONURIA} \quad \downarrow \quad \text{TYROSINOSIS} \\
\text{2,5 dihydroxyphenylpyruvic acid} \\
\downarrow \\
\text{homogentisic acid} \\
\downarrow \text{ALKAPTONURIA} \\
\text{Acetylacetic acid} \\
\downarrow \\
\text{Carbon dioxide} + \text{water}
\end{array}
$$

Breaking the chain of reactions at three distinct points brings about three distinct diseases:

1. Alkaptonuria is a complaint of minor importance of which about 200 cases are known. It consists of the excretion in the urine of homogentisic acid or alkapton bodies which cause the urine to turn a dark mahogany brown. Sometimes a darkening of the skin and cartilage occurs, and a tendency to arthritis. The normal transformation of homogentisic acid into acetylacetic acid does not take place because the necessary enzyme, present in the normal serum, is absent in the serum of persons with alkaptonuria. The gene of alkaptonuria is responsible for the enzyme's absence. Its normal allele determines the presence of the enzyme which

permits the chain of reactions to be concluded.

The irregular deposit of melanin which darkens the skin and cartilages is easily explained, since tyrosine produces 3,4-dihydroxyphenylalanine which in turn produces melanin.

2. Tyrosinosis is characterized by the presence of tyrosine in the urine. In this case the chain of reactions has been interrupted at the preceding stage, that is when p-hydroxyphenylpyruvic acid is normally oxidized. Again the cause is the lack of a specific enzyme, a lack which is genetically determined. In the only known case of tyrosinosis the symptoms are intermediate between those of alkaptonuria and phenylketonuria.

3. With phenylketonuria or phenylpyruvic oligophrenia, the urine of those with this disease contains phenylpyruvic acid and they suffer from varying degrees of mental debility and even idiocy. They are generally small of stature, with heads smaller than normal and incisor teeth tending to be widely spaced. Their hair is fair and their eyes blue, this hypo-pigmentation being a result of faulty metabolism. The disease is hereditary and recessive. The gene responsible blocks the chain of reactions in its first stages, at the point when phenylpyruvic acid is normally oxidised into p-hydroxyphenylpyruvic acid. The mental deficiency probably arises from a diminished rate of oxidation in the nerve cells. The frequency of the disease varies from one in 35,000 to one in 50,000 according to countries, being a little more common in Norway than elsewhere.

The action of recessive genes which are responsible for the absence of specific enzymes is revealed by the organism's inability to elaborate certain chemical compounds, and by the accumulation of intermediary products which would normally be transformed into these chemical compounds.

Drepanocytanaemia

Drepanocytanaemia or sickle cell anaemia is an hereditary affliction of the blood characterized by acute anaemia and the presence in the blood of drepanocytes or sickle-shaped red blood corpuscles. The gene responsible is dominant and in the homozygous state determines drepanocytanaemia. In the heterozygous state it determines only the presence of abnormal corpuscles, the drepanocytes. These corpuscles are not only different in shape but they contain different haemoglobins. The haemoglobin of normal aa persons behaves like a negatively charged ion. The haemoglobin of homozygous AA sufferers from sickle cell anaemia behaves like a positively charged ion. The haemoglobin of heterozygotes Aa who carry drepanocytes divides into two parts, one reacting like positively charged ions, the other like negatively charged ions.

The biochemical action of these genes thus consists of determining the nature of the haemoglobins: normal haemoglobin, pathological haemoglobin, and a mixture of the two.

The chain of reactions sometimes passes through the endocrine glands and gives rise to hormonal disturbances. A certain gene, whose action is reinforced by the absence of iodine in the organism, stops the secretion by the thyroid gland of thyroxin; the individual's

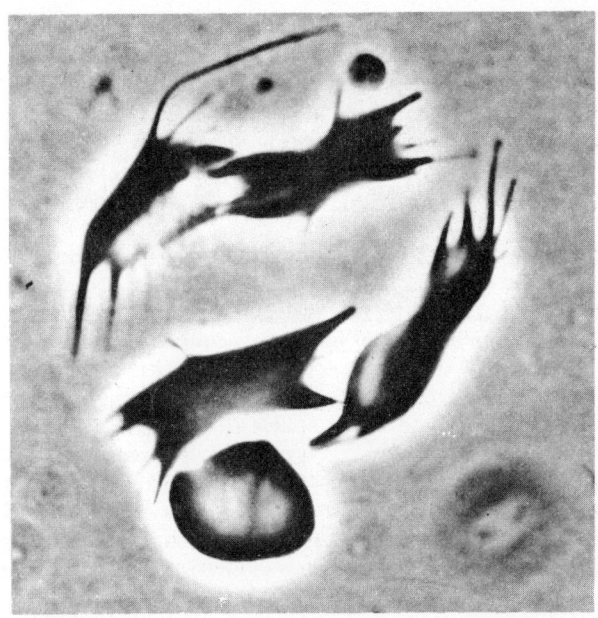

Drepanocytes seen under a phase-difference microscope.

Bessis.

growth and intelligence are arrested and the result is cretinism.

The possibility can thus be envisaged of correcting those undesirable genetic factors which act through the agency of the endocrine glands by injecting the abnormal individual with the deficient hormone. An example of this is the insulin treatment of diabetes, a disease which is often hereditary. The absence of enzymes necessary for protein metabolism could be compensated for by treatment with the intermediary substances normally elaborated by the enzyme. In the case of phenylketonuria the administration of glutamic acid — an amino acid of cerebral metabolism — raises the patient's intellectual level.

Genetic activity is a physico-chemical activity and many stages form the links in the chain of reactions between the gene and the ultimate phenotype. But the surroundings in which these reactions take place, influences both internal and external, play a part which is by no means negligible.

The influence of the environment on the phenotype

The internal environment plays its part in creating the phenotype. Its influence is chiefly apparent in the modifications which the phenotype undergoes during the course of life. The colour of the hair, though genetically determined, sometimes changes from fair in infancy to chestnut or dark brown in adolescence. The genes of pigmentation have one effect on the young organism and another on the older organism.

The striking changes which occur at puberty confirm the importance of the internal environment. The timbre of the voice, genetically conditioned, alters at puberty. The bass voice of the man and the soprano voice of the woman are determined by the same dominant gene; the man's tenor voice and the woman's contralto voice are controlled by the gene's recessive allele. Those who are heterozygotes for this gene will have, if men, baritone voices, and, if women, mezzo-soprano voices. These different voices correspond to the three genotypes AA, aa, and Aa, with incomplete dominance of the gene A in the heterozygote. At puberty voices change, and boys with low voices become high-voiced adults and vice versa. It is well known that boys who have been castrated preserve the high voices of childhood. Similarly a woman who, as a result of adrenal disturbance, manufactures an excessive amount of male hormone, acquires a masculine voice. The influence of the hormones is evident; the level of the voice is controlled by the sex through the agency of the hormones.

The dominance of a gene is sometimes modified by sex: an example of this is baldness. Baldness is determined by a gene situated on an autosome; it affects both sexes but it affects the male sex very much more frequently. That the gene is not sex-linked — that is, that it is located on an autosome — is shown by the fact that bald men transmit the character to half their sons and rarely to their daughters.

Baldness, or calvities, is controlled by a gene C, which is dominant in men and recessive in women. The normal allele of hairgrowth is c, so that the following formulae indicate the three possible genotypes:

cc, a man or woman with normal hair;
Cc, a bald man or a woman with normal hair;
CC, a bald man or a bald woman.

Male homozygotes and heterozygotes for C are bald, while only female homozygotes CC are bald, whence the fact that the character is much more frequently encountered in men.

It would seem that the sex hormones are responsible for this modification in dominance, as the following pathological case suggests: a woman with normal hair developed a tumour of the adrenal cortex which brought about an excessive secretion of male hormone. Secondary masculine sexual characters then appeared: beard and moustache, and also baldness. When the tumour was removed the characters stimulated by the male hormone disappeared.

A hormone treatment for baldness has been proposed which employs female hormone either by injections or by application to the scalp. In some cases the treatment has prevented further hair from falling out and even made hair grow again. It is not, however, without its disadvantages, since it can also endanger a man's virility, diminish his sexual desires, and cause a swelling of his breasts. Even local treatment is no assurance against such secondary effects, for the hormone, absorbed by the scalp, then passes into the bloodstream.

The foetal environment is another aspect of the individual's internal surroundings and also plays its part in the creation of the phenotype. The congenital luxation or dislocation of the hip is a dominant defect which nevertheless is almost exclusively found in women, whereas the frequency of the gene responsible must be roughly the same in both sexes. It arises from a malformation of the hip; the cotyloid cavity is flat and oval and projects slightly. This malformation can be observed in both sexes, being no more marked in one than in the other. But in men the conformation of the pelvis is such that the anomaly does not affect the gait, and only radiography reveals its presence. In women, on the other hand, the shape of the pelvis is incompatible with the malformation, with the result that they walk in a peculiar manner. During intra-uterine life, moreover, the dislocation is further accentuated by the position of the foetal thighs which explains the frequency of the luxation in breech presentations at childbirth. Thus congenital dislocation of the hip, though like all dominant defects it exists in both sexes, remains invisible in sons, but due to the shape of the female pelvis is manifest in daughters.

The maternal environment is also of importance in cases where an infectious disease from which the expectant mother is suffering affects the development of the foetus. German measles, a mild disease in childhood, will, if it attacks a woman in the first months of her pregnancy, often have grave consequences for the new-born baby: deafness, blindness through cataract or retinitis, cardiac malformation, mental backwardness. In a pregnant woman mumps can also bring about foetal malformations. The mechanics of how this occurs are still unclear.

Like the internal environment the external environment also takes part in forming the phenotype. The interaction of the genes and the external environment sets a problem for genetics, for it is far from easy to define the limits of their reciprocal influences. It is the still unresolved problem of 'nature' versus 'nurture'. Its experimental study has been facilitated by a method which offers great possibilities: the comparison of identical twins. Before describing the method, however, we must examine the phenomenon of twins and its genetic importance.

Opposite page
Far right, pop singer Barry Ryan with his twin brother.
Above, the Beverley Sisters, twins Teddy and Babs and elder sister, Joy.
Below, twin sisters Jennifer and Susan Baker.

P. Popper.

Twins

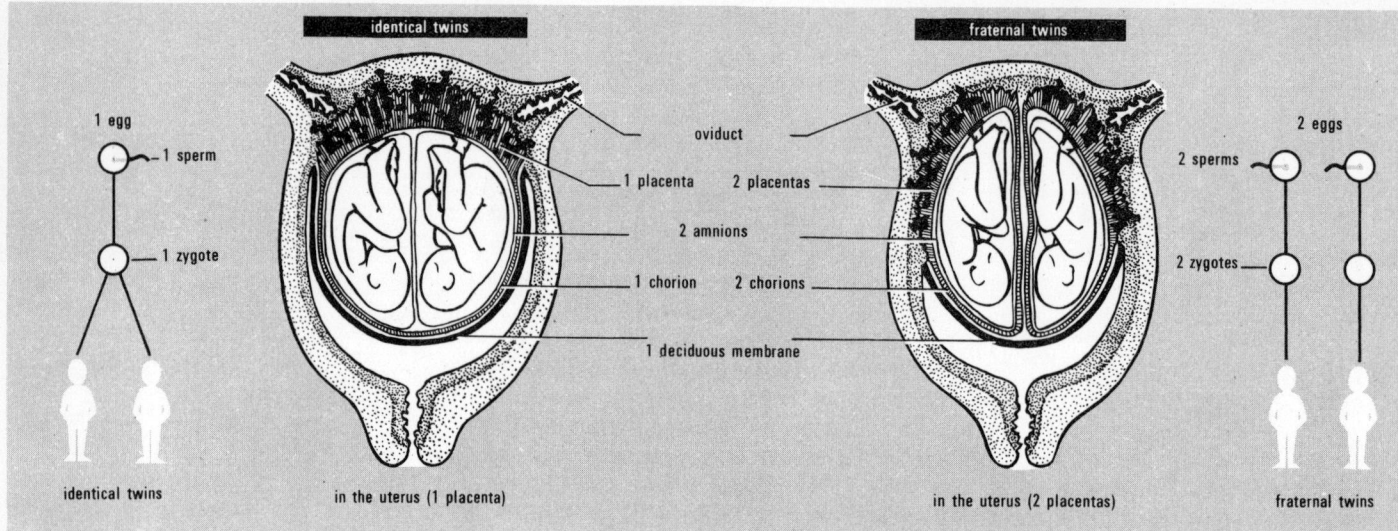

1 egg — 1 sperm

1 zygote

identical twins

in the uterus (1 placenta)

oviduct

1 placenta   2 placentas

2 amnions

1 chorion   2 chorions

1 deciduous membrane

2 eggs

2 sperms

2 zygotes

in the uterus (2 placentas)

fraternal twins

Diagram of univitelline and bivitelline twins.

In man the phenomenon of twins is, as it were, a kind of natural experiment which permits us to distinguish between the influences of heredity and environment, both of which play essential parts in the lives of all organisms.

As a rule women bear one child at a time. The frequency of twin or of multiple births may be found in statistics gathered in various countries. The frequency itself varies and seems to be influenced by the climate, attaining its maximum in cold regions. Twin births represent 1.4 per cent of all births in Scandinavia; 1.25 per cent in Germany; 1.13 per cent in France and Italy; 0.8 to 0.4 per cent in Greece, the Argentine, and Brazil, and 0.5 per cent in Japan. Other surveys give the highest frequency in Belgium (1.79 per cent) and the lowest in South America and Mongolia (0.8 per cent). Roughly speaking, the ratio of twin births to single births is one to eighty.

The birth of three or more children at a time is a rarer event, the ratio of triplets to single births being about one to eighty squared and, as a general rule, the ratio of n multiple births is $1/80^{n-1}$. During the last 500 years only thirty-two authentic cases of quintuplets have been known, and usually the children have not lived. This explains the scientific curiosity aroused by the Dionne quintuplets who have been the object of examinations and observations which still continue.

All children who are born simultaneously to one mother are called twins; but there are two categories of twins. Twins of the first category resemble each other so strikingly that they are known as true or identical twins. The others resemble each other as do brothers and sisters born at different times: these are non-identical twins, and are called false or fraternal twins. The two categories of twins have a different origin. Identical twins arise from a single zygote, the result of one ovum fertilized by one spermatozoon. At the beginning of development the embryo is unique; then it doubles and separates into two independent embryos. Generally the two embryos share a part of the foetal membranes, the chorion and the placenta. Hence identical twins are also called 'uni-ovular' or 'mono-

zygotic' or 'monochorionic' — synonyms which recall that they derive from a single ovum, a single zygote, and share a single chorion. They are always of the same sex. They are of the same genotype and are, indeed, two copies of the same individual. It is a case of human polyembryony.

Fraternal or false twins arise from different eggs which develop simultaneously like those of multiparous mammals which give birth to several offspring in one litter. For example, if two ova are detached from the ovary at about the same time ànd each is fertilized by different sperm the two resulting zygotes will develop side by side. Their foetal membranes will be distinct; they will have two amnions, two chorions, and two placentas. Fraternal twins are therefore known as 'bi-ovular', 'dizygotic' or 'dichorionic' to indicate that they arise from two ova, two zygotes, and have a chorion each. They can be of the same sex or brothers and sisters. Their resemblance to each other is neither more nor less thàn that of ordinary brothers and sisters. They are of different genotypes. It is a case of poly-ovulation.

The simultaneous births of more than two babies can have various origins. They can be monozygotic or the result of monozygotic and dizygotic combinations. For example, triplets can originate in three ways:
1. From a zygote which produces identical monozygotic triplets of the same sex.
2. From two zygotes, one of which produces identical twins while the other becomes simply a fraternal twin — brother or sister.
3. From three zygotes which produce three false or fraternal triplets.

The Dionne quintuplets were monozygotic and arose from the development of a single egg. The Argentine quintuplets were pentazygotic and derived from the development of five separate eggs.

By a statistical method, known as Weinberg's method, it is possible to calculate the relative frequency of the two types of twins. In the white population of the United States 34.2 per cent of all twin births are of identical or monozygotic twins. In the coloured popu-

332

Identical twins: Auguste and Jean Piccard. Auguste Piccard invented the bathyscaphe.

A. F. P.

lation the percentage is 28.9 per cent. The statistics for Germany give roughly the same results, although the frequency of identical twins is higher in towns than in the remainder of the country.

There would seem to be a correlation between the age of the mother and the relative frequency of mono-zygotic and dizygotic births. From the age of fifteen to thirty-nine the frequency of twin births rises slightly for identical twins and appreciably more so for fraternal twins. After the age of forty the frequency of identical twins continues slowly to rise while the frequency of fraternal twins falls.

A woman less than twenty years of age has about the same chance of bearing identical twins as she has of bearing fraternal twins; while a woman between thirty-five and forty has three times as many chances of having fraternal twins.

Twins seem to run in families. Many cases are reported of women who not infrequently have two or even more children at a time. One, who brought thirty-two children into the world, gave birth eleven times, producing three pairs of twins, six sets of triplets and two of quadruplets. Another produced thirty children in twenty-two years by her first marriage, and by her second a further fourteen in the space of three years. Still another had forty-four children in thirty-three years — thirteen pairs of twins and six sets of triplets. Such cases — and there are many others — strongly suggest an individual predisposition to multiple births. Among some women the simultaneous maturing of two or more Graafian follicles or an abnormal formation of multi-ovarian follicles would explain such, exceptional fertility and the resulting production of twins, triplets, etc.

But is a predisposition to have twins hereditary? To answer this question cases have been sought of a repetition of the phenomenon in various generations of a family. Direct and collateral lines are studied so that the frequencies observed may be compared with those given by statistics. The results which Greulich obtained in 1946 seem to indicate that twin births are in fact more frequent in the collateral lines of families with twins than in the general population. Beginning the investigation with couples who had twins, inquiries were made to discover whether the brothers and sisters of such couples also had twins. It was found that in these collateral lines the percentage of twin births was 6.3 per cent, whereas in a random sample of the population it was only 4.5. The difference is significant. In other surveys of a similar nature the frequency not of twins in general but of identical twins and of fraternal twins was studied. It was then found that the frequency of identical twins was not appreciably higher than in the general population, but that the frequency of fraternal twins, on the contrary, was unmistakably higher: 2.32 per cent as against 1.15 per cent of the general population. Hereditary factors would thus seem to be responsible for the tendency to have fraternal twins.

The resemblance of identical twins

Since they have the same hereditary endowment monozygotic twins must be genetically identical and therefore unlike their other brothers and sisters who, though born of the same parents, derive from different eggs — that is to say from different combinations of chromosomes carrying different hereditary factors or genes.

The resemblance of identical twins is a permanent phenomenon which becomes even more obvious during the course of growth. It is unmistakable from the second year of age, and is the essential test in the diagnosis of identical twins, confirming the indications given by the foetal membranes.

At birth identical twins may reveal unequal development, one weighing more than the other. The difference arises from conditions during foetal life, often from differences in placental circulation.

In addition to a general resemblance identical twins have in common a number of physical, psychological, intellectual, and pathological characters. Among their identical physical characters the most significant are their blood-groups (A, B, AB, O — M and N — Rhesus system, etc.), the colour, texture, and form of hair, colour of eyes and skin, the density and distribution of hair on the skin, shape of the nose, ears, lips and chin, the teeth and their irregularities, freckles, shape and size of hands and fingers, fingerprints, and prints of palms and soles of the feet.

Interesting facts are obtained from the study of fingerprints. Bertillon has pointed out 100 characters which differentiate fingerprints. The probability that a given number n of these characters will be identical is $\frac{1}{4}^n$. For instance, the probability that sixteen characters will coincide is $\frac{1}{4}^{16}$, which works out at one chance in 4,294,967,296. Bertillon once examined two individuals with thirty coinciding characters — an astronomical improbability only explained by the fact that the two were identical twins.

These physical resemblances, revealing the identity of hereditary endowment, also appear in physio-logical characteristics — the electrocardiograms are

The Houra quads of Czechoslovakia, two boys and two girls, who were born in 1959.

identical and the pulses beat in the same way for example — and in anomalies, malformations, and hereditary diseases. The following have, for instance, been reported: the same syndactylia in one pair of identical twins, hexadactylia of hands and feet in still-born twins, dental anomalies (median diastema and displaced canines), harelip, inguinal hernia in the same position but on opposite groins.

Such symmetrical disposition is, in fact, fairly often found in identical twins, one being as it were the mirror image of the other. In a pair of 'mirror' twins one will be right-handed, the other left-handed; the hair at the crown of the head grows clockwise in one and anticlockwise in the other; freckles, birthmarks, skin-discolorations on the left side of one's face occupy the right side of the other's; a higher than average frequency of *situs inversus viscerum* is found, that is to say, one of the twins has his heart on the right and his liver on the left, while the other has these two organs in the normal position. Among the Dionne quintuplets Marie and Emilie were mirror twins. The characteristic asymmetry of mirror twins is the result of a delayed separation which takes place only after embryonic development has begun.

The almost total identity of monozygotic twins and the ludicrous confusion to which they give rise has provided themes to playwrights of all ages, from the *Menaechmi* of Plautus, and Shakespeare's *Comedy of Errors,* to the *Invitation au Château* by Anouilh.

The scientific study of twins

The comparison of identical twins enables a distinction to be drawn between those characteristics for which the individual's genetic constitution is responsible and those which are the result of his environment. In short, the study of twins helps to solve the problem of 'nature versus nurture'.

In principle the method is simple: dissimilarities between identical twins who have the same hereditary endowment must arise from the influence of the environment. While discordant characters derive from the environment those in which identical twins agree are the result of hereditary factors. The differences observed between two identical twins reared apart and in different circumstances may thus be attributed to the environment, while their resemblances can safely be assumed to be hereditary.

On the other hand the dissimilarities of fraternal twins arise from both heredity and environment. Differences which are observed between fraternal twins who live in the same family and under identical conditions reveal the effect of two different hereditary constitutions.

As early as 1875 Sir Francis Galton drew attention to the possibilities of this method, although it was only much later that his suggestions were pursued in a scientific manner. The method consists of the statistical study of a given character or disease in the largest possible number of pairs of identical and fraternal twins. Percentages are then drawn up for points of resemblance and difference between each pair of each of these two categories. Whether or not the character or disease is hereditary will then be indicated by the relative frequency with which it is found in identical twins as opposed to fraternal twins.

The most important twin studies have been made in America and in Germany. In the University of Chicago Newman, a biologist, Freeman, a psychologist, and the statistician Holzinger, have examined fifty pairs of identical twins and fifty pairs of fraternal twins of the same sex. They analysed various physical characters, height, weight, length and breadth of hands, cephalic index, colour, texture and manner of hair-growth, colour of eyes and skin, fingerprints, palm-prints, shape of ear, left-handedness, right-handedness; and also psychological characters, intelligence, mental aptitude, and so on. In their work they also studied the resemblances and differences between nineteen pairs of identical twins who had been reared apart, some in widely separated parts of the world. Without going into details of their analyses — and of similar studies

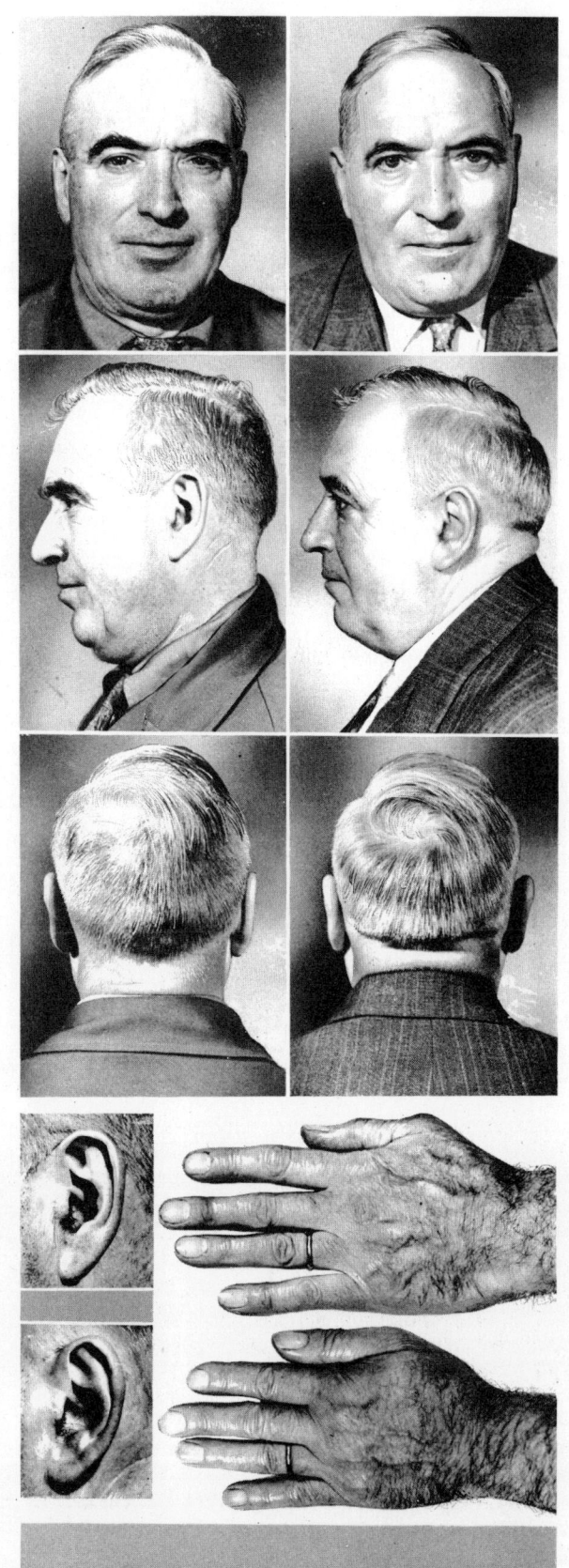

Identical twins born at the beginning of the century and their points of resemblance.

Larousse.

335

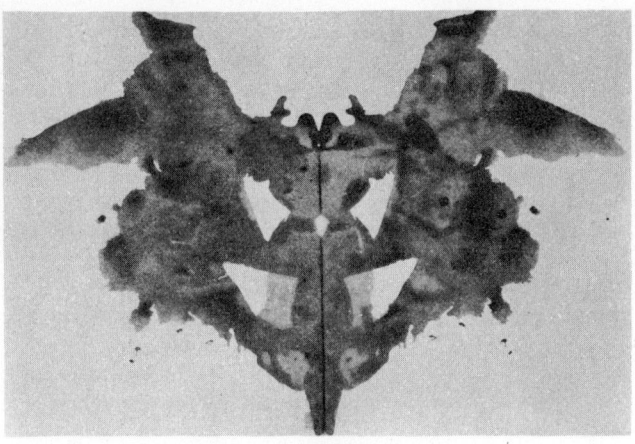

According to Rorschach's 'psychodiagnosis' the person tested can see in this blot of ink:
   a bat
   or
   a butterfly
   or
   two angels with floating robes around the figure of a woman in the centre
   or
   in the centre one angel enveloping two St. Nicholases with his wings.

Verlag Hans Huber, Berne.

Four pairs of twins born consecutively to the same parents. *At the top,* two brothers born at an interval of one year.

A. F. P. Photography.

made elsewhere — a few significant points may be considered.

Do genetic factors play a part in longevity? A correlation undoubtedly exists between the length of the parents' lives and those of their children, and the study of twins contributes interesting facts to this subject. In one case identical twins died of natural causes on the same day at the age of eighty-six. Identical twin sisters, one of whom married a farmer and had several children, while the other was a spinster and a dressmaker, had clinical histories which were extraordinarily parallel: both became totally blind and deaf during the same month and both had a cerebral haemorrhage on the same day, followed by a similar paralysis — hemiplegia. They both died within an interval of twenty-five days at the age of sixty-nine. The physical symptoms of old age are analogous in identical twins; the similarities embrace their general condition, the colour and density of hair, the form of baldness and wrinkles, deficiencies of hearing, sight, and teeth. Identical twins examined at the age of ninety-five were practically indistinguishable: their pulmonary radiographs, the size and shape of their hearts, their electrocardiograms were alike. The sole difference between them was that one spoke English and the other did not, a character obviously not hereditary. Such close resemblances are not found among fraternal twins. It would seem safe to conclude that heredity plays a basic role in the maintenance of relatively good physical health before and during senescence.

Twins studies also tell us whether an illness is or is not hereditary. Multiple sclerosis, a serious affliction of the nervous system, is characterized by the presence in the brain and spinal cord of isolated formations of sclerotic tissue; the various positions of these forma-

tions account for the disease's various symptoms, paraplegia, trembling, scanning speech, visual disturbances. But is the disease hereditary or infectious? Thums conducted a broad enquiry among fifty-one pairs of twins, in whom one twin at least suffered from multiple sclerosis. For various reasons — death principally — only twenty-nine pairs were studied, of whom seven pairs were identical twins and twenty-two fraternal twins, of the same or different sexes. Among the seven identical pairs he found no concordance — that is, one twin had the disease and the other had not. The conclusion is inescapable: multiple sclerosis is not hereditary.

The particularly difficult problem of tuberculosis has been approached in the same manner. Between 1929 and 1936 in Germany 205 pairs of twins were studied. Among the eighty pairs of identical twins fifty-two, that is sixty-five per cent behaved identically with respect to tuberculosis. Among the 125 pairs of fraternal twins only thirty-one — that is twenty-five per cent — behaved identically. The difference is significant: when an identical twin suffers from the disease his fellow twin is in two out of three cases also attacked; when a fraternal twin suffers from tuberculosis his fellow twin is attacked in only one out of four cases. The concordance among identical twins also

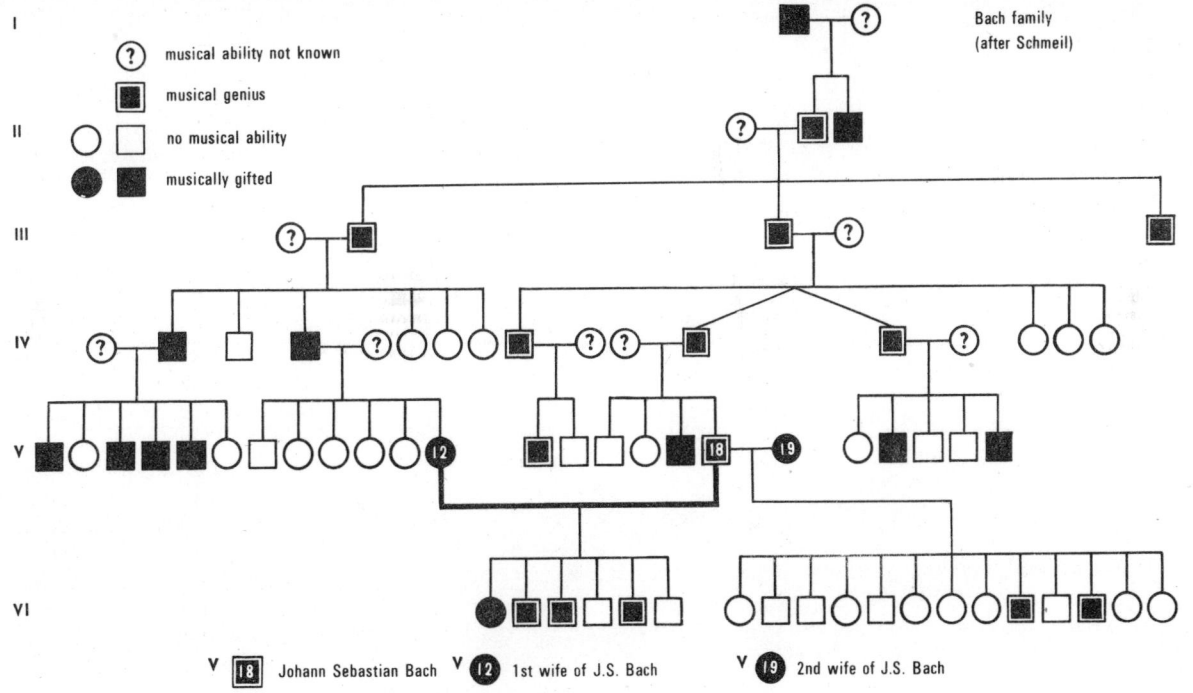

? musical ability not known

▣ musical genius

○ □ no musical ability

● ■ musically gifted

II

III

IV

V

VI

Bach family
(after Schmeil)

V ▣18 Johann Sebastian Bach   V ●12 1st wife of J.S. Bach   V ●19 2nd wife of J.S. Bach

Heredity of aptitudes. The Bach family produced an impressive
number of good and of exceptional musicians. The precise
way in which aptitudes are transmitted remains unknown.

embraces the localization of the disease and the way in which it develops. Both twins will for instance be stricken by tuberculosis of the lungs or of the kidneys or of the bones. In both twins the disease will evolve at the same speed, rapidly or slowly as the case may be. Among fraternal twins — if both have tuberculosis — the similarity of the disease is not so striking. Nor is the discordance among identical twins as marked as it is among fraternal twins. Two examples will illustrate the maximum concordance between identical twins and the minimum concordance between fraternal twins. Identical twin sisters who had long been separated (one lived on a farm and the other in a town) contracted tuberculosis at the same time. Two fraternal twins lived in the same conditions and slept in the same bed: one died of tuberculosis while the other's health remained unaffected. From such results it would seem to emerge that tuberculosis of infectious origin has a hereditary basis; there is a kind of predisposition which takes the form of supplying a favourable field of action for the bacillus. In other words, both factors, predisposition and bacillus, play a part.

Twin studies also throw light on the inheritance of psychological characters. Among identical twins various tests reveal that the mental level is similar and even identical. Among fraternal twins, on the other hand, mental divergence is as marked as it is between ordinary brothers and sisters. Statistically the index of correlation between identical twins is 0.881 and between fraternal twins 0.631. In tests for skill, memory, and scholastic accomplishment the indices of correlation are distinctly higher for identical twins — 0.830 on the average — than for fraternal twins — 0.50. Graphs showing the work capacity of identical twins curve in the same way, rise and dip and indicate the same tendencies to fatigue. Free association tests,

relative progress in learning, and so forth, always reveal the close resemblance between identical twins. An identity of genetic constitution would thus seem to bring about marked resemblances in sensory-motor and intellectual aptitudes.

Personality, character, and temperament follow much the same pattern. The Rorschach test which measures the emotional elements of the personality has been applied to twins. The test consists of interpreting a series of ink blots produced by folding cards in two. The person tested is invited to say what the designs suggest to him. Thirty pairs of twins, of whom eighteen were identical, were tested in this way. By the results the twins could be classified in four groups. In the first group (seven pairs of identical twins) the answers revealed an identity of interpretation. In the second group (nine pairs of identical twins) the answers were very similar. In the third group (five pairs which included two pairs of identical twins) the designs were interpreted differently. In the fourth group (nine pairs of fraternal twins) the interpretations were so unlike that they revealed profound dissimilarities. Answers to Bernreuter's questionnaire, designed as an inventory of personality, gave a correlation index of 0.595 among identical twins, of 0.230 among fraternal twins of the same sex, and of 0.105 among fraternal twins of different sexes. To questions regarding professional ambitions and their future three-quarters of the identical twins had formed identical plans, as against only half of the fraternal twins.

The faculty of association has a higher correlation among identical twins than among fraternal twins, the index being 0.656 for thirty-one pairs of the former as against 0.594 for fifty-three pairs of the latter. In the quality of perseverance the environment seems to be the preponderant factor; for the correlation is reversed:

0.101 for identical twins and 0.465 for fraternal twins.

Lange has studied criminality among twins in the prisons of Bavaria. Thirteen of the convicted had an identical twin and among these identical twins ten had also been in prison for similar crimes. Seventeen prisoners had fraternal twins; only two of them had been charged with criminal activities. In another survey of eighteen pairs of identical twin criminals, eleven pairs had committed the same crime, while of nineteen pairs of fraternal twins only seven pairs were imprisoned for similar reasons. Still other figures give the same picture: among identical twins the same crime in seventy-one per cent of the cases; different crimes in twenty-nine per cent; among fraternal twins the corresponding percentages are thirty-nine and sixty-one. Among identical twins there is also a striking similarity in the manner, gravity and timing of their criminal acts.

It would appear then that certain qualities of character, tendencies and a predisposition to given acts and reactions are hereditarily determined; although a very important part is also played by external circumstances. Psychiatrists report similar findings among schizophrenic, manic-depressive, and epileptic twins.

Further light on the subject is supplied by the examination of multiple births. The Key quadruplets, comprising a pair of identical twin sisters and two non-identical sisters, were brought up in the same fashion and, during childhood and adolescence, their family life was especially close and intimate. They went to the same schools and all received degrees. But in spite of identical environments the non-identical twins differed from each other and from the pair of identical twins. The latter, on the contrary, were remarkably alike in scholastic success, mental ability, and psychological traits.

Among triplets and quadruplets one individual, in spite of profound resemblances, will often stand out as the leader. This conduct, which reveals a difference in personality, acts on the group and introduces an element of dissimilarity. The leader is usually brighter and more lively than the others. The Morlok quadruplets, born in the United States in 1930, resemble each other physically but with a slight difference in height. The smallest of the four is intellectually less bright than her sisters whose lead she always follows. A similar situation exists among the Dionne quintuplets who, being identical, developed on parallel lines. Nevertheless Yvonne, always the strongest, was also the most intelligent. Mary, the 'baby' and the smallest, always revealed an intellectual inferiority connected with her physical backwardness due to unfavourable foetal conditions. Intra-uterine life thus exercises an influence which is often most apparent in multiple births. When foetal development is impaired later growth, physical and mental, will suffer correspondingly.

Other surveys designed to study the nature versus nurture problem have been concerned with identical twins who have been brought up apart and submitted to different influences. Identical twin sisters, married at the ages of nineteen and twenty-one, had since that time lived totally different and separate lives. At the age of forty-three both were divorced; afterwards, at an interval of a few days, both entered hospital to be treated for hypochondria. Tests revealed that their mental level was identical. Two men of thirty, identical twins who had been separated at the age of two weeks and brought up in contrasting social environments, were of the same high intellectual level although their personalities were very different. Identical twins who were adopted and brought up, one in the family of a commercial agent and the other in the family of a journalist, met again at the age of eighteen. Both were boxing champions; tests of their personality and I. Q.s gave the same results. Some differences appeared in their scholastic aptitudes and in their deportment, one having anti-social tendencies. Their tastes also varied in that one preferred music and the other drawing. Identical twin sisters reared in totally unlike environments — one rural and the other urban — met again at college when they were eighteen. Various tests of intelligence, knowledge, and personality resulted in answers which were very similar and at times identical.

The behaviour of identical and fraternal twins brought up in the same environment and that of identical twins separated for a part of their lives provide evidence that mental and sensory-motor capacities are a function of the individual's genetic constitution. Heredity thus exercises an unmistakable psychological influence. Nevertheless this psychological influence is not quite so powerful as the physical and pathological influence of heredity. Holzinger estimated that heredity is responsible for seventy-four to ninety per cent of the individual's physical characters and for sixty-four to eighty per cent of his psychological traits.

The genotypic identity of monozygotic twins finds a still further application in the realm of experimental medicine. In therapeutic research, for instance, one twin can serve as the subject of experiment, while the other acts as a control.

Teratological twins

The phenomenon of monstrous or teratological twins is merely a special case of normal twinning. The two budding embryos which derive from the single egg do not, as with normal twins, remain separate and evolve into normally formed individuals, but instead become fused in varying degrees. In this way twin monsters can be born which are totally doubled, coalescent, or partially doubled. Doubling may be incomplete so that the forward or the rear section of the embryo remains single. Or one of the embryos may develop fully while the other does not, and the double monster will then be composed of one entire individual, the *autosite,* and a secondary component, the *parasite,* which may itself be reduced to a region of the autosite, or even to an organ or a cyst.

The bewildering variety of double monsters, their strangeness and rarity, have long excited curiosity. They were studied as early as the sixteenth and seventeenth centuries, but it was at the beginning of the nineteenth century that Etienne Geoffroy Saint-Hilaire inaugurated a new branch of science, teratology, which dealt with monsters in general. He attempted to explain how they arose, classified them, and provided

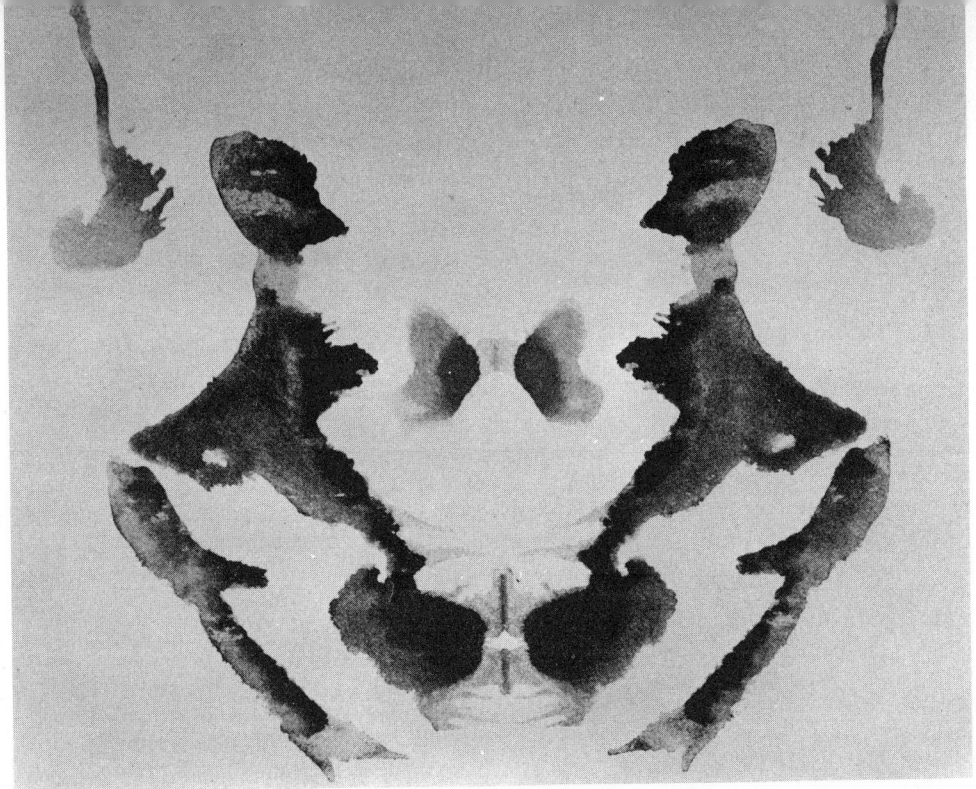

This ink blot evokes various images:
two clowns
or
two waiters bearing between them a champagne bucket
or
two dandies in tail coats, top-hats in hand, who bow to each other with exaggerated ceremony
or
two juggling monkeys, face to face, balancing something on their heads.

Verlag Hans Huber, Berne.

A further ink blot which may suggest:
a bat
or
(when viewed upside down) a butterfly
or
a dancer on tip-toes in a moment of impassioned movement
or
(when viewed from the side) a bent old woman with two umbrellas under her arm.

Verlag Hans Huber, Berne.

technical terms. His work was continued by his son Isidore Geoffroy Saint-Hilaire. More recently teratogenesis has become the subject of experimental research. Chemoteratogenesis, or the creation of monsters by means of chemical substances, has produced results which are of interest and value.

Some cases of celebrated double monsters have been mentioned in an earlier chapter. The Siamese twins Chang-Eng, who lived in the United States, were rather loosely attached by the lower part of the sternum; they could walk, run, and swim without difficulty. They married two ssiters and lived for sixty-three years (1811—1874). Both of them had several children, all of whom were quite normal. Chang, who drank too much, was stricken with hemiplegia in 1872 and died in 1874. His death led to the death of Eng a few hours later. The Chinese brothers Tsang-Sing were fused face to face at the base of the thorax. Attempts have been made to separate pairs of girl twins, but the fusion of their livers in the shared part of the body and the communication of their blood

vessels always make the success of such operations doubtful.

Attachment by the lower part of the trunk is generally compatible with life. The Czech sisters Rosa-Josepha Blazek (1878—1922) were fused in the region of the pelvis. Their viscera were separate except for the large intestines and the urino-genital organs which were united in their lower regions. Anus and urino-genital orifices were shared. One of the sisters had a normal child; during her pregnancy the other sister's organism functioned regularly but when the child was born she formed milk. The brothers Lucio-Simplicio Godena, born in Marseilles, were joined back to back at the lower part of the trunk. The fusion permitted them a certain liberty of movement; they married twin sisters. The physical resemblance of such 'Siamese' twins is often less striking than that of normally separated identical twins.

Other types of double monsters suffer from abnormalities of the respiratory and circulatory systems — atrophied lungs, cardiac malformation — and rarely live. Individuals fused end to end in the region of the cranium or pelvis usually survive for only a few hours. Dicephalous monsters with a single trunk, two arms, two legs, and two heads, may live for several months. Ira-Galya, born in Moscow in 1938, lived for a year in a clinic; they succumbed to an infection of the respiratory tract. Their spinal columns and nervous systems were distinct, and they reacted differently to stimuli. One would laugh while the other cried. They would sleep at different times. Autopsy showed that their viscera and skeletons were fused in the region of the pelvis.

A dicephalous male couple with two legs and two chests lived for twenty-eight years at the court of James IV of Scotland; both spoke several languages and were skilled musicians.

The morphology of monsters is determined by their degree of separation and the orientation of their axes, that is to say, their two germinal streaks. The forward convergence of the two axes and their divergence at the rear give rise to posterior duplications. In man the phenomenon is rare, though Blanche Dumas, born in 1860, had one head, a single thorax and four legs, one of which was rudimentary. Her urino-genital orifices were double, as were her menstruations.

Conversely, the rear convergence of the two axes and their divergence in front give rise to anterior duplications. An extreme example of this type is when the trunk is single and only the two heads are distinct. The case is known of an apparently normal woman who died in Montpellier at the age of fifty-three in whom the doubling of the head was indicated only by a double nose, by a third rudimentary eye placed between the two noses, and by the abnormal width of the mouth. The autopsy revealed a third and reduced cerebral hemisphere situated between the two cerebral hemispheres.

In a third possibility the two axes are arch-shaped and fuse to form an X, so that they diverge both in front and behind. This fusion, which gives both individuals more freedom, is the most viable.

Triple monsters are extremely rare. One case occurred in Catania. It consisted of a single trunk with two legs, three arms and two necks, one of which bore a head while the other, larger, bore two heads. The autopsy revealed three larynges, three tracheae, three oesophaguses, two pairs of lungs, two hearts and two spinal columns one of which bifurcated towards the upper end. The liver, the pancreas and the urino-genital organs were tripled. This monster resulted from a subdivision of one of the two primitive embryos.

Such monstrosities are not hereditary, but arise from accidental anomalies of the zygote's development. In cases of extremely unequal embryonic development the undeveloped parasite is, in a sense, grafted on to the normally developed autosite. A celebrated example is that of Lazzaro Collondo, born towards 1625 in Genoa, who was examined by Doctor Bartolin and drawn by Licetus in 1665 wearing a court costume of the period. To the ventral surface of the trunk was affixed the parasite, Jean-Baptiste, who had a head, a left leg and two arms; the hands had only three fingers each. Another case was that of the thirteen-year-old Hindu boy who appeared at the colonial exhibition of Marseilles in 1906. The parasite had no head; its hands were misshapen, but it possessed external genital organs and could urinate. At times the parasite is reduced to a single member, deformed and more or less inert. A girl named Anna Maria Przesomyl was born with a tumour on the coccyx which developed and three years later erupted, revealing a leg which continued to grow larger. When she was sixteen the leg was surgically removed: it was in reality double, two fused femurs being attached to the pelvis. The parasite is sometimes enclosed in a tumour — a teratoma — which can occupy various regions, external or internal, of the autosite. The teratoma encloses whole sections of a body, organs or tissues, hairs or teeth. Such partial formations arise in a similar fashion: during the early doubling of the individual embryo one of the resulting divisions is incomplete and remains fused to the other. It acts in a sense like an embryonic graft.

The study of twins, both normal and teratological, is closely linked to fundamental biological problems of special importance to embryology, genetics, and the study of the individual's constitution and behaviour. The progress of the various branches of biology enables us to give a scientific interpretation to the phenomena of twins and finally to dispense with the extravagant and often supernatural hypotheses of the past.

Eugenics

The science of eugenics was founded in 1883 by Sir Francis Galton, a cousin of Charles Darwin. The idea itself is very ancient and can be found in the works of Plato and the sages of antiquity.

Inasmuch as many diseases, defects, and abnormalities are hereditary and some are regularly transmitted by parents to children, it was not unreasonable to propose that children unfit for life should not be brought into the world. This, of course, amounts to preventing persons who are seriously defective from reproducing.

Legislation to sterilize such persons exists in many countries. In the United States the first law of this kind was adopted in 1907 by the State of Indiana. By 1940 similar laws existed in thirty states, being more strict in some than in others. In all thirty states mental deficiency is subject to sterilization; in twenty-nine mental diseases are liable to the same treatment. In twenty-three states there are laws controlling epilepsy. In other states sterilization can be applied in certain cases of physical malformation and nervous diseases. From 1907 to 1950 more than 50,000 such operations were performed, two-fifths of the sterilized being men. It is still difficult to weigh the results of these eugenic experiments which have not in all cases been based on sufficiently solid genetic and medical grounds. Similar legislation has been proposed in certain provinces of Canada, in Tasmania, and in New Zealand.

In Europe sterilization laws were first introduced in 1929 in the Swiss canton of Vaud, and in the same year in Denmark, Norway, Sweden, Finland, and Iceland. In Germany and in Estonia the law, after being blindly applied, has been repealed. Laws have been proposed, without being passed, in England, Holland, Hungary, Czechoslovakia, and Poland. At the moment such laws are retained only in the Scandinavian countries and in some Swiss cantons.

The Scandinavian laws provide that the patient himself can ask to be sterilized and that the operation cannot be performed without his freely given consent. During the period 1929—1950 more than 6,000 operations took place in Denmark, two-thirds of those sterilized suffering from mental deficiency, and two-thirds of these being women. In Sweden, for the period 1935—1948, some 15,650 people were sterilized, women again outnumbering men. The number of those sterilized continues to rise and Professor Kemp estimates that in Denmark today there are 600 operations a year, distributed equally between the mentally deficient and those suffering from other afflictions. At this rate an increasingly rapid elimination of mental deficiency may be anticipated. Denmark not only has laws providing for sterilization but also others covering legal abortion, the prevention of certain marriages, and eugenic measures such as isolation, the obligatory declaration of disease, etc.

Of course these various measures encounter moral, social, and religious opposition, and the question arises whether their efficacy is genuine. The answer will vary according to the way in which the gene responsible for the affliction is transmitted.

The elimination of a dominant gene

In the case of a dominant defect sterilization is effective. By eliminating only half the individuals affected in each generation those suffering from the defect will in three generations be reduced to $(\frac{1}{2})^3$ or 1/8th. The almost total disappearance of the defect may be accomplished in an easily calculated number of generations. Naturally the elimination of the defect may be frustrated by spontaneous mutations, leading to its sudden reappearance.

The elimination of a recessive gene

A recessive gene is much more difficult to eliminate, since heterozygotes, apparently healthy, cannot be

Some human anomalies and malformations which are hereditary.

1. Hand with seven fingers; one of the earliest documented cases of human polydactylia.

After Morand, 1770.

2 and 4. Acrocephalosyndactylia — or acrocephalia (a cranial distortion) associated with syndactylia of fingers and toes.
3. Bonnevie-Ullrich's syndrome.
5. 8 and 9. Malformation of the feet.
6. Malformation of the hands.
7. Hereditary oedema of the legs — Milroy's disease.
10. Aplasia of the ear.

McGuire and Pearl Zeek (7); Lamy (2 – 6, 8 – 10).

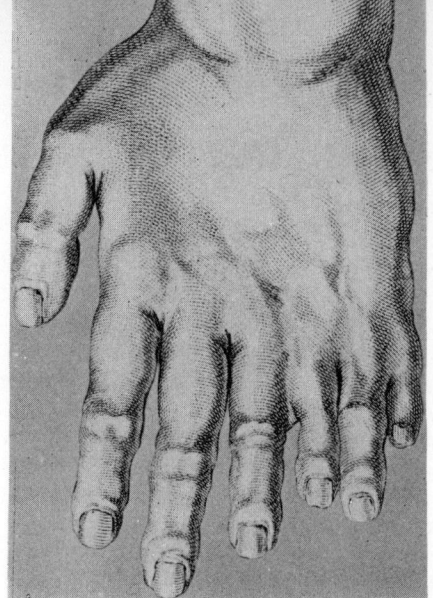

1

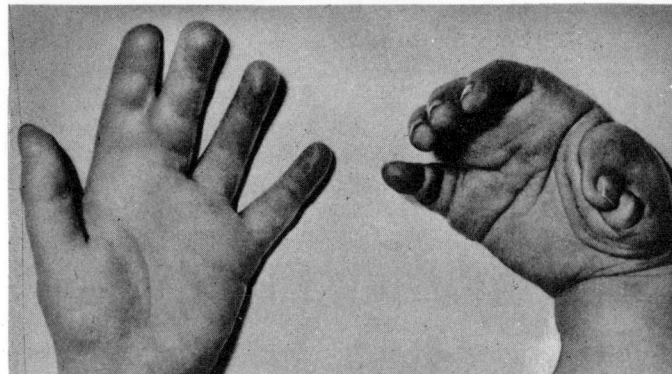

2

detected readily. They are nonetheless carriers of the harmful gene and spread it unwittingly when they have children. Only the elimination of recessive homozygote gene combinations is possible, and the efficacy of this measure is very slight. Calculations reveal that the total elimination of defective recessive homozygotes would have to be repeated for twenty generations (that is, for about six centuries) before the frequency of the defect would drop from one per cent to one per thousand. To reduce the frequency to one in 10,000 would require no fewer than seventy generations. Thus it would be eugenically valuable if heterozygotes for a recessive defect could be recognized.

The elimination of a sex-linked gene

The elimination of a gene carried on the Y-chromosome is simple, since the gene always manifests its presence in men.

The elimination of a recessive gene carried on the X-chromosome is less simple. By preventing a defective man from reproducing, a third of the genes — namely, those carried on the X-chromosomes of his sons — would be eliminated. But two-thirds of the genes carried on the two X-chromosomes of women would persist. It would thus be necessary to forbid the daughters — and indeed the sisters — of defective men to marry. Even then girls who had received the gene with the maternal X-chromosome would not be affected by the measure. At best the results would be indecisive.

From this brief survey it would appear that sterilization measures would not solve the problem and that the elimination of a defect, or of the gene responsible for it, is impracticable. Only the elimination of a dominant defect is a reasonable possibility, and dominant defects are much rarer than recessive defects.

Eugenics may perhaps be most efficiently applied in the form of consultation before marriage with a doctor whose knowledge of genetics makes him competent to give advice. Certain marriages he will consider desirable, while others he will judge to be reckless, or even disastrous.

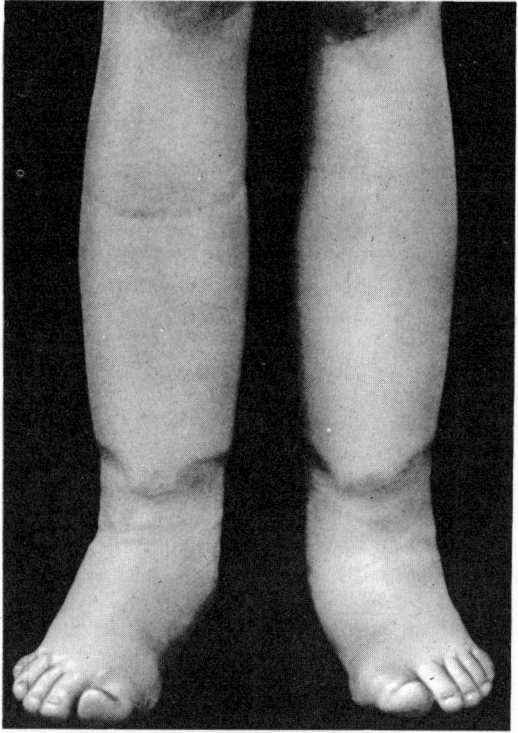

3

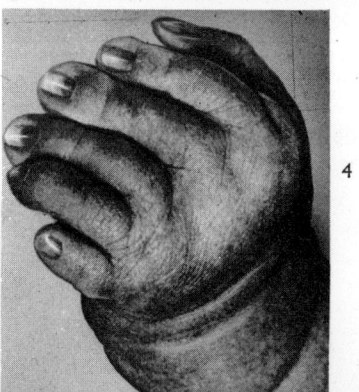

4

342

Eugenic consultation

Eugenic consultation is usually based on a few questions, the following being those most frequently asked:

Is it dangerous to marry a person in whose family one or more members is defective? Should a person who is himself defective marry? If a defective child is born does this mean that subsequent children run the risk of being defective? Should the marriage of first cousins be advised against or forbidden? What should the competent and conscientious doctor reply to such questions — especially as his advice may not be taken? In most countries the doctor has no legal right to forbid marriages on eugenic grounds.

But in answering such questions the nature of the defect must first be clear: is it or is it not hereditary? Some well-known hereditary taints present no difficulty, but in doubtful cases it is less easy to speak with certainty. If the tainted individual has a large family free from the defect, or if particular circumstances accompanied his birth (maternal German measles, influenza, threatened miscarriages, transfusions of incompatible blood), and if there is no suggestion of consanguinity in the proposed marriage, then the likelihood of the blemish being inherited by his children is slight. If the defect is in fact hereditary the manner in which it is transmitted will determine the consultant's advice.

If the defect is dominant and its penetrance complete, no individual unaffected by it can carry the gene responsible. Thus the unaffected cannot transmit it to their offspring. Nevertheless two precautions are necessary: first, the individual must have passed the age at which the flaw appears in the family, for some hereditary afflictions are manifest only at a relatively mature age. Second, the dominance must have complete penetrance. Whether this is so is not easy to decide, especially if the individual has a brother or sister who is tainted, or if the defect is apparent in one of his close relatives.

If a healthy individual has a parent who carries a recessive defect it is probable that he will himself be heterozygous for the responsible recessive gene. This probability is $^2/_3$ if the defect is visible in his brother or sister. It will be remembered that the offspring of two heterozygotes Aa for a recessive gene will be composed of $\frac{1}{4}$ AA individuals who are healthy, $\frac{1}{2}$ Aa individuals who are healthy but carriers of the recessive gene, and $\frac{1}{4}$ aa individuals who are tainted. A person who is not aa thus has two chances out of three of being Aa and one chance out of three of being AA.

In order to engender a defective child our heterozygote must marry a heterozygote for the identical recessive gene. If the frequency of the defect is one in 10,000 the frequency of heterozygotes is one in fifty.

Finally, the marriage of two heterozygotes Aa will give rise to three healthy children, one AA and two Aa, and one recessive homozygote aa who will be tainted. Hence the probability of a defective child being born is $\frac{1}{4}$.

The combined chances are thus $^2/_3$ times $^1/_{50}$ times $\frac{1}{4}$ which equals 1/300 — which means that the candidate

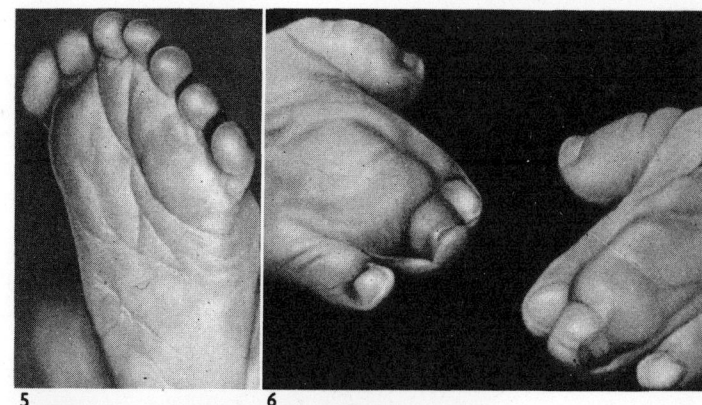

5   6

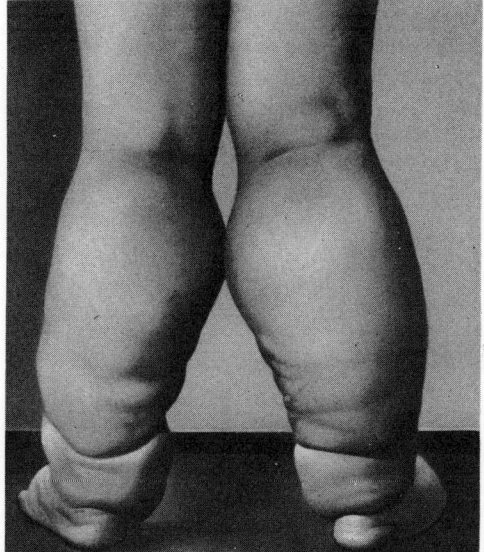

7

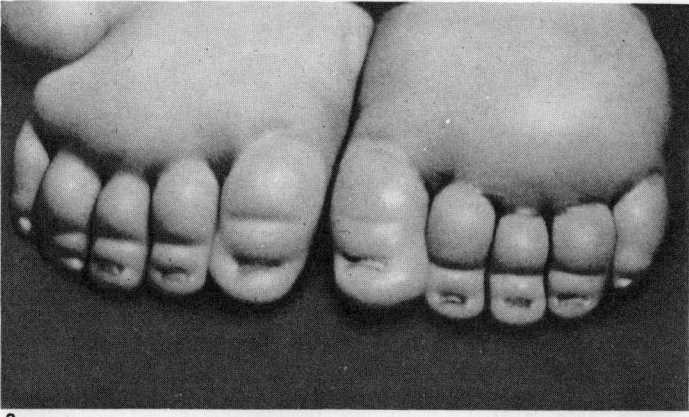

8

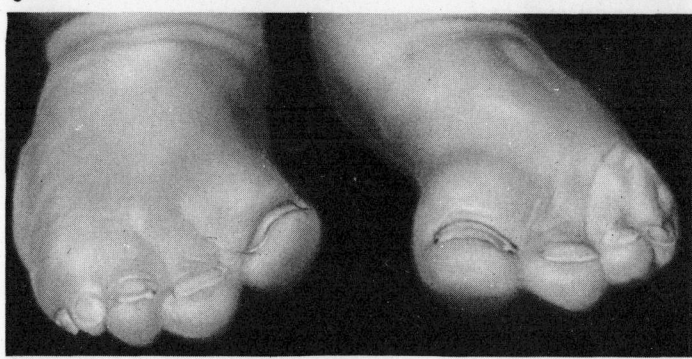

9

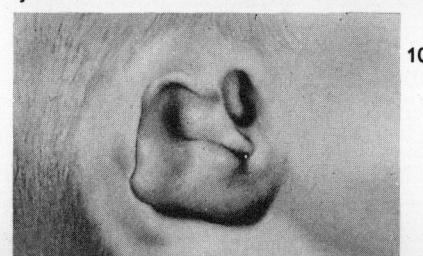

10

343

for marriage who marries the brother or the sister of an individual suffering from a recessive defect with a frequency of 1/10,000 (albinism, for example) has one chance in 300 of having an albino child. This probability is relatively slight, but it is considerably greater than that of a marriage between two persons picked from the population at random. If we take the same example — namely, albinism — the frequency of heterozygotes is 1/50. Thus the chances that two heterozygotes for the gene of albinism will marry is 1/50 times 1/50 or one in 2,500. The probability that this couple will have a recessive homozygote for the defect is $\frac{1}{4}$ — which makes $\frac{1}{4}$ times 1/2,500 or 1/10,000. In other words, in a random marriage the chance of an albino child being born is only one in 10,000 — which is a very much smaller risk than in the marriage first discussed. To sum up, the healthy person who is related to someone who suffers from a recessive defect has a more than average chance of engendering a child with that defect.

There is a third eventuality: the healthy individual is related to a person suffering from a sex-linked recessive defect. There is then the risk that this healthy individual is heterozygous for the gene of this defect; the probability of this being so depends on the individual's sex and degree of relationship to the tainted person.

If it is a man he cannot be a heterozygote. If it is a woman she can be heterozygous if she is related to the tainted person through the female line. In this case marriage is risky. Marriage to the sister of a man suffering from haemophilia or Daltonism cannot be advised: there is a fifty per cent chance that she will be a carrier and that half of her sons will be afflicted with haemophilia or Daltonism. If the defect is dominant and he or she marries a healthy person there is one chance in two of engendering a similarly tainted child, assuming that he or she is heterozygous which is most likely to be the case.

If the defect is recessive, half the offspring will be tainted if the person married happens to be heterozygous for the recessive defect. Let us again consider albinism. The frequency is about 1/10,000. To produce albino children the albino *aa* must marry a heterozygote for albinism *Aa*. Half the children will be heterozygotes *Aa* and the other half will be albinos *aa*. The probability of an *aa* being born is $\frac{1}{2}$. The albino's risk of marrying a heterozygote for albinism is 1/50. The combined chances are thus $\frac{1}{2}$ times 1/50 or 1/100. In other words, when an albino marries the chances are that one child in 100 will itself be an albino.

If the defect is recessive and sex-linked the afflicted person, if a man, will transmit the faulty gene to all of his daughters, who will be carriers, and half of his grandsons will be affected. The doctor's advice in this case will depend on the gravity of the defect.

If the defect is dominant the probability of having a defective child is $\frac{1}{2}$. If it is recessive, sex-linked or not, the probability is $\frac{1}{4}$. But it is essential to grasp the nature of possibility. The concept is only valid when dealing with large numbers and in each birth it acts without taking previous births into account. Although the probability of having a defective child is $\frac{1}{4}$ the birth of such a child is no guarantee that the three following children will be normal. Defective children may be born twice, three times or four times in a row, as in roulette the colour red may turn up ten times or even twenty times running.

The marriage of blood relations is another problem. It has had its partisans and its opponents; and the solution of this insidious problem is not always agreed upon.

Historically such marriages have rarely been encouraged, but the Incas and the Pharaohs practised marriage between brothers and sisters in order that the royal blood should remain unadulterated. For three generations the Pharaohs of the XVIII dynasty provided a good example of uninterrupted consanguineous marriages from which some remarkable sovereigns issued.

Most religions forbid the marriage of blood relations; Catholics require a special dispensation before they can marry even first cousins once removed. In the United States marriages between niece and uncle, and between nephew and aunt, are illegal, while in some third of the states marriages between first cousins are forbidden.

The injurious effects of consanguinity were suspected long before the science of genetics was developed. In 1862 Boudin pointed out the dangers inherent in the marriage of blood relations. In 1883 A. G. Bell noted the frequency of marriages between first cousins in the ancestry of deaf-mutes. It was also known that the number of deficient children was relatively high in small communities which were isolated through lack of communications or by natural barriers such as islands or mountain valleys. Cut off from the world the inhabitants inter-married, as the frequency of identical last names in such communities still indicates.

Darwin recommended that the results of consanguineous unions be attentively studied. When the principles of reproduction and heredity were better understood we should, he felt, no longer hear ignorant law-makers refusing with disdain to consider a plan designed to verify by a simple method whether consanguineous marriages were or were not harmful to man. Like most naturalists of his period Darwin believed that the marriage of blood relations was harmful.

Although many facts seem to support the soundness of this belief there are others which contradict the assumption that consanguinity is an evil which enfeebles the race. Marriages between blood relations have produced totally normal individuals and indeed exceptional individuals. Lincoln and Darwin himself are excellent examples. No less significant is the case of Fort-Mardyck, whose population furnishes a living experiment in natural history.

Fort-Mardyk is a small village near Dunkirk, founded in 1670 by Louis XIV. The king, being in need of mariners, decided to create a community of sailors and fishermen who, from father to son, would supply crews to the nation. To each family that settled at Fort-Mardyk he allotted 2,200 square metres of land, a parcel which was individual, inalienable, and could be neither sold nor seized. Four families, comprising some thirty people from the neighbourhood of Etaples, founded the village. Young married couples,

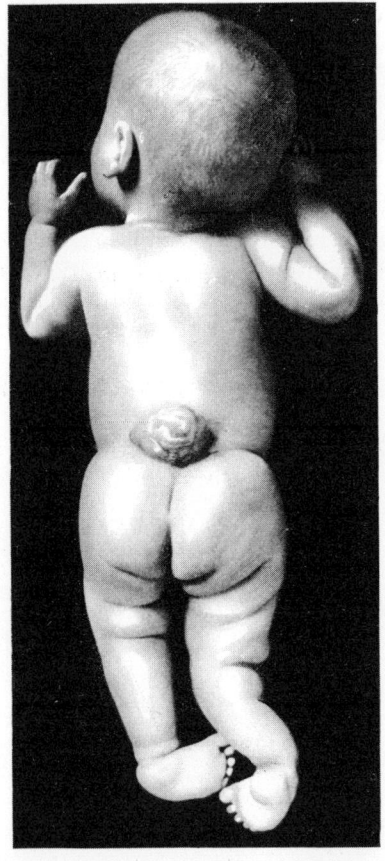

Spina bifida. A developmental anomaly characterized by a defect of closure of the bony spinal canal, sometimes leading to hernia of the meninges and of the spinal cord.

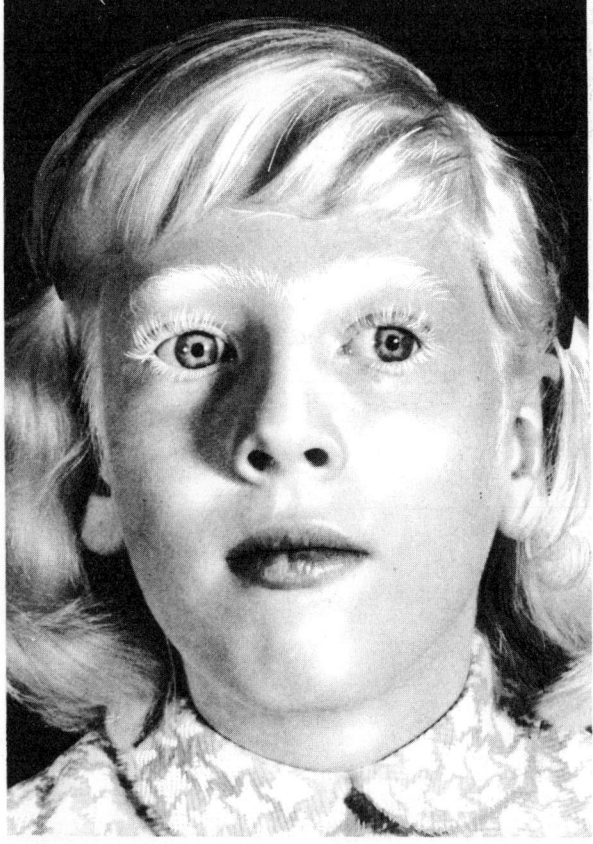

Albinism in a girl of eight born to first cousins, the paternal and maternal grandmothers being sisters. The hair, the eyebrows, the eyelashes, and the iris are all unpigmented.

I. Dainow, Geneva.

like their parents, received their 2,200 square metres of land and the right to fish, on condition that the husband enlisted in the navy. The demographic, sanitary, medical and social history of this happy village has since its creation been carefully recorded. Its population has continued to grow: there were 615 inhabitants in 1861, 1,375 in 1881, 1,960 in 1938. (During the war the village suffered.) Almost all the adult inhabitants are parents and marriages between first cousins are frequent. Between 1882 and 1886 twenty-four per cent of those married were blood relations, a very high ratio when compared with the rest of the country in which the figure varies between one and four per cent according to the region. Examination of the health of the children born of these consanguineous unions has shown that such marriages provide a kind of physiological test: though inbred, healthy stock produced healthy children; tainted stock produced tainted children.

Let us examine what modern genetics contributes to the problem of consanguinity. Sutter undertook a wide statistical survey among the 282,861 inhabitants of the three French departments of Morbihan, Finistère, and Loir-et-Cher. In this area he found that 13,455 marriages had been contracted during the years 1919—1929, and that 8.1 per cent of these marriages were consanguineous. The results of his investigation

showed that in marriages between blood relations the percentage of sterility was higher than in other families, though the accuracy of this finding is doubtful: that male births were no more frequent in one type of union than in the other; and that in consanguineous marriages infant mortality was always greater, but proportionately members of such families lived to the same age as those of other families. Abnormalities such as albinism, insanity, deaf-mutism, and harelip were commonly found in interbred families. Such are some of the general findings of Sutter's enquiry.

What is chiefly dangerous in the marriage of blood relations is the increasing threat that recessive defects will appear in their children. Two individuals who are related are more apt to be heterozygous for the same recessive gene than two individuals chosen at random from the general population. The chance that a heterozygote Aa will find a first cousin heterozygous for the same recessive a gene is one in eight. Consanguinity thus increases the probability that both parents though outwardly healthy will be heterozygous for the same recessive defect.

The concrete example of albinism, with its frequency of 1/50 for heterozygotes Aa, may again be considered. In the general population there is 1/50 times 1/50 or one chance in 2,500 that a heterozygote Aa will encounter another heterozygote for the same recessive gene;

but for two first cousins the probability is 1/8 times 1/50 — or one chance in 400. The difference between one chance in 2,500 and one chance in 400 is not negligible, and therein lies the danger of consanguinity. By increasing the possibility of a marriager between heterozygotes consanguinity encourages the appearance of the recessive homozygote *aa,* who will be tainted.

The chance marriage of *Aa* to *Aa* will engender an albino once in 1/2,500 times ¼ or 10,000 births. The consanguineous marriage of *Aa* to *Aa* will engender an albino once in 1/400 times ¼ or 1,600 births. There is still an appreciable difference between one albino child in 10,000 and one in 1,600.

If an albino marries a member of the general population the risk of an albino child is, as we have seen, one in a hundred. If an albino marries a first cousin this risk is increased about sixfold. This ratio of one to six may be generalized: the rarer the defect the higher the ratio will be. This is not illogical, for if in the general population the defect is very rare the heterozygote has a relatively greater chance of finding another heterozygote in his or her own family, the incriminating gene being derived from a common ancestor. In the ancestry of diabetics, for instance, the ratio of consanguineous marriages is almost normal because diabetes is a common defect for which heterozygotes can easily be encountered. On the other hand the ratio of consanguineous marriages in the ancestry of individuals suffering from pigmentary retinitis is fifteen to seventeen per cent. In amaurotic familial idiocy it is fourteen to fifteen per cent, while in deaf-mutism it is from nine to twenty-three per cent.

Thus the marriage of first cousins must be advised against if a recessive defect exists in the direct or a collateral branch of the family. Consanguinity, endogamy — or the union of gametes with the same genetic ancestry — encourages the appearance of recessive homozygotes.

Although from the point of view of the individual or of the family consanguinity is undesirable it may, on the contrary, be genetically beneficial to the community. For each time a recessive *aa* defective individual dies two harmful genes die with him, and in this manner genes responsible for blemishes tend to be eliminated. On the other hand, marriage outside the clan — exogamy — tends to form heterozygotes *Aa* who preserve and as it were shelter the harmful gene *a.* In the generations which follow the threat of a defective *aa* individual being born is always present.

Due to the shift in populations and the present ease and rapidity of travel few communities are now isolated, and consanguineous marriages have become less common. Thus homozygotes *aa* are also less common, while heterozygotes *Aa* are increasing — which in the long run may once more produce a greater number of *aa* defective homozygotes.

To sum up, the consultant must explain to those who ask for his advice before possibly dangerous marriages what the risks are. Whether consanguineous marriages are good or bad will depend entirely on the genetic patrimony of the family itself. The couple contemplating marriage may, when they grasp the possible consequences of their union, hesitate. Even if they choose to run the risk, they will understand the dangers involved and be fully aware of their responsibilities towards their unborn children and society. Coercive measures, laws, and state interference can, we may hope, be effectively replaced by eugenics based on education and persuasion.

Urgently required is an inventory of hereditary diseases, with their manner of transmission accurately defined and their frequency in the population established on a sound statistical basis. It is also desirable that methods be found to identify healthy heterozygotes who are nonetheless carriers of harmful recessive genes. Then, when the qualities of the offspring can be foretold with something approaching certainty, the doctor can encourage or discourage a proposed marriage by referring to genetic information which is based on known and recognized facts.

Species

Difficulty of definition

The widely employed word 'species' fulfils a practical necessity and represents a concrete reality. The living organisms which compose the species must be given names. The hunter, the fisherman, the farmer, the stock-breeder have names for the animals and plants with which they deal. Zoologists and botanists are more or less in agreement when they classify the fauna and flora of the area or region they are studying.

Most species have an ordinary English name: the common or grass frog, the leopard frog, the greater celandine, the lesser celandine, etc. The scientist replaces these names, which are unfamiliar to those who speak other languages, by Latin names which are universally recognized. The Latin name is composed of two words, the first of which designates the genus to which the species belongs, while the second designates the species itself. Thus frogs belong to the genus *Rana;* the common frog is *Rana temporaria,* the leopard frog is *Rana pipiens.* The greater celandine belongs to the genus *Chelidonium* and is called *Chelidonium majus.* The lesser celandine is called *Ranunculus ficaria* which shows that it belongs to another genus, being closely related to the buttercup. In scientific works the name of the species is often followed by the initial or name of the man who first described the species. For example, *Draba verna* L. signifies that the species *Draba verna* (in English whitlow grass) was first described by Linnaeus.

When he wishes to determine the identity of a species the naturalist resorts to specialized works filled with detailed descriptions and supplemented by illustrations. Sometimes the specimen compared is exactly like the species illustrated. But often, though most characters are exactly like those described, there are others — height, colour — which depart to a lesser or greater extent from the description. Identification of the species then becomes less certain.

The tendency of some authorities to appreciate and of others to ignore minor variations greatly complicates the identification of true species. A definition of species seems to be required.

Cuvier supplied one of the best definitions. 'The species,' he said, 'is a collection of all organisms born to each other or to common parents, and of those organisms which resemble them as closely as they resemble each other.' The species would accordingly be an entity characterized by a permanent morphology. The permanent form of a species makes it possible to identify drawings of prehistoric animals. The various species represented in the friezes which decorate the walls of the grotto of Lascaux are instantly recognized by the visitor. The silhouette of a prehistoric horse, of a deer, of a bull, is unmistakable. And yet the sole criterion of morphological identity is soon found to be insufficient, as the case of sexual or seasonal dimorphism illustrates. Indeed it is by their different appearance that we often distinguish the male from the female of the same species. Secondary sexual characters are common in a large number of vertebrates. Similar difficulties arise from seasonal dimorphism: the form of the butterfly in spring and the same insect's form in summer could easily suggest that two different species were involved.

On the other hand, some animals which are identical in shape belong in reality to different species. *Ascaris lumbricoides,* which inhabits the human intestine, and *Ascaris suum,* parasitic in pigs, are so alike that they can only be identified when their host is known.

Buffon had grasped the inadequacy of morphological identity as the sole criterion when he pointed out that to compare the resemblance of individuals was only an accessory means of identification, and often had nothing to do with the essential continuity of the species by breeding. 'For,' he wrote, 'the ass resembles the horse more closely than the poodle resembles the

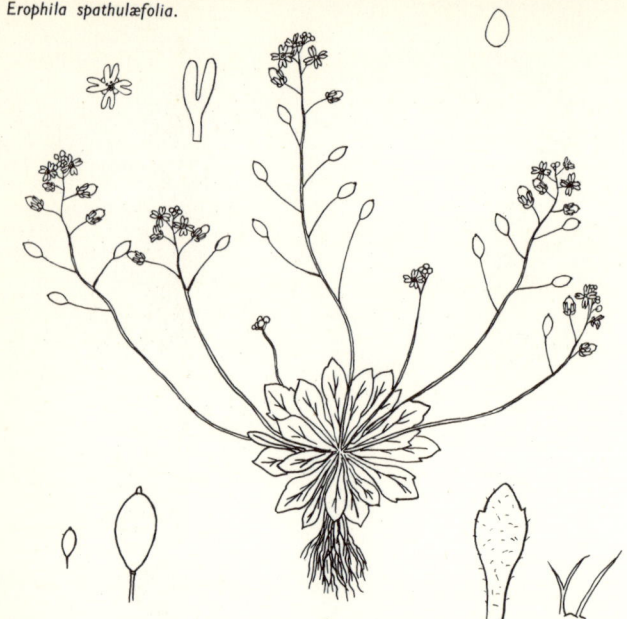

Erophila spathulæfolia.

greyhound, yet the poodle and the greyhound are of the same species since together they produce individuals which in turn produce others. The ass and the horse, on the other hand, are certainly of different species since the individuals they produce together are sterile.'

Thus to the morphological criterion was added what might be called a 'mixiological' criterion. Belonging to the same species were individuals, more or less like each other, who in space and time were linked by their capacity to interbreed. This classic formula gave satisfaction for many years, and numerous biologists phrased similar definitions. In 1942 Mayr considered species as groups of natural populations which did or potentially could interbreed and, from the point of view of reproduction, were isolated from other groups.

Such definitions, based on the two criteria of mor-

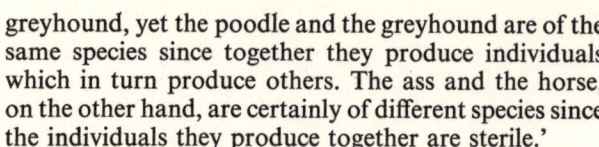

Various true-breeding strains or 'elementary species' of *Erophila verna* or whitlow grass.

Erophila subnitens.

Erophila hirtella.

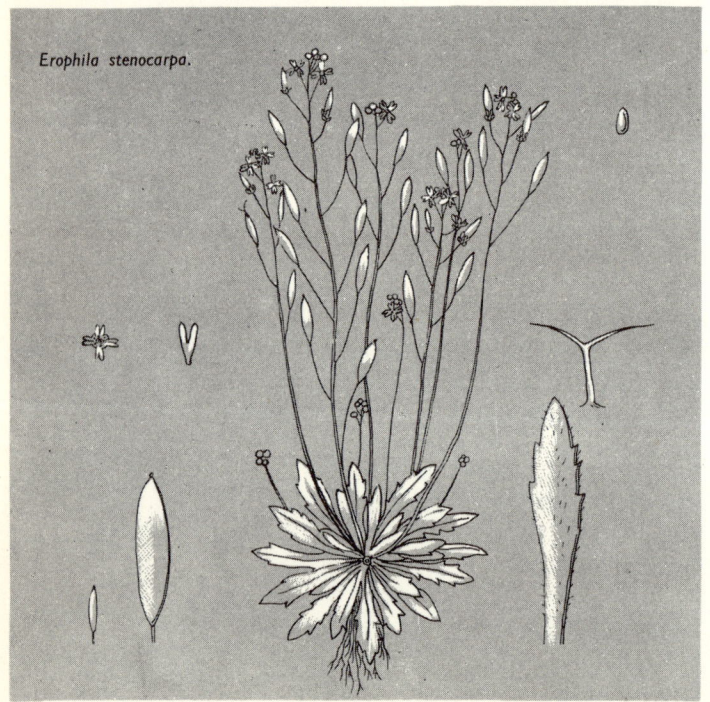

Erophila stenocarpa.

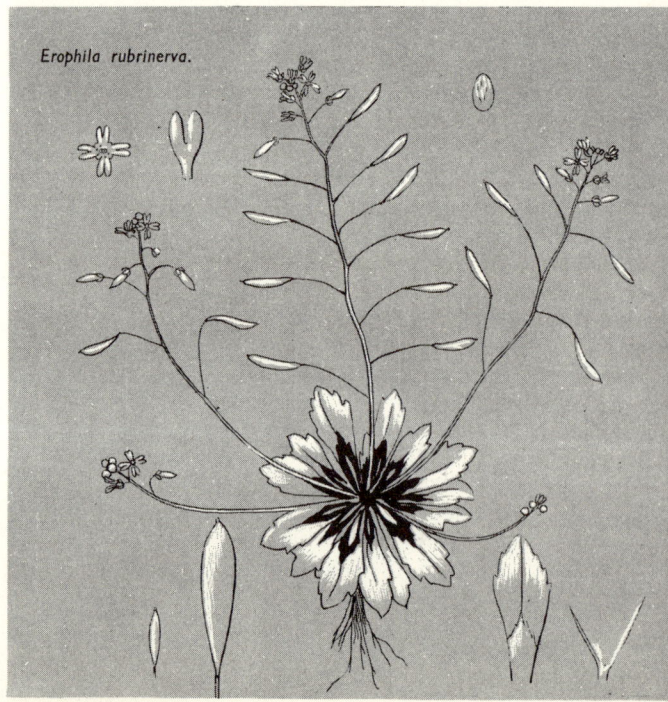

Erophila rubrinerva.

phology and fertility, are valid for restricted populations, for local flora and fauna. But when applied to extended areas difficulties begin to appear. For example the common mouse *Mus musculus* L. inhabits houses in Europe and southern Asia, while another mouse, *Mus spicilegus* Petenyi lives in the fields and forests of the south of France, central Europe, the Balkans, and Spain. Finally a third mouse, *Mus bactrianus* Blyth frequents the houses of south-eastern Russia and central Asia. Among the three species morphological differences exist: size, weight, length of tail, colour, smell. But experiment has shown that they are all perfectly able to interbreed and that their hybrids are fertile. Thus they are not true species but geographical subspecies which belong to the same *syngameon* or great collective species.

Another example is that of the golden pheasant, *Chrysolophus pictus,* and Lady Amherst's pheasant, *Chrysolophus amherstiae,* which are very different in appearance and in geographical distribution. The former, from the north-west of China, has a silky golden-yellow crest and a red belly; in the latter, found in Tibet and Burma, small green quills cover the brow and top of the head and from the back of the head rises a little red plume; its belly is white. In captivity the two species cross and the hybrids they produce are equally fertile. In fact it is rare to find a pure Lady Amherst pheasant, totally uncrossed with a golden pheasant. First generation hybrids are intermediate, and also those of the second generation. The characters which differentiate them are so numerous that it is impossible to obtain a reversion to the parental types. Nevertheless the abnormally high mortality rate of their eggs and chicks, and also the excessive birth of

Golden pheasant and, *below,* Lady Amherst's pheasant.
Larousse.

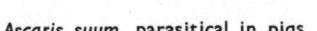

Ascaris suum, parasitical in pigs.

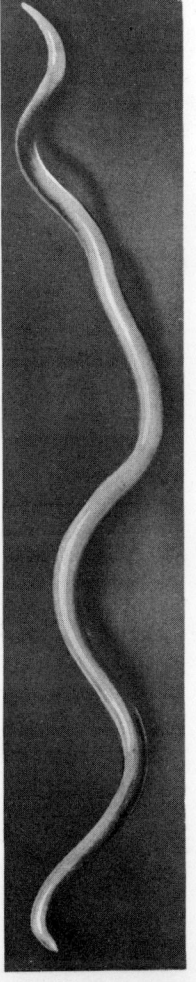

Ascaris lumbricoides, parasitical in man.
d Aguilar.

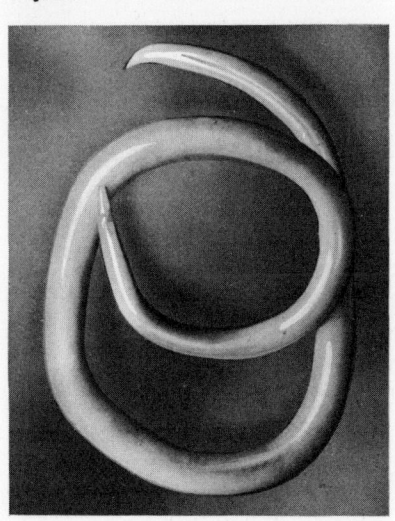

Alsatian.
H. Dimont.

North American wolf.
N. Y. Zoological Society.

Jackal.
Atlas-Photo.

Coyote.
W. W. Photography.

349

Salix caprea

Larousse.

Salix viminalis.

males, reveal that the two hereditary endowments are not perfectly homologous. The two pheasants represent geographical sub-species which together form a great collective species.

The various polar and tropical members of the canine tribe, wolves, jackals, coyotes, dogs are all different morphologically, but they can interbreed and their hybrids are fertile.

The ability to interbreed may be described more exactly as gametic fertility, that is to say that an ovum can receive a foreign sperm and fertilization can give rise to a viable egg that is capable of developing into an adult which, in its turn, will reproduce. Many species which in a wild state have no connection with each other will mate in captivity. Numerous factors — lack of attraction, repulsion, difference in size, slight variations in their periods of reproduction — prevent sex relations between certain species, while gametic fertility between them can be experimentally induced. It is very difficult to ascertain if two species are gametic-ally incompatible or not: it is still unknown whether a cross between the dog and the fox, or between the hare and the rabbit, can or cannot produce hybrids. In plants the question is even more complex because among neighbouring species many degrees between total fertility and total sterility exist. Identical forms may prove sterile, notably among species of *Viola*. In brief, a definition of species based on the two criteria of morphological resemblance and capacity to inter-breed is still not sufficient. For, from a practical viewpoint, it is obviously necessary to recognize specifically a wolf, a dog, a jackal, a coyote, a golden pheasant, a Lady Amherst's pheasant, etc.

To avoid the difficulties we have encountered, a def-inition of species based on three criteria has been proposed. A true species indisputably comprises a collection of related individuals having the same hered-itary morphology and the same physiological char-acters, who lead the same kind of life, and inhabit a given geographical area. A species is separated from neighbouring species by a barrier, psychical, structural, or geographical. The individuals of a species are fertile among themselves and their offspring also interbreed. Three pairs of characteristics, then, define a true species: morphology (M) and physiology; ecology (E) and distribution; fertility among themsel-ves and sterility (S) with other species. Thus the formula of a true species may be expressed as MES, and doubt-ful cases would be those in which the formula lacked one or other of these letters. Forms which are distinct as far as morphology and ecology are concerned but can interbreed would have the formula ME. Such are the jaguar and the panther, the golden pheasant and Lady Amherst's pheasant, the willows *Salix viminalis* and *Salix caprea*. These forms maintain their status as species because their morphological dissimilarity is great and also because the hybrids do not supplant their parents.

More rarely, two forms with identical morphology occupy habitats which make it impossible for them to meet. Thus they are sexually separated, and have the formula E S. If no intercourse between them is possible they are considered to be two autonomous species. The genus *Drosophila* provides several examples, notably the two species *D. pseudo-obscura* and *D. persimilis* which for many years formed only the single species *pseudo-obscura*. Morphologically the two spe-cies are so alike that they cannot be distinguished but physiologically they reveal some differences, the most common being the speed at which their wings vibrate. Both live in the Rocky Mountains, but in distinctly different areas: *pseudo-obscura* in the east and the south where the climate is warm and continental; *persimilis* in the north and the west where the climate is even, damp, and cool. Where the two regions overlap they meet, but in such zones the two species of *Drosophila* remain genetically separate. It seems that sexual attraction between them is lacking. If mating does take place the males of the first generation are sterile and the females only in part fertile — facts which have been ascertained in the laboratory. In view of their genetic isolation the two forms have been separated and now constitute two species: *pseudo-obscura* and *persimilis*. Similar twin species, though difficult to recognize, must be fairly common: examples have been reported in various insects including mosquitoes and flies, and among crustaceans.

In difficult cases in which one criterion is lacking, the character given priority is sometimes morphology and sometimes sexual sterility. The species is then re-assigned to its classic status, the status given in earlier classifications.

Finally an additional criterion has been suggested, namely that of chromosome complement. This seems to be justified by the chromosome formulae of *Drosophila*. In six species of *Drosophila* the chromosomes differ in number, in dimensions and in the way they are grouped. But many other species, such as *D. melanogaster* and *D. simulans,* have the same chromosome complement. Matthey's work on the chromosome formulae of various groups suggests that these formulae are of little help in distinguishing species. They are of much more value in establishing affinities and zoological relationships.

With the progress of biology we can no longer think of a species as a static entity. It is not immutable; it reveals variations which necessitate divisions into geographical and physiological sub-species, the former having a precise geographical distribution, the latter a particular physiological behaviour.

The statistical methods of biometry and Mendelian genetics have led to a new conception of species. Observations are relatively easy with plants which fertilize themselves or reproduce parthenogenetically. In most species it has been possible in this way to isolate the more or less numerous characters which are transmitted by heredity. Whenever a mutation occurs the mutant will no longer cross with the parent plant, but between the various mutants differences involving minute details are observed only by the trained eye of the specialist. In this case it is agreed that the parent species and its mutants constitute a kind of collective species; and each identifiable mutant, of stable type, is a small or elementary species.

A classic example of a collective species is that of the crucifer *Erophila (Draba) verna,* or whitlow grass, a common European plant which grows in fields and on old walls from March to April. In height it varies from 3—15 centimetres and almost all its organs are subject to variation. The leaves may be small or broad, linear or elliptical, of a handsome green or of a glaucous tint, covered in varying degrees with simple or stellate hairs. The breadth of the heart-shaped petals also varies, and they can be deeply or less deeply incised. The stems, squat or slender, carry few or many flowers, which bloom earlier or later; the fruit consists of elliptical pods, linear or rounded.

Each region has its own particular population.

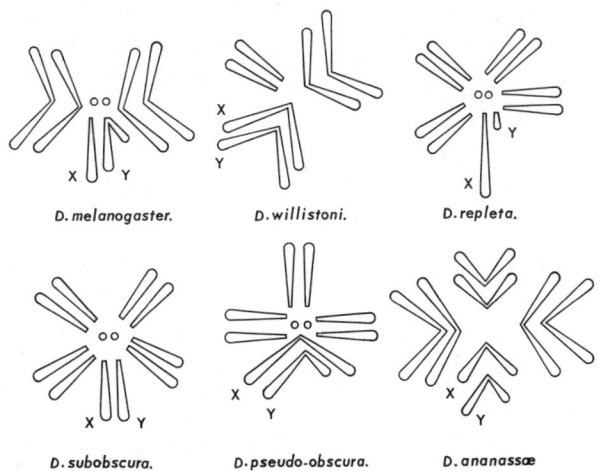

D. melanogaster.   D. willistoni.   D. repleta.

D. subobscura.   D. pseudo-obscura.   D. ananassæ

Male chromosomes of six species of *Drosophila*. X and Y indicate the sex chromosomes.

After Petterson, Stone, etc.

In Europe and Asia more than 200 varieties have been recognized.

The cultivation and breeding of pure strains

Among plants pollinated by the wind or cross-fertilized by insects and among animals the genetic constitution of the individual is supplied by two parents, and such organisms are naturally more or less heterozygous. By various artifices, however, homozygotes can be obtained. If for a sufficient number of generations brothers and sisters are crossed, or young back-crossed with parents, the resulting chromosome formula will in time become homozygous. If care has been taken to eliminate lethal or semi-lethal characters these homozygotes will be hardy and fruitful. In practice one or more such harmful genes are inevitably retained, which can bring about a decrease in vigour when the organisms are compared with the heterozygous stock from which they are descended.

One of the most ambitious experiments of this sort was carried out in the United States with guinea-pigs. Special strains of guinea-pigs were obtained by careful in-breeding and have continued in existence for more than thirty years. Apart from a general decrease in vigour each strain has its own type of markings, its own physiological characters such as manner of growth, and its own psychic characters such as nervousness and spontaneous muscular activity. Each strain is subject to its own type of abnormal development which leads to the birth of typical monsters.

Pure strains of mice, rats and hamsters are also reared in laboratories, being particularly valued by physiologist and geneticist for their homogeneity.

In the vegetable kingdom pure strains are often found in nature. Among animals, however, homozygous purity is maintained only under the strict supervision of man.

The concept of species is, then, singularly complex and fluctuating. It is difficult to apply the many terms at our disposal with accuracy. In view of its historical origin the laciniate celandine may be called a mutant, or, in view of its propagation by seeds, a homozygote of pure line. It can also be a variety or race, if one considers its resemblance to the greater celandine, or, in consideration of its autonomy by self-pollination, a species. The microbiologist, the botanist, the zoologist, the geneticist, and the palaeontologist each has his own, rather special conception of species. Some points are, however, commonly agreed upon: a species is recognizable by certain morphological, physiological, and chromosomal features. Above all, the individuals which compose a species reproduce among themselves or are potentially capable of interbreeding in normal conditions. The limits of a species are thus marked by a genetic barrier. The word 'potentially' in this genetic definition of species implies that the obstacles which prevent interbreeding are imposed by external conditions, the process of isolation or sexual separation.

Beaver.

X Photography.

Effects of isolation

For a new species to arise it must be separated from its original stock and from neighbouring species. 'Without isolation, no species,' said Moritz Wagner. Isolation, before or after variation, comes about in diverse ways.

Geographical isolation

The most important of these ways is geographical isolation. The unbroken area which a species inhabits may be curtailed by the upheaval of a natural obstacle like the Isthmus of Panama which separated the marine fauna of the Atlantic and the Pacific, or by the breaking up of a continent into several islands. Isolation is maybe virtually complete in certain living spaces — or biotopes — to which access is particularly difficult. Such biotopes are provided by Alpine lakes, oceans, oceanic islands, caverns, forests, deserts, deep mountain valleys, and high mountain ranges. For a variety of reasons divergencies are accentuated by geographical isolation. Isolated populations can be composed of different genotypes or mutations can appear. Then by means of selection a new standard type may emerge.

Geographical species are true species in a state of emergence: they form the raw material from which new species arise. It may be remembered that it was the different types of finch that inhabited the different Galapagos Islands which so struck Charles Darwin, and were, as he wrote, 'the origin of all my views'. Darwin's theory of evolution may be said to have dated from that year, 1835, when H. M. S. Beagle visited the Galapagos Islands.

The separation of North America from Eurasia by the two great oceans divided the living areas of several animals, which then evolved separately and are today considered to be separate species. The European beaver, *Castor fiber,* and the beavers of North America *Castor canadensis* and *C. subauratus,* have a common origin, but their differentiation is marked. The Euro-

pean beaver is twice as big as the American beavers, its fur is not so dark and its skull is less developed. But the three species have retained the same external parasite, a blind coleopteran, *Platypsyllus castoris.* The European bison, *Bison bonasus,* and the American bison, *Bison bison,* are morphologically very similar. When crossed their hybrids are always fertile. The panther, once European but today African and Asiatic, and the American jaguar have been separated for thousands of years. They have, however, changed little except in their habits, and in captivity they engender fertile hybrids.

When the Isthmus of Panama rose from the sea it cut in two a marine fauna which was doubtless homogeneous; even today the two faunas, Atlantic and Pacific, are much alike. There are some 600 twin species of fishes on the two sides of the isthmus; the Atlantic and the Pacific forms of the two differ in only slight details. More than thirty genera of sea-urchins are common to both oceans and many species are alike, which betrays their close relationship.

Island fauna provide an excellent example of the role played by isolation. The innumerable islands, small and large, of the Mediterranean and the Adriatic shelter countless wall lizards which vary little in outward form although they exist in many different colours. A large number of geographical species and sub-species has been described by these variations in colour. A crag off Capri, known as the 'Faraglione', is inhabited by a lizard with a black back and deep blue flanks; on another rock there is another type with a dull blue back and a pale blue belly. On other rocks between Capri and Amalfi there are handsome blue-green lizards with blue flanks. On the rocky isles of the Tyrrhenian Sea and the Adriatic still other coloured varieties are found. Insular existence imposes complete geographical isolation. All these island lizards derive from a common ancestor which was widely distributed during the Pliocene when the various islands were connected. Later, isolation and physiological variations gave rise

European bison in the forest of Bialowieza, Poland.

Bertin.

Canadian bison born in the Hamburg Zoo.

J. de Beaupré.

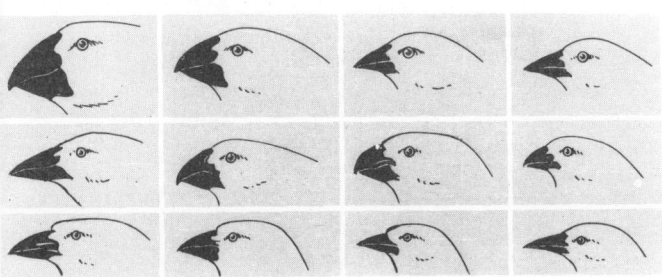

Beaks of many shapes exist among the finches of the Galapagos Islands; these beaks correspond to the many and varied ways in which different types of finches obtain food.

After Lack.

Giant turtle from the Galapagos Islands.

Lizard from the Galapagos Islands.

J. Dorst, Atlas-Photo.

to independent species.

A phenomenon of the same type is illustrated by the giant tortoises which are found in the Galapagos Islands. Each island has its own species with the exception of one which, because it has been recently formed by the merging of three islands which were formerly separate, has three species.

Isolation by modification of behaviour

The adoption by a parasite of a new host can result in a separation of the innovating parasite from the remainder of its species. The false acacia, *Robinia pseudo-acacia,* a tree of American origin, was introduced into Europe at the beginning of the seventeenth century in the form of seeds. Thus it brought no American parasites with it. Since 1879 an insect has ravaged the false acacia plantations in various countries. An examination of the insect by specialists showed that a new species was responsible; the species was then named *Leucanium robiniarum.* Later Marchal showed that *L. robiniarum* was none other than *Leucanium corni,* from which it differs in size, colour, and rugosity. In structure, however, they are identical. *L. corni* is polyphagous and lives on numerous plants. Marchal succeeded in contaminating a false acacia with *L. corni* taken from a peach tree. The larvae born and nourished on the false acacia gave rise to insects which, when completely developed, could be identified as *L. robiniarum.* The reverse experiment, namely the infection of a peach tree by *L. robiniarum* produced no result.

Sometimes the divergence is more accentuated and the new species refuses to lay eggs on the former plant host. Such is the case of the homopteran *Psylla moli,* a jumping plant louse which has split into two species of slightly different size. One — *P. moli* — lays its eggs on the pear tree; the other — *P. peregrina* — on the hawthorn. *P. peregrina* will no longer lay eggs on the pear tree.

The bug, *Cimex lectularius,* which is parasitical on

353

Durmast oak, *Quercus sessiliflora*

Larousse.

Common oak, *Quercus pedunculata*

R. H. Noailles.

Quercus lanuginosa.

Larousse.

Man and that which infests pigeons, *Cimex columbarius,* are morphologically identical but each adheres to its host. In the laboratory they interbreed.

Psychic isolation

What might be called psychic or emotional isolation is generally equivalent to a refusal to mate. Different habits, lack of sexual attraction, or even enmity, prevents the possibly fertile union of species. All species display a marked tendency to homogamy; in other words, the male seeks out a female of his own kind.

Among Hemiptera mutations which affect the length of wing, producing brachypterous insects with small wings and macropterous insects with large wings, result in sexual segregation. In practice when the two types of insect are reared together homogamous coupling — brachypterous with brachypterous, and macropterous with macropterous insects — clearly predominates. In thirty matings observed twenty-seven were of this type while only in three cases did small and large winged insects mate. Among *Hydrometra stagnorum* the cross between small and large winged insects is physically impossible: although the copulating organs are not modified the size of the thoracic segments is affected by the atrophy of the wings. Homogamy is also observed in the two neighbouring species of Hemiptera, *Notonecta glauca* and *N. obliqua,* which differ only in size. In captivity crosses between them are fertile.

Isolation through differing periods of sexual maturity

An advance or delay in the period of sexual maturity, entailing a change in the date of laying eggs, necessarily gives rise to isolation although gametic fertility remains unaltered. This difference in times of mating is often observed in twin species which are morphologically identical, but are in practice mutually sterile. In the neighbourhood of Berlin the edible frog, *Rana esculenta,* mates in May and lays eggs at the beginning of June. *Rana ridibunda* lays eggs from the first of May to the twentieth. Between the two frogs there are only the slightest morphological differences. Their areas of distribution are different, but when they happen to coincide the fact that the two species lay eggs at different times prevents them from producing hybrids.

The same phenomenon is encountered in the vegetable kingdom. In Lorraine, for example, three kinds of oak grow: the durmast oak *(Quercus sessiliflora),* the common oak *(Quercus pedunculata),* and the pubescent oak *(Quercus lanuginosa).* The three species are inter-fertile and are pollinated by the wind. Hybrids between *Q. sessiliflora* and *Q. lanuginosa* are common, but those between *Q. sessiliflora* and *Q. pedunculata* are rare. This is because in the latter tree the yearly appearance of leaves and flowers is later than in *Q. sessiliflora.* The difference varies from five to fifteen days according to the individual tree and the climate; but the last *Q. sessiliflora* to flower has always finished before the earliest *Q. pedunculata* has begun. In zones where *Q. pedunculata* predominates this seasonal difference has given rise to the so-called

June oak, *Quercus pedunculata,* variety *tardissima,* which does not form hybrids even with its parent species.

Mechanical isolation

In certain species the sole obstacle to mating is an exaggerated discrepancy in physical dimensions. Such marked differences in size occur among some breeds of domestic animals: toy dogs and St. Bernards, Shetland ponies and percherons, dwarf fowl and fowl of larger breeds. Copulation is also impossible between dextral and sinistral gastropods whose shells coil in opposite directions.

Cellular incompatibility

In addition to the causes of segregation which have been discussed there is a final obstacle to hybridization, namely cellular incompatibility. In innumerable cases the fertilization of one species by another is impossible because the male gamete cannot penetrate the ovum. This may simply be a question of the size of the spermatozoon or, more often, of its being unattracted by the ovum. In the latter event the physico-chemical state of the egg's membrane may be artificially changed so that it will receive the foreign sperm. In this manner the unfertilized eggs of the sea-urchin, treated with alkalinized sea-water, can be fertilized by the sperm of a starfish or even of a mollusc.

Among higher plants a disproportion between the length of the style and the length of the pollen tube obstructs fertilization. A plant with a long style cannot be fertilized by the pollen of a plant with a short style; but this can be artificially brought about by cutting the longer style.

Many seed plants are auto-sterile, that is to say the pollen of a given plant cannot fertilize the same plant's ovules. They must be cross-pollinated by the wind or by insects.

Even when all other conditions are satisfactory, chemical incompatibility can prevent hybridization. A cross between two species of iris will be successful only if the carbohydrates elaborated by the two parent plants are identical. For instance, crosses between the three species *Iris germanica,* which manufactures starch, *Iris pseudoacorus* which forms fructosan of the irisin type, and *Iris foetidissima* which produces starch and two fructosans that are unrelated to irisin, are impossible.

A disparity between the chromosomes of the parents, or between the ovular cytoplasm and the chromosomes, is one of the chief causes of sterility, for the delicate mechanism by which the reproductive cells are formed is disturbed if the parents have unequal numbers of chromosomes. As an example let us consider the cross between cultivated maize and its wild relations. The haploid number of cultivated maize is ten; that of perennial teosinte is twenty and of annual teosinte ten. The two species of teosinte can be fertilized by the pollen of maize. The annual teosinte-maize hybrid, with the diploid number twenty, forms balanced and fertile gametes; but the perennial teosinte-maize hybrid, with thirty chromosomes, is sterile, for gamete-forming reduction division cannot take place normally, one set of ten chromosomes being unpaired.

Pyrenean dog and chihuahua.

Atlas-Photo.

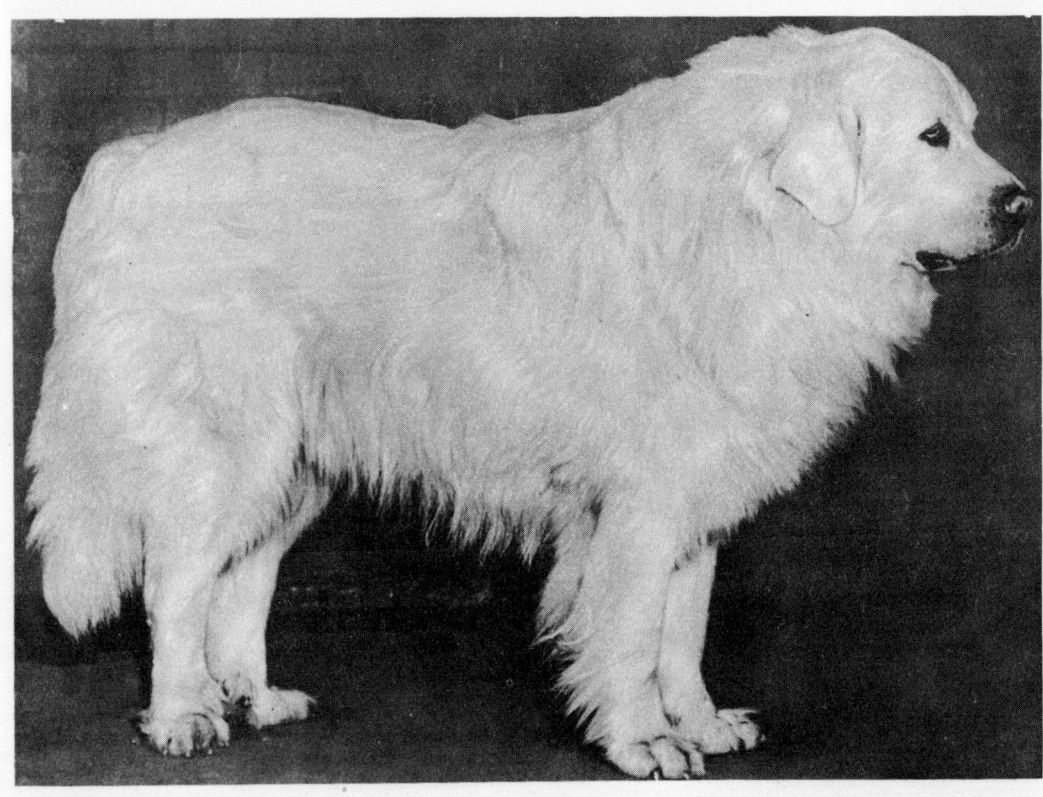

355

It is not, however, enough for two species to have the same number of chromosomes in order to produce a fertile hybrid; between their chromosomes a more or less perfect homology must also exist. Gooseberries with the haploid number eight produce fertile hybrids when their chromosomes are of the same size. But two species, *Ribes sanguineum* and *R. aureum,* although both possess sets of eight chromosomes and are morphologically very similar, produce a completely sterile hybrid, the chromosomes of the former species being larger than those of the latter.

In animals similar phenomena are found. The cross between *Drosophila melanogaster* and *D. simulans,* both with sets of four chromosomes, is achieved only with difficulty. The males mate with the females of both species, but the females prefer males of their own species. An incompatibility is evident, for the cross between a female *melanogaster* and a male *simulans* gives hybrids in ten to forty cases out of 100, while the reverse cross is successful only two per cent of the time. All of these hybrids are sterile, the ovaries and testes being atrophied. A marked disproportion of sexes is also observed, depending on the way the insects are crossed: the female *melanogaster* and the male *simulans* produced 3,698 females to 440 males, while the female *simulans* and the male *melanogaster* produced 408 females to 4,204 males. The proportions given by the first cross are in accordance with Haldane's rule that in hybrids the heterogametic sex is the less viable. Such inequalities of sex are observed in crosses between species and sufficiently different pure strains.

Chromosome discordance can give rise to viable hybrids of which one of the sexes is sterile while the other is fertile when back-crossed. The sterile sex is always the heterogametic sex, among birds the female, among mammals the male. A still greater discordance produces sterile hybrids of both sexes. Such is the case when crossing the lion and the tiger, the male leopard and the lioness, the horse and the zebra, the male zebra and the mare, the ass and the mare, the horse and the she-ass, the Muscovy drake and the common duck.

Maximum discordance gives hybrids which are no longer viable: development may begin, but sooner or later it stops. The male *Rana temporaria* and the female *Rana arvalis,* very similar in form, cross easily and their tadpoles at metamorphosis become adult frogs; but the reverse cross has no result. The female minnow, *Phoxinus loevis,* fertilized by the goldfish, *Carassius auratus,* produces eggs which hatch, but the hybrids die when they are eleven or twelve days old.

The formation of species

Several ways in which species are formed are known to biologists. In plants and in animals the mechanisms are different.

Species formed by mutation

If a mutation occurs in a plant the mutant, if self-fertilized or parthenogenetic, at once forms a pure strain, that is to say it is homozygous. It is isolated from the normal plant with which in most cases it will not form hybrids. Among animals mutation does not, except in the case of self-fertilizing animals like gastropods, give rise immediately to a new species because, to reproduce, the mutant must cross with other individuals. Thus a viable mutation leads to the heterozygous state which is frequently found in free-living animals. To obtain a new stable form man must intervene by crossing the mutants in such a manner that a true-breeding line of pure homozygotes is obtained.

Species formed by hybridization

In plant hybridization, the cross between two different species or even different genera gives rise in the first filial, or F_1, generation to hybrids which are, if viable, uniform and usually of intermediate phenotype, sometimes even to the smallest detail. But the hybrid may resemble one parent more closely than the other, due to dominance or to the number of chromosomes contributed by each parent. The latter influence is seen when tobacco or *Nicotiana tabacum* (twenty-four chromosomes per set) is crossed with *Nicotiana sylvestris* (twelve chromosomes per set). In appearance the hybrids of female *N. tabacum* and male *N. sylvestris* recall the maternal species which provided two-thirds of their chromosomes. The hybrids may be fertile or sterile and of every degree between the two. Their sterility derives from the formation of unbalanced gametes due to the difference in structure between the two sets of chromosomes. Only some of the chromosomes pair. During meiosis (reduction division when gametes are formed) it is possible to observe on the equatorial plate pairs of homologous chromosomes as well as single unpaired chromosomes: the homologous pairs undergo normal disjunction while the single chromosomes are distributed at random. The resulting gametes thus possess varied and irregular sets of chromosomes.

Sterile hybrids can, however, he maintained by vegetative or asexual propagation. In this way a new chromosomal equilibrium may be established, usually by the doubling of the number of chromosomes and the formation of allotetraploids. The hybrids then become fertile. There are numerous examples of polyploidy having re-established the fertility of such hybrids. A great many flowering plants have arisen in this manner.

The extremely varied appearance of the next, or F_2, generation of plant hybrids is the result of Mendelian segregation, the action of multiple factors, and anomalies in the behaviour of the chromosomes. Certain new combinations can be stabilized by means of self-pollination.

In agriculture and horticulture the crossing of species is widely employed to obtain new forms. Cultivated plants are almost all hybrids of more or less complex species. Even in nature the formation of such hybrids occurs: thanks to polyploidy and self-pollination these natural hybrids may survive by displacing another pre-established species or by spreading to sparsely populated areas. Genetic phenomena are thus responsible for the birth of a new species, whether by mutation or by hybridization.

The hybridization of maize. Three crosses.

357

*From left to right, Nicotiana tabacum with pink flowers
(N = 24), Nicotiana sylvestris with white flowers (N = 12),
hybrid of Nicotiana tabacum and Nicotiana sylvestris (2N = 36).*

Panyalakshana.

*From left to right, Primula floribunda, Primula verticillata,
Primula kewensis, hybrid of P. floribunda and P. verticillata.*

Chopinet-Vilmorin.

In the animal kingdom hybridization is less common;
when encountered it is usually among echinoderms,
insects, amphibians, and birds. It is rare in mammals.
Sometimes among hybrid animals only one sex is
viable, generally the homogametic sex; for, as we have
seen, the heterogametic sex is frail.

Geographical sub-species can interbreed, but inhabit different regions; fertile hybrids appear only
where their areas of distribution overlap and representatives of the two sub-species can cross. Many such
crosses give rise to sterile hybrids which, having no
offspring, rapidly become extinct. Less fortunate in
the struggle for existence than sterile plant-hybrids
they are unable to survive by vegetative reproduction
and have no opportunity of re-establishing their
chromosomal equilibrium by tetraploidy.

Doc. Chopinet-V

Species formed by fragmentation

As a result of geological or climatic accident a primitive polymorphous species of wide distribution may be split up, so that it then occupies numerous smaller areas which become the cradle of future and possibly vicarious — or closely related — geographical sub-species. It is believed that all the present-day beeches of Europe, America, and Asia were in this way derived from *Fagus feronioe,* a polymorphous species of wide distribution during the Tertiary period. A similar origin is attributed to the Mediterranean firs which are clearly related and, in parks and 'arboretums', still form hybrids.

The mechanism, which relies essentially on isolation, also encourages the formation of animal species. Indeed geographical segregation, like all other kinds of segregation, plays an important role in the formation of species. Once separated the populations which formed the primitive species are capable of differentiation, and in this manner sub-species, morphologically or physiologically different, arise.

It would appear that speciation takes place very slowly, as the slight differentiation between the fishes on either side of the Isthmus of Panama seems to demonstrate. Britain, separated from the continent for more than 120,000 years, has not yet produced an endemic species of moth. Sub-species, however, are sometimes rapidly formed. For instance, the dark form of the peppered moth, an illustration of the phenomenon of 'industrial melanism', has developed during the last 100 years.

Since mankind is able not only to create new species of plants and domestic animals but also to synthesize wild species, it would seem that the problem of the genesis of species is, at least in broad outline, understood.

The creation of new species

The botanist, working with material which lends itself to such experiments, has succeeded in creating genuinely new species which are fertile and breed true, species morphologically separated from their parent stock and, because of their altered genetic constitution, incapable of cross-breeding with it. Such creations — more difficult to achieve among animals — are simplified by certain characteristics which are peculiar to plants. Plants, being immobile, are often self-fertilized. Polyploidy occurs in nature and can be easily brought about by treatment with colchicine which interrupts mitosis by suppressing the spindle without preventing the cleavage of the chromosomes. Plants, even when sterile, can be perpetuated asexually, by cuttings, by off-sets, by bulbils and by apomixis. Finally the absence of psychic or mechanical obstacles to interbreeding makes it possible for plants to be fertilized by foreign pollen borne by the wind or by insects.

The following are a few examples of plant creations.

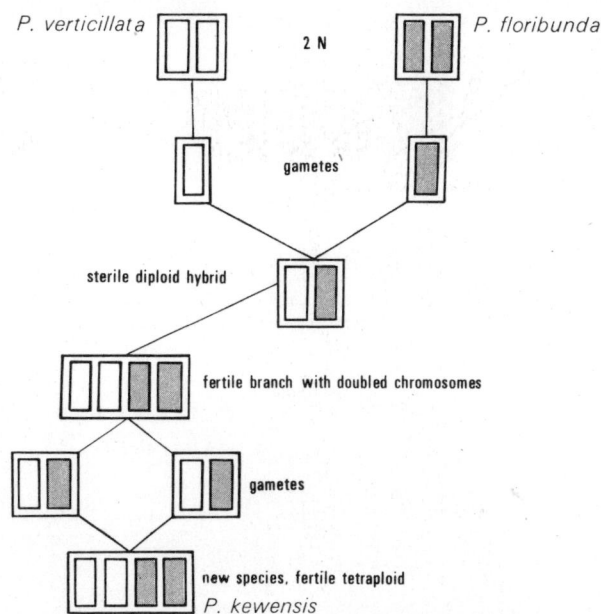

Diagram showing the formation of a new tetraploid species, *Primula kewensis,* by mutation of a bud growing on a sterile diploid hybrid.

Alter Hurst.

A primula, Primula kewensis

In 1900, in a sowing of *Primula floribunda,* a plant native to Afghanistan and west of the Himalayas, there appeared in Kew Gardens an abnormal stool which was assumed to be an accidental hybrid between *P. floribunda* and *P. verticillata,* a plant from south-western Arabia and Abyssinia. In order to verify this assumption Coutts fertilized *P. floribunda* with the pollen of *P. verticillata,* and obtained a hybrid which displayed a mixture of parental characters, a hybrid similar to that which had already appeared by chance. Coutts' hybrid had the same diploid number of chromosomes, eighteen, as the two parent plants. When self-fertilized it produced no seeds. The pollen grains were small and for the most part empty. The diploid hybrids were preserved and multiplied by vegetative reproduction. In 1905, in the greenhouses of Veitch, a mutation of a bud occurred which gave rise to a tetraploid shoot which produced fertile seeds. The same mutation occurred again in 1923 at Kew, and in 1926 at the John Innes Horticultural Institution. The new plant was named *Primula kewensis.* It was grown from seeds and is now found in all greenhouses. It is a hardy plant, larger than its parents, perfectly fertile and breeds true. Its somatic cells possess 35, 36 and 37 chromosomes; the cells of *floribunda (FF)* and of *verticillata (VV)* are doubly represented during meiosis when the homologous chromosomes *(FF or VV)* pair. The non-homologous chromosomes *FV* rarely pair. The Kew primula is sexually separated from its parents.

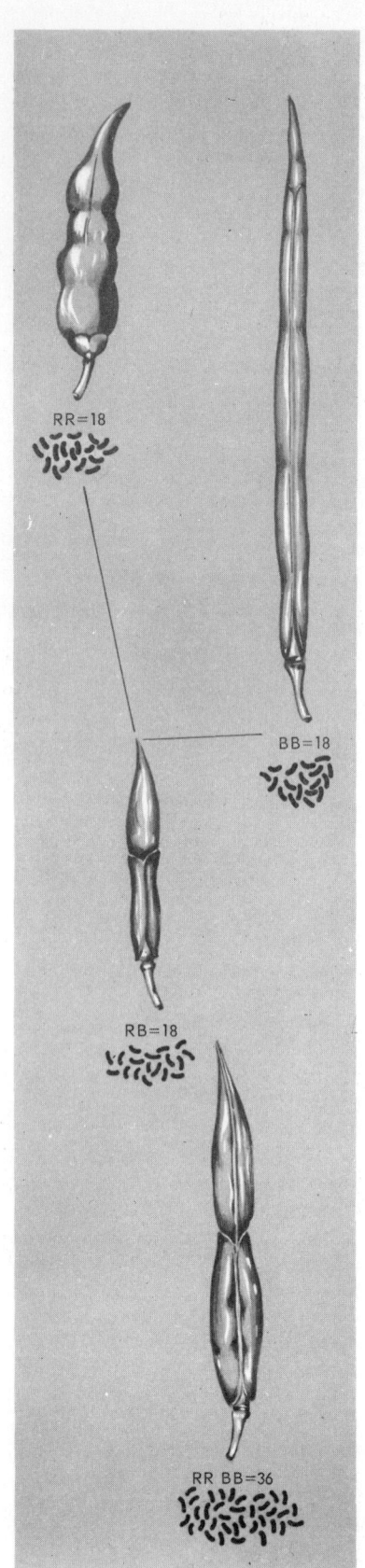

RR=18

BB=18

RB=18

RR BB=36

The fruit (or pod) and the chromosomes

of the radish *Raphanus sativus*, RR, *(top left)*

of the cabbage *Brassica oleracea*, BB, *(top right)*

of the hybrid RB *(centre)*

and of the new tetraploid species *Raphanobrassica karpechenkoi*, RR BB, *(botton)*.

After Karpechenko.

Raphanobrassica karpechenkoi, a new tetraploid species.

J. Vincent.

The cabbage-radish, Raphanobrassica karpechenkoi

The radish *(Raphanus sativus)* and the cabbage *(Brassica oleracea)* are very dissimilar species, but both have the same number of chromosomes, namely eighteen. By crossing the two plants Karpechenko succeeded in obtaining hybrids, also with eighteen chromosomes. The flowers and fruits of the hybrids were intermediate in character. The plants, sometimes very hardy, had giant, glabrous leaves like those of the cabbage. They flowered abundantly but were sterile among themselves and with their parent plants. But certain stools (nineteen out of 123) produced a few seeds by natural pollination, and these seeds in F_2 gave rise to tetraploid plants with thirty-six chromosomes which in other respects were identical to those of F_1. The doubling of the hybrid chromosome complement had taken place during the first reduction division of meiosis. The eighteen chromosomes, instead of dividing and passing half to each daughter nucleus, remained grouped in a single large nucleus. The second division was then normal, and each gamete received eighteen chromosomes. The hybrid, named *Raphanobrassica karpechenkoi*, differs in character from its grandparents. In the new species meiosis is normal; the plant is fertile, stable, and will cross, though with difficulty, with its stock. When fertilized with the pollen of *Raphanus* it produces triploids.

Iris hoogiana.

Plant fairly hardy, early blooming, bifloral inflorescences, unbranched.

Green spathes which are not, or are only slightly, scarious at the top; seed provided with an aril.

Iris macrantha.

Purplish spathes totally (or in the upper three-quarters) scarious; seed without aril.

Plant very hardy, fairly early blooming, multifloral inflorescences, four or five times branched.

Iris autosyndetica, derived by crossing *I. hoogiana* and *I. macrantha.* Left, plant hardy, rather early blooming, inflorescences bifloral or multifloral, single or branched from one to three times; *right,* spathes purplish, the upper half slightly scarious; seed provided with a small aril.

Simonet. Larousse.

An iris, Iris autosyndetica

This species, created by Simonet, is the result of crossing two tetraploid irises, *Iris hoogiana* (twenty two chromosomes per set) and *Iris macrantha* (twenty-four chromosomes per set). The diploid number of the hybrid thus obtained is forty-six, and its characters are a mosaic of those of its parents. Its meiosis is normal, the twenty-two maternal chromosomes pairing among themselves and the twenty-four paternal chromosomes pairing among themselves. Like most irises it cannot fertilize itself because, for structural reasons, the pollen tube is unable to make contact with the ovules. Its gametes, however, are viable.

A hawksbeard, Crepis artificialis

Many species of hawksbeard (of the Compositae family) cross in nature and in botanical gardens. A cross between *Crepis setosa* (haploid number four) and *Crepis biennis* (haploid number twenty) will in spite of the numerical disproportion of their chromosomes produce a hybrid. The hybrid with twenty-four chromosomes — ten pairs of which are bivalent and four univalent — is hardy and in the main resembles *C. biennis.* But there are differences. The hybrid blooms in its first year while *C. biennis* is biennial. The hybrid is not sterile and when self-pollinated produces a few

361

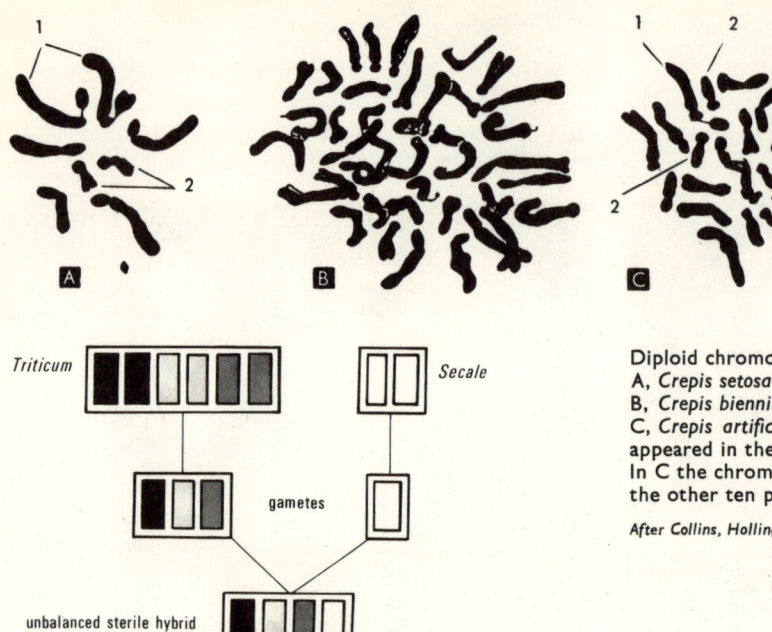

Diploid chromosome content of
A, *Crepis setosa* (2N = 8);
B, *Crepis biennis* (octoploid, 2N = 40);
C, *Crepis artificialis,* a new and stable species (2N = 24) which appeared in the fourth generation of the hybrid *setosa-biennis.* In C the chromosomes marked 1 and 2 are derived from *setosa;* the other ten pairs are *biennis* chromosomes.

After Collins, Hollingshead, and Avery.

Diagram of the formation of a new octoploid species, *Triticale saratoviense,* by the hybridization of a hexaploid wheat (*Triticum vulgare*) 2N = 42 chromosomes, and diploid rye (*Secale cereale*) 2N = 14 chromosomes.

After Hurst.

seeds, one of which grew into a fertile plant — the parent plant of a stable strain with twenty-four chromosomes. The new species, *Crepis artificialis,* displays mixed characters: its achenes recall those of *C. biennis;* and its denticulate leaves those of *C. setosa.* Its twenty-four chromosomes comprise twenty contributed by *biennis* and four by *setosa.* Its meiosis is normal, for the twenty *biennis* chromosomes pair among themselves, as do the four *setosa* chromosomes. Its gametes, with ten *biennis* chromosomes and two *setosa* chromosomes, are thus balanced.

A wheat-rye, Triticale saratoviense

When the soft wheat *Triticum vulgare* (hexaploid, with twenty-one chromosomes per set) is fertilized with the pollen of rye (*Secale cereale,* diploid, with fourteen chromosomes per set) it engenders tetraploid hybrids (diploid number twenty-eight) which are very hardy but generally sterile. Some of the hybrids, in which the gametes have failed to reduce the number of their chromosomes during meiosis, give rise to an octoploid F_2 with the diploid number fifty-six. These second generation plants are fertile, stable, and share the characters of wheat and rye, that is to say of two different genera — hence the name of the new genus: *Triticale.*

Colchiploids — or polyploids created by colchicine

In the preceding examples we have seen the importance of polyploidy in the formation of new species. By doubling or by multiplying the number of chromosomes polyploidy makes normal meiosis possible, for each chromosome can then find a homologue with which to pair. A hybrid which is sterile because its chromosome sets are unbalanced is transformed by polyploidy into a fertile hybrid.

In nature plant polyploidy is not infrequent, but it can be brought about experimentally by treating plantlets or seeds with dilute solutions of colchicine or similar 'stathmocinetic' substances, indol-3-acetic acid, for example. These substances inhibit the formation of the spindle during mitosis, but do not prevent the longitudinal cleavage of the chromosomes. In this manner the number of genomes — or chromosome sets — is multiplied. The plantlet which has just germinated is daubed with a jelly containing thirty per cent agar to which an equal volume of a two per cent solution of colchicine has been added. The treatment results in foliar abnormalities and the death of numerous individual plants, but among those that survive tetraploids will be present. Their pollen grains are double those of normal diploids. Such mutations are characterized by gigantism and their fertility is

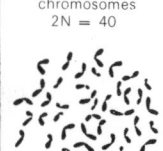

The horse chestnut, *Aesculus hippocastanum*. Large white flowers, tinted with red; large fruit with spiny envelope.

Larousse.

The red buckeye *(Aesculus pavia)*. Flowers a pale red; fruit with spineless envelope.

Larousse.

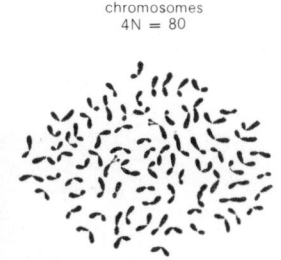

Pavia carnea, the hybrid octoploid, 2N = 80, of *Aesculus hippocastanum* and *Aesculus pavia.*

Larousse.

Prunus cerasifera, 2N = 16 chromosomes.

Prunus spinosa, 2N = 32 chromosomes.

Larousse.

unaffected. Tetraploid forms exist in a diversity of plants: cauliflowers, aubergines, Brussels sprouts, flax, soya beans, beetroot, strawberries, petunias, etc.

The synthesis of natural species

The botanist has not only succeeded in creating valuable new species but, with his knowledge of the mechanism by which species arise, he has also been able to synthesize species found in nature.

Synthesis of Galeopsis tetrahit

Galeopsis is a genus of the Labiatae family and contains the hemp-nettles which grow in fields and waste places. *Galeopsis tetrahit,* the common hemp-nettle, has always been considered a true species. In reconstructing it, however, Müntzig has demonstrated its hybrid origin, its two parent species being *Galeopsis pubescens* (haploid number eight) and *G. speciosa* (also eight chromosomes per set). By crossing *pubescens* and *speciosa* (whose chromosomes have been doubled by colchicine) *Galeopsis tetrahit* is produced. It is a tetraploid with, in all, sixty-four chromosomes;

and it is the first example of the artificial synthesis of a Linnaean species.

Synthesis of the red horse-chestnut, Aesculus carnea

The ordinary horse-chestnut, *Aesculus hippocastanum,* a native of south-eastern Europe and central Asia, has white flowers and attains a height of some 90 feet. *Pavia rubra,* from North America, has red flowers and is much smaller, measuring only some 25 feet. Both have sets of twenty chromosomes and are almost certainly tetraploids. Characteristics of both species are found in *Aesculus carnea* which is frequently cultivated in parks and gardens: its flowers are coloured like those of *Pavia rubra* while its leaves resemble those of *A. hippocastanum;* its fruits and its height (60 feet) are intermediate. It has sets of forty chromosomes and it is thus an octoploid. *A. carnea* has, like *Primula kewensis,* spontaneously appeared on several occasions. Back-crossed with *A. hippocastanum* it gives rise to a hexaploid (thirty chromosomes per set) which is fertile and breeds true.

Synthesis of the plum, Prunus domesticus

Prunus domesticus is hexaploid (twenty-four chromosomes per set) and is the result of a cross between *Prunus cerasifera* from Asia Minor (diploid, eight chromosomes per set) and the blackthorn or sloe *Prunus spinosa* (tetraploid, sixteen chromosomes per set). A doubling of the hybrid's chromosomes, perhaps the result of an encounter between two unreduced gametes, has produced *Prunus domesticus.*

Synthesis of the turnip, Brassica napus

This synthesis has been achieved by crossing a cabbage, *Brassica oleracea,* with a rape, *Brassica rapa.*

There are further examples of plant synthesis which could be given.

The formation of animal species

Among animals man has, by carefully selected mutations and controlled crossing, created new breeds which are sometimes very different from their origins and would, if found in nature, be described as true species. These creations have been chiefly brought about among aquarium fish, birds, and domestic mammals. In the case of fur-bearing animals efforts have been made to produce new types with pelts of different qualities and colours. In this way numerous colour mutations have been achieved in mink. Some thirty of these mutations are stable and by appropriate cross-breeding a wide selection of colours is commercially available, of which the most valuable are silver-blue, pastel, and white.

Less advance has been made in the creation of new animal species than in that of plant species; but among certain animals the mechanism of speciation has been analysed with great precision. As examples we may

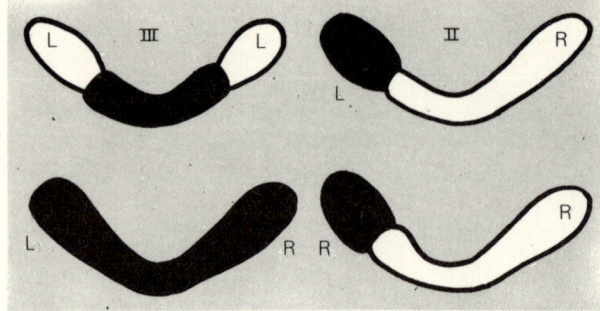

Chromosomes II and III of *Drosophila artificialis.* A stable type which arose after complicated translocations in a strain of *D. melanogaster.* In black, chromosome III or portions of it. In white, chromosome II or portions of it. L, left; R, right.

consider *Peromyscus, Drosophila,* and two birds, the gull and *Zosterops.*

Peromyscus are small North American rodents which somewhat resemble European field-mice. They differ from each other in colour, dimensions, especially length of tail, and habits. Variations are so numerous that more than seventy species and sub-species have been described in the United States and Canada. Certain species — *leucopus, truei, maniculatus* — are true species which differ morphologically and refuse to interbreed either in nature or in captivity. Sexual separation is less complete between the two species, *truei* and *nasatus,* which occupy different habitats but will sometimes interbreed. In nature the absence of intermediate types suggests total sexual separation, but in captivity the two species can, with difficulty, be made to cross. The resulting male hybrids are sterile, while the female hybrids are fertile with their male parents. Between the two species *leucopus* and *gossipus* the separation is still less complete: the two species resemble each other but *gossipus* is larger. They normally occupy different habitats although some areas are shared. In these areas the two species do not, from lack of sexual attraction, mate. But in captivity they interbreed and the hybrids of both sexes are fertile. Thus gametic infertility is not responsible for their sterility in wild life. A third case is provided by the two species *maniculatus* and *polionotus* which have a strong morphological resemblance but are very different in size. As hundreds of miles separate their areas of distribution sexual isolation is complete. But in captivity they mate with ease. The small *polionotus* females, fertilized by *maniculatus* males, usually die during parturition; but the reverse cross, *maniculatus* female and *polionotus* male, produces good results and the hybrids of both sexes are fertile.

Of the great species *maniculatus* there are numerous geographical sub-species, scattered in different localities and differing from each other in colour, length of tail, feet, and ears. Laboratory rearing shows that these variations are fixed characters, independent of external conditions — like the characters of a pure strain. In captivity mutations affecting colour occur and even produce albinism. The other great species, *polionotus,* also embraces a number of sub-species, some of which are characterized by reduced pigmenta-

The formation of species in gulls. *Larus fuscus*, the lesser
black-backed gull of Britain, gives way in Finland to the
Scandinavian *Larus*, which in turn yields to the *Larus* of Siberia,
then to the American *Larus* and finally to *Larus argentatus*, the
British herring gull. Between the two apparently distinct
species, *fuscus* and *argentatus*, the transition can thus be
followed.

After a wall display in the British Museum, Natural History.

tion which seems to be connected with their habitat.
Leucocephalus, for instance, has a pale grey back and
lives on white sand. It appears to be an ecotype, in
other words an ecological pure strain.

The crossings described above demonstrate that the
sexual isolation of the various types of *Peromyscus*
does not derive exclusively from gametic infertility,
but more often from other causes: geographic or
ecological isolation, differences in size, and emotional
aversion.

Speciation in the genus Drosophila

Thanks to the giant chromosomes of its salivary
glands the chromosomal constitution of *Drosophila*
is well known, and its mechanism of speciation
clearly demonstrates the importance of gene variations
and variations in the structure of the chromosomes.
The effect of gene variations, even when accumulated
in a single individual, is superficial, but structural
variations in the chromosomes can bring about sexual
separation and thereby affect evolution. *Drosophila
virilis* and the sub-species *D. americana* are crossed

with difficulty and produce few offspring. The hybrids
are partially fertile among themselves or with their
parents. The genomes of the two species differ: that of
virilis (diploid number twelve) comprises five pairs of
rod-shaped chromosomes and one pair of micro-
chromosomes; that of the female *americana* (diploid
number eight) comprises four V-shaped chromosomes,
one rod-shaped pair, and one pair of micro-chromosomes, while the male *americana* has three V-shaped
chromosomes, two pairs which are rod-shaped and one
pair of micro-chromosomes.

The cross between *Drosophila melanogaster* and
D. simulans, when successful, gives rise to sterile
hybrids. The two species nonetheless have the same
haploid number of chromosomes, namely four. Their
chromosomes are similar and have in common many
genes which in both species occupy the same loci. But
differences in their chromosomal structure have been
observed, differences involving one large inversion,
five small inversions, four small changes at the extremities of the chromosomes, and fourteen regions where
the chromosomes refuse to pair.

One race of *D. melanogaster,* characterized by translocations between two chromosomes, has been renamed *D. artificialis*. One of its number III chromosomes has lost both its extremities, right and left, while its two number II chromosomes have each lost the left extremity. This exchange of extremities is shown in the diagram below. The chromosomal mutant is a fertile and true-breeding species and is sexually separated from its original stock.

Speciation among sea-birds

In Western Europe there are two neighbouring species of gull: *Larus argentatus* or the herring-gull, and *Larus fuscus* or the lesser black-backed gull. The two differ not only in the colour of their plumage and feet but also in their habits. One reproduces in sandy moorlands and migrates in winter; the other builds its nest in cliffs and disperses to a different area. Living side by side they do not interbreed and appear to be two distinct species sexually isolated. But if one traces a circular course around the north pole, following the arctic shores of America and Asia, a continuous chain of gull populations will be found, populations displaying intermediate characters. In the regions of passage no sexual isolation between the two species exists. The *fuscus* of the British Isles with its black back and yellow feet forms a link with the black-backed gull of Scandinavia which then joins the Siberian gull of lighter colour and flesh-coloured feet. This, in turn, joins the silvery American gull which itself forms a link with the British *L. argentatus,* with its lighter plumage and pink feet. These various gulls constitute a continuous series of sub-species. The two terminal sub-species — *fuscus* and *argentatus* — have been separated for such a long time that genetic differences have brought about their sexual isolation.

Speciation in another island bird

A similar phenomenon occurs in Norfolk Island which lies some 600 miles off the east coast of Australia. There, three species of birds of the genus *Zosterops* are found: *Z. norfolkensis, Z. tenuirostris,* and *Z. albogularis*. These species are endemic to the island and all three are related to the species *Z. lateralis* which lives on the Australian continent. To account for the three different insular species it is assumed that each is the result of a successive migration from Australia. Between migrations a sufficient time elapsed for each group to become genetically isolated from its Australian stock and thus to form a new and independent species. This occurred three times.

It seems highly probable that in nature today speciation — in other words the formation of new species — obeys the general laws we have discussed. In whatever manner a species is formed it is essential that a given population ceases to interbreed with neighbouring populations, and that it accumulates hereditary variations by means of gene and chromosome mutations. Animal speciation, like plant speciation, requires long periods of time — which Linnaeus himself pointed out when he said that 'the species are the daughters of Time'.

A species thus makes its first appearance in a localized region of origin, sometimes called its cradle. From this cradle the species gradually spreads — if conditions are favourable, if its fertility is high, and if no obstacle intervenes. Little by little it gains ground although there are, of course, examples of sudden and rapid expansion. The extension of the area of a species' dispersal is often a measure of the species' age. The contraction of its area announces its approaching extinction. Sometimes it will survive for a while in isolated patches, then in a single outpost, the 'tomb' of the species, for instance, the isle of Eldey for *Pinguinis impennis,* or an islet off the coast of New Zealand for the iguana-like tuatara or *Sphenodon*.

The question arises: can a species appear independently in different regions? Several identical mutations have, as we know, occurred at different times and places, such as the copper beech, the Kew primula and the late-flowering common oak. Thus a species can have more than one cradle, but the phenomenon seems to be rare, and a single cradle remains the general rule.

Finally, an understanding of how species arise is provided by genetics; for the hereditary differences which separate two species are of the same order as the variations which occur within a single species. Speciation is only an aspect of evolution. Every useful change constitutes a kind of beach-head from which the species can push onwards and continue its evolution.

Evolution

In all epochs the origin and the past of living beings have engrossed the mind of man. Since earliest antiquity two rival views have been held of the genesis of plant and animal species: the theory that they are fixed and unchanging, and the theory that they evolve.

Immutability

The living world is composed of distinct, individual, and identifiable species. In the opinion of Aristotle and other ancient philosophers these species were eternal and immutable. In the late Middle Ages schoolmen of the Church christianized the philosophy of Aristotle and adopted this opinion. Living creatures were separately created, each according to its kind, an explanation in accordance with what was written in the Book of Genesis. At the origin of each species there was a special intervention on the part of the Creator. Successive species were created in a determined order and succeeded each other progressively, Man being the last to appear. Each species was endowed with every mechanical structure essential to its mode of life. Thus the problem of adaptation was at once solved, in its entirety and in advance. Creatures of the air were provided with wings, creatures of the deep with fins and so on.

Then palaeontology revealed the existence, in various strata of the Earth, of animals which were different in form from contemporary animals. The older they were the more they differed from living organisms. To explain the awkward presence of such ancient flora and fauna Cuvier suggested that geological upheavals and events like the deluge — believed to be the last of such catastrophies — had destroyed all living creatures. He assumed that there had been separate creations for each of the great zoological groups. Only a few recent mammals, who had taken refuge on high-lying points

Left, Cuvier by David d'Angers in the Louvre. *Giraudon.*

Right, Buffon by Houdon in the Louvre.

of the Earth, had returned to the flooded areas.

The theory of evolution

The idea that organized beings evolved and that they diverged and developed from a common ancestry, is as old as Greek philosophy. From lack of documentation, however, it remained within the realms of speculation. Many Greek writers already foreshadowed the principles which Lamarck and Darwin were later to propound. Anaximander of Miletus (610—547 B.C.) and Empedocles of Agrigentum (495—435 B.C.) both

367

interpreted the world in evolutionary terms and pointed out that living creatures adapted to their environment. Theophrastus of Eresos (370—285 B.C.) noted the existence of organs like male nipples which were useless, or of those like the stag's exaggerated antlers which actually endangered their possessor. Lucretius (95—53 B.C.), a disciple of Empedocles, was fully conscious of the struggle for existence. 'All creatures which breathe the life-giving air have had, since their origin, either finesse or vigilance or agility, to defend themselves and preserve their kind . . . Lions triumph by their courage, foxes by their cunning, stags by their swiftness in flight . . . But those to whom nature has given neither independent means to live, nor qualities which make them useful to us, nor any other title to our care and protection, all such, exposed to the rapacity of others, helpless in the bonds which fate has forged for them, must perish to the last individual — by order of nature.'

Until the eighth century the Fathers of the Church, including St Gregory of Nyssa and St Augustine, were evolutionists: the world evolved as a function of the powers which God at the creation had attributed to it. Special interventions were no more needed to explain the origin of life, which derived from organic matter, than to explain the origin of species, which were created with the power to succeed each other.

During the Renaissance scholars were familiar with the evolutionary theories of the Greeks. Although Vanini (1586—1619) believed in the spontaneous generation of plants and animals his scientific intuitions were remarkable. The changes in plants brought about by cultivation suggested to him that one species could evolve into another; the striking resemblances between Man and the ape gave him the idea of their common derivation. Like Democritus and Epicurus, he believed that the universe was not the product of divine spirit, but a fortuitous combination of atoms. Such audacity was not forgiven by the Inquisition; he was found guilty of atheism, tortured, and burnt at the stake.

The seventeenth century made little contribution to evolutionary theory. In 1672 Swammerdam, noting how animals resembled each other, wondered if one might not 'in a certain measure affirm that God had created a single animal only, an animal diversified in an infinite number of kinds or species.'

But with Buffon and the *Encyclopédistes,* the eighteenth century prepared the literate public to accept the idea of evolution. Diderot above all was a precursor of the movement, though Voltaire remained aloof from it and even hostile. In this age of enlightenment *Encyclopédistes,* naturalists, philosophers, all contributed to freeing the minds of their contemporaries from pseudo-scientific superstition and fables. To identify species John Ray in England and Theodor Klein in Germany drew up inventories of flora and fauna, work which had begun at the end of the seventeenth century. But it was above all Linnaeus (1707—1778) who put order into the classification of species and created the binomial nomenclature which is still used. Linnaeus believed in the fixity of species but his work encouraged evolutionary ideas.

The variability of contemporary living organisms

Lamarck, by Tardieu.
Freuler.

Charles Darwin in 1860.
Nadar.

and the derivation of one species from another were concepts with which the eighteenth century was familiar. The first great evolutionist was the immensely learned Buffon. Buffon at first held that species were fixed. As he acquired more knowledge, however, he changed his opinion until his view of the world was distinctly evolutionary. Since the sixteenth century, fossils had been thought to be the remains of living creatures. Buffon confirmed this belief with sound arguments. Often the fossils belonged to totally extinct species and Buffon correctly interpreted the fossil remains of mammoths and extinct rhinoceroses. In his theory of the Earth he proposed a succession of flora and fauna which differed from those then known. These revolutionary ideas provoked a formal condemnation, in sixteen propositions, from the Sorbonne in 1751. Buffon retracted and thus made his peace with the Faculty of Theology. But in 1778 he advanced his ideas more forcibly, and observed that the Earth, the Earth's crust and the Earth's climate, were subject to change, and that sedimentary layers had been deposited in chronological order. He placed Man among the animals which 'he resembles in all material ways'. Even the instincts of animals seemed to him 'to be more trustworthy, perhaps than Man's reason, and their industry more admirable than Man's arts.' He glimpsed the unity of relationship between Man and the vertebrates, suggesting that four-handed creatures 'filled in the great interval which separates Man from the quadrupeds.' He also noted animal and plant variations which he attributed to multiple causes, the influence of nourishment, climate, changes of habit and domestication. In spite of the divergence between American and Eurasian fauna he believed that they had a common origin. 'Even without reversing the order of nature it is not impossible,' he wrote, 'that all the animals of the New World are basically the same as those of the Old World from which they formerly derived. Then, having been separated by vast oceans or impassable land, they could, with the passage of time, have received the imprint of, and been affected by, a climate which itself had changed in quality for the very reason that brought about the separation in the first place. In consequence — and always with the passage of time — such animals might be reduced in size, denatured, and so forth. But all this need not prevent us from regarding them today as animals of different species. From whatever cause the differences arise, be it the lapse of time, the climate and the land, or if indeed they date from the Creation, they are no less real. Nature, we must admit, is in a continual state of flux; but it is enough for us to seize upon her in our own moment of existence, and to cast a few glances forward and behind in order to glimpse what formerly she could have been and what, in the future, she may become.'

From this it would seem that Buffon could have formulated a theory of evolution, but he valued his tranquillity too highly perhaps to risk exposing his thoughts in their entirety.

The views of Maupertuis (1698—1759) were more speculative than those of Buffon, and he hit upon certain ideas which anticipate our present knowledge of mutations. Maupertuis grasped the importance of hereditary variations and of selection. Though climate and nourishment played their parts in the variation of the individual, a much greater part was played by modifications of the seminal fluids. Those variations which could, he wrote, 'characterize new species of animals or plants tend to die out. They are Nature's discards.' Variations which appear by chance are at once submitted to the arbitrament of selection. The only individuals which survive the test 'are those in which order and fitness are found; and the species which we see today are but a minute part of those which blind destiny has produced.' He sought in this way to account for the genesis of species. 'Could we not thus explain how from two individuals a multiplicity of the most varied species might have ensued? Their origin could be due solely to a few fortuitous offspring in which the elementary parts had not retained the order and arrangement which prevailed in the parent animals. Each time such an error occurred a new species would arise, and by force of repeated discards the infinite variety of animals we see today could have been produced . . .'

Towards the end of the century (1794) Erasmus Darwin (1731—1803) published *Zoonomia* or *The Laws of Organic Life,* in which he obliquely foreshadowed both the doctrine of natural selection formulated by his more illustrious grandson and that of the inheritance of acquired characteristics advanced by Lamarck.

Lamarck, a disciple of Buffon, was the first to propound a coherent theory to explain evolution. His work aroused widespread interest, although less perhaps than it deserved. Insufficiently based on demonstrable facts his conclusions were vigorously attacked, notably by Cuvier. Cuvier (1769—1832), a confirmed enemy of dawning evolutionary theory, nevertheless furnished his opponents with solid arguments based on well-established facts in his own valuable work in comparative anatomy, zoology, classification and, above all, palaeontology.

Finally Charles Darwin (1809—1862) arrived on the scene to restore order. His powerful influence on evolutionary thought — described in more detail in a later chapter — was decisive. Darwinism, neo-Darwinism, and mutation-theory lead us into the present period.

Such, in brief, is a historical summary of evolutionary ideas. But before examining evolution itself it is essential to distinguish clearly between the 'fact of evolution' and the theories which have been advanced to explain it. Evolution recognizes the continuity of the living world and the derivation of animal and plant forms from related ancestral species. No serious biologist has ever been gnawed by secret doubts as to the fact of evolution itself; but disagreement arises when scientists attempt to explain its causes, and the mechanism by which changes in the living world come about. That is why, periodically, people speak of a crisis in evolutionary theory. The nature of the 'crisis' should be kept in mind: the fact of evolution itself is indisputable; but exactly how it occurs is still largely obscure.

The fact of evolution and its proofs

The problem of evolution is the central problem of biology itself. Living organisms are what they are because that is what they have become. Change, not immutability, is the rule. In the Universe, moreover, everything evolves; nothing is stable. The Universe is in a state of incessant becoming, and living things are no exception to this general law. Evolution is the history of the changes undergone by living matter since its appearance on Earth, some 1,000—2,000 million years ago. Evolution implies not only change but also continuity and linked descent.

For the fact of evolution and the proofs of its existence we rely upon the evidence of many sciences. Palaeontology, embryology, comparative anatomy, biochemistry, all shed light on how life has developed during the ages.

The fossil evidence

The angle from which the palaeontologist approaches the problem is different from that of the biologist. He studies variation through vast stretches of time. The evolution of living creatures unfolds before his eyes like a film. He sees, as Osborn put it, 'the movement of characters' in a series of inter-related beings. Palaeontology supplies signposts in time which sometimes enable us to retrace the historical evolution and ancestry of living forms back to remote ages, periods which elapsed hundreds of millions of years ago.

The first question which arises is how long has life existed on our globe, and to answer it we must first discover at least the approximate age of the Earth itself, or rather the age of the first elements that formed the crust which solidified when the original gaseous sphere had become sufficiently cool. It is possible to estimate this age by consulting the radioactive clock of Uranium ores: in other words by examining the state of disintegration of the radioactive elements found in rocks.

It is known that uranium, an element widely dispersed through the Earth's crust, disintegrates at a

1   2   3

Below, Walchia schlotheimii. Autunian subdivision of Permian period, from the slate quarry of Lodève (Hérault). Spermatophyta, subdivision Gymnospermae, family Araucariaceae (extant representatives include the Chile pine or monkey puzzle tree).

Muséum d'histoire naturelle. Larousse.

Palaeozoic plant fossils. 1. *Pecopteris polymorpha:* from the coal measures of the Tarentaise; a pteridophyte of the order Filicales, i. e. ferns. 2. *Sphenophyllum emarginatum,* from the middle carboniferous layers of Saarbruck: an extinct pteridophyte related to horsetails. 3. *Odontopteris rotundata* from the upper carboniferous layers of la Péronnière, a fossil fern.

Muséum d'histoire naturelle. Larousse.

Mischoptera nigra, an insect found in the Carboniferous deposits of Commentry.

Muséum d'histoire naturelle. Larousse.

rate which is precise and unvarying. Of 1 gram of uranium present at a given date only half a gram will remain after the lapse of 4,500 million years. If during this interval the rock in which the uranium is found has undergone no profound upheaval this remaining half-gram will be accompanied by the end-products of its disintegration; namely, helium and lead-206, which is one of the isotopes of ordinary lead, itself a mixture of leads 206, 207, and 208. From the moment at which the gram of uranium was incorporated into the rock the proportion of lead-206 to uranium slowly but constantly increases. To determine the age of the rock, that is to say the time which has elapsed since its formation, all that is needed is a sample containing lead-206 and uranium, and to measure the ratio of one to the other. According to such measurements the Earth's crust is at least 3,000 million years old. The age of certain pre-Cambrian uranites is estimated to be 1,985 million years, while other pre-Cambrian rocks are 2,000 — 2,400 million years old.

By the same technique maximum and minimum durations of geological eras and periods have been approximately established.

A table of geological time (minimum durations)

Time elapsed since end of Quaternary era to present: 8,500 to 10,000 years.
Quaternary: 1 million years

		Duration in millions of years
Tertiary or Cainozoic (Beginning 60 million years ago)	Pliocene	6
	Miocene	12
	Oligocene	16
	Eocene	25
Secondary or Mesozoic (Beginning 230 million years ago)	Cretaceous	60 to 80
	Jurassic	20 to 35
	Triassic	27 to 35
Primary or Palaeozoic (Beginning 600 million years ago)	Permian	35 to 45
	Carboniferous	70 to 80
	Devonian	40 to 50
	Silurian	
	Gothlandian	30
	Ordovician	70
	Cambrian	90
2,000 million	Pre-Cambrian . . .	1,400

It will be observed that pre-Cambrian times ended 600 million years ago and that the pre-Cambrian itself extended over a period of time which is more than double that which has elapsed since then until our own days. It can with confidence be asserted that life existed in pre-Cambrian times, for certain deposits of the era contain authentic remains of plants (Cyanophyceae and other algae) and of animals (sponges and polyps). If such remains are rare it is because pre-Cambrian rocks, being extremely ancient, have been subjected to so many strains and stresses that the fossils they contained have been largely destroyed. The rarity of the fossils may also be plausibly explained by the small number of species which had in pre-

Cambrian times evolved a skeleton or a shell. Rare and incomplete though they are, the fossils reveal that in this remote era complex and differentiated organisms already existed. From the Palaeozoic era the earliest plant and animal fossils are sufficiently well-preserved for the palaeontologist to describe their structure with accuracy. From an examination of these findings a few general conclusions may be drawn.

1. *The various plant and animal types appeared successively and in strict sequence, from the simplest to the more complex.* The oldest marine or fresh-water plants are basically algae: some of them had a thallus which measured 1 metre in diameter. The first terrestrial plants, the Psilophytales, are found in the middle and upper Devonian: these are the vascular cryptogams, that is, plants with fluid-conducting vessels, plants which attained their zenith in the Carboniferous period. In the middle Carboniferous appear the Phanerogamia (Spermatophyta) represented by the cordaites, then by Cycadaceae and the conifers. The ancestor of the ginkgos appears in the Permian. By the end of the Palaeozoic era the flora is already highly diversified.

In the Mesozoic era the vascular cryptogams declined. A few Pteridospermae (or Cycadofilicales with fern-like leaves, and reproducing by seeds) still persisted, but the gymnosperms multiplied. The Cycadaceae, very abundant until the Jurassic, then began to diminish. In the Cretaceous the conifers formed the dominant terrestrial flora. At the beginning of this period the angiosperms or flowering plants, appeared. From the end of the Mesozoic era the angiosperms are composed of species which resemble those found today, and have divided into monocotyledons and dicotyledons.

From this rapid panorama we can see that plants of simple structure, considered as primitive or archaic, did in fact appear before those whose structure is more complex. The great phyla or divisions coincide with the various morphological or structural stages reached during the course of time.

The fossil record similarly reveals the successive appearance of diverse animal groups. In the Cambrian many groups, already differentiated, are represented: sponges, coelenterates, annelids, molluscs, brachiopods, crustaceans, trilobites, and so on. But no trace of a vertebrate has yet been discovered from this period.

During the next period the Silurian groups already present are perfected. The trilobites (extinct aquatic arthropods allied to ancestors of the crustaceans) and the nautiloids become more and more diversified. In the upper Silurian the vertebrates make their appearance. The most primitive would seem to be *Jaymoytius* discovered in Scotland in 1946. It belongs to the extremely primitive class of the Agnatha or jawless vertebrates and is a sort of *Amphioxus,* 18 centimetres long, with naked skin, no lower jaw, with a notochord and a continuous fin along its flanks. Then come the bizarre, armoured fish known as the 'shell-skinned' or ostracoderms, also of the class Agnatha.

In the Devonian the Merostomata abound. New types of ammonites appear and also the placoderms, or archaic jawed fishes with cartilaginous skeleton.

371

Section of *Nautilus* showing the chambers, the shell-walls and syphon.

Below, Melonites multipora, a carboniferous fossil sea-urchin, from Saint Louis, Missouri.

Muséum d'histoire naturelle. Larousse.

In the upper Devonian the shark-like *Selachii* appear.

In the Carboniferous and the Permian fossils of myriapods, insects, and arachnids are abundant in continental and coastal sites. The first four-legged creatures, the Stegocephalia, primitive amphibians descended from Devonian Crossopterygii (or lobe-finned fishes), diversify and, in the next period, the Permian, give rise to the first reptiles adapted to terrestrial life. The stegocephalians disappear while the small reptiles increase.

The Mesozoic era is characterized by the extraordinary development of the reptiles which achieve vast size and rule the land, the seas and the air. The higher crustaceans (the Decapoda) and insects capable of complete metamorphosis appear. The ammonites reach their zenith while the stemmed echinoderms, the Merostomata and the trilobites decline or disappear. The first mammals, related to the reptiles, are observed in the Jurassic; they are still small and very primitive. Placental mammals first arose in the following period, the Cretaceous. Also related to reptiles, the oldest known birds, *Archaeopteryx* and *Archaeornis,* date from the upper Jurassic.

The Cainozoic era is marked by the disappearance of the giant ruling reptiles and the extraordinary success of the mammals. In this era all the orders known today arose.

Finally, in the Quaternary, Man appeared.

In the scale of time flora and fauna arise and evolve progressively, beginning with the simplest and proceeding towards the more complex. Organic progression is manifest. A strict concordance exists between the time a given group appears and the position or rank which its organization confers upon it. For example, the jawless Agnatha precede the true fishes which in turn precede the reptiles, themselves followed by the birds and the mammals. The placental mammals are preceded by the simpler aplacental types. The same concordance between time of origin and complexity is observed in the lesser taxonomic — or classificatory — divisions, the classes and orders.

Hind leg of *Ichthyostega* embedded in the rock where it was found.

After Piveteau.

2. *The intermediate forms.* Organization and structure, then, become progressively more complicated. The genetic continuity of the successive types is affirmed by intermediate forms which link two or several types of organization.

These intermediate forms reveal a structure comprising characteristic traits of several types, and for this reason are known as 'synthetics'. The transition from an earlier to a later type of organization is not effected by forms which are exactly intermediate, but by the progressive substitution of new characters for those of the earlier stock. The classification of these intermediate types is obviously arbitrary and will depend on the criteria adopted to determine whether an organism belongs to this class or to that. Broadly speaking, if in the intermediate forms between classes A and B the characters of the former remain commoner and of greater structural importance the organism will still be classified as A; when the characters of B predominate, then it will be classified as B.

We know much less about the transitional forms of the invertebrates than we know about those of the vertebrates. This does not, of course, imply that such forms do not exist. The great invertebrate groups are represented from the Cambrian period onwards; but the primitive pre-Cambrian forms have been largely destroyed in metamorphosed rocks. In addition many invertebrates leave no remains which can be fossilized.

Among the vertebrates, however, the intermediate forms or the manner in which transition occurred have been more successfully analysed. A few examples follow.

The Ichthyostegidae discovered in the Devonian of Greenland link the crossopterygian lobe-finned fishes of the extinct order Rhipidistia with the stegocephalian amphibians of the Carboniferous. *Ichthyostega* has a pelvic girdle and developed legs of the amphibian type, but the low-domed skull is that of a crossopterygian fish. The skull is pierced by two eye sockets and the pineal orifice. The arrangement of the bones is like that of the crossopterygians: there is a single bone in front of the nasal cavities, a single bone behind the parietal bones, and a preoperculum. Again as in crossopterygians the roof of the palate has the orifices of external and internal nostrils, while the external nostrils (ventral) and the sensory canals of the Ichthyostegidae and the Crossopterygii are disposed in the same way. In fact the cranial bones of these two groups are homologous.

The amphibian *Protobatrachus massoti,* discovered in the Triassic sandstone schists of northern Madagascar, is an ancestral form of the tailless amphibians. It is naked-skinned and has a skull which greatly resembles that of the tailless amphibians, the sole difference being a more complete ossification of the auditory region. The spinal column has fewer vertebrae; it forms no urostyle and a few caudal vertebrae persist. The hind legs are of the tailless type and the forelegs of the walking type. *Protobatrachus* thus resembles a frog with a small tail. Its structure seems less perfectly adapted to movement on land and in water than the frog's. The anterior arrangement of the tailless amphibian's lymph heart would seem to have existed in the ancestral form of the early Triassic.

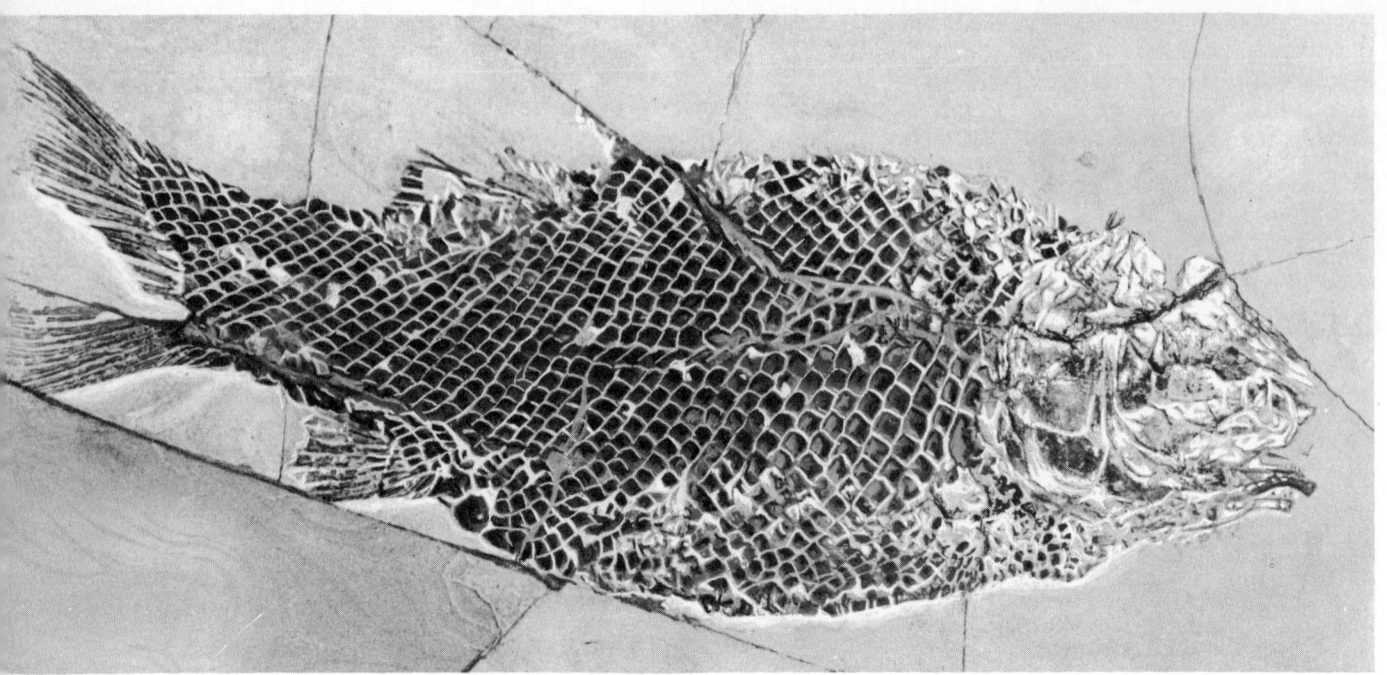

Lepidotus elvensis. An armoured fish from the end of the
Palaeozoic era, 40 centimetres long, discovered in the upper
Lias of Holzmaden.

Paris-Match. Hubert de Segonzac.

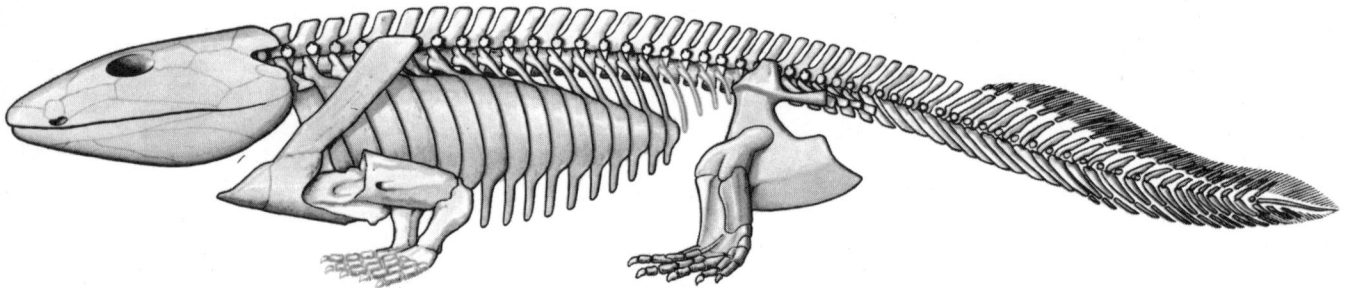

Ichthyostega: the oldest-known four-footed animal (Devonian
of Greenland). It may have been a metre long and is a link
between the crossopterygian fishes and the stegocephalian
amphibians.

After Jarvick.

Skull of *Cynognathus crateronotus.* Fossil therapsid reptile,
from the Trias of Karroo in South Africa. The skull, 40
centimetres long, with its differentiated teeth — incisors,
canines, and molars — resembles the skull of a large mammal.

Larousse.

Seymouria. Cast of a skeleton found in the lower Permian of Texas.

Muséum d'histoire naturelle. Larousse.

Protobatrachus also offers certain skeletal characteristics of the stegocephalians, especially of those belonging to the order Labyrinthodontia. Frogs and toads of the very ancient order Anura are thus apparently linked to the more primitive stegocephalian amphibians.

Forms which link the amphibians with the reptiles are also known. *Conodectes* (or *Seymouria*) of the lower Permian of Texas is sometimes classified among the stegocephalian amphibians because of its tadpole stage, and sometimes among the primitive reptiles of the order Cotylosauria because of its many reptilian characters.

Mammals derive from small reptiles of the extinct order Therapsida which flourished in the Triassic. Among these mammal-like reptiles there was already differentiation of teeth into from three to five incisors, one canine, and from seven to nine molars, preceded probably by milk-teeth. This dental formula recalls that of marsupials which have five incisors, one canine, and nine molars. The skull and part of the skeleton reveal numerous mammalian characters; the posterior bones of the mandible are reduced and the dentary bone enlarged; articulation is still effected by the articulare and the angulare bones which are greatly reduced. The single median articular process characteristic of reptiles has atrophied and given way to two occipital condyles. The quadrate bone has considerably diminished in preparation for its future entrance into the middle ear. The third — or pineal — eye has disappeared. The temporal fossa is confined from above by the parietal bone as in mammals and, as in mammals, a secondary palate has developed in the roof of the mouth. The nostrils have moved to the back of the mouth. The two vomers have fused into a single bone. This, the ploughshare bone, is also single in mammals. There are twenty-eight pre-sacral vertebrae in the spinal column of which six or seven are cervical. (Among mammals the corresponding figures are twenty-seven and seven). The first vertebra of the neck, the atlas, resembles that of mammals. The conformation of the ribs suggests the presence of a diaphragm. There is a small bony outgrowth or process on the scapula. Between the os pubis and the ischium there is a large opening, the foramen obturatum, found in mammals, while the number of the phalanges (2, 3, 3, 3) is again mammalian.

Therapsid reptiles probably had facial muscles and therefore flexible skin. The brain comprises a large cerebellum with a median lobe, like that of mammals, and denotes an aptitude for rapid and coordinated movements. Such basic resemblances between these reptiles and mammals have given rise to some confusion, and one Triassic species from the deposits of Karoo which was first classified as a reptile has since been placed among the mammals. Coincidence can hardly account for the many and striking similarities; plainly they bear witness to the reptilian ancestry of mammals. Mammalian structures are built up with elements already present in reptiles. This is conclusively demonstrated by the middle ear of mammals in which the ossicles can be shown to be homologous with certain reptilian bones. The 'stirrup' is homologous with the reptile's columella auris, the 'anvil' with the quadrate bone, and the 'hammer' with the reptilian articulare.

Archaeopteryx, discovered in the schists of Solenhofen, is the classic example of a transitional form between reptiles and birds. In general structure, appearance and feathers it is a bird; but it reveals numerous reptilian characteristics, notably the arrangement of its cranial bones, the presence of a broad lachrymal fossa separated from the nostrils by a thin partition, teeth in its upper and lower jaw bones, absence of horny beak, long tail with twenty caudal vertebrae, and lack of the pygostyle which forms the rump of most birds. The wings are bird-like but the three fingers which support them still bear claws. Such

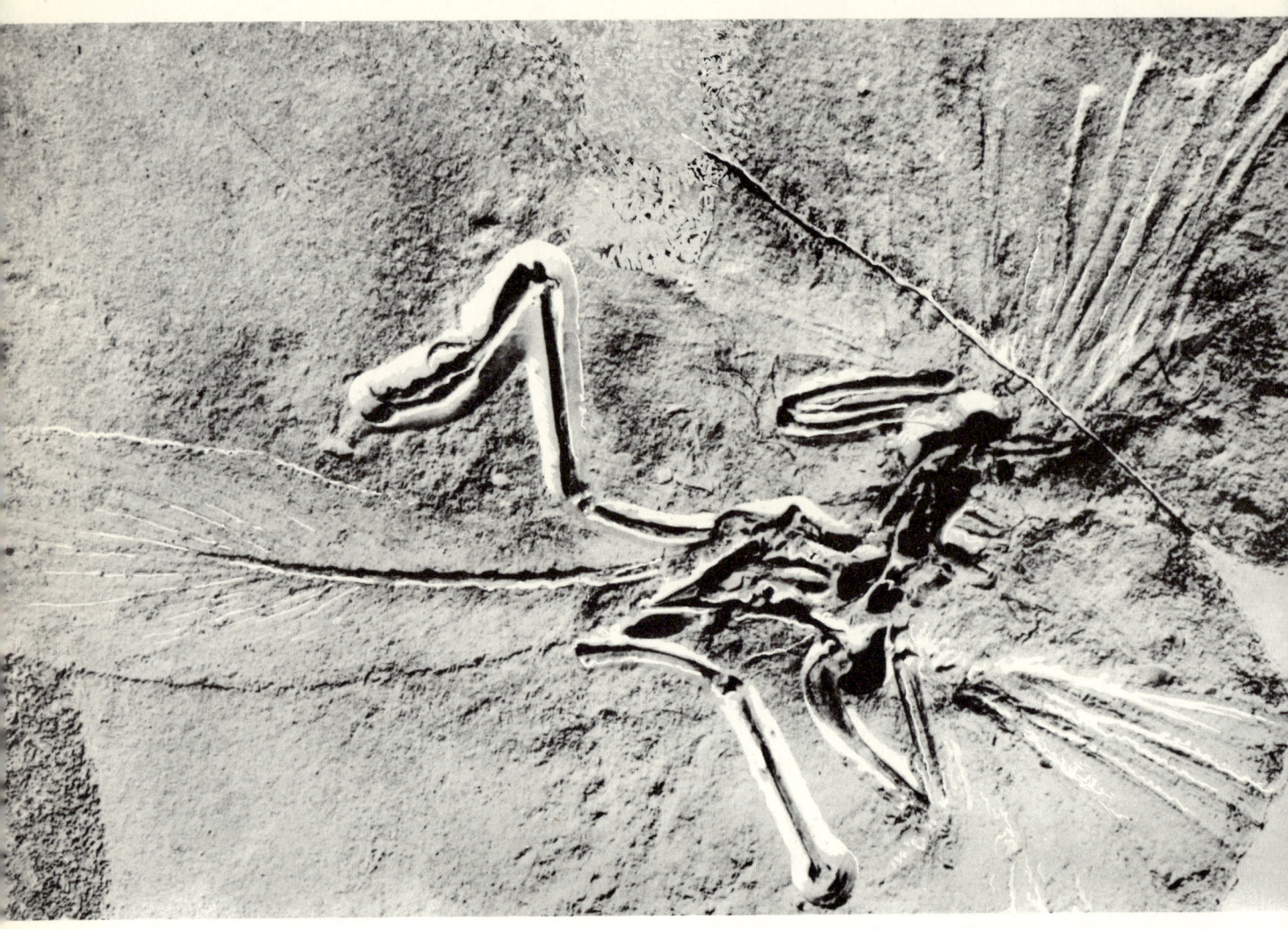

Archaeopteryx lithographica.

Sir Gavin de Beer.

characters came of necessity from reptilian ancestors, probably the pseudosuchians.

Archaeopteryx was a true bird, although its anatomical structure did not allow it to fly as successfully as modern birds. The flexibility of its backbone and the fragility of its pelvis were ill fitted for the jolts and strains of landing. The pelvis of contemporary birds is solidly constructed and distributes the shock of landing over the widest possible area of the body. The habits of *Archaeopteryx* may rather have resembled those of the bat, and the long tail, useless for flight, could have been employed in climbing.

Jaymoytius, Ichthyostega, the therapsid *Cynognathus, Archaeopteryx* — are they primordial species, archetypes from which all extinct and extant forms are derived? Probably not, but they stand out as rare and valuable links between types of animal organization which, on the examination of extant forms alone, would appear to have little connection.

3. *Continuous series.* In certain well-defined regions sedimentary deposits have, without change of climate or nature, been laid down continuously over considerable periods of time. Thus on the south coast of England the chalk cliffs of Margate — of the Turonian and Senonian ages — are formed of from 120—150 metres of sediment which must have been deposited in shallow and calm seas over a period of from 3—4 million years. At all levels the cliff formation contains the fossils of a sea-urchin of the genus *Micraster* which lived in shallow waters. The sea-urchin's larva does not inhabit the sea bottom — its life is free-swimming and pelagic — and the fossil animals are therefore not directly descended from each other. Nonetheless they furnish a well-defined record of the genus *Micraster* over an immense stretch of time. The fossils were gathered with great care by A. W. Rowe, a doctor from Margate. He collected 2,000 specimens and noted, inch by inch, the level at which each fossil was found. An overall picture of evolution was thus evident, involving the general form of the animal, the adornment of the shell, the aspect of the alimentary grooves, the displacement of the mouth towards the periphery. Species based on certain characters have been described, as well as numerous varieties linking well-defined types. In this way all the Margate fossils form a continuous series. They have evolved slowly and without interruption. Their general structure has not been modified; they are always species of the genus *Micraster,* larger or smaller, with more or fewer spines. As a continuous series they offer a vivid panorama of evolution.

Further examples of the same nature are provided

by certain European molluscs *(Paludina)*, and by the snails *(Planorbis)* of Steinheim. During the Pliocene, sediments some 100 metres in thickness were deposited in the great fresh-water lakes which covered the plain of Hungary. There lived *Paludina,* a gastropod of the class Prosobranchia, and fossil shells of the molluscs are found in the sediments. At the lowest level of the deposits the shells are smooth and differ very little from each other. In the higher layers the shells become increasingly ornate; first they acquire one carina or keel, then two; then on the two keels appear tubercles which later grow together. Each layer of sediment shelters several species of *Paludina.* All originated from a single species from which they later diverged into distinct lines and species.

Various lakes in the Chinese province of Yunnan contain Quaternary and contemporary *Paludina* which also exist in the smooth, keeled and ornate forms found in the same molluscs of the Pliocene Hungarian plain. *Paludina* seems to have a tendency to evolve towards nodulous and spiny forms in certain regions. Its evolution always follows the same course: the simple smooth forms are succeeded by the keeled forms and then by those with tubercles.

Planorbis of the Miocene lake of Steinheim in Württemberg furnish an analogous series. In the sedimentary floor are found snails with scalariform conical shells. The middle layers contain transitional types, while in the upper layers the normally coiled *Planorbis* is found. Such uninterrupted series of species in deep and continuous sediments are found among various groups: ammonites, scallops, sea-urchins and brachiopods. The millions of years required for the sediments to be laid down bears mute witness to the slowness of evolution's pace.

The embryological evidence

Many facts of ontogeny, or the development of the individual organism, can be explained only in evolutionary terms.

It has long been known that the embryonic stages of an individual's organs offer striking resemblances to the same organs in other groups. For example, the heart of the young human foetus is composed of a single curved tube divided into an auricle, a ventricle, and a bulb, a disposition which corresponds exactly to the heart of a fish. All vertebrates at one stage in their development have the notochord of *Amphioxus* and the Agnatha; without the notochord the development of the spinal column is impossible. In the region of the neck the embryos of reptiles, birds, and mammals all have gill-like slits like those of fish. Branchial arches are present in all vertebrate embryos. They are functional in the adult stage only in Agnatha and higher fishes; but from amphibians to mammals they take part in the formation of the respiratory apparatus (the Eustachian tube, etc.), the lymphoid organs (the tonsils, thymus, etc.), and the cervical region. The crinoids (sea-lilies, etc.) are normally sedentary and stalked; the adult *Comatula,* the only free-living form of the Atlantic coasts, has sedentary larvae which recall the adults of other crinoid species.

Such phenomena can be explained only if it is agreed that organisms which in the course of their development possess the same organs and pass through the same stages have a common origin. Haeckel in 1866 expressed the matter succinctly in a phrase which is often quoted: 'Ontogeny recapitulates phylogeny'; (phylogeny being the evolution of the race or group), by which he meant that the animal in the course of its development passes rapidly through all its ancestral stages. Haeckel believed that he had propounded a fundamental biogenetic law.

His law at first aroused enthusiasm, for it offered an explanation of the transitory organs of embryos. The foetus of the whale develops rudimentary teeth which do not pierce the gums and atrophy without serving any purpose. They bear witness, however, to the fact that the whale's ancestors had teeth, which the modern whale has replaced by flattened plates of whalebone. Nudibranchiate molluscs have no shells when they are adults; but the transitional shell they possess in the course of their development indicates that they are descended from shelled gastropods. The development of the Tunicata (sea-squirts, etc.) which were once considered to be molluscs, has shown their affinities to the vertebrates. At certain stages their larvae have a structure analogous to that of *Amphioxus*.

Today Haeckel's fundamental biogenetic law is less respected than the more exact laws which Karl von Baer enunciated in 1828. Revived by English embryologists these laws state:

1. In the course of embryonic development general characters appear earlier than specific characters. For instance, a dog, in the course of its development, is a vertebrate before becoming a mammal, and a mammal before becoming a carnivore.
2. Less generalized structures derive from more generalized structures, and so on until the formation of the most highly specialized characters.
3. The embryo of a given animal is always different from the embryos of other kinds of animal.
4. Fundamentally, the embryo of a higher animal never resembles the adult form of a lower species, but only its embryonic form. The gill-slits of embryonic reptiles, birds, and mammals do not at all resemble the gill-slits of an adult fish, but they are altogether analogous to those of an embryonic fish.

It is untrue to say that 'ontogeny recapitulates phylogeny', for the law applies to organs in particular and not to the entire organism. The human embryo possesses gill-slits, a notochord and a heart, each similar to the corresponding organ of an embryo fish; but at no time does the human embryo have the organization of an adult fish. Each organ follows its own development, slowly or rapidly, a phenomenon known as heterochronism or, literally, 'irregularity in time of occurrence'. It is, however, reasonably certain that the ancestors of mammals had gill-slits, a notochord, a tubular heart, etc.

Haeckel was, in fact, well aware of the differences between ontogeny and phylogeny, and distinguished between structures which do and those which do not have phylogenetic value. The former or palingenetic characters are ancestral characters, while the latter or cenogenetic characters appear afterwards and are

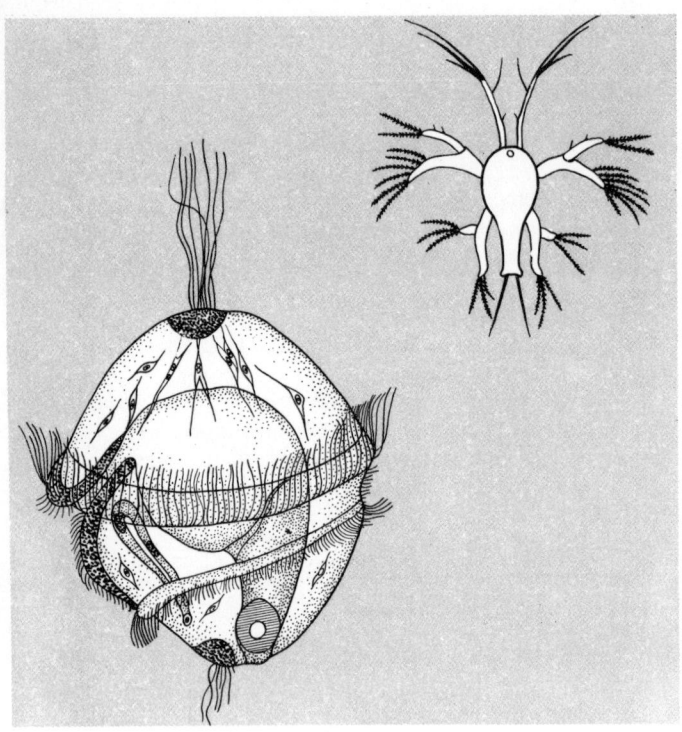

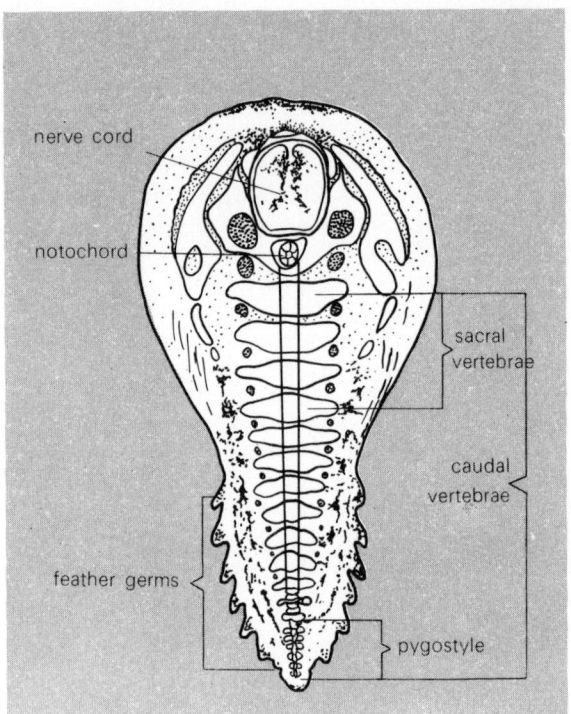

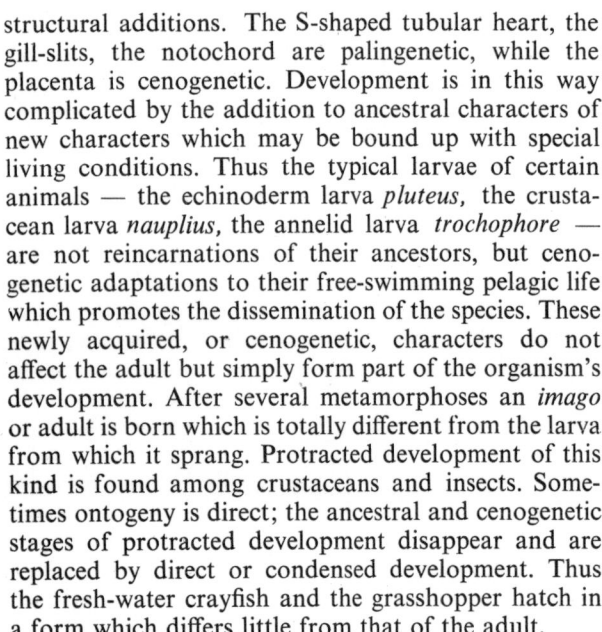

Left, trochophore, the larva of Polychaeta, an order of annelid worms. A ring of cilia divides the larva into two hemispheres, the episphere and the hyposphere. *Right*, nauplius larva.

After Woltereck.

Horizontal section of the tail region of an embryo bird *(Syrnium aluco)* showing its archeopterygian aspect.

structural additions. The S-shaped tubular heart, the gill-slits, the notochord are palingenetic, while the placenta is cenogenetic. Development is in this way complicated by the addition to ancestral characters of new characters which may be bound up with special living conditions. Thus the typical larvae of certain animals — the echinoderm larva *pluteus,* the crustacean larva *nauplius,* the annelid larva *trochophore* — are not reincarnations of their ancestors, but cenogenetic adaptations to their free-swimming pelagic life which promotes the dissemination of the species. These newly acquired, or cenogenetic, characters do not affect the adult but simply form part of the organism's development. After several metamorphoses an *imago* or adult is born which is totally different from the larva from which it sprang. Protracted development of this kind is found among crustaceans and insects. Sometimes ontogeny is direct; the ancestral and cenogenetic stages of protracted development disappear and are replaced by direct or condensed development. Thus the fresh-water crayfish and the grasshopper hatch in a form which differs little from that of the adult.

The persistence of ancestral stages allows us to define the position of special groups which are hard to classify. *Sacculina* and *Peltogaster,* the two chief genera of the Rhizocephala — a degenerate division of Cirrepedia, are parasites of the crustacean decapods, the first living on crabs and the second on hermit-crabs. Both are so profoundly modified by parasitism that they are no longer recognizable except in their larval stage. The discovery of their larval forms demonstrated not only that they belonged to the crustaceans but also showed their affinities with the Cirrepedia. Nevertheless *Sacculina* and *Peltogaster* in no way resemble the

typical Cirrepedia, which is the barnacle. Usually even the most degenerate parasites continue to produce larvae which are more or less identical to the larvae of closely related free species. Parasitical life, which so deeply alters the adults, has no effect on the larval phases. The gametes, the embryos, and the larvae seem to be more stable than the adult forms.

Resemblances during embryonic development also reveal the connection between classes which are in appearance very different. The first stages of development of segmented worms or annelids and of molluscs are so alike that their common origin is betrayed. *Amphioxus,* or the lancelet, is a primitive chordate which forms a link between invertebrates and vertebrates; it still retains invertebrate characters (skin composed of a single layer of cells, colourless blood, and a nephridium like the annelids) while its embryonic development is of vertebrate type. Indeed, certain arrangements of the *Amphioxus* embryo are found exclusively in vertebrate embryos. Such embryonic resemblances are not accidental, but reveal a common origin or, at least, bonds of relationship.

Evolution does not involve a modification of ancestral adult characters, but rather a rearrangement of the hereditary endowment which itself modifies the organism's development. For example, all vertebrates hold in common a fund of genes which determine the vertebrate structure. Their action is supplemented by the action of further genes which determine less generalized structures, and so it proceeds until specialized characters are formed, characters typical of the phylum, the class, the order, the genus. Genetic change may take place earlier or later in the organism's development, and until it occurs recapitulation is observed.

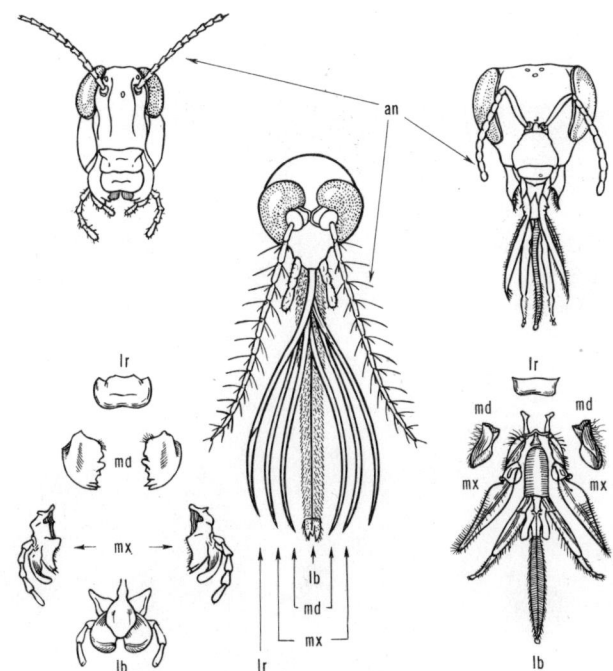

The buccal elements of three types of insect, showing homology.
Left, a grinder, the grasshopper; *in the centre*, a pricker, the
mosquito; *right*, a licking-grinder, the bee.
an, antennae; *lb*, labium or lower lip with feelers; *lr*, labrum
or upper lip; *md*, mandibles; *mx*, maxillae with feelers.

After Dobzhansky.

Ontogeny then recalls phylogeny. Thus *Comatula*, formerly sedentary like other, older crinoids, broke free from its stalk and became a swimmer. The bird's embryo seems to pass through the *Archaeopteryx* stage with a long tail and no pygostyle; the short tail of the contemporary bird is composed of four caudal vertebrae and a single bone, the pygostyle: which is hidden under the skin, the tail feathers forming a fan. In one phase of development, however, the embryo bird has four sacral vertebrae and some fifteen caudal vertebrae, most of which later fuse to form the pygostyle. The phase recapitulates an ancestral state.

The anatomical evidence

Supplementary evidence of value is also furnished by anatomy and in particular by comparative anatomy.

Embryology shows us that organs which differ in form and function can nevertheless have a common origin. Such organs are said to be *homologous*. In addition to their common origin these organs have a fundamental similarity of structure and position relative to other organs. This anatomical similarity sometimes allows us to observe homologies which would otherwise escape notice. The swim or air bladder of the fish and the lungs of the mammal are homologous organs and both result from the evagination, or turning inside out, of the anterior digestive tube. Among lungfish the air bladder folds, forms alveoli, and acquires the structure of a lung. The greatest complexity is found in *Protopterus*, the African lungfish, in which the very large air bladder has become

a true functional lung. The homology between the ossicles of the middle ear in mammals and certain reptilian bones has already been mentioned. There are seven vertebrae in the neck of the giraffe; all mammals, with rare exceptions, have seven cervical vertebrae whether their necks are long like the giraffe's or non-existent like the whale's. The buccal structure of the stinging insect and of the chewing insect is composed of the same elements. Crabs have a symmetrical abdomen provided with paired appendages. In the exterior aspect of its dorsal surface *Lithodes* resembles a crab, but its reduced abdomen, definitely asymmetrical, is provided with only the left ovigerous legs. Morphology shows that *Lithodes* is not a crab but a hermit-crab. Now hermit-crabs habitually hide their soft abdomen in a spiral shell, and abdominal asymmetry is connected with the presence of such a shell. The right side is pressed against the axis of the shell so that ovigerous legs can appear on the left side only. *Lithodes,* a hermit-crab without a shell, has retained this asymmetry, a legacy from its ancestors.

The cranium of mammals, though extremely varied in form, is always composed of the same bones. In the course of numerous evolutionary changes the same plan of organization and the same homologous organs are found again and again. The connection between bones and bones, or between cranial bones and sense organs, tells us when they are homologous. Among the aquatic crossopterygians and their terrestrial descendants, the stegocephalians, Westoll has been able to demonstrate the homology of the cranial bones by pointing out that the pineal orifice always opens between the parietal bones. Topographically the parietal bones of the crossopterygians occupy the place of the frontal bones of the stegocephalians. The large-sized bones once thought to be parietal are in reality post-parietal. The relative proportions of the two animals' craniums are different; the anterior region of the crossopterygian's cranium is reduced and the posterior region is elongated, while the opposite is true of the stegocephalians.

Rudimentary organs are under-developed organs which serve no purpose. In highly organized animals they are frequently found. They are homologous with organs which are normally developed in more primitive groups. Such 'anatomical relics' are not uncommon in man: the half-moon fold of skin of the eye is the remains of a nictitating membrane or reptilian third eyelid which is still functional in some mammals. The vermicular appendix is a vestige of the distal portion of the caecum. The atrophied muscles of the ear are the souvenir of a movable ear, while the non-functional musculature of the coccyx is evidence of muscles which once moved a tail.

Transitional organs exist only in embryos or in very young individuals. The embryonic slow-worm has rudimentary limbs whereas the adult is legless. The embryos of various birds have jaws with tooth-buds which rapidly disappear.

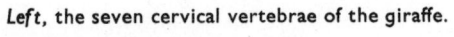

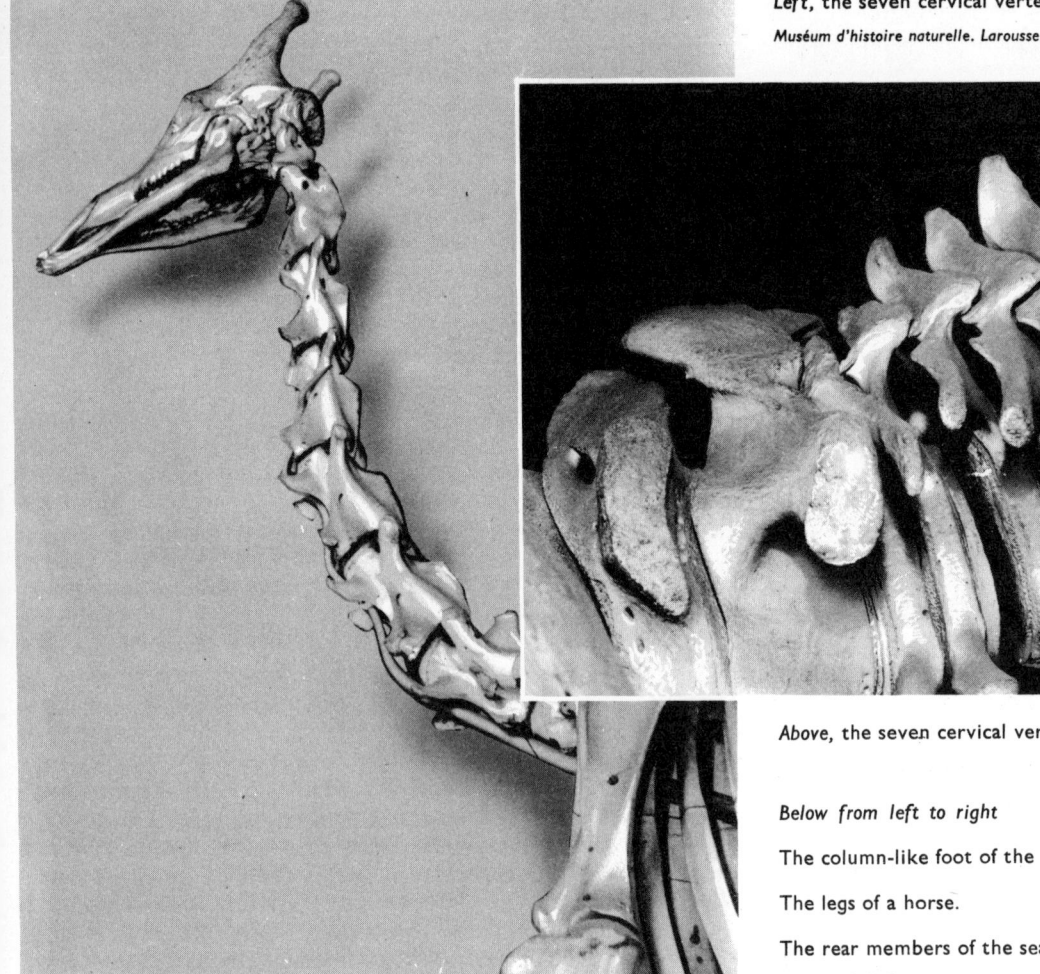

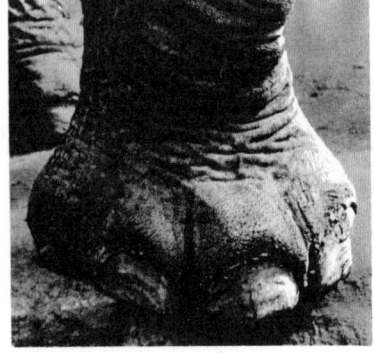
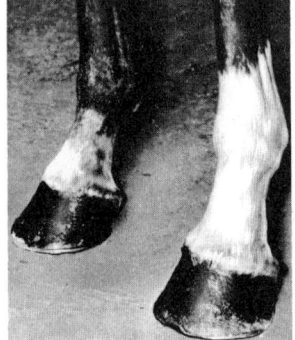

The chemical evidence

Biochemistry, like comparative anatomy, bears witness to the unity of living creatures. Closely related forms often have chemical substances in common. All arthropods contain chitin. Haemoglobin is found in all vertebrates. The Compositae contain inulin, while the Cruciferae and related families contain various compounds with a sulphur base. But unrelated forms, too, can possess the same chemical substance: inulin also exists in the Campanulaceae, while rubber is found in the Euphorbiaceae, the Urticaceae, the Apocynaceae, and the Compositae. Haemocyanins or haemerythrins are present in worms,

molluscs, and crustaceans. The haemoglobin of species belonging to large genera form crystals of the same type. The genus *Canis,* which includes dog, wolf and jackal, the genus *Vulpes* (fox), and the genus *Felis* produce isomorphous haemoglobin crystals.

The antigens present in the blood of higher mammals, determining the blood-groups to which individuals belong, show how close the relationship is between Man and the anthropoid apes. The same fact is confirmed by reactions to agglutination. When an animal receives a blood transfusion with the blood of another species its serum acquires the property of clumping or agglutinating the foreign red blood corpuscles and also the red blood corpuscles of related species. Thus the rabbit, which receives human red blood-cells, elaborates an anti-human serum which clumps the cells of men and anthropoids alike. But it has no effect on the red blood-cells of the lower primates. An anti-bovine or an anti-ovine serum has a maximum clumping effect on the blood of cows, sheep and goats, which are all ruminants, but much less on the blood of pigs, horses, dogs, or Man.

By the technique of agglutination, genuine affinities can be revealed. In this way Nuttall has shown that the king-crab *(Limulus)* which in form is very unlike the arachnids (which include spiders, scorpions, mites, etc.) belongs in fact to the class Arachnida. This confirms the findings of palaeontology and of anatomy, which classify *Limulus* among the arachnids. By the same method of agglutination, forms which are very similar can be separated into species, an example being the two lobsters, *Homarus vulgaris* and *Homarus americanus.*

The evidence from the various sources we have briefly examined establishes the fact of evolution without question. Palaeontology reveals that living creatures have appeared in a given and hierarchical order during the course of geological time. The arguments of palaeontology are reinforced by the findings of embryology, anatomy, and biochemistry. In animal and vegetable kingdom alike development and structure evolve in a connected manner. The skeleton and the head of the true vertebrate can only be explained as arising from the head of a jawless or agnathous vertebrate. The birth of a mammal requires the previous existence of reptiles whose structure contains the basic elements of mammalian structure. Comprehension of the organic world is, indeed, impossible without the concept of evolution.

Having seen how the fact of evolution is proved we must now turn to the manner in which it operates.

The laws of evolution

Palaeontology shows us the unwinding reel of evolution. As the ages pass before our eyes the simplest living things first appear and are followed by organisms of growing complexity. The primitive Thallophyta precede plants with roots, stem, and leaves; plants without flowers are followed by plants with flowers. The invertebrates arrive before the vertebrates, and the most primitive invertebrates precede those of more advanced organization. Mammals, still more complex, are the last to arrive. No exception is found to this first law, the law of increasing complexity. It is verified with each fresh discovery of palaeontology; in the hierarchy of life and in the scale of time each new fossil falls into the position which evolutionary theory predicts.

The evolutionary history of each group forms a pattern which recalls the development of the individual. The group is born in a more or less explosive fashion, spreads and diversifies; then it diminishes and finally becomes extinct — except in the case of certain species of limited dispersal and restricted membership which have survived to the present day and form what may be called relics or 'living fossils'. This phenomenon is repeated in every group and comprises a preparatory phase, a phase of evolutionary climax, and a phase of senescence which ends in death.

The preparatory phase

The phase of preparation is generally long, and is the phase of which least is known. During the Carboniferous and Permian periods the reptilian structure was elaborated; the entire Mesozoic era was required for the appearance of mammalian organization. The great groups of mammals which are reasonably well known (carnivores, ungulates, and primates) all began as modest, undemanding species, known as generalized, synthetic, or composite species. This phenomenon is so general that it may be considered to constitute a second law, Cope's law.

The common ancestor of the mammals must have been a small terrestrial animal, an insectivore, walking on the sole of a five-toed foot, and able to survive in many environments. From such non-specialized organs specialized structures could be derived, structures adapted to special tasks. An ordinary foot with five toes could give rise to the prehensile hand of the arboreal ape, the cat's paw with its retractile claws, the mighty foot of the elephant, the mining tool of the mole, the foot of a jumping rodent like the jerboa, the bat's wing, the two-toed or one-toed hoof of running animals and the flipper of the whale.

The phase of evolutionary climax

The preparatory phase is succeeded by the critical phase of evolutionary climax during which life branches out into a diversity of forms and variations on a single basic type. Evolution is now essentially occupied with differentiation and diversification. The phase marks the zenith of groups rich in individuals and types of structure. It is a conquering phase and tends to invade all environments. Life appeared in the waters, but soon the conquest of the land was undertaken by various groups with varying degrees of success. Even groups which are essentially aquatic have their terrestrial representatives: unicellular organisms which inhabit the soil, terrestrial ribbon-worms and round-worms, oligochaetes, terrestrial leeches, pulmonate molluscs such as snails, isopods such as woodlice, land crabs, lungfish and some of the Teleostomi group of fish. Next, the major groups become land-dwellers, arthropods, and four-legged vertebrates. The Earth becomes the normal habitat except for some families which remain aquatic like the crustaceans and the turtles, or else return to the sea like the whales, the Sirenia (sea-cows, etc.) and the Pinnipedia (seals, walruses). Then the terrestrial forms invade the air. The higher insects, Hymenoptera, Lepidoptera, Diptera, the last to appear, are the best fliers. The air has been conquered on several occasions by the vertebrates: first by the reptilian pterodactyls, then by the birds and the bats. Other mammals (some marsupials, some rodents and flying lemurs) glide through the air by means of membranes stretched like a parachute between forelegs and hind legs.

Increased specialization, however, limits future possibilities. Specialized species can no longer abandon

The Dodo, or *Didus ineptus*, of Mauritius.

American Museum of Natural History.

In Mauritius the Dutch pigeon was still common in 1770. Much prized for its flesh it was hunted down and remorselessly killed off, until in 1826 the last specimen was slain. Only three examples of this handsome bird still exist, all stuffed.

Library of the Muséum d'histoire naturelle. Larousse.

environments to which they have become too perfectly adapted. They are slaves to their feeding habits and have reached a point of no return. When existing forms spread to fresh territory an organic innovation may aid them to occupy the new environment. Thus the paired fins of certain crossopterygian fishes were transformed into the legs of the stegocephalians, the ancestors of all four-legged land animals. The passage from fin to leg constitutes an organic revolution which, like the acquisition of air-breathing lungs and an appropriate circulatory system, was indispensable for the occupation of a terrestrial environment. The reptiles, laying their large eggs on land, were finally emancipated from an aquatic existence and, with their dry, scaly skin, were fitted to take over the warm regions of the Earth.

The phase of senescence

The phase of reptilian expansion was of limited duration. It was followed by the phase of senescence, the beginning of which was announced by various signs. The number of genera and species diminished as though 'vital potentiality' was on the wane. Individual reptiles grew to immense size. Now gigantism is a serious handicap when the surrounding environment changes or when a newly arrived occupant offers competition. Large animals have the disadvantages of requiring large amounts of food, and they are slow to grow and reproduce. Thus smaller creatures, less demanding, were able to eliminate them. In the vegetable kingdom similar phenomena are found: the giant tree forms of the horsetails and the fern-like *Selaginella* gave way to herbaceous members of the same groups.

The extinction of groups

The ultimate causes of the extinction of a group remain unknown. Changes of environment and food restrictions tend, however, to lessen fertility, while epidemics and the depredations of carnivores play parts which are probably secondary.

In any case species and individuals become fewer in number and their area of distribution diminishes. Finally the group becomes extinct and vanishes. A certain number of species are known to have disappeared within historical memory. By indiscriminate hunting, by forest and swamp clearance, by the introduction of new animals and so forth, man himself has hastened the natural extinction of too many species, in particular of certain island birds. The dodo (*Didus ineptus*), the foolish and flightless bird, was exterminated by the settlers of the island of Mauritius between 1679 and 1693. On the same island a similar fate overtook eleven other species of birds as well as the giant tortoises. *Pezophaps,* which was larger than a swan, disappeared from Rodriguez Island at about the same time. These two birds laid no more than a single egg and that perhaps only once a year. Such restrained fertility, though sufficient in the absence of enemies, explains the rapidity of their extinction. The gigantic *Dinornis* of New Zealand became extinct towards 1770, *Aepyornis* of Madagascar towards 1650 and the great penguin of the North Atlantic isles towards 1844. The passenger pigeon *Ectopistes canadensis,* once common in North America, began to die off towards 1860; the last wild flocks were observed in the last years of the century. The Labrador duck *Camptorhyncus labradorius,* once common on the Atlantic

382

The crossopterygian fish, *Eusthenopteron*.

After E. Jarvik.

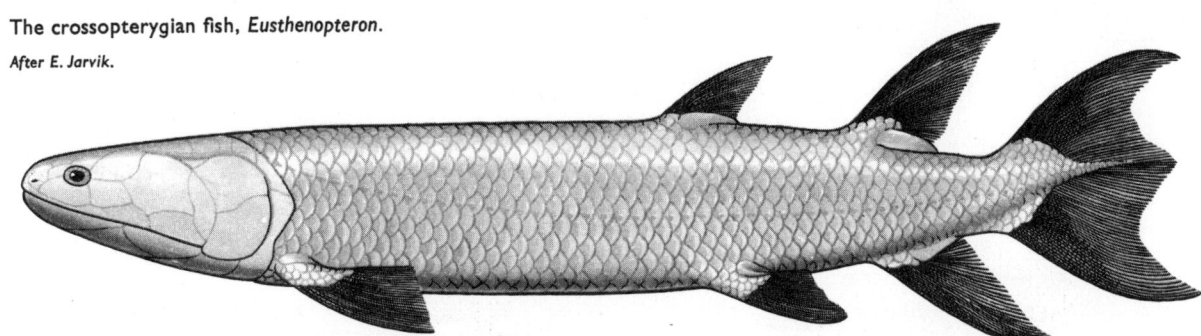

coast of Labrador, has been extinct since 1875. The last aurochs died in a Polish forest in 1927. The European bison and the American bison have passed through a critical period, but it has been possible to build up new herds by rearing them in protected reserves and parks. The saiga antelope is almost extinct. The chinchilla of the Andes only survives on farms maintained for that purpose. Two large ungulates, the blue antelope *(Hippotragus leucophaeus)* and the quagga *(Equus quagga)* have disappeared, the former towards 1799, the latter in 1858. Aquatic mammals like fur-bearing seals, the dugong, the manatee and whales are threatened with extinction. Steller's seacow *(Rhytina)* which inhabited the islands of the Bering Strait disappeared in 1768, twenty-seven years after its discovery. The South American toothless

mammal, the armadillo-like *Glyptodon,* and the giant lemurs of Madagascar have also vanished.

Relics or living fossils

Some groups have not yet totally disappeared and, represented by rare specimens of archaic aspect, persist to this day. *Sphenodon punctatus,* a reptile from New Zealand, is the last survivor of the Rhynchocephalia which abounded in the Triassic and Jurassic of Europe and Africa. It is 75 centimetres in length and resembles a long lizard. It has taken refuge in an island north-east of North Island where it inhabits the nesting place of the petrel, eating part of the food which the bird brings to its young. *Sphenodon,* or the tuatara, is now hunted and eaten by a large falcon which has

Quaggas *(Equus quagga)* were a kind of zebra whose heads and necks alone were striped. Great numbers of them lived in South Africa where the Boers hunted them to extinction. The last specimen died in captivity in Amsterdam in 1883.

Library of the Muséum d'histoire naturelle. Larousse.

383

Skeleton of the *Glyptodon*.

Muséum de Paris. Larousse.

become naturalized on the island.

When, not many years ago, a coelacanth *(Latimeria chalumnae)* was captured the event was greeted as the twentieth century's most sensational discovery in the field of natural history. The coelacanth is a representative of that great group of fish, the Crossopterygii, which flourished in the Devonian and was thought to have disappeared at the end of the Cretaceous, in other words 60 to 70 million years ago. Today it seems to be localized in the deep waters bordering the Comoro Islands in the Indian Ocean. The surviving species scarcely differs from its Devonian ancestors which go back for more than 350 million years. It has retained archaic structures which throw light on the processes of evolution. Among the interesting anatomical features of this living fossil we find a lung which is evidence of air breathing among primitive fishes, a heart which is much simpler than that of other adult vertebrates, a brain which has developed very slightly and a skeletal axis composed of a fibrous and homogeneous notochord. These large fish, measuring from 1.3—1.8 metres in length, have seven fins: a pair of pectoral, a pair of pelvic, and three fins which are unpaired, namely an anterior dorsal, a posterior dorsal, and an anal fin. Only the anterior dorsal has the structure of normal fins. The others, paddle-like, are carried on scaly lobes which are thought to adumbrate the limbs of air-breathing vertebrates.

Another interesting living fossil is the mollusc *Neopilina galathaea* which is the only surviving member of a group which was previously represented only by Cambrian and Silurian fossils.

Such living fossils demonstrate the stability of living matter, once the critical period of evolutionary climax has ended. Their remarkable longevity is, however, only a special case of protracted senescence.

The abyssal mollusc *Neopilina galathaea*. Dorsal view of the 'shellfish' and ventral view of the anterior part of the body, showing the mouth in the middle of the contracted disc, surrounded by the rhinophores.

H. Lemche, Zoological Museum of Copenhagen.

Caninia gigantea. Tetracorallum of the Carboniferous of Tournay.

Muséum de Paris. Larousse.

Sphenodon punctatus, a New Zealand reptile, the last
representative of the Rhynchocephalia.

Mertens.

The coelacanth, *Latimeria chalumnae*.

Detail of the jaws. *(right)*.

J. M. Baufle.

385

Megatherium, fossil skeleton preserved in the Muséum d'histoire naturelle.

Rhamphorhynchus gemmerigi and *Pterodactylus*.

Bernissart's iguanodon, *Iguanodon bernissartensis*, a dinosaurian herbivorous reptile of the Wealdian Cretaceous. Height: 5 metres. Length: 10 metres.

Triceratops calicornis, Upper Cretaceous of Colorado.

Larousse.

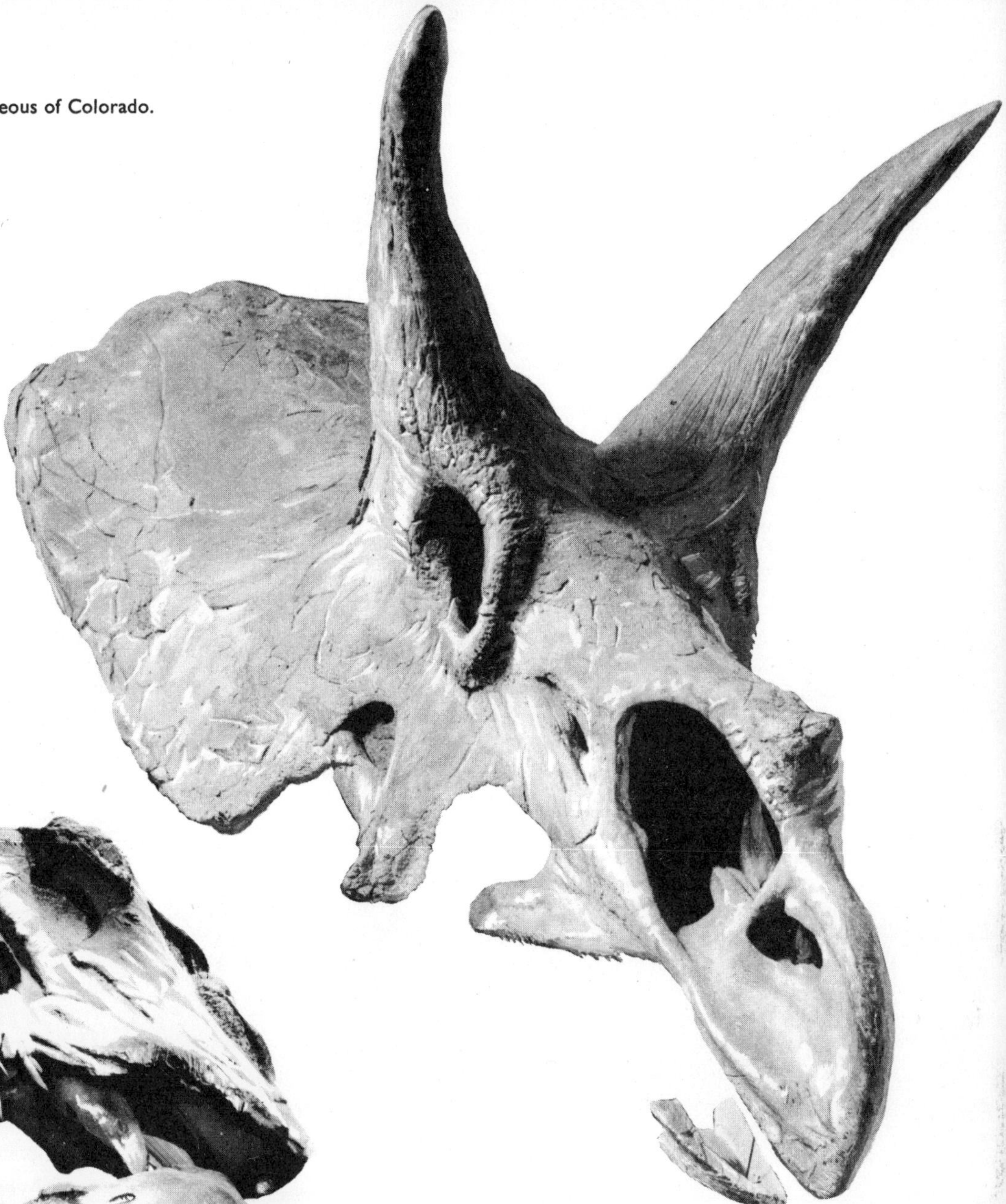

The skull and impressive denture of a *Tyrannosaurus* which measured 15 metres in length.

Muséum d'histoire naturelle. Larousse.

Skeleton of *Ichthyosaurus*.

Muséum d'histoire naturelle.

The law of succession

The evolutionary cycles which we have examined, the birth, zenith and death which is the lot of all great groups of living organisms, implies that earlier groups are succeeded by or, as in a relay race, relieved by newcomers. After its period of success and expansion the group becomes static and decay sets in. It is then replaced by a fresh group which, in its turn, becomes dominant, degenerates, and is replaced. This rise and fall of reigning dynasties is characteristic of evolution and might be called the law of succession. Examples abound.

The Tetracoralla and the Tabulata which in the Palaeozoic era constructed coral reefs were replaced in the Triassic by the Hexacoralla which still live. With the exception of the Cidaridae, Palaeozoic sea-urchins are extinct, being replaced in the Triassic by our present-day forms. Many groups of fishes have been replaced. The Creodonta, carnivorous mammals found in Eocene and Oligocene formations in Europe and North America, included species which were more powerful than lions and tigers; one, from Mongolia, had a skull 1 metre long. The Creodonta succumbed to their descendants who totally eliminated them. The flying reptiles of the Mesozoic era were succeeded by the birds.

Among terrestrial vertebrates the same process is clearly evident. Until late Devonian times vertebrates remained essentially aquatic, inhabiting the seas, fresh-water streams and lakes. Towards the end of the Devonian an important modification appeared: the transformation of fins into limbs among the first tetrapods, the stegocephalians, who still led a more or less amphibious life, crawling out at times from their lakes and pools. From the Permian period the adults became terrestrial; only the larvae remained aquatic. Their expansion took place in the Permian and Triassic; then they waned, leaving after them only certain small forms which have survived until today.

In the Permian-Triassic the first reptiles appeared; during the entire Mesozoic era — the age of the reptiles — they multiplied, diversified, and took possession of all environments, land, sea, and air. At first the group consisted of small quickly growing reptiles, but giant forms soon arrived on the scene. The increase in size

Catarrhinian anthropoids.
From left to right, the Asiatic gibbon, *Hylobates lar;* the Indonesian orang-utan, *Pongo pygmaeus;* the African chimpanzee, *Pan troglodytes;* the African gorilla, *Gorilla gorilla.*

was probably caused by overactivity of the pituitary body. When skulls of the great reptiles are examined it is found that the sella turcica — a depression of the sphenoid bone which contains the hypophysis cerebri or anterior pituitary lobe — is particularly large. Among the Ceratopsia the rate of growth has been determined: at the beginning of the upper Cretaceous *Protoceratops* of Mongolia was about 2 metres long; *Menoclonius* from the middle of the upper Cretaceous attained a length of nearly 6 metres, while the final forms, like *Triceratops,* exceeded 7 metres in length. Increase in dimensions was much more rapid in the beginning; gigantism was quickly attained, and size then became more stable.

The largest and strangest of the giant reptiles, the dinosaurs (Greek *deinos,* terrible, and *sauros,* lizard) were land-dwellers. *Iguanodon* was a vegetarian biped, 4—5 metres high and some 10 metres long, somewhat resembling a kangaroo. *Stegosaurus* was 7 metres long and its back was armed with a double row of triangular bony plates while in the region of its tail there were four pairs of great sharp spikes. *Triceratops* (Greek *treis,* three, and *keras,* horn) had the aspect of a rhinoceros with a skull extending backwards in a broad bony plate to protect the neck. Two solid horns rose above the eye sockets, and an axe-like protuberance surmounted the nose. *Triceratops* was a dangerous assailant; powerful muscles moved its head and the horns were redoubtable weapons.

Diplodocus, Brontosaurus, Brachiosaurus and *Titanosaurus,* were herbivorous quadrupeds which, with their long necks and tails, attained fantastic lengths. Their limbs seem to have terminated in huge rounded pads with claws. *Brachiosaurus,* giant among giants, had an estimated weight of some 30 tons. It must have eaten at least 600 pounds of food a day — in other words it must have eaten almost continually. In much the same way the Asian elephant can work for a few hours only because of the time it must spend in feeding itself.

In the seas lived the swimming reptiles: *Ichthyosaurus, Plesiosaurus, Mosasaurus.* Ichthyosaurs (or 'fish-lizards') were from 1—10 metres long; their huge heads contained from 100—200 conical teeth. Like fish they had four pairs of fins constituted, however, like the anterior fins of whales. They lived on fish, ammonites, belemnites, and crustaceans. Fossils have been found with foetuses which prove that they were ovoviviparous. The plesiosaurs, with their broad, flat and inflexible trunks, paddled themselves along somewhat like turtles. The mosasaurs, abundant in the

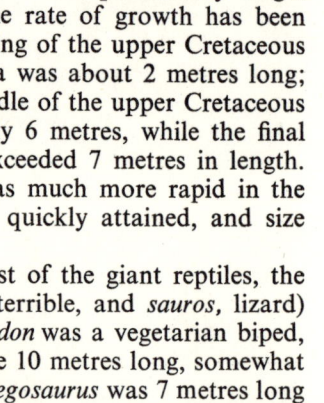

Diplodocus.

Muséum d'histoire naturelle. Larousse.

Skeleton of *Dimetrodon*. This enormous carnivorous 'lizard' was 3 metres long and lived 240 million years ago in North America.

Paris-Match. Hubert de Segonzac.

Cretaceous, reached lengths of 10—15 metres.

The air was ruled by the pterosaurs (Greek *pteron,* wing, and *sauros,* lizard) or flying reptiles, *Rhamphorhynchus,* the pterodactyls, and *Pteranodon. Rhamphorhynchus,* a creature about 1 metre in length, had a long tail ending in a small 'rudder'. Its jaws were armed with fine, pointed teeth. In one specimen indications of sebaceous glands and a hairy covering are thought to be visible. The much smaller pterodactyl was about the size of a sparrow. Its tail was short and its head, in comparison with the rest of its body, large. *Pteranodon* of America grew to great size, having a wing-span of as much as 9 metres, which no bird has ever approached. Its long beak was toothless and a huge crest rose from the back of its head.

Mammals were present from the beginning of the Jurassic, but throughout the age of the reptiles they remained small in size and relatively undifferentiated. The disappearance of the great continental dinosaurs left immense empty spaces on the earth's surface. Birds attempted to occupy these free areas by adopting a terrestrial manner of life. *Gastornis,* for example, recalls the ostrich in its gait and appearance. The attempt, however, rapidly failed and the mammals established their supremacy. By the end of the Cretaceous the marsupials had appeared, as well as archaic placental mammals which became extinct without leaving direct descendants. Finally, in the Tertiary or Cainozoic, placental mammals which are ancestors of existing types began to spread. From the Oligocene we can follow the line of tree-dwelling insectivores

Hippurites, a bivalve mollusc of the order Rudista, showing the valves separated.

Muséum d'histoire naturelle. Larousse.

from which the primate group is descended. They diverge from common forms like the mole and the shrew, and their anatomy begins to assume the organization of the lemurs. Today they are represented in Asia by the *Tupaia* or tree shrew and in Africa by the *Macroscelides* or elephant shrews. Monkeys appear in the Eocene and by the end of the Tertiary the primates were flourishing. The Catarrhines, or Old World monkeys, replace the primitive primates. The anthropoid apes attain their zenith in the Miocene and the Pliocene. From the Quaternary they regress, and their present day representatives — gibbon, orang-utan, chimpanzee, and gorilla — are the last survivors of a group which prospered during the second half of the Tertiary.

Man appeared in the Quaternary and did not, as we shall see in the chapter devoted to human evolution, escape from the law of succession.

As in the animal kingdom so in the vegetable kingdom the same law was enforced. At the end of the Palaeozoic era the horsetails, the tree-like club mosses and ferns give way to the conifers. In the lower Cretaceous the *Cycadales* and the *Ginkgoales* disappeared, leaving only a few 'living fossils' like the maidenhair tree, while the flowering angiosperms multiplied and spread.

Increase in size

In many groups the evolution of a species systematically entails a progressive increase in size. Such growth often leads to gigantism and to an exaggerated development of external organs like horns, tusks, and armour which thus become hypertelic (not explainable on grounds of utility). The phenomenon is clearly observed among vertebrates. We have already seen that the giant reptiles of the Mesozoic era were heavily armed monstrosities sometimes 20 metres in length.

Small sharks of the genus *Carcharodon* which appeared in the Eocene vanished in the Pliocene with species which had reached a length of 20 metres. The stegocephalians of the Triassic included large forms, *Mastodonsaurus* having a skull 1 metre long. Three lines of Proboscoidea ended in giant forms: the mastodons of the Miocene and Pliocene, *Dinotherium gigantissimum* of the upper Miocene, and the great elephants *Antiquus* and *Primigenius* of the Quaternary. The titanotheres (American perissodactyls or odd-toed ungulates) began in the Eocene as small creatures and ended in the lower Oligocene as the enormous *Titanotherium*. Rodents, usually small in size, have one representative the size of a rhinoceros, namely *Megamys* of the Argentinian Pliocene and Quaternary.

Among the invertebrates the tendency to grow larger is less frequently encountered. Nonetheless it exists in sea-urchins, the bivalve molluscs Rudistae, and the ammonites. The last ammonites of the genus *Pachydiscus* of the Cretaceous measured as much as 2 metres in diameter. From the Permian to the present day kingcrabs have progressively enlarged, the carapace having grown from 1 to some 30 centimetres. Even the protozoans have giant representatives such as the nummulites (7—8 centimetres in diameter) and the foraminifers, among which the first orbitolites measured a few millimetres in diameter and the last some 3—6 centimetres. Gigantism, as a rule, affects the last members of a race and marks the beginning of senescence and the approach of extinction.

Orthogenesis

The progressive stages by which one form of organization is replaced by another of increased complexity suggest persistent evolutionary trends. Palaeontology shows that evolution steadily pursues a given course, as though to achieve a more and more specialized type

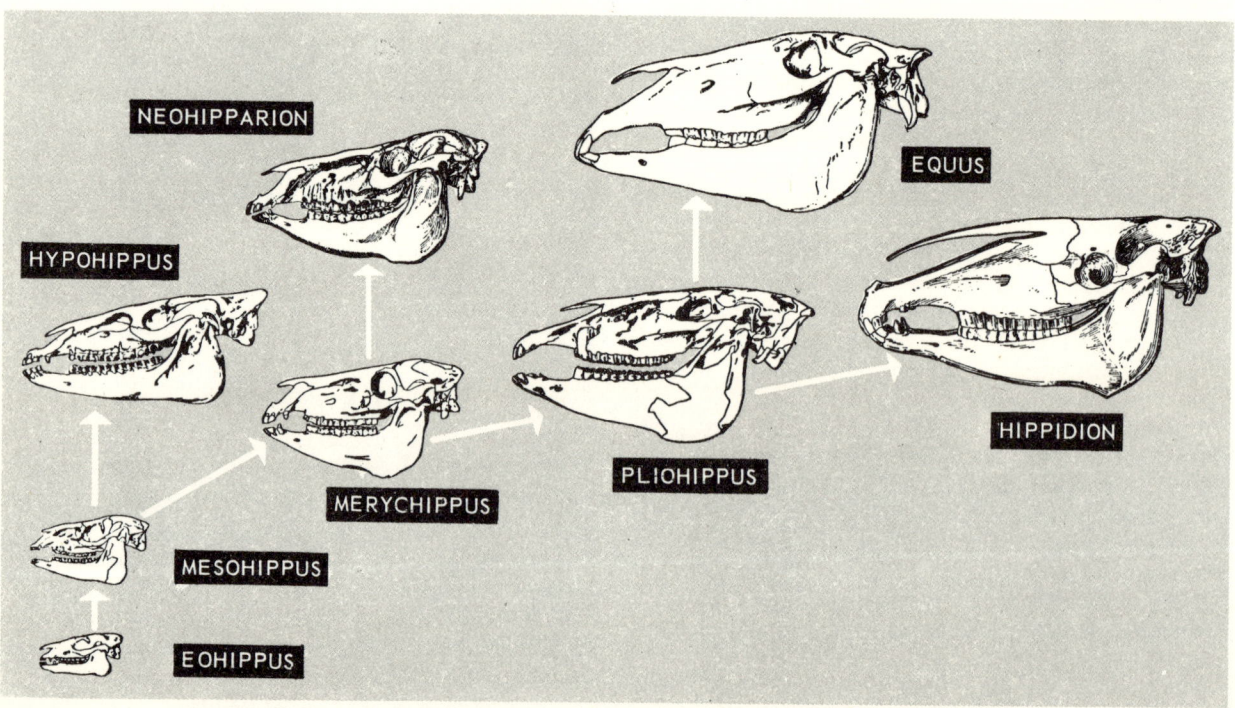

The evolution of the Equidae (after Arambourg)

		AMERICA	ASIA	EUROPE
Quaternary:	Single-toed.	Equus. Hippidius	Equus	Equus
Upper Pliocene:	Single-toed. Hypsodont teeth.			
Lower Pliocene:	Size of a pony. Hypsodont teeth. Rudimentary lateral toes.	Pliohippus	Hipparion	Hipparion
Upper Miocene:	Hypsodont teeth with abundant cement. Reduced lateral toes. Ulna fused with radius.	Protohippus	Hipparion	Hipparion
Middle Miocene:	Moderately hypsodont teeth with abundant cement. Lateral toes non-functional. Ulna and fibula regressing.	Merychippus	Anchitherium	
Lower Miocene:	Molars with traces of cement and cusps or prominences. Two lateral toes non-functional.	Parahippus		Anchitherium
Upper Oligocene:	Size of a tapir. Brachyodont molars with two crescent-shaped cusps. Three toes, the central toe predominating. Developed ulna and fibula.	Miohippus		
Middle and Lower Oligocene:	Size of a large dog. Premolars similar to molars with six cusps set in oblique crests. Three functional toes, with vestige of the fifth. Bones of the forelimb (ulna and radius) separate.	Mesohippus		
Upper Eocene:	The two last premolars molarized. Forelegs with four toes, the third prominent and the little or 'fifth' reduced.	Epihippus		
Middle Eocene:	Cusped or tuberculated brachyodont teeth. Last premolar molarized. Four-toed front feet.	Orohippus		
Lower Eocene:	Size of a fox. Tuberculated brachyodont teeth, the premolars differing from the molars. Front feet four-toed and hind feet three-toed. Ulna and fibula well developed.	Eohippus		Hyracotherium

Opposite, the evolution of the horse's skull, drawn to scale.

After G. G. Simpson.

of organization. In the process certain characteristics become progressively more complicated while others are progressively simplified. Such a persistent trend in a given direction is known as orthogenesis (from the Greek *orthos,* straight, and *genesis,* generation). Examples of the phenomenon are numerous, and indeed the history of every group may be studied in the light of the concept.

The ancestors of the horse

The classic example quoted by those who defend orthogenesis is that of the Equidae whose anatomy includes many characteristic traits. From the Eocene to the Pliocene the equine form progressively evolved. Two lines have been distinguished by palaeontologists, one developing in America, and the other in Eurasia. The American line, whose stages are well marked by fossils, has been thoroughly studied. As the diagram and table show, the evolution of the horse involves its size, skull structure, shape and dimensions of the teeth, size and shape of legs, and number of toes. It begins with a creature the size of a fox and ends with the contemporary horse. The face grows longer in comparison with the brain case so that the eye-socket recedes towards the occiput. The teeth, too, grow longer: teeth with short crowns and no cement, that is, brachyodont teeth which cease to grow after the roots are formed, give way to hypsodont teeth with high crowns and cement, teeth with complicated folds of enamel which continue to grow until old age. The legs lengthen: the ulna and the fibula, at first normally developed, regress. The toes, originally five in number, are reduced to the single toe of the modern horse.

From *Eohippus* (the 'dawn horse') to *Equus* the race evolves towards a running animal, an objective which is achieved by shortening the femur, slightly lengthening the tibia, considerably lengthening the metatarsus, and above all by the fusion of the toes into a single unit. Man has attempted to continue these evolutionary trends by demanding of the racing thoroughbred more and more speed at an ever earlier age.

The titanotheres

The titanotheres were perissodactyls — odd-toed ungulates — belonging to the same order as the horse, the tapir, and the rhinoceros. Their evolution took place during the Eocene and the Oligocene in North America and Eurasia, their fossil remains being especially abundant in the western part of the United States where they have been the object of careful study. In size they rapidly increased: the first Eocene forms were, like *Eohippus,* the size of a fox, while Oligocene titanotheres like *Brontotherium* and *Brontops* resembled an enormous rhinoceros some 2½ metres high. In the earliest forms the head is slender and tapering, the face being twice as long as the cranium. Then, little by little, the face diminishes while the skull grows bigger. *Brontotherium* has a short muzzle bearing two great branching horns. The two horns have fused into a single mass in the gigantic *Embolotherium* of Mongolia. The cranial profile, almost flat in the small early forms, becomes more and more concave. Modifications of the limbs are of minor importance except for the lengthening of the thigh-bone or femur and the reduction of the foot. The number of toes remains unaltered, four in front and three behind. The molars undergo little change. The race of the titanotheres evolved towards a ponderously walking type which was achieved by lengthening the femur to nearly half the total length of the leg, shortening the tibia to somewhat more than a quarter, and above all the foot to somewhat less than a quarter of the total length of the leg.

The camels

Of the even-toed ungulates the sub-order Tylopoda is today represented by the ruminant camels and llamas. Camels and dromedaries live in Asia and Africa, while the llamas inhabit South America. But North America was the original home of both animals, and their ancestral fossil remains form an interesting series. In the North American upper Eocene we find the species *Protylopus* which was no bigger than a large hare. On the forefeet the second and fifth toes are smaller than the third and fourth toes which, in the hind feet, have atrophied. *Poebrotherium* in the Oligocene is the size of a gazelle and already camel-like. The front of the muzzle has sharpened, and both front and hind feet have only two toes. The orbit is still incompletely formed, while the jaws are provided with teeth of all categories. The metacarpal and the metatarsal bones are fused at their proximal extremities. The forms which now follow assume the characteristics of the family Camelidae more and more clearly: the orbits are complete; spaces between the teeth are formed by the disappearance of the two incisors from the upper jaw; the molars become hypsodont; the metatarsals fuse into a cannon bone; the metacarpals remain partially free but finally also fuse into a cannon bone. In the Pliocene the animals have completed their evolution, and those which emigrate to South America become llamas. *Camelus* of North America produced giant Pliocene and Quaternary forms. In the Pliocene *Camelus* spread into China and India; in the Quaternary Africa and Europe were reached. Then the decline began. Apart from this main branch which led to the camels and to the llamas there were lateral branches which gave rise to Tylopoda such as *Alticamelus* which looked like a giraffe and the gazelle-like *Stenomylus.* The evolution of the camel family led, like that of the Equidae, to a race of runners.

The elephants

The evolution of the horse family, of the titanotheres, and of the camel family was relatively simple because we have considered only the chief line; lateral branches of little importance have been ignored. But sometimes secondary groups acquire great importance and, while retaining the same general form, include special and well characterized types. This is the case when we examine the Proboscidea or elephant group, the evolution of which is complicated by an abundance of forms, differing chiefly in the variety of their molars and incisors or tusks. All the proboscidians with the exception of the primitive form are particularly distinguished by their ponderous build, well known to everyone from the group's only living representative, the elephant. Proboscidians first appear in Egypt, in the middle Eocene and Lower Oligocene. The deposits at Fayum reveal an archaic form, *Moeritherium,* the size of a bear. Its skull is normal and its teeth comprise three upper and two lower incisors, an upper canine, and six upper and lower molars. The dental formula for the half-jaw is thus I 3/2, C 1/0, P 3/3, M 3/3. The second upper and lower incisors are well developed and meet, as they do in rodents. There is no trunk. From the incomplete and limbless skeleton it would appear that *Moeritherium* led an aquatic or amphibious life.

From this primitive form three branches diverge: the mastodon, the elephant, and *Dinotherium.* All three have common tendencies: to grow larger, to develop a trunk, to lengthen the second upper and lower incisors — or tusks — and to reduce the number of molars. But as well as these general tendencies each branch acquires its own special characteristics. The mastodon race is perhaps the most interesting, as it is the richest in secondary branches. The mastodons (as the Fayum deposits suggest) arose in Oligocene Africa, but rapidly conquered all continents except Australia. From Africa they passed, in the early Miocene, into Europe and Asia. Then, reaching North America in the middle Miocene, they spread southwards as far as the Argentine and Bolivia. They diversified into numerous and strange lines. The primitive *Palaeomastodon,* the size of a rhinoceros, had short upper and lower tusks and probably a small trunk. Then the *Tetrabelodon* branch was formed,

On the left the forelegs and *on the right* the hind legs of the camel family. Note the decrease in the number of toes and the formation of a cannon bone as the earliest forms give way to the most recent.

After Scott.

1. *Protolypus,* upper Eocene.
2. *Poebrotherium,* Oligocene.
3. *Procamelus,* upper Miocene.
4. *Lama,* of today.

Evolution of the titanotheres.

After the restoration by Osborn.

Brontotherium of the lower Oligocene with massive horns.

Protitanotherium, of the upper Eocene.

Manteoceras, of the upper Eocene.

Eotitanops, of the lower Eocene, without horns.

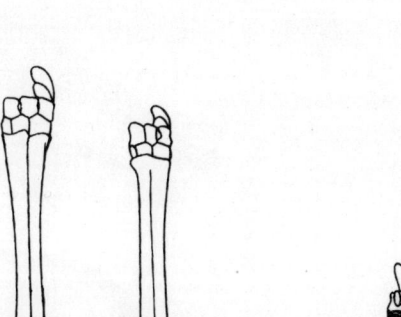

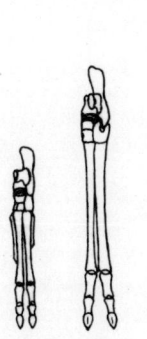

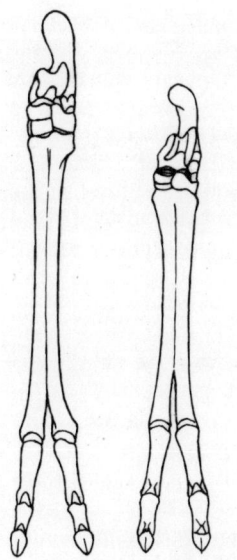

393

composed of species with four tusks, the lower tusks being larger than those of *Palaeomastodon*. As a result the mandible was lengthened. In another branch *Amebelodon* and *Platybelodon* developed mandibles of increasing flatness, provided with two lower tusks widened like a shovel which must have served to dig in the mud or earth in search of roots. In the mastodon line itself the lower tusks were atrophied or missing. The mastodon had two upper tusks which could reach a length of 2½ metres. The disappearance of the lower tusks was at times accompanied by the transformation of the lower jaw bone into an immense spatula 1½ metres long, reminiscent of the beak of *Ornithorhynchus*. It may have been covered with a horny casing and used to extract buried roots. The mastodon's molars were furnished with cusped ridges.

The elephant branch, characterized by grinding molars and two upper tusks, appeared in the Old World during the Pliocene. These early elephants, usually called mammoths, spread to North America, and then to South America with the mastodons. At the end of the Quaternary both mastodons and mammoths disappeared, when *Elephas primigenius,* the last mammoth, became extinct. Today these powerful, varied and once numerous creatures are represented by two species, the Indian elephant *(Elephas indicus)* and the wilder, more primitive African elephant, *Loxodonta africana*. The former is derived from *antiquus* and the latter from *atlanticus*. Both are huge and of massive build, with a thick and almost hairless hide, limbs like pillars, a prehensile trunk and upper incisors transformed into tusks. Their four molars — one for each half-jaw — are powerful grinders, cross-ridged with enamel crests and alternate valleys of cement. When one tooth is ground down to the root it falls out and is replaced by a larger and newly formed molar.

The third branch of the group, *Dinotherium,* appeared suddenly in Europe during the lower Miocene. It gave rise to an elephantine form with a low cranium, very unlike the high skull of contemporary elephants. It had no upper tusks. Only the lower incisors persisted, to become strong tusks which curved downwards. The five molars of each half-jaw had transverse crests and their archaic character recalls the teeth of *Moeritherium*. The size of the dinotheres increased considerably. In the middle Miocene some species measured 2.70 metres, while in the lower and middle Pliocene *Dinotherium gigantissimum* was nearly twice that size.

The concept of orthogenesis is well illustrated in the evolution of these proboscidians in which inherent and persistent trends affect the entire group, while in lateral lines secondary trends, without masking the overall orthogenesis, confer on each branch of the family its special characteristics.

Series which are still extant illustrate the same concept. The shell of the Fissurellidae, limpet-like gastropod molluscs, has an opening into the mantle cavity which varies in position according to the species. *Hemitoma* has a slight indentation on the upper edge while *Emarginula* has a marginal cleft or slit. *Macroschisma* is intermediate between *Emmarginula* and *Rimula (*or *Cranopsis)* in which the slit has become a

hole halfway between the edge and the apex. In *Puncturella* the hole has not quite reached the apex, while in *Diodora* its position is completely apical. Now, in the embryonic development of *Diodora* the orifice shifts from the edge towards the centre exactly as it does in the orthogenetic series beginning with *Hemitoma* and ending with *Diodora*.

The mechanism of orthogenesis

No complete lines of species exist to illustrate these evolutionary trends. The cases we have considered are gathered at random and merely furnish signposts, indicating directions. There would seem to be three possible aspects of orthogenesis:

I. It proceeds steadily in a given direction and ceases to operate when the form achieved appears to have reached equilibrium, for example, the single toe of the horse, the two toes of the ostrich and the ruminants, the eye of the primates, the circulatory system and constant body temperature of mammals, the trunk of the elephant. Such orthogenesis is progressive and

Elephas, contemporary.

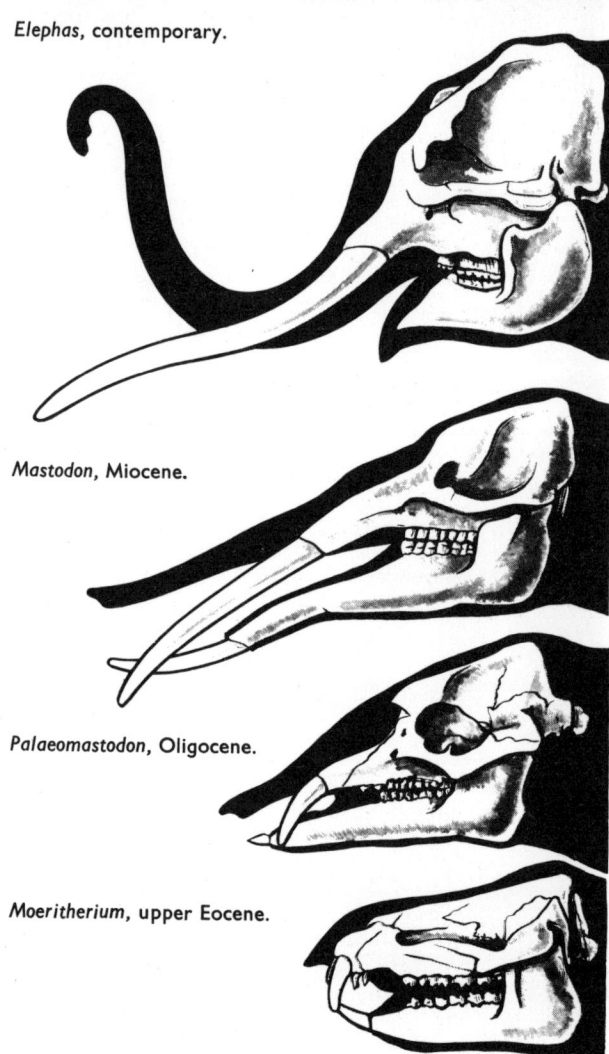

Mastodon, Miocene.

Palaeomastodon, Oligocene.

Moeritherium, upper Eocene.

Transformation of the head of the Proboscidea from the earliest form to the most recent.

After Boule and Piveteau.

The mammoth, *Elephas primigenius.*

X *Photography*

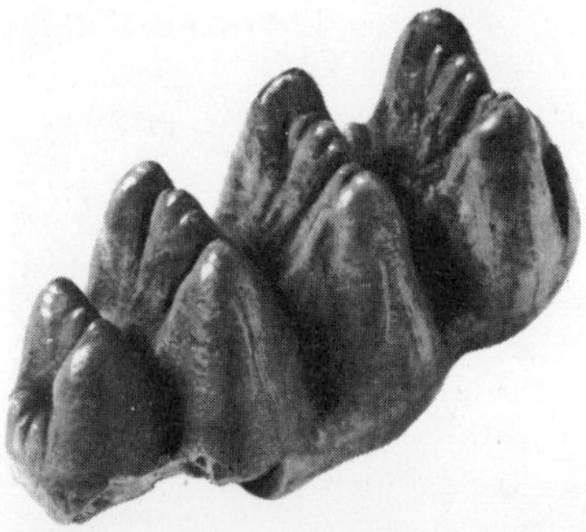

Above left, molar of *Mastodon turicensis; above centre,* molar of the mammoth, Le Raincy; *above right,* lower molar of *Elephas antiquus,* from the river deposits of Chelles.

Muséum national d'histoire naturelle. Larousse.

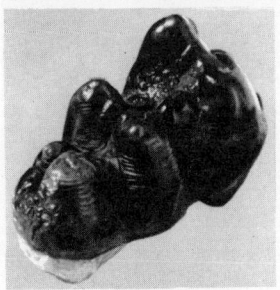

The molar of *Palaeomastodon,* Oligocene of Fayum, Egypt.

Muséum national d'histoire naturelle. Larousse.

Restoration of *Platybelodon* of the upper Miocene of Mongolia. These elephants presumably used their flattened incisors as a spade to dig in the earth.

After Osborn.

leads to the organism's physical improvement. The horse is an excellent runner and the elephant's trunk, just long enough to brush the ground, compensates for the animal's great height.

2. Instead of stopping when physical perfection is reached progressive evolution may continue until the result is exaggerated development or hypertely, always an encumbrance and sometimes a danger to the species. Instances are the inward curving of the mammoth's tusks, the extravagant horns of certain members of the deer family, notably the antlers of *Megaceros* of the Quaternary which had a span of 3 metres, the gigantic canine teeth of *Machairodus, Smilodon,* and *Babirussa,* the heavy armour of the stegocephalians, the enormous spines of certain sea-urchins and the outsize legs and antennae of certain insects.

3. Finally orthogenesis can be retrogressive and lead to rudimentary organs without function. Such are the pineal eye of contemporary lizards, the vestigial nictitating membrane of the human eye and the atrophied muscles of the ear. It can also lead to the total disappearance of an organ: amphibians and reptiles without legs, whales and sea-cows without posterior limbs.

In practice these three aspects are interwoven: the evolution of the Equidae involved not only progression for the middle toe and the animal's size, but also regression for the second and fourth toes and the total disappearance of the first and fifth toes. Hypertely, which affects only certain organs like horns, canine teeth, and tusks, has little evolutionary importance. The opposite is true when orthogenesis acts in a co-ordinated and harmonious fashion on various organs and gives rise to a new and more specialized form capable of living in another environment or of leading a new kind of life. In the course of a group's evolution, innovations appear, that is to say organs which are new to the race: for example, the varied types of grinding molars, the dorsal fin of the ichthyosaurs and the dolphins. But orthogenesis does not explain the birth of new types of organization, or the appearance of new phyla or sub-kingdoms. Nonetheless systems of classification are based, at least implicitly, on the concept of orthogenesis.

What is the mechanism of orthogenetic series? Huxley proposed an explanation based on the allometric or differential growth of the organs. It is, in fact, established that certain organs have a different rate of growth from that of the organism of which they form a part. Differential growth thus results. The ratio between the growth of the organ and the growth of the organism can be expressed as a mathematical formula. An organ whose rate of growth is different from that of the organism is known as allometric. If the two rates are the same the organ is called isometric. Such relations explain hypertely, for as the total organism grows the allometric organ will acquire excessive dimensions. But in coordinated orthogenetic series the rates of growth remain in harmony so that organ and organism achieve a balanced and increasingly specialized form.

To some philosophers these persistent evolutionary trends imply predetermination, and for them the word orthogenesis assumes a mystical connotation. In this sense the concept is not valid. The facts at our disposal show that the course which evolution pursues is not straight but wavering, and full of twists and turns.

The law of irreversibility

Of all the laws of evolution the law of irreversibility is perhaps the most general and important. Edgar Quinet grasped its implications in 1870 when he wrote: 'Nature does not turn back; what she has destroyed she does not re-create. She does not return to the millstone she has broken. In the infinite number of combinations which the future has in store you will never twice behold the same person, the same flower, the same creature.' In 1893 Dollo amplified the statement. When an organ begins to regress it can no longer return to its former condition or recover what it has lost. Ungulates whose toes have been reduced in number to three, two or one, have never engendered species with a greater number of toes, although they have always been preceded by four- or five-toed species. Since man has five toes it follows that all his ancestors from the earliest Devonian tetrapod possessed five toes. No four-legged creature having fewer than five toes participated in his lineage; human pentadactylism is a primitive character. The thirty-two human teeth are a secondary character, for it is certain that the number of human molars has been reduced and that ancestors of man had a greater number of molars. Forms having fewer than thirty-two teeth are, however, excluded from man's possible ancestors.

Animals which return to a mode of life similar to that of their ancestors develop analogous organs serving the same functions, but such organs have another origin. There are a number of land vertebrates which have re-adapted themselves to an aquatic life: reptiles like *Ichthyosaurus* and *Plesiosaurus;* crocodiles, turtles, and serpents; mammals like the Cetacea (whales, porpoises, and dolphins), seals, sea-cows, etc. In none of them, however, have gills and fins of the fish-type reappeared. Although the transverse caudal fin of the Cetacea may recall that of a fish its structure is peculiar to itself. The insects, which are terrestrial arthropods, breathe with the aid of tracheae. The larvae of certain insects, reverting to aquatic existence, do not breathe with gills but with modified tracheae.

Parasitism, the result of regressive evolution, leads to a structure which cannot recover its earlier form. Groups composed exclusively of parasites — Trematoda, Cestoda, Acanthocephala, fleas, and lice — derive from free-living ancestors; but no parasite abandons its mode of life to regain its independence.

The law of irreversibility applies not only to organs but also to the history of animal and plant groups. The evolutionary climax occurs once only; the phase of senescence never includes an evolutionary resurgence. And no extinct group has ever reappeared.

The facts we have reviewed thus bring to light a certain number of rules or laws which preside over the course of evolution: the law of increasing complexity, the law of primitive synthetic forms, the law of succession, the law of orthogenesis, and the law that regressive evolution is irreversible.

The pace of evolution

Although evolution is an observable phenomenon it proceeds with extreme slowness. To bring about a change of minor importance millions of years may be required. Osborn estimated that twenty million years were necessary to produce the finished version of the trilophodont mastodon's third molar. In the animal's Eocene ancestor this tooth had seven tubercles; in the last of the *Trilophodon* the number had risen to thirty-seven. Thus twenty million years were needed to acquire thirty tubercles, which works out at roughly one every 660,000 years! The student of nature, during his brief life, finds it impossible to grasp processes which evolve so slowly; the impression he has is one of stability. Nevertheless he sometimes observes the appearance of clear and swift variations: namely, mutations, of which we have seen numerous examples. What connection exists between mutations, which more or less slightly modify individuals, and the extremely slow variations which transform structures and types of organization?

At times the pace of evolution seems more rapid than at others: the rise of the reptiles, for instance, was more rapid than that of the mammals. The passage from aquatic life to life on land entailed a complex organic and physiological revolution which required immense stretches of time and was achieved in successive stages. Lungs developed among crossopterygian fishes which then possessed both lungs and gills. The transformation of fins into limbs capable of walking allowed the stegocephalians to lead an amphibious existence and become the first animals able to live, at least temporarily, on land. The disappearance of aquatic larval life — that is, of the tadpole stage — and the acquisition of an impermeable skin made it possible to abandon the water permanently.

Within the confines of the great groups it would seem that the more recent the group is the more rapidly it evolves. Thus it is generally agreed that the evolution of the bony fishes required between 350 and 375 million years, while that of the birds was achieved in about 150 million years. The great diversity of bird species, moreover, is thought to have arisen only during the last fifty million years. All the mammals were differentiated during the Tertiary in some twenty to forty million years. The pace of primate history appears to be even further accelerated. The lemurs and *Tarsius* (a hopping, rat-like creature, not yet a monkey) date only from the Eocene and the Oligocene. The more highly developed monkeys then appeared. The point at which the human line emerged is still debated, but it can scarcely be more than a million years ago. Contemporary man goes back only for some thousands of years.

Has evolution come to an end? Today all the great groups appear to be static; for thousands of years no new family has arisen. But in the scale of geological time the human species is still very recent. Will it persist indefinitely to become, as it were, a 'living fossil' or, subject in its turn to the law of succession, will it be replaced by a group whose existence is still unsuspected?

Megaceros hibernicus, elk of the peat-bogs.

Muséum national d'histoire naturelle. Larousse.

Left, the Ring-tailed Lemur, *Lemur catta*.

Above, the Tarsier, *Tarsius* spp.

Bruce Coleman.

Adaptation and its Problems

The basic problem of evolution is that of adaptation. The word is derived from the Latin *adaptare,* itself composed of *ad,* 'to' or 'towards', and *aptare,* 'adjust' or 'arrange'. As its etymology indicates, adaptation means an adjustment of the organism to all conditions of its environment, external and internal. In definite conditions and in a given environment this adjustment permits the organism to live, to endure, and to reproduce itself. Adaptation may be considered in relation to the environment under consideration, environment in the widest sense of the word, or in relation to a mode of life, to the organism's activities. The concept implies fitness rather than usefulness. Adaptation is a general biological phenomenon and a fundamental attribute of living organisms. When we speak of adaptation we envisage the living organism in its functional responses to its environment. Since the word 'adaptation' is often employed in other senses it is important to define our meaning.

Individual adaptation

When we say that a European has become adapted to a tropical climate we mean that after a certain time he has learned to support the conditions of life in a tropical climate without inconvenience. He has accommodated himself. Accommodation occurs when an individual, man, animal or plant, successfully lives in an environment different from its normal environment. Such accommodation can give rise to modifications of form or physiology, but if the individual returns to its normal life and surroundings these modifications more or less rapidly disappear.

The organism responds to physical, chemical, and mechanical factors by reacting in a manner calculated to neutralize their effect. Appropriate reactions allow the organism to live and to adapt itself to new conditions. The mammal responds to cold by growing more fur and to heat by growing less. When settled in a mild climate the woolly goats of the high plateaux of Asia lose their hair while their skin thickens. General metabolism, upset by unfamiliar cold or heat, causes the organism to react in such a way that the hair either grows or falls out. Such adaptation takes place unconsciously, without voluntary participation on the part of the individual organism, which simply supports environmental changes within its capacity and becomes adapted with ease or with difficulty. Life in high altitudes, where atmospheric pressure is low and oxygen scarce, demands increased circulation and more red blood-cells to distribute oxygen through the body. The body's reaction to cold brings into play a series of immediate defences: a constriction of the surface blood vessels, a rising of hair or ruffling of feathers to lessen the loss of heat, the secretion of adrenalin and shivering to increase the production of heat.

These individual adaptations which protect men and all living creatures against a new environment are particular cases of a general and normal phenomenon: regulatory adaptation or, as W. B. Cannon called it, *homeostasis.* Homeostasis, or maintenance of the internal environment's equilibrium, affects both matter and mechanisms. Materially it involves the storing up of reserves and the excretion of waste; mechanically it coordinates the functioning of the organs with the requirements of the organism. These requirements may in certain conditions be sharply increased. A sudden need for oxygen entails increased activity on the part of both respiratory and circulatory systems, for the blood must be more rapidly aerated and oxygen supplied at a higher than normal rate of pressure.

In the adult homeostatic mechanisms are, in normal circumstances, effective: a given change in the internal or external environment is always followed by the same appropriate homeostatic response. But as the organism ages, its capacity for keeping the internal

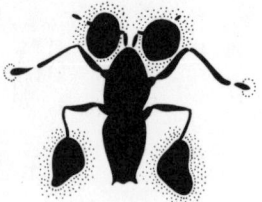

Shadow of *Velia* in repose.

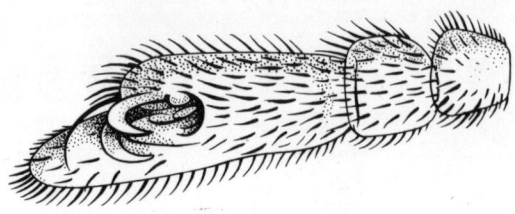

The tarsus of the foreleg.

Shadow of *Velia* on the bottom of a shallow stream.

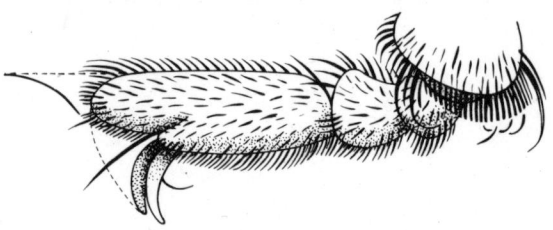

The same tarsus in the meniscus or depression it has made on the surface of the water, a depression roughly equal in depth to the thickness of the tarsus.

Shadow of *Velia* washing itself; the left foreleg has moved.

medium constant in composition begins to fail. In the old, homeostasis as it affects temperature, the glucose content of the blood, etc., is frequently defective.

Certain diseases in man appear to result from a breakdown in the hormonal mechanism of adaptation; some are caused by lack of adaptation, others by excessive adaptational reactions.

Specific adaptation or naturalization

Let us now consider not the individual but a group of individuals, capable not only of existing but also of reproducing in conditions that differ from those to which the group is accustomed. With the aid of Man this can occur in specially protected environments and is then known as 'acclimatization'. Obvious examples are garden and greenhouse plants, domestic animals, and animals reared in zoos.

Acclimatization, however, does not always require human intervention. The plant or animal may adapt itself, and thereafter form part of the community's natural flora and fauna. The species is then said to be 'naturalized'. Capable of living and of reproducing, it illustrates specific adaptation. Every species which forms part of an association, or natural vegetation unit, is specifically adapted to the environment in which it lives.

The bivalve molluscs *Dreissensia polymorpha,* native to the brackish waters of the Pliocene Ponto-Arabo-Caspian basin, became naturalized to fresh-water in the eighteenth century when they entered the Volga. Then, by way of rivers and canals, they reached Germany, were carried to England, reached Holland, Belgium, and France. They now abound in canals.

The Colorado beetle, which causes such havoc among potatoes, was discovered in Colorado in 1824 and from there set out on the conquest of America. In 1859 it invaded Nebraska and by 1874 had reached the Atlantic coast. It arrived in Bordeaux in 1919, probably introduced with merchandise imported from the United States. From Bordeaux it spread throughout Europe. Only England, by drastic measures, was able to keep the pest under control.

The Chinese crab *Eriocheir sinensis* is, as its name suggests, of Chinese origin. It lives along the coasts, in rivers, and in irrigation ditches. It was first introduced into the Elbe or the Weser in Germany in 1912 by accident. It spread through rivers and along the coast until it reached the coasts of Denmark. It has penetrated the Baltic sea, reached Sweden and even the Gulf of Finland. It has also invaded Holland, Belgium, England, and France. The crab is a pest since it devours the food of the native aquatic population. In addition it damages fishing nets and, worse, undermines river-banks by its tunnelling operations.

Many other examples, some happier, could be cited of animal and plant adaptation. The orange tree, a native of southern Asia, is acclimatized to the regions surrounding the Mediterranean but it is perfectly naturalized in, for instance, Florida and Paraguay.

Species which seem to be comfortably naturalized

Adaptation to walking on water. Certain insects are adapted to life on the smooth surface of water. They are supported by the minute depressions — or menisci — which their water-proof tarsi make in the water's surface. These depressions are only a tenth of a millimetre in depth, but are made visible by the optical effects they produce:
1. Light waves, refracted by the concave surface of the menisci, throw shadows on the bottom of shallow streams or pools;
2. Furthermore the colour effects due to the interference of light waves enable the curvature of the depressions to be measured.

Velia. A fresh-water hemipterous insect measuring 7.5 millimetres and its shadow on the bottom of a stream. The insect is supported by its front and hind legs only; thus the middle pair throws no shadow. It is used as a pair of oars.

Baudoin.

Velia on the surface of a stream. Interference caused by different degrees of light wave refraction makes the menisci visible.

Françon and Baudoin.

Aepophilus, a marine, hemipterous insect which lives in tidal basins and is found on the surface of the water when the tide is in. The insect rests on six menisci which are roughly of the same depth, and is thus supported by all three pairs of legs.

Françon and Baudoin.

401

Spread of the Colorado beetle.

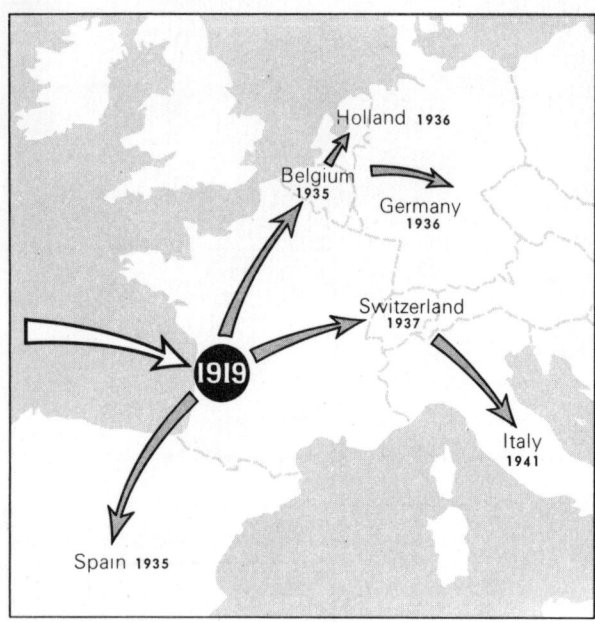

The Colorado beetle, *Leptinotarsa decemlineata*.

R. H. Noailles.

Distribution and limits of extension of the Chinese crab in the river basins of France.

Marc André.

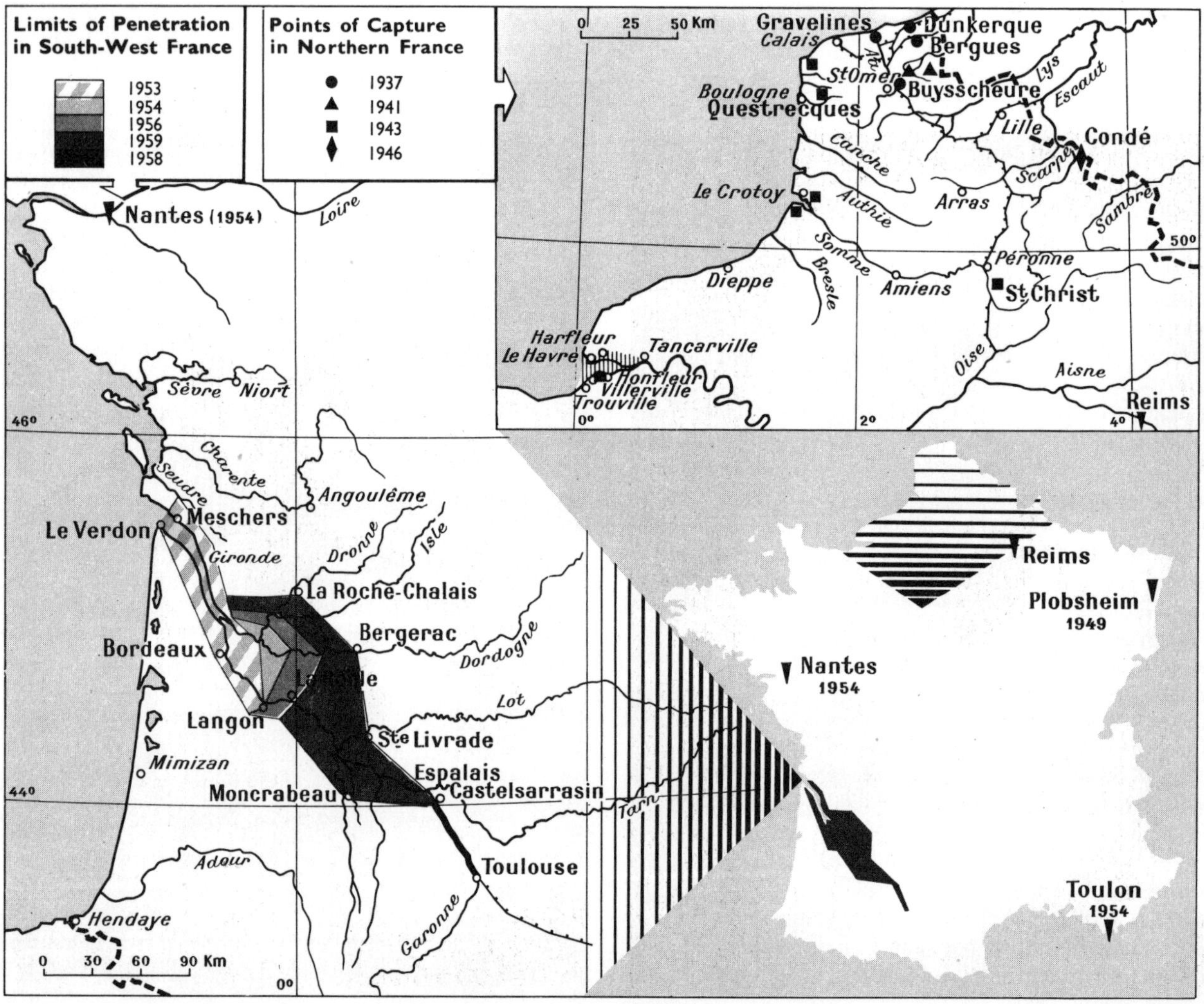

The Chinese crab, *Eriocheir sinensis*. Note the thick hairy covering of the claws.

Marc André. Larousse.

Limits of Penetration in South-West France	
	1953
	1954
	1956
	1959
	1958

Points of Capture in Northern France	
●	1937
▲	1941
■	1943
▼	1946

may disappear as a result of exceptional conditions. Severe winters, notably the winter of 1879—1880, killed trees and shrubs which had belonged to the local flora for many years. In the north of France the Norway pine resisted the cold while the maritime or sea-pine, which had been naturalized for nearly a century, succumbed.

Naturalized plants, like naturalized animals, can be pests. Oidium or vine-mildew originated in America while the oidium which attacks the oak probably came from Portugal. *Elodea,* the common water weed of Canada, was known as 'Babington's curse' in England where for ten years it obstructed inland navigation.

Ethological adaptation

Ethological adaptation, or adaptation to a particular mode of life, is found in animals which, though totally unrelated lead similar lives. Among aquatic animals, for instance, fins, flattened tails and webbed feet are frequent. The modified forelegs of the mole-cricket (an insect), of the mole-rat (a rodent), and of the marsupial mole *Notoryctes,* recall those of the mole itself (a burrowing mammal) and suggest that all these animals are excavators. The familiar face of the owl leads one correctly to suppose that the Malayan tarsiers, and the lemurs, which also have great protruding eyes in a flat face, are, like the owl, animals which are active at night and sleep during the day. Fins are so common in aquatic animals that their presence in a fossil is evidence that the fossil lived in the water. Such evidence though great is not, of course, conclusive; for some aquatic animals have no fins while land animals are occasionally found with appendages which are analogous to fins. It is also true that *Palmatogecko rangei* which lives in the coastal sand dunes of southwest Africa has webbed feet. Similarly, the feet of many burrowing animals are not mole-like. Nevertheless the great frequency with which these ethological adaptations are found in special environments is indisputable, and Cuénot has suggested that they be called 'statistical adaptations'.

Whether an animal is well or ill adapted to a given mode of life will depend on the number and importance of its ethological adaptations. The wharf rat, the water vole, the otter, the seal, and the whale illustrate increasingly successful adaptations to aquatic existence. The wharf rat, a terrestrial rodent, frequents damp places, wharfs, and drains, and is able to swim across rivers. The water vole, a good swimmer, lives at the water's edge where, though omnivorous, it feeds chiefly on fish and frogs. The otter, an excellent swimmer and diver, has webbed feet and lives on fish. The seal, with its spindle-shaped body and flippers, lives in the water and comes to land only for the breeding season. The whale is an aquatic mammal which is no longer able to live out of water. This transition of terrestrial mammals to aquatic mammals is effected in graduated stages and as the adaptations become more characteristic they increasingly limit the animal's possibilities. An animal or plant whose ethological adaptations are different from those which are normal to its family is called 'specialized'. The whale is specialized to aquatic

Shell of the bivalve mollusc *Dreissensia polymorpha.*
Muséum national d'histoire naturelle. Larousse.

life and dependent on its marine environment which it cannot leave. A few examples of ethological adaptations follow.

Webbed feet

Many aquatic animals possess webbed feet; these include the otter, the desman, the duck-billed platypus, and web-footed birds. Statistically the great majority of aquatic animals have webbed feet or, as in the case of coots and grebes, possess lobes along the sides of their toes. *Sorex palustris,* the marsh shrew has developed flattened paws edged with stiff hairs while water snakes and crocodiles have transversely flattened tails.

Among aquatic birds the majority have webbed feet and a correlation between webbed feet and aquatic habitat is apparent. This correlation becomes even more obvious when we examine the manner in which these birds live. The unwebbed or slightly webbed genera live on river banks or in damp fields and are not truly aquatic. All the web-footed genera are, on the contrary, fully aquatic and swim with ease. Complete webbing does not exist in any bird which is truly terrestrial. It may be pointed out that indications of webbing exist in certain breeds of domestic poultry, in turkeys, wood-pigeons' and pheasants. These, however, are only vestiges, similar to those found in herons or storks. The domestic goose, though a land-dweller, has webbed feet; but it is descended from the wild goose which is a good swimmer.

Although webbed feet are connected with an aquatic environment they are not indispensable to swimming. They are an advantageous specialization which encourages swimming, especially in animals otherwise fitted for or attracted to life in the water. When the swan swims its webbed feet act like the blades of an oar.

In mammals, as in birds, palmated or webbed feet are associated with aquatic life. The beaver and the nutria-bearing coypu have palmated hind feet while the mole-like desmans of Eurasia have, like otters, two pairs of palmated feet. All, including the seal, are skilled swimmers. Others, like the African *Potamogale,*

Above, *left to right,* cuscus or *Phalanger,* an Australian marsupial; *Galago,* an African lemur; *Loris,* a lemur from Ceylon; screech owl.

Larousse.

Webbed feet. *From left to right,* of swan, of duck, of goose.

Jean.

a short-legged, otter-like insectivore, are not palmated, but have a long flat tail which enables them to swim rapidly.

The amphibian Urodela (newts and salamanders) use their tails for swimming and are generally unpalmated, while the lack of a swimming tail among the amphibian anurans is in the case of the frog compensated for by the broad webbing of the hind feet, and the more aquatic the frog the more highly developed its webbing. From this rapid survey of palmation certain facts emerge: webbed feet are an adaptation which is not necessarily connected with aquatic life. Thus Lamarck was incorrect in his assumption that a bird, when it wished to swim, spread the toes of its feet so that the skin acquired the habit of stretching, and that in this manner palmation little by little developed. Webbed feet also supply one answer to the swimming problem, the other being an undulating spindle-shaped body and a transversely flattened tail. The first solution seems to have been adopted by thick-bodied animals like frogs, birds, and mammals, while the second was adopted by slim and elongated animals like reptiles, newts, and salamanders. These arrangements regress if the aquatic animal becomes a land-dweller.

Adaptation to diving

Diving and swimming under water require special adaptations. Many birds dive and, although the speed with which they immerse is not great, they may remain under water for as long as six minutes. The penguin can, as it were, 'fly under water', for it propels itself with short and powerful wings. For this reason it has well-developed muscles. The great size of the coracoid (lower bone of the shoulder-girdle) prevents the dorso-ventral flattening of the body. The shoulder-blade, which transmits the vigorous movements of the pinions to the rest of the body, is exceptionally long and particularly broad in the distal region. The sternum is like a powerful breastplate which protects the ventral region from shocks incurred in diving.

Aquatic mammals also are able to dive, and can remain under water much longer than terrestrial species. For example, a white rat can remain immersed for some three minutes, a muskrat for twelve and a beaver for fifteen; while Sirenia (manatees, dugongs), Pinnipedia (seals, walruses), and Cetacea can stay under water as long as eighty minutes and, in the case of the whale, for an hour and forty minutes. These animals, moreover, descend to great depths.

Air-breathing vertebrates capable of surviving so long without oxygen and of returning to the surface without embolism must obviously possess a respiratory equipment specially adapted to the physiology of land animals which have returned to the sea. For some years it has been known that the lungs of aquatic animals are no larger than those of land animals. The volume of a man's lungs, for instance, is exactly the same as that of a seal or of a manatee. The oxygen-bearing capacity of human blood is roughly the same as that of the rorqual and the dolphin. In addition to these negative findings certain positive facts have emerged:

Tarsius spectrum, from Malaya.

X Photography.

The swimming flipper of the seal.

Larousse.

every time a diving mammal dives it almost completely renews the air in its lungs. In the whale ninety per cent of the air is renewed whereas in man the percentage is fifteen to twenty. Moreover the respiratory centre of diving mammals is remarkably insensitive to the increased ratio of carbon dioxide in the blood. This peculiarity is found not only in skilled divers like the manatee, seal, and porpoise, but also in lesser divers like the beaver and muskrat. Thus the two mechanisms

The gecko, *Ptychozoon kuhli*, seen through a pane of glass so that its adhesive discs may be observed.

Tweedie-Rapho.

tend to augment the quantity of oxygen at the beginning of the dive and to reduce the stimulating effect of carbon dioxide on the respiratory centre. Supplementary mechanisms allow the consumption of oxygen to be reduced and restrict its employment to essential activities. As the dive begins the heart beats more slowly. In the grey seal the number of heartbeats is reduced from 100 to 10 per minute, and the slower rate continues during 15—20 minutes of submersion. The electro-cardiograms of dolphins diving more or less freely have shown that at each dive the rate of the animals' heartbeat is on average reduced by fifty per cent. The same is true of the Florida manatee. Thus there is a minimum expenditure of available oxygen. During the first minute of its dive the seal consumes only a fifth as much oxygen as it consumes during a minute of repose. On the other hand the blood pressure of the large arteries remains normal; it rapidly falls, however, to the venous pressure, while in the small arteries the pulse ceases altogether. In the mesenteric blood vessels vasoconstriction occurs though the cerebral circulation remains normal. Only those exchanges required by essential vital processes take place.

The problem arises of how diving cetaceans escape the embolism which constitutes the great danger to human deep-sea divers. In men embolism results from returning too quickly to the surface. Nitrogen, which is dissolved in the blood in great quantity during the descent, can no longer remain dissolved when pressure diminishes; instead it is released in the form of gas bubbles which cause discomfort, illness, and

405

Young cape penguin.

L. Rollet.

A cape penguin.

France-Reportage.

King penguins, *Aptenodytes patagonicus*, Adélie Coast.

C. N. R. S. Antarctic subcommittee, Jacques, Masson.

sometimes fatalities among deep-sea divers. It was long thought that the blood of whales contained some micro-organism capable of using the nitrogen dissolved in the blood. This proved to be untrue; the mechanism is in fact quite simple. In the first place the cetacean is not, like the deep-sea diver, in an atmosphere which is constantly renewed by surface pumps. In its lungs it carries a limited amount of nitrogen. Also, while diving, pressure on the animal's flanks compresses the pulmonary alveoli and drives most of the free gases into the mouth and nasal cavities. Nitrogen therefore dissolves in the plasma at a slower rate. Finally the animal's quickened heartbeat as it rises towards the surface facilitates the elimination of the nitrogen which has been dissolved. This mechanism is not infallible, for a seal submitted to an experimental dive of 300 metres suffered from nitrogenous embolism when it came up again. It is highly probable that Cetacea also come to the surface by reasonably slow stages.

Adaptation to climbing: tree-dwellers

Some climbing birds have the faculty of going up or down vertical tree trunks. Their toes are arranged in a special manner: the fourth toe is turned back in such a way that it joins the first toe which in this case is well developed. The foot thus forms a shape resembling a pair of pincers with four claws opposed in pairs. The claws are strongly curved and their sharp points are noticeably perpendicular to the plane on which the extended foot is placed. In this way the foot is, as it were, embedded and cannot be torn from its support without tearing away a part of the foot itself. The bird's grasp is sufficient to assure a stable position. In

order to avoid fatigue the bird presses its body against the tree, thus reducing torque, and closing the 'pincers' as tightly as possible. Generally speaking climbing birds are small and slight.

The tree-frog is, as its name implies, an arboreal amphibian, and its front and hind legs are adapted to its habitat. The extremity of each toe is furnished with an adhesive disc which permits the animal to affix itself to smooth and vertical surfaces.

Lizards of the gecko type move across the ceiling by means of enlarged toes. The undersurface of the toes is provided with minute transverse lamellae, flexible and overlapping, which enable the animal to maintain its reversed position. Pressure against a smooth surface stretches and distends these lamellae which expel the air contained between them. Then a partial retraction of the muscles restores them to position. A vacuum is thus produced which is sufficient to support the gecko.

The ethological adaptations of the tree-dweller are largely designed to enable the animal to remain affixed to surfaces or to leap from branch to branch. Many arboreal creatures have a prehensile tail (the chameleon, New World monkeys, certain marsupials, the pangolin or scaly anteater), a prehensile tail with a scaly underside *(Spilocus)*, opposable hands or claws (lemurs, monkeys, parrots, woodpeckers), powerful claws (three-toed sloths), adhesive pads *(Tarsius,* tree-frogs), a parachute (for example, the wing-membrane of the flying lemur, *Anomalurus,* flying squirrels and marsupials, the flying lizard or flying dragon *Draco*). Animals with parachutes seem to predominate in Holarctic, African and Australian forests, while animals with clinging tails such as

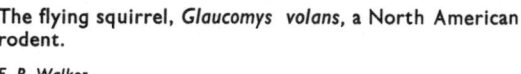
The flying squirrel, *Glaucomys volans,* a North American rodent.

E. P. Walker.

Anomalurus pelii, an African rodent.

G. Cansdale.

The tree-frog, *Hyla arborea meridionalis.*

J. Six-Atlas-Photo.

monkeys, the small South American opposum, and the quica are exclusively New World.

Climbing plants are, like tree-dwelling animals, adapted to an arboreal existence. Their adaptations include the ivy's aerial clinging roots, tendrils which entwine like spiral springs (vetches, bryony, convolvulus), branched tendrils with adhesive discs (Virginia creeper), sensitive hooks like those of *Artabothrys* which thicken after grasping a support.

Other adaptations could be described as ethological or fitting the organism to its mode of life: the disproportionate lengthening of appendages such as the legs and antennae of insects and cave-dwelling crustaceans; the melanism of numerous mountain and island animals, etc.

Adaptation to dryness

The adaptations of the xerophytes, or plants adapted to inadequate water supply, are designed chiefly to reduce surfaces and thus limit evaporation. The cactus and related forms have succulent green stems enclosing an abundance of tissue composed of thin-walled cells called *parenchyma* which act as a water reserve. These stems, bulbous, thickly jointed, or cylindrical, single or branched, smooth or grooved, are leafless; and thus the surface exposed to the air is greatly reduced. As a rule the roots of xerophytes are long and highly developed and draw moisture from deep in the earth.

Ethological adaptations are also evident in the structure of water plants. The stems are thicker and softer than aerial stems and are modified in two essentials: the more highly developed bark encloses large air-filled spaces, and the supporting tissues of wood and sclerenchyma are less developed.

Examples of ethological adaptations could be multiplied indefinitely; for all living organisms inhabiting a definite environment reveal anatomical and physiological modifications which are useful, if not essential, to their survival.

Convergence

In the development of certain adaptations the phenomenon of convergence is evident. Convergence is the term applied to those striking resemblances between individuals leading the same kind of life or between organs performing the same functions. Such resemblances do not imply any degree of relationship, but are simply due to similar responses to needs or stimuli, or to identical living conditions. We might think of convergence as a consequence of accumulated ethological adaptations.

In both the animal and vegetable kingdoms the phenomenon of convergence is encountered. Among vertebrates that swim well fish-like convergence is observed: the shark, a selachian fish, and *Ichthyosaurus,* an extinct reptile, and the dolphin, a cetacean mammal, are all three adapted to rapid swimming and are all three similar in shape. Their slender, spindle-shaped bodies are provided with a triangular dorsal fin. But these fins evolved independently and differ in structure. In the shark the fin is armed with bony or cartilaginous rays, while in *Ichthyosaurus* and the dolphin it is made of connective tissue and skin. In form, however, it is much the same; the convex forward edge which cuts the water is reinforced, while the concave rear edge is slimly tapered. The fin is situated

at the summit of the arching back. Its presence is not indispensable for it is lacking in *Plesiosaurus, Mosasaurus,* seals, the Sirenia and other coastal cetaceans which are far from poor swimmers. Like the centre-board of a boat it presumably acts as a stabilizer and prevents drifting.

Mole-like convergence is seen in burrowing mammals: the mole itself, an insectivore, *Myotalpa,* a rodent, and *Notoryctes,* an Australian marsupial. All three are blind or have much reduced eyes; their more or less misshapen forefeet are furnished with strong claws; short and silky fur and a short or missing tail are adaptations to their mode of life; and their burrowing instinct leads them to dig holes in which to live.

We have already seen that certain tree-dwellers have a 'patagium' or wing-membrane which enables them to glide; Australian phalangers, flying squirrels, *Anomalurus,* flying lemurs, and the flying lizard *Draco.*

Bird-like convergence is found in various vertebrates capable of flying: *Pterosaurus* (reptile), birds, and bats (mammals). The wing has evolved four times with the same function, but with an entirely different structure: the insect's wing with its stiffening tubular

Agave (Amaryllidaceae).

Atlas-Photo.

Tendrils of the white bryony. Aerial roots of the ivy.

R. H. Noailles.

Convergence in marine vertebrates which belong to three distinct classes, all adapted to rapid swimming. *Top,* the shark, a selachian fish; *centre, Ichthyosaurus,* an extinct reptile; *bottom,* the dolphin, a cetacean mammal. The dorsal fin serves to maintain equilibrium.

In the flying reptile the wing-membrane is supported by the four greatly elongated phalanges of the fourth digit, articulated on a powerful external metacarpal bone. The other three metacarpals, more fragile, bear two, three, or four phalanges terminated by claws. In the bird the wing skeleton comprises a humerus, a radius, a cubitus, a carpus with four bones, three metacarpals, and three digits. The first metacarpal bears the reduced first digit or thumb; the second and third are fused and bear the second and third digits, digit II being distinctly longer. (In the diagram digit III cannot be seen.) The remiges or quill feathers are inserted into the digits, the metacarpals, the carpus, the cubitus, and the humerus. In the bat the wing membrane is supported by the metacarpals and digits II to V which are long and disposed like a fan. The thumb is short and can move in any direction.

veins; the bat's wing ribbed like an umbrella; the feather wing of the bird; and the pterodactyl's wing supported by a single, immensely elongated finger.

Certain air-breathing animals which lead an aquatic life are organized in such a way that when they are in the water their nostrils and eyes emerge. This arrangement is found in an amphibian (the frog), in a reptile (the crocodile), and in a mammal (the hippopotamus).

Convergence also occurs in internal organs. The big glands connected with the digestive tube of molluscs, crustaceans, and arachnids recall the liver of the vertebrates. They are, indeed, called by the same name, but in origin they have nothing in common with the vertebrate liver. *Amphioxus* has organs of excretion of a type found in many worms and annelids. Vertebrates, certain molluscs, scallops, and the cephalopods have eyes with a crystalline lens. The heart with four chambers has appeared in four groups: Dipnoi or lungfish, Crocodilia, birds, and mammals. The placenta, which unites the embryo with the maternal organism, is found in some protochordates, some fish of the Teleostomi group, some reptiles, and in all placental mammals. In these cases convergence

skeleton of wings

digits   metacarpals

of pterosaur

digit IV

radius   ulna

thumb   carpals   humerus

digit II

of bird

metacarpals II and III fused

of bat

thumb

II

III   IV   V

The heads of a frog, a crocodile, and a hippopotamus. Nostril and eye alone appear above the water level: an example of convergent adaptation to amphibian life.

After Hesse.

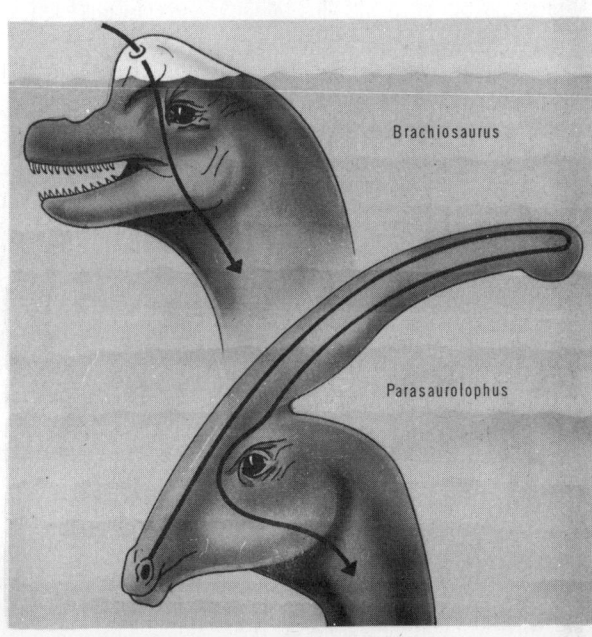

Brachiosaurus

Parasaurolophus

Brachiosaurus. A small part of the head emerges from the water. The nostrils are situated on top of the head, and the arrow indicates the route which the air takes to reach the animal's lungs.
Parasaurolophus, a dinosaurian, has a long nasal canal which runs through a cephalic appendage and in this way forms an air reserve which enables the animal to remain submerged for long periods.

After E. H. Colbert.

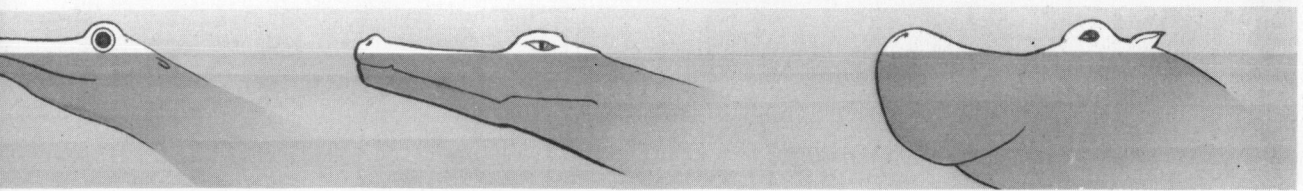

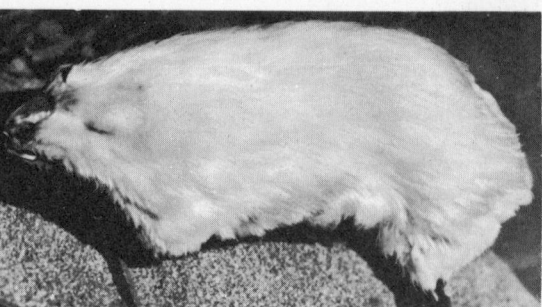
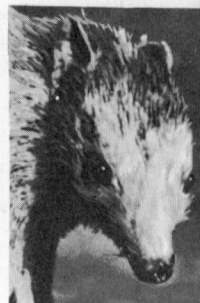

The mole, *Talpa europaea.*

R. H. Noailles.

The marsupial mole, *Notoryctes typhlops*, of Australia.

X Photography.

Above right, the marsupial bandicoot (Peramelidae.)

Australian Embassy, Paris.

Adaptation to a hot dry climate of three plants belonging to three different families. *Above, Cereus,* of Mexico, Cactaceae family. *Below on the left, Stapelia desmeoiana* of South Africa, Asclepiadaceae. *On the right, Euphorbia bothae* of South Africa, Euphorbiaceae.

J. Vincent, Larousse.

appears as the repetition in different animal series of an invention or new structure.

Ecological convergence is not uncommon in plants growing in similar surroundings. Succulent or fleshy plants store water in their tissues and in this way can live in tropical deserts where water falls only occasionally. They tend to assume the same cactus-like appearance, although they may belong to different families: the Cactaceae *(Cereus)*, the Euphorbiaceae *(Euphorbia bothae)* and the Asclepiadaceae *(Stapelia)*. Convergence also affects plant posture, the formation of leaves and fruit among other things.

Not only ecological but also ethological convergence — that is to say, convergence concerned less with surroundings than with behaviour — is found in different species. Thus the trap constructed by the larva of *Vermileo,* a dipteran, is exactly like that which is built by the ant-lion, family Myrmeleontidae. Both animals dig a conical pit in fine sand and lie in wait at the bottom for their prey.

Convergence occurs not only among individuals belonging to totally different families but also among balanced populations, fauna or flora, which occupy specific and well-defined regions. The vast continent of Australia, which was separated from the rest of the world before the Cretaceous when placental mammals developed, possesses a rich marsupial fauna that gave rise to types which parallel those evolved on other continents by the mammals. There are herbivorous marsupials (kangaroos) which are analogous to the ruminants; rodent marsupials (wombats) which recall the marmot; marsupial insectivores *(Notoryctes)* which resemble the mole; carnivorous marsupials (the thylacine or Tasmanian wolf, and the Tasmanian devil — *Sarcophilus*); fruit-eating marsupials with prehensile tails (phalangers); arboreal marsupials (the so-called koala 'bear'); and marsupials with the aspect of lemurs, three-toed sloths, and so on.

An analogous phenomenon, though less diversified, is found among the lemurs and the insectivores of Madagascar, and in certain South American rodents.

In the Mesozoic era the ruling reptiles, having little or no competition, also evolved a wide variety of types.

410

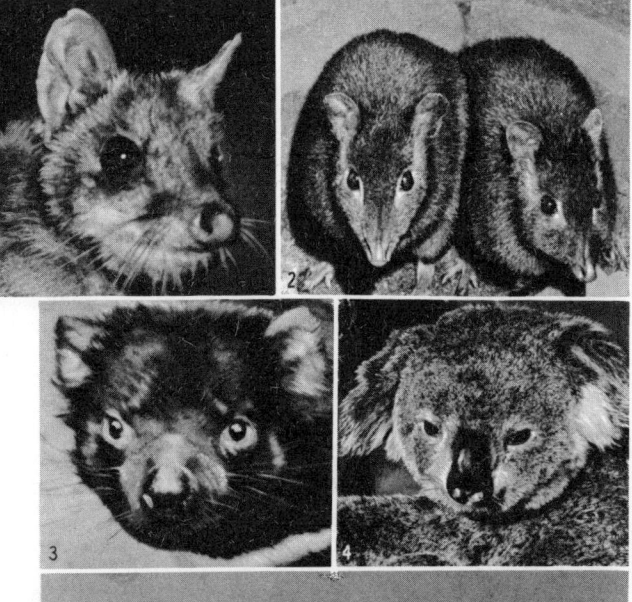

Marsupials: 1. Marsupial cat, *Dasyurus*. 2. Kangaroo-rat. 3. The Tasmanian devil, *Sarcophilus harrisi*. 4. The koala, *Phascolarctos cinereus*. 5. The red kangaroo, *Macropus rufus*. 6. The opossum or sarigue, *Didelphis*. 7. The wombat, *Vombatus ursinus*.

Australian Embassy, Paris.

Organic and functional adaptations

Complete harmony exists between the structure of an organ and the function it fulfils. Without this adaptation neither the digestive tube, the lungs, the kidneys, the heart, nor indeed any other organ could do its job properly. But the same harmony is also observed in numerous arrangements which perform no vital function; such arrangements may seem to be of secondary importance, but they bear witness to profound adaptation.

Means of defence

Means of defence are many and varied, though their defensive value is sometimes difficult for us to judge, owing to our tendency to interpret them from a human point of view. Only experiment can test their efficiency, efficiency which must not be confused with invulnerability; for one single individual saved is evidence that the defensive mechanism is effective.

Flight and autotomy

The instinctive reaction to an attack is flight; a hasty retreat, an abrupt taking wing, a leap backwards, a precipitous dive, and the threatened animal is beyond the reach of the predator.

Try to capture a lizard. You may attempt to prevent its escape by seizing its long tail, but the tail on the slightest contact breaks and the lizard, thus released, scurries away. Similarly the grasshopper, seized by one of its long jumping legs, leaves the leg in your hand and flees. Crabs, harvest-spiders, daddy-long-legs with their exaggerated limbs, and the Ophiuroidea which are for this reason called brittle-stars, also illustrate the same phenomenon which is known as autotomy. Autotomy is a reflex brought about by sudden pressure. If the pressure is insufficient nothing occurs. The rupture is the result of a rapid muscular contraction

411

which snaps off the appendage at its weakest point. The structure of the appendage itself makes autotomy possible; it is as it were prepared in advance for the contingency. At regularly spaced intervals the tail, for example, is divided by planes or sections of autotomy which pass through tissues and vertebrae; in front of each plane of autotomy the caudal artery is furnished with a sphincter while the vein, too, is able to contract. Thus the break is not followed by haemorrhage. In many species regeneration destroys this mechanism and autotomy can occur only once. A new tail or new legs are regrown fairly quickly. The grasshopper does not in fact regenerate its self-amputated leg, but the species is annual.

Some rodents like the field-mouse and the garden dormouse *(Eliomys quercinus)* have the faculty of abandoning the furry skin of their tails. The long tail of these rodents is easy to seize; but at the slightest contact the cutaneous sheath is detached and the animal liberated. Left without covering, the vertebral axis withers away and falls off. After healing the tail is simply shorter. The detachment takes place in a zone of loose or slack connective tissue.

Weapons

Many species capture their prey or above all defend themselves with varied and sometimes complicated weapons. Their defensive and offensive armoury contains hatch-like lids, perforating devices, stylets, spurs, teeth, poisonous pincers, spines, stinging organs, and so forth.

Perforating devices in the form of sharp spines, rough protuberances, pointed quills and knife-edged scales, protect the test of the sea-urchin, the shell of the mollusc, the carapace of the crustacean, the chitinous surface of the insect, the bodies of fish, lizards, snakes, and mammals. One of the best armed

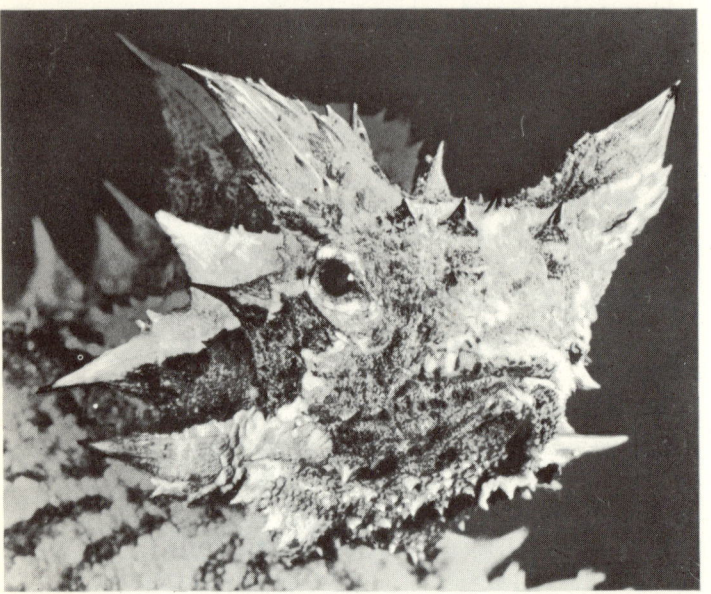

Top right and above right, Dasypus novemcinctus.

Colyann.

Above left, the head of Moloch horridus.

J. Warham.

Left, hedgehog rolled up in a ball.

Vienne-Atlas.

412

lizards, *Moloch horridus,* is a living pin-cushion; sharp horns adorn its head, great spikes rise from its neck, spines and polygonal scales cover its legs and body, while its tail is armed with further spikes. The spurs of the cockerel and the cock pheasant are formidable weapons. In many animals the spines are fixed, but in others they are erectile and can be raised at will. When roused, the hedgehog, the porcupine, the ant-eater, the porcupine fish will raise their quills. Such devices seem to suggest classic measures of military defence, *chevaux de frise* or barbed-wire entanglements. They are, indeed, redoubtable arms and the mechanical nature of the damage they inflict is sometimes rendered more dangerous by accessory poison, the spine or stinger being connected with a gland which secretes venom into the wound. Poisonous spines are frequent in fishes like the weeverfish, the scorpionfish, and the whip-tailed stingray. A complicated stinging apparatus exists in Hymenoptera such as wasps and bees. In some fishes and lizards it is the bite which is dangerous. The bite of the Mexican *Heloderma horridum,* a close relation of the gila monster, is said to be as venomous as that of the rattlesnake.

Poisonous hairs are found not only in plants belonging to the Urticaceae or nettle family but also in animals, and especially in caterpillars with barbed and stinging hairs. Processionary caterpillars are well known for the itching caused by their long, rigid hairs, barbed and brittle. The Cnidaria (jelly-fish, etc.) are provided with complicated stinging organs known as nematocysts. The sting of the medusa can bring about a rash with fever, inflammation and symptoms of poisoning. The powerful nematocysts with which the long tentacles of the Portuguese man-of-war *(Physalia arethusa)* are armed can cause intense pain. The poison ejected by the Portuguese man-of-war is, it would seem, specific in its action; for certain fish living in close association with the animal are able to touch the

The porcupine fish.

Above, the Portuguese man-of-war, *Physalia,* a cnidarian coelenterate, which seizes its prey with filaments whose sting can be dangerous. The animal floats on the surface of the sea by means of a bladder.

D. P. Wilson.

Above left, the skunk, *Mephitis mephitis,* a North American carnivore.

J.-M., Baufle.

Left, mountain thistle.

Rose Nadau.

Means of defence. *Caligo prometheus* of Brazil. *Above*, the upper and *below* the lower surface of the butterfly. The lower surface of the rear wings resembles the face of an owl.

Muséum national d'histoire naturelle. Larousse.

formidable filaments without danger.

Plants that are furnished with spines, such as *Eryngium campestre,* cactuses, and certain euphorbias, are left alone by the ruminants. In deserts the only plants which persist are thorny bushes, respected by the herbivores.

Armour

Many animals are enclosed in protective armour. The skin of the armadillo and the tortoise has hardened into a carapace. The thick cuticle of the arthropods is often calcified. Molluscs enclose themselves in a resistant calcareous shell which if damaged is soon repaired. Armoured protection is sometimes supplemented by the animal rolling up into a compact ball. Such is the practice of the three-banded armadillo, *Tolypeutes tricinctus,* the millipede *Glomeris,* the terrestrial and marine isopods *Armadillidium* and *Sphoeroma.*

Defensive secretions

Many animals have special glands which secrete foul-smelling and sometimes poisonous substances. Among the small carnivores which defend themselves by this means the striped skunk, *Mephitis mephitis,* has an unsavoury reputation. Within its anus two muscular-walled perineal glands open and eject a mephitic secretion for distances of several yards. The secretion, rich in sulphurous products, infects the atmosphere for a considerable time. Teledu, the Malayan stink badger, ejects a sticky liquid which is as malodorous as that of the American skunk. The European polecat, *Putorius putorius,* can also spray its enemy with a secretion of nauseating smell.

Some millipedes have cutaneous glands which secrete liquids containing quinone (the genus *Julus*) or prussic acid. The acidic and odorous secretion emerges from the animal's dorsal pores.

Many ants are supplied with a gland which under pressure emits the formic acid it contains. The large red ant which inhabits pine forests squirts formic acid to a height of half a yard, and those who gather the nymphs of this species — much sought after by pheasant breeders — suffer from skin which reddens and peels in consequence.

Reflex bleeding is observed among certain beetles, for instance Cantharidae and Coccinellidae or ladybird beetles. The Spanish fly, *Lytta vesicatoria,* when touched by an enemy, falls on its side as though dead. Drops of pale yellow blood ooze from the tibio-femoral joints of its six legs; the drops contain a powerful toxic substance, cantharidin. The Spanish fly waits for a few minutes, then gets to its feet and, having no more to fear from its assailant, crawls off. A cockchafer smeared with the blood of the Spanish fly is no longer eaten by its customary enemy, the carabid beetle.

Plants also employ chemical means of defence. Plants protected by a disagreeable taste, often a warning that they are poisonous, thrive in pastures. Animals refuse to browse on poisonous buttercups, bitter gentians, and the toxic *Veratrum,* or false hellebore. Cattle avoid the autumn crocus, leaves and flowers

The Australian collared lizard *Chlamydosaurus kingii* as it normally appears. *Below,* in its attitude of defence. When its collarette is fully extended it forms a large and highly coloured ruff in the centre of which gape jaws, terrifying in appearance.

Zoological Society of London.

alike, with its bitter flavour. An Indian tree, *Melia azedarach,* which is naturalized in Algeria, contains a repellent substance and is left untouched by locusts. Extracts of this substance are successfully used as a spray to protect crops from insects.

The secretion of sticky substances also serves to protect an organism from its enemies. *Holothuria forskali,* a sea-cucumber found off the coasts of England and France, is known as the 'cotton-spinner' because when irritated it abruptly ejects from its cloaca some half a dozen long, white sticky filaments which adhere to all they touch. The animal's prey, crab or fish, becomes entangled in these clinging filaments and the more it struggles to free itself the more it becomes enmeshed.

The slug, when seized, quickly contracts and exudes a glistening, sticky slime; and generally speaking carnivores leave sticky creatures alone.

Certain soldier termites are provided with an enormous gland with an outlet at the extremity of a prolongation of the head; when this gland contracts a sticky liquid is ejected.

415

Oedipoda coerulescens on bauxite.

'Measuring-worm' on a cystus twig.

The phasmid, or stick insect, *Clonopsis gallicus,* on a grass.

Oedipoda coerulescens on quartose gravel.

Oedipoda coerulescens on a blue mineral background.

The caterpillar *Poecilocampa populi* on an alder.

Satyrus hermione on quartose rock.

Feigned death

Some animals defend themselves by pretending to be dead. At the slightest alarm certain insects collapse and remain totally motionless. These tactics are usually rewarded with success: predators like frogs and lizards who attack moving objects are duped, while birds no longer see prey which has disappeared in the grass or the undergrowth.

Bravado

Certain animals, when attacked, assume a terrifying aspect; like blustering bullies they attempt to intimidate their adversaries. Everyone has seen the cat when angry, the flattened ears, the hair on end, the arched back and lashing tail. The ferocious appearance of the mantis in the presence of a lizard is characteristic: the mantis faces its enemy, spreading, its wings and wing-sheaths, while the jerky rubbing of its abdomen against the wings produces a menacing sound; its predatory legs are drawn back before its head to display the black patches on the underside of its haunches. When prepared to defend itself the East African mantis, *Pseudocreobotra wahlbergi,* raises its wing-sheaths or elytra to show the great ocellus with which each is adorned. The impression given is that of two enormous eyes belonging to some strange and terrifying animal.

The Australian lizard *Chlamydosaurus* startles its enemy by raising its head and ruffling out the disproportionately large and brightly coloured encircling membrane.

In repose the caterpillar of the hawk noth, *Leucorhampha ornatus,* resembles a branch covered with creamy white lichen. When frightened it twists to show the underside of its forward segments which are adorned with an olive green band. Its minute legs, pressed against the body, are invisible. As its thoracic segments swell two spots shaped like eyes lend it the aspect of a little snake.

To the natives of the African savannah the small lizard *Psilodactylus caudicinctus,* allied to the geckos, is the lizard with two heads. It has, in fact, a tail which retracts and swells and assumes the exact shape and size of the head; both extremities have the same markings and in appearance are amazingly similar.

When disturbed the caterpillar of the moth *Stauropus fagi,* which attacks the beech tree, raises the forepart of its body and takes on the terrifying pose of a spider. Its legs lengthen and quiver as though on the point of seizing its victim. The rear of its body is also raised until, with its two appendages suggestive of antennae, it curves forward above the head.

Coloration

Adaptive coloration

The general term adaptive coloration designates the various means by which an animal, in order to escape its enemies, remains inconspicuous in its normal environment.

The colour and markings of animals which live in well-lighted regions are varied, while animals which frequent caves or darker habitats are dull or colourless. In a ploughed field the hare crouching in the furrow is practically invisible; the colour of desert animals merges into the general shade of the surrounding sands. In the snow polar bears and foxes are white.

This great variety of protective coloration may take several forms which we shall now examine.

Cryptic coloration

Coloration which conceals an animal by its resemblance to the environment is known as cryptic coloration. It may take the form of brightening the colour of the animal which lives in bright surroundings or of darkening the animal which, though a daylight creature, inhabits darker surroundings. This phenomenon is common in fishes, frogs, lizards, and crustaceans. It may take a more advanced form and result in a quasi-identity of colour between the animal and its environment. The green tree-frog lives in green-leafed trees. Various insects, grasshoppers for instance, are the same colour as the grass or foliage they frequent. Identity of colour may be permanent or changeable. In the latter case colour alters rapidly to harmonize with that of the animal's situation. The case of the chameleon whose colour varies from greenish to brownish is proverbial. Other lizards, certain flatfish, cephalopods, and crustaceans also have the faculty of modifying their colour.

Similarity of colour is sometimes reinforced by a further phenomenon, namely similarity of form. The animal's body then copies not only the colour of its surroundings but also their shape and apparent texture. Thus it will resemble a twig, a piece of bark, a dead leaf, a bit of seaweed, or a pebble. At times the animal's posture may accentuate the resemblance. In this way the measuring worm which has stiffened into total immobility and the phasmid or stick insect attached to the stem of a plant look exactly like small lateral twigs.

By these varied devices cryptic coloration helps the animal to remain inconspicuous. In contrast there also exists the phenomenon of aposematic conspicuous coloration.

Conspicuous coloration

Brilliant and gaudy colours in designs calculated to catch the eye naturally make an animal particularly visible. Such conspicuous coloration is seen in ladybirds, the harlequin frogs of tropical America, fishes which live in coral reefs, red slugs, yellow wasps ringed with black, red and black wood-infesting bugs, certain snakes ringed with black and red or black and yellow. These animals, diurnal and slow to move, do not hide themselves. In general they are provided with effective weapons of defence: stings, repellent or poisonous glands, mucus, and so forth. They are thus relatively inedible. Now predators learn by trial and error to recognize whether prey is or is not edible, and if inedible species are flagrantly conspicuous the predator can all the more easily identify them. Conspicuous, or aposematic, coloration thus acts as a warning and has been aptly compared with the red label on bottles containing poison. Wallace suggested that the connection is so close that it is possible to predict that a species of striking colour is not edible and that it possesses adequate means of defence.

Certain animals are dull and inconspicuous in repose. When resting their coloration is cryptic; but when they move they suddenly reveal regions of their bodies which are of a brilliant scarlet, orange, yellow, blue. The phenomenon is known as 'flash-colour'. Flash-colour is not uncommon in Lepidoptera. *Catocala nupta* has greyish, cryptic forewings as it rests flat against a wall; when it flies away its hind wings, red with two broad black bands, abruptly appear. The eye, struck by the flash of red, loses all trace of the moth as it once again alights and hides its coloured patches. The same phenomenon is seen in the acridians or grasshoppers. Many exotic insects combine the two types of coloration, cryptic and conspicuous.

Some animals are adorned with brightly coloured ocelli which give an impression of large eyes. Such ocellate spots are found, for example, on the fins or flanks of certain fishes, on the tail feathers or wings of birds, notably peacocks, on the elytra and wings of insects, and on the forward segments of certain caterpillars. On the hind wings of *Sphinx ocellata* there are handsome black and blue ocelli on a pink background; in repose they are invisible, but appear when the moth spreads its wings. We have observed the same occurrence in the mantis *Pseudocreobotra wahlbergi*. The ocelli are designed to frighten the assailant or to distract its attention from vital regions of the animal. They are situated on the extremities of the wings, in other words at a distance from the animal's body; and it is not uncommon to find evidence in the ocelli of butterflies that birds have pecked at them. The ocelli, in fact, serve to misguide the enemy, to deflect him from his target.

Mimicry

Mimicry is the outward resemblance of one species to another to which it may be entirely unrelated. The resemblance has no connection with the two species' mode of life, but is simply a protective measure on the part of the mimicking species, which is generally edible and unprotected. It makes no effort to avoid detection; on the contrary it makes itself conspicuous so that the possible predator will confuse it with the species mimicked, as a rule a species brilliantly coloured and provided with properties, defensive or repellent, which render it inedible. In this way the mimic gains protection which it would otherwise lack. Predators, deceived by false warning signals, refrain from attacking it. The coloration of the well-defended animal is, it will be remembered, known as aposematic. By analogy the coloration of the mimicking animal is called pseudo-aposematic.

Certain elements are characteristic of mimicry. In the first place the mimicking species inhabits the same areas as the species mimicked. It is ill-protected or unprotected, while its model defends itself actively against predators by poison and the warning coloration which announces poison. The mimicking species is much rarer than the species it imitates and differs from normal members of its group only in a superficial appearance designed to create illusion. In reality its outward appearance indicates no affinities or connection with the animal mimicked. Its enemy, however, is misled into leaving it alone. Numerous examples of protective mimicry are found among the butterflies of South America, Africa, and Asia. Butterflies of .the Ithomiidae and the Heliconiidae groups serve as models frequently copied by other butterflies. The most extraordinary mimic belongs to the family Pieridae a family which is otherwise uniform and in colour generally white or lemon-yellow. Only by the closest examination can *Dismorpha fortunata* be distinguished from its model, *Leucothyris quintina*, family Ithomiidae. *Papilio paradoxa,* which has lost the characteristic markings of its group, copies *Euploea diocletianus,* family Danaidae, with its odorous organs. Wasps are frequently imitated by butterflies, by Coleoptera, Diptera, and even by Orthoptera. The moth *Aegeria apiformis* provides a classic example of a moth mimicking a hornet. The colour of its thorax and abdomen, and the shape of its abdomen, exactly copy the hornet; even its flight is unlike that of a moth and recalls that of the Hymenoptera. In addition, these moths plunder flowers, behaviour which adds to the confusion between model and copy. There are also families of exotic butterflies which mimic the wasp or other hymenopterans with astonishing exactness.

The Asilidae (robber flies, hawk flies, etc.) are vigorous and voracious Diptera. Normally grey or yellow with transparent wings, there are also darker varieties with bands of white bristles and bluish-black wings. These latter, *Hyperechia bifasciata,* resemble large carpenter bees of the genus *Xylocopa,* and furnish an example of a dipteran mimicking a hymenopteran.

Ants are frequently mimicked by various insects and also by spiders. Ladybirds, which are not eaten by insectivores, are copied by other coleopterans, by certain cockroaches and even by one spider.

Mimicry is uncommon in vertebrates. The most famous example is that of the venomous coral snakes of the genus *Elaps* which are imitated by non-poisonous snakes. *Homalocranium semicinctus* has the same appearance as *Elaps corallinus;* both are brilliantly coloured and ringed in red and black. The resemblance is so great that the innocuous snake, mistaken for the poisonous snake, is left severely alone by its potential enemies. *Ophiphthys colubrina,* an eel of the muraenoid family, is an exact copy of the very poisonous snake *Laticauda colubrina,* and even an expert ichthyologist cannot at first sight say which is which.

The cases we have considered in which the predator is deceived by spurious warning coloration are examples of 'Batesian mimicry', so called because the great English naturalist H. W. Bates was the pioneer in this field of study.

Protective colouring.   *Saga pedo* in foliage.

J. Vanden Eeckhoudt.

The phasmid, *Bacillus rossii* on *Calycotome spinosa*.

The tree-frog *Hyla arboraea* on ivy leaves.

The gecko, *Tarentola mauritanica,* on mica-schist.

Mimicry.

Senckenberg Museum. Mertens.

Micrurus frontalis, the poisonous coral snake of Brazil which, when in danger, raises the tip of its tail.

Simophis rhinostoma, also Brazilian but not poisonous.

Another type of mimicry was established by Fritz Müller. In 'Müllerian mimicry' a general resemblance of colour and shape is observed between species belonging to the same group, all of which possess warning coloration and are well-defended. Three families of South American butterflies, the Danaidae, the Neotropinae, and the Heliconidae, are avoided by predators because of their disagreeable taste. Gaudy colours and markings of the same type give them an air of family resemblance which is in fact spurious. But their apparent similarity results in fewer losses during the apprenticeship of their predators; for young birds must learn to distinguish insects which are edible from those which are not.

Finally there is a type of mimicry for which there is no very clear explanation, namely parasitical mimicry. It takes the form of an astonishing resemblance between the host and its commensal associate or its parasite. A rove beetle, *Mimanomma spectrum,* an ant-associate of the Cameroons, not only lives with the blind ant *Anomma nigricans,* but looks like an ant; only by examining the tarsi and the buccal elements can the beetle be recognized. The spider *Myrmarachna formicaria* also resembles an ant; its two raised front legs vibrate like antennae and, like an insect, it walks on six legs.

Camouflage

Before concluding our survey of protective coloration and the various forms it takes, the question of camouflage in general must be considered. The object of camouflage is, by one means or another, to make the animal as indistinguishable as possible from its background.

Conspicuous markings break up the animal's silhouette and in this way make it hard to see. An animal's outline is accentuated when the contour is repeated by its coloration, for coloration then acts as an additional outline or frame. But if it cuts the contour perpendicularly the outline is broken and the animal camouflaged. The zebra with its stripes is strikingly visible from a short distance, but in the savannah at a distance of some fifty yards it vanishes, and game-hunters report that its perpendicular black and white stripes merge into the landscape. This is true of all those animals which, like the tiger, have their contour broken up by perpendicular stripes. Contrasting blotches of black on a white background similarly conceal the animal. The larger snakes have various markings in which contrast is emphasized by white edges bordering dark patches. The chicks of numerous coloured species which, like the ringed plover, the woodcock, the quail, and the bittern, are hatched on the ground profit from this principle of broken contours and in consequence are hard to see against the earth or sand.

Protective coloration may also serve to disguise an animal's appendages. A good example of this is provided by a small frog of the lower Zambesi, *Megalixalus fornasinii.* The frog, very conspicuous, has a brown band separated by two white bands in the middle of its back and on its flanks; there is also a brown band and a white band on its legs. When the frog is resting the bands of the legs form a prolongation of the bands of the body, so that the familiar silhouette of a frog is no longer noticeable. Coloration similarly designed to mask the animal's shape is not uncommon among insects, fish, and reptiles.

The vertebrate eye, with its brightness and the regular form of the pupil and iris, is generally very visible, and efforts are made to camouflage it. Often it is lost in a black irregular band which seems to form part of a design — for example in fish, frogs, snakes, birds, and antelopes. Often a coloured band is disposed in such a way that it crosses the eye and tends to make the black depths of the pupil disappear. Many vertebrates instinctively close their eyes in full daylight or when in danger; the lid which protects the eye is often camouflaged and exactly resembles the region above the eye-sockets. The eyelids may be covered with scales or with feathers. In the Australian lizard *Moloch* the eyelids are not only scaly but are crossed by a coloured band perpendicular to the palpebral fissure.

The presence of an inconspicuous animal may be betrayed by its shadow, which is sometimes more obvious than the animal itself. Thus many crepuscular moths which spend the daylight hours on tree trunks flatten themselves against the bark to avoid casting a shadow. The Satyridae, in repose, raise their wings and press them together so that only the dull under-surfaces can be seen. But as well as this they face in the direction of the sun so that the thin line of their shadow is as inconspicuous as possible. Many lizards lie flat against the ground and the shadow they throw is masked by lateral fringes.

An animal's presence is betrayed not only by its outline but, being three-dimensional, by its relief. Most invertebrates and almost all fishes, amphibians, reptiles, and mammals have more highly coloured backs than bellies. The more intense lighting to which the back is normally exposed reduces this contrast of pigmentation and, when the animal is seen at a certain distance, creates the illusion of a flat surface. In this way small marine animals, swimming over a sandy bottom, are invisible. Sometimes this colour scheme is reversed: in certain animals which swim upside down the belly is more highly pigmented than the back. A Nile catfish, and the pelagic Nudibranchiate mollusc *(Glaucus atlanticus)* have silvery backs on which they swim, and dark blue bellies which are visible from above. The caterpillar of *Sphinx ocellata* lives on a willow; its back is pale green and its belly a dark greyish green. But very frequently the position the caterpillar takes up is the reverse of normal. It then seems to merge into the leaf. Its rear pair of legs is fixed to a branch, and its pale back is turned towards the ground, its darker belly towards the light; its oblique lines give an impression of the play of light and shade on the veins of a leaf.

Coloration designed to lower the animal's relief sometimes takes the form of a gradation of colours. The cheetah, seen from a distance, appears to be greyish and flat, thanks to a yellow skin dotted with black patches which are dense on its back and increasingly spaced out as they approach its underside. The

Commensalism and symbiosis.

P. Pesson.

The sea-anemone *Anemone sulcata*.

Symbiosis of the hermit crab *Eupagurus bernhardus* and the sea-anemone *Sagartia parasitica*.

brilliant colours of most coral-reef fishes are superimposed on a tinted background in shaded designs.

The biological value of adaptive coloration

At the beginning of the century certain biologists saw protective colouration and mimicry everywhere. Such exaggeration roused other biologists to affirm that cryptic coloration and mimicry were figments of the imagination and had no significance whatsoever. Both of these extreme positions are, as we might expect, false. An objective answer to the question can be arrived at only by experiment. Varied and repeated experiments have, in fact, revealed the protective value of cryptic coloration, of warning or aposematic coloration, and of mimicry.

Protective coloration is of course of value only when predators can see their prey: fishes and cephalopods in the sea, amphibians, reptiles birds, and mammals on land. But some predators are guided by other senses, smell and feeling, and colour schemes cannot interest

cryptic coloration and resemblance to surrounding objects are found in the sea and in fresh water: *Histrio histrio* and *Hippolyte acuminata* of the Sargasso Sea, *Phycodurus eques* of Australia which resembles algae, *Monocirrhus polyacanthus,* sometimes called the 'leaffish' of the Amazon. On the other hand, only one case of mimicry is known. But camouflage of the type unobserved by predators can have no protective function.

On land, questions of colour are certainly of no interest to crepuscular or nocturnal animals. On the other hand birds and the primates with their acute eyesight can be deceived by camouflage, and it is among these animals that the majority of experiments designed to verify the protective value of cryptic coloration and mimicry have been undertaken.

Isely chose locusts gathered from three types of terrain, pale, dark, and red. The locusts, whose coloration normally matches their habitat, were whitish, blackish, and reddish respectively. They were then anaesthetized or otherwise affixed to terrain which was sometimes matching and sometimes contrasting in coloration; there they became the prey of various birds, with the following results.

Predator	Colour of Prey and Terrain	Prey Eaten	Prey Uneaten	Total Prey
Cocks	Matching	81 (44%)	104 (56%)	185
	Contrasting	157 (85%)	28 (15%)	185
Wild birds	Matching	39 (34%)	75 (66%)	114
	Contrasting	96 (84%)	18 (16%)	114
Turkey cocks	Matching	37 (46%)	43 (54%)	80
	Contrasting	73 (91%)	7 (9%)	80

them. Fish can see, but poorly; they are much more aware of movement. Thus true protection in the water depends on absolute immobility, on rapid flight, or the sudden leap or dive. And yet numerous examples of

In an experiment of the same type the locusts were left free to move at will over a matching or contrasting background before the arrival of hens and turkeys. The results were even more striking:

Predator	Colour of Prey and Terrain	Prey Eaten	Prey Uneaten	Total Prey
Hens and Turkeys	Matching	9 (30%)	21 (70%)	30
	Contrasting	30 (100%)	0	30

The gastropod mollusc *Sipho curtus* with, on its shell, a commensal, the sea-anemone *Allantactis parasitica* and, between its shell and the sea-anemone, a ribbonworm, *Nemertopsis actinophila*.

After P. H. Winther.

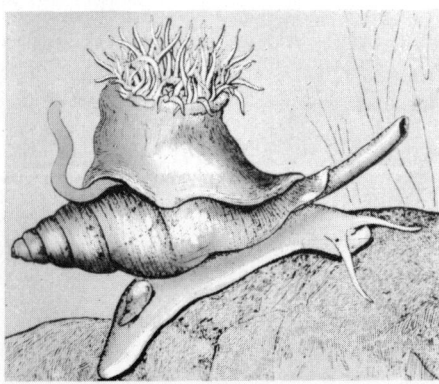

Eupagurus bernhardus entering a vacant shell.

X Photography.

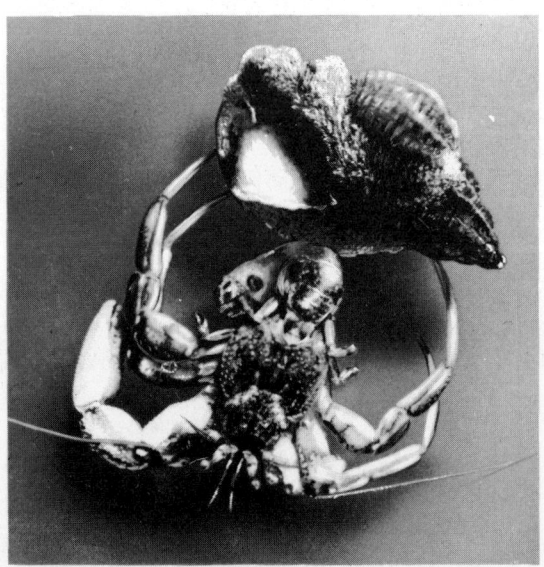

Some animal tools.

1. Predatory feet of a water scorpion (Nepa rubra).

Six.

2. Siphon of a water scorpion (Ranatra).

Six.

3. The 'parachute' of the flying lemur (Cynocephalus).

Beamish-Polunin.

4. The diving-bell of the water spider.

R. H. Noailles.

5. The suckers of the octopus (Octopus vulgaris).

Six.

1

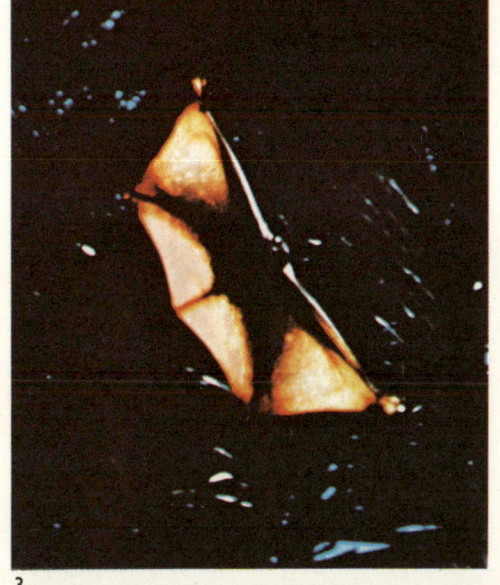

2

3

5

4

425

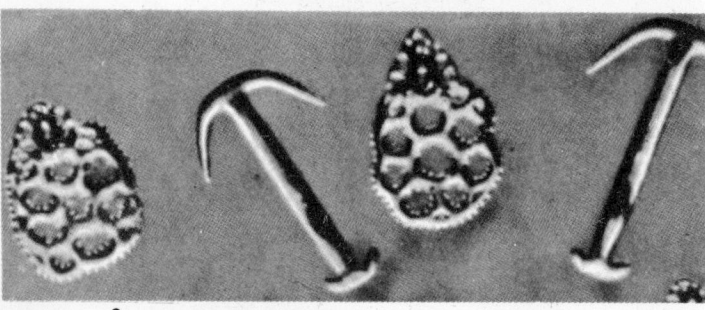

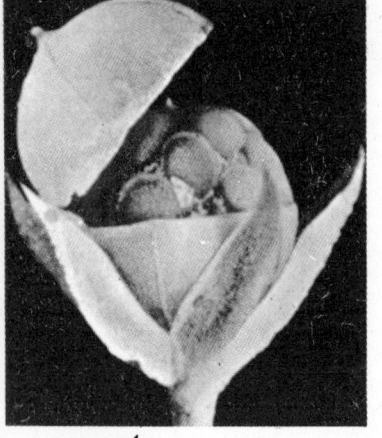

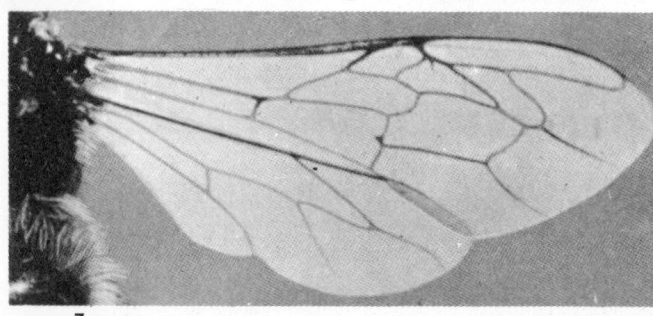

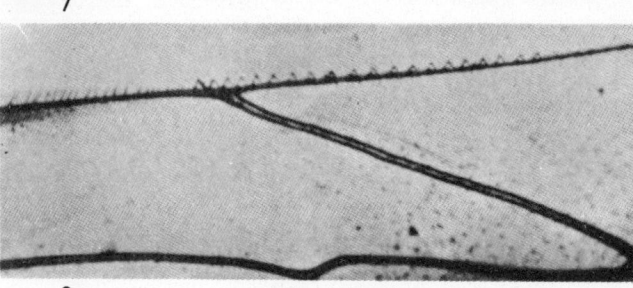

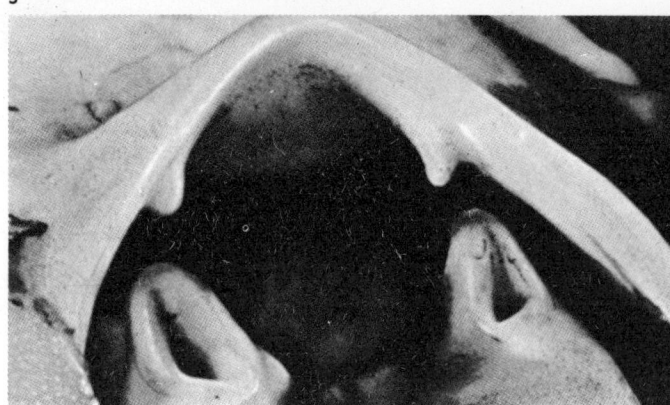

1

2

3

4

5

6

7

8

9

Some hooking devices in plants and animals.

1. Fruit of the agrimony (Rosaceae).

R. H. Noailles.

2. Anchors and dermal plates of *Synapta* (Echinodermata).

L. Plouvier.

3. Tentacles of the octopus with suction discs.

Larousse.

4. Opening of the capsule or pyxidium of the scarlet pimpernel, *Anagallis arvensis.*

R. H. Noailles.

5. Foot of a male diving beetle.

R. H. Noailles.

6. The press-stud of a cuttlefish.

Larousse.

7. and 8. The wings of a bee. The insect's front and rear wings are held together by minute hooks known as hamuli.

Larousse.

9. Radula or 'rasp' of a snail, enlarged 450 times.

R. H. Noailles.

426

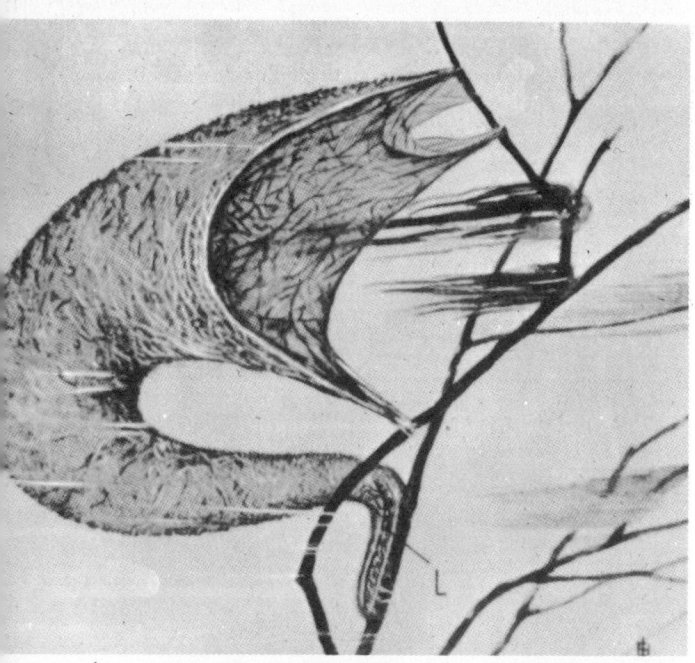

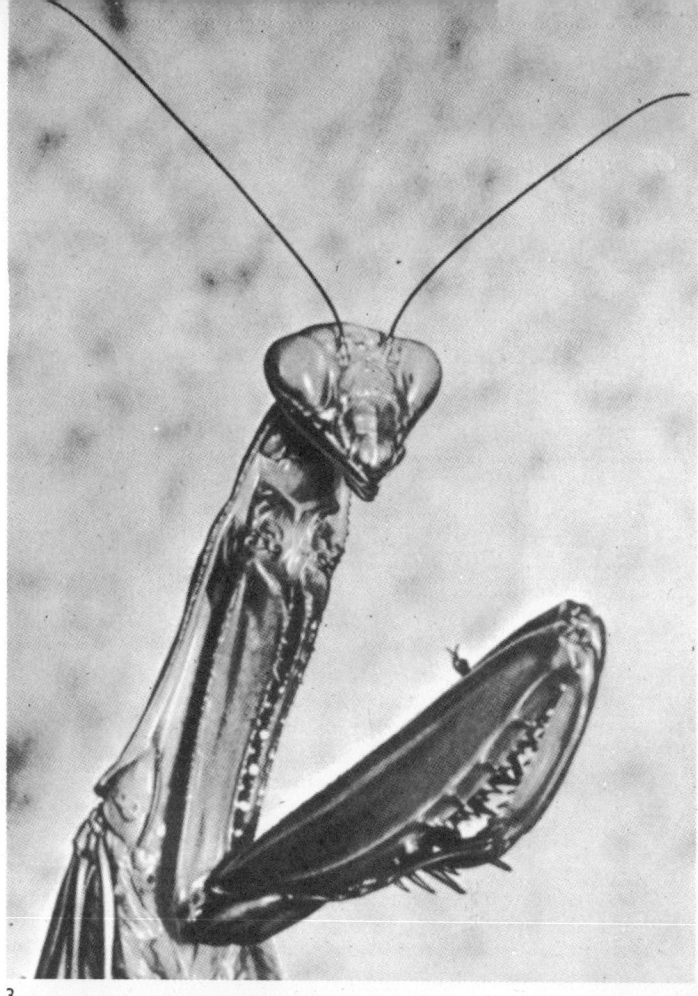

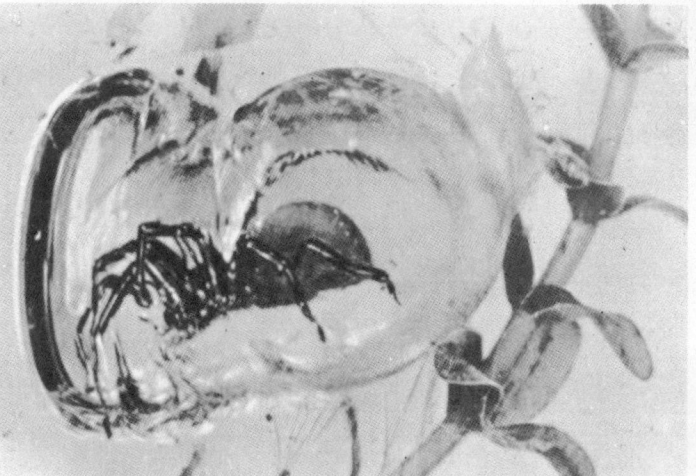

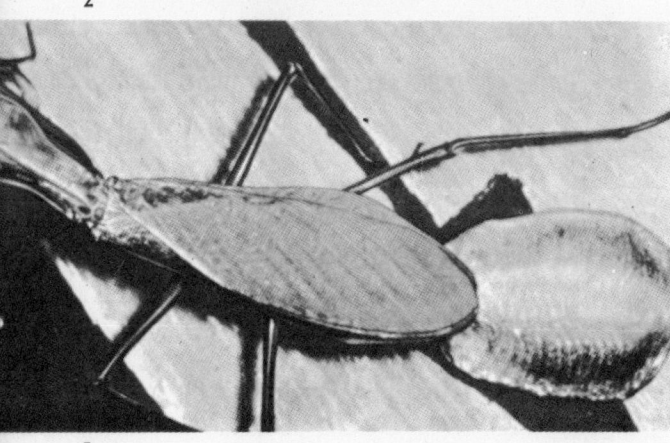

1. Snare of the larva of *Neureclepis bimaculata*.

2. *Nepa* or water scorpion with a pair of predatory legs which recall those of the mantis.

R. H. Noailles.

3. Head and predatory forelegs of a female praying mantis.

J. Vanden Eeckhoudt.

4. The water spider *Argyroneta aquatica* in its diving-bell.

R. H. Noailles.

5. The African mantis *Sphodromantis* having just deposited its ootheca or egg case.

P. Pesson.

More recent experiments show that *Oedipoda coerulescens* and *Acrida turrita,* members of the acridian family which includes locusts and grasshoppers, can discern colours; this faculty is lost when their eyes are painted over. These insects actively seek out backgrounds which merge with their own coloration. Chameleons introduced into a cage full of Acrida ate every insect which was not on a background of its own colour. Similarly in an environment of its own colour, the water-bug *Corixa* is three times as safe from its normal enemies, the diving beetle *Dytiscus,* larval dragonflies and fish, as it is in surroundings of contrasting colour.

More numerous experiments have been made to test the protective value of aposematic or warning coloration, and research has been carried out among predators of various kinds: mammals, birds, reptiles, amphibians, and fish. An examination of 244 animals captured by long-tailed African monkeys of the genus *Cercopithecus* showed that of 143 aposematic species 120 were avoided by the monkeys and twenty-three eaten. Of 101 species of dull or cryptic coloration eighty-three were consumed and eighteen were not. Generally speaking, other monkeys, bats and lemurs also devour cryptic species while refusing those with warning coloration. In this respect birds, reptiles and amphibians behave in the same way. When dull-coloured insects and aposematic insects were fixed alive to branches near the nests of wrens, willow-warblers, and hedge-warblers eighty-six per cent of the inconspicuous species were eaten as against only 4.9 per cent of the conspicuous.

Experiments also confirm the protective value of mimicry. The lizard *Anolis* never eats Lycaenidae of the genus *Thonalmus,* brilliantly coloured in red, orange, blue. *Thonalmus* is mimicked by other butterflies which live in its neighbourhood, and the lizards will eat neither model nor mimicking species. For six days Darlington observed that *Thonalmus* and its mimics were left strictly untouched while six other totally dissimilar species were devoured with relish.

The results of various experiments thus agree in confirming that cryptic coloration, warning coloration, and mimicry are of real protective value. But it is by no means certain that all camouflage is designed as a protection against predators. Camouflage could also have a connection with light, heat, or other forms of radiation. If this is so, certain apparently meaningless camouflage may in fact serve a purpose which is still unknown.

Commensalism and symbiosis

Commensalism or close association with a species which is non-comestible or particularly well armed is a defensive measure which is often found in nature. In addition to the protection afforded an exchange of food sometimes takes place. Association of dissimilar organisms to their mutual advantage is known as symbiosis.

The most sought-after protectors are the Cnidaria or jellyfish, animals particularly respected for their batteries of nematocysts which act like poisoned arrows. Next in popularity come the inedible sponges, the echinoderms with their tough or spiny coverings, and the bivalves protected by their shells. Most actinians (sea-anemones) give shelter to a pair of fish; without such protection both male and female fish are devoured by predators. Hence the commensals do not in their search for food venture far from the sea-anemone, and at the slightest signal of alarm speedily regain its sheltering tentacles. It is possible that the fish contribute a little to the sea-anemone's diet. The mollusc *Sipho curtus* of Greenland carries an associate or commensal on its shell, the actinian *Allantactis parasitica;* and in twenty cases out of 100 a ribbon-worm, *Nemertopsis actinophila,* lodges between the shell of the mollusc and the foot of the sea-anemone, deriving its nourishment from the gastral cavity of the latter.

Hermit crabs with their soft abdomens take shelter in empty shells. One hermit crab, *Eupagurus bernhardus,* often affixes to its adopted shell one or more sea-anemones of the species *Sagartia parasitica.* When it changes shells it removes its sea-anemones and re-attaches them to its new lodging. In other regions the hermit crab's borrowed shell is covered with a colony of hydroids. Hydroids and sea-anemones, well supplied with nematocysts or stinging organs, provide the hermit crab with protection against its enemies, in particular against the octopus. But in the case of *Eupagurus bernhardus* commensalism is not obligatory, for it sometimes inhabits bare shells.

On the contrary the association — or in this case symbiosis — is permanent between another species of hermit crab, *Eupagurus prideauxi* of the Mediterranean, and the sea-anemone *Adamsia palliata.* The hermit crab shares its food with *Adamsia.* When young the two species are capable of living independently, but later their association becomes indispensable. Separation is then fatal to both animals, for the hermit crab is rapidly devoured while the sea-anemone is unable to feed itself. Still other hermit crabs live in association with inedible sponges.

In these associations, commensal and symbiotic, the protected does not bring the protector's defensive weapons into play. Thus the sea-anemone's commensal fish do not release its stinging organs; the hermit crab *E. prideauxi* is almost totally immune to the poisonous nematocysts of *Adamsia.*

The organs and tools of living creatures

Living organisms possess certain organs of primary importance which fulfil essential functions, organs which might be compared to tools or industrial equipment. It is a commonplace to say that the heart works like a pump, a lift-and-force pump, that the kidneys act as a filter, or that the eye resembles a camera. The Greek word for 'organ' and the Latin word for 'tool' mean exactly the same thing. An organism is thus, literally speaking, a box of tools,

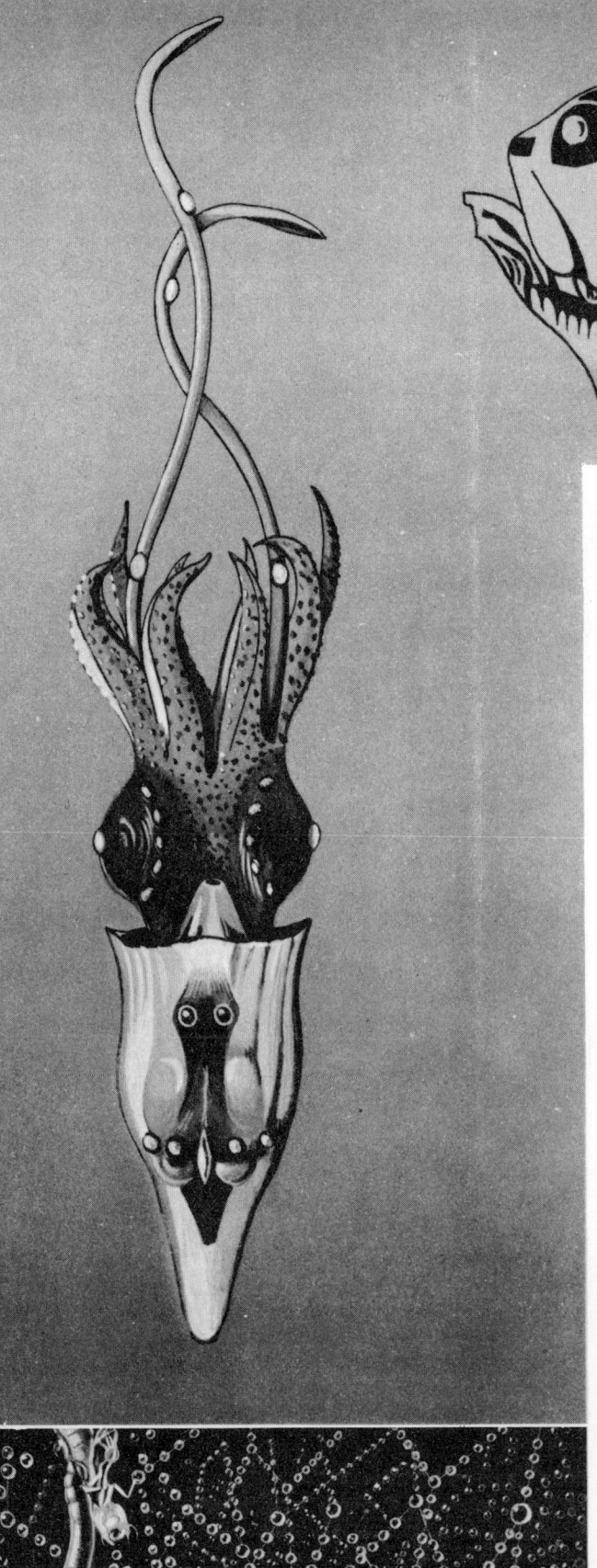

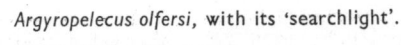

and the living creature has a collection of implements of graded importance.

In addition to the indispensable organs the organism is supplied with a number of secondary arrangements which are not essential to life but are nevertheless of interest and value. Thus in animals and plants we find a wide selection of such devices as press-studs, jack-knives, and powder-boxes, various hooking contrivances — hamuli, suckers and hooks of all kinds, tongs, trap-doors, nets, flanges, snares, gliders, parachutes, floats, lamps, electric batteries, ejectors, thermos flasks, arms of all kinds, and even of musical instruments — violins, cymbals, and xylophones.

The form and structure of such appurtenances have been widely studied and described. They fall into two categories, one comprising structural units often of great complexity and the other two-part devices which fit together with astonishing precision. To the latter category belong the press-stud, the jack-knife, and the hook or hamulus.

The press-stud

The press-stud, by means of which two distinct parts of the body are united, is found in certain arthropods, both insects and crustaceans, and also in some cephalopod molluscs. Aquatic Hemiptera (bugs) of the Hydrocorixidae family, such as *Nepa, Ranatra, Nototecta, Corixa* or water boatmen, possess the typical press-stud. It unites the outer edge of the hemelytron (or thickened front wing) to the thorax. In *Notonecta* the thorax carries a protruding knob, hard and chitinous, which fits into a circular depression on the outer edge of the hemelytron. Adherence is reinforced by the scaly nature of the two surfaces in contact. As the insect flies away its wings are freed and the click of the press-stud as it comes apart can be heard. The aquatic life of these insects means that the device

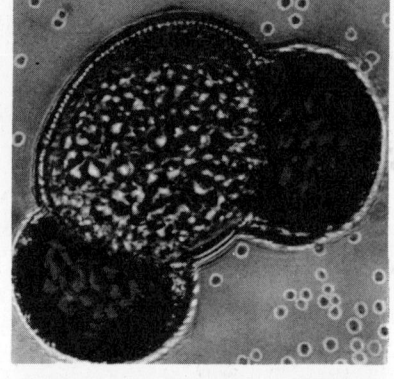

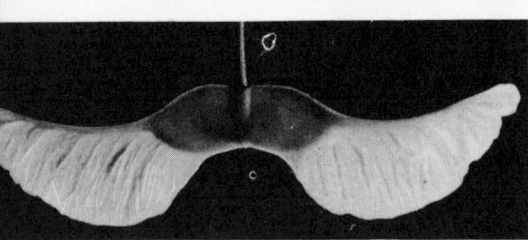

Top left, a human implement copies a plant implement: salt-sprinkler and the capsule of a poppy.

Larousse, R. H. Noailles.

Top right, grain of pine pollen with its two tiny air-filled balloons.

R. H. Noailles.

Above left, samara, or winged achene, of the maple.

R. H. Noailles.

Above right, the winged fruit of the birch.

R. H. Noailles.

Right, fruit of the salsify (*Tragopogon pratensis*) lands with a parachute.

R. H. Noailles.

is particularly valuable. *Notonecta,* when deprived of its press-studs, dives with difficulty; since the hemelytra are no longer affixed to the thorax air accumulates in the intervening spaces. This air acts as a float and the insect tends to rise to the surface. If the hemelytra are artificially stuck to the thorax the insect can again dive normally. The mutilated insect will, however, swim like a normal insect after a few hours have elapsed, though the period of time it stays under water is shorter. The press-studs of the other Hydro-corixidae have the same general structure with certain modifications of detail.

The press-stud is commonly found among cephalo-pod molluscs in which it fastens the lower ventral edge of the 'funnel' to the internal surface of the 'mantle'.

Its position is fixed, but its structure varies in different families. The best developed is that of the cuttlefish which fits tightly and allows the animal to swim backwards. Water in the mantle cavity, forcibly ex-pelled by the cuttle-fish, can escape only by the narrow mouth of the funnel, and the jet thus produced shoots the animal backwards like a torpedo. The press-stud relies on the active participation of the muscles and ceases to function with the animal's death. In repose, the elasticity of the muscles keeps the device closed and its solidity is further increased by the muscular contraction which accompanies swimming. The cephal-opods which swim most strongly are, in fact, endowed with the most efficient press-studs, reinforced by cartilaginous elements.

Hamuli

The flight of insects with two pairs of wings requires an automatic coupling of forewings with hind wings, for without coordination of wing-movements the single plane which makes for efficient flight could not be maintained. Such coupling is achieved in various ways. Among the Hymenoptera (wasps and bees) it is assured by hamuli (Latin *hamulus,* little hook). The rear margin of the forewing is grooved while the front

Fruits with parachutes.

R. H. Noailles.

Left, clematis *(Clematis vitalba)*; above, coltsfoot, *(Tussilago farfara)*; right, *Anemone pulsatilla*; below, dandelion, *Taraxacum officinale.*

edge of the hind wing is provided with hamuli with two bends, one at the base and one at the extremity, so that the hook is not too sharp. When not in motion the wings are superimposed, but as they spread to fly the grooved forewing engages the hooks of the hind wing. The second bend of the hooks helps to secure this contact and enables the hooks to slide in the grooves. The number of hamuli varies from between two to fifty according to the species, but in a given species it is always the same.

Different devices for coupling wings are found in Hemiptera or bugs, in Trichoptera or caddis flies, in Mecoptera or scorpion flies, and in Lepidoptera.

Jack-knives

The predatory legs of some insects, including Hemiptera, Diptera, and Mantidae, may be likened to jack-knives. A predatory leg is a modified leg designed to seize prey. Gripping devices of a jack-knife type are formed by the folding back of one or more terminal segments (comparable to a blade) against a thicker and immediately anterior segment (comparable to a handle). The blade is formed by the tibia, or the tibio-tarsus, or a tarsal joint. The handle is formed by the femur or a tarsal joint. Sometimes the lengthening of a joint,

usually the haunch, increases the reach of the predatory leg. That of the water-scorpion is typical; it consists of a blade composed of the tibia and the tarsus, and a handle composed of the femur. In repose the blade rests snugly against the handle, the concavity of the tibia exactly housing the convexity of the femur. Other curious structural details contribute to this morphological adjustment, strengthen the adherence of the two parts, and cushion their contact.

Other tools

Like the tools we have already examined others, equally abundant and varied, fulfil tasks in conformity with their structure. The 'searchlight' of the fish or the insect, composed of a chemical source of light, a lens and a reflector, obeys the laws of optics. The grasshopper's violin, the cicada's cymbal, the xylophone of *Phrynus,* produce sounds in accordance with the rules which govern vibrating strings or percussion instruments. The organs with which electric fish are provided are electrochemical generators comparable with a battery. A wide variety of traps, snares, nets, nooses, pitfalls, lures and decoys attract and capture prey. The many systems of attachment are in general efficient: the hooks, the hamuli, the tendrils, the burrs, the grappling devices, the coils, the suckers, the sticky substances, the press-studs, the parachutes and gliders, the floats and sinkers, follow the principles enunciated in text books of physics and mechanics. The ootheca or egg case of the mantis is a thermos flask; the central chamber containing the eggs is enclosed in solidified froth of light and airy texture which forms a poor conductor of heat. This frothy envelope isolates the eggs thermically and protects them against changes of temperature. The powder-sprinkler of the poppy is an authentic prototype of the pepperpot on the restaurant table. The sugar tongs on the tea-tray copy the shape and function of small forceps devised by the sea-urchin, technically its tridactyl pedicellaria. The many organs designed for piercing, cleaning, brushing, and sweeping perform their allotted tasks.

The genesis of tools
and of adaptations

For many biologists the importance of these natural tools resides in the fact that they seem to offer proof of the existence of purposeful finality. This proof will not bear analysis. Minor tools, like organs of importance, simply accomplish functions compatible with their form and structure; they are merely evidence of factual finality. In normal conditions every organ which assures the equilibrium and well-being of the organism accomplishes its function, and any derangement it suffers entails a derangement of the organism itself. Less well known than the essential organs which anatomists have thoroughly dissected, nature's lesser tools have the attraction of novelty and awaken curiosity. But the evidence they provide of purposeful finality is neither greater nor less than that provided by the other organs.

It is essential to keep in mind the contrasting characteristics of organic invention and human invention. The former is a phenomenon of which the organism is unaware; it is accompanied by nothing analogous to human consciousness. It results from the extremely slow work of unspecialized protoplasm in a species, a genus, or even a phylum. Human invention is, on the contrary, a mental act, individual and rapid.

But how could these organic inventions, these small tools, appear? It seems most improbable that a single mutation could have given rise simultaneously to the various elements which compose, say, a press-stud or hooking device. Several mutations must therefore be assumed, but this implies the further assumption of close coordination between different and distinct mutations. Such indispensable coordination is a major stumbling block, for no known mutations occur in this way. Neo-Darwinism recognizes only fortuitous and always isolated mutations.

In addition, the fossil record shows that the musical equipment of the cricket, for instance, and the hooking apparatus of insect wings, evolved and slowly improved. Successive mutations, not being simultaneous, must have been coordinated in order to achieve those adjustments without which the devices could not function. It is extremely difficult to imagine such an unlikely mechanism. The genesis of tools, in fact, remains an enigma.

Equally unsolved is the problem of how ethological adaptations arose. Many hypotheses have been suggested, but none is satisfactory. Adaptation appears as a structure which fulfils a function. The problem of its genesis thus comes to the same thing as the problem of the genesis of the functional organization and its indispensable correlations. Minor tools, essential organs and adaptations are all genetically conditioned. Their origin raises the question raised by the development from a single cell — egg or zygote — of every coordinated organism. Vital processes act as though they amplify probabilities, achieving in effect a coordination whose spontaneous appearance would be highly improbable. In its nature the problem to be solved is thus essentially embryological.

The neo-Darwinists do not subscribe to this opinion; for them adaptation results from the interplay of natural selection and the competition or struggle for existence. For them the organism passively supports the birth and development of the adaptation without in any way taking part in the process.

In spite of all the efforts of experimental embryology the genesis of adaptations is still a mystery. We know nothing about the origin of adaptations to walking, swimming, flying, except that their achievement required vast stretches of time, running into tens of millions of years. The difficulties are great and experiment is impossible. It is probable that the structure, transformations, and innovations of living organisms are conditioned by the properties of life itself. It is in this sense that certain authors, notably Cuénot and Vandel, have spoken of 'organic invention'. Organic invention would be a property of living matter acting very slowly at the specific level and at the level of the phylum. Life's faculty of invention would be merely an extension of its power of organization. 'Organic invention' is an unhappy expression, for it automatically suggests the characteristics of human invention from which it fundamentally differs. Human invention is personal and rapid. It is a mental phenomenon. Organic invention is unconscious, slow in the extreme, and the outcome of non-specialized matter.

Theories of Evolution

The fact of evolution is solidly established; proofs of its existence are many and incontestable. The mechanism of evolution, however, remains unknown. Of the various hypotheses which have been advanced none seems altogether sufficient. A satisfactory theory must explain not only the origin of species but, more important, why living organisms have evolved in a given direction; it must further explain how organisms adapt themselves to their environment and organs to their functions.

Lamarckism

Jean Baptiste de Monet de Lamarck (1744—1829), author of the theory which bears his name, was a naturalist distinguished for his contributions to classification and to natural history.

Faced with the difficulty of isolating species, Lamarck came to the conclusion that they merge into each other; species are not individually created one by one, but have descended one from another. Animals form a progressive series, ascending from the Protozoa to the mammals. 'The species exists so that we can stick labels on it.' It possesses a temporary and apparent stability which is dependent on the stability of the environment. 'Change in the circumstances of habitat, exposure, climate, food, and way of life is attended by a corresponding change in the animal's size, shape, proportions, colour, character, agility, and industry.' Changed environment means changed requirements; animals thus contract new habits which 'last as long as the needs which created them.' Changed habits produce changes in the animal's physical actions. Its movements are no longer the same; a certain organ will work harder, and increased work will strengthen and transform it. On the other hand, an organ which is not used will atrophy and may even disappear. According to Lamarck 'it is not the animal's

Jean-Baptiste de Monet, Chevalier de Lamarck (1744—1829).

organs which have given rise to the animal's habits and particular faculties; on the contrary, it is the animal's habits, way of life, and the circumstances which surround the individuals from which it derives that have in the course of time formed the shape of the animal's body, the number and state of its organs and finally the special faculties at its command.' A change of environment or of habits must, then, always precede changes in form which, usually take place only in subsequent generations. The theory can be summed up as follows: a change of circumstances entails a change in habits, itself entailing a change in action which, in its turn, brings about a change in form.

Lamarck illustrated his theory with numerous examples. Thus the giraffe, obliged to browse off the leaves of trees, makes an effort to reach them. The result of this habit, followed over a long period of time by all members of the species, is a pair of forelegs which have grown longer than the hind legs and a neck

433

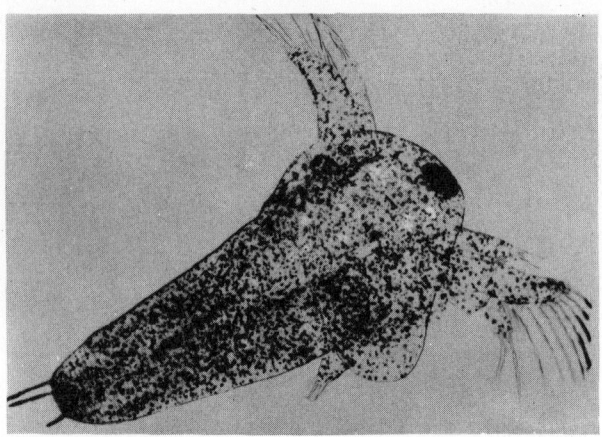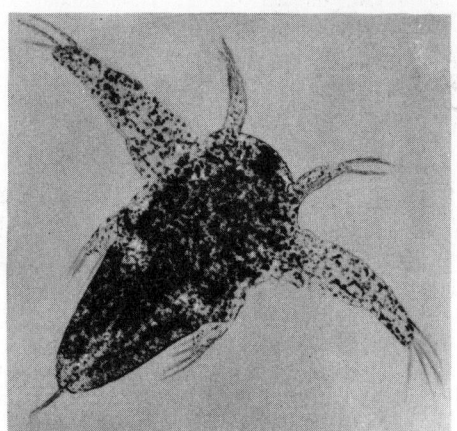

In water which is slightly salty *Artemia salina* has long, hooked bristles;

and, in very salty water, short bristles.

Mugard.

so elongated that it can reach branches at a height of some 20 feet. The bird which ventures into the water in search of food spreads its toes when it wishes to swim; in this way the skin between its toes acquires the habit of stretching: and in time the inherited effect of this exercise, repeated during the course of many generations, will be the webbed foot of aquatic fowl. Water-side birds on the other hand attempt to spread and enlarge their feet to avoid sinking into the mud. They become waders and their necks lengthen as a result of their efforts to fish without wetting their bodies.

Useless organs atrophy and disappear. Many vertebrates which have taken to an underground life have lost their eyesight, having no further occasion to use their eyes.

The theory can be summed up in two rules: the first is that need creates the required organ; use strengthens and enlarges it. Lack of use is followed by the atrophy and final disappearance of the useless organ. Secondly, a character acquired under the influence of environmental conditions is transmitted from one generation to the next. The acquired character is therefore hereditary.

Lamarckism thus rests on two postulates: first, to a change of environment or habit the organism responds by supplying exactly what is required. The examples chosen by Lamarck demonstrate this faculty of the individual's auto-adaptation, a faculty always to the individual's advantage. There is no doubt that using an organ strengthens it, and that with disuse organs tend in a certain degree to atrophy. In this way oarsmen develop powerful biceps, stevedores broad shoulders, cyclists large calves, and so forth. It follows that the inheritance of these acquired characters — the second postulate — is indispensable to the theory.

Objections to Lamarckism

It is precisely the two postulates mentioned above which are questioned by the critics of Lamarckism. That the environment has an effect on the organism is indisputable. In new surroundings the organism re-

sponds by physiological changes which may or may not be accompanied by changes in form and appearance. The spines of a prickly plant like gorse *(Ulex europaeus)* grown in a shady and humid position tend to diminish or even disappear; but if gorse is grown in dry, sunny conditions the spines increase in size and number. The plant's response is valuable because it helps to lessen transpiration by diminishing leaf-surfaces.

The response of the plant to a change of environment is an adaptive response. Adaptive responses are well known among animals: a few examples may be given. Exposure to ultraviolet light increases the pigmentation of the skin, because the pigment forms a protective screen which prevents burns. Continued friction will bring about the formation of a callus which prevents ulceration of the skin. It is a commonplace that athletes tend to develop powerful muscles. Healing, compensatory hypertrophy, and regeneration are adaptive responses to wounds. The same is true of inurement to gradual changes in external conditions such as temperature, salinity and atmospheric pressure; of tolerance to poisons (mithridatism); of the production of antibodies against toxic microbes. Such valuable responses by the organism are examples of the general faculty of functional regulation. It is well known that when a man's kidney is removed the remaining kidney will enlarge until it tends to weigh as much as two kidneys. After a fracture muscles and bones, being unused, regress. Regulatory adaptation is manifest in various useful and individual reactions to changes in external conditions: it is an aspect of morphosis, wich means the process of an organ's formation.

Lamarck believed that the organism invariably responds to environmental changes or dangers by a useful modification. Now such modifications are, in reality, often trivial and devoid of value. The little crustacean *Artemia salina* lives in waters which vary greatly in salt content. As salinity increases the bristles which adorn its posterior extremity become shorter and in high concentrations disappear altogether. The seasonal variations of *Daphnia* (affecting shape of head and length of appendages) and of

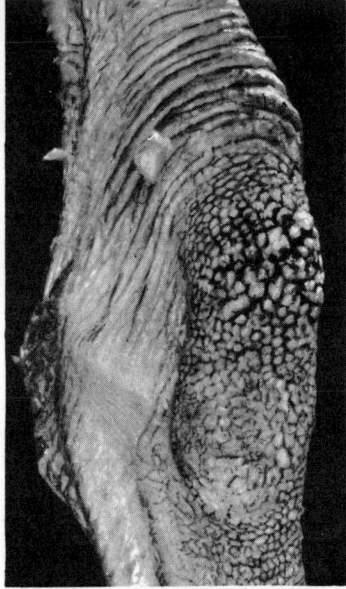

A callosity on the leg of the ostrich.
Jean.

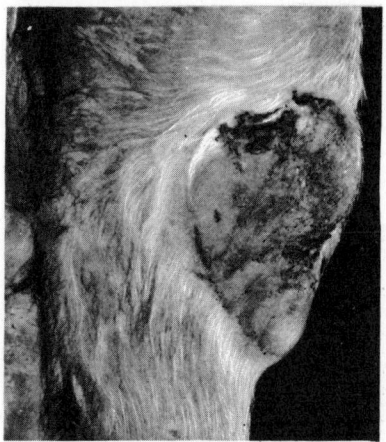

A callosity on the leg of the dromedary.
Jean.

rotifers (affecting dimensions, ornamentation of the cuticle, and number of appendages) are determined by external factors.

The question of the inheritance of acquired characters has been discussed earlier in this book. It will be remembered that experiment has shown that modifications which take place in an organism as a result of changed environment are not hereditary. They influence the somatic cells only and leave the germ-cells unaffected. All properly controlled experiments designed to establish the transmission of acquired characters have failed. This alone destroys Lamarckism. The fact that adaptive responses are not transmitted at once deprives them of all evolutionary significance. Partisans of Lamarckism, however, argue that the constant failure of experimental proof is due to the fact that the experiments have not lasted long enough to allow the time factor full play. If the cause of a variation operates over a period of thousands of years the physiological or morphological response may perhaps become hereditary. To this it could be objected that if centuries are necessary before the germ-cells are prepared to transmit acquired characters, then there is plenty of time for other mechanisms, like selection, to intervene.

A certain number of facts are disconcerting and hard to explain. For instance, one maritime plant, cultivated far from the sea, retains adaptive characters like hairiness and succulent tissues, while another, in identical circumstances, will lose them. In the first plant the characters form part of its hereditary endowment while in the second they do not. They have affected only the somatic cells. But why the difference? The question remains unanswered.

Biological features of adaptation

In spite of criticism, refutation, and experimental failure, Lamarckism has not totally disappeared. Many naturalists persist in believing that the influence exerted by the environment on the living organism is direct,

instead of indirect — that is, acting through the agency of selection. Although a new character may not be experimentally acquired, is this always so in nature? It must be admitted that if Lamarckism were true it would give a simple explanation of many awkward biological facts: hereditary callosities, the blindness of cave-dwellers, the abdominal curvature of hermit crabs living in adopted shells, and so on.

As everyone knows pressure and constant rubbing will thicken and harden the skin. Calluses are common among workers of many kinds: they are developed on the ankles of tailors who sew with crossed legs, on the thumb and third finger of the barber manipulating his scissors, and on the ball of the thumb and edge of the left forefinger of the violinist. Metal engravers, riveters, and even writers have their characteristic callosities, while sometimes bursae, or sac-like cavities filled with viscid fluid, develop among rag-and-bone-men, dockers and bootmakers.

Certain vertebrates are provided with cutaneous thickenings or specific callosities on parts of the body in frequent contact with the ground. Such callosities are characteristic of the species and are found in all its representatives. Ostriches, dromedaries, goats, wart-hogs, and monkeys have specific callosities related to the positions they assume when crouching. A fatty pad or thick layer of connective tissue is interposed between the bone and the horny epidermis. Kneeling dromedaries and camels rest on seven callosities formed by a thick horny layer which covers thickened connective tissue. The sternal callosity of the dromedary is plainly evident in the foetus, though it is still tender and cartilaginous. When the animal is about four years old it becomes hard and horny.

The case which has been most thoroughly studied is that of the African wart-hog which with its fleshy snout and tusks digs in the earth for the roots on which it lives. To do this it supports itself on bent-back wrists while the hind legs buttress and drive it forward. Due to this frequently adopted position the animal has carpal and sub-calcaneal callosities. The callosities are hairless, the skin thick, wrinkled, and folded. They are more marked in older animals. A section of the skin reveals hypertrophy of the epidermis and all the classic

Darwin's house at Downe in Kent.
Unit.

'Darwin and the squirrels' by Meredith Nugent.
Library of the Muséum national d'histoire naturelle. Larousse.

symptoms of chronic irritation of the skin. The callosities are perfectly visible in very young embryos; they are smooth, glabrous surfaces with folded epidermis which is three times as thick as the normal surrounding epidermis. In other words, the callosities are prepared before birth, *in utero,* as Leche has observed in an embryo 18 centimetres long. In older embryos it can be seen that the structure has developed, although there has as yet been no contact with a hard surface. The callosity or its preparation thus forms part of the wart-hog's hereditary endowment. It would appear to be a recent acquisition, since it does not exist in the wild boar or the river hog, *Potamochoerus*, from which it is derived. It seems to be connected with the instinct which makes the wart-hog kneel while rooting in the ground in search of food.

Disciples of Lamarck have, of course, their explanation of such developments. The wart-hog, or its ancestor, fell into the habit of kneeling on its wrists to dig for roots. The reaction to constant rubbing was the normal development of a callosity. Then, after a sufficient number of generations, the acquired epidermal thickening was inscribed into the animal's hereditary patrimony. But how? It has been suggested that 'messengers' could have been sent out by the somatic cells to impress the appropriate genes. With the passage of time a germinal tendency to localize over-activity of the skin would thus develop. The callosity would appear independently of initial cause and without any stimulation of the epidermis in the dromedary and the goat almost immediately after birth, and would be evident in the embryo of the wart-hog. This explanation is far from satisfactory, for the existence of 'messengers' is purely hypothetical. Phenomena whose mechanism remains unknown do not help us to solve the problem of the transmission of acquired characters, a premise which may itself prove to be false.

Lamarckism, or rather neo-Lamarckism, was not, however, without its enthusiastic supporters at the turn of the century and, as we have seen, it became the official dogma of Soviet biologists during the Stalinist era.

Darwinism

Charles Darwin (1809—1881) joined H.M.S. *Beagle,* as naturalist, on its famous voyage in 1831 to South America and certain islands of the Pacific. The voyage lasted five years and had a decisive influence on Darwin's thinking. Like all naturalists of that epoch he had, until then, believed in the immutability of species.

He began to think about the *Origin of Species* as

Brachiopods of the genus *Lingula*, which have existed since Cambrian times, and now frequent the coastal waters of warm seas, in particular the Pacific as far as India and Japan. The shells are attached to flexible and extensile tubular stalks which burrow in the muddy sand at the level of the lowest tides.

Left, Thomas Huxley (1825—1895).

Centre, August Weismann (1834—1914).

Right, Ernst Haeckel (1834—1919).

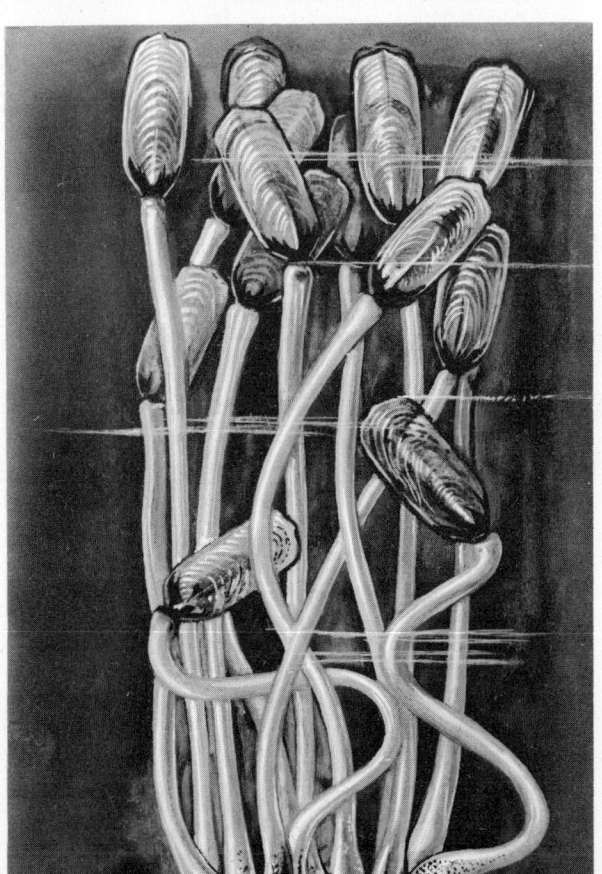

early as 1837 and when the volume appeared in November 1859 it caused an immediate sensation.

During the voyage of the *Beagle* certain facts had struck Darwin, notably the following: the substitution from north to south of species plainly allied to each other; the relationship between species inhabiting the islands near South America and those inhabiting the continent; the variety of fauna found on the various Galapagos islands; the connection between living toothless mammals (Edentata) and the fossils of extinct species found in the pampas. None of these facts, which he observed with great accuracy and attention, was compatible with the principle of immutability. They could be explained only on the assumption that animal forms had evolved gradually. Darwin sought to discover the mechanism by which such evolution was possible. Domestic animals and cultivated plants showed him the importance of variations and the advantage man could derive from them by artificial selection. His ideas were also much influenced by the *Essay on the Principle of Population* which the political economist Malthus had published

in 1798. Malthus had shown that population tends to increase much more rapidly than the means to sustain it. The unavoidable result was an increasing struggle for food in which victory belonged to those with advantages over their competitors. From these two sources, artificial selection and competition for food, arose the concepts of natural selection and the struggle for existence.

Darwin's theory of evolution may be summed up as follows: changes in the environment such as climate, nourishment, and so on, bring about variations in living organisms. Darwin made a distinction between definite variations, which affect all modified individuals in the same way, and indefinite variations, which differ from one individual to another and are generally slight. Due to the growth of populations a struggle for existence is inevitable. In a given area each individual competes with its kind for food and for living space. Every unfavourable variation will be condemned and destroyed. Every advantageous variation, morphological, reproductive, or in habit, will subsist and be handed on to the progeny of those who possess it. The best types are, as it were, sorted out and preserved by natural selection. Thus natural selection leads to the survival of the fittest. The giraffe owes its origin to the preservation, during periods of famine, of the tallest individuals which were able to browse among the highest branches. In addition the animal's great height helped to protect it against carnivores.

Evolution is thus a function of variability and of competition, and the various combinations of these two factors explain numerous phenomena. The absence of the two factors explains why a genus like *Lingula* has remained practically unchanged for nearly 500 million years. Slight variation, combined with fierce competition, results in the extinction of a species. Finally great variation together with much competition assures the evolution and diversification of species. Widely distributed species rich in individuals established in sufficiently different surroundings, thus providing an important impetus to variation, give rise to new species capable of occupying new environments. Rare and specialized species are, on the contrary, more liable to be wiped out. In other words, the more plastic a species is the greater its evolutionary potentiality. Natural selection, varying in intensity but acting over vast stretches of time, must in the long run appreciably affect vast populations.

Darwin admitted that environment played its part, though a somewhat secondary part. Towards the end of his life he expressed regret that he had not attributed greater importance to it. To his theory of evolution he afterwards added the concept of sexual selection. Among many arthropods and vertebrates there is a striking difference between the sexes; the males possess better arms, odorous glands, and musical ability; often they are also brilliantly adorned. They fight among themselves for possession of the females, and the stronger and better armed, winning the contest, set their rivals aside and alone procreate. The females choose the more handsome, highly coloured, and vigorous males. In this way a rather special selection is achieved, namely sexual selection.

Darwin's theory had an enormous impact on biological and, indeed, on all branches of thought. Clear and logical, it apparently accounted for all the facts by the inescapable play of blind variation and death which, with the collaboration of time, sorted out the unfit. But the *Origin of Species* and even more the *Descent of Man* aroused bitter opposition as well as enthusiastic support. In England Thomas Huxley and in Germany Ernst Haeckel spread and amplified Darwin's conceptions. Thanks largely to them evolutionary ideas won ground while the arguments of the opposition slowly convinced fewer people.

The Ultra-Darwinists

Darwin's principles were carried to the extreme by those who were known as the ultra-Darwinists, and principally by Weismann and Wallace who accepted everything in Darwinism apart from its element of Lamarckism. In natural selection they recognized the sole effective factor and conferred upon it absolute power. To be selected, a character must be of value. A. R. Wallace, who had discovered the principle of natural selection at the same time as Darwin, strove to demonstrate it and in doing so formulated views which were almost as excessive as those of Bernadin de Saint-Pierre who argued that nature's sole purpose was the achievement of Man. Weismann reduced the importance of the environment still further, considering it, in fact, non-existent. He contended that all multicellular organisms are divided into body-cells or soma and germ plasma or germen totally independent of each other. All variation arises from an internal cause localized in the germen. Somatic modifications are without evolutionary value, being incapable of action on the germen and hence of becoming hereditary. He explained the persistence of useless and of rudimentary organs: while the harmful organ is eliminated by selection, the useless organ is ignored by selection. Selection ceases to operate and the useless organ will either degenerate or become rudimentary. He called this cessation of selective action *panmixia*.

Let us see how the Darwinists and the neo-Darwinists explain the development of webbed feet in aquatic fowl and the acquisition of transversally flattened tails

by aquatic vertebrates. Among the various individuals of a given species there will occur slight and random variations in the folds of the skin between the toes; in the same way the shape of the vertebrate's tail will vary. During infancy the mortality rate is considerable, since many are born and comparatively few become adults. Among aquatic animals the mortality rate will be especially high in mediocre swimmers that are slower in pursuit of prey and in flight from danger. Those individuals in whose swimming organs advantageous variations have by sheer chance occurred will survive. This sorting out of the fittest, repeated by natural selection for generation after generation, brings about the slow and continuous progress of those features useful in swimming. Thus webbed feet and the flattened tail developed automatically by the survival of the best swimmers.

Objections to Darwinism

Variations form the point of departure of Darwin's theory. But most variations are somatic and therefore not hereditary. Only mutations are hereditary. This confusion between what is hereditary and what is not narrows the scope of the theory, for Darwinism still depends on the same inheritance of acquired characters which brought about the downfall of Lamarckism. In addition the Darwinian hypothesis relies on the advantage conferred by a random variation. Now it is a delicate matter to estimate such advantages; an adaptation which is useful in one situation can be a handicap in another, and the true balance may be impossible to decide. The bird with webbed feet swims with ease and perhaps more rapidly, but it walks with difficulty over rough ground. Palmation, an advantage in the water, limits possibilities on land. The criterion of usefulness is thus hard to establish.

But as a logical theory Darwinism collapses because the keystone of the system, selection carried out by death, is an illusion. The famous 'struggle for existence' is, of course, a familiar phenomenon. In nature species are extremely prolific and tend to increase in number. But massive destruction of eggs, larvae, and even of adults maintains an equilibrium between their number and the available food supply. The idea of active struggle, however, suggests a false image which is slightly too anthropomorphic. Equilibrium is a function of a large number of factors, and the least disturbance upsets it. Certain species are destroyed to the benefit of others. Living organisms passively submit to the action of climatic factors or of parasites. Such sufferance, such passive resistance, has little resemblance to 'struggle' and victorious combat. One species is successful for a period of time, and then gives way to another.

So do we really know whether organisms which persist actually possess advantages of which those species that disappeared were deprived? Is there, in fact, a genuine difference between those that live and those that die? Does massive mortality strike at random or does it pick and choose?

For Darwin the answer would have been yes: those who die are not identical to those who do not die, although, as he said in 1844, less than a grain in the balance would determine which individuals live and which perish. It is of course evident that in each generation monstrous eggs, abnormal larvae, misshapen embryos, malformed young, and adults which are seriously handicapped, are eliminated. Every creature devoid of a certain minimum viability disappears; but selection of this sort affects a relatively small number of individuals. The remainder are eliminated by a chance deficiency of some kind, or as the result of a combination of unfortunate circumstances, like travellers killed in an aeroplane accident. Is the struggle for life really a battle won by the superior, the stronger? Is death, in fact, selective?

Only experimental analysis can furnish data on which an attempt to answer the question may be made. Frequency curves of the various characters of certain species have been established and then compared with corresponding curves of members of the same species who were either killed as a result of unfavourable conditions or of members who survived periods of high mortality. For instance, the characteristics of 136 sparrows wounded or killed in a winter storm were compared with those of living sparrows. Certain cocoons of *Bombyx* of the ailanthus *(Philosamia cynthia)* neither hatch nor produce moths; they have been measured and compared with normal cocoons. A thousand cocoons gave the following results: five per cent of the cocoons contained dead caterpillars before the pupa stage, sixty-four per cent died in the course of metamorphosis, and thirty-one per cent hatched normally. The dimensions of the organs of the dead and of the living revealed that individual differences of length and form had no connection with survival; thus these characters have no bearing on viability. Experiments have been made to determine whether there is a difference in the weight of peas and beans which germinate and of those which do not germinate. The results have not been of great interest: difference in weight between seeds which grow and those which do not are, when they exist, slight. It would seem that the average type survives, while the extreme types are eliminated. In certain cases it even occurs that the less well-armed individuals of a species win the 'struggle'.

Let us consider the life cycle of an ordinary species, the common frog, *Rana temporaria.* The frog lays from some 1,200 to 4,000 eggs which in normal conditions of temperature arrive at the hatching stage. Abnormal eggs and defective embryos are removed by selective action: the elimination of the unfit. The tadpoles are born and begin their free life. Many fall prey to predators, especially to aquatic insects. The hazards of the weather, a hot spell in which ponds dry up, may make living conditions impossible, and in this way large numbers of tadpoles perish. But the most difficult period of their lives is that of metamorphosis. Young frogs emerging from their pools find difficulty in procuring suitable insects and worms to eat, and many die at this stage. Once this critical phase has passed the mortality rate becomes normal. The frogs succumb in the ordinary way to predators, to parasites, and to accidents. Mortality, it will be observed, is highest in the younger stages of the animal's life; and in this connection there is a correlation between the number of eggs laid and the amount of protection which is afforded them. Frogs and toads lay thousands of eggs enclosed in a jelly-like secretion and abandon them. The female of an equatorial frog, the leptodactyl *Ceratohyla,* on the other hand, lays only nine eggs, but they are protected and carried on the female's back. A 17 pound carp lays about 800,000 eggs which it abandons in pond water, whereas during its life the stickleback lays only once — or at most twice — some 200 eggs in a nest closely guarded by a fighting male.

Death, after the first eliminations it makes, has no further selective value. It is above all the young that are destroyed and, since the young have not yet reached full development they leave no offspring and from the evolutionary point of view have no importance. Adults handicapped by abnormalities or malformations are, moreover, not uncommon in nature: fish with swimming restricted by curvature of the spine, grasshoppers deprived of their jumping legs, small mammals with fractures indifferently mended. In a homogeneous species living in a more or less stable environment it is observed that selection plays no part. The survival of the fittest is an illusion; for in practice the average type is preserved by the elimination of the blemished, the infirm, and extreme types of any kind. Death thus appears to be conservative, preserving rather than differentiating.

That the action of selection is not automatic and unrelenting is visibly apparent; for otherwise all discords and anomalies would have long since been eliminated. All that man judges in nature to be imperfect, dangerous, useless or excessive ought no longer to exist. Darwin himself saw and attempted to resolve these difficulties. The gorgeous spread of the male peacock's tail could be explained by sexual selection, but it is also a hindrance in flight.

Since death does not differentiate, one of the basic assumptions of Darwinism is untenable. But what of selection and its influence?

Natural selection

Natural selection is not comparable to artificial selection. Nature's methods of eliminating the weak or the blemished have little relation to the techniques of the stock-breeder. Races and varieties can be perpetuated by breeding which would, in nature, inevitably perish. Natural selection eliminates deviations from the norm while artificial selection preserves them. Artificial selection attempts to concentrate in a single individual the genes which determine a desired character, genes which in a normal population are scattered among various individuals. Its aim is to obtain a pair of homozygotes from which a pure line may issue. Natural selection creates nothing; its action is quantitative rather than qualitative. Within certain limits it intensifies characters which already exist. It intervenes in the case of mutants which live in association with normal members of their species and again with mutants which change their environment. In the first case the phenomenon of 'substitution' is said to take place, in the second that of 'pre-adaptation'.

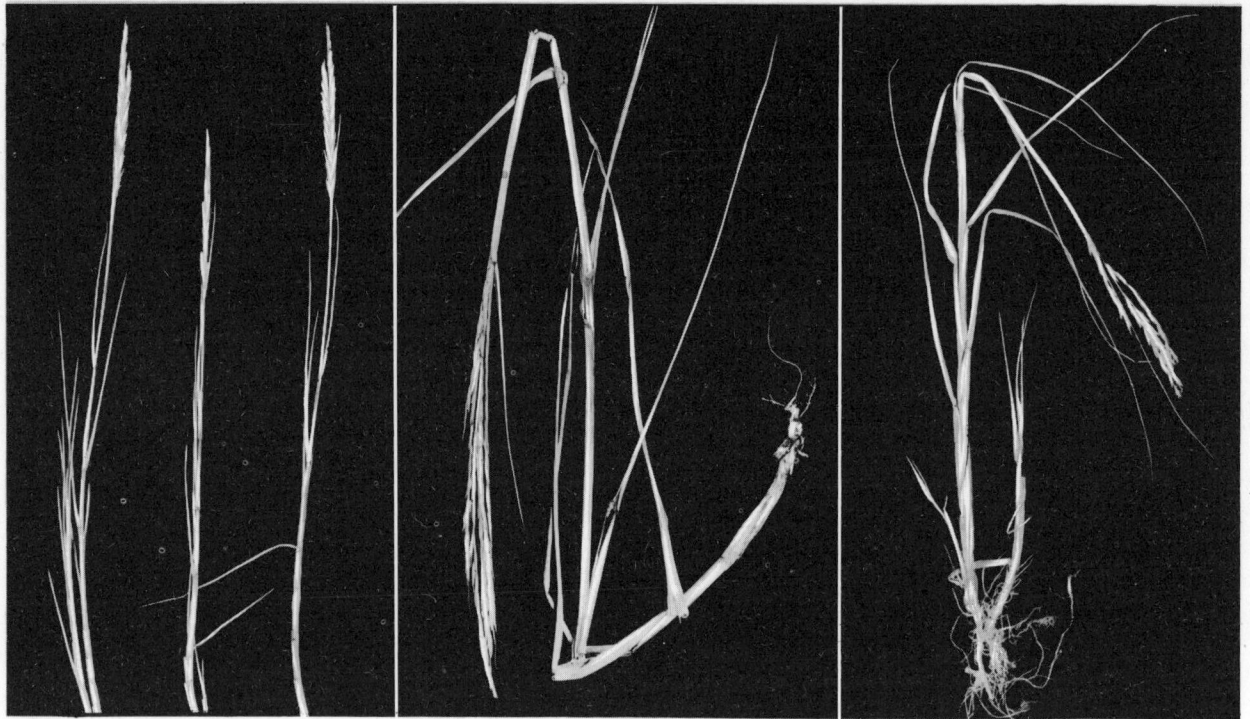

In the centre, Spartina townsendii, between its two parent
species, Spartina maritima and Spartina alterniflora.

Parriaud.

Substitution

When a mutation occurs in nature the mutant may
simply disappear by chance. If it survives and crosses
with normal members of its species heterozygotes
will be born. If the mutation is dominant natural
selection will then operate; if the mutation is recessive
selection will play no part except in the case of homozygotes, that is to say the progeny of two mutants. If
the individuals affected by the mutation are well-formed,
if they resist disease, and if they are particularly fertile,
they have an excellent chance of being selected and of
increasing in number.

Geneticists have studied selection as it takes place in
the miniature world of the laboratory. Into a large
container supplied with abundant and regularly renewed nourishment a population of *Drosophila* of known
genetic composition was placed. For several months
the fruit flies were thus reared under controlled
conditions. In the experiment the wild type of the
insect was opposed by a mutant or by two mutants,
one a lethal mutant, the other non-lethal. In this way it
is possible to appraise what G. Tessier calls the value
of selection under experimental conditions. Some of the
results obtained were as follows.

Drosophila melanogaster and *Drosophila funebris*
were put together in the receptacle; no matter what the
initial proportion of the two species was, at the end of
a few months *melanogaster* overwhelmingly predom-
inated. *Melanogaster* was more fertile and its development more rapid; only some two per cent of the
funebris remained. At a temperature of 25 °C twice as
many *melanogaster* eggs as *funebris* eggs will hatch;
but at the temperature of 15° this proportion is
reversed. This explains why in nature *melanogaster*
lives in temperate and *funebris* in cooler regions.

In a population of pure *bar* mutants (or ebony
mutants) with their dark coloration a few individuals
of normal *melanogaster* species were introduced. At
the end of twenty months the wild type was clearly
dominant; of the bar mutants only one per cent remained, and of the ebony mutants fourteen per cent,
the latter defending themselves more successfully than
the former. In nature there are three possible mutant
situations.

Firstly, a mutant may appear which is inferior to the
typical member of the species, which in normal

Ladybird.

Bisserot.

Helix nemoralis with its peristome or grooved shell-lip and two
broad dark bands.

Larousse.

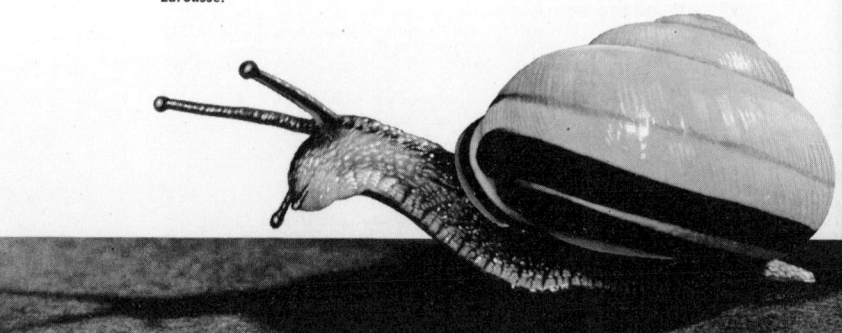

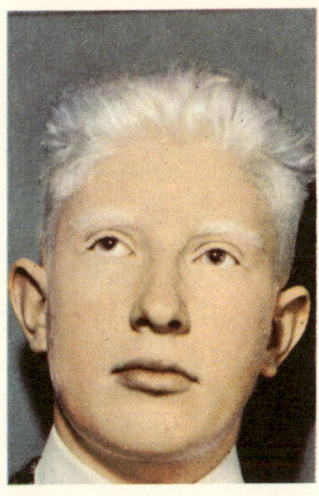

Albinism

Far left, mutant adult albino *Rana temporaria;* examples are very rare.

After J. Vasserot. Larousse.

Left, young man, born in 1946, revealing complete albinism: hair and eyebrows colourless, cheeks pink, pupils with the characteristic glint of red.

Klein, Geneva.

Below left, group of adult albino guinea-pigs from strain K; skin and fur completely unpigmented.

Larousse.

Below, albino tadpoles and young frogs *(Rana temporaria).*

After J. Rostand. J. Vincent.

Bottom left, pair of albino mice. ('Swiss' strain.)

Bottom right, hypodactyl mutant of albino rat.

Laboratory at Gif-sur-Yvette. Larousse.

441

The peppered moth, *Biston betularia:* the normal and the melanic forms as they appear against pale bark.

S. Beaufoy.

conditions is less viable than the average representative. It does not prosper. A good example is that of albinism which exists in all vertebrates from fishes to Man. Albinos are always fairly rare, and in nature no race of albino vertebrates has ever been formed. They usually die before having reproduced. Although they live and multiply with ease in captivity, in nature they are unable to compete with wild types. Even in species which form societies normal individuals often refuse to tolerate albinos and drive them from the group.

Secondly, there can be a mutant equal to the typical member of the species that does not differ essentially from the standard type, particularly in physiology. The two cohabit and can even cross. The species is then said to be 'polymorphous'. The ladybird, *Adalia bipunctata,* has a black spot on each of its red wing-covers or elytra. Ladybird populations are composed of the *bipunctata* type mingled with *quadrimaculata,* dominant mutants which have two red spots on each black elytron, and with *sexpustulata* mutants which have three red spots on each black elytron. In a given region the proportions of the two mutants are roughly constant, which means that selection is inactive; but the proportions change from region to region and sometimes from season to season. The black forms, commoner in autumn, are more resistant to heat, while the red forms, commoner in spring, support the winter better and the heat of summer less successfully.

An analogous phenomenon is encountered in colonies of the wood snail *Helix nemoralis* and the garden snail *Helix hortensis.* Colonies with well-defined boundaries number on an average from 1,000—2,000 adult individuals, while exceptional colonies may attain a population of 20,000 snails or more. The snail is very polymorphous, chiefly in characteristics affecting the shell — which may be yellow or pink, and may or may not lack one or more of its five typical bands. Again these bands may in varying degrees coalesce, and their pigmentation may be normal, incomplete, or absent. The peristome, or lip of the shell, may also vary in colour. It is known that the pink colour at the base of the shell is dominant

(genetically) to the colour yellow; but as both colours are very common in natural populations it is not known which should be considered normal and which mutant. The same is true of the presence or absence of bands. Absence of bands is dominant to presence of bands, and it is supposed that the five-band form represents the original type. Albinism of the bands and of the peristome is determined by a fairly rare recessive gene. The great number of combinations arising from the absence of one or more of the five parallel bands, the fusion of from two to five of them, and the varying coloration of the shell result in a polymorphism which is general to the different snail colonies of all regions. Although no two colonies are the same in composition this extreme diversity is common to them all.

All the various forms continue to exist side by side; selection plays no part and panmixia is total. Crosses between the different forms take place by pure chance. No mating between like types, no gametic correlation can be detected.

Another type of mutant is that which is superior to the typical member of the species. If the mutant succeeds in replacing the normal type it is, by definition, superior. In several localities the purple gentian *Gentiana campestris,* has given way to a white mutant. *Spartina townsendii,* known as rice grass, appeared as recently as 1870 on the coast near Southampton. It was thought to have arisen from a cross between *S. stricta,* of American origin and the native S. *maritima.* The hybrid, very hardy, thrives in salt marshes from which it has eliminated its two parent species wherever it has cohabited with them. Rice grass has spread far from its centre of origin, being dispersed by the currents during high tides.

Another case of substitution is that of the peppered moth *(Biston betularia)* which, until 1850, existed in England in its normal colouring, a speckled or peppered grey which resembles the lichen or bark of the trees on which the moth commonly rests. In 1850, however, dark, melanic forms of the moth appeared near Manchester, mutants which were named *Biston insularia* and *B. carbonaria.* The former still retained some white, the latter were totally black. Both mutants and especially *carbonaria* stand out vividly when seen against pale lichen or bark. The two mutations are independent. *Carbonaria* depends on a single dominant Mendelian gene and is slightly more hardy and fertile than the standard species. Mutant moths live on the same plants as normal moths, cross with normal moths, and rapidly replace them. Near Huddersfield, the black mutant, which first appeared between 1860 and 1870, has totally eliminated the normal type. In Newport towards 1870 the normal moth and the mutant *carbonaria* coexisted in equal numbers; a few years later only *carbonaria* remained. In the mining district of Birtley in Durham the melanic mutant alone is known. In France and Belgium the melanic mutation appeared towards 1900; six years later the mutant *insularia* was as common as the normal *Biston* in Belgium; near Lille and in the Ardennes two-thirds of this genus were melanic by 1913. On the Franco-Belgian frontier, by 1926, five-sixths were *carbonaria.* In 1934 near Nancy two-thirds were melanic while in 1936 *carbonaria* were as common in the Bois de

Hugo de Vries examining the evening primrose plants of which he made a particular study.

After Cleland.

The evening primrose, *Oenothera lamarckiana* with which Hugo de Vries began his experiments.

Underside of the leaves: *on the left,* the diploid *Oenothera lamarckiana; on the right,* the tetraploid *Oenothera lamarckiana gigas.*

After Hugo de Vries.

Boulogne in Paris as the pale type; in Flanders the normal type has become a rarity. In Holland the melanic mutant was still rare in 1914. In Germany the first black mutant was seen in Hanover towards 1875. In 1884 it was rare in Crefeld, but eleven years later it composed half the *Biston* population. The mutants were, on the other hand, still unknown in Scotland and in Brittany. In the United States, however, the melanic mutant has appeared near Pittsburg.

Research indicates that among the Lepidoptera the moth family Geometridae *(Biston, Boarmia,* and *Selenia)* displays a marked tendency to melanic mutation. This replacement of the original pale moths by melanic moths raises many questions. It is, of course, impossible to ignore the fact that the dark mutants appear and subsequently multiply in industrial regions. The centre of their dispersal coincides with some town in a 'black country' where forges, coal mines, munition factories, or steelworks abound. For this reason the phenomenon has been called 'industrial melanism'. Is there a cause and effect relationship between industrial centres and melanic mutation? Food has been suggested: metallic salts contained in the smoke of factories are deposited on the plants which nourish caterpillars and, it is argued, these salts could bring about mutation. But experimental verification of this hypothesis has yielded negative results. It must also be remembered that melanic moths have been captured in forests far from any industrial town.

What effect has natural selection on the phenomenon? When the *carbonaria* mutation first appeared it was eliminated because against light grey tree trunks the dark moths were easily seen by birds. It persisted only because the mutation occurred again and again. Then the environment was profoundly altered by the intense development of industry. The air was polluted by coal smoke and many pale lichens were destroyed. The trunks and branches of trees were blackened by deposits of soot. These new conditions altered the circumstances of the life of *Biston.* The normal pale form was no longer, as originally, at an advantage. The black mutant, on the other hand, found the new conditions more favourable than the old. This fact is demonstrated by an examination of the moths captured by birds and the mutants' rates of survival in different localities. The survival rate of the mutant *carbonaria,* for instance, is ten per cent higher in a polluted area and seventeen per cent lower in a region which is not polluted. A hundred years ago the black mutation formed less than one per cent of the population; today it forms ninety-nine per cent. Selection has also intensified melanism and mutants are even darker now than formerly.

Since the fertility and vitality of the melanic forms are higher than those of the pale forms why did not dark moths replace pale moths sooner? Some factor must have changed, and the factor must have been biological. The transformation of an agricultural region into an industrial region is accompanied by a whole

443

series of changes, notably in the bird and bat populations, these animals being the chief consumers of moths. Thus, in addition to the moth's protective coloration, there may also be fewer predators to fear.

Sometimes a normal species, driven out and replaced by a mutant form, will survive in a neighbouring region in which climatic conditions are slightly different. In the U. S. S. R. the hamster *Cricetus cricetus,* normally two-coloured, produced a melanic mutation which, at the end of the eighteenth century, thrived in the damp forests of the eastern steppes. There it supplanted the typical hamster which took refuge elsewhere.

Selection can account for the replacement of the original type by a mutant type, and certain experiments seem to show that this, in fact, occurs. But another factor plays a part of undoubted importance, namely that of 'differential fertility'. A fertile race, characterized not only by a high birth rate but also by the number of adults which reproduce, will displace a less fertile race. This would seem to be a general law applicable to all living creatures. If fertility is F and the rate of elimination is E, then the ratio of F to E is vital; for if E is greater than F the species is on the road to extinction. Mutations of the greatest survival value will increase the number of eggs produced, or produce better protection for the egg and the early stages of its development or, in general, improve any device which insures more certain fertility, more straightforward development and better protection for the young.

The theory of mutation

The first modification of orthodox Darwinism had been made in 1880 by the ultra-Darwinists. A second deviation took place some twenty years later when the theory of mutation was propounded. The new theory soon superseded both orthodox Darwinism and the neo-Darwinism of Wallace and Weismann. It might be defined as Darwinism deprived of its two chief tenets: the inheritance of acquired characters and the omnipotence, down to the minutest details, of natural selection. It is, however, a theory which, thanks to its alliance with genetics, is perhaps more solidly founded than Darwinism.

Curiously enough, the mutation theory arose from an erroneous interpretation of indisputable facts. It was first suggested by the Dutch botanist Hugo de Vries.

De Vries cultivated on a large scale a plant of unknown origin which grows in the dunes of Holland, *Oenothera lamarckiana* or the evening primrose. Among the descendants of *O. lamarckiana* he found slightly different forms which he named *O. gigas, lata, nanella,* etc. He concluded that they were new species which had suddenly appeared by mutation. In reality the cause of the phenomenon was not mutation, but a mechanism which was understood only later. *O. lamarckiana,* it was afterwards found, is a polyhybrid heterozygote; when crossed, disjunctions of various hereditary characters occur in the hybrids, and the hybrids thus obtained — wrongly thought to be mutations — are simply throwbacks to one of the hybrid's ancestors.

Lamarck and Darwin thought of variation as being continuous and continuously acted upon by natural selection. De Vries replaced this concept with the concept of 'discontinuous variation', in other words 'mutation', which at once becomes part of the organism's hereditary endowment. Examples of discontinuous variation, of new viable or semi-viable forms appearing abruptly, are not uncommon in nature. Darwin was aware of them and called them 'sports', but thought them of slight evolutionary importance.

The theory of mutation was confirmed by the rediscovery of Mendel's laws of heredity, which showed how mutations were transmitted and could combine to produce novel forms. By judicious crossing it proved possible to combine desirable qualities or eliminate defects. Characters could be so grouped as to resist parasites, to raise the yield of, for example, wheats sugar-beet, and potatoes, or to create new shades of colour in fur-bearing animals.

The mutation theory totally rejects the inheritance of

Variations of the stickleback, *Gasterosteus aculeatus.*
The *trachurus* form, whose flanks are completely covered with some thirty bony plates.
The fresh-water *gymnura* form, with its five or six plates near the head.
The form found in the saline pools of Lorraine, large in size and with ten lateral plates.

Oposite page

1. Colony of serpulids, *Mercierella enigmatica* — sedentary annelids of the Polychaetea — affixed to a stem of *Phragmites* found in a canalized river.
2. The calcareous tubes of *Mercierella enigmatica* with their collarettes and plumed tufts. Commensalism with the lamellibranchiate mollusc *Dreissensia polymorpha* (in the centre) covered by the polyzoan *Membranipora* — from the Caen canal.

Rullier.

characters acquired by the body (soma) under the influence of the environment, of usage, or of non-usage. In this respect it disagrees with Lamarckism and shares the opinion of the ultra-Darwinists. In addition to the concept of mutations imprinted in the hereditary patrimony and transmitted during crosses, the theory admits that selection tends to preserve the average type, selection which plays a much less important role than that attributed to it by Darwin. Finally, a new hypothesis advocated in 1902 by Cuénot was incorporated into the theory of mutations. This hypothesis, 'pre-adaptation', may be considered as a special aspect of natural selection.

Pre-adaptation

Let us consider the case of species living in an environment which is subject to abrupt and violent changes, for example to a particularly bitter winter or an exceptionally dry summer, to the arrival of unfamiliar parasites or predators, or, if water-dwellers, to altered salt concentrations. Such sudden catastrophes will result in the immediate or slow death of individuals incapable of supporting the change, and only those which are adapted in advance to resist the new conditions will survive. In this case death does, indeed, differentiate. Selection operates, but its concern is with the entire organism, the total collection of characters, anatomical, physiological, and chromosomal, which happen to be suited to the new conditions. In this way, 'pre-adapted forms' are selected. At the same time correlative characters, height and longevity for instance, are indirectly selected. A new type is then established which differs from the old not only anatomically and physiologically but sometimes in outward appearance. Selection of this sort gives rise to 'ecotypes' or local races.

Many examples of the phenomenon could be given, the following being a few instances chosen at random. Douglas firs of varying origin, which grew side by side in the forest of La Joux in the Jura, were almost entirely destroyed during severe winters. But sixty per cent of those which had been grown from seeds collected in the Vosges survived. Darwin mentions that in his garden after a bitterly freezing night only twelve out of 390 stools of *Phaseolus multiflorus* (scarlet runners), of the same age and similarly exposed, survived. Four days later it froze even harder and only three plants resisted. These three were pre-adapted to intense cold. Sometimes resistance or susceptibility to cold will affect a certain feature of an animal. For instance, the combs and wattles of a cock will freeze at temperatures lower than minus 12 °C, and breeds which live in cold climates do not have large combs. Bareskinned mammals, like Mexican hairless dogs, are confined to warm countries.

Certain fishes are capable of supporting different degrees of salinity. The three-spined stickleback, *Gasterosteus aculeatus* lives in both fresh and in seawater. When sticklebacks caught in fresh-water streams are suddenly plunged into salt water many of them die within twenty-four hours; others resist indefinitely. Exactly the same result is produced by the opposite experiment. Differences in ability to support the change are not accompanied by any consistent morphological differences in the fish. The stickleback's tolerance to varying saline concentrations permits it to live in the sea, in brackish waters and, for preference, in fresh water.

A mutation which is a drawback in a certain environment may prove to be of value if the environment changes. In an experiment, L'Héritier, Neefs, and Tessier demonstrated how the wind is the agent of selection among insects flying in particularly windy places. A culture of *Drosophila* composed of a mixture of normal winged fruit flies and *vestigial* mutants incapable of flight was exposed to the wind at Roscoff on the coast of Brittany. A second group of flies, of identical composition, was enclosed in a room. Thus protected, the normal winged type, more fertile and longer lived, rapidly multiplied and replaced the vestigial mutant. But in the open air the normal winged type, swept away by the wind, was at a disadvantage. The wingless mutant resisted the wind and rapidly became the dominant type. In this environment where

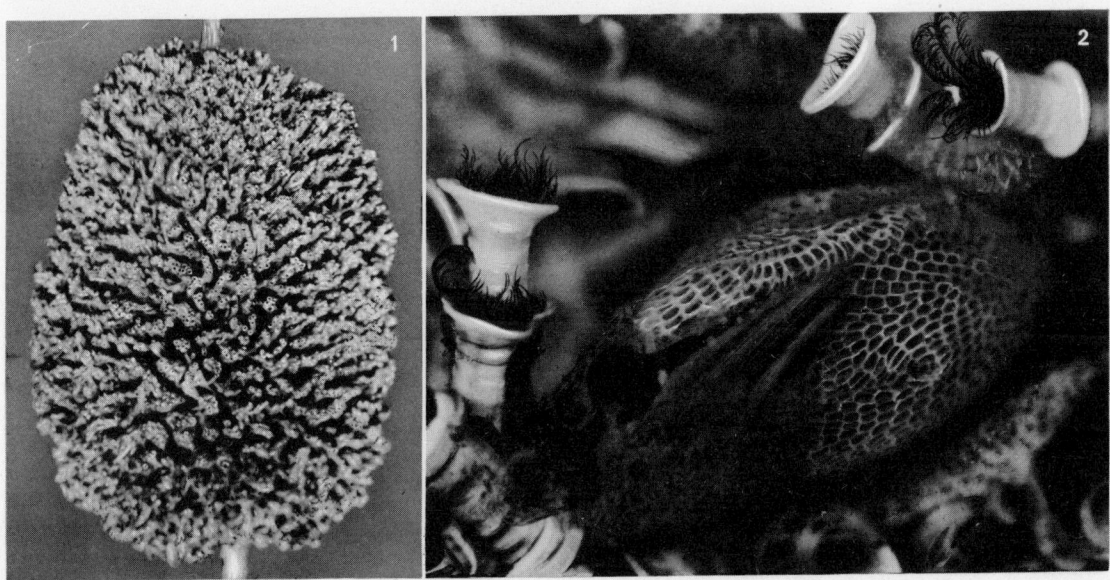

the wind raged violently the wingless mutants established themselves with relative ease.

Pre-adaptation, as defined by Cuénot, is the presence in a species of those neutral or semi-useful characters which are susceptible to becoming definite adaptations if the species adopts a new habitat or acquires new habits, changes made possible precisely because these pre-adaptations are present. The concept has been more or less implicitly accepted by many biologists, though it has sometimes been twisted, especially by the anti-evolutionists, to imply predestination.

The peopling of empty spaces

Pre-adaptation supplies a feasible explanation of how an animal is able to withdraw or escape from an environment or from a way of life for which it is specialized, and adopt a new environment or take up a new way of life. Pre-adaptation thus plays a part in the peopling of empty spaces, an occurrence which could be called one of nature's experiments. By an empty space we mean a new environment offering conditions which are compatible with life, an environment which appears as a result of natural or man-made geographical changes, such as the building of a dam resulting in the formation of a lake, a sand-bank or an island emerging from the waters for some reason or other, the digging of a canal or of a mine shaft, a forest fire, the evaporation of land-locked waters forming salty pools or even seas. All these create empty spaces ripe for colonization.

The canal from Caen to the sea, dug between 1844 and 1857, provides a good example of an empty space colonized by animals. It is almost 9 miles long and its maximum depth is 6½ yards. It begins at Caen where it is fed by the river Orne and reaches the sea at Ouistreham where a lock prevents the inrush of the tide. Within 2½ miles of the sea the salinity of the canal is 2—3 grams per litre; further inland it decreases progressively. The fauna includes some fifty species, all highly tolerant to salinity and often cosmopolitan. Among them are fresh-water species, brackish-water species, and marine species which enter when the locks are open or have been brought in by ships. Some species are particularly interesting: a little worm, the serpulid *Mercierella enigmatica,* which thrives on bankside plants and submerged objects, is a recent arrival; of Indian origin it has been carried all over the world by ships. The mollusc *Congeria cochleata,* of West African origin, has been common since 1898. *Cordilophora caspia,* a hydroid from the Caspian Sea, is perfectly at home in the canal.

Empty spaces are more or less rapidly invaded by animals from similar environments in the neighbourhood. Animals which are attracted by the new living space and are prepared for the conditions it offers will establish themselves and multiply: a kind of infiltration of neighbouring species takes place. As many marine species are incapable of living in brackish water they cannot move into fresh water. Very few aquatic groups can support life on land, even marshy land. Many more varieties live in the sea than in

fresh water, and many more in fresh water than on land. Animals and plants survive only in environments compatible with their structure and physiology. The kind of adaptation which is essential before establishment in an empty space is of necessity a pre-adaptation.

Pre-adaptation and the occupation of empty spaces may be an important element in evolution. Any natural process which totally alters living conditions, such as a glacial period, a general rise or fall of temperature, creates a new and possibly immense ecological void in which only those pre-adapted mutants of the former population will survive. The pre-adapted will extend both in space and, since they persist, in time.

The succession in time of the various groups seems to be connected with the presence of empty spaces. This may be seen in the rise and spread of the vertebrates. The Cyclostomata (lampreys and hagfish) inhabited the shallow coastal waters while the Selachii (sharks, etc.) lived in deeper waters. The Crossopterygii (of which the only known living representative is the coelacanth) and the Dipnoi (or lungfish) peopled the rivers, lakes and swamps. Here an innovation of capital importance occurred: the birth of an animal with four legs. The first tetrapods, with the skull of a crossopterygian, a dorsal fin and walking limbs, were still aquatic; but thanks to air-breathing lungs and feet they could move overland from pool to pool. The Stegocephalia (fossil Amphibia) and the Urodela (newts and salamanders) were satisfied with an environment which was merely humid. The reptiles then occupied dry regions. A certain number of pre-adaptive characters, including hard skin, toes armed with claws, internal fertilization, and large eggs which developed without need for an aquatic phase, enabled them to install and maintain themselves in empty spaces where conditions were new. The birds, originally runners, next prospered in the air which was left free to them by the disappearance of the last flying reptiles. Meanwhile the mammals vegetated so long that the reptiles still occupied regions of all kinds. The eventual rise and spread of the mammals was contemporary with the extinction of the reptiles. In the vegetable kingdom the appearance of monocotyledons and dicotyledons provided an important source of nourishment which encouraged the expansion of herbivorous mammals, themselves eaten by the carnivores. Evolution is progressive; in the competition to occupy new environments structures and organization become increasingly complex.

Adaptive radiation

A species suitably adapted to a sufficiently large empty space will multiply enormously. Then, sooner or later, it will differentiate into distinct forms further adapted to occupy the various 'ecological niches' of the habitat. This expansion of the primitive form is known as 'adaptive radiation'. The phenomenon is observable chiefly in islands, lakes, mountain tops, and, generally speaking, in isolated areas.

The Galapagos Islands shelter a family of finches, the Geospizinae. The primitive form resembles a warbler; other species subsist on insects, berries, fruit, seeds. The morphology of the beak is suited to its diet. One species lives like a wood pecker and perforates the bark of trees to obtain the larvae of insects; lacking the extensible tongue of the woodpecker this finch holds in its beak a cactus spine or a small twig to winkle out the larvae from their holes.

Objections to the mutation theory

Certain aspects of the mutation theory are based on well-established facts: mutation itself, the transmission of mutations, the non-inheritance of characters acquired by the soma, pre-adaptation. The theory does much to explain what is sometimes called 'micro-evolution', that is to say the evolutionary changes within a population, and the formation and isolation of new species. But many facts it can only partly account for, nor does it pretend to be as comprehensive as previous theories. It rejects the idea of a direct causal connection between a given character and the environmental conditions to which it must respond. It also denies the automatic selection of useful characters in a species by the differentiating action of death. When a habitat is changed or an empty region occupied innumerable combinations are possible; among thousands of attempts one will from time to time succeed and the mutant finds a niche which is unoccupied or ill defended. The all-inclusive physiological fitness of a certain hereditary type — that is, its pre-adaptation — conditions success much more than any structural detail. Every organism which lives and multiplies has found by chance the environment which suits it. If the environment changes, the organism adapts itself to the change, emigrates, or dies. Mutation theory thus accepts an overall selection of the pre-adapted. This selection eliminates the species which are unfitted to given conditions and allows only those with the necessary qualities to survive. In the hot, wet climate of sub-tropical regions where cold winters are unknown, trees with large leaves retain their foliage permanently. In temperate climates with cold winters, or in tropical climates with a dry season, trees lose their leaves during the winter or during the dry season, periods in which water is hard to obtain. By losing their leaves the trees avoid desiccation.

Chance plays a large part in mutation theory. The mutation itself occurs by chance, and the discovery of an adequate environment is also a matter of chance. Since overall success is thus a random occurrence the theory offers no explanation of the origin of adaptations or of orthogenesis. It is difficult to imagine chance mutations giving rise to orthogenetic series, to grasp how random variations could pursue the same course and direction over vast periods of time, or indeed produce such complicated organs as an eye or an ear. Then again, a species does not always evolve in a manner which is to its own advantage; the growth of organs is sometimes so exaggerated that they become a hindrance. Such anomalies are not infrequent. Why, then, have they not been eliminated by selection?

The very principle of the theory may be criticized.

If all evolution is accomplished by transmissible variations, in other words by mutations, the evolutionary value of mutations must be carefully scrutinized. Mutations usually subtract from the organism; they rarely add. Above all they add nothing new; they are not constructive. All the mutants of *Drosophila* are still unmistakably fruit flies. Numerous characters are simply modified in varying degrees: the shape, size and veining of the wings, the form and texture of the eye, the colour of the eyes and body, the number and dimensions of the bristles, the form of the head, abdomen and feet, fertility and vitality. These remarks apply to all known mutations, animal or plant. Evolution does not merely modify characters which already exist, but brings about the appearance of new characters, new plans of organization. A bird differs from a reptile by the acquisition of feathers, of wings, of a gizzard, of warm blood, and by the transformation of various organic equipment. In an animal there are countless possible combinations of genes, but none can make a new organ appear. Many mutations are, moreover, recessive. In *Drosophila* ninety-four per cent of the mutations which are not lethal are recessive, and therefore do not appear until a homozygote is born. Before this occurs the heterozygote which carried the recessive mutation has behaved exactly like the normal type and populations are composed very largely of heterozygotes. In addition, many mutations are lethal and result in the death of the individual. In *Drosophila* two out of five mutations prove to be lethal. The only mutations which can have a certain minor evolutionary importance are dominant mutations or recessive mutations in homozygotes.

Thus three great theories, Lamarckism, Darwinism in its original form, and the mutation theory, have in turn attempted to explain the mechanism of evolution. None of them is totally satisfying and all are open to serious objections. Lamarckism fails because acquired characters are not transmissible; the two other theories contain certain elements which are of genuine explanatory value. Each theory is a slight step forward and cannot be entirely dismissed. Lamarckism affirmed the fact of evolution, the modifying action of the environment on the individual, and the effect of the use or disuse of an organ. Darwinism revealed the significance of the struggle or rather the competition for existence, of natural selection and of adaptation. The mutation theory emphasized mutations, the formation of new species and the non-inheritance of acquired characters.

The future theory of evolution will be slowly constructed: it must attempt to explain not only adaptation but also how the major groups (the phyla) arose, the paths, the directions, which evolution follows, and the origin of complex organs like the eye or the cephalopod's two part press-studs.

The synthetic theory of evolution

What is the attitude of contemporary biologists to evolution? Most English-speaking biologists are very firm advocates of a synthetic evolutionary theory which is, in fact, an amended version of neo-Darwinism, or rather a synthesis of mutational and of neo-Darwinist theories.

The brilliant American palaeontologist G. G. Simpson followed this school of thought. He postulated the concept that life appears to be a strange mixture of the directed and the non-directed, of the systematic and the non-systematic. Adaptation is the guiding force of evolution, and the mechanism of adaptation is supplied by natural selection acting on the genetic structure of populations. Amended neo-Darwinism is simply the 'extrapolation' or the extension to all geological times and to all organisms, of the mechanisms of transformation as analysed by the genetics of populations, that is to say by the science which studies the genetic structure of populations and its variations from generation to generation. The direction which evolution takes is, then, the result of an interaction between the genetic constitution, which controls the mechanisms of development, and natural selection, which serves as guide. Mutations supply the possibilities, selection chooses the path.

Evolution carries out no pre-arranged plan; the essence of its character is 'opportunism'. It explores all potentialities, and what can occur, does occur. Opportunism is limited only by the means which organisms have at their disposal to profit by the possibilities which the environment offers. Adaptation to these new possibilities is the guiding factor. Evolutionary material is composed of the genetic systems of populations and their mutants. Evolutionary changes are the outcome of new genetic combinations arising from crosses, of the nature and rate of mutations, and of natural selection.

The size of populations is an essential factor in evolution. In large populations evolution, normally slow, is accelerated by selection. It is adaptive in rough proportion to the intensity of selection. In small populations selection plays a lesser part; the effects of chance are apparent and become predominant. Medium-sized populations occupying an extensive geographical area are, in general, those most favourable to rapid and sustained evolution.

Selection conceived in this way is not exactly the same as Darwin's natural selection which recognized a differentiating death to eliminate the unfit. This kind of selection is more active and creates combinations and genetic systems which could not have been brought about by Mendelian mechanisms alone. The driving role of selection is no longer limited to 'canalization in a restrictive sense'; it is also creative in that it gives rise to a progression in a definite direction. Selection may be thought of as a directed magnitude, a vector, characterized by its intensity and direction. Intense selection in an adapted population will slow down or arrest an evolving modification; but in an unadapted population it will, on the contrary, quicken evolution. Its direction, which varies according to conditions, has three separable components: one which tends to concentrate the population around a standard type, another to break it up into diverse types, while under the influence of the third the standard population tends to slide in one direction.

Selection acting on small and completely isolated populations leads to 'quantic evolution'. This is the rapid achievement by an unbalanced population of an equilibrium different from its ancestral equilibrium. Between the two states of equilibrium there is an interval, the biological 'quantum', during which the system is unstable. Three solutions are then possible: a return to the old conditions, extinction, or the achievement of a new equilibrium. In the latter event quantic evolution will occur. Quantic evolution comprises three phases. Firstly there is an inadaptive phase in the course of which the group loses its equilibrium. This is followed by a pre-adaptive phase, marked by the intense pressure of selection driving the group towards a new equilibrium. Finally, an adaptive phase appears in which the new equilibrium is reached. Quantic evolution can take place only if pre-adaptation for a possible new way of life exists.

Simpson took the example of the horse and the evolution of the animal's size, its hypsodont or high-crowned cheek teeth, the design of its dental crowns, and the structure of its foot. In the Eocene period the ancestors of the horse inhabited forest regions and ate leaves. Variation was intense and the acquisition of new characters was to favour the subsequent change in diet. The growth of the dental crown was a pre-adaptation to the hypsodont molars indispensable to an animal which feeds on grass. Reduction in the number of toes was a pre-adaptation to running across treeless hills. The lengthening of the face was necessitated by the development of hypsodont teeth which require more space. This sequence of pre-adaptations enabled mutant animals to gain the grassy savannahs. At that time they were imperfectly adapted both to a diet of leaves and to a diet of grass, but they were pre-adapted to grazing. Those with the highest-crowned teeth digested the grass most efficiently and thus obtained most nourishment. These animals, grouped around the ancestral species *Merychippus,* evolved fairly quickly towards forms which were totally herbivorous, the Equinae. Under the pressure of selection the pre-adaptations had become post-adaptations, while the organism as a whole underwent organic transformations in accordance with the new diet.

Quantic evolution would, it is argued, be the essential process by which families, orders, and classes arose; it would also participate in the formation of the great groups, the phyla. The less important taxonomic units such as species and genera would derive from speciation or from evolution within the phyla.

Sub-species, species, and even genera would result from speciation or specific differentiation. Speciation is brought about by adaptation to local ecological conditions and by chance segregation. It is shifting and changeable like a net, sometimes closely, sometimes loosely, meshed. It acts on average populations, affecting secondary morphological characters: colour, number of scales, fur, slight changes of height, proportions, shape.

Evolution within the phylum implies the continuous and orientated development of a population's average characters. It is the domain of post-adaptation. It produces changes of greater importance than speciation and creates groups of wider scope — genera, subfamilies, and families. Its operation is unceasing and parallel to that of speciation, affecting populations of all sizes but especially those of large or medium size.

The development of the extinct ungulates, litopterns and notoungulates peculiar to South America demonstrates the combined action of the three forms of evolution which Simpson envisages. At the beginning of the Tertiary their ancestors consisted of very small forms, widely dispersed and provided with a variety of pre-adaptations. They invaded numerous unoccupied areas where their numbers rapidly increased by quantic leaps. The fauna of Casamayor the oldest known mammalian fauna of South America, displayed extravagant variation. This exuberance died down in the fauna which followed. Transitional types disappeared and the fauna became stabilized in a limited number of ecological zones. Quantic evolution had more or less come to an end. Then evolution within the phylum took over; surviving races pursued their evolution and more or less rapidly became stereotyped. Meanwhile, within the races, specific differentiation came into play.

Thus pre-adaptation, which in the theory of mutation is basic, is far from negligible in the synthetic theory of evolution; though in the synthetic theory the decisive factor is the action of selection on the pre-adaptive structures at its disposal.

Objections to the synthetic theory of evolution

The synthetic theory of evolution is an original construction, supported by sound reasoning and facts. Many of its aspects are valuable and positive, and it opens out new approaches to the problem. Mutations, the genetics of populations, and selection, are indisputably elements of importance in evolution.

The genetics of populations, analysing genetic structure statistically, is a science which grants wide scope to mathematical speculations. From one generation to the next, factors of four kinds are capable of modifying genetic structure: namely, the statistical laws which govern the encounter of gametes, the forces of selection which favour certain genotypes, mutations, and the immigration of individual carriers of new genotypes, and, finally, the factor of chance. These four factors can be represented by mathematical formulae. Situations assumed to be typical are chosen and then, by working out the mathematical problems involved, an answer is reached, and the modifications which genetic structures undergo can be expressed in figures.

While admitting that the evolutionary value of mutations is great, we have no unimpeachable proof that the variations of fossil forms were identical to those variations of present-day forms which we can produce experimentally. The variations observed in fossil forms surely involved greater and more profound rearrangements of genetic material, and the mechanism itself may well have been different.

Within populations which exist today new species do, indeed, arise; but no profound change modifying basic structure has yet been observed. It is hard to believe that an accumulation of the kind of mutations we are familiar with can bridge the gulf between phyla or the major groups into which plants and animals are classified.

Again, is mutation the only manner in which variations today occur? The study of directed mutations has scarcely begun. Nucleic acids determine the characters of the cell which contains them. When liberated can these nucleic acids make the same characters appear in another cell? If the answer proves to be yes, then the influence of somatic cells on germ-cells would no longer seem impossible.

The order in which mutations appear is purely random. The paths which evolution follows, the apparent sense of direction, is difficult to explain by the simple interplay of random mutations and selection.

If it is denied that evolution is an orientated process, then how must we explain the phenomenon of convergence leading to similar organs in animals which are unrelated? If, of course, pure coincidence is invoked no problem arises.

How are we to explain the genesis of those devices or organic tools which are formed by the reciprocal adjustment of two independent parts of the body? The two parts develop without contact, with nothing external to mould them. Hence the character is inscribed in the genotype and depends on a precise morphogenetic territory. When the foreleg of a stick insect (phasmid) is amputated a regenerative bud is formed; if the bud is transplanted to the territory of the second leg it will produce a leg with a femur curved exactly like that of a foreleg. Even if we agree that two-part devices are susceptible to selection, a difficulty remains: how could complementary mutations appear simultaneously and with the coordination required to create a precision instrument as exact as the hooking apparatus of the wings of the bee, which functions correctly from the moment the insect first flies? We come up against the same difficulty when we consider complex organs with highly coordinated structures, like the eye, the ear, and the human brain. Complex organs are not merely simple organs enlarged. More than quantitative factors are involved; in complex organs new qualities appear, new coordinations, new and different plans and organization. Numerous mutations, occurring just at the right moment, would thus have been indispensable.

The synthetic theory of evolution, like classic Darwinism, is based on the principle of utility. Now, all that exists in living creatures is not necessarily useful; and that which is useful to the individual may not be of value to the species. Sexual selection favours the most attractive and vigorous males; but are these always the most fertile?

The principle of utility implies a judgment of value which is incompatible with the purely mechanistic interpretation of evolution which the neo-Darwinists defend.

The synthetic theory of evolution offers a satisfactory explanation of many facts; but, as P. P. Grassé remarked, it harps too much on one string to give a general explanation of evolution. The mechanisms of evolution are many and their causes are no less numerous.

Prehistoric paintings.

Above, the grotto of Lascaux in the Dordogne; the so-called chamber of 'the Bulls'.

Skira.

Left, ibex, in the grotto of Niaux (Ariège).

Languepin-Rapho.

Human Evolution

Before considering the origin of Man we must attempt to place him in his zoological framework. Man is a primate and although his constitution differs in various ways from that of the other primates it would be incomprehensible if treated in isolation. The group or order of the Primates is divided into two sub-orders: the Prosimii and the Anthropoidea. The Anthropoidea are further subdivided into the Platyrrhina, or South American monkeys, and the Catarrhina, which include the Old World monkeys, the great apes (Pongidae), and the Hominidae, the group to which Man belongs.

Characteristic of the human race is its lack of excessive specialization; archaic structures persist side by side with more highly developed structures. The four limbs are still pentadactyl, and the human foot is more like that of the first four-legged vertebrates than it is like the foot of anthropoid apes. The design of the teeth, the four-cusped molars, had already been achieved by mammals in the Tertiary. A portion of the tympanum reveals the persistence in the middle ear of the tympanic membrane of the theriodont reptiles from which the mammals themselves arose. The human digestive tube is not specialized. On the other hand, the erect posture of the biped, the projecting nose, the straight, high forehead, the prominent chin, and above all the complex and highly differentiated brain are all evolved structures which appeared at the end of the Quaternary. Man made his appearance as a primitive being with generalized characters and therefore susceptible to further structural evolution. But he became capable of progress without changing his structure, for he invented, made, and used artificial organs — tools. The emergence of Man was marked by somatic changes which were slight and by psychic changes which were profound. The first men differed among themselves more by their activities and industries than by their anatomical structure.

There are two phases of human genesis. The first, before the emergence of true Man, is the phase when, among the primate branches, that branch which was to terminate in Man became distinct. The second, after the emergence of Man, began with the first manifestation of his arrival and has continued ever since. Between these two phases Man was born. We must, then, try to discover those sub-human forms which led to the appearance of Man, and then analyse the paths by which modern Man has reached his present anatomical and psychological condition. The task is complicated by conflicting hypotheses, dissenting opinions, and unanswered questions. We shall present the point of view of the palaeontologist, Professor Piveteau.

Phase one: the Hominidae before the appearance of Man

Walking on two legs, which required a modification of the locomotor apparatus, entailed anatomical changes which mark the initial divergence between the man-like hominids and the anthropoid apes. But which diverged from which? The fossil record suggests hypotheses but provides no answer. The numerous resemblances between the hominids and the anthropoid apes, and between these two primate families and that of the Old World monkeys, makes highly probable the existence of a common ancestral group (the Protocatarrhina) from which by increasingly marked divergencies the three families (Old World monkeys, great apes, and hominids) successively branched off.

Over a stretch of millions of years palaeontology supplies only two signposts: one, *Oreopithecus,* which dates from the Tertiary or more exactly from the upper Miocene or Vindobonian; the other, *Australopithecus,* from the Quaternary or lower Pleistocene.

Oreopithecus

A few remains — teeth, skull fragments, almost complete jawbones, the upper part of the cubitus, the articular head of the radius and the femur — were discovered in 1872 by Gervais in the Vindobonian lignite deposits of Monte Bamboli in Tuscany. They were studied by Gervais and later by the Swiss Hürzeler. In 1954 Hürzeler published the results of their combined findings and came to the conclusion that *Oreopithecus* was a hominid. His conclusions were disputed and widely denied. The question of *Oreopithecus* again arose when, in 1958, an entire skeleton was discovered in the same lignite deposits. Its study will no doubt furnish valuable information and establish whether it has a place among the ancestors of Man or belongs to a lateral branch.

Among its characters of particular interest are the morphology of the canine teeth, the bicuspid structure of the anterior premolar, in other words the presence of two tubercles on the crown, the vertical disposition of the incisors, the shortness of the jawbone, and the beginning of a zygomatic arch above the final premolar. These features suggest a shortening of the face which is characteristic of modern Man.

Oreopithecus marks a stage in the very ancient individualization of the human branch. As fossil anthropoid apes are already highly specialized in the lower Miocene, the group from which both anthropoid apes and hominids descended must have existed before the lower Miocene.

Australopithecinae

Numerous fossils of man-like apes — *Australopithecus, Plesianthropus, Paranthropus* — were discovered by Dart and Broom in the limestone quarries of South Africa and not, as the name *Australopithecus* ('southern ape') might suggest, in Australia. They date from the Quaternary or more exactly the lower Pleistocene.

Their relationship to Man is revealed by a mixture of simian and modern characters. Archaic characters include certain peculiarities of the skull and teeth, the developed sagittal crest, the great size of the premolars and molars, the molarization of the first lower premolar. Modern characters are the erect posture, the structure of the palatine arch which is analogous to that of Man, the bicuspid design of the first lower premolar, and a mandible without the simian plate.

These 'southern apes' are probably not direct ancestors of Man, for they are of too recent date. Their localization in the southern part of South Africa suggests a refuge, a place of asylum, rather than an area of expansion. It is likely that they form a small lateral and fairly independent branch of the family. Their final forms, powerful and provided with extraordinarily large teeth, emphasize their differences from the human branch. They reached an impasse and became extinct. But it is possible for the ancestors of *Australopithecus,* living elsewhere, to have been the direct ancestors not only of the 'southern apes' but also of Man.

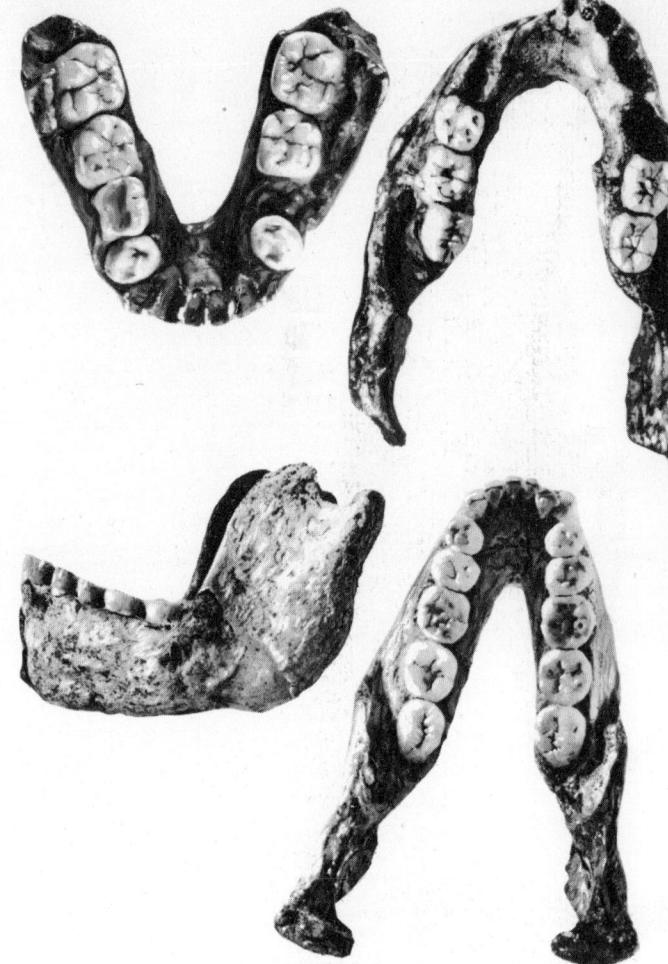

Top left, lower jaw of *Australopithecus promethus.*
J. Oster.

Top right, lower jaw of *Telanthropus capensis* (from South Africa). *Above,* two pictures showing the lower jaw of female *Paranthropus crassidens* (from South Africa) viewed from above and in profile.

Collection Musée de l'Homme.

The emergence of Man

Developments pointing towards Man were taking place very slowly in *Oreopithecus* and in *Australopithecus.* It is, however, almost always impossible to determine when a threshold has been crossed and a new branch begun. Two innovations were to occur: the liberation of the hand and the elaboration of the brain. The hand was freed as a consequence of the upright posture. The complicated structure of the brain would take longer to achieve, but between its increasing complexity and the increasing dexterity of the hand a certain correlation may well have existed.

True Man is normally identified by his capacity to make artificial tools. Thus the oldest worked flints bear witness to the existence of men. During the course of the Tertiary period no member of the Hominidae crossed the dividing line to become true Man; that is to say no manufactured tool of that period is known to

Skeleton of *Oreopithecus*, discovered by Hürzeler in 1958 in a lignite deposit, Tuscany, Italy.

Rothpletz.

Above, skull of *Paranthropus crassidens*; below, skull of female of the same species (South Africa).

Collection Musée de l'Homme.

exist. But from the beginning of the Quaternary period the activity of Man can be observed in Europe, in Africa, and probably in Asia. Already in this period Man had acquired a certain psychic development; but, as in all genealogical trees, the stem, the base of the human branch, has left no visible traces.

The coming of Man was accompanied by no important organic changes or structural novelty, but by the appearance of new modes of activity, born of mental reflection. Life from its very origin is a psychic process, but with the arrival of Man, a threshold is crossed. This psychic break-through is linked anatomically with increase in size of the brain and increase in the brain's structural complexity.

Palaeontology limits its investigations to the structure of the skull; the enlarged volume of the brain is indubitable, but its mechanism remains unknown. It is generally agreed that the number of nerve-cells in the brain of an anthropoid ape would require relatively little multiplication to equal that of the human brain.

Actually the cerebral cortex of an anthropoid ape contains some 3,000 million cells, derived from thirty-one successive cell-divisions. The human cerebral cortex contains about 9,200 million cells, derived from thirty-three cell-divisions. Thus two extra mitoses would seem sufficient to turn an anthropoid ape's brain into a human brain. This line of reasoning assumes, of course, that the anthropoid brain is the brain of a hominid arrested in its development — an assumption which is very probably false. The anthropoid brain is of different structural type, a type which terminates an evolutionary branch, that of the family Pongidae, the great apes.

The emergence of Man is thus characterized by the appearance of mental reflection in an organism elaborated during the course of millions of years. The mechanism by which this psychic innovation came about remains totally unknown.

453

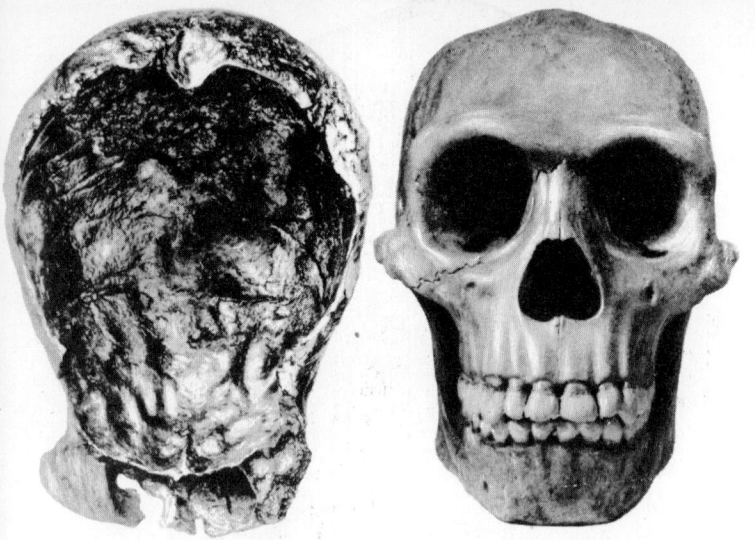

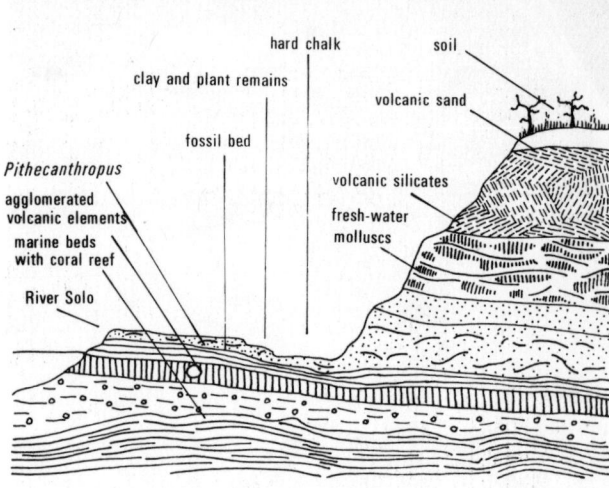

Left, partial skull of *Pithecanthropus erectus* viewed from above.

Right, reconstruction of skull of *Pithecanthropus*.

After Weinert. Collection Musée de l'Homme.

Cross-section of the beds where *Pithecanthropus erectus* was found in Trinil (Java).

Collection Musée de l'Homme.

Phase two:
the protoanthropic stratum

After the sub-human forms we have considered we reach forms that are incontestably human. These do not occur as a homogeneous group, but rather in successive waves. The human branch is generally believed to have spread in the following manner: first, the oldest part, dating from the beginning of the Pleistocene, consists of a protoanthropic stratum comprising remains discovered at the extremities of the Old World: *Pithecanthropus* of Java, *Meganthropus* and the pithecanthropes of China — known as *Sinanthropus* — the child of Modjokerto. The etymology of these names is significant, for *Pithecanthropus* means 'ape-man' while *Sinanthropus* signifies 'man of China'. Protoanthropic forms must have been much more widely distributed; they also existed in Africa as we know from the recent discovery at Ternifine in the province of Oran of the pithecanthrope *Atlanthropus mauritanicus*. These latter remains are associated with a stone industry composed of primitive tools: hand-axes, flint 'bifaces'. Such signs of industry are not found among the 'ape-men' of Java, and indicate the emergence of true tool-making men. At the side of *Sinanthropus* hearths and cinders are found, together with chipped flint scrapers and perforators, and some other implements made of antler horn. *Sinanthropus* thus knew and used fire; the instrumental phase had been reached. In addition, the thighbones of *Sinanthropus* are found broken as though the marrow had been extracted, while the occipital holes in the skulls have been enlarged to allow the brains to be more easily removed. This practice is still encountered among Melanesian head-hunters, and a reasonable conclusion would be that *Sinanthropus* was the first of the cannibals.

A comparison of the skeletons of *Pithecanthropus* and modern Man reveals a series of modifications, the chief of which are an enlargement of the brain case and a reduction of the jaws and teeth. The bones of the legs and arms and their muscular attachments also alter in structure as the erect posture and biped gait become habitual.

We have already seen that it is a rule of evolution that 'nature does not turn back', that tendencies are irreversible; thus *Pithecanthropus* had a larger brain than his ancestors, but his jaws and teeth were smaller.

The Palaeoanthropic stratum

The early wave of protoanthropic forms was followed by the appearance of palaeoanthropic forms which held the stage for more than 100,000 years and comprised several human types that were contemporary with the Mousterian culture of the middle Pleistocene.

The oldest known of these types, dating from the end of the lower Pleistocene, is *Homo heidelbergensis,* Heidelberg man, represented by a jaw found in 1907 near the little village of Mauer. Then, during the relatively recent Riss-Würm interglacial period, lived the Ehringsdorf group. Its representatives, found in central Europe, on the shores of the Mediterranean, and in Palestine, include the fossils of Ehringsdorf (near Weimar), the Taubach teeth, the Steinheim skull (a little older, probably Rissian), the men of Krapina in Yugoslavia, the man of Saccopastore in Rome, the child and the adult skull of Gibraltar, and Palestine men (the Galilee skull, the men and children of Mount Carmel, the skeletons of Djebel Kafzeh). The human remains of Arcy-sur-Cure and the Montmaurin jaw should, though more doubtfully, be added to the list. To the same group belong two fossils which are sometimes considered to be pre-sapiens, that is to say more or less direct ancestors of *Homo sapiens.* These are Swanscombe man, found in 1936 in Kent, and Fontéchevade man found near Montbron in France. The first is known by a brain-pan, the second by a skull.

These different populations, widely dispersed, were highly varied and in many ways distinctly modern.

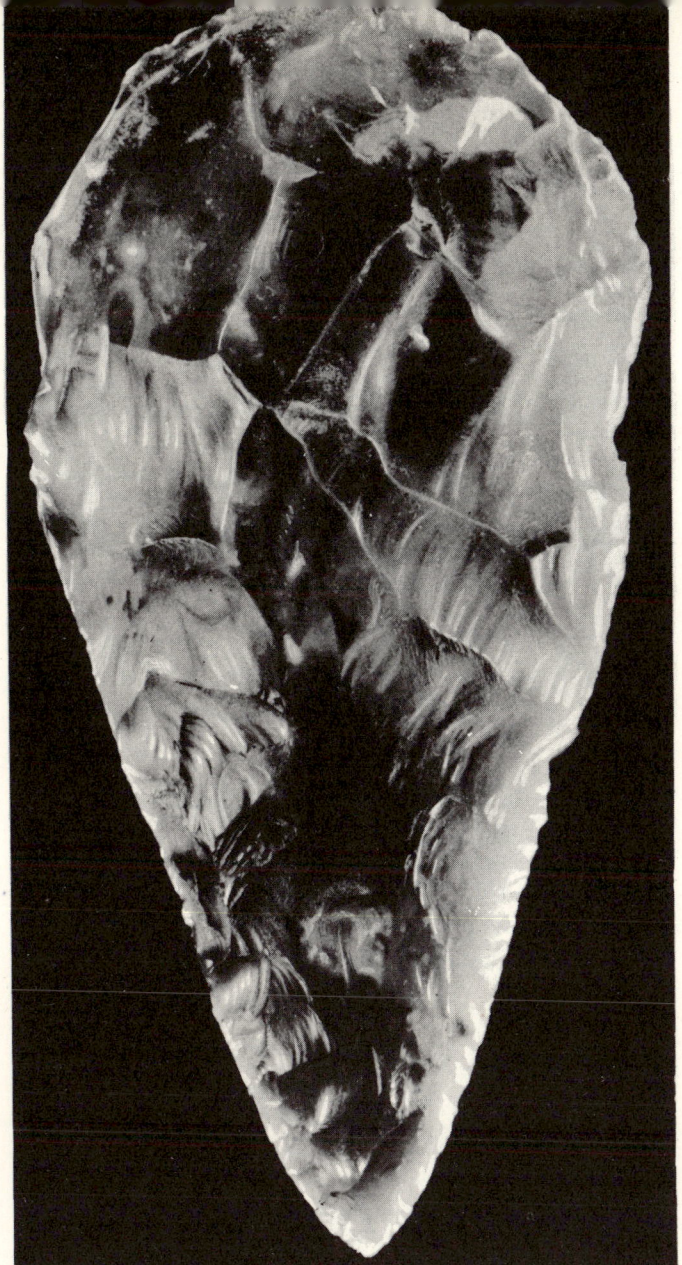

Skull of female *Sinanthropus pekinensis*.

Collection Musée de l'Homme.

Acheulian flint hand-axe (Lower Paleolithic).

Collection Musée de l'Homme. Larousse.

Reconstruction of skull of *Sinanthropus pekinensis* (frontal view) by F. Weidenreich.

Collection Musée de l'Homme. Tracol.

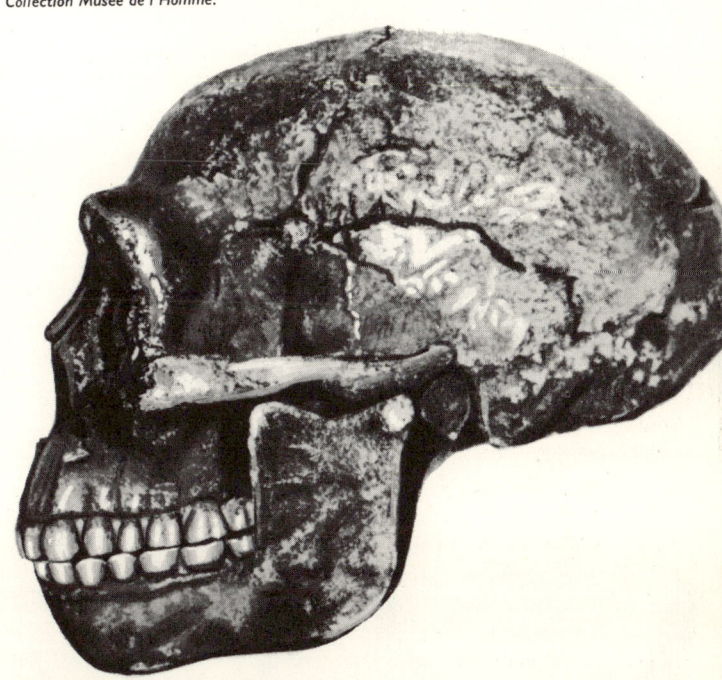

Skull of *Sinanthropus pekinensis*.

Collection Musée de l'Homme. Tracol.

Lower jaw of *Atlanthropus mauritanicus*.

Collection Musée de l'Homme.

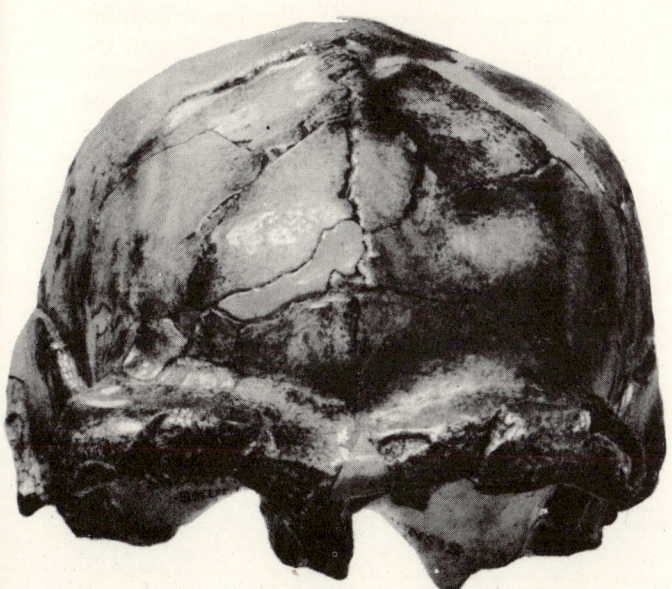

Their relationship to the earlier protoanthropic types raises many unanswered questions. But the Ehringsdorf group appears to be of fundamental importance and probably constitutes the common stock from which diverged the two branches *Homo neanderthalensis* and *Homo sapiens.*

Stratographic section of the Quarry at Mauer where the lower jaw of Heidelberg man was discovered (Germany).

After Shoetensack. Collection Musée de l'Homme.

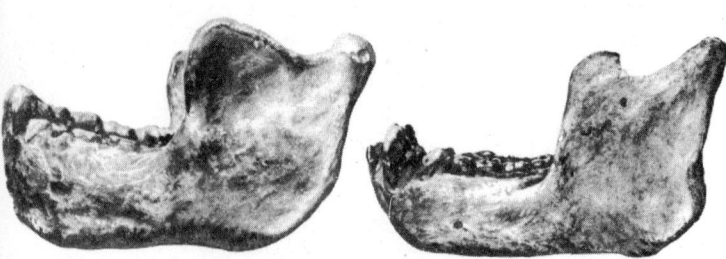

Heidelberg man   Chimpanzee

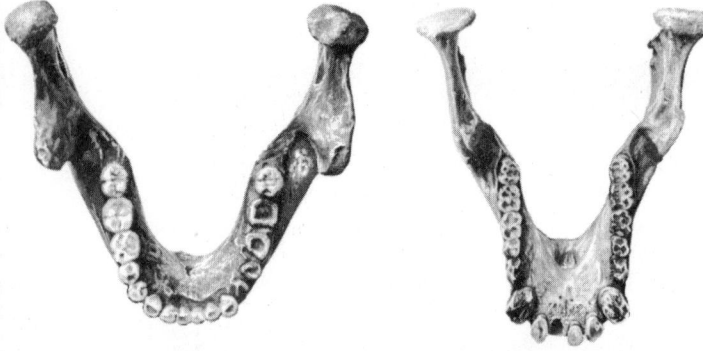

Heidelberg man   Chimpanzee

Comparison of the lower jaws of Heidelberg man and chimpanzee; *above,* the mandibles in profile; *below,* viewed from above.

Collection Musée de l'Homme.

Neanderthal man

The first branch, that of Neanderthal man, *(Homo neanderthalensis),* shows certain signs of regressive evolution. Neanderthal men were relatively homogeneous and lived during the Würm glaciation period in the middle Pleistocene. They have left numerous remains throughout western Europe, in France, Belgium, the Rhineland, and Italy. In France they are abundant in the department of the Charente and particularly so in the Dordogne. The man of La Chapelle-aux-Saints, discovered in Corrèze, is a typical Neanderthal representative.

Other well-known deposits and fossils include the men of La Ferrassie (near Eyzies), the man of Moustier (in Vézère), the man of Petit-Puymoyen (Angoumois), the jaw of Marillac, the remains of adults and children of La Chaise, the jaw of Malarnaud, the man of Neanderthal (Rhineland), the jaw of La Naulette (Belgium), the men of Spy (near Namur), the child of Engis (near Liège) and the skull of Mount Circeo in Italy. All these men had corpulent and powerful bodies, thick bones, a skull with a continuous bony brow-ridge and powerful jaws. The volume of the cranium was fairly large, being 1,540 cubic centimetres. The differences in the skeletons of Neanderthal man and modern Man are pronounced.

Neanderthal man lived in caves and was a hunter. He used fire and made flint tools which were superior in workmanship and finish to earlier examples. The improved technique of his Mousterian culture resulted in better yield. Thus the old 'bifaces' required some 2 pounds of flint to produce 8 inches of cutting edge in the finished article. In Mousterian culture 2 pounds of flint yielded some ten splinters which represented a length of over 6 feet of cutting edge. Improvement was brought about not only by more skilful stone cutting but also by adapting the tool to the job for which it was intended. Some objects reveal a certain aesthetic sense. Progress was, however, extremely slow. It is hard to believe that these artifacts were simply copied; the skills required must have been taught — which implies the existence of language. Neanderthal man practised the cult of the dead; there were funeral rites, and dead bodies were treated with particular care.

The Neanderthal branch seems to have been confined to the western regions of the European peninsula; it was cut off from the rest of the world by the vast Alpine and Scandinavian glaciers which then covered so much of Europe. This isolation may be responsible for the branch's regression and final extinction.

Homo sapiens

The other branch which sprang from the Ehringsdorf group is that of *Homo sapiens.* It is characterized by a general refinement of physical structure and is sometimes called the neanthropic branch. Anthropologists are more or less in agreement that all contemporary races are variations of the single species

Homo sapiens. From the upper Pleistocene human fossils are numerous. During Aurignacian times we find remains of the Cro-Magnon race in the Dordogne, of the Grimaldi race in Italy, of the Chancelade race near Périgueux. Anatomically these men differ little from modern men.

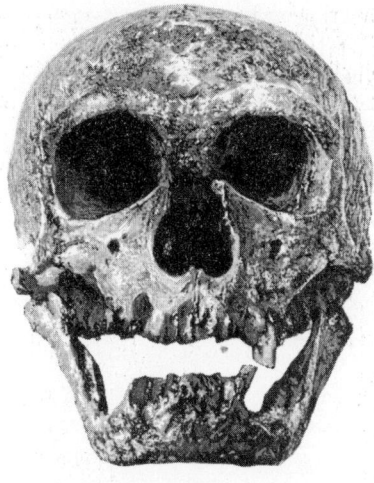

Skull of La Chapelle-aux-Saints man.

Collection Musée de l'Homme.

Skull of Monte Circeo man discovered in 1939, in the Guattari Cave, San Felice Circeo, Italy.

Collection Musée de l'Homme.

Skull of La Chapelle-aux-Saints man.

Collection Musée de l'Homme.

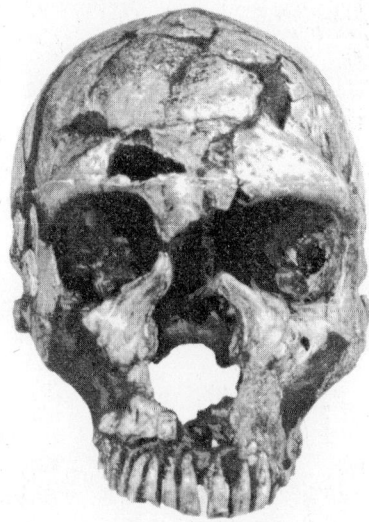

Skull of La Ferrassie man. *Below*, skull of La Quina man.

Collection Musée de l'Homme. J. Oster.

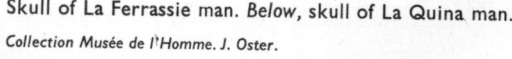

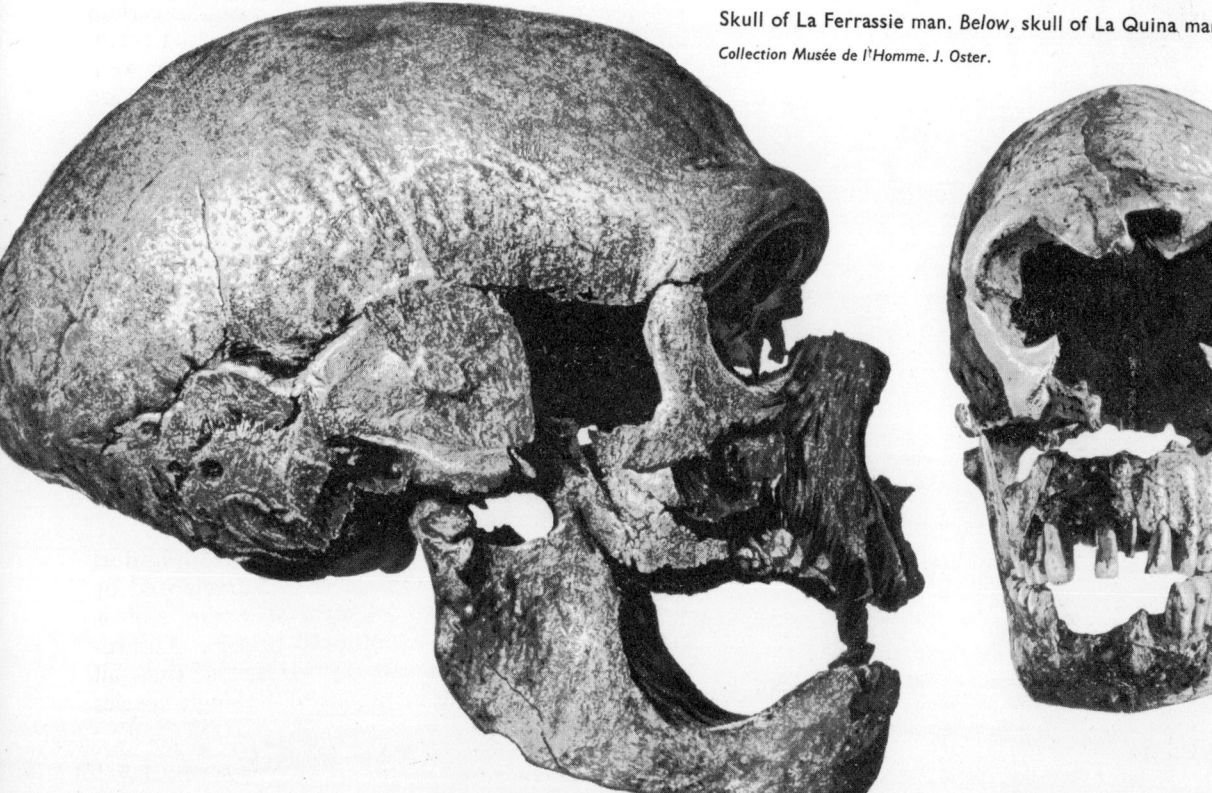

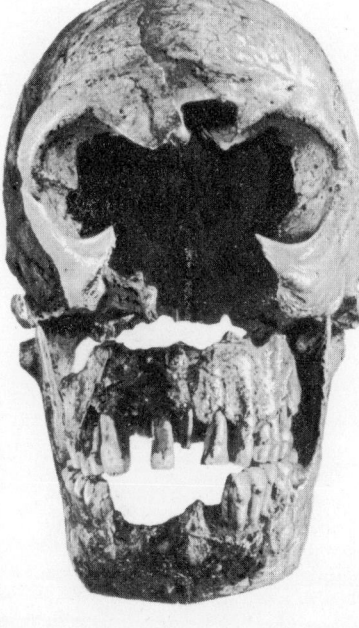

457

The Cro-Magnon race was widely distributed throughout France, Belgium, Germany, and Moravia. Cro-Magnon man was tall, 6 feet at least, and his skull, dolichocephalic, had a capacity of 1,590 cubic centimetres. The Grimaldi race was smaller and the capacity of its dolichocephalic skull was 1,580 cubic centimetres. The Chancelade race was also rather short, about 5 feet 3 inches, but its skull capacity was considerable — 1,710 cubic centimetres, which is more than the average volume of contemporary European skulls.

The neanthropic or proto-Europeans still lived in caves and were great hunters. The reindeer, the horse and the bison provided their chief source of game. They also fished in the rivers. They practised the cult of the dead. Their tools were numerous, varied, and well adapted to the use they made of them. The grotto of La Vache (Ariège) has yielded a large number of flint arrowheads, bone arrows, more or less complicated spearheads, barbed harpoons, etc. Industry had reached a high degree of perfection. Propulsive devices and traps of various kinds were made. Art which had begun to develop in Aurignacian times reached astonishing heights during the Magdalenian period. Artists engraved animals on bone, on pebbles, in clay, and on the walls of grottoes. The carving which decorates the various implements and the celebrated paintings in the grottoes of Lascaux, Niaux, and Altamira, are the work of artists of the first rank. One can, indeed, speak of 'schools of art' from this time; for the similarity of the work found in widely separated regions can only be explained by the existence of 'studios' producing itinerant artists.

The men of Rhodesia

The Rhodesian group, comprising types which are on the whole archaic with a few more modern traits, belongs to the palaeoanthropic stratum we have already discussed. Some — Rhodesian man, the Saldhana skull, *Africanthropus njarensis* — have been discovered in eastern and southern Africa. Others — Solo man — have been found in Java. Anatomically this group is slightly aberrant and lies outside the evolutionary line. It presumably forms a lateral branch which may have given rise to such primitive races as the bushmen and the Australian aborigines.

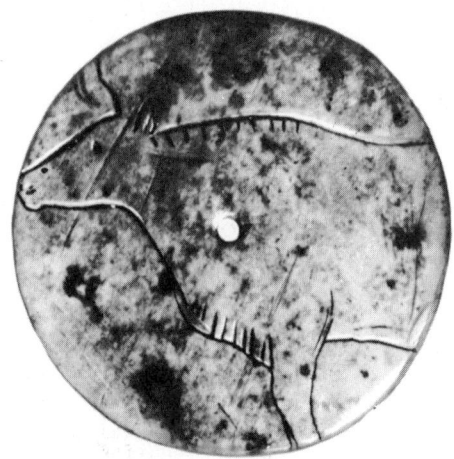

From top to bottom
Solutrean laurel-leaf blade (Volgu, Saône-et Loire); Magdalenian bone harpoon (Laugerie-Basse, Dordogne).

Musée des Antiquités nationales, Saint-Germain-en-Laye.

Bone amulet decorated with a cow, diameter 45 mm (1/5 in.), Magdalenian (Mas-d'Azil Cave).

Collection Péquart.

Spearthrower in the shape of a fawn and a bird, total length 32 cms (13 ins). (Mas-d'Azil Cave).

Collection Péquart. X Photography.

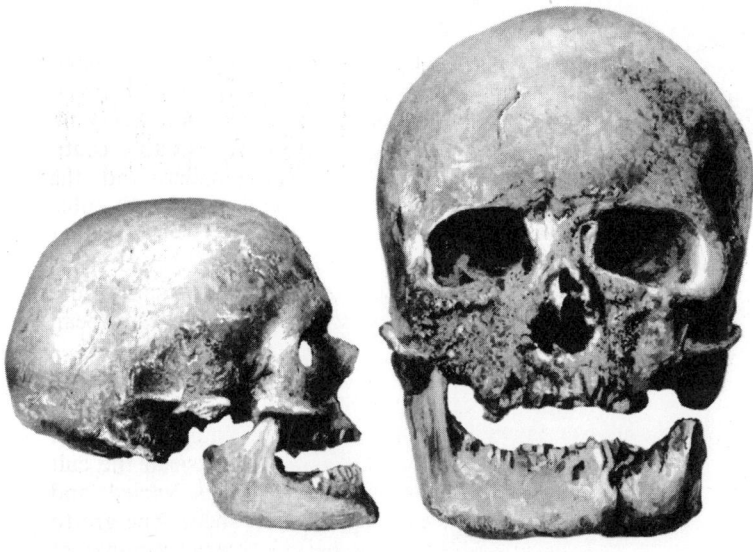

Skull of Cro-Magnon man, viewed in profile and full face.

After Boule and Vallois. Musée de l'Homme.

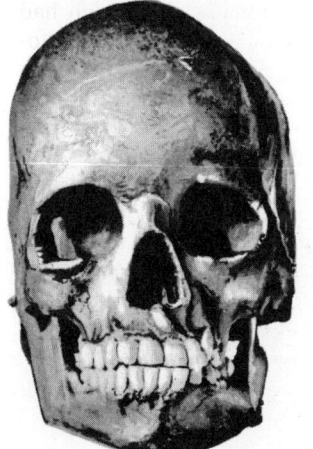
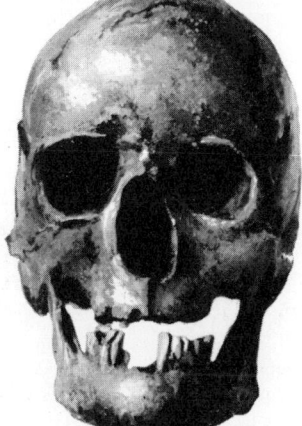

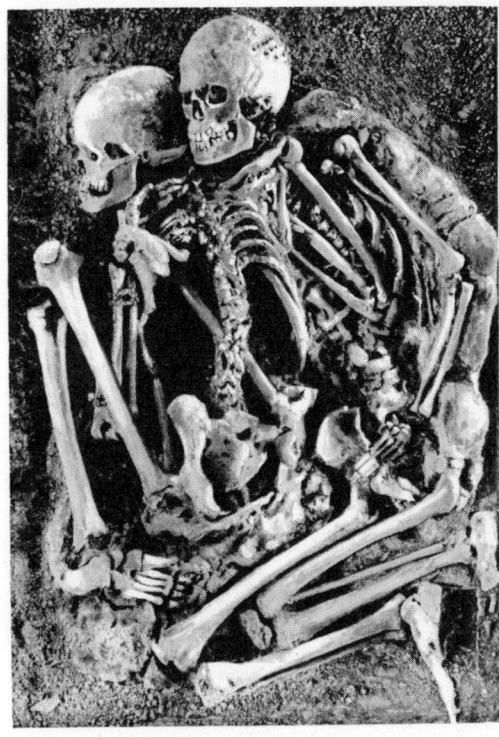

Two skeletons of Grimaldi man discovered in the Grotte des Enfants.

Monaco Museum of Anthropology. After Verneau, in Boule and Vallois.

Far left, skull of adolescent Grimaldi Negroid.

After Verneau, in Boule and Vallois.

Left, skull of Chancelade man.

After Boule and Vallois.

Anatomical changes

The branch of the Hominidae is thus marked out by numerous and varied fossil remains, although the outline is not sufficiently detailed to give an exact picture of how types arose and succeeded each other. From the most ancient men to Man today a series of anatomical transformations is apparent, the most characteristic of which involve the structure of the skull and the teeth.

From *Pithecanthropus* to modern Man the occipital region of the skull gradually alters; changes occur in form, in proportions, and in the inter-relationship of various bones. There is also a tendency for the brow ridges to disappear and the occipital crest to be reduced. In modern Man the presence of a brow-ridge is exceptional, while in modern primitive races the median occipital crest not infrequently persists. The skulls of the Cro-Magnon, Grimaldi, and Chancelade races are dolichocephalic, that is to say elongated. Even today an evolution of the skull from dolichocephalic to brachycephalic can be observed; the elongated skull tends to become more and more round.

The causes of this phenomenon are obscure. Round-headedness has no connection with increased cerebral content; it may, perhaps, lead to a better balance of the head on the spinal column.

After the disappearance of the palaeoanthropic group the skeleton becomes lighter and less robust.

The chin was late to appear; it is absent in the first hominids. Its formation is probably bound up with the withdrawal of the face and the reduction of the teeth.

The size of the brain-pan continues to increase all along the line, and especially in the *Homo neanderthalensis* and *Homo sapiens* branches. Its structure becomes much more complex.

The evolution of the dentition is also characteristic. It is marked by reduction and simplification, both connected with the reduction of the masticating apparatus. This results in less powerful jaws and the recession of the face. As the teeth diminish in size the crowns become shorter, the roots less deep and thick; the number of denticles is reduced while the folds of the enamel become less distinct.

The Future of Man

The broad outlines of human evolution, which we have rapidly glanced at, reveal many characteristics of evolution in general. The usual tendency to ramify is manifest. From the central axis leading to *Homo sapiens* lateral lines, some short, some long, spring out to confer the typically branching aspect of a genealogical tree. The stems of these various offshoots have not been preserved and therefore their origins remain unknown. We re-encounter the same evolutionary laws: Cope's rule that animal groups begin as generalized, undifferentiated species, the irreversibility of regressive evolution, the law of succession by which dominant groups are replaced. Nevertheless there is a profound difference between the genesis of Man and the genesis of the animal phyla. Human evolution involved no anatomical revolutions; no new structure made its appearance. Man himself created new structures, that is, tools, and invented new techniques which are the homologues of structural modifications in lower animals. Today, as in the distant past, the human species must include strains, lines, races; but their evolution is of a psychic rather than a somatic nature. Such races are differentiated essentially by their cultures, their ideologies. Man has reached a position of eminence and become in some respects responsible for the future course and development of evolution. The world of the future will be what Man makes it. Of all living beings it is incontestably Man who occupies the highest rung of the organic ladder.

It has sometimes been argued that criteria of superiority are relative, and that if some other animal had classified the animal kingdom a very different hierarchy would have been established. But, as G. G. Simpson has remarked, such reasoning is merely a form of paradox. Avoiding the anthropocentric viewpoint and setting aside all 'human chauvinism', the fact remains that among all the animals living on earth today Man is the one which has the most complex and delicately constructed brain. He is the only animal that possesses the faculty of reasoning, the power of abstract thought, and a capacity to use symbols with its major consequence: an aptitude for conceptual language. These qualities establish his legitimate claim to be the first and sovereign animal of our planet.

Justified though the claim may be, a question arises: will Man's supremacy remain undisputed? Among the multitude of creatures which share the surface of the globe with him are there none, even though inferior, capable of offering such competition that his survival is endangered?

Howard has spoken of the 'insect menace', and it is true that the voracity and ability to proliferate of many insects cause havoc in our forests and ravage our crops. In this unhappy pursuit the Colorado beetle, the cockchafer, and the locust are all too famous. Millions are spent annually in coping with them. But this is a far cry from concluding that insects can actually threaten our food supply. In the endless battle against insect pests Man is, moreover, constantly armed by science with more effective weapons, some (like D. D. T.) supplied by the chemist, others (like the introduction of appropriate predators or parasites) by the biologist.

Far more dangerous than the insect's threat to our food supply is its capacity to carry disease. It is well known that the plague is spread by fleas and exanthematic typhus and recurrent fever by lice, while mosquitoes spread malaria, yellow fever, etc.

And this point raises a more general question: will the human race one day be exterminated by the action of some microbe or virus which has either newly appeared or else acquired, through mutation, exceptional and overwhelming virulence?

It is by no means impossible that certain species and even entire groups have, in the past, been eliminated by action of this kind. But with the ever increasing variety and efficacy of the means at its disposal Man should be able to win the fight against infection.

Again it may be asked — for it is not certain that the evolution of life on our earth has reached its limit — if a newly formed species endowed with particularly advantageous characters may not some day challenge the species *Homo sapiens*. To this question, which raises all the unsolved problems of the evolutionary process, biology can obviously give no satisfactory answer.

Finally, does the human species carry within itself the seeds of its own destruction? Some biologists believe that species slowly 'age' under the influence of internal factors which are still ill defined. In this way, they argue, the ruling reptiles, the giant amphibians, etc., long ago disappeared by 'natural' extinction. This hypothesis of the fatal 'senescence' of species is not, in fact, supported by any positive data. On the contrary, all that we know about the mechanism of heredity and of the genetic continuity of races leads us to suppose that a species can persist indefinitely unless it is supplanted by a rival species or destroyed by external changes to which it is unable to adapt itself.

The degradation of the species

If the human species is not threatened by 'senility' it is, however, threatened by a deterioration of its genetic constitution.

Mutations occur continually. They are very numerous and almost always result in variations which are disadvantageous or harmful.

In the past, when living conditions were much harder for the individual than they are to day, natural selection weeded out those who were feeble, malformed, or unfit. In this way the species maintained the level of its genetic qualities in spite of mutations.

In the civilized world of today the situation is no longer the same. Every day the progress of medicine, surgery, and hygiene, together with improved social legislation, which softens living conditions for increasingly large numbers of human beings, more and more powerfully counteracts the purging action of natural selection. The result is the survival of individuals who, formerly, would have left no descendants. Myopia, diabetes, susceptibility to tuberculosis, the narrowness of a woman's pelvis, etc., no longer cause death or sterility; and, in consequence, the genes responsible for these defects are multiplied without restraint. It is almost certain — although statistics may not yet have furnished direct proof of this genetic degradation — that the hereditary patrimony of *Homo sapiens* has badly deteriorated since prehistoric times and that its loss in quality still slowly continues.

There is no escape from the sobering conclusion that civilization undermines the biological assets of Man, lowering the quality of human protoplasm at the same time that it increases the total quantity of human protoplasm on earth.

In theory there would seem to be no other methods of counteracting this danger than those which the science of genetics proposes: namely, to replace natural selection which is no longer effective by artificial selection in the form of negative and positive eugenics and ultimately to improve the quality of genes by the direct action of certain chemical substances derived from the nuclei of cells.

The recommendations of eugenics, however, arouse violent and emotional resistance, while the voluntary modification of genes — that is to say, the practice of 'directed mutations' — is still only a somewhat vague aspiration of biologists.

Over-population

Another danger to humanity — a danger which also derives largely from scientific progress and civilization — is that which must result from the over-population of the globe.

How to feed so many mouths, how to supply all these human beings with an adequate number of calories, is plainly a vital problem. As to the imminence of the peril opinions differ. Some authorities, including Robert Matthey, the eminent professor of genetics at the University of Lausanne, believe that births should here and now be limited by international control. Others argue that in view of the almost unlimited resources of the Earth the threat of famine is remote, although each day this opinion becomes less tenable.

Whatever the truth may be, nations in all parts of the world are preoccupied with increasing their potential food supply. Efforts are being made to obtain higher agricultural yield by the use of better fertilizers and the genetic improvement of cultivated species. The exploitation of new sources of nourishment is being studied, among them being certain microscopic algae like *Chlorella,* leaf proteins, the autolysates of yeast, and animal plankton.

The evolution of Man

It seems unlikely that the human race will allow itself to starve to death. If universal famine ever threatens us with obliteration we are well able to take the necessary steps to limit the number of births. In that event biology will be invaluable in providing means of temporarily sterilizing individuals without harmful after-effects to their health.

We can, then, perhaps reasonably predict a very long future for the human race. Unless, driven by blind hate or ideological fanaticism, we commit some irreparable act of madness, mankind should pursue its course until the time when the Earth is no longer habitable — a time which astronomers assure us lies in the extremely remote future and which science and human technical skills may by then be able to postpone still farther.

During the course of his long and adventurous sojourn on Earth will Man continue to develop, to evolve still further, to become, in fact, 'super-man'?

As we have seen he could by selection — or more precisely by the methods of positive eugenics — not only arrest his decadence but also raise his genetic level. Progress of this kind, however, would soon reach its limits. More promising are the techniques of 'directed mutation', if it ever proves possible by such means to bring about changes of major importance.

461

We can also envisage the possibility of directly modifying the human embryo; but in this case the modification, being purely somatic, would not be inherited and would have to be renewed with each generation.

The most tempting project is to increase the number of cells in the brain, for it is known that the degree of intelligence is largely linked to this number, which is constant in members of the same species. Very interesting experiments were made along these lines by the biologist Zamenhof, while in several laboratories in America such experiments are now in progress. Zamenhof succeeded in increasing the number of cerebral cells in frogs and even in rats. When, during the final period of gestation, pregnant female rats were injected with pituitary or growth hormone, he found on examining the brains of the new-born rats a very marked — about forty per cent — increase in the total number of cortical cells. The increase persisted until the animals reached maturity. Unfortunately this increase in cerebral capital was not accompanied by an increase in intelligence, intelligence being measured by the 'maze test' but it is of course possible that this test is not sufficiently sensitive to detect delicate changes in intellectual aptitude.

In order to obtain similar results in human beings it would be necessary to intervene before the fourth week of pregnancy, since the multiplication of nerve-cells in the human brain then ceases.

Again, it may be that other hormones or chemical substances which are unknown today will prove to be more effective than the pituitary growth hormone. For example, it has recently been found that certain snake poisons very powerfully affect the growth of the nervous fibres of the sympathetic ganglia in embryo chickens.

Whatever process Man may employ to evolve into 'super-man', his self-improvement will require the guidance of some ethical criterion. Biology itself may well supply the proper criterion; for, without resort to transcendental realities and relying solely on the knowledge of how normal Man best functions, biology can define with some precision what is good for Man, that which makes him more *human,* and, at the same time, point out what is bad for Man, that which 'dehumanizes' him.

This 'biological morality' was the concern of Lecomte du Nouÿ, Teilhard de Chardin, J. Grasset and Carrel; today it is important to be aware of the responsibilities which Man, as the sole possessor of a highly organized brain, must assume. This brain is something more than an organ of intelligence and will, capable of giving Man self-control and mastery over his own behaviour; it is also an organ of feeling and emotion. Man is not making full use of this marvellous 'instrument for being' which is his brain, unless he exploits all its potentialities, its capacity to understand, its will-power, and its faculties of altruism, devotion, and love.

Biology, at this point, merges with what psychology tells us of the normal evolution of the instincts during the course of building up the individual's personality. Beginning as purely egocentric, these instincts pass through an infantile stage of reaching out for what is desirable to achieve in maturity, the capacity for devotion and even self-sacrifice.

Finally, this aspect of biology is emphasized by certain findings of psychiatry which show us that many neuroses and mental deficiencies associated with a state of immature or arrested emotional development are characterized by an inability to take an interest in others. A certain degree of altruism — of consideration for the welfare of others — seems to be indispensable to the proper functioning of the brain and to the full development of the individual.

The principles of biological ethics might be briefly summarized as follows: to be as fully *human* as possible, to develop in oneself that which is 'proper to Man', and thus to become less bestial, less infantile, and less neurotic.

Such a conception is far removed from the doctrine that might is right and the morality of Nietzsche which was once claimed to be based on biology. It is nearer to the teachings of traditional morality. The philosopher Guyau, already speaking in the name of science, once wrote:

'The individual is not sufficient unto himself; men with the richest lives are also those who are most apt to lay themselves out to please, to sacrifice themselves within reason, and to share themselves with others. From this it follows that the most perfect organism will also be the most sociable. The anti-social being is a departure from the human type just as, on the physical level, the hunchback is a departure from the normally formed man. Pure egoism, instead of being an affirmation of the self, is in reality a mutilation of the self.'

For the moment it is difficult to assert that 'biological ethics' occupies an established position among the other systems of 'scientific ethics'; but it would seem that very serious efforts are being made to approach the subject objectively and to give it a positive content which is independent of metaphysical concepts and religious beliefs. It is true that certain of its particular applications, urged by over-zealous biologists, are open to dispute; but at least we can retain its fundamental principle: namely, that biology can help us to understand the manner in which Man must conduct himself if he is to accomplish the task of being human and earn full rights to his title *Homo sapiens*.

If biology is capable of distinguishing what, in Man, is the most truly human, it naturally follows that it can also guide us in preparing the way for 'super-humanity'.

Thanks to biology we shall be able more or less clearly to discern the 'direction of human evolution' and this knowledge will serve as a criterion on every occasion when we must pronounce judgment on the 'moral' value of Man's intervention in the life of Man.

Index

465

467